Fourth Edition

STATISTICAL PROCESS CONTROL AND QUALITY IMPROVEMENT

Gerald M. Smith
Cayuga Community College

Upper Saddle River, New Jersey
Columbus, Ohio

Library of Congress Cataloging-in-Publication Data
Smith, Gerald, 1939–
Statistical process control and quality improvement / Gerald M. Smith.—4th ed.
p. cm.
ISBN 0-13-025563-7
1. Process control—Statistical methods. 2. Quality control—Statistical methods. I Title.

TS156.8 S618 2000
658.5'62—dc21

00–037481

Vice President and Publisher: *Dave Garza*
Editor in Chief: *Stephen Helba*
Executive Editor: *Debbie Yarnell*
Production Editor: *Louise N. Sette*
Production Supervision: *Clarinda Publication Services*
Design Coordinator: *Robin G. Chukes*
Cover Designer: *Alan Bumpus*
Cover Art: © *Alan Bumpus*
Production Manager: *Brian Fox*
Marketing Manager: *Chris Bracken*

This book was set in Times Roman by The Clarinda Company. It was printed and bound by R. R. Donnelley and Sons Company. The cover was printed by Phoenix Color Corp.

10 9 8 7 6 5 4 3 2
ISBN 0-13-025563-7

To Julie

CONTENTS

7 Additional Control Charts for Variables 237

8 Variables Charts for Limited Data 301

9 Attributes Control Charts 337

PREFACE

Comments and suggestions from users of the previous editions have been very helpful in the preparation of this fourth edition. The following changes have been made:

1. The answers to all the odd exercises have been provided in the back of the text.
2. Lab exercises have been developed for each chapter.
3. Definitions and formulas have been boxed to highlight them.
4. Chapter 6 has been split into two chapters. The introduction to control charts and the work with $\bar{x}$ and R charts have been retained in Chapter 6. The $\tilde{x}$ and R, $\bar{x}$ and s, and the small and short run variations have been moved into a new Chapter 7.
5. New exercises have been added.
6. The discussion on linear regression has been expanded to include the application of the linear regression mode on statistical calculators.
7. The discussion on Walter Shewhart's contribution to control charts has been expanded to provide a better understanding of the concepts.
8. A new section on Six Sigma quality concepts has been introduced.

NEED

Statistical process control (SPC) is not a new topic in industry: It has been used off and on since its development in the 1920s. However, since the 1970s it has become an extremely important tool. A new economic age is developing in which the demand for quality is rapidly increasing, with a resulting global competition of companies striving to provide that quality. The detection system of final inspection, a costly method of quality control, is giving way to a prevention system that uses in-process inspection and SPC to build quality into a process. This change requires extensive training in SPC. Also, for the most effective application of SPC, management must coordinate a team effort in which everyone in the work force can contribute meaningfully to the quality effort.

PURPOSE

This book was written with the following goals:

1. To provide an understanding of basic statistical concepts.
2. To present a management philosophy for successful application of statistical process control.
3. To give the student a solid foundation on control charts: setting scales, charting, interpreting, and analyzing process capability.
4. To teach the student the quality concepts and problem-solving techniques associated with statistical process control.
5. To provide a readable source of SPC topics that the student can refer to as the on-the-job need arises.

FLEXIBILITY

The book is designed for use in two-year and four-year colleges, as well as industry. The order of the chapters features a low-level mathematics approach so that anyone with a basic mathematics background can learn the control chart concepts in Chapters 1 through 10 and the problem-solving concepts in Chapter 11.

The book is mathematics-friendly:

- Only the needed mathematics is presented.
- The mathematics knowledge that is required for each topic is reviewed at the introduction of the topic.

The entire book contains enough material for a three-credit-hour course. The mathematics prerequisite for someone studying the entire book should be elementary algebra.

The recommended sequence for college is Chapters 1 through 13, with the basic algebra in Appendix A reviewed at the beginning of Chapter 3. The probability section in the Appendix is optional and can be taught with the introduction to probability in Chapter 5 or with the applications of probability in Chapters 8 or 10. One possible variation in the sequence would be to teach Chapter 10 after Chapter 5. Then all of the out-of-control patterns would be available for analyzing the control charts presented in Chapters 6 through 9. The book sequence introduces a few basic out-of-control patterns for use in the presentation of control charts, followed by a more comprehensive analysis after all the control charts have been introduced.

The recommended sequence for industry is Chapters 1 through 11. Chapters 12 and 13 are more job-specific and may be taught to particular groups. The probability in Appendix A can be taught at any time if a more thorough understanding of the probability concepts in chart interpretation or sampling is desired. The basic algebra in Appendix A can be taught at the beginning of Chapter 3.

EXAMPLES AND ILLUSTRATIONS

The examples have been carefully chosen to provide a thorough understanding of the concepts involved. A detailed, step-by-step format has been used throughout the book to provide a pattern that can be used effectively, both for the immediate problems and for future reference. The examples feature worksheets and control charts to be filled in by the student and completed worksheets and charts for checking results. Control chart masters have been included at the end of the solutions manual.

ACKNOWLEDGMENTS

I thank the following reviewers for their comments and suggestions: Paul Wright, Paul Wright Consulting Services in Ft. Wayne, IN; George E. Brown, Jackson State Community College; Hank Campbell, Ph.D., Illinois State University; David H. Devier, Owens Community College; and Marty Hodges, Colorado Tech.

1

INTRODUCTION TO QUALITY CONCEPTS AND STATISTICAL PROCESS CONTROL

OBJECTIVES

- Know the definitions of quality.
- Know the statistical signals that are used to improve a manufacturing process.
- Differentiate between the detection model and the prevention model for quality control.
- Identify the goals for using statistical process control (SPC).
- Learn the techniques that utilize SPC.
- Identify several positive effects of SPC.
- Learn the problem-solving model that utilizes SPC for process improvement.
- Describe the important aspects of quality.
- Know the seven tools for SPC.
- Describe how designed experiments are used.

1.1 WHAT IS QUALITY?

Quality can mean different things to different people and can be interpreted in a variety of ways by an individual. Quality may be thought to have two main divisions: the quality of a manufactured product and the quality of services received. From a manufacturing standpoint quality is simply conformance to specifications. The ultimate customer could describe quality as fitness for use. When trying to edge out the competition, quality can be interpreted as producing the very best product or providing the very best service. In some industries a set of classifications have been established by *design quality.* For example, several levels of design quality exist in the automotive industry, from top-of-the-line luxury models down to economy cars. At each level, however, the buyer would expect good *conformance quality.* In fact, auto manufacturers encourage in-class comparisons to show that they have the best conformance quality in their class. Buyers who are not pleased with the overall quality of a specific model car are encouraged to "step up" a class or two (for more money, of course!). In the service sector, the hotel industry provides a good example of differences in design quality. All hotels and motels provide a place to sleep, but many features of design quality, such as services available, comfortable to luxurious surroundings, exercise rooms, pools, and hot tubs, separate the bargain hotel from a five-star hotel. Companies that produce products at the higher levels of design quality and companies that produce products for a market that has primarily a single level of design quality would be more inclined to use the combination of the two categories which stresses excellence in the quality definition.

Quality can also be linked to customer satisfaction. Some companies have used that definition for years, but there is now a broad move toward defining quality as total customer satisfaction. To use that definition, a company must know its customer, and in the multilevel markets, it must know the customers at each level for which it produces. The customer is becoming the driving force for quality.

Many companies that initially aimed at improving the quality of their products found that to satisfy the final customer, it was necessary to satisfy a whole sequence of internal customers. Each person involved in the manufacturing process received a partially completed product, performed the assigned operation(s), and passed it on to the next person. At each step in the process the internal customer had to receive a quality product and pass on a quality product. Away from the manufacturing area, those responsible for the order entry, shipping, and billing were also involved in achieving total customer satisfaction. Again, the final customer cannot receive a quality product unless the associated internal customers receive a quality product as well. This awareness led to the concept of total quality, that there are no exceptions to producing quality work. Everyone in a manufacturing environment or in the service sector has customers, internal or external, and must maintain total customer satisfaction.

One more recent development in the definition of quality is that of exceeding the customer's expectations. When the service is so good that the customer feels "special," when a product has an outstanding feature, or when the combination of product, service, and delivery leaves the customer truly amazed, the customer's expectations have been exceeded. Exceeding customer expectations has been extremely effective in building a loyal customer base. It is estimated to be five to seven times more costly to attain a new customer than it is to retain a current one, so it makes a lot of sense to go that extra step.

Quality is

1. Fitness for use
2. Conformance to specifications
3. Producing the very best products
4. Excellence in products and services
5. Total customer satisfaction
6. Exceeding customer expectations

Total quality in an organization means simply that quality work is expected in every job. There are no exemptions. When something is done, it should be done right the first time. When a product is made, it should be defect-free. When a service is provided, the customer should be pleased with the result. Total quality has evolved as a necessary process for delivering a quality product or service. (Figure 1.1).

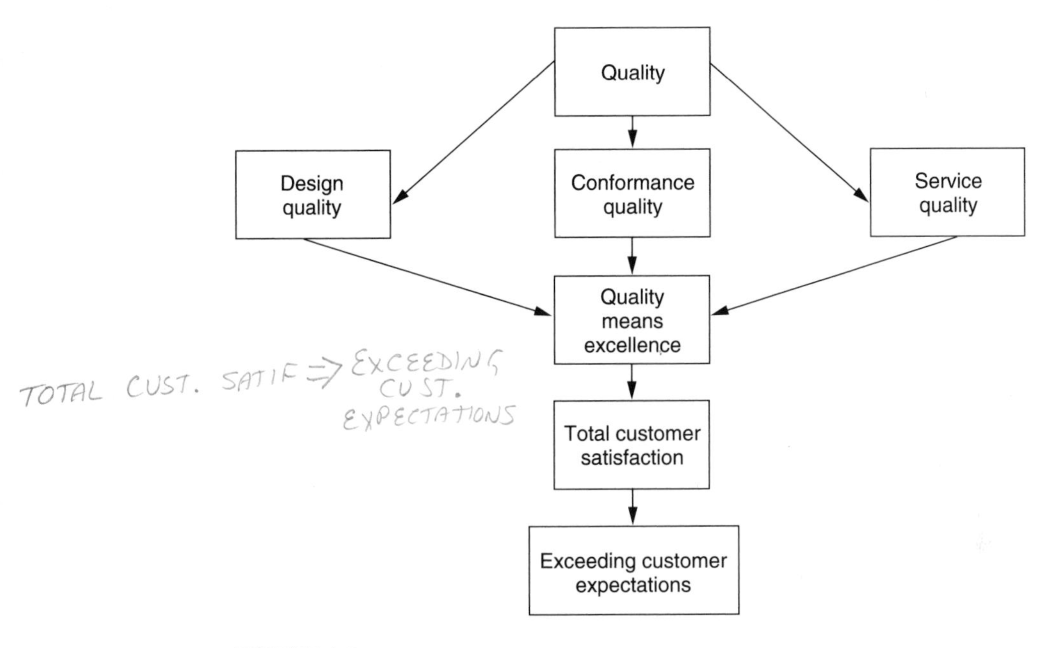

FIGURE 1.1
Quality definitions.

1.2 THE NEED FOR SPC

Dedication to constant improvement in quality and productivity is needed to prosper in today's economic climate. Yesterday's standards are not good enough. A company's product has competition from companies throughout the world because modern communication and transportation have created a world marketplace. The quality of a product has to be *world class,* as good as the best in the world, in order to compete. Consumers are looking for the best combination of price and quality before they buy. The "Made in America" label is no longer enough to sell a product. If U.S. companies want to succeed in a world market, their products must be competitive. Today each company employee must be committed to the use of effective methods to achieve optimum efficiency, productivity, and quality to produce competitive goods.

When a company produces a product or service it utilizes many interralated processes and each process involves several to many steps to accomplish a specific task. There may be several sources of data available as the task is undertaken; measurements that have to be within specified limits or outcomes that can be judged acceptable or not acceptable. All the different processes are combined to yield the final product or service.

> *Statistical Process Control (SPC)* is a procedure in which data is collected, organized, analayzed and interpreted so that a process can be maintained at its present level of quality or improved to a higher level of quality.

- *Collection of data:* The measurement data from a process is usually collected in small samples. Large samples are generally used to determine the proportion of non-conforming items, or services or to analyze shipments from suppliers.
- *Organization of data:* The data is organized in tables, charts and/or graphs.
- *Analysis of data:* Average values are calculated and the spread of the data is determined.
- *Interpretation of data:* The graphs and charts are interpreted to determine if the process is acceptable. The first check is to see if the process is in statistical control; i.e. the process is running as good as can be expected, given the prescribed procedures. The second check is to see if the product or service meets the specifications or prescribed quality level.

SPC can be applied wherever work is being done. Initally it was applied to just production processes, but it has evoloved to the point where it is applied to any work situation where data can be gathered. As companies work toward a total quality goal, SPC is used in more diverse situations.

SPC involves the use of statistical signals to identify sources of variation, to improve performance, and to maintain control of processes at higher quality levels. The statistical concepts that are applied in SPC are very basic and can be learned by everyone in the organization. All workers must know how SPC applies to their specific jobs and how it can be used to improve their output. Supervisors must be aware of the ways SPC can be used in their sections, be prepared to help their production workers utilize SPC, and be receptive to suggestions for improvements from the workers who are effectively using SPC. Managers must know how SPC can be used to improve quality and productivity simultaneously. When supervisors recommend process changes based on SPC analysis and interpretation, managers must understand the SPC concepts in order to make knowledgeable decisions. They must create and maintain a management style that emphasizes communication and cooperation between levels and between departments. Their goal must be to develop a working atmosphere that maximizes everyone's contribution to the production of competitive products.

1.3 PREVENTION VERSUS DETECTION

One of the major problems in manufacturing today is that some companies' version of quality control is simply to find defective items after they are made and remove them before shipment to the customer. When mistakes are found in service industries, documents are redone or corrected. If the mistake involved the customer, apologies must be made. These are examples of trying to achieve quality through the *detection* method. The quality of the system has not improved, even though some inferior products were weeded out or some mistakes were found and corrected. SPC, on the other hand, leads to a system of *prevention*, which will replace the existing system of detection. Statistical signals are used to improve a process systematically so that the production of substandard materials is prevented or the procedure that led to mistakes is improved.

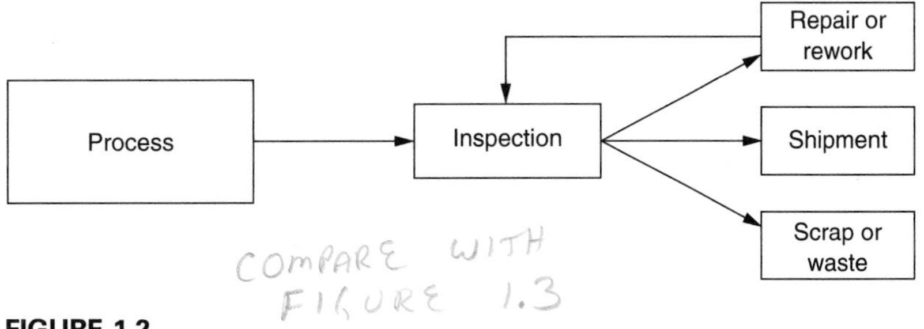

FIGURE 1.2
The detection model.

Detection models, as Figure 1.2 illustrates, usually rely on a corps of inspectors to check the product at various stages of production and catch errors. This quality control method is inadequate and wasteful. Money, time, and materials are invested in products or services that are not always usable or satisfactory. After-the-fact inspection is uneconomical and unreliable. Inspection without analysis and subsequent action on the process neither improves nor maintains product quality. Inspection plans cannot find all the defective items, and the waste is appalling! The company pays an employee to make the defective item and then pays an inspector to try to find it. If the inspector finds it, the company pays another employee to fix it. Also, defective products that are not found lead to warranty costs, reputation damage, and canceled orders. Unless action is taken to correct a faulty process, the percentage of the output that is defective will remain constant.

EXAMPLE 1.1

How good are your inspection skills? In the following paragraph, treat the letter *f*, whether capitalized or lowercased, as a defective item. Inspect the paragraph and count the number of defective items.

> The study of SPC can be both fun and rewarding for everyone. When you find out that the fundamental ideas of statistics are fairly easy to learn, you will discover that your efforts result in a great deal of satisfaction. If you treat a production problem as a puzzle, the application of SPC provides clues for its solution, and when the puzzle is finally solved, the feeling of satisfaction is very fulfilling. Puzzles can be frustrating, but their final solution is fun.

Solution

There are 23 defective items. Did you find them all the first time? The second time? This example illustrates one of the problems with inspection: It is easy to miss defective items. The results can actually be worse when two inspectors are used, because the first inspector expects the second inspector to find the missed defectives and the second inspector assumes that the first inspector found all the defective items. Shared responsibility can mean that no one takes responsibility.

One of the statistics lessons taught later in this text emphasizes that unless improvements are made on the manufacturing process, the percentage of defective products that are produced now, next week, and next year will remain substantially the same. That is why it makes sense to avoid producing the defective item in the first place! This is the basis for the prevention model.

The prevention model uses statistical signals at appropriate points in the process to improve the procedure and to maintain control at the improved level. Statistical signals provide an efficient method for analyzing a process to indicate where improvements should be made to prevent the manufacture of defective items and to improve the quality of the items produced. This is illustrated in Figure 1.3.

Prevention avoids waste! If the product is not flawless the first time, fix the process so the products will be right the next time. Monitor the process so that needed adjustments can be made before quality suffers.

Statistical process control is becoming the core for both quality improvement and quality maintenance. Important decisions from optimum adjustment time decisions made at the shop-floor level to process change decisions made by management involve SPC. Statistical methods and techniques, such as control chart analysis of a process or its output, are now being used extensively to make economically sound decisions. The process analysis leads to appropriate actions for achieving and maintaining a state of statistical control and for reducing process variability.

A major obstacle to achieving high quality is product variability. Design quality may differ between products—for example, the Lincoln Towncar has design quality superior to that of the Ford Escort—but the demand for quality exists within each design category. All cars should have consistently high conformance quality regardless of their competition, and that quality can be achieved only by decreasing the variability of all component parts.

Statistical process control can improve quality by reducing product variability and can lead to improvements in production efficiency by decreasing scrap and rework. It can be used to monitor a process to determine when substandard items are about to be produced so that adjustments can be made to prevent the production of defective items. However, SPC is primarily a trouble indicator. For each statistical application, such as a control chart or histogram, there is an expected form or pattern, and when the actual form or pattern differs from the expected, it is usually a signal that a problem exists. The

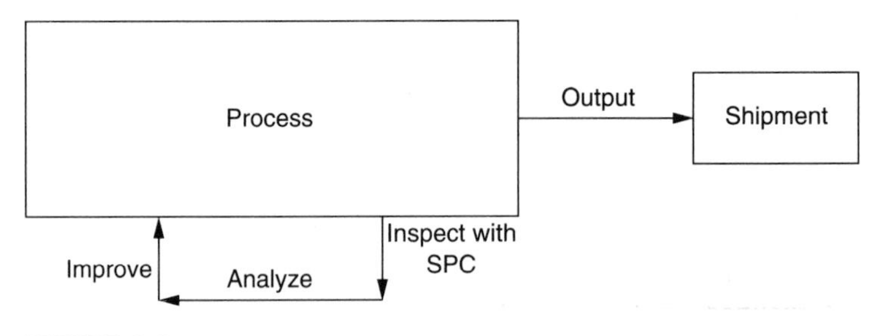

FIGURE 1.3
The prevention model.

potential problem must then be investigated and eliminated. So, SPC by itself will not improve quality; only the appropriate action to its signals can improve and/or maintain quality.

1.4 SPC GOALS

The following are the primary goals of SPC:

1. Minimize production costs. This is accomplished with a "make it right the first time" program. This type of program can eliminate costs associated with making, finding, and repairing or scrapping substandard products.
2. Attain a consistency of products and services that will meet production specifications and customer expectations. Reduce product variability to a level that is well within specifications so the process output will match the desired design quality. This consistency leads to process predictability, which benefits the company by helping management meet quantity targets.
3. Create opportunities for all members of the organization to contribute to quality improvement.
4. Help both management and employees make economically sound decisions about actions affecting the process.

Statistical process control can be used by both management and production people because it includes statistical methods that utilize the expertise of all employees of the company for problem solving. Management can use SPC as an effective tool for reducing operating costs and improving quality by using its methods for organizing and implementing the quality effort. The various processes can be charted and product flow can be analyzed at any point of the process. Bottlenecks and quality problems can be pinpointed for management planning and solutions. The capability of the processes to attain desired quality levels could be analyzed statistically for management decisions regarding overhaul or replacement of machinery. Analysis of control of charts can lead to training, retraining or re-assignment decisions regarding production workers. Management decisions on recommendations for process changes can be based on a combination of statistical analysis and cost analysis. The entire process becomes predictable so managers can achieve better planning for quantity targets. SPC creates a new management philosophy: Lines of communication are opened among all employee levels for the betterment of the company and the product.

Statistical process control also works for production employees. These employees can use it to develop effective tools to work more efficiently (not necessarily harder). When employees learn SPC, they work more intelligently. They know from their control charts when they're doing a good job. SPC gives them the opportunity to influence work operations and be responsible for their jobs. It can help increase employees' pride in their work by allowing them input into the production process. Production workers are often the most qualified employees to determine what is right or wrong with their particular step in the process. As contributing members of a process control team, they can help with quality improvement.

Both management and employees in the service sector can take advantage of SPC techniques to analyze processes and procedures. Processes may be streamlined to save employee hours. Procedures that lead to mistakes may be changed so that the incidence of mistakes is reduced or eliminated. Employee involvement in the use of charts and check sheets can lead to valuable input in improving the service.

1.5 THE BASIC TOOLS FOR SPC

The following tools are described briefly. They will be presented in detail in Chapter 4.

1. *Flowchart:* The entire process is diagrammed from start to finish with each step of the process clearly indicated. All involved in the process should know their position(s) on the flowchart and at least a partial upstream and downstream trace from their position(s). All should know who their suppliers are and who their customers are in the process flow.
2. *Pareto chart:* The number of occurrences or the costs of occurrences of specific problems are charted on a bar graph. The largest bars indicate the major problems and are used to determine the priorities for problem solving.
3. *Checksheet:* A data-gathering sheet is prepared that categorizes problems or defects. Checksheet information may be put on a Pareto chart or, if a time analysis is included, may be used to investigate problem trends over time.
4. *Cause-and-effect diagram:* A problem (the effect) is systematically tracked back to possible causes. The diagram organizes the search for the root cause of the problem. A similar diagram can be used to systematically search for solutions to a problem.
5. *Histogram:* A bar graph shows the comparative frequency of specific measurements. The shape of the histogram can indicate that a problem exists at a specific point in a process.
6. *Control chart:* A broken-line graph illustrates how a process or a point in a process behaves over time. Samples are periodically taken, checked, or measured, and the results plotted on the chart. The charts can show how the specific measurement changes, how the variation in measurements changes, or how the proportion of defective pieces changes over time. Control charts are used to find sources of special-cause variation (variation that is caused by specific, fixable occurrences), to measure the extent of common-cause variation (variation that is inherent in the process), and to maintain control of a process that is operating effectively.
7. *Scatterplot:* Pairs of measurements are plotted on a two-dimensional coordinate system to determine if a relationship exists between the measurements.

1.6 STATISTICAL PROCESS CONTROL TECHNIQUES

Essential techniques in SPC include the use of

1. Process control charts to achieve and maintain statistical control at each phase of the process

2. Process capability studies that use control charts to assess process capability in relation to product specifications and customer demands
3. Statistical sampling as part of a self-certification plan for vendors
4. Gauge capability studies
5. The seven SPC tools for problem solving

Statistical process control is an important tool and leads to many process improvements and positive process results, such as

- Uniformity of output
- Reduced rework
- Fewer defective products
- Increased output
- Increased profit
- Lower average cost
- Fewer errors
- Higher quality output
- Less scrap
- Less machine downtime
- Less waste in production labor hours
- Increased job satisfaction
- Improved competitive position
- More jobs

The preceding list is interactive: When SPC helps create a product that is low in variability and well within specifications, the output is more uniform and of higher quality. That automatically means that there are fewer defects to rework and less scrap, so output and profit increase. The use of SPC by production personnel can show the way toward refining the process and ironing out mistakes. More work time and machine time can be geared toward making good products instead of repairing faulty ones, so costs decrease. This leads to a lower average cost, an improved competitive position, and more jobs because demand for the company's product will increase.

CASE STUDY 1.1

Consultant Ray Milton[1] reports how EIS Brake Parts, a division of Standard Motor Parts Incorporated of Middletown, Connecticut, introduced SPC to its operation.

> The key to management's application of SPC was starting small. When the first results showed positive, it was easy to expand the program a step at a time to other areas. The first application of SPC took place in the wheel cylinder department. The scrap rate of one of the parts, a master cylinder, was reduced from 30 percent defective to zero defects in 10,000 pieces inspected. The overall scrap rate for the department was reduced from a high of 1.1 percent to 0.3 percent over a seven-month span.

[1]Ray Milton, "SPC: A Search for Zero Defects," *Modern Machine Shop* (May 1988), 60-67.

The application of SPC helped to identify faulty parts, processes, and machines. In one case it revealed that a turning machine required a new set of jaws. SPC slowly began to improve the individual products and the manufacturing process itself. It resulted in reduced scrap, higher quality, increased productivity, and lower costs.

Statistical process control must be adopted as an integral part of a long-term policy for continuous improvement in product quality and productivity. If SPC is limited to the use of control charts only, positive results will be rather limited. There is no quick fix for quality problems. SPC charts and techniques will show where problems exist and provide hints of the problem cause. Management must establish a responsive action process. SPC can be applied in any area where work is done: It is used to solve problems in engineering, production, inspection, management, service, and accounting. To be truly effective, however, SPC must become an important segment of corporate life as part of a total quality control program. It is an integral part of a new philosophy for doing business. Management has to change its traditional top-down approach and create, through proper training, a structure that is fully intralevel cooperative. New lines of communication must be formed, responsibility must be realistically assigned where the work is done, and a spirit of full cooperation for the good of the company must prevail.

1.7 APPLYING SPC TO AN EXISTING MANUFACTURING PROCESS

When SPC is introduced to a process, the initial thought should be about the quality characteristics. What are they and what are their associated measurement requirements? In most situations a statistical quality control (SQC) analysis would be done first by identifying the quality characteristics of the product or service and determining how those characteristics can be controlled. As an example, when an automobile transmission or a four-wheel drive transfer case is made, one of the quality characteristics is the noise (or lack of it) at various speeds. The SQC analysis of the noise at specific speeds could then lead to several SPC charts of critical measurements in the manufacture of the gears that would contribute to the noise factor.

The following steps briefly outline the SPC procedure. A more complete and detailed set of steps are presented in Chapter 6.

Step 1. The initial step toward applying SPC is diagramming and analyzing the process to decide where control charts may best be applied. Brainstorming sessions with representatives from all levels—from the shop floor to management—can be effectively used to create a thorough picture of the process. Useful tools for this procedure include flowcharts (the step-by-step process flow), cause-and-effect diagrams (an organized list of potential problem causes), and SQC analysis results. If different types of defects occur, Pareto charts can be used to prioritize them. The critical measurements to be charted should be identified.

Step 2. Decrease any obvious variability in the target process.

(CUT OBVIOUS VARIABILITY)

Step 3. The third step involves statistically testing the gauges using a gauge capability study. This must be done before measurements are taken for control charting. The variation that shows up on the control charts must reflect the process variation that needs to be reduced. If gauge variation is present in excessive amounts (there will always be some gauge variation), causes of the process variation cannot be isolated. Gauge capability is presented in Chapter 12. (GAUGE CALIBRATION)

Step 4. Make your sampling plan. Determine the size of the sample and when the samples are to be taken. (FORM A SAMPLING PLAN)

Step 5. The fifth step dictates the use of control charts to get the process in statistical control. There are two main sources of variation: One is embedded in the system and is referred to as *common-cause* variation, and the other, which should be eliminated at this step, is called *special-* or *assignable-cause* variation and can be eliminated by the process control team. The team uses the control chart to find an out-of-control situation, evaluates what happened at that specific time to cause it, and then works to prevent that cause. This procedure continues until the control chart indicates that there are no more special-cause variation problems. By this time the process is running as well as it possible can without process modifications and is said to be in *statistical control.*

(USE OF CONTROL CHARTS)

Step 6. The sixth step is to put the operator in charge. This step and step 5 actually occur simultaneously because the operator should be doing the control charting and attaining statistical control with the help of the process control team. The operator must know what this step entails:

- The operator must be taught what is "right."
- The operator needs the means for determining if that phase of the process is right.
- The operator must be able to change what is wrong in order to achieve statistical control.
- The operator is *responsible* for that specific phase in the process.

Meeting these four criteria is referred to as *ownership.* When operators are given ownership of their part of the process, a sense of pride and responsibility may be engendered that often results in higher quality work.

(OPERATOR OWNERSHIP)

Step 7. The seventh step is applied once the process is in statistical control following steps 5 and 6. This step is used to determine how capable the process is according to product specifications and customer expectations. A process capability study will measure the extent of common-cause variation, or the inherent variability in the process, for comparison with the allowable variation given in the product specifications.

(PROCESS CAPABILITY)

Step 8. The eighth step is designed to improve the process. Eighty-five percent or more of the process problems are handled at this stage, according to quality consultant W.

Edwards Deming.[2] Changes in the process require management action. Teamwork and brainstorming sessions involving representatives from all levels of the operation can determine probable causes of excessive variation in the process. Process changes can be analyzed on control charts either singly or in variable interaction studies for signs of process improvement. Designed experiments may also be used in the search for improvements. When improvements are found, management must follow through and see that the appropriate changes are incorporated in the process *without backsliding.*

Step 9. The ninth step calls for a switch to precontrol, a monitoring technique that compares a measurement with target and warning measurements, when the process is in control and capable. This is a much simpler system and is quite effective for monitoring a process. Precontrol charts are presented in Chapter 8. An occasional return to control charts can provide a check on process capability. Another alternative at this step is to continue using control charts but increase the time interval between samples.

Step 10. Quality improvement is a continuous process. Two things should be done at this step. First, continue to look for ways to improve the process at hand and, second, return to step 1 for the next critical measurement.

Historically, many companies did not begin using SPC until they were forced to. Either they could see their competitive position diminishing or they were obliged to meet their customer's requirement that contracts would not be awarded until their workforce was trained in SPC. Unfortunately, some companies just met the minimum requirement of providing basic SPC training and discovered that it wasn't good enough. A follow-up was needed to show how the various SPC techniques would apply to specific jobs. The supervisors needed more intensive training because they needed to know how SPC could be used at each job in their area of responsibility. They also had to provide "coaching" or reminders to the people in thier area on SPC applications. Workers recommendations for process changes go through the supervisors and they both have to thoroughly understand the SPC interpretations in order to present the recommendation to management. The larger companies now have on-going SPC training because of worker turnover: both new workers coming in and workers moving to new jobs within the company need specific SPC application training.

1.8 DESIGNED EXPERIMENTS

Many different variables can affect an output measurement. For example, a machining operation has to cut a groove in a cylindrical shell. Important characteristics would include the position of the groove on the shell and the depth and width and the consistency of the depth and width of the groove. Some of the variables that could affect the outcome would include machine speed, tool quality and the time since the cutting tool was sharpened, oil flow and temperature control, tool pressure, chip control, quality and

---

[2]W. Edwards Deming, *Out of the Crisis* (Cambridge, MA: Massachusetts Institute of Technology, 1986).

consistency of the bar stock for the shell, and fixture consistency for a multispindle machine. Often a change in one variable will cause a second variable to change. This is referred to as a cause-and-effect relationship between the two variables. Cause-and-effect relationships are difficult to identify, especially when several variables are involved. An excellent method to help identify cause-and-effect relationships is called designed experiments. A designed experiment would find the cause-and-effect relationships and would determine which variables are controllable and which combination of variable measures would yield the best results. Various combinations of speed, tool pressure, oil flow, and tool brand, for example, would be tested statistically to do this. SPC could be applied to the best combinations to estimate longevity of tool wear and the potential process capability (how good it will be with respect to the specifications). SPC can also be applied to a spindle analysis to determine the machine's consistency.

The best use of designed experiments occurs in the development of a process. Then production pressures do not get in the way of process design. Also, thorough planning with Design of Experiments (DOE) will give the best chance of producing a product of superior quality without costly process corrections and changes.

Design of experiments was mentioned in step 8 of the problem-solving process as a way to improve an existing process. It is rather difficult from a production point of view to use the DOE on an existing process because the process output can change for the worse. If the output is going to the customer, 100 percent inspection may be needed to ensure that no substandard pieces are included. If production is temporarily halted while the experiment is run, then it is not a problem and the pieces produced are for analysis only. This application of the DOE is really a "better late than never" approach to quality. However, companies that are just getting started in the quality improvement process will find themselves in this situation.

The goals, techniques, and results of SPC have been discussed briefly in this introduction. The various tools mentioned, such as diagrams and charts, are explained in detail in the following chapters.

EXERCISES

1. What does *ownership* refer to?
2. What is the detection model for quality control?
3. What is statistical control?
4. What is the prevention model for quality control?
5. What is common cause variation?
6. What are the primary goals of SPC?
7. What is special cause variation?
8. What techniques does SPC involve?
9. How does reduced rework increase profit?
10. What are some of the improvements that can result from using SPC?
11. Why do managers have to understand SPC concepts?
12. What ten steps are followed when SPC is used in manufacturing?
13. Does the use of SPC guarantee high quality? Why?
14. Give six definitions of quality.

15. Why doesn't the detection model provide good quality?
16. What are the seven basic tools for SPC?
17. Why is the detection method more costly than the prevention method of quality?
18. Discuss the interaction between the various process improvements that can occur when SPC is used.
19. Why has the total quality concept evolved in most companies?
20. Why is the "complete customer satisfaction" definition of quality an all-inclusive definition?
21. Why isn't just basic SPC training enough for the workforce?
22. What are designed experiments?
23. Why is it necessary for supervisiors to have the most thorough SPC training?
24. Why is it better to use designed experiments when a process is being developed than on a process that is in operation?
25. Why is on-going SPC training needed?
26. Choose seven items you have or would like to have and give a definition of quality for each.
27. What does process capability refer to?
28. Choose five specific service organizations that you are familiar with (stores, restaurants, etc.) and evaluate them as follows:
    - **a.** Where do you expect quality service? (There may be several situations within each establishment.)
    - **b.** Define what quality service is in each organization.
    - **c.** What could each organization do to exceed your expectations?
29. What is a checksheet?
30. Give an example of an internal supplier and an internal customer in
    - **a.** A manufacturing environment.
    - **b.** The service sector.
31. What is a flowchart?
32. How does SQC differ from SPC?
33. Why does the prevention method lead to better quality?
34. Give an example of how "ownership" can improve quality:
    - **a.** In manufacturing.
    - **b.** In service.

2

STRIVING FOR QUALITY: MANAGEMENT'S PROBLEM AND MANAGEMENT'S SOLUTION

OBJECTIVES

- Discuss the historical background to management's problem.
- Know Deming's 14 points for management.
- Know Deming's Seven Deadly Diseases.
- Learn Crosby's symptoms of trouble.
- Define Crosby's absolutes of quality.
- Discuss Crosby's 14 steps to quality improvement.
- Identify the major changes needed to institute a quality process.
- Know the major pitfalls that must be avoided when starting the quality process.
- Understand total quality management (TQM).
- Know the total customer satisfaction concepts.
- Know how the multiplier effect works with customers.
- Know the concepts, purpose, and criteria of the Baldrige Award.
- Know the function of the five basic parts of the ISO-9000 series and the importance of the standards.
- Know the extent of the service sector in America's workforce.

2.1 MANAGEMENT'S PROBLEM

During the 1970s, U.S. companies noticed a dramatic loss of market share to Japanese competitors in many different product areas. The most noticeable losses occurred in the electronics and automotive industries. Comparative product analysis and consumer surveys indicated that the Japanese products were superior in quality and lower in price.

Visits to various Japanese companies revealed that they were basing decisions on statistical signals, most notably SPC charts. Consequently, many U.S. companies instituted crash courses is statistical process control for their employees. The technical literature in the 1980s contained many SPC success stories for U.S. industries, but it also showed many failures. Since the remarkable success of the Japanese had been attributed to their widespread use of statistical signals, why didn't that application work for *all* the U.S. companies that tried it?

Comparisons have been made between Japanese and U.S. schools, families, workers, and management techniques to try to pinpoint critical differences and discover the secret to Japanese success. The answer that has slowly emerged is that SPC and other basic applications of statistics are absolutely necessary to achieve top quality, but they must also be combined with the appropriate management methods.

2.2 MANAGEMENT'S DILEMMA

American management techniques have been developed by 40 years of minor competition, quantity pressures to meet a seemingly unquenchable, worldwide demand, and a secondary regard to quality. Companies took a "weed out the bad ones, but keep that production rolling" attitude.

Management–employee relationships have been primarily adversarial in the United States. This relationship stems from a history of big business exploitation of workers and the reactionary pressures of unions. Many companies have become accustomed to a constant power struggle between union and management. There is so much distrust that confrontations over contracts have led to plant closings and lost jobs.

An additional attitude problem hinders labor–management relations. Artisan skills long ago gave way to repetitive assembly line work. Workers were given relatively little training to do a very specific job. They were not credited with much knowledge of the process in which they were involved. The difference in education levels of management and the work force, combined with this process of deskilling jobs, has brought about a general underestimation of employees' intelligence and abilities. In many companies the worker has become the quality scapegoat. Management is convinced that poor quality stems from the faults of inadequate workers.

The educational lines that lead to management positions include primarily business and engineering. Neither curriculum, in all its variations, has traditionally included coursework on quality achievement or quality control (QC). Quality and QC were always relegated to checkers and inspectors to identify the bad products, both incoming and outgoing. Quality and productivity were thought of as opposing attributes. Slow careful work was thought to lead to high quality and low productivity, whereas fast work was associated with high productivity and low quality. The concept that high quality may be accompanied by high productivity and low production costs is completely foreign to many Chief Executive Officers (CEOs).

U.S. managers, with their heritage of disregard for quality, their noncommittal attitudes toward employees, and their adversarial management techniques, are now in a quandary. They are being economically defeated by companies that have a sense of fairness in their management style, respect and admiration for the abilities of their employees, and a firm commitment to competitive quality. Changes must be made in order for U.S. companies to claim their share of the top in an emerging global economy.

Management can no longer pass the buck and blame poor quality on workers. According to W. Edwards Deming, only about 15% of the quality problem can be attributed to the worker.[1] The other 85% of the problem is embedded in the manufacturing process, and nothing can be changed with the process unless it is management directed.

Another favorite management dodge is the claim that U.S. companies cannot compete with foreign companies because the U.S. wage scale is too high. The wage discrepancy between the U.S. industries and their foreign competitors can often be counterbalanced by a combination of better technology, a better educated work force, and better management methods all focused on producing a top-quality product. Some U.S. companies have already turned things around: Their quality has improved, and they have beaten the competition. Some U.S.-run plants have been closed, bought by Japanese firms, and reopened successfully with Japanese top management and the same American workers. Sharp Electronics of Japan did that with a company in Tennessee, and that plant has produced well over a million TV sets and microwave ovens since the change. A turnaround can be accomplished in U.S. industry, but management must lead the way.

2.3 LEADERSHIP BY MANAGEMENT

The importance of the leadership role that management must take cannot be overstated. The commitment to making quality products and/or providing quality service can't be delegated; top management must lead the way. There is such a dramatic change in the way a company is run when it makes a commitment to quality that anything less than total dedication will eventually erode the quality effort.

Statistical thinking must permeate all levels of management.[2] Managers in the company must realize that all the work done in their area and in related areas forms a series of interconnected processes. Those processes are subject to variation. It is the manager's job to become aware of the sources of variation in these processes, to be able to classify them as special-cause or common-cause variation, and to treat them as opportunities of improvement. Managers must gather data from their processes, analyze them with the appropriate statistical tool(s), and base their decisions on that analysis. If any level of management is omitted in this total quality format, managers will eventually be making decisions counterproductive to the company's quality effort.

The features of statistical thinking are listed as follows:

- All work is part of interconnected processes.
- Variation is always present in a process.

---

[1]W. Edwards Deming, *Out of the Crisis* (Cambridge, MA: Massachusetts Institute of Technology, 1986).

[2]Lynne B. Hare, Roger W. Hoerl, John D. Hromi and Ronald D. Snee, "The Role of Statistical Thinking in Management" (*Quality Progress,* February 1995).

- More than 85% of the problems are due to the variation in the processes.
- Process decisions (quality improvement) must be based on appropriate data from the process.

Good managers base their decision on hard information i.e. relevant data. When they use statistical thinking they identify what sources(s) of data is (are) most relevant. The data is organized, analyzed, and the variability in the data must be classified as common cause variation or special cause variation. The special cause variation leads to a specific source of variation that can be eliminated. Common cause variation, however, can't be eliminated, only reduced by improving the process. If the type of variation is misclassified, time, money and effort is wasted looking for a source of variation that doesn't exist, or quality may suffer by making unnecessary changes in the process when a source of variation could be found and eliminated.

(pg. 50)

2.4 DEMING'S CONTRIBUTION TO QUALITY

W. Edwards Deming is perhaps the consultant most responsible for reawakening U.S. industry to the need for higher quality. The initial quality breakthrough was created by Walter Shewhart in 1924 when he developed the concept of the control chart while working for Bell Labs, the research division of AT&T. Western Electric, the manufacturing division of AT&T, was the first to use control charts. Additional use of control charts and other statistical methods developed slowly in the 1930s and then received a big push during World War II. Deming became closely associated with Shewhart during the 1930s and was instrumental in the nationwide training program of teaching statistical methods to industries involved in the war effort. After the war, however, American industry was intact while much of the rest of the world's industry was devastated. The resulting demand for goods put quality on the back burner. Companies resorted to produce, inspect, and ship. Some managers who initially developed their management skills and techniques during this period are still resisting the change to a quality emphasis.

In 1950 Deming was invited to Japan by the Union of Japanese Scientists and Engineers (JUSE) to present seminars and lectures on statistical quality control. JUSE had initiated research on quality control, but had trouble adapting statistical theories to industrial applications. Deming's analysis of his impact was as follows:

> Management learned in 1950 what [it] must do to achieve continual improvement of quality. Hundreds of Japanese engineers learned the methods of Walter A. Shewhart. Quality became at once in 1950, and ever after, everybody's job, company-wide, nationwide. . . . Improvement of quality leads to decrease in cost of production, and captures the market. This is the fundamental chain reaction that Japanese industry learned and adopted as a way of life in my visit to Japan in the summer of 1950 and in many subsequent visits. . . . The achievement of Japanese industry toward ever better quality at reduced cost constitutes lessons in management for the whole world.[3]

[3]Jungi Noguchi, "The Legacy of W. Edwards Deming" (*Quality Progress,* December 1995).

Deming's lectures on quality control in 1950 were transcribed and printed and are still in use today. In 1980, at the age of 80, Deming appeared on NBC in the TV special "If Japan Can, Why Can't We?" From 1980 to 1993 he worked with many of America's major industries.

The Ford Motor Company was one of the first American companies to change to a quality-first system of manufacturing. In the following case study, Donald Katz[4] reports on W. Edwards Deming and his efforts to arouse American industry to the need for competitive quality.

CASE STUDY 2.1

Faced with drastic losses, Ford Motor Company became one of Deming's first U.S. customers in 1980, according to Katz. Deming helped to reorganize the entire process that Ford used to make cars. Senior managers attended Deming's seminars, and statistical control methods were used in every department. Deming broke through departmental barriers and had machinists and assemblers work together to determine critical measurements. When the machinists made the parts, the assemblers were then better able to install them. Quality drastically improved.

Team members responsible for building the $3.25 billion production lines for the Taurus and Sable held seminars with suppliers regarding the necessary quality of the incoming parts. They collected 1400 suggestions from line workers and used 550 of them. When the two cars hit the market in December 1985, they were instantly successful. They became an integral part of Ford's remarkable comeback. The company's profits shot way up, to $4.6 billion in 1987.

Walter Shewhart's development of the control chart and his theories on variation, presented in his 1931 publication *Economic Control of Quality of Manufactured Product,* formed the base of Deming's management philosophy. The message that Deming expounded in his four-day seminars, until his death at age 93, was that management must understand how variation is present in every process. There are two types of variation: (1) special-cause variation, which can be attributed to a specific source and can be eliminated completely, and (2) common-cause variation, which is imbedded in the process that management created. Common-cause variation can never be eliminated; there will always be some variation in every process. But, variation is the enemy of quality; whenever variation is decreased, quality is increased. It is management's never-ending job to constantly investigate processes, identify process procedures that contribute to variation, and improve the process so that variation is decreased.

Deming's message emphasizes that quality starts at the top of the corporate structure and pervades every phase and level all the way to the shop floor. Without top management's unequivocal commitment and constant pressure for higher quality, a company's efforts will usually be wasted.

Deming contends that management style has to change. Companies must work strategically to beat the competition. Each employee in the company must be a contribut-

---

[4]Donald R. Katz, "Coming Home" (*Business Month,* October 1988), 57–62.

ing member of the company team, dedicated to making top-quality products. Many companies have an operation structure in which supervisors and engineers are responsible for making production improvements. However, these workers are usually so involved in solving the company's never-ending sequence of daily problems that they can devote very little time to improvements. Companies that initially thrived because of technological improvements in a product are stalled when it comes to new innovations and advancement in technology. Instead of falling behind the competition, these companies should get more people involved in process improvements. In fact, *everyone* involved in a process should take part in process improvements. Companies need a new teamwork structure that will actively involve all employees. This *is* possible, and it has already happened in some companies, but management has to lead the way. Management must organize and orchestrate the company team. Deming has outlined a 14-point management plan to coordinate a company-team approach to top quality and high productivity.

2.5 DEMING'S 14 POINTS FOR MANAGEMENT[5]

Point 1. Create a constancy of purpose to improve quality and service, to become competitive, and to stay in business. Many companies become trapped in a constant sequence of short-term solutions and become overinvolved in immediate problems. It is important to budget for the future. Long-term planning is a necessity and must be based on an unshakable policy of high quality. Continually investigate the possibility for new products, new markets, and new ways to compete efficiently in the present markets. Plan carefully for the necessary training, equipment, and production. Allocate funds for ongoing education, design improvements, and equipment maintenance. Both management and production workers should constantly seek technological improvements of processes. Major technological breakthroughs do occur, but a continuous sequence of small gains in efficiency and minor improvements over a period of two or three years can often be the best path toward beating the competition.

Point 2. Adopt the new philosophy. There is a new economic age of competition in a global economy. Previously accepted delays, mistakes, poor quality, and poor workmanship are no longer tolerated. Every company should be striving to eliminate substandard products. Defects and defective items are not free: The total cost of producing, finding, and disposing of or repairing a defective item exceeds the cost of producing a good one.

> Few companies have grasped the enormity of the task that faces them. Quality management . . . is a holistic philosophy that must be adopted in its entirety if it is to work at all . . . companies spend millions retraining their

---

[5]Dr. Deming's commentary paraphrased by author. Reprinted from *Out of the Crisis* by W. Edwards Deming by permission of MIT and The W. Edwards Deming Institute. Published by MIT, Center for Advanced Engineering Study, Cambridge, MA 02139. Copyright 1986 by The W. Edwards Deming Institute.

> workers, but neglect to educate managers about their role in the process . . . Deming's philosophy has been widely hailed throughout corporate management, yet in the U.S. it has rarely been adopted in its entirety because of the magnitude of change it requires.[6]

Point 3. Cease dependence on mass inspection. Inspection, by itself, is too late and too costly. Scrap, downgrading, and rework do not correct the process: The process will keep producing the same proportion of defective items until process improvements are made. Basic applications of statistics are needed at every phase of a process in order to make necessary improvements or to maintain an acceptable level of quality and production. Inspection must be built in as an integral part of the improvement–maintenance procedure, not as the single control of quality.

Point 4. End the practice of awarding business on the basis of price tag alone. Learn to use meaningful measures of quality as well as price when deciding on purchases. Eliminate suppliers who cannot provide statistical evidence of quality. Both government and industry are being cheated when they follow rules that award business to the lowest bidder without considering quality. A large fraction of the problems that lead to poor quality and low productivity are due to poor-quality incoming parts and materials and low-quality machines and tools.

Purchasing managers should be trained to judge quality based on appropriate statistical evidence. Specifications have to be thorough so that incoming parts and material will blend properly in production assembly. The critical measurements on all incoming parts and material must be known. In some situations it may be necessary to follow a sample of incoming material through the whole production process to be sure that it blends adequately.

To qualify for the contract, suppliers should meet the following standards:

- Their management should be actively involved in the 14-point program.
- There should be evidence of sustained use of SPC. Any vendor who cannot provide evidence of quality will either have higher costs or be sacrificing quality.
- The aim of suppliers should be to improve quality and decrease costs to the point at which customers search them out.
- A dependable source, responsive to needs on a long-term arrangement, is more important than initial price. Economies should result. If two or more suppliers are used on an item, a cutback to one who satisfies these criteria will reduce costs. Quality will improve, too, because variation owing to different suppliers will be eliminated.

Point 5. Constantly and forever improve the system of production and service. Continually search for problems, reduce waste, and improve quality in every phase of the process. Statistical leadership will be required to separate special-cause and common-cause variation and to design and analyze tests for reducing variation.

---

[6]Andrea Gabor, *The Man Who Discovered Quality* (New York: Times Books/Random House), p. 29.

Quality must be built in from the design stage. Teamwork among the designers, manufacturers, and customers is necessary to ensure that the final product does what it was intended to do. Quality control teams should be formed to utilize the expertise from all phases of management and production associated with the process. Combined team effort can be very effective in increasing both quality and production.

Point 6. Institute modern training methods on the job. One of the greatest wastes in the United States is the failure to use the full abilities of the work force. Training must include a thorough understanding of the entire process and an individual's part in it. All workers should know who their customer is and what their customer requires. Their customer is often the next person in the production process, and it is just as important to satisfy that customer as it is to satisfy the customer that eventually receives the final product.

Statistical methods must be used to determine when training is completed. A trainee's output can be sampled and tracked on a control chart. When that chart is in statistical control, the training is completed. An analysis of the trainee's capability can be determined from the control chart and compared to the required (or historical) capability for that job. If the trainee's capability is not good enough, then either a better training process has to be developed or the trainee should be switched to another job. The training results have to be assessed because a variable standard for acceptable work is a big problem in both training and supervision that must be overcome.

Managers must be trained as well. Their work is also part of a process and subject to variation. They should understand how SPC and other basic statistical tools are used to effectively decrease variation and increase quality. They should specifically know how statistical thinking applies to every process they are involved with.

Point 7. Institute modern methods of supervision. Modern supervisors must be regarded as leaders and facilitators, not as overseers. They must emphasize quality first and look for improvements in production quantity within the quality framework. The improvement in quality will automatically result in some improved productivity.

Supervisors must report to management on all conditions that are barriers to quality and need correction, and management must be prepared to respond. Supervisors should help the workers perform quality work.

Point 8. Drive out fear. The economic loss due to fear is appalling. Fear of asking questions, expressing ideas, and reporting trouble can lead to problems with quality and lagging improvements. Employees must develop confidence in management in order to participate fully in the company's quality process. Workers cannot perform their best unless they feel secure. Fear of change, of new knowledge, and of new responsibilities has to be confronted.

> Fear puts an upper limit on improvement. Fear of failure, fear of supervisors, fear of voicing an opinion . . . are the chief enemies of timely action and flexibility. Until fear is minimized or eliminated, experimentation with new concepts will not become commonplace. Information hoarding rather than sharing will be the modus operandi. Delaying tactics will become the norm.

Indecisiveness or safe decisions will prevail. Politics will be a hallmark throughout the organization.[7]

Ken Kivenko, President and CEO of Canada Macaroni Company, lists several indicators of fear in an organization[8]:

- Low scores on employee attitude surveys
- Few employee suggestions
- A chronic lack of questions at meetings
- Multiple approval signatures
- Memos that justify actions
- Padded estimates

Kivenko also suggests several ways to reduce fear in an organization:

- Improve two-way communication at all levels.
- Actively seek comments and suggestions and provide positive feedback.
- Recognize good work.
- Define roles and responsibility, especially with empowerment.
- Emphasize training and retraining as needed.
- Have an open-door policy.

Point 9. Break down barriers between staff areas. People in research, design, purchasing, production, and sales must work as a team to anticipate production problems. A rush into production can often cause costly delays that could have been avoided by a team approach. Process problems can best be solved by quality control teams that have members in all areas associated with the process.

CASE STUDY 2.2

Motor Trend's 1988 Special Issue[9] is devoted to Ford's big turnaround. In the three years previous to 1987, Ford had lost $1 to $1.5 billion per year. One of the major contributors to Ford's new success was reported to be teamwork. The old sequence of design to engineering to manufacturing to assembly, in which each phase was primarily finished before the next began, had been changed. Now there was teamwork and consultation between the various divisions. The result was fewer surprises and impossible situations planned by one division for another to accomplish. The new procedure eliminated many problems and effected a big improvement in quality.

Point 10. Eliminate numerical goals for the work force. Eliminate targets, slogans, posters, and unrealistic goals such as zero defects. That approach makes management appear to be dumping their responsibilities on the work force. It implies that

---

[7,8]Ken Kivenko, "Improve Performance by Driving Out Fear" (*Quality Progress,* October 1994).

[9]*Motor Trend Magazine,* Special Issue, 1988.

properly motivated production workers can accomplish zero defects, higher quality, and higher productivity and ignores the fact that the bulk of the problem lies with management. Instead, offer a plan that has realistic goals.

Statistical methods, such as SPC, can be used to analyze a process. If the process is stable, setting arbitrary (wishful thinking) goals is useless. The only way to achieve a new goal with a stable process is to change the process, so any plan has to include a management-directed process change.

Point 11. Eliminate work standards and numerical quotas. These guarantee inefficiency and high cost. Quotas encourage sporadic work habits such as working fast until the quota is almost met and then easing off for the remainder of the shift, which is detrimental to the quality effort. Piecework is even worse. People are either paid to make defective items or penalized for something that could be the system's fault if defect penalties are involved. What really caused the defect, the worker or the process? There is no piecework in Japan!

A quota is a fortress against improvements in quality and productivity. Work standards, rates, incentive pay, and piecework are manifestations of an inability to understand and provide appropriate supervision. Workers have no incentive to help management improve the process because any process improvement would be followed by an increase in the quota. Incentives can sometimes lead to a decrease in quality. If the quota appears to be out of reach, workers develop a don't-care attitude and careless work may be the result.

> When we give a quota or goal without an in-depth look at the plan for achieving it, we are not really managing . . . Management by objective (MBO) has become an adversary game in which we focus on results. . . . People . . . will perform only [well] enough to look good to their supervisor. They will avoid doing more. If they greatly exceed targets set for them, MBO requires a raise in next year's targets. Workers understand variation better than their managers. They know that results will vary up or down from one measurement to the next. They want to keep their quota as low as possible. In effect, MBO sets a ceiling on performance.[10]

A better alternative to MBO is management by planning (MBP) which focuses on improving the process with the cooperation of the workers.

Point 12. Remove barriers that hinder the hourly worker. Nothing should interfere with the worker's ability to do a good job. Job instructions should be thorough, concise, and understandable, and problems with inspection or gauges should not reflect unfairly on the worker. The supervisor should listen to the worker's observations regarding process problems and report back on management's response.

---

[10]William J. Latzko and David M. Saunders, *Four Days With Dr. Deming* (Reading, MA: Addison Wesley, 1995), 110.

Point 13. Institute a vigorous program of education and training. An organization needs not only good people, but also people who are improving themselves with education. Encourage education: Eliminate the fear of layoffs. As quality and productivity increase, fewer people will be needed on some jobs, but education and retraining can move displaced workers to another department. Education in simple but powerful statistical techniques are required of all people in management, engineering, production, design, quality control, finance, purchasing, and sales.

Point 14. Create a structure in top management that will push every day on the above 13 points. Agree on the direction you want to take. Have the courage to break with tradition and explain to employees via seminars and other means such as newsletters and meetings why the change is necessary and why everyone must be involved. Management will require guidance from an experienced consultant who can teach statistical methods and develop in-house teachers.

Methods for improvement of quality are transferable to different problems and circumstances. Problem-solving principles are universal in nature. There is no quick fix, however. Sufficient education, effort, and time are needed for the 14-point program to pervade and be followed by all levels of management.

In Deming's commentary on point 13, he emphasizes that the fear of layoffs should be eliminated. That's one of the big differences in management style between Japanese and U.S. companies. In lean times, when product demand drops, U.S. companies lay off workers. In Japan, there is a set pattern for lean times:

1. Corporate dividends are cut.
2. Salaries and bonuses of top management are cut.
3. Salaries for middle management are cut.
4. Workers' salaries are cut. Any work force reduction occurs naturally through attrition or voluntary discharge.

Japanese companies also take advantage of low production periods to build for the future. They place more emphasis on long-term projects. For example, when steel production dropped drastically worldwide, the Kawasaki steel plant had the work force build a new-generation blast furnace that would give them an edge when the lean times were over.

Deming's 14-point program has not been widely adopted by U.S. industry. Top managers in many companies simply do not believe it. No more piecework? Eliminate quotas? "We've got production schedules to meet!" Quality? "If the workers would do their jobs right, we wouldn't have that problem!" The trouble is really with management and the production process? "Don't be ridiculous!"

The 14 points are definitely idealistic, and even Deming agrees that to get a company to follow the program effectively can take years. U.S. executives have relied on short-term goals and solutions for so long that they can barely conceive of a 10-year plan that will gradually and fully adapt the 14 points to their company. Too often, companies in economic trouble will shake up their top-management structures or institute new programs and then eagerly look for results in the next quarterly report.

How should a specific company adapt the 14 points? Every company has its existing management style and track record of improvement changes. The 14-point program requires such a drastic change in attitudes and interpersonal relationships that many companies are stymied at the very start. The plan calls for the development of communication skills: people dealing with people. Within-level communication, between-level communication, and interdepartmental communication all have to develop in order for the 14 points to work. Group dynamics and teamwork are needed for planning instead of the "decree from above."

Deming and other quality consultants insist that changes have to start at the top-management level. Real improvement cannot happen any other way. If major changes were initiated at any intermediate level, they would eventually meet with disapproval at some higher management level. Training starts at the top as well; all managers need to understand the concepts and reasons for the recommended changes that will be made when the work force is trained. Top management also has to initiate and promote the new management style and ensure that it is adopted at all levels.

2.6 DEMING'S SEVEN DEADLY DISEASES[11]

1. Lack of constancy of purpose, failure to plan ahead. It is management's job to set the future direction for the company and to communicate that direction to everyone in the company. Management must coordinate the resolution of both the problems of today and the problems of tomorrow. Today's problems encompass

- Production levels
- Budget, sales, profit
- Employment, safety
- Service
- Public relations

Tomorrow's problems include

- New products and new services
- New materials required
- New methods of production
- New levels of product performance

2. Emphasis on short-term profits. "Wall Street creates heavy pressure for short-term, quarter-by-quarter gains. These gains ignore variation, and each quarter they must go up. So managers rely on creative accounting, mergers, acquisitions, tax schemes, for-

[11]William J. Latzko and David M. Saunders, *Four Days With Dr. Deming* (Reading, MA: Addison Wesley, 1995), 118–132.

eign currency swaps, and all sorts of finagling to boost the short-term profit. This disease, unchecked, will ruin our economy."

3. Evaluation of performance, merit rating, or annual review.

$$X + (XY) = \text{Results}$$

"Let X equal the performance of the individual. Let Y equal the performance of the system. Let XY represent the interaction of individual and system. Here's the catch: X is unknown, and Y is unknown. He who can solve a single equation with two unknowns can rate people. It can't be done."

4. Mobility of management. This is the "white knight syndrome." A manager takes on a job in which things are a mess, makes some changes by tampering with the system, and achieves some short-term gains. Then the manager collects the "rewards" and leaves before the long-term problems show up.

5. Running a company with visible figures alone. Visible figures are important, but as Deming quotes Lloyd Nelson, "The most important figures needed for management of any organization are unknown and unknowable." Who has numbers on the multiplying effect of a happy customer or on the opposite effect of an unhappy one? Who has numbers on improvement of quality and productivity from teamwork between engineers, production, sales, and marketing departments? Study the process, not just the results alone.

6. Excessive medical costs. The effect of variation on the necessary time for patient care is often ignored. The result is excessive expense when the average time is not needed and cutbacks in patient care when more than the average time is needed.

7. Excessive legal costs. The combined effect of excessive medical and legal costs can inflate the final costs of products and services to the point where they are no longer competitive.

2.7 CROSBY'S APPROACH

Philip Crosby is probably the most commercially successful quality consultant in the United States. He was the vice president in charge of quality operations for 14 years at ITT before leaving to form his own corporate consulting firm, Philip Crosby Associates. As an integral part of his consulting service, he and his associates run a quality college in Winter Park, Florida, for seminars on various quality topics. Crosby's concepts agree for the most part with Deming's 14 points, but there are a few notable exceptions. The main thrust of the Crosby method is that it provides more of a step-by-step plan for management to follow.

Symptoms of Trouble[12]

How does a company tell if it is heading for trouble? It may surprise some managers that what Crosby lists as a trouble indicator is standard operating procedure in many companies. The following are symptoms of organizational trouble:

1. *The outgoing product or service normally contains deviations from the published, announced, or agreed-upon requirements.* There always seems to be another crisis or problem cropping up. Brush fire management and off-specification waivers are common.
2. *The company has an extensive field service or dealer network skilled in rework and resourceful corrective action to keep the customers satisfied.* The company has a large rework department, a customer relations department, or both. Customer engineers may "customize" the product to customer requirements, but that may be an admission of on-site finishing and debugging. Service organizations also have hot lines and other trouble-handling procedures for system failures.
3. *Management does not provide a clear performance or definition of standard quality, so the employees each develop their own.* The traditional routine follows a pattern that emphasizes a schedule first, cost second, and quality third. Sloppy definitions of quality allow for regular exceptions and a "close enough" attitude.
4. *Management does not know the price of nonconformance.* Generally nonconformance with standards amounts to 20% or more of the sales dollar in production industries and 35% or more of the operating cost in service industries. When the costs of appropriate education and training, with the subsequent savings that always accompany doing things right, are compared with the losses, it is always astounding that action was not taken sooner.
5. *Management denies that it is the cause of the problem.* Executives fail to realize that inadequate attention to quality is the main problem in their company and that quality improvement will eliminate what they currently perceive as the problem.

Crosby believes that U.S. businesses have had problems with quality because they do not take quality seriously. They have to be as concerned with quality as they are with profit. When management respects the rights of customers as much as they do the rights of owners, banks, and stockholders, then consistent quality will be achieved. Management must believe that there is absolutely no reason to deliver a nonconforming product or service. The chief executive officer must be dedicated to customer satisfaction, and that satisfaction has to be measured in a way that can lead to corrective action. Management should establish a companywide emphasis on defect prevention and use that as a basis for continuous improvement. Few managers understand how the SPC techniques work, so it is a top priority that they learn the concepts involved. Statistical signals are used throughout the operation, and in order for managers to provide the necessary leadership, they have to know the

---

[12]Reprinted from Philip B. Crosby, *Quality Without Tears* (New York: McGraw-Hill, 1984), 1–5.

potential applications and reasons that these applications are so important to the quality effort.

Crosby's Four Absolutes of Quality[13]

1. The definition of quality is conformance to requirements. Be sure the requirements are clear, understandable, and, when necessary, accepted. Emphasize a "do it right the first time" attitude. Management must insist that there is absolutely no reason for selling a faulty product to a customer.

2. The system of quality is prevention. The prevention method features corrective changes in the process when problems occur with the product. SPC is used as an integral part of the prevention system. Think ahead. Look for opportunities for error and take preventative action.

3. The performance standard is zero defects. If you do not insist on zero defects from your suppliers, you are telling them in effect that it is acceptable for them to send nonconforming parts and materials. If you do not insist on zero defects from your workers, you're also telling them that it is all right to produce nonconforming products. People are conditioned to accept the concept that to err is human and therefore believe that all humans will err, which is faulty logic. Errors occur in some situations, but not in others; it really becomes a function of importance. Crosby claims, "Mistakes are caused by two factors: lack of knowledge and lack of attention." Education and training can eliminate the first cause, and a personal commitment to excellence (zero defects) and attention to detail will cure the second.

Nebulous standards, such as the Acceptable Quality Level (AQL) for incoming products, have a detrimental effect on concepts of excellence and high quality, but specific standards like zero defects or "do it right the first time" will lead the way to problem prevention.

4. The measure of quality is the price of nonconformance. Quality, as a management concern, has not traditionally been taught in management schools. Quality has been considered to be a technical function, not a management function, because it has not traditionally been evaluated in financial terms. Crosby states that "the cost of quality is divided into two areas—the price of nonconformance (PONC) and the price of conformance (POC)." *PONC* is the total cost of doing things wrong: It is the sum of all the costs that are unnecessary when the product is made correctly the first time. By Crosby's estimate, PONC is approximately 20% of the sales dollar in a product industry and 35% in a service industry. *POC,* alternatively, is the sum of the costs associated with the quality effort: prevention measures and education costs. POC is about 3 to 4% of the sales dollar.

---

[13]Crosby, *Quality Without Tears,* 59–86.

Crosby's 14 Steps to Quality Improvement[14]

Step 1. Management commitment. Senior management holds the key. It has the mission of changing the culture of the company. It has to initiate a hassle-free management style and insist on an attitude of strict conformance to requirements. Management's credibility is usually low at first because of its history of short-term, short-lived solutions. There are several ways to change that attitude. First, a firm, clear statement that emphasizes the company policy on quality should be issued. Second, quality should be the first item on all meeting agendas. Third, top-management officials should compose clear messages on quality and deliver them to everyone. The basic policy message should make these points:

- "We will deliver defect-free products to our customers, on time."
- Our company policy is conformance to requirements, and "do it right the first time" is our method.
- Complete customer satisfaction is the goal.

Step 2. The quality improvement team. The team needs a clear direction and a good leader. Form the team to guide, coordinate, and support the quality process. Its members should consist of people who are good at clearing roadblocks and who represent all parts of the operation.

A committee consisting of a full-time quality coordinator, the team chair, top-management official(s), and a quality consultant sets the overall strategy for the team. The team members all need the same educational base in quality improvement. The team functions include the following:

- Setting up appropriate educational activities for all involved
- Methodically creating procedures and actions
- Learning more about quality improvements through continued work with the improvement process

Step 3. Measurement. Measure your progress. The team needs to know how it is doing as the quality process evolves. The team is continuously involved in sequences of input, process, and output; these aspects of the quality process can be measured.

Step 4. The cost of quality. Maintain an ongoing cost of quality. The company comptroller can work out a cost analysis of the quality effort. The initial PONC should be determined before the quality effort begins so that POC and savings can be realistically reflected.

Step 5. Quality awareness. Create a total quality awareness. If a communication device such as a regular newsletter exists, use a part of it to keep all employees aware of the effort and successes associated with the quality process. Start a quality newsletter if

[14]Crosby, *Quality Without Tears,* 101–124.

there is not one initially. Emphasize management's commitment, the quality policy, and PONC.

Step 6. Corrective action. Use SPC and problem-solving techniques to identify problems, find their root cause, and eliminate them. If suppliers have quality problems, meet with them and discuss requirements and ways of reducing nonconformities. Crosby emphasizes that "the real purpose of corrective action is to identify and eliminate problems forever."

Step 7. Zero defects (ZD) planning. "A zero defects commitment is a major step forward in the thrust and longevity of the quality management process." Start planning for ZD day early. Invite speakers from government, customer companies, management, and unions. The quality team should carefully plan this first public commitment of the company's new quality process.

Step 8. Employee education. See that everyone receives the same education on the quality process.

Step 9. ZD day. This is the day when management makes its official commitment to quality in front of everyone. The public commitment ensures that management is serious about quality. ZD day is important because it is a deterrent to management backsliding on the quality process.

Step 10. Goal setting. Set goals for the quality team. "Major goals should be chosen by the team and put on a chart for all to see." The chart should show a progress indicator.

Step 11. Error-cause removal. Ask people to submit statements about problems that they are aware of so that solutions can be developed. The response to this request is usually overwhelming, so be sure to set the error-response procedure ahead of time:

- Decide upon an initial response to the submitting person.
- Choose a method of analysis and action for each response.
- Make a concluding response to the submitting person.

Step 12. Recognition of good work in the quality process. The creation of a recognition program for both management and employees is an important part of the quality effort. Do not rush into it, though; make sure it is meaningful and well deserved. The program should do the following:

- Recognize hard-working people who are valuable to the quality effort.
- Ensure that those recognized are chosen by their peers.
- Provide a clear picture of what quality work is.
- Provide living "beacons of quality" for others to emulate on a daily basis.

Step 13. Quality councils. "Bring the quality professionals together and let them learn from each other."

Step 14. Repetition. "Do it all over again." Quality improvement is an ongoing process. Choose a new team with perhaps a one-member carryover from a previous team. The new-team approach provides a fresh look at the quality process and the quality problems.

2.8 A COMPARISON OF DEMING'S 14 POINTS AND CROSBY'S 14 STEPS

Both Deming and Crosby emphasize a total commitment to the quality process by top management, and both strongly promote the prevention system. They both stress education and the concept that striving for quality is a never-ending process. Deming is more explicit with his 14 points, but Crosby gives more direction by providing specific steps to take. Crosby urges a hassle-free management style, and Deming encourages managers to drive out fear and remove barriers that come between the workers and their ability to do a good job. The language may differ, but the message is similar.

The two do differ on some concepts. Some of Deming's points, such as the need for eliminating quotas and work standards and using modern methods of training and supervision, are not in Crosby's approach at all. Crosby believes that it is important to know PONC, keep an ongoing POC, maintain an awareness of the quality effort, and have a recognition program. Deming opposes slogans, posters, and "unrealistic" goals such as zero defects. He believes that they are a management responsibility dodge that implies that quality problems are due to poor work by the work force. Crosby, on the other hand, likes the slogans and posters as part of a quality campaign and strongly believes in the zero-defect concept.

2.9 WHICH WAY TO TOP QUALITY?

The preceding comparison of the approaches to high quality by two of the best quality consultants in the country shows enough difference in style and content to raise the question, "Is there a best way?" Recent literature on quality seems to emphasize that a melding of statistical and management methods is needed to bring U.S. industries into world-class competition. The use of statistical methods by everyone involved in the manufacturing process is really the third phase of the industrial revolution, following mechanization and mass production. All of a company's employees need some knowledge of statistical methods for SPC and problem solving. Statistics courses are often the most feared courses in any curriculum of study, but by focusing on practical statistics and applying the procedures to in-house data, employees can learn the important concepts in a meaningful way. People who have been life-long math dodgers can deal effectively with on-the-job data when they are shown the basic techniques and given a calculator. Deming emphasizes the belief that simple statistical concepts form a powerful tool for industrial applications.

Management methods must improve! Management's major goal should be to maximize the quality effort of the entire company work force. This can be realistically achieved only by fostering a dedication to quality that permeates every level, and the generating power has to be at the top level. Furthermore, the management method has to be hassle free, as Crosby calls it. People do not work in a dedicated manner when they are hassled. Deming refers to this as removal of fear and barriers. This ideal, positive approach promotes teamwork:

- All channels of communication are kept open.
- Everyone is treated with respect.
- All employees are well trained.
- All employees have a clear concept of the goals, both broad and specific, and of their potential contribution toward achieving those goals.
- All workers and managers pursue those goals in a coordinated manner.

CASE STUDY 2.3

The total commitment of the Industrial Specialities Division (ISD) of 3M to the quality process is reported in the July 1988 issue of *Quality Progress.*[15] ISD's program featured a combination of outside consultants and inside champions to help implement SPC and the quality effort. The quality function at ISD played a major role in quality training. Quality was becoming everyone's responsibility: it was not just the job of the quality control department.

When the quality training process began, outside consultants were used to train the management team and to help the quality function put together a training program. Top management then helped train others in the division. The managers became better quality coaches after their participation in the training program, and they also understood the concepts more thoroughly because they had to explain them to others. A strong sense of internal control and direction was maintained by adapting the materials to the unique division's operations.

ISD's quality training was implemented in five phases that featured the teachings of some of the top quality consultants. Philip B. Crosby's explanation of quality was used, and everyone in the division received a copy of his book, *Quality Is Free.* Joseph Juran's set of 16 videotapes was used to teach corrective and preventive action techniques. W. Edwards Deming's book, *Quality, Productivity and Competitive Position,* was used extensively to present his principles. The company is developing designed experiments materials based on the work of Genichi Taguchi. Designed experiments feature an analysis of planned changes in a process for the optimum path to process improvement. The last phase in the present quality plans is development of advanced designed experiments using reference materials by George E. P. Box.

---

[15]Michael J. Bulduc and Kimberly S. DeGolier, "The Expanding Role of Quality in Specialized Training" (*Quality Progress,* July 1988), 34–38.

2.10 PITFALLS IN THE QUEST FOR QUALITY[16]

Companies should avoid the following situations as they introduce their new strategies:

1. Instant results. Quality consultants emphasize that there is no quick fix to the quality problem. Remember, the Japanese were first introduced to the concepts of applying statistical methods in 1950, and their dominating influence on world markets was not really felt until the late 1970s. Fortunately, U.S. industry does not have as far to go as Japanese industry did following World War II.

2. Lack of commitment by management. Do not dump the burden on the work force by providing SPC training and nothing else. The quality effort is a coordinated process that involves everyone, and management has to lead the way.

3. Lack of long-term planning. Too often management is looking for results in time for the next quarterly report. Short-term impatience leads to a waste of time, effort, and money. A haphazard approach involves no meaningful change in the system.

4. Limited application. If only one part of a process uses SPC, no improvement in quality will be realized. GIGO: garbage in, garbage out! When SPC is started in an industry in a limited way, it should be applied to a complete process, not just a part of a process, from start to finish. Then it should be established in another complete process, and another, until the entire company uses SPC.

5. Overdependence on computerized QC. Thorough training with the hand-held calculator on the basic concepts is necessary for complete understanding. Computerized outputs can then be analyzed properly by those at the workstation and by the process control team when necessary.

6. No market research. The customer is the most important part of the process. Management must know the level of quality that is expected by customers.

7. Lack of funds committed to the quality process. Do not "pinch pennies." The initial monetary outlay in education costs and in production time lost for on-the-job training may seem extensive, but keep that long-term plan in mind. Remember, PONC is about 20% or more of sales, and POC is about 3 to 4% of sales.

8. Underestimating the work force. The best competitive effort can occur only when everyone in the organization is actively contributing to the quality process. Keep avenues of communication open and make sure that all employees are aware that their input is important, is wanted, and is expected. Utilize the knowledge and skills of everyone associated with the process.

---

[16]Richard McKee, "The ABC's for Process Control" (*Quality,* August 1985).

9. Failure to acquire a statistician/consultant. Hire a consultant to train the work force. After people learn the statistical concepts and techniques, they have to be trained to apply them.

10. Failure to involve the suppliers. The quality of incoming goods must equal the quality of your products. Garbage in, garbage out (GIGO)!

2.11 TOTAL QUALITY MANAGEMENT (TQM)

Virtually all quality consultants have been promoting the concept of a top-down quality process, that top management must lead the way and provide the constant pressure to develop quality processes in their companies. It is important that quality is instilled and improved in every process. One section of a company cannot have different quality standards or different work standards from another. Consequently, companies are now using or are moving toward a TQM system. When a company uses TQM, quality pervades every job, every operation, and every process, and the key measure of that quality is total customer satisfaction. There are two aspects to total customer satisfaction. First, the external customer, the one who ultimately receives the company's product or service, must be totally satisfied. Second, the internal customers must be totally satisfied. The internal customer is the next person in any sequential process, so the manager is the customer of the secretary and the next person in a production line is the customer of the previous person. The second aspect above leads to the first because when total customer satisfaction is accomplished internally, the end result will satisfy the external customer as well. All levels of management have to be involved because the application of quality principles must occur in every process.

The introduction of TQM in a company or the first change toward TQM must be preceded by training. All who will be involved must be trained in quality concepts and shown how those concepts apply to their specific jobs. Definitions of quality must be clearly established for every job. All must answer the question "What is quality work for my job?" and they have to compare their answers with their customer's answers (internal and/or external) to determine their specific definitions of quality. Quality tools such as process flow charts, check sheets, and appropriate performance charts can then be used to measure the present status of quality in the specific jobs. SPC charts can be used to determine if there is any special-cause variation that has to be tracked down and then used to determine the extent of common-cause variation and capability. Teams can be formed to formulate mission statements and to decide on what procedures should be used to achieve the goals set in the mission statements. Management has the task of providing the leadership and coordinating the action. Management has to maintain a very positive approach to eliminate the fear of honestly reporting a poor status quo and has to emphasize that any evidence of lack of quality represents an opportunity for improvement and that additional training will be provided, as needed, to achieve the quality goals.

Adopting TQM principles involves organizational change within a company. According to John K. Hawley, a quality consultant in El Paso, Texas,[17] six important facts about organizational change apply.

[17]John K. Hawley, "Where's the Q in TQM?" (*Quality Progress,* October 1995).

Fact 1. Change can be accomplished, but it is difficult.

> Most people have tried to either lose weight or quit smoking, generally with limited success. If individual change is that difficult, why should anyone think that changing an entire organization will be any easier?

Fact 2. Imposed change will be resisted.

> Impending change produces considerable anxiety. . . . [O]nly a few are going to go along with it without some explanation of why the change is necessary. . . . [S]imply firing those who resist . . . will traumatize survivors and virtually guarantee that the change will fail.

Fact 3. Full cooperation, commitment, and participation by all levels of management is essential.

> For any change program to work, every organizational level must be committed and involved. Each level has its own lessons to learn and tasks to perform.

Fact 4. Change takes time.

> For organizations that have achieved success with TQM, it usually takes years to get results. . . . W. Edwards Deming, Joseph M. Juran, and others first went to Japan in the late 1940s, and it took decades for their efforts to really bear fruit.

Fact 5. You might not get positive results first.

> Change is disruptive and will often result in reduced levels of organizational performance before positive effects become apparent.

Fact 6. Change might go in directions you didn't intend.

> One of the tools that is regularly used during TQM implementation is employee empowerment. . . . It can, however, be very dangerous . . . in areas where management has not yet let go. . . . [It] must therefore be tempered with explicit bounds on what employees can and cannot address.

Hawley goes on to say that a company can improve its odds for success if it has realistic expectations, if it starts slowly with small projects, and if it sets objectives, but does not dictate how they must be achieved.

Robert J. Masters,[18] Dean of the School of Business at Southeastern Oklahoma State University, in an extensive review of the literature, has concluded that the following eight barriers to success with TQM occur the most often:

---

[18]Robert J. Masters, "Overcoming the Barriers to TQM's Success" (*Quality Progress,* May 1996).

1. Lack of management commitment
2. Inability to change organizational culture
3. Improper planning
4. Lack of continuous training and education
5. Incompatible organizational structure and isolated individuals and departments
6. Ineffective measurement techniques and lack of access to data and results
7. Paying inadequate attention to internal and external customers
8. Inadequate use of empowerment and teamwork

2.12 THE MALCOLM BALDRIGE NATIONAL QUALITY AWARD[19]

The Malcolm Baldrige National Quality Award was established in 1987 as a response to public findings that American companies were lagging behind their competitors in production growth. Quality improvement was recognized as the key to increasing competitiveness, and strong leadership from management along with increased emphasis on customer satisfaction was seen as forming the backbone of any quality improvement program. The award is designed to recognize world-class quality management and to help raise national quality standards by providing a framework for the development of a TQM system. Each year there is a maximum of two award recipients in each of three categories: manufacturing, service, and small business. As of 1995, a total of 24 companies have received the Baldrige Award. The award appears to be living up to its design purpose. Each year the number of people investigating the award criteria is growing (more than 1 million by 1995), the number of companies using the criteria is increasing, and the number of companies applying for the award is increasing as well.

The award criteria are built on 10 core concepts:

1. Customer-driven quality. Both customer retention and the acquisition of new customers involve defect and error elimination, satisfying the customer's expectations, and anticipating market changes and opportunities.

2. Leadership. Top management must be actively and personally involved in leading the company toward excellence in quality. They should be able to demonstrate how their day-to-day activities support the quality values and are customer focused.

3. Continuous improvement. The company should be actually involved in a continuous improvement process. The improvements should be both "breakthroughs" and incremental improvements in

- New and improved products
- Error, defect, and waste reduction

---

[19]Current copies of the Baldrige Award Criteria may be obtained by writing to Malcolm Baldrige National Quality Award, National Institute of Standards and Technology, Route 270 and Quince Orchard Road, Administration Building, Room A537, Gaithersburg, MD 20899.

- Responsiveness and cycle time
- Productivity and efficiency

4. Employee participation and development. The company should invest in improving its workers'

- Skills and dedication
- Involvement in problem solving and decision making
- Diversification (cross training)

5. Fast response. The company should decrease cycle times for new and improved products and services. Decision-making processes should be streamlined.

6. Design quality and prevention. The company should have a prevention-oriented system. Preventing problems at the design stage costs much less than correcting the problems when they occur in the production stage. The design-to-product cycle times should be minimized.

7. Long-range outlook. The company should be involved in long-term planning and commitments and should not sacrifice long-term goals for short-term gains.

8. Management by fact. Management and process decisions must be based on reliable information, data, and analysis.

9. Partnership development. Internal partnerships would include labor–management pacts or cooperative agreements with unions. External partnerships would be with customers, suppliers, and educational institutions and could also include strategic alliances.

10. Corporate responsibility and citizenship. This encompasses business ethics, waste management, and public health and safety and environmental considerations.

There are seven specific examination categories, and these are broken down into subcategories and particular areas to address. The total package constitutes a comprehensive TQM outline that any company can use. Coaching is available as well, since past recipients of the award are expected to share information on their quality strategies. The individual items reflect the procedures and strategies of the more successful companies. They also echo the recommendations of Deming and other quality consultants. Highlights from the award criteria are shown matched with the corresponding concepts from the quality consultant advice of Deming and Crosby:

Baldrige Award Criteria	Consultant Advice
1. *Leadership:* Top management should develop and maintain an environment of quality excellence and integrate quality values into day-to-day operations.	Deming: point 14 Crosby: steps 1, 5
2. *Scope and management of quality and performance data and information:* The company should be using	Deming: point 4 Crosby: step 3

information and data for planning, management, and evaluation of quality and for financial and competitive quality comparisons including benchmarking.

3. *Strategic quality and company performance planning process:* The company should have a strategic planning process for both the short term and long term for process improvement, quality integration, and customer satisfaction. Specific goals should be cited.

Deming: points 1, 2, 5
Crosby: steps 7, 10

4. *Human resources development and management:* The company's human resource plans should be integrated with its quality and performance goals. Employees should receive both quality and job-specific training and be empowered to use that training. The company should maintain a work environment conducive to the well-being and satisfaction of the employees and also have a performance and recognition program.

Deming: points 6–8, 12, 13
Crosby: steps 2, 8, 10, 12

5. *Management of process quality:* The company should have a policy that specifies how customer requirements are translated to design requirements and how quality is integrated into the design-to-delivery process. The company's business and support services should have quality requirements. Quality and operational performance should be continuously improved in every company department and operation. The company should also have a policy on supplier quality and quality improvement.

Deming: points 1–5, 9
Crosby: steps 2, 6

6. *Quality and operational results:* The company should know its quality levels, operational performances and trends in all areas. These should be compared to those of competitors and/or benchmarks.

7. *Customer focus and satisfaction:* The company should have a procedure for determining both short- and long-term requirements and expectations of customers. There should be a proclaimed commitment to customers and a procedure for measuring and managing customer satisfaction. Comparisons should be made to competitors and/or benchmarks.

Crosby: step 1

2.13 TOTAL CUSTOMER SATISFACTION

Principles of TQM and the Baldrige Award criteria have a very strong emphasis on total customer satisfaction. Why is that so important when instilling quality into every phase of a company's operation? First, it is a measure of quality. If a customer sets the specifications for a product and is completely satisfied, then quality has been achieved. A consumer has specifications vis-à-vis competitive products and their associated prices, and if that consumer is completely satisfied, then quality has been achieved.

A second reason for aiming for total customer satisfaction is that it makes sense economically. Pete Babich,[20] owner of Total Quality Engineering, did a study of the economics involved. His study showed that it is five times as costly to recruit a new customer than to keep a current one. Also, dissatisfied customers generally tell 8 to 20 people about their dissatisfaction while satisfied customers generally tell only 3 to 5 people. His mathematical model based on these facts showed that a company with a 95% customer satisfaction rate would eventually end up with more than twice the market share over competitors with a 90% rate. In another illustration, he showed that a company with a 99% customer satisfaction rate would end up with more than five times the market share of a competitor with a 95% rate. His study has reinforced his perception "that continuously improving customer satisfaction is the best way to achieve business success." The study also indicates that it is important for a company to know its competitor's customer satisfaction rate as well as its own.

When quality is customer driven it is absolutely necessary to know your customers. What are their priorities with respect to your product or service? Do you really know or are you using what management thinks are the priorities? What are the critical aspects that determine customer loyalty for your product or service? Do you measure customer satisfaction as often as you measure other financial data? What are the sources of your customer satisfaction information? How are customer complaints handled? Are they treated as important data that identify opportunities for quality improvement? Are complaints channeled to the right place for response and corrective action? A five-step process that systematically deals with these questions has been suggested by John A. Goodman, Pres., Gary F. Bargatze, Sr. Vice Pres., and Cynthia Grimm, Vice Pres., all of Technical Assistance Research Programs, Inc. of Arlington, VA.[21]

Step 1. Inventory sources of data. List all the functions that receive customer complaints, inquiries and comments. . . . Functions that handle routine transactions should also be included.

Step 2. Determine the number of problems. List the frequency of reported problems by type of problem from each function. Use a multiplier to project the actual number of problems in the marketplace. Research indicates that the number of problems reported by customers to headquarters is a small percentage (0.1 to 5%) of the actual number of problems in the marketplace. . . . Through research, a company can determine the appropriate multiplier. . . . For example, . . . if 5% complain . . . the multiplier is [$1/.05$] or 20. . . . Thus, if headquarters received 40 complaints, it should assume that 800 customers had similar problems.

---

[20]Pete Babich, "Customer Satisfaction. How Good Is Good Enough?" (*Quality Progress*, December 1992).

[21]John A. Goodman, Gary F. Bargatze, Cynthia Grimm, "The Problems With TQM" (*Quality Progress*, January 1994).

Step 3. Identify what is important to the customer. [A] survey should also be conducted to estimate the loss of loyalty that these problems are causing (i.e., market damage).

Step 4. Develop a focal point and methodology to reconcile the data. A single function should be charged with collecting and reconciling internal quality data, data on customers' problems, and survey results. . . . The goal is to identify the key drivers of customer satisfaction—that is, those product and service attributes that, if done poorly, are most likely to cause a loss of customer loyalty.

Step 5. Track impact results monthly. Most companies . . . depend on monthly internal operational and cost data to evaluate quality initiatives. . . . But [that does] not indicate whether quality initiatives have an impact on the marketplace. . . . Once the key drivers of customer satisfaction and loyalty and their impact on revenue have been identified using effective baseline customer research, these key attributes should be measured monthly.

EXAMPLE 2.1 A company surveys its customers and the survey indicates that 4% of the customers would complain to the company about their dissatisfaction with the company's product. The survey also revealed that the loyalty factor was 20% (20% of the dissatisfied customers would switch to a competitor).

(a) Determine the multiplier effect.
(b) If the company received 30 complaints last month, how many customers actually were dissatisfied with the company's product?
(c) How many customers did the company lose last month?

Solution

(a) Since 4% complain, the multiplier effect is $1/.04 = 25$.
(b) 30 people complained, so 30×25 (the multiplier effect) $= 750$.
There were 750 dissatisfied customers.
Check: 4% of the $750 = 30$ complaints.
(c) 20% of $750 = 150$ customers switched to a competitor.

2.14 ISO-9000

Another important outcome of the global quest for quality has been the development of international standards of quality called the ISO-9000 series. Initially published in 1987, ISO-9000 has had a big impact worldwide. It has been adopted as the national standards in more than 60 countries, including all the developed countries.[22] It is rapidly becoming

[22] Ian G. Durand, Donald W. Marquardt, Robert W. Peach, and James C. Pyle, "Updating the ISO-9000 Quality Standards: Responding to Marketplace Needs" (*Quality Progress,* July 1993).

the basis of quality system requirements in the global marketplace. Over 40,000 supplier companies are registered ISO-9001, ISO-9002, and ISO-9003, and that registration is often sufficient for classification as a certified vendor for many customer companies. It is generally believed that in order for any company to compete internationally, it will have to be registered in the ISO-9000 series. Supplier companies have to be able to prove that they are supplying a quality product. ISO-9000 series registration will be either sufficient proof or a necessary part of the proof, depending on the customer company.

There are five parts to the series:

- ISO-9000 — Quality management and quality assurance standards: This provides guidelines for the selection and use of the standards.
- ISO-9001 — Quality systems: The model for quality assurance in design, development, production, installation, and servicing. This is the most inclusive of the external quality assurance standards (9001, 9002, and 9003).
- ISO-9002 — Quality systems: The model for quality assurance in production, installation, and servicing.
- ISO-9003 — Quality systems: The model for quality assurance in the final inspection and testing. A company may start by registering as a 9003 or a 9002 and then work toward the more inclusive 9001 registration as it builds quality into its other functions.
- ISO-9004 — Provides guidelines for quality management and quality system elements. This assists an organization in developing and implementing a quality system. It also helps to determine which of the 9001, 9002, or 9003 is applicable.

The ISO-9000 series (Figure 2.1) is heavy on documentation. It is up to individual companies to determine how the quality standards apply to them, but their interpretation must be carefully documented. The official investigative group then certifies that quality.

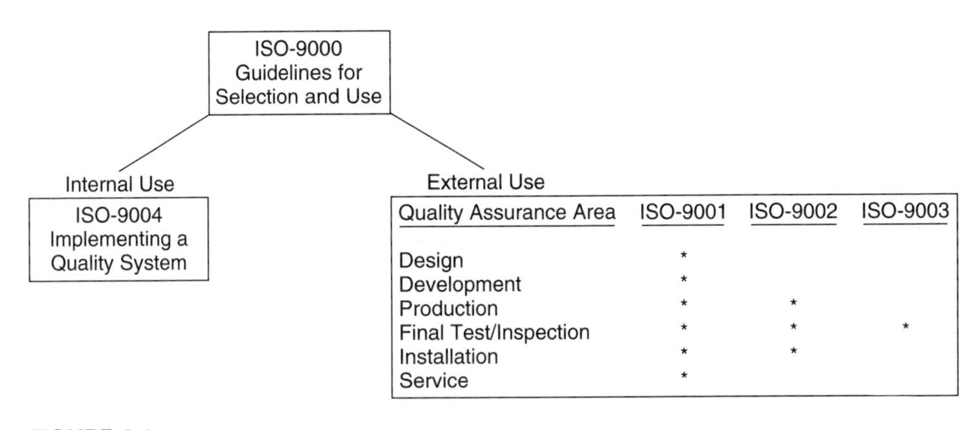

FIGURE 2.1
ISO-9000 series.

Once certified, it cuts down on customer interaction. Instead of going through a certification process with each customer, the ISO-9000 certification is usually sufficient. In addition, registration may very well be a marketing advantage for the short term, but it could also become a requirement for doing business in the long term.

ISO-9000 registration is really an ongoing process. First, each registrant undergoes well-documented periodic audits so that no backsliding can occur. Second, the ISO-9000 standards are reviewed and updated every five years.

CASE STUDY 2.4

A prime example of a company's long-term commitment to quality and the elimination of defects was reported by Basile A. Denissoff, the senior project quality engineer manager of the AVX Corp. in Raleigh, NC.[23]

In 1982 the AVX Corp. changed from a final lot inspection procedure to an application of SPC on sublots at their Raleigh, NC, plant. This led to improved quality, reduced scrap and rework, and a decrease in both inspectors and rework operators. In 1983 they upgraded their SPC by applying it more effectively to the manufacturing processes. The result was that fewer inspectors were needed and they gained more control of their product's characteristics. In each situation, displaced inspectors and rework operators were transferred to manufacturing duties as the process was streamlined. In 1984 everyone in the company underwent training based on the concepts of Crosby, Deming, and Juran. They emphasized principles of satisfying requirements, error-free work, and managing by prevention. Everyone became involved in the total quality program. They upgraded the use of statistics in the manufacturing engineering department in 1985 and then implemented operator self-inspection in 1986. The operators were given ownership of their operation and the roving inspectors were transferred to manufacturing duties. This resulted in a 17% increase in product output and a savings of $500,000 per year. 1987 saw the advent of module manufacturing where several independent operations were blended into module units. This increased efficiency cut down on between-operations inventory and increased customer service. In 1988, with the SPC charts showing statistical control, an initiative was made to increase the Cpk of the various processes to 1.67 as an immediate goal and then to 2.0. (Cpk is a process capability index and is presented in Chapter 6.) The company shifted its focus in 1991 to total customer satisfaction in both products and services and they used the Malcolm Baldrige Award guidelines for self-assessment. Then in 1992 the company began the ISO 9002 certification program. They achieved certification and registration in January 1993.

Through constant efforts, the AVX Corp. has shown that the quest for quality is a never-ending process.

---

[23]Basile A. Denissoff, "War With Defects and Peace With Quality" (*Quality Progress,* September 1993).

2.15 THE SERVICE SECTOR

W. Edwards Deming, in his four-day quality seminar, explained that "Figures published by the Census show that 75 out of 100 people are employed in service organizations. If we add to these figures the ones in manufacturing industries who provide support services, we find that 86 in 100 are engaged in service. This leaves only 14 out of 100 to make items that we can drive, use, misuse, drop, or break, and these 14 include those in agriculture."[24]

With 86% of the workforce in the service area and with Crosby's estimate that the price of nonconformance is more than 35% of the sales dollar, there is an extensive opportunity for savings, nationwide, by applying quality-control concepts to service. The concepts will be the same, but the applications may differ. The service sector has its own set of interlocking processes within each service area. All processes are subject to variation and variation is the enemy of quality. Reduce the variation and quality will increase and costs will decrease. Service processes have measurable results or outputs that can be tracked on control charts to identify sources of special-cause variation. When the special-cause variation is eliminated from a service process, the process is stable (in statistical control) and its capability can be measured. Further improvements in quality may be achieved by investigating the process with flow charts, check sheets, and other quality tools to uncover new opportunities for improvement.

EXERCISES

1. Name three management problems that are causing some U.S. companies to fall behind their competitors.
2. List Deming's 14 points for management.
3. Basically, what two things are needed to achieve top quality in a company?
4. What new philosophy does Deming mention?
5. What is the main reason for the adversarial relationship that exists between labor and management in some companies?
6. Why is mass inspection a poor approach to quality?
7. What quality criteria should be expected of a supplier?
8. Who should be involved in the quality process?
9. What is meant by modern methods of supervision?
10. Why does fear cause a company economic loss?
11. How can high quality, high productivity and lower production costs be achieved simultaneously?
12. Why is it important to eliminate departmental barriers?
13. Why do you suppose work quotas have been used? Why should they be eliminated?
14. What are some of the barriers that prevent a worker from doing a good job?
15. Why is it necessary to eliminate the fear of layoffs when starting the quality process?
16. Why does the quality process have to start at the top-management level?
17. According to Deming, what percent of the quality problem is attributed to the workers and what percent is attributed to management?

[24]William J. Latzko and David M. Saunders, *Four Days With Dr. Deming*. (Reading, MA: Addison Wesley, 1995), 184.

18. List Crosby's symptoms of trouble.
19. When was the control chart concept developed and who invented it?
20. What are Crosby's 14 steps to quality improvement?
21. What are POC and PONC?
22. What are the five criteria for promoting teamwork?
23. How may inspection within a process lead to improved quality?
24. What are the main pitfalls in the quality process?
25. Why is it important to include middle management and support areas in TQM?
26. What are the 10 core concepts of the Baldrige Award?
27. If you buy a new tool, why isn't it a good idea to just buy the least expensive one?
28. What are the seven Baldrige Award criteria?
29. Why is total customer satisfaction so important?
30. Why is ISO-9000 registration becoming so important?
31. What is the main purpose of the Baldrige Award?
32. What are four features of statistical thinking for management?
33. What main points about variation did Deming emphasize in his four-day seminars?
34. How does fear limit process improvement?
35. Why should the pursuit of quality be continuous?
36. Give five indicators of fear in an organization.
37. How can fear be reduced in an organization?
38. How does MBO put a ceiling on performance?
39. Why is it useless to set higher goals for the workers when the process is in statistical control?
40. What are Deming's Seven Deadly Diseases?
41. If a company wants to improve the quality of its product, why should it initiate TQM?
42. List six barriers to TQM.
43. What is a "top-down" quality process?
44. What important facts about organizational change have to be considered when a company initiates TQM?
45. Suppose your company surveys its customers and concludes that only 6% would complain about a specific problem.
    a. What is the "multiplier"?
    b. If 40 complaints are received on the problem, how many customers actually had that problem?
    c. If the company's customer research reveals that the loyalty factor is 15% (fifteen percent of the people who have a problem of this type would switch to a competitor), how many customers would the company lose if 40 complaints were received?
46. What five steps should a company take to deal with customer information such as complaints or other comments about the product or service?
47. What percentage of the American workforce is engaged in providing services?
48. Consider a service area you are familiar with (restaurant, hotel, retail business, etc.). What procedure does it use that is subject to variation and how does that variation cost the company money?
49. A random survey revealed that only 3% of a company's dissatisfied customers complained.
    a. What is the "multiplier"?
    b. If 24 complaints on the problem were received in a month, how many dissatisfied customers actually had that problem?
    c. If the loyalty factor is 40%, how many customers were lost?

3

INTRODUCTION TO VARIATION AND STATISTICS

OBJECTIVES

- Understand that variation occurs in all measurements and that quality control translates to control of variation.
- Know the round-off rules.
- Distinguish between two types of variation: special cause and common cause.
- Know what distributions are and how they are used with SPC.
- Calculate the mean, median, mode, range, and standard deviation for a set of numbers using formulas and using a statistical calculator.
- Draw a histogram for a set of numbers.
- Find $\bar{x} \pm 3s$ for a set of data.
- Know the relationships between a population distribution and a sample distribution.

The concept of measurement variation is introduced in this chapter, as are some statistical methods for measuring and describing that variation. Any large set of measurements will form a graphical pattern when the frequency of each measurement is charted; that pattern is called a *distribution*. The concept of distributions is important because all the SPC techniques use the distribution of a limited number of measurements to imply the "true" distribution of all measurements.

3.1 MEASUREMENT CONCEPTS

It is important to understand that no two products or characteristics are exactly the same. Their differences may be large or imperceptibly small, but they are always present. Any measurement is only as good as the measuring device and the person's reading of the measuring device.

Measurement Error

The concept of error in a measurement is the same whether the device is a ruler or a micrometer.

> *Definition:* The *accuracy* of a measurement is the smallest unit of measurement on the measuring device.

> *Definition:* The *maximum error* of a measurement is half the accuracy.

A measurement of 2.34 centimeters, for example, is accurate to the nearest hundredth of a centimeter, and the maximum error is 0.005 centimeters. If the width of a table top is measured to the nearest inch, any width between 41.5 and 42.4 inches will be called 42 inches and the maximum error will be .5 inches. Likewise, if a bearing is measured to the nearest thousandth of an inch, a bearing measuring between 0.7475 and 0.7484 inches will be given the dimension 0.748 inches and the maximum error will be .0005 inches. If two bearings are thought to measure "exactly" 0.748 inches, a measuring device accurate to the nearest ten thousandth or hundred thousandth may be needed to show a difference, but the difference will be there.

A measurement is an approximate number because it has been *rounded off* to the accuracy implied in the measurement (the nearest measurement mark on the measuring device). Calculations with measurements must be properly rounded off so that the correct accuracy is implied in the calculated values.

EXAMPLE 3.1 A bar has been marked into two sections and each length measured. A meter stick marked to the nearest centimeter was used to measure the first part, and a ruler marked to the nearest millimeter was used to measure the second part. The two measurements were 96 centimeters and 15.3 centimeters. What is the length of the bar?

Solution

The answer is *not* 111.3 centimeters because that measurement implies that the true length of the bar lies between 111.25 and 111.34 centimeters:

Minimum Possible Measurements	Maximum Possible Measurements
95.5 cm	96.4 cm
+ 15.25 cm	+ 15.34 cm
110.75 cm	111.74 cm

The simple sum, 111.3, implies more accuracy in the true length of the bar than is warranted. The calculations with the minimum and maximum possible true measurements show the actual possible range in the true length of the bar.

The correct calculation method is to add the measurements, then round off the answer to match the accuracy of the least accurate number used in the calculations. The simple sum of 111.3 centimeters should be rounded off to 111 centimeters. This measurement has an implied accuracy of 110.5 to 111.4 centimeters, which more closely reflects the actual range of measurements determined by using the minimum and maximum values.

Round-off Rules

If the number to the right of the desired place value is half or more of that place value, round *up* to the next digit:

$23.\underset{\wedge}{4}72$ to the nearest tenth is 23.5.

$4.14\underset{\wedge}{5}6$ to the nearest thousandth is 4.146.

$5\underset{\wedge}{6}92$ to the nearest hundred is 5700.

$23.\underset{\wedge}{4}5$ to the nearest tenth is 23.5.

If the number to the right of the desired place value is less than half of that place value, *truncate* to that place value:

$23.\underset{\wedge}{4}14$ to the nearest tenth is 23.4.

$4.14\underset{\wedge}{5}4$ to the nearest thousandth is 4.145.

$5\underset{\wedge}{6}48$ to the nearest hundred is 5600.

> *Definition:* *Tolerance* is an acceptable range of measurements on a specific dimension.

The tolerance is usually set at the design stage of the product, but it may be changed at a later time, depending on how critical the measurement is, how easily or economically the part can be made, and/or how the customer gauges the resulting quality. The tolerance is given as a target measurement plus or minus (±) the variation that is acceptable. The target measurement is usually centered in the tolerable range, as implied by the plus or minus, but the tolerable range may occasionally be unsymmetric, such as $2.450^{+0.001}_{-0.003}$ inches. Any product that has its measurement beyond the tolerable range is said to be unacceptable or defective.

> *Definition:* A *distribution* is an ordered set of numbers that are grouped in some manner. The distribution may be in a table form, a graph form, or a picture form.

Suppose that ball bearings were to measure 0.745 ± 0.002 inches and that a large number of them were produced, measured to the nearest 0.0005 inches, and stacked

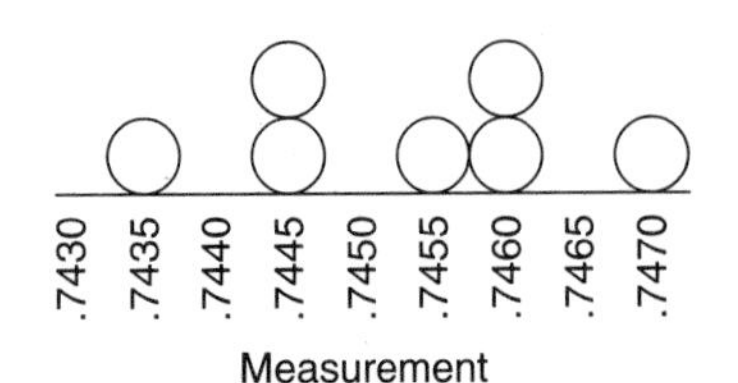

FIGURE 3.1
The first few bearings.

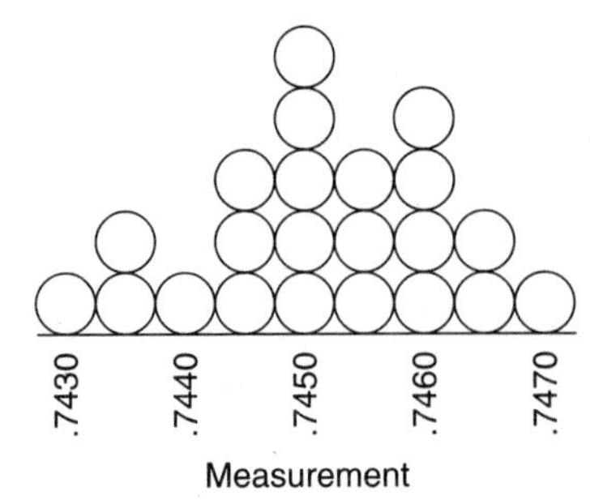

FIGURE 3.2
More bearings accumulate.

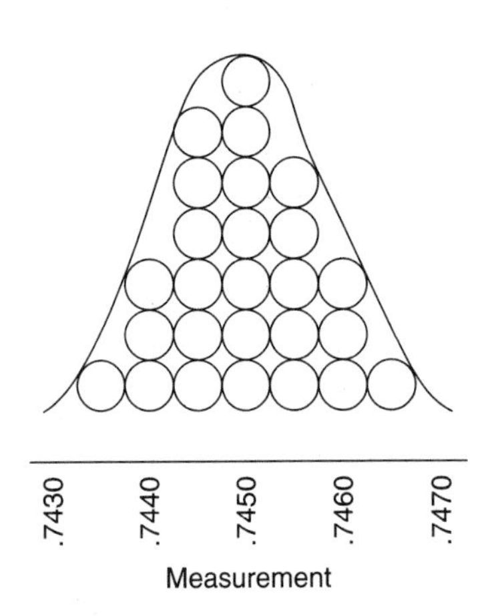

FIGURE 3.3
The day's accumulation forms a pattern shown by the curve.

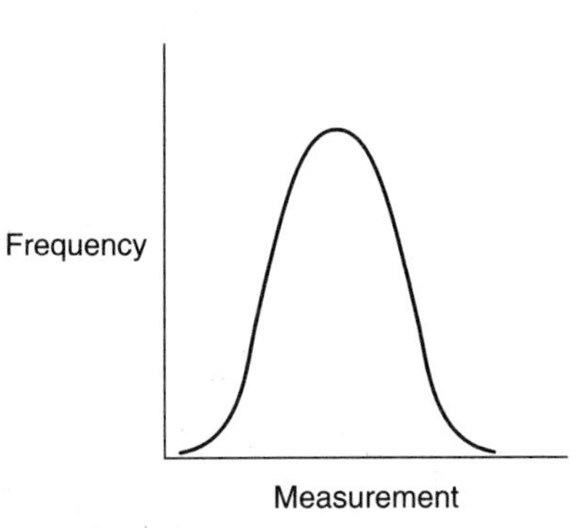

FIGURE 3.4
The pattern, or distribution, by itself.

according to size. The slots at each measurement between 0.743 and 0.747 inches would fill in a random manner, but as more were measured and stacked, the bearings would form a pattern that is called a *distribution.* This process is shown in Figures 3.1 to 3.4. The shape of the distribution can be described by a mathematical curve and the percentages in each section predicted using the branch of mathematics called statistics. Normally, most of the bearings will fall in the slots closest to the target measurement with the numbers per slot decreasing as the slots approach the tolerance limit. However, other distribution shapes are possible as well.

3.2 SPECIAL-CAUSE AND COMMON-CAUSE VARIATION

Two types of variation must be dealt with in SPC. The first is called *special-cause variation,* or assignable-cause variation. It affects the process in unpredictable ways and can be detected by simple statistical techniques. It can be *eliminated from the process* by the worker or process control team in charge of that particular segment of the process; this is referred to as *local action.* Approximately 15% of all process problems can be handled by local action. When all the special-cause variation is eliminated, the process is said to be in *statistical control.*

The second type of variation, called *common-cause variation,* is inherent in the process. When the special-cause variation has been eliminated, the process is working as well as it possibly can without modifications. Approximately 85% of all process problems are due to common-cause variation. The only way to decrease common-cause variation is to make improvements in the process. The extent of common-cause variation can be measured statistically and compared to specifications; if improvements are needed, action on the process is necessary. *Management action* is needed for any process changes.

A major error that can occur when trying to improve quality is to incorrectly classify the variation type. If a common-cause variation is misinterpreted as special-cause, extensive work with local action will be wasted with efforts to eliminate the source of variation. Common-cause variation cannot be eliminated because it is part of the process itself and local action does not involve changes in the process. The worker frustration level will increase as well because of the impossible task undertaken. Also, management will be incorrectly blaming the workforce for the poor quality. If a source of special-cause variation is incorrectly classified as common-cause variation, changes to the process will be undertaken that may be expensive, will change the quality of the output (often for the worse), and will not lead to the elimination of the problem. One of the main functions of SPC charts, which are studied in Chapter 6, is to indicate when special-cause variation is present. SPC charts provide hints to the source of special-cause variation and show when local action has eliminated it.

CASE STUDY 3.1

The Ace Machining Company experienced excessive variation among products from one of their machines. They decided that, since it was an old machine, it must need an overhaul. It was taken out of service, torn down to replace worn parts, and reassembled for in-service operation. The overhaul costs exceeded $10,000 in worker hours and lost production time. They did find one bearing that seemed to be worn and they replaced it. When the machine was back in service, the excessive variation in its output was virtually the same. The real problem in that case was in the process. In a companywide downsizing change, the operators on each shift had been laid off and their responsibilities for running the machine were now shared by two other operators on each shift. The two operators were instructed to check the machine each half hour in turn. One checked it on the hour and the other checked it on the half-hour. They were instructed to adjust the machine if the critical measurements on the part being produced were more than .005 inches from the target measurement. Each operator silently blamed the other each time they adjusted the machine and, consequently, the machine was adjusted every half hour. Overadjustment was the true source of the excessive variation and was due to the process that was established by management as part of the company downsizing. Without the use of control charts to identify the type of variation, it was misdiagnosed as special-cause variation. The mistake cost the company a large amount of money and a morale problem with the machine operators.

Production quality is directly linked to the amount of variation in the product measurements. For example, quality is a factor in the closeness of fit of the various parts of an automobile, and quality is visible in large parts such as doors, hoods, and finish. The hidden quality lies in the moving parts of the engine and drive train, where top quality means years of trouble-free driving. In the manufacture of all products, there are critical measurements for which adherence to the right tolerances is necessary for high quality.

Service quality *may* be related to a stated standard of the company providing the service. Suppose that a company guarantees delivery of an item within three days of when the order is received. Quality service, in that case, would be receiving the ordered item within the three-day limit. One of the things that makes service quality more difficult to achieve than production quality is that service quality is often subject to individual expectations or individual interpretations of what the quality feature should be. For example, a hotel chain advertises clean and comfortable rooms as part of its quality service, but what is clean and comfortable to one guest may not be clean and comfortable enough for another.

3.3 THE VARIATION CONCEPTS

Individual measurements are different, but as a group, they form a predictable pattern called a distribution. The distribution can be pictured as a statistical curve. The curve may be easier to conceptualize if imagined as stacks of like measurements within it. For example, if the curve in Figure 3.5 represents the distribution of a day's production of some specific product (dimension), how can it be interpreted? The horizontal scale in Figure 3.6

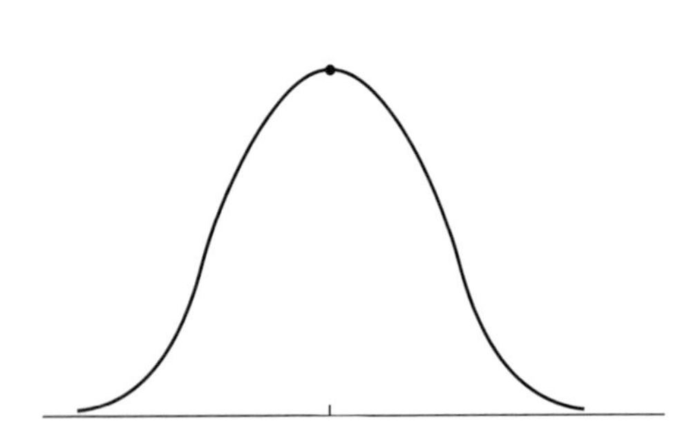

FIGURE 3.5
The distribution of a day's production.

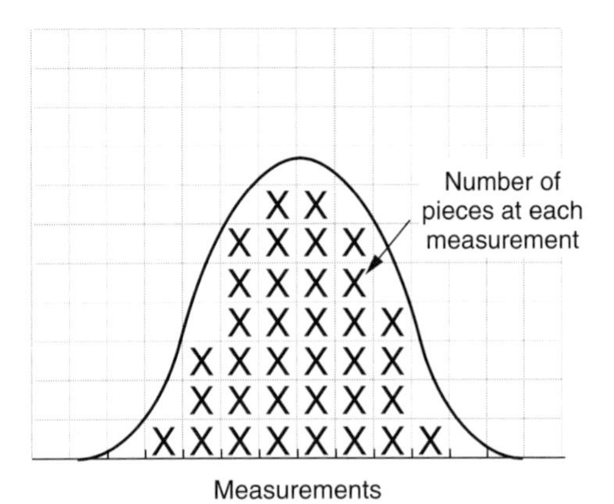

FIGURE 3.6
The day's distribution curve with tallied measurements.

is marked off in measurements. The columns above each measurement represent the stacked height, or frequency, of each specific measurement (the column of X's).

> Every distribution has the following measurable characteristics:
>
> - *Location:* The middle value, or average value, may differ for different distributions. Figure 3.7 shows three distribution curves that differ only in their middle, or average, value.
> - *Spread:* The width of the distribution is a measure of the extent of variation from one extreme to the other. Figure 3.8 shows three distribution curves that differ in spread, or range.
> - *Shape:* The way the measurements stack up can cause different-shaped distributions. Figure 3.9 shows three distribution curves of different shape.

Important information about the process can be determined from the distribution pattern of the measurements being tracked. Suppose the target measurement is .510. Figure 3.7 indicates that the distribution on the left, with a middle value of .498, is producing pieces whose measurements are too small. An adjustment of the machine (local action for a special-cause variation problem) can change the position of the next batch of pieces to the middle position—on target. Likewise, a distribution pattern centered at .519 can be adjusted smaller so that the resulting pattern is centered at the target value. The spread of the distribution, as shown in Figure 3.8, indicates the amount of variation in measurements. The figure on the right shows the best distribution because it has the smallest amount of variation. The poorest distribution is in the middle, with the widest pattern indicating a large amount of variation in the measurements. It should be noted that since all three distributions in Figure 3.8 are on target, the narrowest distribution indicates

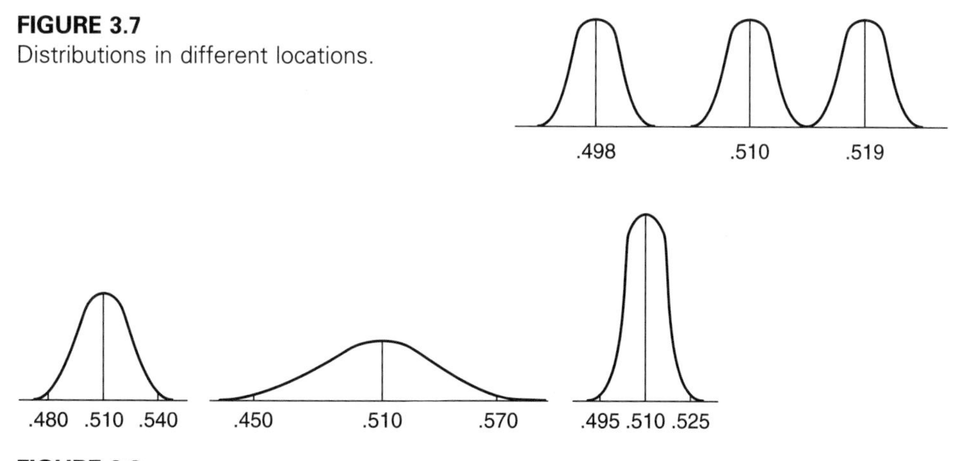

FIGURE 3.7
Distributions in different locations.

FIGURE 3.8
Distributions with different amounts of spread.

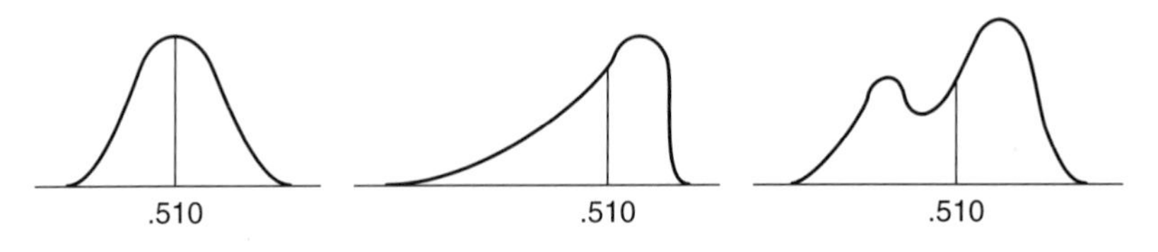

FIGURE 3.9
Distributions with different shapes.

the one with the smallest amount of variation and the highest quality. The shapes of the distribution will also indicate the presence of special-cause variation. The expected shape of a distribution is usually a normal pattern, as shown on the left of Figure 3.9. The middle shape is skewed left (the long tail section of the curve is on the left side), indicating a problem with some excessively small measurements. When the problem is found and eliminated, the shape of the distribution should return to a normal pattern. The shape on the right has two distinct high peaks and is referred to as bimodal (bi indicating two and modal indicating high-frequency values). This also indicates the presence of special-cause variation, which can be eliminated by local action. When the cause of the two separate (merged) distributions is determined and the problem fixed, the distribution pattern should become normal.

A basic concept of distributions, when applied to a process that is in statistical control (no special-cause variation), is that the distribution pattern will remain fixed as long as nothing changes in the process. As soon as some change is made, such as an adjustment on the machine, a new distribution pattern will result. In the case of a simple adjustment, the change should be just in the location of the distribution pattern. When a machine is overadjusted, as occurred in Case Study 3.1, it results in each distribution pattern superimposed over the previous patterns. Excessive variation results. This is shown in Figure 3.10.

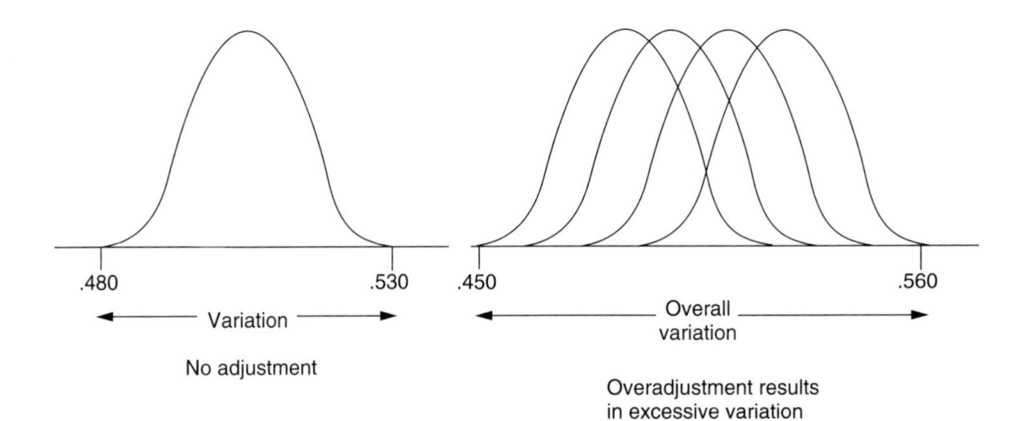

FIGURE 3.10
The effect of overadjustment.

3.4 DISTRIBUTIONS AND SPC GOALS

Figure 3.11 illustrates day-to-day production when special causes of variation are present. The special-cause variation generally affects the process in unpredictable ways. What will tomorrow's output be? No one knows because of the erratic behavior of special-cause variation! Most processes fit this pattern before a company introduces SPC because most processes have some degree of special-cause variation.

Special-cause variation has been eliminated in the distributions shown in Figure 3.12. The process is now predictable, so tomorrow's product distribution is known as well as next year's. Only common-cause variation is present and the process is said to be in *statistical control.*

The next step is to compare the distribution of the *individual* measurements with the specification limits. The specification limits along the measurement scale are shown in Figure 3.13. The *lower specification limit* (LSL) and the *upper specification limit* (USL) are determined by applying the tolerance to the target measurement. The distribution curves in Figure 3.13 show that day after day, a constant percentage of product is out of specification. The area under the curve corresponds to percentage of product, so the area in the shaded sections represents the percentage of product that is out of tolerance, or out of specification. It shows that unless the process is improved to decrease common-cause variation, the manufacturer will have to accept this constant percentage of defective products. The goal, therefore, is to decrease variation.

The application of SPC methods can achieve the goal illustrated in Figure 3.14. The spread, or range, of the product's distribution pattern narrows when the amount of common-cause variation in the process decreases so that the range of measurements is

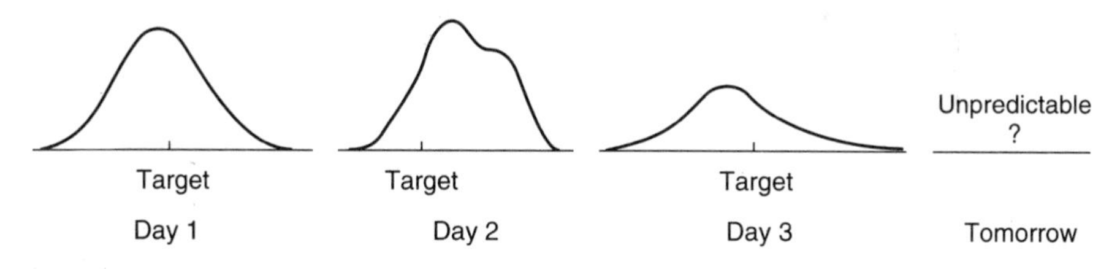

FIGURE 3.11
Special-cause variation is present. The product distribution is therefore unpredictable.

FIGURE 3.12
The process is in statistical control, so only common-cause variation is present. Production is predictable.

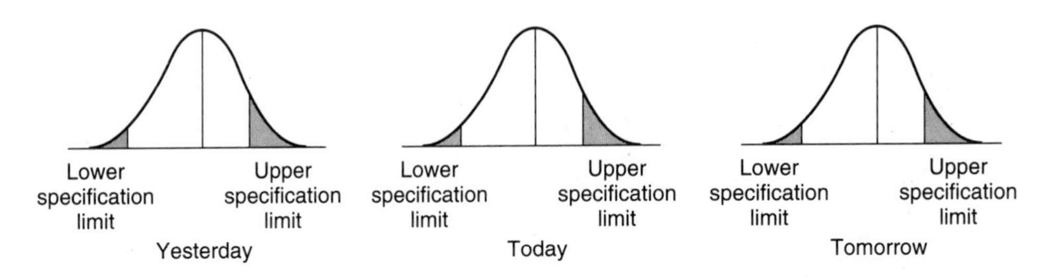

FIGURE 3.13
The process is predictable, but not capable. The shaded areas represent a constant percentage of product that exceeds specification. The common-cause variation is excessive.

well within the specification limits and virtually no out-of-specification products are produced.

A strict engineering approach to the variation problem is to decrease variation to the point where the distribution is narrow enough to stay within the specification limits. However, W. Edwards Deming taught that just staying within the specification limits is not good enough. The level of quality stays constant. If quality is going to improve, the amount of variation must decrease. The true goal then should be to continuously improve the process so that the distribution stays on target and the width of the distribution is continuously decreasing. For many industries, the first step is to get the distribution inside the specification limits. The next goal is to decrease variation to the point where the distribution utilizes three-fourths of the tolerance. The ultimate goal is to use just half of the tolerance spread. The Japanese approach would be to just continuously work at decreasing variation.

One important point to consider when concentrating on specification limits is that tolerance and the corresponding specification limits are a calculated hedge against quality. An engineer designs a product with specific measurements. Then, either the engineer, the manufacturer, the customer, or some combination of the three *estimates* how much variation can be tolerated without sacrificing quality (too much).

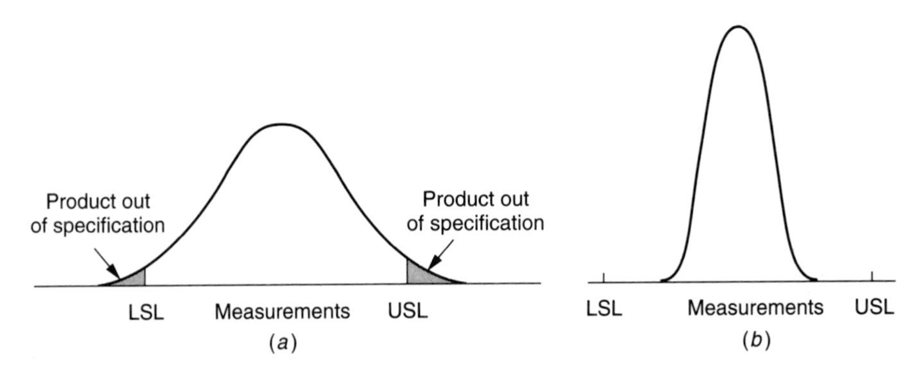

FIGURE 3.14
The goal of SPC is to change each day's production pattern from (*a*) to (*b*) by reducing product variability.

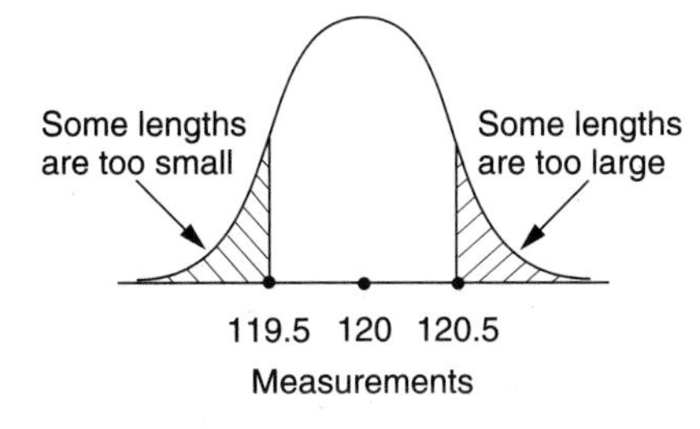

FIGURE 3.15
Too much variation. Some lengths are not within the specifications.

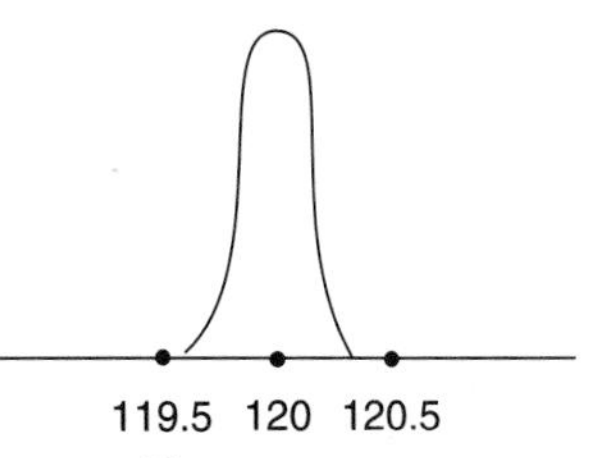

FIGURE 3.16
Variation has been decreased. All measurements are within the specifictions.

EXAMPLE 3.2

A steel company is making steel reinforcing bars and the length is supposed to be 120 inches ± .5 inches. A large sample of 200 bars has a smallest length of 119.2 inches and a largest length of 120.9 inches. The middle of the distribution pattern is located at 120 inches and the shape of the distribution is normal (bell shaped). Is this acceptable? What should be done?

Solution

It is NOT acceptable. The location of the distribution is good and the shape of the distribution is good, but there is too much variation. The tolerance is from the Lower Specification Limit (LSL) of 119.5 inches to the Upper Specification Limit (USL) of 120.5 inches. There are lengths in the sample smaller that 119.5 and there are also lengths larger than 120.5. (Figure 3–15) Variation must be reduced so that all lengths are between 119.5 inches and 120.5 inches, i.e. within the specification limits. (Figure 3–16.)

3.5 BASIC STATISTICAL CONCEPTS

Statistics is the science of data handling. The application of the statistical concepts normally involves using sample information to make decisions about a population of measurements:

Definition: A *population* is the set of all possible data values of interest.

Definition: A *sample* is a subset, or part, of the population.

The four phases involved in the application of statistics:

1. *Collection of data.* The data collected can be all the data in the population or, as is usually the case, a sample of the data in a population. When a sample is used, the sample must fairly represent the population because conclusions about the population depend on the information from the sample.
2. *Organization of data.* The data collected must be organized so that the information they contain can be understood. Tables, charts, graphs, and other pictures are used for the organization.
3. *Analysis of data.* The analysis of data involves concise numerical measures of the data. The measures fall into two main classifications: a middle value, called a measure of central tendency, and a data spread indicator, called a measure of dispersion.
4. *Interpretation of the data.* Conclusions are made about the population of measurements based on the charts, graphs, and statistical calculations that are applied to the samples taken.

The concepts of inferential statistics have been developed to predict population values from the information gained by sampling the population. The samples must represent a true picture of the population in order to accurately predict the population values.

Definition: A *random sample* is one in which every element in the population has an equal chance of being chosen for the sample.

Every sampling method used in statistics is based on the random sample concept. All the statistical theories and formulas that use samples to estimate population values require random samples. An extensive amount of work has been done on sampling methods. These methods all deal with different ways to draw a random sample from the population.

Definition: A *biased sample* is any sample in which some elements of the population are more likely to be chosen for the sample than others.

A biased sample should never be used to make inferences about a population. Each piece produced in a manufacturing process has a specific number of dimensions, and each dimension has its own population of measurements. If all the measurements of a dimension are stacked up according to size, the distribution will usually take on a *normal* shape. In a normal distribution, which is discussed more thoroughly in Chapter 5, most of the data cluster close to a central or middle value. The height of the stacks diminishes rapidly as the measurements get farther from the central value. The shape of the normal distribution is a bell-shaped curve (Figure 3.17).

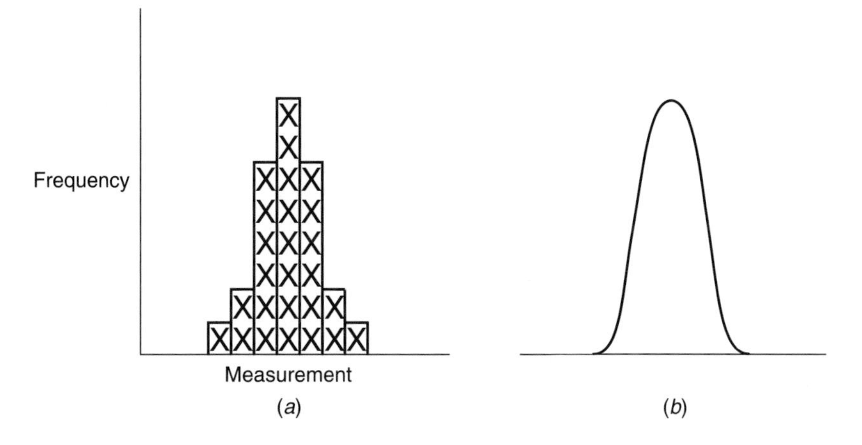

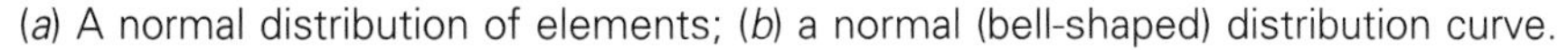

FIGURE 3.17
(*a*) A normal distribution of elements; (*b*) a normal (bell-shaped) distribution curve.

When a random sample is taken from a population and the pieces are stacked according to size, the shape of the developing distribution can take almost any form. However, as more pieces are added, the shape of the sampled distribution will closely resemble that of the population's distribution. This concept was shown in Figures 3.1 to 3.4, and in that case, the population distribution will nearly match Figure 3.4. A large sample, usually defined as a sample with about 100 pieces, is needed for this to occur. The larger the sample, the more closely the sample distribution will resemble the population distribution.

The application of statistics provides a way to determine population measures from specific sample measures. This is pictured here for a specific dimension. If the frequency distribution for a large random sample was compared to the frequency distribution of the population, the results would look similar to Figure 3.18 and 3.19. The expectations for the three measurable distribution characteristics are also shown:

- The center of the sample distribution should be very close to the center of the population distribution.

FIGURE 3.18
Expected relationship between population distribution and large-sample distribution.

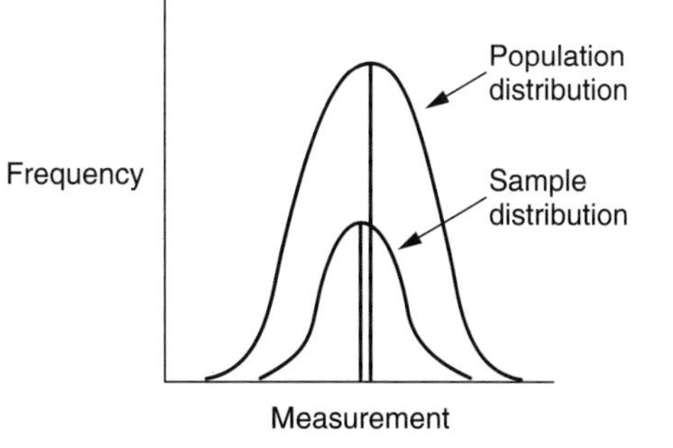

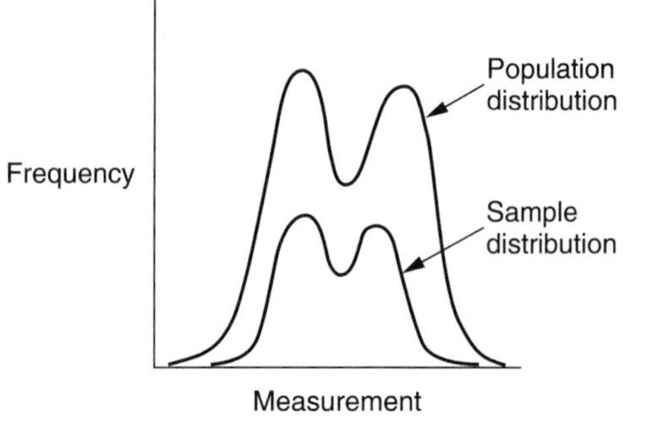

FIGURE 3.19
Expected relationship between population distribution and large-sample distribution.

- The spread of the sample distribution should be slightly smaller than, but close to, the spread of the population distribution.
- The shape of the sample distribution should mimic the shape of the population distribution.

For example, a machine produces cylindrical shell blanks for making the YF-42 spark plugs. If the measurement of interest is the diameter of the shell blank, the population would consist of the diameter measurements of all the YF-42 shell blanks produced. As in most manufacturing situations, this is an ever-growing population. If five shell blanks are chosen and their diameters measured, those five diameter values would be a sample. The collection of data for the shell blanks will consist of taking ten samples at random times throughout a shift. The samples of data are then organized using a control chart and a histogram. The data are analyzed by calculating important numbers that describe the position and the spread of the distribution of data values. The data, charts, graphs, and numerical analysis values are then interpreted and conclusions about the population of all YF-42 shell blanks are drawn.

Data Collection

Data collection can sometimes be the most difficult part of a statistical application. The sample chosen must be random, so the sample requires careful planning. In a *random* sample, each data value in the population has an equal chance of being used in the sample. All the formulas in statistics that relate to samples assume the use of random samples. If the sample is not random, then it is *biased* and some elements of the population are more likely to be chosen for the sample than others. *Biased samples lead to false conclusions regarding the population.*

A haphazard sample is often thought to be random, but unexpected biases may sneak in. For example, pieces chosen haphazardly on an incoming inspection may have a large proportion of the "easy to get at" pieces that can be carefully checked and placed by the vendor. The resulting inspection will imply that the quality of the incoming shipment is higher than it actually is. For another example, an operator given some leeway in taking production samples may subconsciously wait until things seem to be going well before taking the sample. Again, the biased sample may imply that the pieces produced are better than they actually are. Use of random numbers in planning a sample can be the best defense against a haphazard bias.

EXAMPLE 3.3

Use a random-number generator on a statistical calculator to plan the random times for six samples during an 8-hour shift that begins at 7:00 A.M. Plan how the random-number digits will be used before starting. The calculator generator has three-digit numbers.

Solution

Let the digits represent hour:minutes from 0:00 to 7:59, and add the hour digit to the 7:00 A.M. beginning of the shift. Omit any random numbers that do not apply.

Random Digits	Sample Time	
370	—	(does not apply, 70 > 59)
297	—	(does not apply, 97 > 59)
078	—	(does not apply, 78 > 59)
679	—	(does not apply, 79 > 59)
644	1:44 P.M.	(7:00 + 6:44)
541	12:41 P.M.	(7:00 + 5:41)
825	—	(does not apply, 8 > 7)
115	8:15 A.M.	(7:00 + 1:15)
046	7:46 A.M.	(7:00 + 0:46)
343	10:43 A.M.	(7:00 + 3:43)
193	—	(does not apply, 93 > 59)
538	12:38 P.M.	(7:00 + 5:38)

If the time that it takes to gather the sample pieces exceeds three minutes (if, for example, the sample times at 12:38 P.M. and 12:41 P.M. overlap), take two consecutive samples starting at the earlier time (12:38 P.M.). This procedure is quick and should be done for each shift taking samples each day. As a last step, the sample times should be listed sequentially.

Data Organization

In tables, data can be listed in order from smallest to largest. They can also be condensed into a frequency distribution. Data can be shown in charts, graphs, pictures, and diagrams. Individual data values can be charted on a run chart as they are produced. Either individual values or sample averages can be charted on a control chart. If the data occur in pairs of measurements, the paired values can be graphed on a scatterplot. A bar graph or histogram can be used to picture the distribution of the data or the frequency distribution. The tables, charts, and graphs are presented in Chapter 4.

Measures of the Center of a Distribution

The three characteristics of a distribution were presented in Figures 3.7 to 3.9. The first, location, is described using measures of the distribution's center. Mean, median, and mode are three measures used to determine the center.

Definition: The *mean* is the average value.

Mean Given a set of numbers, to find the average, or mean, add all the values and divide by the number of values that you added. The symbol for the sample mean is $\overline{x}$. The statistical notation or formula is $\overline{x} = (\Sigma x)/n$. It is read, "$x$ bar equals the sum of the x values divided by the number of values, n." The formula can also be written

$$\overline{x} = \frac{x_1 + x_2 + x_3 + \cdots + x_n}{n}$$

where subscripts are used to indicate the different x values that are being averaged. Figure 3.18 illustrates the statistical concept that the mean of a large random sample distribution should be approximately the same as the mean of the population distribution. These two measures have different symbols. Generally, in the study of statistics, population measures are indicated with a Greek letter and sample measures are indicated with an alphabetic letter:

The population mean = μ (mu):

$$\mu = \frac{\Sigma x}{N} \qquad N = \text{the number of measurements } (x) \text{ in the population}$$

A sample mean = $\overline{x}$

$$\overline{x} = \frac{\Sigma x}{n} \qquad n = \text{the number of measurements } (x) \text{ in the sample}$$

In SPC, a large sample mean, $\overline{\overline{x}}$ (x double bar) is calculated from a control chart and is used to approximate μ:

$$\overline{\overline{x}} \approx \mu$$

EXAMPLE 3.4 Given the sample measurements 8, 15, 12, 9, and 6, find the sample mean and demonstrate the different formulas.

Solution

$$\overline{x} = \frac{\Sigma x}{n}$$

$\overline{x} = \dfrac{8 + 15 + 12 + 9 + 6}{5}$ Following the directions of the formula, the five measurements are added first.

$\overline{x} = \dfrac{50}{5}$ Their sum, 50, is then divided by the number of measurements, 5. The mean $\overline{x}$ is 10.

$\overline{x} = 10$

Also, if we designate $x_1 = 8$, $x_2 = 15$, $x_3 = 12$, $x_4 = 9$, and $x_5 = 6$, then

$$\overline{x} = \frac{x_1 + x_2 + x_3 + x_4 + x_5}{5}$$

$\overline{x} = \dfrac{8 + 15 + 12 + 9 + 6}{5}$ Add the five measurements and then divide by 5. The mean $\overline{x}$ is 10.

$$\overline{x} = \frac{50}{5}$$

$$\overline{x} = 10$$

NOTE: A bar over a variable indicates an average for all the values of that variable. For example, R is the variable we use for the range; $\overline{R}$ is the average for all the R values, and $\overline{R} = (\Sigma R)/N$ for N range values.

EXAMPLE 3.5 The control charts that will be studied in Chapter 6 consist of a sequence of small samples of size n ($n = 5$ is used most often). The initial calculations with a control chart are as follows:

(a) Find the sample mean $\overline{x}$ for each sample.
(b) The range values R have been calculated. Find $\overline{R}$
(c) Calculate the average of the $\overline{x}$ values, $\overline{\overline{x}}$.

Do these calculations for the partial control chart in Table 3.1.

TABLE 3.1

Sample number	1	2	3	4	5	6	7	8
Measurements	15.1	14.8	15.2	15.0	14.9	14.5	13.8	15.2
	14.9	15.3	15.0	14.6	14.7	14.9	14.3	14.5
	15.3	14.9	14.6	14.6	14.2	14.9	14.3	14.7
	15.4	15.0	14.8	14.9	14.2	14.1	14.6	14.8
	14.8	14.5	14.4	14.4	15.0	15.1	14.5	14.8
Range	.6	.8	.8	.6	.8	1.0	.8	.7

Solution

(a) Column $\Sigma x = 75.5$ 74.5 74 73.5 73 73.5 71.5 74

$\bar{x} = \dfrac{\Sigma x}{n} = \dfrac{\Sigma x}{5} = 15.1$ 14.9 14.8 14.7 14.6 14.7 14.3 14.8

(b) $\bar{R} = \dfrac{\Sigma R}{N} = \dfrac{.6 + .8 + .8 + .6 + .8 + 1.0 + .8 + .7}{8} = \dfrac{6.1}{8} = .7625$

(c) $\bar{\bar{x}} = \dfrac{\Sigma \bar{x}}{N} = \dfrac{15.1 + 14.9 + 14.8 + 14.7 + 14.6 + 14.7 + 14.3 + 14.8}{8} = \dfrac{117.9}{8} = 14.7375$

EXAMPLE 3.6 Find the mean for the set of measurements in Table 3.2.

TABLE 3.2

2.340	2.339	2.352	2.348	2.344
2.342	2.342	2.337	2.343	2.346
2.351	2.345	2.350	2.344	2.345

Solution

$$\bar{x} = \frac{\Sigma x}{n}$$
$$= \frac{35.168}{15}$$
$$= 2.3445$$

One disadvantage of using the mean as a measure of the center of a set of data is that if extreme data values are present, the average is no longer in the middle. Extreme values adversely affect the mean. To illustrate this effect, suppose there is one more data value in Table 3.2 that has the value of 2.460. The mean is now

$$\bar{x} = \frac{\Sigma x}{n}$$
$$= \frac{35.168 + 2.460}{16}$$
$$= 2.35175$$

This number is larger than all but one of the original 15 values. The 2.35175 is the mean, but it is no longer in the center of the data.

The Median The median represents the data value that is physically in the middle when the set of data is organized from smallest to largest, i.e. ordered. The symbol for the median is $\tilde{x}$. The median can occur in two possible ways:

1. If there is an *odd* number of data values, there will be just one value in the middle when the data are ordered, and that value is the median. If there are n data values, the median will be the $\frac{n+1}{2}$ data value in the ordered set. For example, if $n = 3$, $\frac{3+1}{2} = 2^{nd}$ data value will be the median. If $n = 7$, $\frac{7+1}{2} = 4^{th}$ data value will be the median.
2. If there is an *even* number of values, order the values and average the two values that occur in the middle. If there are n data values, the median will be the average of the $\frac{n}{2}$ data value and the next one in the ordered set. For example, if $n = 4$, $\frac{4}{2} = 2^{nd}$. The median will be $\frac{2nd + 3rd}{2} = \tilde{x}$. If $n = 10$, $\frac{10}{2} = 5^{th}$. The median will be $\frac{5th + 6th}{2} = \tilde{x}$.

EXAMPLE 3.7 Find the median for the set of data in Table 3.3.

TABLE 3.3

4.231	4.238	4.234	4.227	4.225	4.230	4.225

Solution

Step 1. Order the data values:

4.225, 4.225, 4.227, 4.230, 4.231, 4.234, 4.238

Step 2. There are seven values (an odd number), so the median is the $\frac{7+1}{2} = 4^{th}$ one when they are ordered. The median $\tilde{x}$ is 4.230.

EXAMPLE 3.8 Suppose one additional data value, 4.150, is included in the data of Table 3.3. Find the median.

Solution

Step 1. Order the data values:

4.150, 4.225, 4.225, 4.227, 4.230, 4.231, 4.234, 4.238

Step 2. Average the two in the middle, $\frac{n}{2} = \frac{8}{2} = 4^{\text{th}}$.

$$\tilde{x} = \frac{4^{\text{th}} + 5^{\text{th}}}{2}$$

$$\tilde{x} = \frac{4.227 + 4.230}{2}$$

$$= \frac{8.457}{2}$$

$$= 4.2285$$

Even though an extremely small data value was included, the value of the median was not adversely affected; it is still a measure of the middle.

Mode The mode is usually included in a discussion of measures of central tendency, but it is not necessarily a measure of the center. The mode represents the data value that occurs the most or the class that has the highest frequency in a frequency distribution. It is often used in a descriptive manner with distributions.

EXAMPLE 3.9 Find the mode for the set of data in Table 3.4.

TABLE 3.4

2.9	2.5	2.8	2.5	2.9	2.8
2.7	2.1	2.9	2.9	2.4	2.9
2.5	2.6	2.9	2.2	2.5	2.7
2.5	3.0	3.1	2.5	2.9	3.0

Solution

Order the data:

2.1, 2.2, 2.4, 2.5, 2.5, 2.5, 2.5, 2.5, 2.5, 2.6, 2.7, 2.7, 2.8, 2.8, 2.9, 2.9, 2.9, 2.9, 2.9, 2.9, 2.9, 3.0, 3.0, 3.1

The mode is 2.9 (the measurement that occurs the most). The distribution is bimodal because it has two distinct high frequencies at 2.5 and 2.9.

Three major control charts for variables can be used in process control. Two of them feature the mean and the other uses the median. The mean is the more traditional middle value in statistics; most of the statistical formulas use it. The median has more appeal in process control for some companies because it is easier to find in a set of measurements.

The median is a more stable measurement than the mean because it is unaffected by extreme data values. The two groups of numbers, 10, 13, 14, 17, and 21 and 10, 13, 14,

17, and 36, have the same median, 14, but different means. The mean of the first group is 15 and the mean of the second group increased from 15 to 18, which does not seem to represent the middle. For symmetrical distributions, the mean and the median will have virtually the same value.

EXAMPLE 3.10 The sample data in Figure 3.20 has been organized in a frequency distribution. There are two measurements of 3.0, 5 measurements of 3.1, etc.

1. Calculate the mean.
2. Find the median.
3. Find the mode.
4. Draw the expected shape of the population distribution based on the given sample data.

Solution

1. The repeated additions in each column can be done by multiplication:

$$\bar{x} = \frac{\Sigma x}{n}$$

$$= \frac{3.0\times2+3.1\times5+3.2\times8+3.3\times12+3.4\times7+3.5\times4+3.6\times2+3.7\times10+3.8\times6+3.9}{57}$$

$$= \frac{195.4}{57} = 3.43$$

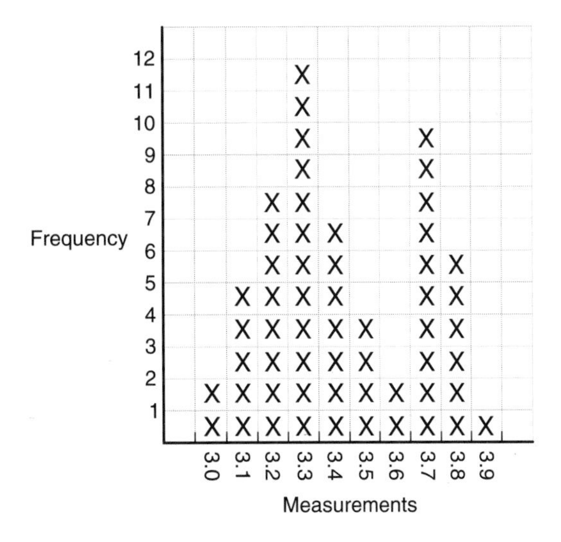

FIGURE 3.20
A bimodal distribution.

FIGURE 3.21
The expected shape of the population distribution.

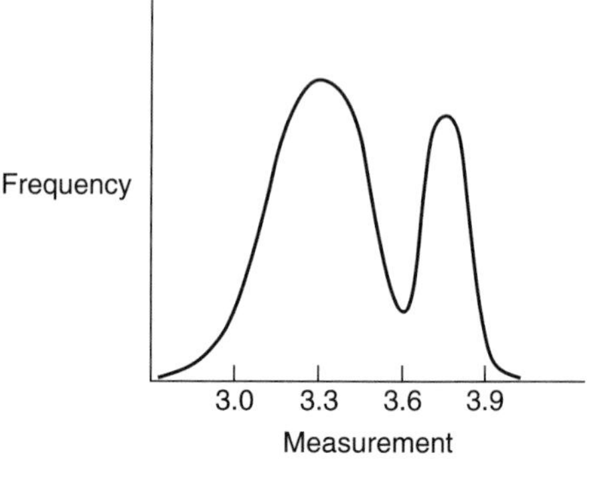

2. The median of 57 values is the 29th value, (57 + 1)/2.
   a. Order the values: The frequency distribution has them ordered.
   b. The middle value is the 29th value from the left:

$$\tilde{x} = 3.4$$

3. The mode is 3.3, the measurement with the highest frequency.
4. We would need a larger sample ($n > 100$) for more assurance, but the shape of the population distribution should mimic the bimodal shape of the sample (Figure 3.21).

Measures of the Spread

The second characteristic of a distribution, the spread, was pictured in Figure 3.8. The spread of a distribution of measurements illustrates the extent of variability in those measurements.

Range The *range, R,* is an easily calculated measure of the spread and is used extensively in process control. It is the difference between the largest and the smallest data values:

Definition: Range = largest value − smallest value

EXAMPLE 3.11 Given the measurements 8, 15, 12, 9, and 6, find the range.

Solution

Scan the data to find the largest and smallest data values. Then, subtract them to find the range.

$$\begin{aligned} R &= 15 - 6 \\ &= 9 \end{aligned}$$

Standard Deviation One of the needed features of a measure of variability is a way of determining how the data values are distributed within the range of values. The measure that does this is called the *standard deviation* and it is used extensively in statistics. The work in Chapter 2 emphasized that quality can be improved by reducing variation in the process. The two types of variation, special-cause variation and common-cause variation, both rely on the standard deviation for analyses. Since the application of statistics deals primarily with making decisions about a population based on the information in a sample, there are two versions of the standard deviation, one for the population and one for the sample.

The population standard deviation uses the symbol σ (sigma, the lowercase Greek letter. Remember, the uppercase sigma, Σ, indicates "the sum of . . .").

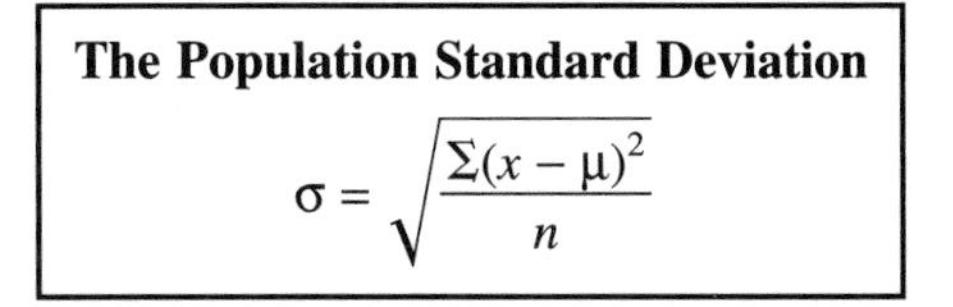

The Population Standard Deviation

$$\sigma = \sqrt{\frac{\Sigma(x - \mu)^2}{n}}$$

Generally, the population mean μ is not known and the population standard deviation cannot be calculated. The measure used most often is the sample standard deviation. The symbol for the sample standard deviation is s:

The Simple Standard Deviation

$$s = \sqrt{\frac{\Sigma(x - \bar{x})^2}{n - 1}}$$

The procedure for calculating the standard deviation is illustrated in the following example.

EXAMPLE 3.12 Given the measurements 8, 15, 12, 9, and 16, calculate the standard deviation with the formulas

$$\sigma = \sqrt{\frac{\Sigma(x - \mu)^2}{n}} \qquad s = \sqrt{\frac{\Sigma(x - \bar{x})^2}{n - 1}}$$

Solution

The mean is $\Sigma x/n = 60/5 = 12$. If the five data values represent the entire population, then $\mu = 12$ and the population standard deviation σ can be calculated with the formula on the left. If the five data values are a sample from a population, their mean is $\bar{x} = 12$ and the sample standard deviation s can be calculated with the formula on the right. The procedure for the two is primarily the same, and the steps follow the order of operations for the formula.

Step 1. Find the differences between the data values and the mean to calculate $(x - \bar{x})$ or $(x - \mu)$:

$$8 - 12 = -4$$
$$15 - 12 = 3$$
$$12 - 12 = 0$$
$$9 - 12 = -3$$
$$16 - 12 = 4$$

Step 2. Square the differences to find $(x - \bar{x})^2$ or $(x - \mu)^2$:

$$(-4)^2 = 16$$
$$(3)^2 = 9$$
$$(0)^2 = 0$$
$$(-3)^2 = 9$$
$$(4)^2 = 16$$

Step 3. Add the squared differences:

$$\Sigma(x - \bar{x})^2 = 16 + 9 + 0 + 9 + 16 = 50$$

or

$$\Sigma(x - \mu)^2 = 50$$

Step 4. Divide by n for the σ formula and by $n - 1$ for the s formula:

$$\frac{\Sigma(x - \mu)^2}{n} = \frac{50}{5} = 10 \qquad \frac{\Sigma(x - \bar{x})^2}{n - 1} = \frac{50}{4} = 12.5$$

Step 5. Find the square root (Use the $\sqrt{\ }$ button on the calculator):

$$\sigma = \sqrt{\frac{\Sigma(x - \mu)^2}{n}} = \sqrt{10} = 3.162 \qquad s = \sqrt{\frac{\Sigma(x - \bar{x})^2}{n - 1}} = \sqrt{12.5} = 3.536$$

In the 1920s, when Walter Shewhart developed his control chart concepts at Bell Laboratories, the calculation of s for 100 or more sample values was a tedious calculation. To make the calculations easier for the average worker, he developed a simpler formula:

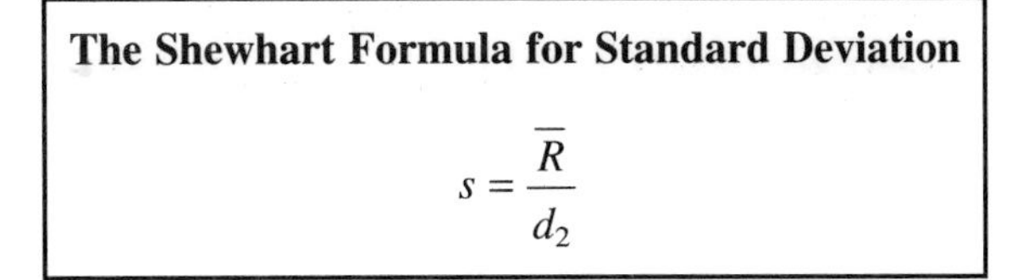

based on the average of the sample ranges $\bar{R}$. Shewhart's control charts were designed to give good information about a population of measurements based on 20 to 25 small samples. The sample sizes were usually $n = 3$ to $n = 5$, and the sample averages $\bar{x}$ were charted along with the sample ranges R. The range is the simplest measure of variability, and Shewhart's formula, $s = \bar{R}/d_2$, gives a good estimate of the population standard deviation. However, the sample range is very sensitive to the sample size. The range values for larger samples will generally be larger than those for small samples. For example, the R values for $n = 5$ will be larger than the R values for $n = 3$. The d_2 constants, dependent on n, are designed to give consistent estimates of σ regardless of the sample size. Table B1 in the Appendix lists the d_2 values with their sample size n.

The calculation of s can be easily done on a statistical calculator, but Shewhart's formula is still used with control charts because it is faster to calculate $\bar{R}$ for 25 samples than to calculate s directly from 125 data values.

The standard deviations in the preceding square-root form are the most complicated statistical calculation we have encountered thus far. The simpler range method using Shewhart's formula $s = \bar{R}/d_2$ is the one used in SPC. The square-root method is included to illustrate the idea that, conceptually, the standard deviation is something like an average difference between the data values and the mean. When s is used on a control chart, as shown in Chapter 6, a statistical calculator should be used.

EXAMPLE 3.13 Twelve samples of five measurements each have the following range values: 2.7, 3.2, 1.1, 2.7, 2.5, 0.4, 3.0, 2.2, 1.8, 2.6, 1.3, and 1.6. Calculate the standard deviation of the measurements using Shewhart's formula, $s = \bar{R}/d_2$.

Solution

Calculate $\bar{R}$:

$$\bar{R} = \frac{\Sigma R}{k}$$

$$= \frac{25.1}{12}$$

$$= 2.092$$

From Table B.1, $n = 5$, so $d_2 = 2.326$.

Second, calculate s:

$$s = \frac{\overline{R}}{d_2} = \frac{2.092}{2.326} = .899$$

Note: The symbol that represents the number of values may differ in statistical formulas. One common convention is to use n for the sample size, N for the population size and K for the number of samples. In Example 3.13 the sample size is $n = 5$, there are $K = 12$ samples and the total number of measurements is $N = 60$.

Variance The population variance is an average squared difference between the individual measurements and the mean. The symbol for the variance uses the Greek letter sigma squared, σ^2:

The Population Variance

$$\sigma^2 = \frac{\Sigma(x - \mu)^2}{N}$$

NOTE: The square root of the variance gives the standard deviation.

There are statistical techniques that use variance directly, but for the general use in statistical process control, the variance is one step away from the standard deviation. As was noted with the standard deviation, the population mean is usually unknown, so the population variance cannot be calculated. Usually the sample variance is used. The symbol for the sample variance is s^2 and the formula is

The Sample Variance

$$s^2 = \frac{\Sigma(x - \bar{x})^2}{n - 1}$$

The square root of the sample variance gives the sample standard deviation.

EXAMPLE 3.14 Given the data 2.71, 2.92, 2.67, 2.78, 2.84, and 2.82, calculate the mean, median, range, sample variance, and sample standard deviation s.

Solution

Put the data in a column format, in order (for the median), in column x and calculate Σx:

x	$x - \bar{x}$	$(x - \bar{x})^2$
2.67	$2.67 - 2.79 = -.12$	.0144
2.71	$2.71 - 2.79 = -.08$	.0064
2.78	$2.78 - 2.79 = -.01$	.0001
2.82	$2.82 - 2.79 = .03$	.0009
2.84	$2.84 - 2.79 = .05$	.0025
2.92	$2.92 - 2.79 = .13$	.0169
$\Sigma x = 16.74$		.0412

Calculate the mean:

$$\bar{x} = \frac{\Sigma x}{n} = \frac{16.74}{6} = 2.79$$

The 2.79 value is used to calculate $x - \bar{x}$ for each data value x.

Calculate the median:

The data are ordered in the first column for the median. There is an even number of values, so the median is the average of the middle two:

$$\tilde{x} = \frac{2.78 + 2.82}{2} = \frac{5.60}{2} = 2.80$$

Calculate the range:

$$R = \text{largest value} - \text{smallest value} = 2.92 - 2.67 = .25$$

Calculate the variance:

The x values were put in a column format and the mean was calculated to be 2.79. Calculate the differences $x - \bar{x}$ in column 2. Square the differences in column 3. Add the squared values in column 3 for $\Sigma(x - \bar{x})^2 = .0412$. Substitute the values found in the preceding steps into the formula and calculate the final answer:

$$s^2 = \frac{\Sigma(x - \bar{x})^2}{n - 1}$$

$$s^2 = \frac{.0412}{5}$$

$$s^2 = .00824$$

Calculate the sample standard deviation:

To calculate the sample standard deviation s take the square root of the variance:

$$s = \sqrt{.00824}$$

$$s = .091$$

3.6 DISTRIBUTIONS AND THREE STANDARD DEVIATIONS

The standard deviation is one of the most difficult concepts in the introduction to statistics. Its importance as a measure of the variability in a set of data can be illustrated by a few numerical examples. First, if the data values are all clustered close to their average value, the standard deviation will be small and when the data values are spread farther away from their mean, the standard deviation will be large.

EXAMPLE 3.15 Find the sample standard deviation s using a statistical calculator. (Your calculator may use any of the following symbols instead of s: σ_{n-1} or σ_{xn-1}.)

(a) For the data 7, 7, 7, 8, 8, 8, 8, 8, 8, 9, 9, 9

Solution

$$\bar{x} = 8, \quad s = .74$$

(b) For the data 7.9, 7.9, 7.9, 8, 8, 8, 8, 8, 8, 8.1, 8.1, 8.1

Solution

$$\bar{x} = 8, \quad s = .074$$

(c) For the data 1, 2, 3, 6, 7, 7, 8, 8, 8, 8, 9, 9, 10, 13, 14, 15

Solution

$$\bar{x} = 8, \quad s = 3.93$$

The second standard deviation concept that is relied on in data analysis is that in most distributions, virtually all the measurements lie within three standard deviations of the mean. All the control charts have their control limits three standard deviations from the mean. All the charted points are expected to fall between the control limits and when they do not, trouble is indicated. When an in-control process is analyzed to determine how good it is relative to the process specifications it is called a capability analysis. The capability analysis relates the three standard deviation spread from the mean to the tolerance. If the three standard deviation spread is well within the tolerance, virtually all the measurements in the population will be too. The following example indicates how the individual data values relate to the three standard deviation spread.

EXAMPLE 3.16 For the following sets of data determine the $\bar{x} \pm 3s$ spread and the relationship of the data values to it:

1. For the three data sets in Example 3.15 (Figures 3.22–3.24)

Solution

(a) $\bar{x} \pm 3s = 8 \pm 3 \times .74$

$= 5.78 \text{ to } 10.22$

(b) $\bar{x} \pm 3s = 8 \pm 3 \times .074$

$= 7.78 \text{ to } 8.22$

(c) $\bar{x} \pm 3s = 8 \pm 3 \times 3.93$

$= -3.79 \text{ to } 19.79$

2. For the skewed distribution of data (Figure 3.25)

7, 7, 7, 7, 8, 8, 8, 8, 8, 8, 8, 8, 9, 9, 9, 9, 10, 10, 12, 13, 15, 20, 25, 26

Solution

$$\bar{x} \pm 3s = 10.8 \pm 3 \times 5.4$$
$$= -5.4 \text{ to } 27$$

3. For a large data set with one outlier (Figure 3.26)

Solution

X	**Frequency**
7	10
8	15
9	10
12	1

$$\bar{x} \pm 3s = 8.11 \pm 3 \times 1.01$$
$$= 5.10 \text{ to } 11.14$$

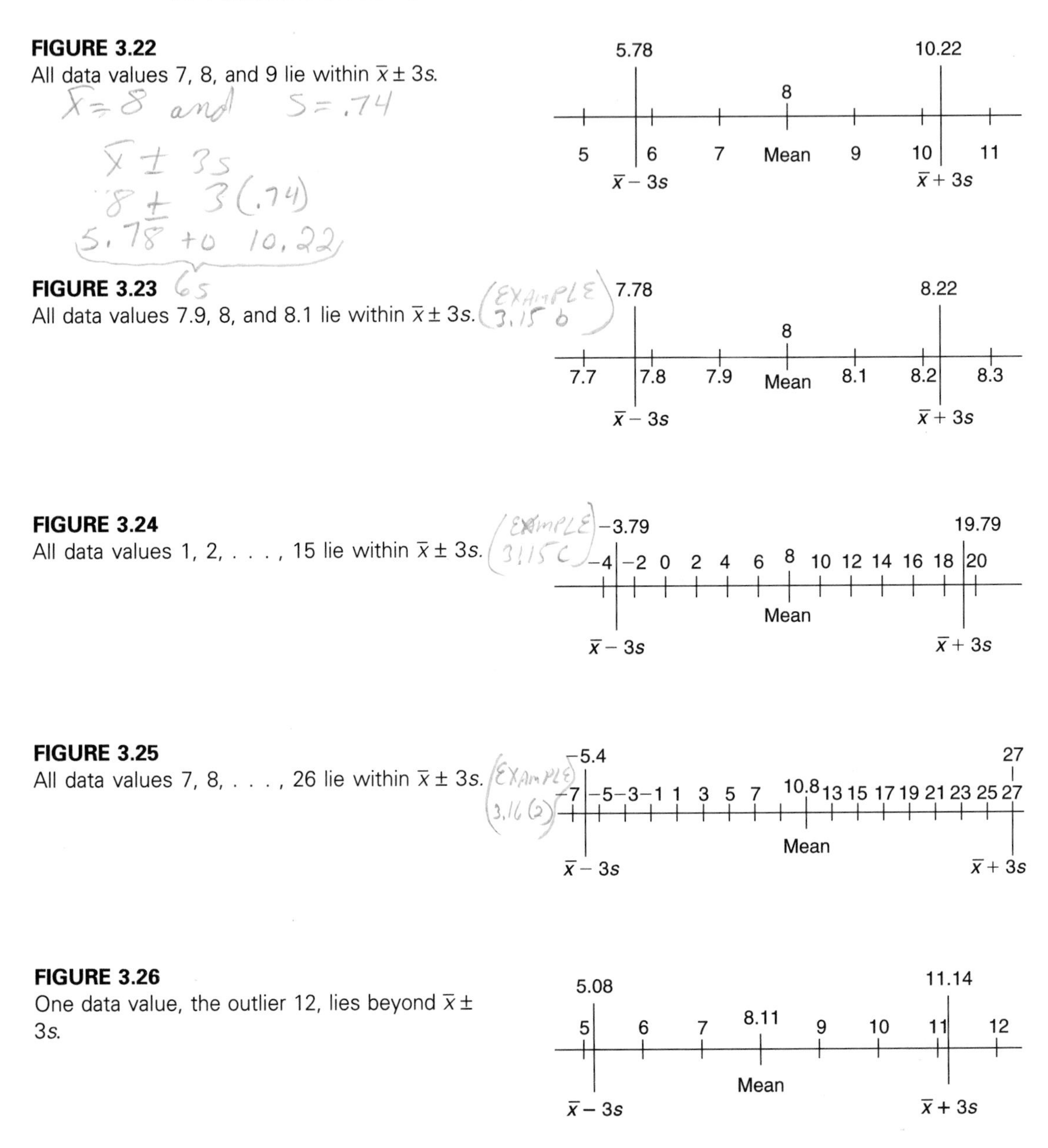

FIGURE 3.22
All data values 7, 8, and 9 lie within $\overline{x} \pm 3s$.

FIGURE 3.23
All data values 7.9, 8, and 8.1 lie within $\overline{x} \pm 3s$.

FIGURE 3.24
All data values 1, 2, . . . , 15 lie within $\overline{x} \pm 3s$.

FIGURE 3.25
All data values 7, 8, . . . , 26 lie within $\overline{x} \pm 3s$.

FIGURE 3.26
One data value, the outlier 12, lies beyond $\overline{x} \pm 3s$.

The introduction to distributions in this chapter indicated that all distributions have three measurable characteristics. The measure of the *location* of the distribution is generally the *mean.* The measure of the *spread* or *variability* is usually either the *range* or the *standard deviation.* The measure of the *shape* of the distribution is usually done informally with a picture analysis using a *histogram.*

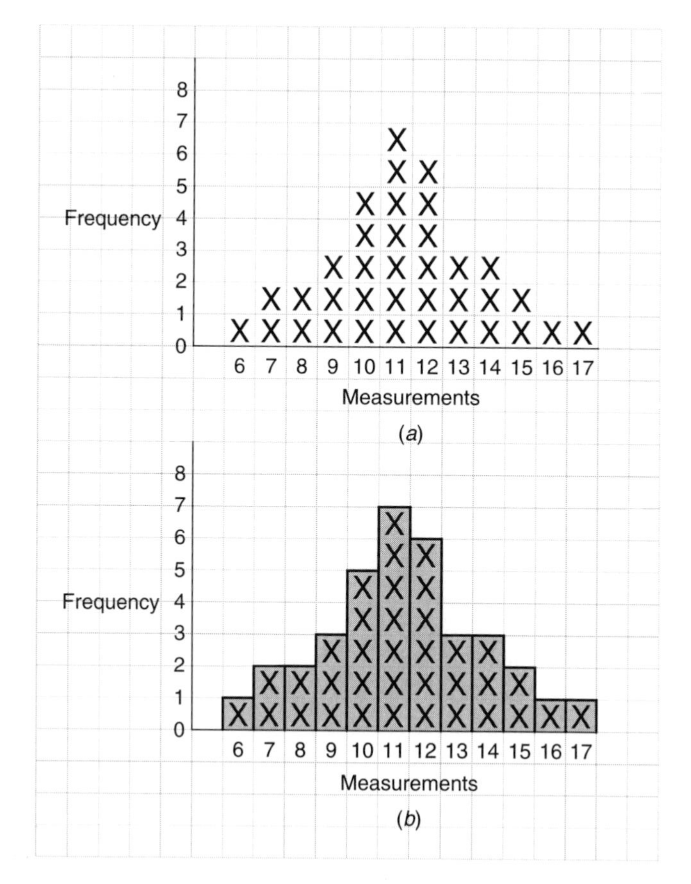

FIGURE 3.27
(*a*) The tally form; (*b*) the histogram.

The Histogram

A *histogram* is a picture of the data that shows the comparative frequency of the measurements. Figure 3.27 is a histogram drawn from the measurements in Table 3.5. The measurements range from 6 to 17, so the horizontal axis is marked off with these values. The initial histogram is in tally form: As we go through the set of data, we put an "X" above the appropriate value on the horizontal axis. The frequency naturally builds. The frequency value can be read from the vertical scale for each column.

TABLE 3.5

12	11	6	17	10	14	11	12	14
11	7	9	11	12	9	10	16	8
11	13	12	7	13	15	14	10	12
10	12	15	8	10	11	9	11	13

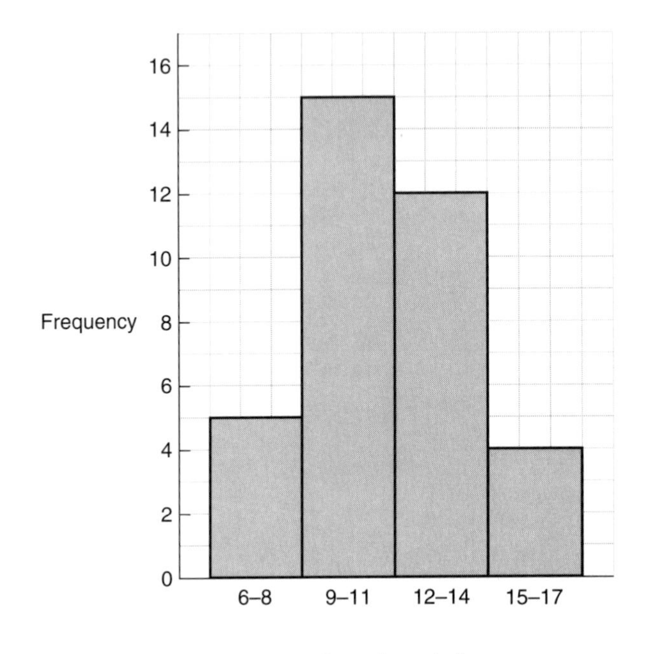

FIGURE 3.28
A histogram for grouped data.

The tally histogram is formalized to create the official histogram in bar graph form: Make a bar over each data value. The bars should touch and have the same one-unit width.

Some situations have too many different individual values for this type of histogram to be meaningful. The histogram should have a distinct shape similar to the shape of the population distribution. If the 36 data values given for the histogram in Figure 3.27 ranged from 6 to 17 and were accurate to the nearest tenth with values like 7.2, 12.3, and 13.1, there would be too many different values to chart individually. More than 100 measurements would be on the horizontal axis, if it was marked off in tenths, and the chart would consist of many columns with heights of 0, 1, or 2. This would not represent the shape of the population distribution. Grouped data should be used when this occurs (see Figure 3.28). Groups of 6 to 8, 9 to 11, 12 to 14, and 15 to 17 can be formed and all the variates totaled in each to create a histogram for grouped data. It is important that all the group widths be the same. Chapter 4 provides a more detailed approach to making the grouped data histogram.

Exercises

1. For the two measurements, (1) 1.4982 inches and (2) .631 millimeters, answer the following:
   a. What is the accuracy of each measurement?
   b. What is the maximum error of each measurement?
   c. Round off each measurement to the nearest 10th.
   d. Round off each measurement to the nearest 100th.
   e. Round off each measurement to the nearest unit.
2. Round-off exercises:
   a. Round off the following to the nearest thousandth:

   2.41349 2.4157 2.4198 2.4173

   b. Round off the following to the nearest hundredth:

   4.245 4.242 4.235 4.2349
   4.246 4.2476 4.2451

3. Three workers measure three different shafts with three different measuring devices. The measurements are 4.8 inches, 4.76 inches, and 4.764 inches.
   a. Use the appropriate round-off rules to find the sum of the measurements.
   b. Apply the maximum error to each measurement to find the minimum possible sum and the maximum possible sum. Does your answer from (a) lie between them?
4. What are the two types of variation in SPC?
5. What is the tolerance of a measurement?
6. If the curve in Figure 3.4 represents the distribution of a day's production, how can it be interpreted in terms of frequency?
7. Every time a measurement deviates from the target value, the worker adjusts the machine toward the target value. Why does this conscientious approach result in poor quality?
8. What are the three basic ways that distributions can differ?
9. A supplier sends your company a shipment of parts and one of the critical measurements has the specification of $3.5 \pm .3$ inches.
   a. What is the target value for the measurement?
   b. What is the tolerance?
   c. The distribution of a sample of measurements from the shipment is shown in Figure 3.21. Are these parts acceptable? Why or why not?
10. If special-cause variation is present, can a future distribution of product measurements be predicted?
11. What are the four phases involved in the application of statistics?
12. If common-cause variation is the only variation present, can a future distribution of product measurements be predicted?
13. What are the two main classifications for the measurements in the analysis of data?
14. Evaluate the following algebra expressions using the values $x_1 = 4$, $x_2 = 8$, $x_3 = 10$, and $y = 9$: (A review of Appendix A may be helpful).
   a. Σx
   b. yx_2
   c. $x_3 - x_1$
   d. $7y$
   e. $\frac{x_3}{x_1}$
   f. $x_2 + 5x_1$
   g. y^2
   h. $\sqrt{x_1}$
   i. $\sqrt{x_2}$
   j. $\frac{x_1 + x_3}{3}$

15. Use the following random numbers, taken in order by column, to plan the times for a random sample of 12 samples for the 3:00 PM to 11:00 PM shift. If it takes 4 minutes to collect the data for one sample, organize the sample times sequentially.

765	628	079	227	468	391
618	044	278	448	651	452
671	406	860	011	806	652
492	310	936	550	284	850

16. For the following statistical calculations, complete the table:

x	$x - \bar{x}$	$(x - \bar{x})^2$
2.4		
1.6		
.8		
1.2		

a. What is the symbol for the sample mean?
b. What is the symbol for the range?
c. What is the symbol for the median?
d. What are the standard deviation symbols and what do they represent?
e. Calculate $\bar{x}$.
f. Calculate R.
g. Calculate $\tilde{x}$.
h. Calculate σ assuming that $\mu = 1.5$.
i. Calculate s.

17. Evaluate the following algebra expressions using the given values. (A review of Appendix A may be needed.) $x_1 = 8$, $x_2 = 18$, $x_3 = 5$, $x_4 = 9$, $y_1 = 13$, $y_2 = 6$

a. $\Sigma\, x$ **c.** $x_1 - y_1$ **e.** $\dfrac{x_2}{y_2}$ **g.** $\sqrt{x_4}$

b. $x_2\, y_1$ **d.** $4y_2$ **f.** $\Sigma\, x - \Sigma\, y$ **h.** $\dfrac{x_2 + x_3}{2}$

18. Given the table of measurements

Table 3.6						
24	26	30	28	29	28	29
29	25	30	26	29	27	31
32	28	28	27	27	26	27
25	29	27	30	28	27	28

a. Make a tally histogram.
b. Make a histogram for the individual measurements.
c. Make a grouped histogram with the first group 24 to 25, the second group 26 to 27, etc.

19. Evaluate. (A review of Appendix A may be needed.)

a. $5 + (-8.2)$ **d.** $-6.7 + 4.2$ **g.** $-2.3 - (-3.5)$ **j.** $-(2.3)^2$

b. $-9.3 + (-7.5)$ **e.** $\dfrac{-2.8}{.7}$ **h.** $\dfrac{-18.3}{1.5}$

c. $-4.2\,(-3.5)$ **f.** $-.9.4 - 4.3$ **i.** $(-2.8)^2$

20. Evaluate: (A review of Appendix A may be needed.)

a. $2 + (-9.4)$
b. $-6.5 + (-4.1)$
c. $(9.2) + (-2.8)$
d. $-3.4(-5.1)$
e. $\frac{-6.5}{-5}$
f. $-8.7 - 3.4$
g. $(2.4)(-3.6)$
h. $4.6 - (-3.8)$
i. $\frac{8.4}{-1.4}$
j. $(-2.3)^2$

21. Complete the table and do the statistical calculations.

x	$x - \bar{x}$	$(x - \bar{x})^2$
4.8		
5.2		
6.4		
7.3		
7.8		

a. $\Sigma x =$
b. $\bar{x} =$
c. $\tilde{x} =$
d. R =
e. σ, assuming $\mu = \bar{x}$
f. $s =$
g. Population variance =
h. Sample variance =
i. Identify (b) through (f) by name.
j. Identify (g) through (h) by symbol.

22. Use a statistical calculator to check the results for $\bar{x}$, σ, and s in the following:

a. Exercise 16
b. Example 3.12
c. Example 3.14
d. Exercise 21.

23. A control chart has 9 samples of size $n = 4$. The Range values are given below. Use Shewhart's formula and Table B1 in the appendix to find the value of s.
Range values: 1.23 .86 1.02 1.14 .72 1.05 .63 .86 .32

24. For the data in Figure 3.27 calculate:

a. $\bar{x}$
b. $\tilde{x}$
c. the mode
d. s
e. σ, assuming $\mu = \bar{x}$

25. Use a statistical calculator with the data values in Exercise 18 to find

a. $\bar{x} =$
b. Population standard deviation, assuming $\bar{x} = \mu$
c. Sample standard deviation
d. σ^2
e. s^2

26. Which type of variation can be eliminated from a process?

27. Find the median and the mode for the data in Excercise 18.

28. Which type of variation contains most of the process problems?

29. Use the results from Exercise 25 to find the following values for the data in Exercise 18.

a. $\bar{x} \pm s$

b. $\bar{x} \pm 2s$

c. $\bar{x} \pm 3s$

d. What percentage of the data values in Exercise 18 are more than two standard deviations from the mean?

30. Why is it a problem if common-cause variation is mislabeled as special-cause variation?

31. For the measurements in Table 3.7, each column represents a sample of size $n = 4$.

Table 3.7

4.2	3.7	3.5	3.9	4.1	$n = 4$ data values per sample
3.8	4.8	3.6	4.2	4.4	$k = 5$ samples
4.1	3.6	3.8	4.5	4.5	$N = 20$ data values
4.3	3.9	3.9	4.2	4.6	

a. Calculate the sample mean for each sample.

b. Calculate the Range for each sample.

c. Calculate $\bar{\bar{x}}$ two ways: $\bar{\bar{x}} = \dfrac{\Sigma \bar{x}}{k}$, $\bar{\bar{x}} = \dfrac{\Sigma x}{N}$

d. Calculate s using Shewhart's formula.

32. Why is it a problem if special-cause variation is mislabeled as common-cause variation?

33. Make a tally histogram for the data in Table 3.7 using the groups 3.5 to 3.7, 3.8 to 4.0, etc.

34. Why is service quality sometimes more difficult to achieve than production quality?

35. Use the results of Exercise 31 to do the following:

a. Calculate $\bar{\bar{x}} \pm 2s$

b. Calculate $\bar{\bar{x}} \pm 3s$

c. What percentage of the data values is *within* 3 standard deviations of the mean, $\bar{\bar{x}}$?

d. What percentage of the data values are *more than* 2 standard deviations from the mean, $\bar{\bar{x}}$?

36. When a company's goal is to stay within specification and it does, why does that eventually cause problems for the company?

37. What is the mode for the data in Table 3.7?

38. How does a biased sample cause problems?

39. What is the modal class or modal group for the histogram in Exercise 33?

40. What relationships are expected between the population distribution and a large sample distribution?

41. When a process is in statistical control, what do the daily distributions of a specific measurement look like relative to each other?

42. Calculate the standard deviation from the range values given in Example 3.5 using the Shewhart formula $s = \bar{R}/d_2$.

43. Why is it always important for a sample to be a random sample?

44. Find the $\bar{x} \pm 3s$ spread of values for the data values in the histogram in Figure 3.27.

45. How are specification limits determined?

46. Find the $\bar{x} \pm 3s$ spread of values for the data in Exercise 18.

4

ORGANIZATION OF DATA: INTRODUCTION TO TABLES, CHARTS, AND GRAPHS

OBJECTIVES

- Organize data with stemplots, split stemplots, and tally charts.
- Make a frequency distribution.
- Make a histogram.
- Interpret histograms.
- Make a Pareto chart.
- Make a flowchart.
- Create a storyboard.
- Make a cause-and-effect diagram.
- Use a cause-and-effect diagram to search for root causes.
- Make a checksheet.
- Draw a scatterplot.
- Interpret a scatterplot.

4.1 STEMPLOTS

When data are gathered in the usual rows and/or columns of numbers, they are usually impossible to comprehend. When data are gathered over a time sequence, the order in the sequence is important and the variation over time is analyzed with control charts. Control charts are studied in Chapter 6. However, the data on the control chart must be analyzed in two basic ways: the time sequence analysis (Chapter 6) and an overall distribution analysis. The data must be reorganized from the control chart to do the distribution analysis. Incoming parts and subassemblies have to be inspected, so when the sample data are collected they must be organized. Data are usually put in order from smallest to largest or grouped smallest to largest.

The *stemplot* is an efficient data organizer in a table format that has the first measurement digits ordered in a column. Then as the data are gathered, the last measurement digit is put in the row formed by its first digits.

EXAMPLE 4.1 Make a stemplot for the data in Table 4.1.

TABLE 4.1

2.36, 2.39, 2.51, 2.43, 2.56, 2.72, 2.45, 2.83, 2.75, 2.27, 2.63, 2.92, 2.88, 2.44, 2.57, 2.61, 2.58, 2.81, 2.81, 2.43, 2.62, 2.96, 2.48, 2.31, 2.67

Solution

The *stems* consist of the column of numbers 2.2 to 2.9 (Table 4.2).

Work through the data just once. For each measurement, put the last digit in the appropriate stem row.

2.36 – put the 6 in the 2.3 row

2.39 – put the 9 in the 2.3 row

2.51 – put the 1 in the 2.5 row

Continue through the data in this manner. Keep the columns of digits straight so the distribution pattern can be seen.

TABLE 4.2
The stemplot

2.2	7				
2.3	6	9	1		
2.4	3	5	4	3	8
2.5	1	6	7	8	
2.6	3	1	2	7	
2.7	2	5			
2.8	3	8	1	1	
2.9	2	6			

TABLE 4.3
The ordered stemplot

2.2	7				
2.3	1	6	9		
2.4	3	3	4	5	8
2.5	1	6	7	8	
2.6	1	2	3	7	
2.7	2	5			
2.8	1	1	3	8	
2.9	2	6			

In this stemplot, the data are ordered by tenths (with the stems). If the data must be completely ordered, just order the hundredths digits in each row (Table 4.3).

If there are just a few categories of first digits, then a split stemplot can be used. For example, if the data in Table 4.1 just had measurements from 2.40 to 2.69 there would be only three rows or categories in a regular stemplot. A split stemplot would double that to six rows or categories by putting the last digits 0 to 4 in the first 2.4 row and the last digits 5 to 9 in the second 2.4 row. This is illustrated in the following example.

EXAMPLE 4.2 Make a split stemplot for the data in Table 4.4.

TABLE 4.4

.488, .492, .491, .482, .496, .501, .499, .486, .498, .505, .491, .487, .503, .480, .495, .490, .496, .507, .491, .495, .484, .488, .490, .492, .494

Solution

The stems are the numbers .48, .49, and .50. Use each one twice for split stems (Table 4.5).

TABLE 4.5
The split stemplot

.48	2	0	4					
.48	8	6	7	8				
.49	2	1	1	0	2	1	0	4
.49	6	9	8	5	6	5		
.50	1	3						
.50	5	7						

The first digits are referred to as the stem and in the split stemplot the stem is used twice, first for the last digits 0–4 and second for the last digits 5–9. As before, the last digits may be ordered if necessary. When the variation between measurements is quite small and the data are to be organized, the use of individual measurements may result in too many categories. For example, a set of measurements with the specification .750± .006 would have 13 categories if the individual measurements were used: .744, .745, ..., .756. A normal split stemplot would have just 4 categories and that would be too few if we want a shape anaysis as part of our organization. A special split stemplot with two units per stem would provide a good alternative. There would be 7 categories:

.74	4's and 5's
.74	6's and 7's
.74	8's and 9's
.75	0's and 1's
.75	2's and 3's
.75	4's and 5's
.75	6's and 7's

The stemplots are used as a first-step data organizer and they also show the distribution pattern of the data in a form that is similar to that of a histogram.

The introduction to distributions in Chapter 3 emphasized that every distribution has three measurable characteristics that can provide important information about the population from which the sample of values was taken. The position of the distribution can be seen informally by observing where the center of the stemplot is relative to the target value for the measurements. The spread of the distribution, which shows the amount of variation, can be compared to the specification limits of the measurements. If the stemplot lies well within the specification limits, then the variation is within reason. If the stemplot crowds or overlaps the specification limits, then variation problems are evident and improvements must be made. If the columns of the stemplot are straight then the distribution pattern may be compared to the expected normal pattern. A stemplot that doesn't have the normal shape is another indicator of problems with the process that is producing the data values.

In Example 4.1, the stemplot has:

1. Location: the middle value is between 2.5 and 2.6
2. Spread: The smallest is 2.27, the largest is 2.96 and the range is .69.
3. Shape: there is a bimodal pattern with "high" points at the stems 2.4 and 2.8

In Example 4.2 the stemplot has:

1. Location: the middle is at the first .49 stem.
2. Spread: The low value is .480, the high value is .507 and the range is .027.
3. Shape: there is a normal pattern.

4.2 FREQUENCY DISTRIBUTIONS AND TALLY CHARTS

A frequency distribution, in its simplest form, is a table that indicates all the different values from a set of data and the frequency with which each different value occurred. When the number of different values is so large that many of the values have a frequency of 0, 1, or 2, the values should be grouped: the frequency distribution will then list the groups and the number of data items in them (the frequency of each group).

When putting data in a frequency distribution, the objective is to

1. Organize the data.
2. Condense the data information when there are too many different data values.
3. Obtain a distribution pattern that mimics the population distribution as much as possible.

NOTE: For item 3, the *G* Chart in Appendix B, Table B.7 gives the recommended number of groups for the frequency distribution.

EXAMPLE 4.3 Organize the data in Table 4.6 in a frequency distribution for individual measurements.

TABLE 4.6

2.8	3.2	3.3	3.0	3.5	3.5	3.5	3.4	3.0	3.3	3.1	3.0	3.0	2.9	2.8	2.9	2.6	2.7	2.8	2.8
2.9	3.1	3.6	3.4	3.5	3.4	3.1	3.0	3.4	3.2	3.3	3.4	2.9	2.9	2.8	2.8	2.9	2.7	2.9	2.8
3.1	3.4	3.3	3.6	3.5	3.6	3.3	3.4	3.4	3.0	3.2	3.4	3.1	2.9	3.0	3.0	2.9	3.1	2.6	2.6
2.8	3.1	3.4	3.6	3.5	3.5	3.3	3.4	3.0	3.3	2.8	3.0	3.0	2.9	3.0	2.9	2.8	2.7	2.5	3.0

Solution

An easy way to form the frequency distribution is to make a *tally chart* first. Notice that for this set of data a regular stemplot would not apply. A split stemplot with two units per stem would be an alternative data organizer. Scan the data for the range and list all possible data values. The data range from 2.5 to 3.6. Go through the data once and put a tally mark, x, next to the value on the tally chart (grouped in fives for easy counting). Make the frequency distribution by indicating the tally totals in the frequency column (Tables 4.7 and 4.8).

TABLE 4.7
The tally chart

2.5	x
2.6	xxx
2.7	xxx
2.8	xxxxx xxxxx
2.9	xxxxx xxxxx x
3.0	xxxxx xxxxx xxx
3.1	xxxxx xx
3.2	xxx
3.3	xxxxx xx
3.4	xxxxx xxxxx x
3.5	xxxxx xx
3.6	xxxx

TABLE 4.8
A frequency distribution of individual measurements

Measurement	Frequency
2.5	1
2.6	3
2.7	3
2.8	10
2.9	11
3.0	13
3.1	7
3.2	3
3.3	7
3.4	11
3.5	7
3.6	4
	$N = 80$

The tally chart, with straight columns, shows the shape of the distribution and the basic characteristics of the distribution can be determined:

1. Location: the center is about 3.0.
2. Spread: the smallest value is 2.5, the largest is 3.6, and the range is 1.1.
3. Shape: the distribution is bimodal with "high" frequency values at 3.0 and 3.4.

EXAMPLE 4.4 Organize the data from Table 4.6 in a frequency distribution with grouped measurements.

Solution

Since there are only 12 different measurements, groups of two would be best. The first step would be to make a tally chart for the groups, but since an individual tally chart was made in Example 4.3, we can use that. The frequency distribution is shown in Table 4.9.

TABLE 4.9
A frequency distribution with grouped measurements

Measurement Group	Frequency
2.5–2.6	4
2.7–2.8	13
2.9–3.0	24
3.1–3.2	10
3.3–3.4	18
3.5–3.6	11
	$N = 80$

Information on a population of measurements comes from the data in a sample. The sample data can be organized in many ways, depending on the number of groups or classes used, and each way provides a picture of the population distribution with various degrees of accuracy. For every sample size, there is a preferred choice for the number of classes that will make the sample distribution most accurately mirror the population distribution. The G chart, which is given in Appendix B, Table B.7, shows the recommended number of groups or classes that correspond to specific sample sizes. Sometimes common sense overrides the G chart. There were 80 data values in Table 4.6. The G chart recommends either 9 or 10 groups. In this case that is impossible if we want to preserve the basic shape of the distribution. There are 12 different consecutive measurements. The obvious choices would be to use the 12 different measurements, as shown in Example 4.3, or to use groups of two measurements each for a total of six groups, as shown in Example 4.4. Either frequency distribution would be good here because the bimodal shape of the data is evident in both forms. The frequency distribution in table form is a data organizer and is often followed by a histogram, which is a bar graph of the distribution. When the number of groups in the frequency distribution is too large, the histogram has a large spread and a low height, which does not mimic the shape of the population distribution. If the number of groups is too small, distribution details are lost and again the shape does not fairly represent the shape of the population distribution.

The histogram provides a logical lead into the concept of the normal distribution. In Chapter 3 the histogram was introduced as a picture of the shape of the distribution of measurements. With a large number of different values, measurements have to be grouped in a frequency distribution in order to picture the product distribution from the sample distribution. The more formal procedure for setting up a frequency distribution will be demonstrated here using the data from Table 4.10.

EXAMPLE 4.5 Construct a frequency distribution for the data in Table 4.10.

TABLE 4.10

Sample						Sample					
1	933	897	885	900	879	2	911	898	900	905	862
3	889	915	905	902	873	4	882	913	930	900	871
5	903	930	890	890	900	6	890	940	895	909	915
7	892	912	895	896	902	8	908	920	896	894	906
9	895	920	922	928	926	10	916	890	891	920	915
11	901	892	892	895	898	12	908	895	896	925	933
13	909	904	906	892	927	14	895	902	902	932	932
15	893	906	917	910	925	16	909	907	904	923	888
17	885	892	942	911	916	18	897	904	916	912	920
19	912	896	932	936	913	20	882	894	941	934	917
21	896	912	907	928	926	22	912	909	913	915	928
23	926	903	908	910	885	24	917	917	918	914	925
25	884	889	912	919	898						

Solution

Step 1. Scan the data for the range:

$$\begin{aligned} R &= \text{largest} - \text{smallest} \\ &= 942 - 862 \\ &= 80 \end{aligned}$$

Step 2. Should we use individual values or grouped values in the frequency distribution? Groups should be used in this case because there are so many different individual values. How many groups? The G chart shown in Appendix B, Table B.7, indicates the optimum number of groups for the size of the set of data. For the 125 data values, 11 groups should be used.

Step 3. Find the group range R_G and set up the group endpoints. Divide the number of groups into the range and round *up* to determine the range of each group. *The group ranges must all be the same.*

$$80 \div 11 = 7.27 \qquad \text{Round } up \text{ to } 7.3, \text{ or } 7.4, \ldots, \text{ or } 8.0.$$

The minimum group range is 7.3. Round up to 8 to make the group range easier to work with.

Step 4. Check to see how the group boundaries will fit the set of data. This can be done by either algebra or trial and error. The algebra approach parallels what would be done using trial and error. Start at the smallest value and add 8 (the chosen group range) eleven times (once for each group). This determines the largest data value in the frequency distribution (950):

Trial and Error	Algebra	
862–870 (add 8)	$S + G \times R_G$	S = Smallest data value
870–878 (add 8)	$862 + (11 \times 8)$	G = Number of groups
878–886 (add 8)	$862 + 88$	R_G = Group range
•	950	
•		
•		
942–950		

Subtract the largest data value from the result. The difference is the amount of overlap. If that difference is negative, you either made an arithmetic error or did not round *up* at step 3.

$$950 - 942 = 8 \qquad \text{The overlap is 8 units.}$$

If the frequency distribution starts at the smallest data value, 862, it will end at 950, which is eight units larger than the largest data value, 942. When the overlap exceeds half of the group range, center the data by "backing up" half the number of overlap units. The overlap in this case, 8, is more than half of the R_G value, 8, so divide the overlap number in half and back up four units. The frequency distribution, then, begins at 858 (the lowest value, 862, minus four units). See Table 4.11.

It was mentioned in step 3 that there are several choices for the group range value R_G. This means that there are several possible histograms for the same set of data. The criteria for any grouped frequency distribution are as follows:

- The number of groups either matches the recommended value or is reasonable for the set of data values.
- The smallest data value fits in the first group.
- The largest data value fits in the last group.
- All the groups have the same group range.

Table 4.12 shows one of the alternate sets of group endpoints using the group range 7.3. Notice that the smaller round-up (7.3 versus 8) eliminated the overlap problem.

Step 5. Complete the frequency distribution by determining the number of data values that fall in each group (the frequency) with a tally chart like Table 4.13. The completed frequency distribution consists of a column of group endpoints and a column of frequency values, as shown in Table 4.14.

TABLE 4.11
Group endpoints for the frequency distribution ($R_G = 8$)

858–866
866–874
874–882
882–890
890–898
898–906
906–914
914–922
922–930
930–938
938–946

TABLE 4.12
Group endpoints for the frequency distribution ($R_G = 7.3$)

862.0–869.3
869.3–876.6
876.6–883.9
883.9–891.2
891.2–898.5
898.5–905.8
905.8–913.1
913.1–920.4
920.4–927.7
927.7–935.0
935.0–942.3

TABLE 4.13
The tally chart

858–866	X
866–874	XXX
874–882	XXX
882–890	XXXXX XXXXX X
890–898	XXXXX XXXXX XXXXX XXXXX XXXX
898–906	XXXXX XXXXX XXXXX XXXX
906–914	XXXXX XXXXX XXXXX XXXXX XXX
914–922	XXXXX XXXXX XXXXX XXX
922–930	XXXXX XXXXX XXX
930–938	XXXXX XX
930–946	XXX

TABLE 4.14
The frequency distribution for Example 4.5

Group Boundaries	Frequency
858–866	1
866–874	3
874–882	3
882–890	11
890–898	24
898–906	19
906–914	23
914–922	18
922–930	13
930–938	7
930–946	3

One of the problems that you may have encountered when working Example 4.5 was assigning a measurement to a group when the measurement fell on the group boundary. What group should 882 be put in? 874–882 or 882–890? There are three ways to handle this problem.

1. Use "halfway" boundary values. If the data are accurate to the nearest tenth, the boundary values would be to the five-hundredths ($\frac{.1}{2} = .05$). In this example the data are accurate to the nearest whole number so the boundary values would be to the five-tenths ($\frac{1}{2} = .5$)—857.5 to 865.5, 865.5 to 873.5, etc. This method makes the group selection obvious since a measurement of 858 would go in the first class, a measurement of 866 would be in the second class, etc. The extra

place value on the boundaries, however, makes the frequency distribution or the histogram look more complicated.

2. Use consecutive number boundary values. If the data were accurate to tenths with the smallest value 2.2 and a group range of .6, the boundary values would be 2.2 to 2.7, 2.8 to 3.3, etc. For the data in Table 4.10 the boundary values would be 858 to 865, 866 to 873, etc. With this method an adjustment must be made to the addition number (1 less than the group range value) in order to satisfy the group range and number of groups requirement. There are other little complications that have to be dealt with, such as the numerical scale on the histogram (which number goes on the common boundary of adjacent bars?).
3. Use the same number as the upper boundary of one group and the lower boundary of the next group (common boundaries). This is the method shown in Table 4.14. The advantage of this method is that it is easier than the other two to set up and use. There is just one extra rule that must be followed:

Common Boundary Rule: If a data value coincides with a common group boundary value, assign it to the higher group.

The group boundaries could be written using inequalities to specifically show this rule. Tables 4.11, 4.13 or 4.14 would then show the group boundaries as:

$858 \leq x < 866$
$866 \leq x < 874$
$874 \leq x < 882$ etc.

A data value of 858 would go in the first group (if there were a previous class, its upper boundary would be 858, and by the rule, the 858 would go into the 858 to 866 group), a value of 866 would go in the second class, etc. All three methods, as described, would result in the same frequency distribution (the same frequencies would result for each respective group).

EXAMPLE 4.6 A set of data is given in Table 4.15.

1. Find the recommended number of groups on the *G* chart in Appendix B, Table B.7.
2. Calculate the group range and set up the group boundaries by all three methods.
3. Tally the data in a tally chart with the group boundaries.
4. Make the frequency distribution.

Table 4.15
Data for Example 4.6

4.522	4.510	4.555	4.534	4.550	4.528	4.538
4.544	4.535	4.539	4.517	4.553	4.548	4.518
4.536	4.526	4.546	4.542	4.525	4.537	4.525
4.526	4.536	4.530	4.540	4.531	4.532	4.524
4.529	4.538	4.535	4.533	4.520	4.543	4.535

Solution

There are 35 data values, so either six or seven groups will be appropriate. We will use seven groups in this example. Following the preceding list of steps, we will next calculate the group range, check for overlap, and set boundaries:

$$
\begin{aligned}
R &= 4.555 - 4.510 \\
&= .045 \\
R_G &= .045 \div 7 \\
&= .0064286 \\
&\cong .007 \text{ (always round } up\text{)} \\
S + (G \times R_G) &= 4.510 + (7 \times .007) \\
&= 4.559 \\
4.559 - 4.555 &= .004 \text{ (this is the amount the table overlaps the data)}
\end{aligned}
$$

The overlap of .004 is more than half of R_G, so center the data by starting the first group at 4.508 instead of 4.510. The three methods for setting the group boundaries are shown in Tables 4.16 through 4.18.

Table 4.16
Group boundaries using method 1: "half-way" boundary values.

4.5075–4.5145
4.5145–4.5215
4.5215–4.5285
4.5285–4.5355
4.5355–4.5425
4.5425–4.5495
4.5495–4.5565

Table 4.17
Group boundaries using method 2: consecutive boundary values.

4.508–4.514
4.515–4.521
4.522–4.528
4.529–4.535
4.536–4.542
4.543–4.549
4.550–4.556

Table 4.18
Group boundaries using method 3: common boundary values.

4.508–4.515
4.515–4.522
4.522–4.529
4.529–4.536
4.536–4.543
4.543–4.550
4.550–4.557

When making the frequency distribution, just one method would be used. Whichever one is used, check to ensure that the four main criteria are satisfied:

- The number of groups is reasonable for the given data.
- The smallest data value fits in the first group.
- The largest data value fits in the last group.
- All groups have the same group range.

All three versions of the group boundaries shown in Tables 4.16 to 4.18 will have exactly the same frequencies as is shown in Table 4.19.

TABLE 4.19
The frequency distribution for Example 4.6: using method 3 common boundaries

Group Boundaries	Frequency
4.508–4.515	1
4.515–4.522	3
4.522–4.529	7
4.529–4.536	9
4.536–4.543	8
4.543–4.550	4
4.550–4.557	3

4.3 HISTOGRAMS

A *histogram* is a bar graph with a measurement scale on one axis and a frequency or percentage scale on the other. The adjacent bars share a common side and all bars are of equal width. The histogram is a picture of a frequency distribution and is generally used to show the distribution pattern of a large sample of data.

The following steps are used to make a histogram:

Step 1. Set the scale for the measurement axis using the group endpoints from the frequency distribution or the individual measurements. If the measurements have specifications, include the specification limits on the measurement scale.

Step 2. Set an appropriate scale for the frequency or percentage axis.

Step 3. Make a bar of the proper height for each group or measurement. The adjacent bars should share a common side.

When the bars represent a single measurement, the sides of the bars would be at the "half" measurement. This is illustrated in the following example.

EXAMPLE 4.7 Draw the histogram for the frequency distribution in Table 4.8.

Solution

The highest frequency is 13, so the vertical scale can be marked off in unit values from 0 to 13. The measurement scale represents the individual measurements 2.5, 2.6, . . . , 3.6, so the edges of each bar can be marked 2.45, 2.55, . . . , 3.65 or the middle of each bar could be marked 2.5, 2.6, ... These alternatives are shown in Figures 4.1 and 4.2.

A *tally histogram* can be used to organize the data instead of a tally chart or a stemplot. The tally marks, x, should be evenly spaced so the result easily converts to a histogram. The tally histogram would be organized in the same way as the tally chart for a frequency distribution to determine the group range.

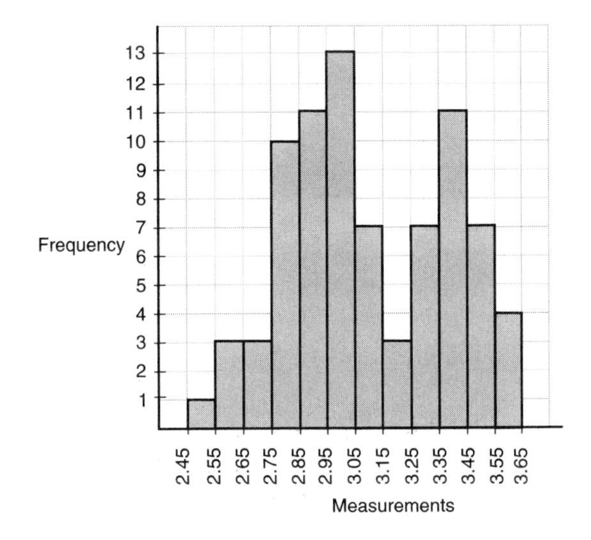

FIGURE 4.1
A histogram for individual measurements with the measurement scale at the "half" measurement.

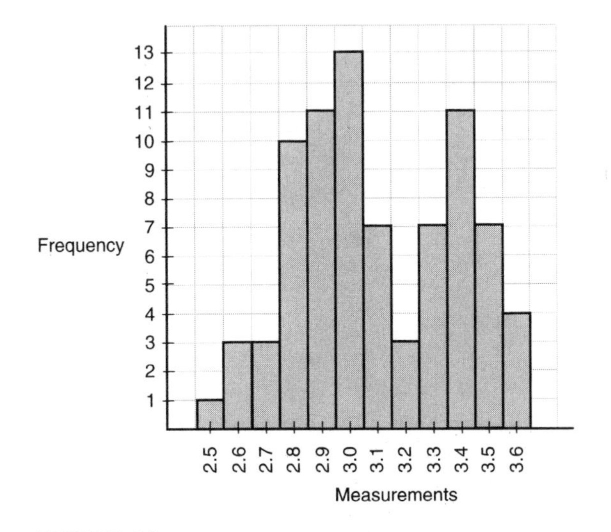

FIGURE 4.2
A histogram for individual measurements with the measurement at the middle of the bar.

EXAMPLE 4.8

1. Organize the data in Table 4.20 using a tally histogram.
2. Draw the histogram.
3. Make the frequency distribution.

TABLE 4.20

.912	.910	.911	.904	.905	.910	.910
.914	.912	.914	.910	.913	.908	.909
.907	.909	.913	.913	.912	.909	.908
.902	.906	.908	.909	.907	.906	.910
.915	.909	.909	.910	.911	.912	.909

Solution

1. The range = .915 − .902 = .013.
2. Thirteen units is too many for this chart of individual values. The *G* chart recommends seven classes.
3. Divide:

$$7\overline{).013} = .001+$$

Round up to .002 units per class.

The horizontal scale for the tally histogram is marked off with the .002 units per group or class, starting with the smallest measurement .902. The vertical scale will be in units 0, 1, 2, . . . ; the height is unknown at this time. Go through the data in Table 4.20 once putting an "x" in the appropriate column for each measurement. Remember to use the "higher class" rule since the consecutive classes share a boundary measurement. The result is shown in Figure 4.3, the frequency distribution is given in Table 4.21 and the histogram is drawn in Figure 4.4. Notice that drawing bars over the tally marks in Figure 4.3 creates the histogram in Figure 4.4. This shows that the organization of the data and a picture of the data distribution pattern can be accomplished in one step.

TABLE 4.21
Frequency distribution using the common boundary method

Class	Frequency
.902 to .904	1
.904 to .906	2
.906 to .908	4
.908 to .910	10
.910 to .912	8
.912 to .914	7
.914 to .916	3

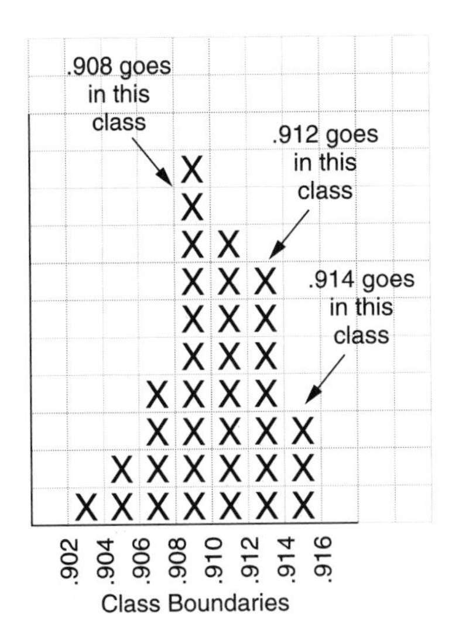

FIGURE 4.3
The tally histogram (common boundary method) for Example 4.8.

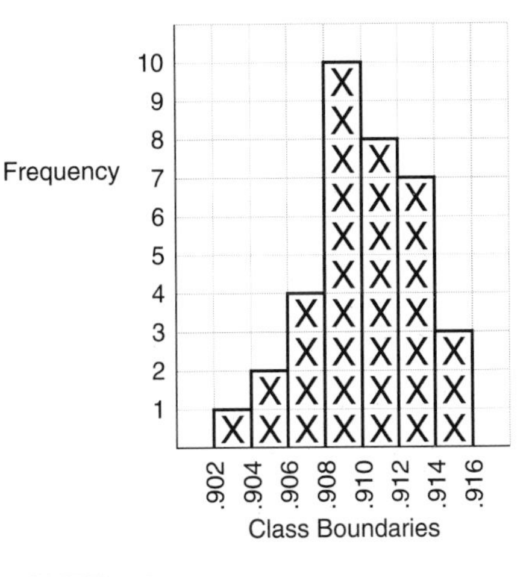

FIGURE 4.4
The histogram.

It is assumed that there is a definite shape of the population histogram at any one time and that the shape of the sample histogram will be quite similar to that of the population histogram. It is important to realize that a tally histogram grows in jumps and spurts. For example, Figure 4.5 shows what the tally histogram from Example 4.8 looks like after 10 entries, 15 entries, 20 entries, and 30 entries. Compare this with the final histogram shown in Figure 4.4.

As the sample size increases, the shape of the histogram becomes stable and the assumption is made that it looks more and more like the true population histogram. In application, a sample size of $n = 100$ or more is usually relied on to set the pattern.

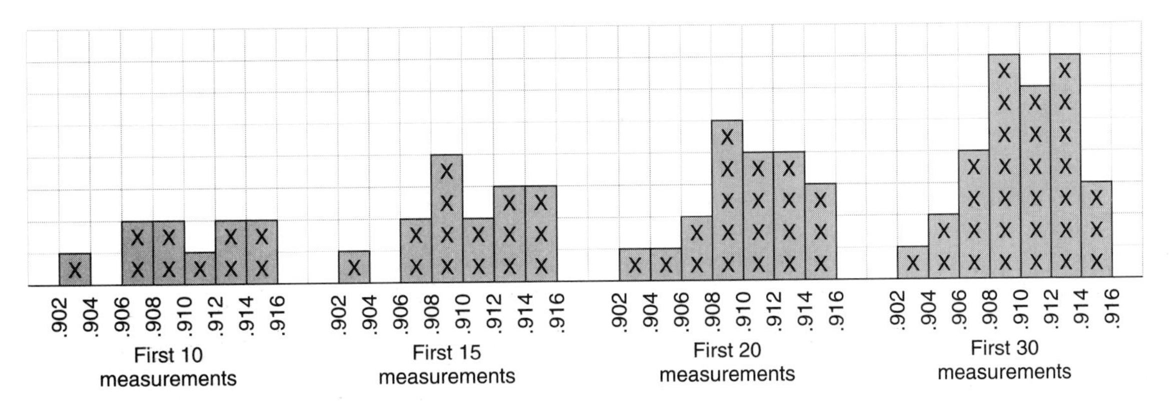

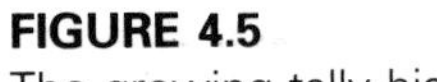

FIGURE 4.5
The growing tally histogram.

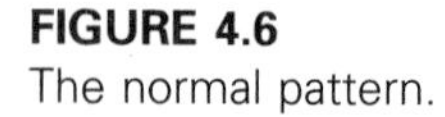

FIGURE 4.6
The normal pattern.

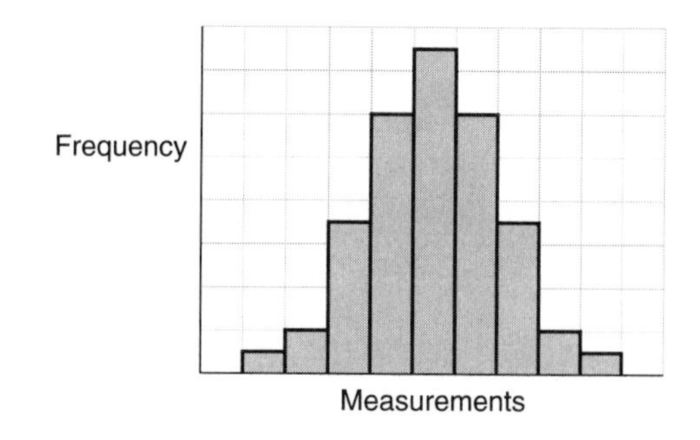

A histogram is usually expected to have a normal pattern, as shown in Figure 4.6, with the highest frequency in the center and the frequencies decreasing symmetrically on the sides as the distance from the center increases. When a data histogram does not have that pattern, something has probably affected the process. The shape of the data histogram, along with other pertinent information, is used to analyze the process and to find the cause of the process problem(s). The histogram is one of the simplest statistical concepts, but it is a very powerful tool in process analysis.

Histograms are very important in SPC because all the usual statistical formulas that are used apply to a *normal probability distribution.* The normal distribution has the bell shape that was illustrated on some previous charts, such as Figures 3.13 and 3.14. The histogram indicates when the distribution of data is normal or nearly so. A normal pattern is necessary for the application of the process capability formulas to give meaningful results.

If a histogram does not have the expected bell shape, then either something in the process is out of control and must be fixed or the shape of the distribution naturally follows some other basic pattern. If a different pattern is evident, the appropriate statistical distribution must than be matched to the shape of the histogram; the statistical formulas that accompany that distribution will then give the most meaningful results.

The previous discussion has introduced several ways to tackle the initial organization of data: stemplots, split stemplots, ordered stemplots, tally charts, tally histograms, frequency distributions, and histograms. Usually the frequency distributions and histograms require the previous organizers be used first. Which one is best or easiest? It depends on the data and the information you would like to gain. If the data have to be ordered, the stemplot followed by an ordered stemplot would be the logical choice. If a strict ordering isn't needed, the G chart can provide assistance by giving the optimum number of groups for your set of data. If the number of stems or split stems is close to that optimum number of groups, the stemplots will generally be the most efficient organizer. If the number of stems were too large, relative to the recommended number of groups, either a tally chart or tally histogram would be used to form the required number of groups.

Histogram Analysis Examples: Shape Interpretation

Production samples or samples of incoming parts from a supplier can be organized into histograms for analysis. Figures 4.7 through 4.12 illustrate different trouble patterns and the possible causes for those patterns. The two peaks shown in Figure 4.7 may indicate that

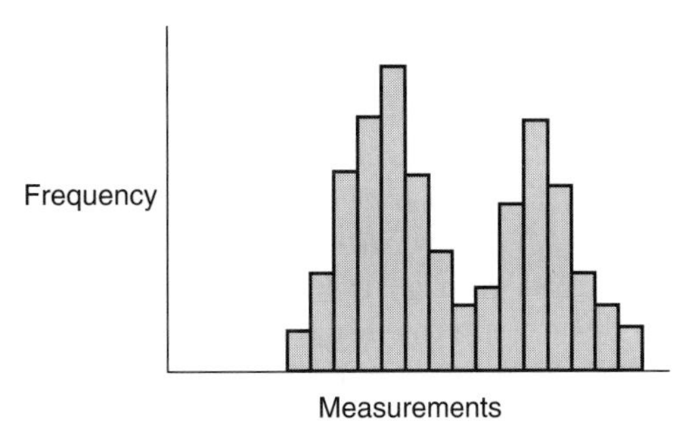

FIGURE 4.7
A bimodal distribution: a mix of two or more distributions.

the variable of interest has not been isolated. Two or more process streams may be feeding into the process point being analyzed. The bimodal pattern can indicate that there are really two different distribution patterns for a critical measurement when it was assumed or expected that there would be only one. For example, two operators may be running one machine and some additional training is needed for one or both of them. Two different shifts may be producing different results or maybe two different suppliers are producing the same part but with different distribution patterns. All of these lead to excessive variation which can be reduced if the different distribution problem is resolved. It can be seen from the histogram that if the two distributions had the same target point at their center, the overall variation would be cut almost in half.

Figure 4.8 doesn't have a normal pattern because the left end appears to be chopped off. Since the missing columns occur at the lower specification limit, it appears as if the pieces with the smaller, out-of-specification measurements have been removed. If this distribution pattern comes from the company's manufacturing section, what is it costing the company to produce unacceptable parts and sort them out? If the distribution pattern is from incoming inspection, who is paying the supplier to make the out-of-specification parts and then paying to have them sorted out? Further, the histogram represents a sample of the incoming parts. What if their inspectors didn't find all the bad parts and they end up in the company's production process? Case Study 4.1 deals with this particular problem.

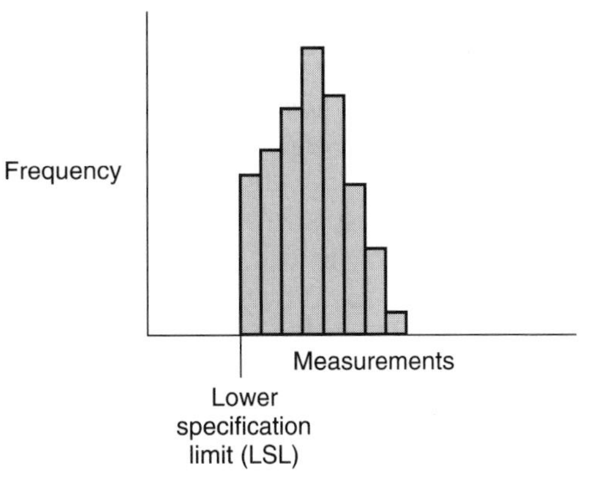

FIGURE 4.8
A distribution with the smaller measurements removed.

CASE STUDY 4.1

A companywide, total involvement in the quality effort occurred at Velcro and was reported in the *Harvard Business Review.* The author of the report, Theodor Krantz,[1] had been the president of Velcro USA since 1984. He stated that the company received a shock in 1985 when General Motors, one of Velcro's important customers, dropped the company's rating as a supplier from the highest level to next to the lowest level. The main problem, according to GM, was that Velcro was inspecting quality into its product instead of manufacturing quality into it. The waste that accompanied Velcro's detection method kept its prices high in an industry in which continual improvement in quality and productivity was needed for cost effectiveness. Velcro was throwing away 5 to 8% of its tape, and that much waste had to be reflected in the cost of its product.

Velcro's initial response to the deadline imposed by GM was to hire a local group of consultants for training and guidance. The training involved teaching SPC and problem-solving techniques to about 500 hourly and salaried personnel. The consultants were helpful because they worked across the traditional hierarchical lines of management and could effectively expose some of the company's problems.

The usual disbelief among both hourly and salaried personnel had to be overcome. Total management commitment was a necessity to convince the rest of the company that high quality and continual improvement was the new, unrevokable policy. Factors obstructing the interaction between supervision and production workers had to be eliminated. Some workers did not want the responsibility for quality placed on their shoulders; it was always easier when the quality control division had that responsibility (detection) further down the line. Some supervisors did not like the change in their job structure; they became defensive and resisted the full utilization of the production workers.

New lines of communication had to be formed. Data from the application of SPC were used by ad hoc teams organized to develop process improvements. Velcro had had about 120 teams over the first three years, and at the time the article was written, about 50 teams were functioning. The teams were initially only in production, but now some of the teams were in the administrative area, too; quality consciousness was spreading.

The results of Velcro's quality effort initially showed up in waste reduction. Waste in 1987 decreased 50% from 1986, and waste in 1988 was down 45% from 1987. The results in 1989 were not as dramatic because the company was at the little-by-little stage of progression. The management realized, however, that there is no end to the quality process; it pervades every department and becomes a way of life.

The histogram in Figure 4.9 has an unusually high stack of measurements at the lower specification limit and is labeled "inspector flinching." This could indicate that the inspector is using a "close enough" attitude on parts that measure slightly out of specification. This could be an indicator of the presence of fear in the workplace, one of the problems that Deming mentions in his 14 points. The inspector could be afraid to "rock the boat," to point out with honest measurements that a problem exists.

---

[1] K. Theodor Krantz, "How Velcro Got Hooked on Quality" (*Harvard Business Review,* September–October 1989), 34–40.

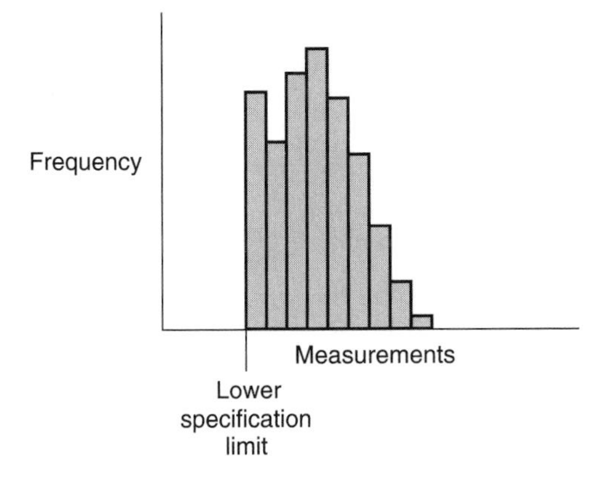

FIGURE 4.9
Inspector flinching at the lower specification limit.

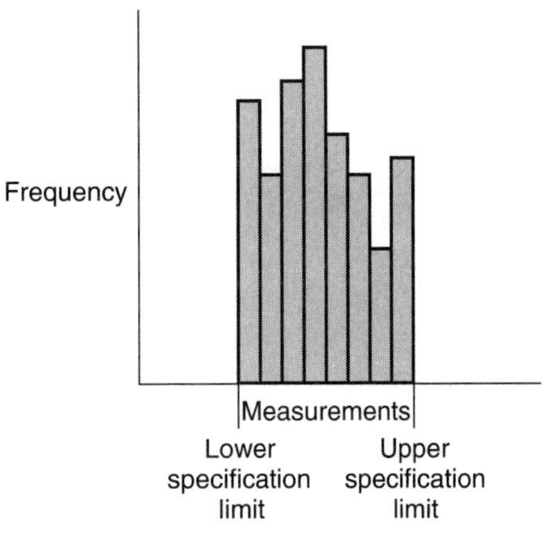

FIGURE 4.10
Inspector flinching at both limits.

Figure 4.10 shows the existence of two problems. One is the inspector-flinching problem indicated by the unusually high stacks of measurements at both specification limits. An even worse problem is the excessive variation that is indicated by the histogram crowding both specification limits and most likely overlapping them if honest measurements were taken.

A distribution pattern as erratic as that in Figure 4.11 can indicate either a faulty gauge or that the person using the gauge needs more training on how to use it. Another possibility is that the distribution has several modes that are due to different spindles, shifts, operators, or machines.

The histogram in Figure 4.12 may be skewed because of excessive tool wear. The shape shows a different statistical distribution, such as the Poisson or the chi-square distribution; a statistical goodness-of-fit test would determine which fits best. If the tool cannot be normalized for a specified product run, a set of histograms from that process point will have to be matched to the statistical distribution that best describes that product distribution. The formulas from that distribution will be used in any capability analysis.

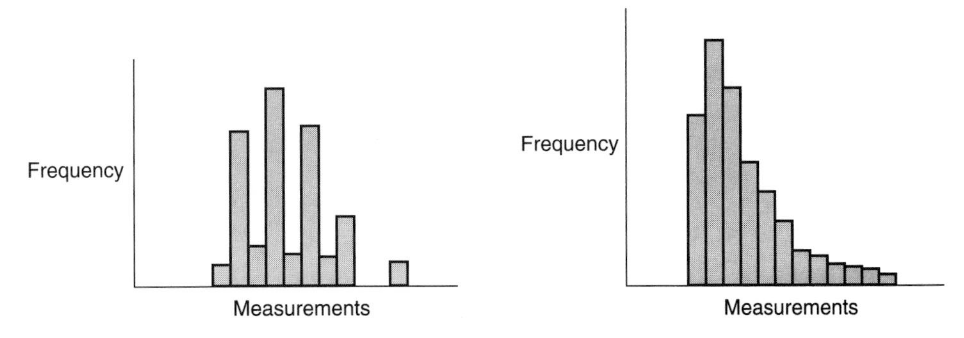

FIGURE 4.11
A faulty gauge or poor training.

FIGURE 4.12
A skewed distribution.

4.4 PARETO CHARTS

The Pareto chart, a bar graph that ranks problems in decreasing order of frequency, was adapted to quality control by Joseph M. Juran,[2] a noted authority and leader in the resurgence of quality in U.S. industry. The Pareto principle, credited to Italian economist Vilfredo Pareto, involves the concept that the comparative distributions of certain economic factors, such as wealth, follow an inverse relationship. Pareto discovered that 80% of the wealth in his country in the early 1900s was concentrated in 20% of the population. Dr. Juran discovered that the 80%–20% split also occurs in quality control. Eighty percent of the scrap is caused by 20% of the problems, and 80% of the dollar loss caused by poor quality is concentrated in 20% of the quality problems. Of course, the 80%–20% split is not exact; the percentages vary quite a bit. The important outcome of a Pareto chart is its assessment of process problem priorities. It separates the vital few problems from the trivial many. Another plus for the Pareto chart is its elimination of *recentivity,* the tendency to overestimate the importance of the most recent problem.

When a problem analysis is done for a Pareto chart, data are gathered that give the number of occurrences for each problem and the dollar loss associated with it. When all the data have been gathered, percentages can be tabulated for both the number data and the dollar loss data.

The procedure for making a Pareto chart is as follows:

1. Decide on the subject of the chart. Usually the need to set data priorities suggests the use of a Pareto chart. Determine what data are to be collected.
   a. Where is the problem?
   b. What are the categories?

   Where should the data be gathered?
   a. Should they come directly off a line?
   b. Should they come from a bin of nonconformities that have accumulated in the specified time period?
2. Be sure the time period for all the categories is the same: Use the number of nonconformities per hour, per shift, or per week.
3. What type of chart is needed? Should you track the numbers in each category, the percentage in each category, or the costs in each category? A cost chart is usually included with either a numbers chart or a percentage chart.
4. Make a table by gathering the data and tallying the numbers in each category. Find the total number of nonconformities and calculate the percentage of the total in each category. Make a cost of nonconformities column and a cumulative percent column.
5. Arrange the table of data from the largest category to the smallest.
6. Set the scales and draw a Pareto chart.
7. Include all pertinent information on the chart. Are the categories clear? Has the time frame been specified?

---

[2]Joseph M. Juran, ed., *Quality Control Handbook,* 3d ed. (New York: McGraw Hill, 1974), 2.16–2.19.

8. Analyze the chart. The largest bars represent the vital few. The cumulative percentage line levels off and emphasizes the trivial many. If the chart does not show a vital few, check to see if it is possible to recategorize for another analysis.

EXAMPLE 4.9 An analysis of nonconforming shirts in a week's production revealed the following causes. The nonconforming shirts were discounted according to defect and sold to Bargain Bin Incorporated and the dollar loss was noted. Make a number-of-defects Pareto chart, a dollar-loss Pareto chart, and a cumulative percentage Pareto chart.

Solution

Defect Category	Number of Shirts	Percentage of All Shirts	Dollar Loss	Cumulative Percentage
Loose threads	2300	46%	$9,200	46%
Hemming wrong	1650	33%	$9,900	79%
Material flaw	300	6%	$2,400	85%
Collar wrong	250	5%	$1,250	90%
Cuffs wrong	200	4%	$1,200	94%
Buttons	100	2%	$ 400	96%
Stitching	100	2%	$ 400	98%
Buttonholes	50	1%	$ 400	99%
Material tear	50	1%	$ 600	100%
TOTALS	5000	100%	$25,750	

The data have been organized in the table according to the number of defects in each category. The percentages were calculated as follows:

1. Add the values in the number-of-defective-shirts column to get the total number of defects.
2. Divide each number of defects by the total from step 1 and multiply by 100 to get the percentage of the total number of defects.

The percentage of defective shirts with loose threads or hemming problems, for example, are calculated according to two steps:

$$\text{Loose threads} \qquad \frac{2300}{5000} \times 100 = 46\%$$

$$\text{Hemming} \qquad \frac{1650}{5000} \times 100 = 33\%$$

The percentages were totaled in the last column:

46%	Loose threads
79% = 46% + 33%	Threads and hems
85% = 79% + 6%	Threads, hems, and flaws
90% = 85% + 5%	Threads, hems, flaws, and collars

The Pareto charts can be formed in a few different ways. Figure 4.13 shows the basic chart with the number of defective shirts on the vertical scale and the different categories on the horizontal scale. Figure 4.14 shows the dollar loss on the vertical axis and the categories on the horizontal axis. Notice that the order has changed among the categories as the charted values go from largest to smallest. Figure 4.15 is a combination of two Pareto charts. The left scale tracks the number of defects per category with the bar graph, and the right scale tracks the accumulated percentage of all defects with a line graph. It is common practice to combine a cumulative percentage chart with a percentage-of-defects chart and also with a dollar loss chart. In all cases, the cumulative percentage line levels off at the trivial many.

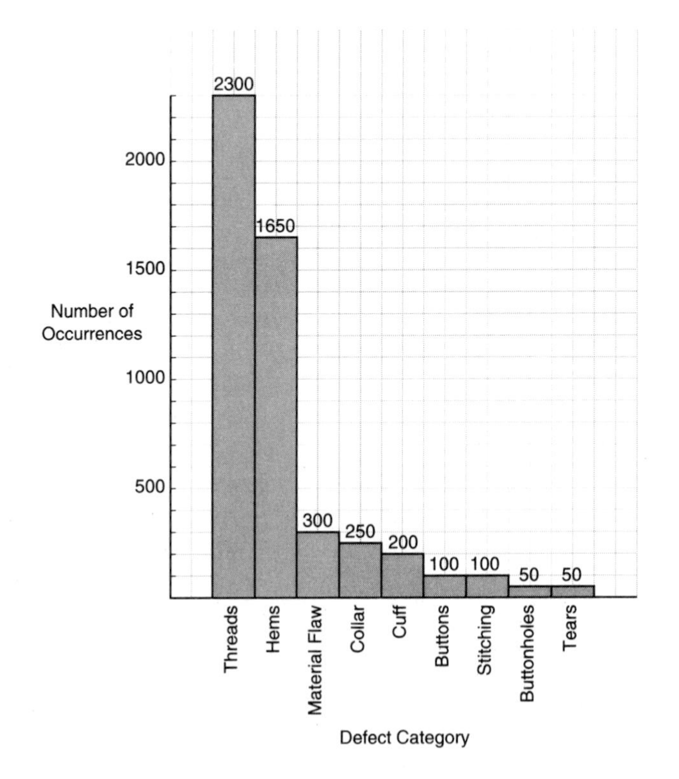

FIGURE 4.13
A Pareto chart for Example 4.9 of the number of nonconforming shirts per week with the number of defects per category per week.

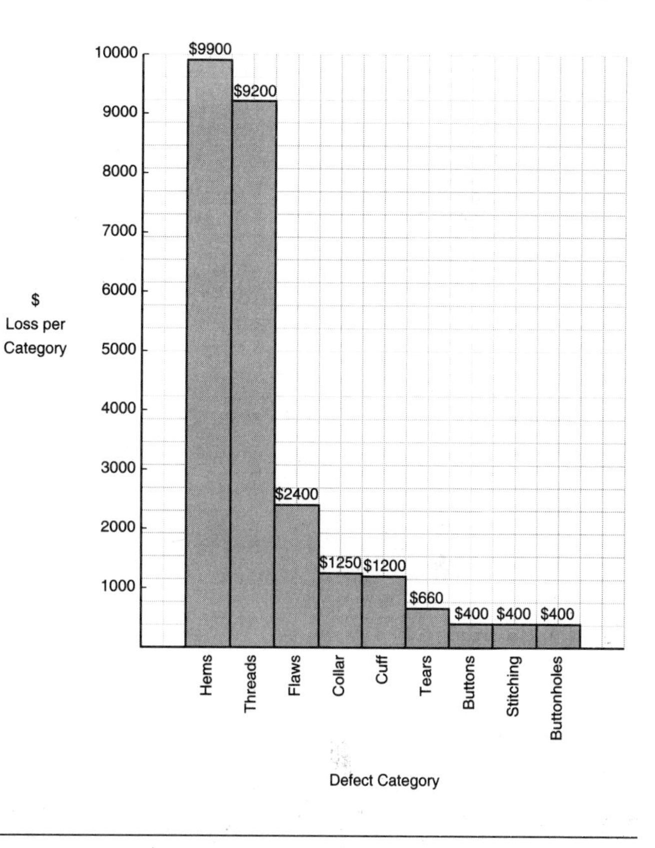

FIGURE 4.14
A Pareto chart for Example 4.9 showing the dollar loss per category per week.

EXAMPLE 4.10 In a customer satisfaction survey at a local fast-food restaurant, the following complaints were lodged:

Complaint	Number of Complaints	Complaint	Number of Complaints
Cold food	105	Poor service	13
Flimsy utensils	2	Food too greasy	9
Food tastes bad	10	Lack of courtesy	2
Salad not fresh	94	Lack of cleanliness	25

Using these data:

1. Make a table with the data ordered.
2. Include columns for percentage of complaints and cumulative percentage. Because of rounding, the cumulative percentage may be slightly different from 100%.
3. Make a Pareto chart for the number of complaints and include the cumulative percentage graph.
4. Check your results with the completed table and chart in Figure 4.16.

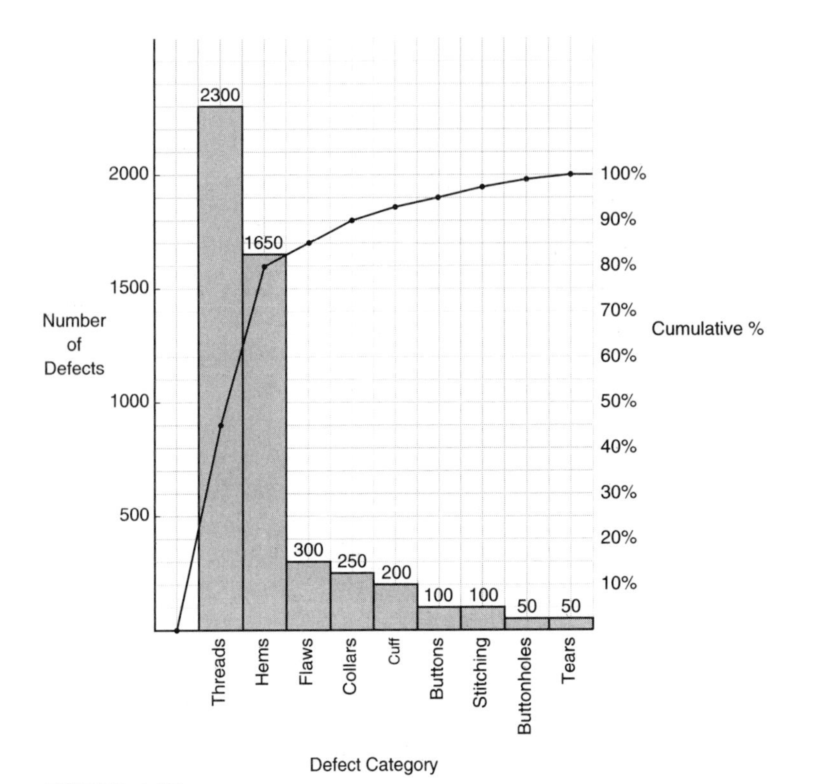

FIGURE 4.15
A Pareto chart for Example 4.9 showing the number of defects and the cumulative percentage.

Solution

Category	Number	Percent	Cumulative %
Cold food	105	$\frac{105}{260} \to 40\%$	40%
Salad not fresh	94	$\frac{94}{260} \to 36\%$	76%
Cleanliness	25	$\frac{25}{260} \to 10\%$	86%
Service	13	$\frac{13}{260} \to 5\%$	91%
Taste	10	$\frac{10}{260} \to 4\%$	95%
Greasy	9	$\frac{9}{260} \to 4\%$	99%
Utensils	2	$\frac{2}{260} \to 1\%$	100%
Courtesy	2	$\frac{2}{260} \to 1\%$	101%
Totals	260	101%	101%

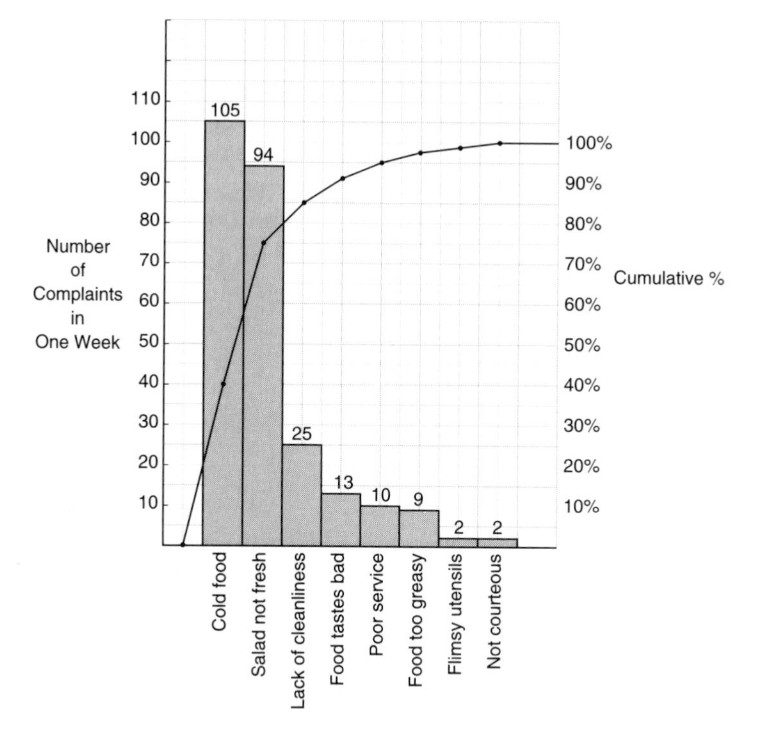

FIGURE 4.16
The Pareto chart for Example 4.10.

4.5 FLOWCHARTS

The charts that have been discussed up to this point are usually done individually. One person has the responsibility to gather the information, organize it, and make the appropriate chart. Flowcharts and a few other types of charts are often developed by a group of people. A useful group technique in making these charts is called *brainstorming*. Brainstorming, which is discussed more thoroughly in Chapter 10, generates a large number of ideas on a specific topic. An important feature of brainstorming is that one person's idea can spark a modification or a new idea from someone else.

A *flowchart* is a diagramming tool that is used to trace a process from start to finish. It can be used for an entire complicated process or for some segment of the process. Different symbols specify what is being done to the product as it progresses from the input stage to the output stage of the process. When problems exist within a process or process segment, the problem-solving team should clearly understand what is being done to the product at the various stages in the process. A completed flowchart should make the step-by-step procedure within the process clear to the entire team.

In the problem-solving sequence, making a flowchart of the process is usually one of the first steps. Brainstorming is very useful in developing the flowchart, and using it ensures that important process details are not omitted.

Completed flowcharts can be helpful in finding the root causes of problems. Brainstorming sessions with a flowchart in hand allow the team to trace the product back and forth in the process until the cause of the problem is found or until several good candidates for the root cause have been uncovered, leading the way to further data gathering and analysis.

Two types of flowcharts will be demonstrated: a straightforward procedure in which symbolism is unnecessary and a more complicated process that uses defined symbols to indicate the action at each step. The first step to take with either one is to brainstorm for steps in the process and for a logical process sequence.

EXAMPLE 4.11 Make a symbol-free flowchart for the process of making coffee in a coffeemaker at work.

Solution

Step 1. Brainstorm for process steps:

Buy the coffee.	Buy the filters.
Clean the pot.	Clean the basket.
Buy the sugar and creamer.	Put in the coffee.
Put in the filter.	Fill the reservoir with water.
Turn it on.	Put the basket in place.
Put the pot in place.	Wait until it is finished.
Pour the coffee.	Add sugar and creamer as needed.
Tidy up.	Drink the coffee.

Step 2. Make the flowchart using the following:

Materials	Process	Controls
	Make sure everything is clean.	
Filter	Put in a filter.	
Coffee	Measure the coffee into the filter.	
Water	Fill the reservoir.	
	Put the basket in place.	
	Put the pot in place.	
	Turn the coffeemaker on.	
	Wait until it is done.	
Cup	Pour the coffee.	
Sugar	Add sugar if needed.	
Creamer	Add creamer if needed.	
	Adjust:	Taste
	Amount of coffee (for next time)	
	Sugar and creamer	
	Tidy up.	
	Drink the coffee.	
	Clean the basket and pot when the pot is empty or at the end of the day.	
	Buy needed materials for tomorrow.	

FIGURE 4.17
Flowchart symbol definition. *Source:* "Problem Solving Tools and Techniques" Saginaw Division, General Motors Corporation, Decatur, AL 20–23.

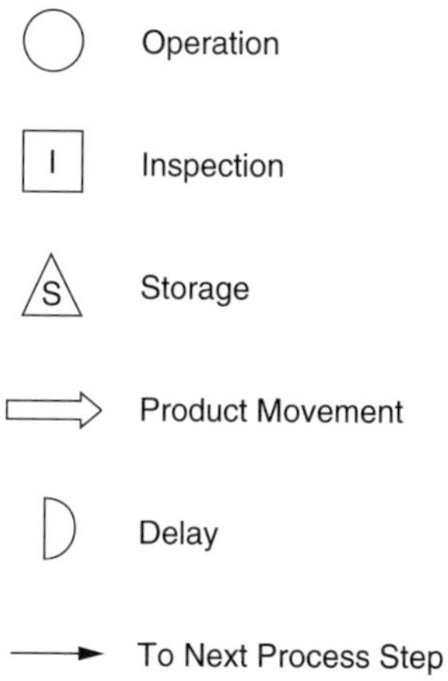

Operation

Inspection

Storage

Product Movement

Delay

To Next Process Step

More complicated processes can be flowcharted with the use of standardized symbols to indicate what is being done to the product. This type of chart requires that everyone using it understand the symbols, which are displayed in Figure 4.17.

EXAMPLE 4.11 Make a flowchart, complete with symbols, for the process of buying a new car.

Solution

Step 1. Brainstorm for process steps:

Read the new-car literature.

Decide on down payment.

Check your budget for maximum payments.

Visit a new-car dealer.

Pick a car you like.

Discuss purchase price.

Find out about the payments.

Wait for the new-car preparation.

Check *Compucar* (gives dealer costs on model and options).

Find out about extendable warranties.

Investigate the length of payment period.

Check on loan rates.

Check the papers on price of your present car.

Narrow your choices of cars.

Look at the cars on the lot.

Drive the car.

Discuss trade price.

Sign the papers and buy the car.

Pick up the car.

Do a final inspection.

Drive the car home.

Step 2. **Make the flowchart, as shown in Figure 4.18.**

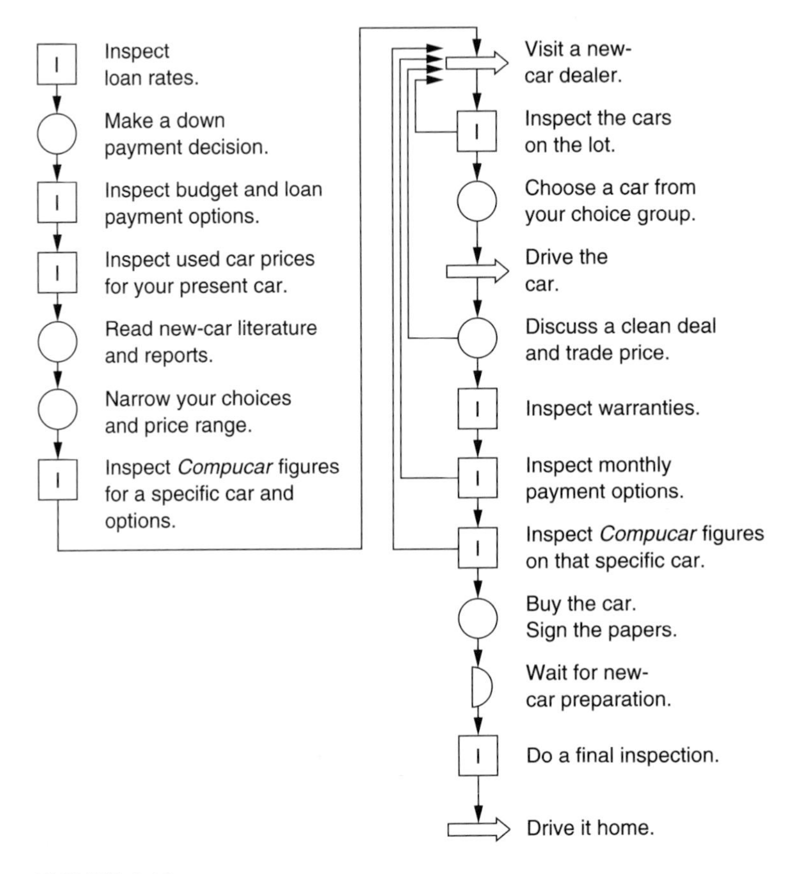

FIGURE 4.18
Flowchart for buying a new car.

Making a flowchart can be difficult because of all the revisions and afterthoughts that can occur. One way to improve the procedure is to hand out Post-it notes.[3] Each team member writes down one specific activity that they are involved in or are aware of on the Post-it note. Continue with a few more, writing one activity per note. Ask for an activity that is central to the process (about in the middle). Post that note in the middle of a wall or a large sheet of paper. Then arrange the other notes in proper order, both upstream and downstream of the starting note. Continue the writing and posting until the activity sequence is completed. Ask for more details, inputs, and outputs, each on their own note. When the Post-it flowchart is completed, the team should be satisfied with it, every process step should be clear, and everyone's part in the process should be completely

[3]Leland R. Beaumont, "Sticking With Flowcharts" (*Quality Progress,* July 1993).

detailed. Make the semifinal drawing with the appropriate symbols and pass copies around for the final check. Make sure that all control charts, quality checks, delays, completion times, and rework are clearly shown. When the final version of the flowchart is handed out, all must realize it is a working tool, not a work of art. They should mark it, write on it, and revise it as process improvement progresses.

4.6 STORYBOARDS (OPTIONAL)

Walt Disney is credited with the development of the *storyboard.* When he made his feature-length animated pictures, he started by outlining the story on a board that ran the length of the wall. He and his associates would then fill in various substories and details. The format allowed them enough flexibility to try different ideas, change their minds, rearrange events, and see how it would logically fit in the storyline. As they worked in the details, the complete story developed on the storyboard.

The storyboard concept works very well in outlining a process for both initial setup and problem solving. A flexible structure such as a corkboard wall is effective, but chalkboards, markerboards, and even scotch tape and chart paper may be used. The storyboard is similar to a flowchart, but its format allows more attention to detail. Subroutines can either be included in the main process structure or be shown separately. Both options are illustrated in Figure 4.19.

The choice of storyboard structure depends on the degree of complication in the process and subroutines. Color coding subroutines in either structure can also be helpful. The goal is clarity; everyone involved should be able to follow the process from start to finish when the storyboard is completed.

When a storyboard is going to be developed, brainstorming and teamwork are necessary ingredients. If a new process is being developed or an existing process is being

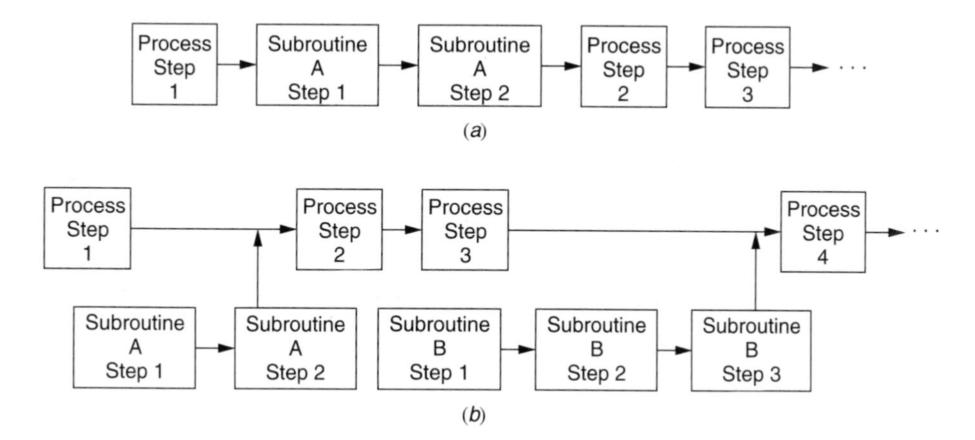

FIGURE 4.19
Storyboard patterns.

outlined for troubleshooting, a diverse team with representatives from all phases of the process should be involved. Just as with the other problem-solving tools, it is important to have ideas and questions from different viewpoints.

Process problems can be tracked on the storyboard from their point of discovery in the process back to their root cause. When the path back is unclear, the storyboard can be helpful in deciding where additional information is needed.

4.7 THE CAUSE-AND-EFFECT DIAGRAM

The *cause-and-effect diagram* is a useful tool in a brainstorming session because it organizes the ideas presented. It is sometimes called a *fishbone diagram* because of its shape or an *Ishikawa* diagram after Professor Ishikawa of Japan, who first used the technique in the 1960s. The basic shape of the cause-and-effect diagram is shown in Figure 4.20.

The diagram is a format for logically aligning the possible causes of a problem or effect. A basic way to organize the "ribs," or main categories, is to assign them the four M's: methods, machines, measurement, and materials. As the ideas are presented, they are inserted as the "bones," or possible causes of the effect, in the appropriate category. The bones can be subcategorized as causes of a cause are presented. The subdividing continues until the root cause to the problem is found. There may not always be a single root cause, but at least a few potential root causes will surface, and a decision for action can be made.

The four M's are generally used as the initial main categories for a cause-and-effect diagram. Other categories specific to the particular process may be added if the team decides they are important. Environment is one example of a possible other category. It may be considered important enough to be a main category of problem causes, or it may be a subcategory in any or all of the other categories, depending on the process being analyzed.

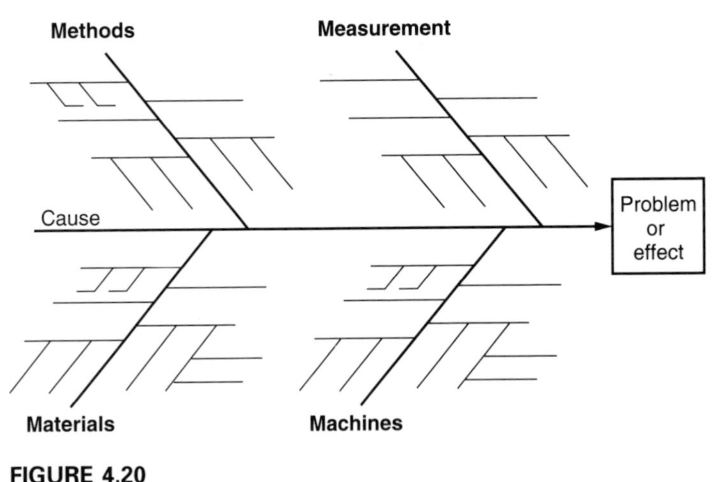

FIGURE 4.20
The cause-and-effect diagram.

When a problem has been isolated, a series of brainstorming sessions may be necessary before it is ultimately solved. Occasionally, for a complex process, the ideas generated in a brainstorming session may come too fast to be entered on the cause-and-effect diagram. If the session leader has to slow things down in order to make sure an idea fits in the right category or subcategory position, the spontaneous flow of ideas may be hindered. Housekeeping details at this point can slow the generation of ideas. This may occur, for example, at the very first session or at the beginning of a session when there is a large initial volume of ideas. The session leader may at that time switch to a list or a roughly categorized list. Later in the session when new ideas are scarce or at the next session, the group can decide where each idea belongs on the cause-and-effect diagram. That organization step of entering the ideas on the diagram can be very beneficial in generating new ideas. As categories and subcategories are arranged, the logical structure often suggests the root cause to the problem or specific phase of the problem.

Sometimes a systematic, directed approach to the cause-and-effect diagram is more effective than the "shotgun" method described above. In this approach a specific category is designated, such as Machines. Then discussion and brainstorming traces potential sources of trouble until the category is completed (no further possibilities at that time). Another category is chosen and the process continues until the diagram is completed and a list of possible causes is developed for investigation.

EXAMPLE 4.13 Make a cause-and-effect diagram for the following problem: The car will not start.

Solution

A group brainstorms the problem using the four M's to create rough categories of possible causes. Enter the causes from the list on a cause-and-effect diagram. If that organization step makes you think of any other possible causes, enter them as well. Figure 4.21 shows a completed diagram.

Measurement	**Methods**
Faulty gas gauge	Flooded engine
	Standard: clutch must be in
	Automatic: must be in neutral or park
	Wrong car
	Wrong key
	Seatbelt not fastened
Machines	**Materials**
Carburetor problem	No gas
Wiring problem	No oil
Computer malfunction	Oil too thick (cold)
Electrical trouble	Dead battery
	Loose battery connection
	Faulty spark plugs
	Plugged gas line
	Air cleaner too dirty

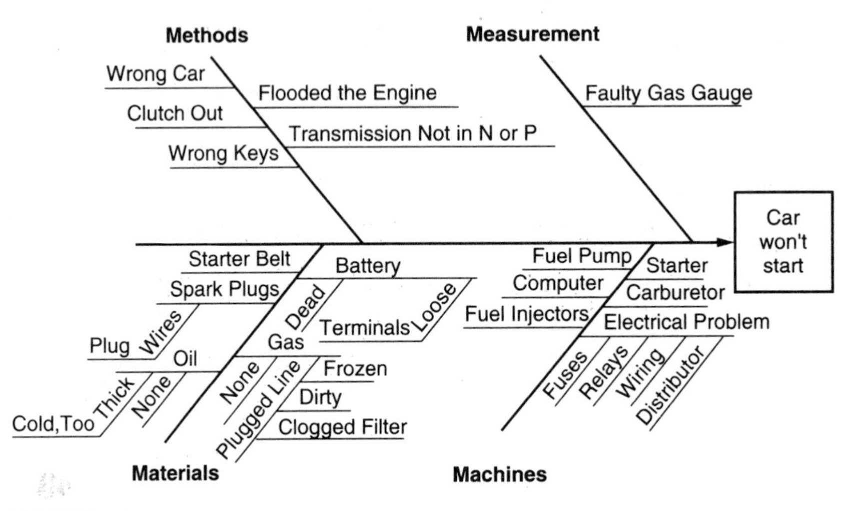

FIGURE 4.21
Cause-and-effect diagram for Example 4.13.

A variation on the cause-and-effect diagram is the tree diagram. This has been used extensively in mathematics to develop a list of logical outcomes. An example will illustrate the use of a tree diagram.

EXAMPLE 4.14 Use a tree diagram to explore ways to improve your grades.

Solution

The three main requirements are attending class, doing assigned work, and studying. These are the first three branches, as shown in Figure 4.22.

Following the top branch, continue with additional branches representing related activities and details. This is shown in Figure 4.23. The best in-class activities include note taking, staying awake, and asking questions. The notes should bring attention to the major ideas that were developed in class and they should be reviewed before the next class.

FIGURE 4.22
Beginning a tree diagram.

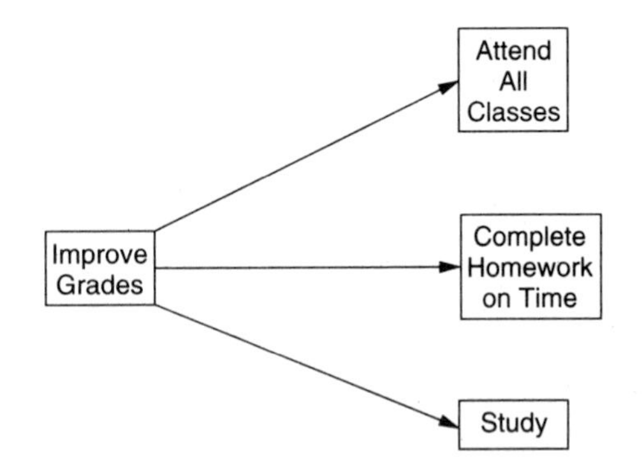

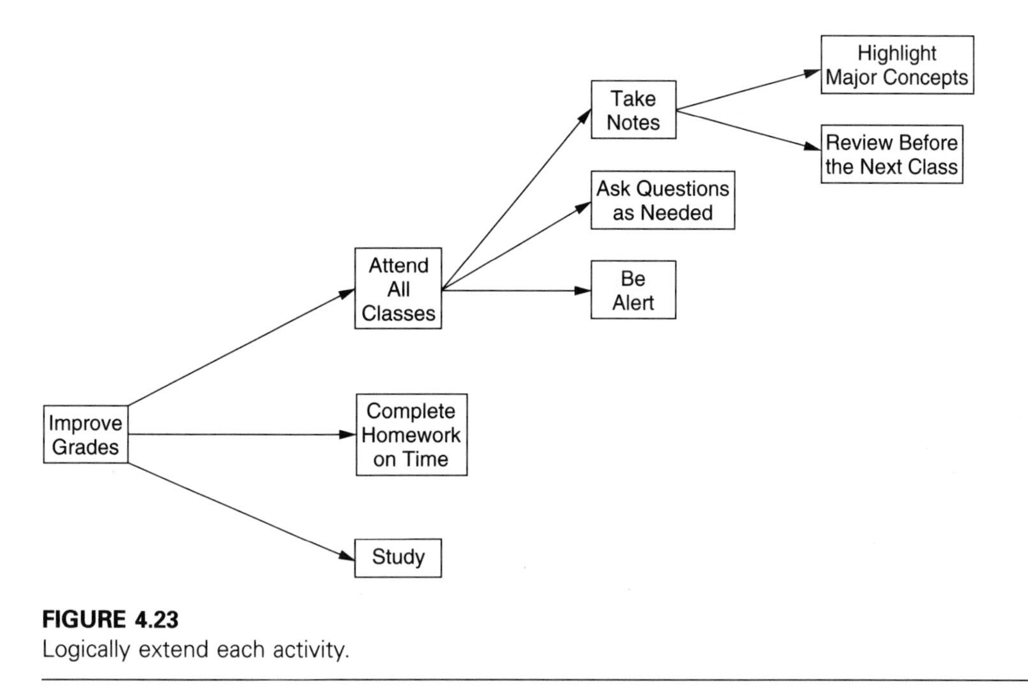

FIGURE 4.23
Logically extend each activity.

Follow the logical sequence for the other main categories in a like manner. Homework results can be charted on an *nq* chart (discussed in Chapter 9), and any incorrect work should be redone. The study sessions should be built around the basic concepts of the course and their applications. The completed diagram is shown in Figure 4.24. There may be other branches that could be added that are more specific to an individual's learning style. Tree diagrams are useful in industry for planning sessions in areas of management, engineering, and quality improvement.

4.8 CHECKSHEETS

A *checksheet* is a data gathering device that consists of a list of the different types of data to be gathered and a row or column in which to put tally marks or brief descriptive remarks. The heading on the checksheet should contain information such as the name of the individual gathering the data, the time frame in which the data are gathered, and any specific information about the source and type of data. The information from a checksheet of this type could be put in a Pareto chart for prioritizing problems.

Another type of checksheet is used in the inspection process. The checksheet contains a list of specific items that have to be checked or it could contain a picture of the item being inspected with inspection areas highlighted or numbered. Some also contain codes that are used to specify the problem area and the type of problem.

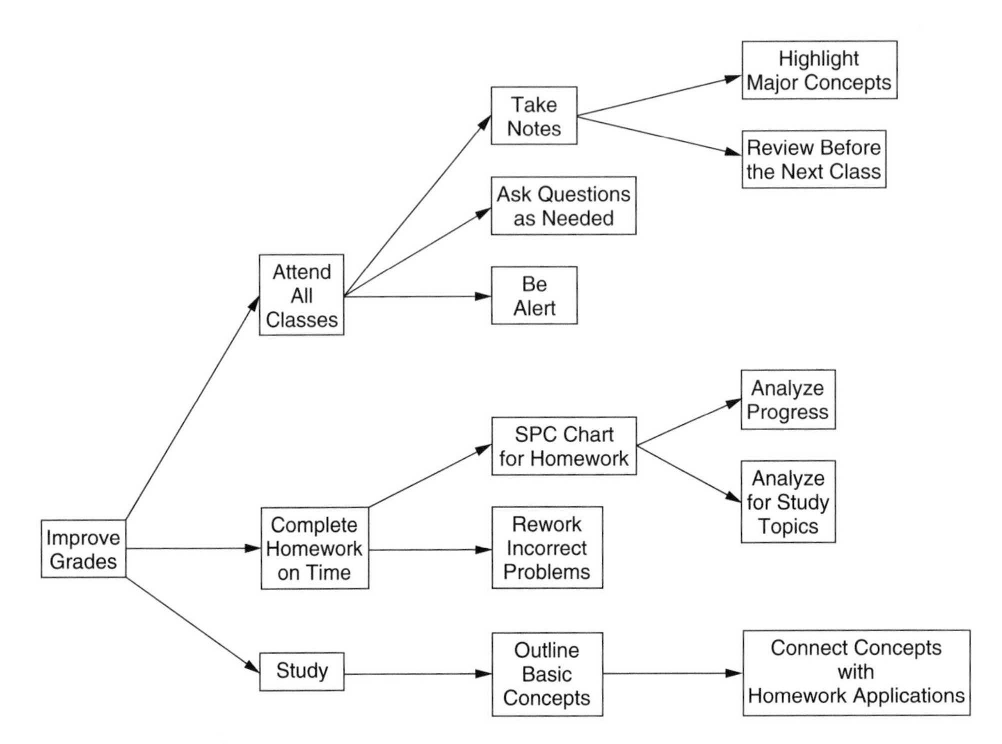

FIGURE 4.24
The completed tree diagram.

4.9 SCATTERPLOTS

The *scatterplot* is a graph of measurement pairs that shows whether there is correlation between the measurements. When correlation exists, changes in one measurement will be accompanied by proportionate changes in the other. If there is *positive* correlation, the changes will be in the *same direction*. When the first measurement increases, the second increases, and when the first measurement decreases, the second decreases. When the correlation is *negative*, the two measurements move in *opposite directions*. If the first measurement increases, the second decreases, and vice versa.

Two relationships may exist between measurements that are correlated. If a *common-cause* relationship exists, then the measurements are similarly affected by the common-cause variation in the process. So, if one measurement is in control, the other one will be too. Likewise, trouble with one can signal trouble with the other. In a *cause-and-effect* relationship a change in one measurement will cause the other measurement to change as well.

Scatterplots can be useful in both quality control and problem solving. If measurements have a common-cause correlation, just one of them has to be tracked on a control chart. If measurements have cause-and-effect relationship, the important measure can be optimized and controlled with the other. Also, if improvements on the process of the first

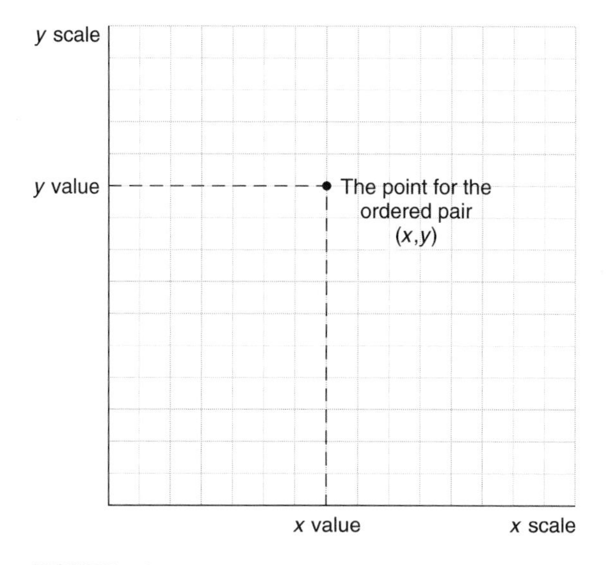

FIGURE 4.25
An ordered pair on a scatterplot.

measurement result in decreased variability, the second measurement will often undergo a similar improvement. When process changes are made, however, the correlation factor should be verified.

The scatterplot is a graph of ordered pairs of numbers (x, y). The first number in the ordered pair, represented by x, is located on the horizontal scale. The second number, represented by y, is located on the vertical scale. A point that aligns with both values is graphed for the ordered pair. This is shown in Figure 4.25.

EXAMPLE 4.15 Graph the following ordered pairs (x, y): (2, 12), (5, 14), (7, 15), (10, 16), (12, 18).

Solution

The horizontal or x scale extends as far as 12 and the vertical or y scale goes up to 18. The points, plotted in Figure 4.26, are almost in a straight line. That indicates that there is a strong linear correlation between the measurements x and y.

If it is suspected that a cause-and-effect relationship exists between the two measurements, it is important to associate the independent variable or measurement with x and the horizontal axis. The dependent value, associated with y and the vertical scale, is the measurement of interest that is to be controlled. When the cause-and-effect relationship is verified, the independent variable x will be held at the optimum value predicted by the graph that will keep the measurement of interest y at the desired value. A *line of best fit* is drawn through the points to approximate the slope and to balance the points above the line and the points below the line. There are statistical equations that will define the line precisely, but for now just an estimated line of best fit will be used to discuss the concepts involved. A statistical calculator with linear regression may also be used to find the line of best fit.

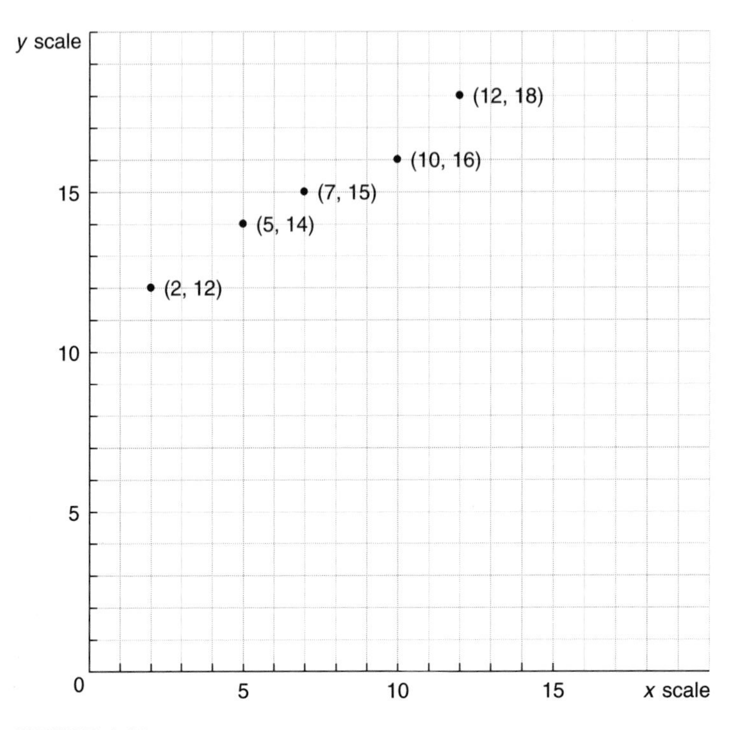

FIGURE 4.26
Ordered pairs on a scatterplot.

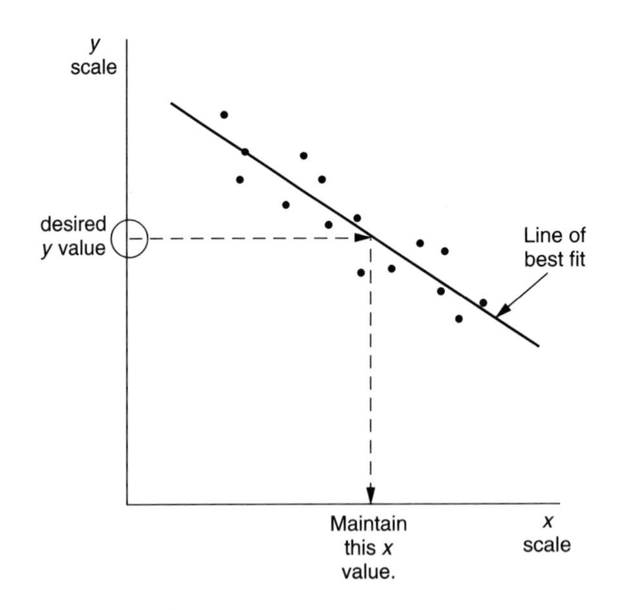

FIGURE 4.27
The line of best fit on a scatterplot with a cause-and-effect relationship.

In a cause-and-effect relationship, the desired *y* value is associated with a specific *x* value using the line of best fit. That *x* value will then be maintained in the process in an effort to keep *y*, the measurement of interest, at the desired level. This is shown in Figure 4.27.

If the two measurements are thought to have a common-cause relationship instead of a cause-and-effect relationship, the specific choice of which variable to graph on the horizontal scale and which on the vertical scale is arbitrary.

Table 4.22 gives pairs of measurements of corresponding internal and external diameter measurements. Each pair of numbers specifies a point on the graph. The measurement pairs are graphed on Figure 4.28. Figure 4.28 illustrates perfect-positive correlation. All points are on a straight line and the measurements increase and decrease proportionately. Perfect-negative correlation is illustrated in Figure 4.29. The points on the graph, given in Table 4.23, are on a straight line. One measurement increases as the other decreases.

The scatterplot is commonly used to calculate process cause and effect on a specific measurement. Variables such as RPM, temperature, pressure, spindle speeds, and voltage can influence process measurements dramatically. Figure 4.30 shows that RPM has a direct effect on measurement A, so control of the measurement requires control of the RPM. There is very good negative correlation in the scatterplot in this figure. The points, which are listed in Table 4.24, lie very close to a straight line.

Figure 4.31 shows that there may be some correlation between measurement C and temperature but that other factors must be affecting the pair of measurements as well. See Table 4.25 for the measurements. Here the scatterplot shows poor positive correlation. The points do not lie close to a straight line.

TABLE 4.22
Measurements for Figure 4.28

Internal Diameter	External Diameter
.212	.512
.210	.510
.213	.513
.215	.515
.209	.509

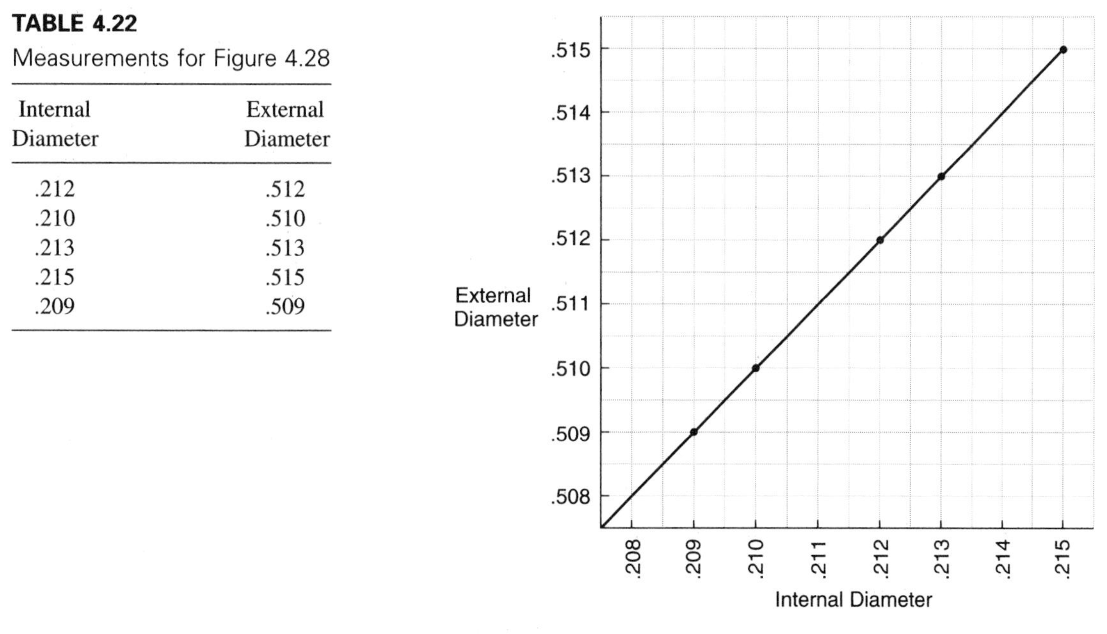

FIGURE 4.28
Perfect-positive correlation.

TABLE 4.23
Measurements for Figure 4.29

Internal Diameter	Weight
.212	.713
.210	.721
.213	.709
.215	.701
.209	.725

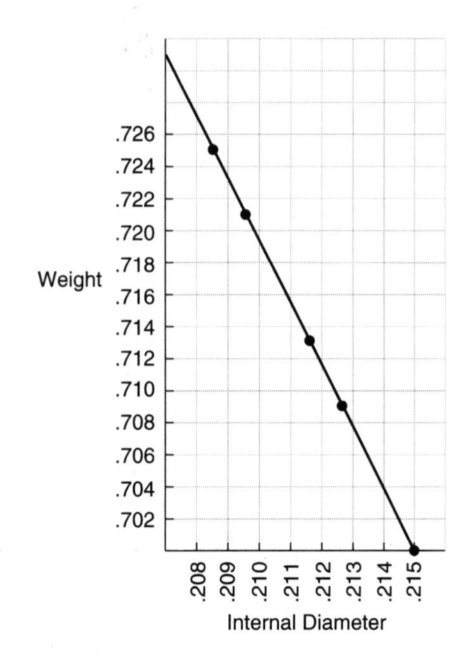

FIGURE 4.29
Perfect-negative correlation.

TABLE 4.24
Measurements for Figure 4.30.

RPM	Measurement A
862	.5025
856	.5035
860	.5028
862	.5024
858	.5032
861	.5028
859	.5030

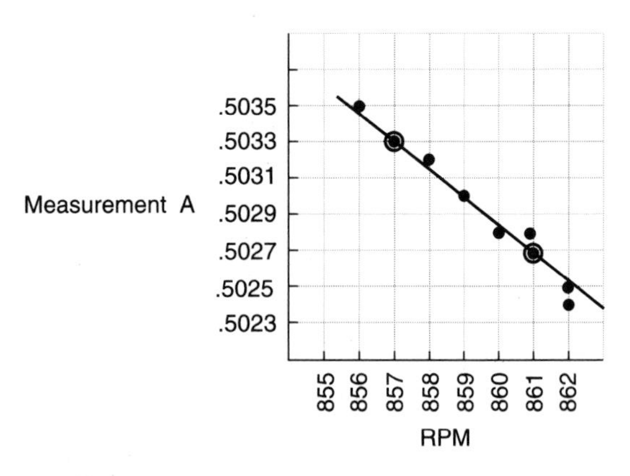

FIGURE 4.30
Very good negative correlation.

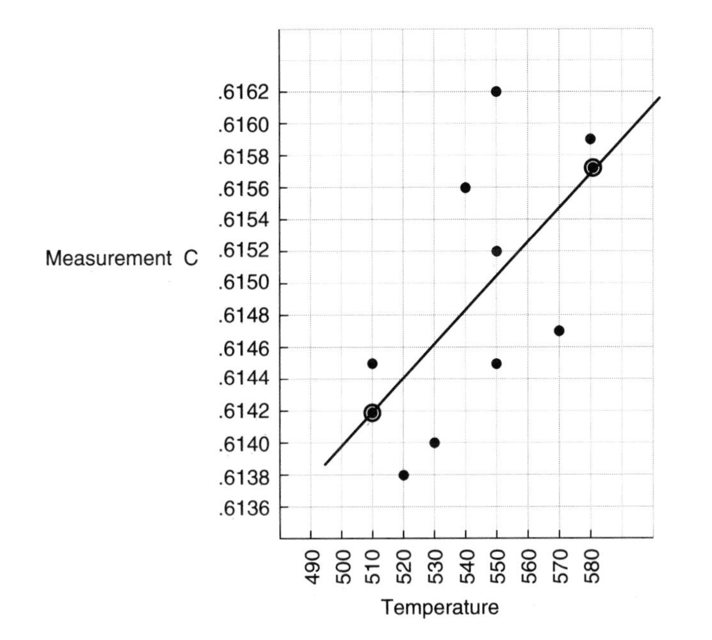

FIGURE 4.31
Poor positive correlation.

TABLE 4.25
Measurements for Figure 4.31

Temperature	Measurement C
550	.6152
580	.6159
510	.6145
530	.6140
540	.6156
520	.6138
570	.6147
550	.6145
550	.6162

A graph that has all the points close to the line of best fit is said to have a strong linear trend and good correlation between the variables. Figures 4.32 and 4.33 show examples of poor correlation between x and y. In the latter figure, there is no line of best fit.

A calculator with linear regression capabilities will provide the information to graph the regression line and it will give the value of the correlation coefficient, r that indicates whether there is good or poor correlation between the variables. The different calculator brands have their own instruction on how to enter the ordered paired data and how to interpret the results. Please refer to your owner's manual.

When the data have been entered, the calculator will give the slope of the regression line, B, and the y intercept of the line, A. In a basic algebra course, the equation of a line, in slope-intercept form, is generally written y = mx+b, where the coefficient of x, m, is the slope of the line and the value b is the y intercept. Following that basic pattern, the equation of the regression line can be written y = Bx+A using the information from the calculator. Usually the slope, B, is given in a decimal form and doesn't translate easily to the graph. A better way to graph the regression line (the line of best fit through the data points) is to enter a x value into the calculator and the calculator will give the predicted y value (predicted by the regression equation). Choose a second x value sufficiently far from the first one and find its predicted y value also. Plot the two ordered pairs and draw a straight line through them. For a visual check, the line should go through the middle of the scatterplot pattern with a good balance of points on either side.

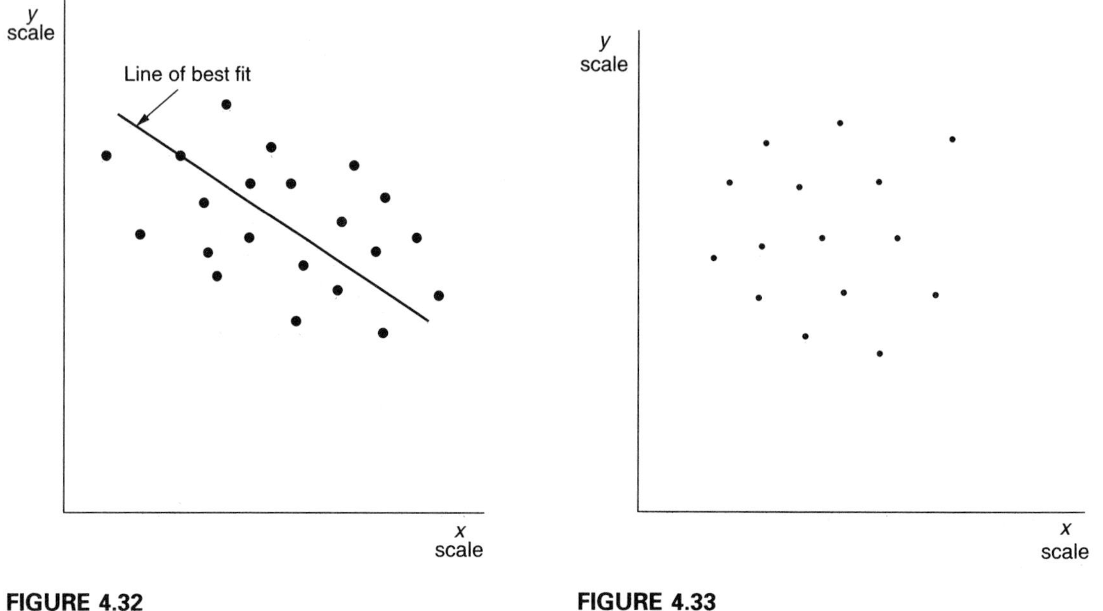

FIGURE 4.32
Poor correlation between *x* and *y*.

FIGURE 4.33
No correlation between *x* and *y*.

The correlation coefficient, r, will have the same algebraic sign as the slope of the regression line and the value of r will always be between ± 1, $-1 \leq r \leq 1$. If $|r| = 1$, there is perfect correlation and all the points will be on the regression line. If $|r|$ is close to 1, the points in the scatterplot will be clustered very close to the regression line. The regression equation, or the calculator, will predict accurate y values for given x values. When $|r|$ is close to .5 there is some linear trend, but the predicted values of y will not be very accurate. If $|r|$ is close to 0, it indicates there is no linear relationship between the x and y values and there is no usable regression line of best fit.

EXAMPLE 4.16 Find the regression line for the data given in Example 4.15.

Solution

Enter the ordered pair data in a linear regression calculator.
Read: slope B = .56
y-intercept A = 10.99
correlation coefficient $r = .99$
The regression equation is y = .56x + 10.99

The scatterplot is shown in Figure 4.26. The regression line can be drawn using any two ordered pairs. If x = 3, the calculator indicates that y = 12.7. You would get the same result if you had substituted 3 for x in the regression equation. Likewise, if x = 11, the

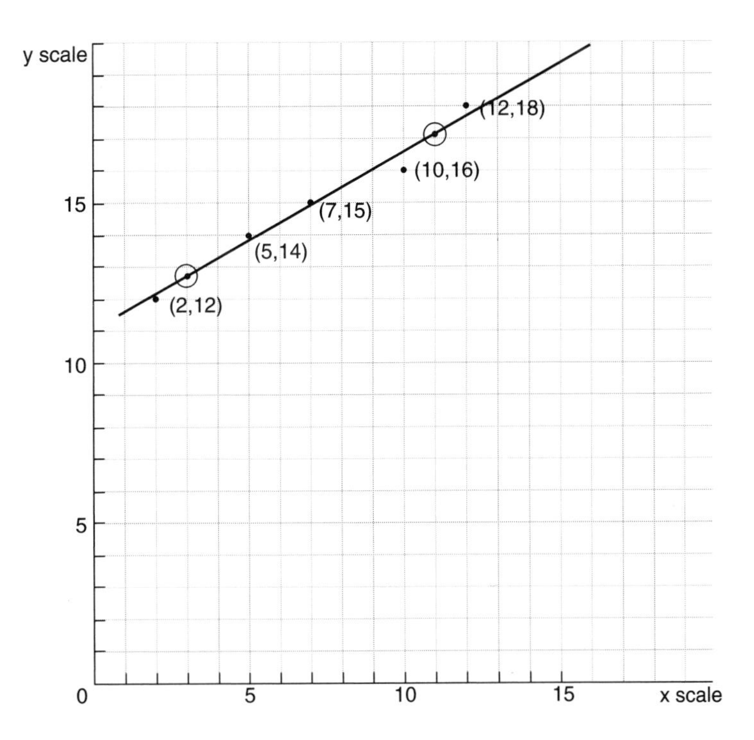

FIGURE 4.34
The line of best fit is drawn through the points predicted by the regression equation, (3,12.7) and (11,17.1).

calculator indicated that y = 17.1. Figure 4.34 shows the scatterplot and the regression line. The excellent correlation, indicated by $r = .99$, is seen on the graph by how close the data points are to the regression line.

EXAMPLE 4.17 Find the regression line for the data given in Table 4.24.

Solution

Enter the data in a linear regression calculator.

Read: slope B = −.00017

y-intercept A = .65

correlation coefficient, r, is $r = -.98$

The regression equation is $r = -.00017x + .65$

Two points for the regression line:

if x = 857, the calculator indicates that y = .5033 and if x = 861, the calculator predicts y = .5027. The excellent correlation, $r = -.98$, ($|r| = .98$) indicates that the points on the scatterplot will be very close to the line and the line will have a negative slope. A visual check in Figure 4.30 shows that this is the case.

EXAMPLE 4.18 Draw the regression line for the data in Table 4.25.

Solution

Enter the ordered pair data in a linear regression calculator.

Read: slope B = .000022

y-intercept A = .6031

correlation coefficient r = .58

Two points for the regression line:

if x = 510, the calculator indicates that y = .6142 and if x = 580, then y = .6157. The correlation coefficient indicates poor correlation and the visual check shows that the line does go through the scatterplot with balance above and below, but there is too much variation and the points are not very close to the line. The regression equation should not be used to predict y values for given x values because of the large amount of variation.

EXERCISES

1. Refer to the *G* chart in Appendix B, Table B.7. How many groups should be used for making a histogram if there are the following number of data values:
   a. 150
   b. 70
   c. 300
2. If your data range from a low of 2.824 to a high of 2.852 and you are making eight groups, what are the group boundaries?
3. A set of data values range from 5.7850 inches to 5.8428 inches. Find group boundaries that could be used to organize the data in a frequency distribution with seven groups.
4. **a.** Make a tally histogram for the data in Table 4.26. Use the group boundaries 1.725 to 1.765, 1.765 to 1.805, 1.805 to 1.845, and so on.

TABLE 4.26

1.81	1.86	1.87	1.85	1.87	1.81	1.85
1.76	1.79	1.82	1.89	1.83	1.79	1.83
1.84	1.93	1.85	1.87	1.86	1.78	1.87
1.73	1.81	1.81	1.98	1.95	1.81	1.86
1.77	1.94	1.74	1.90	1.84	1.86	1.87
1.87	1.74	1.73	1.81	1.74	1.73	1.90

   b. How could the histogram be interpreted if the LSL = 1.73 and the USL = 2.00?
5. Make a flowchart for getting up in the morning and going to work or to school:
   a. Without symbols.
   b. With symbols.
6. Make a flowchart for buying a week's supply of groceries:
   a. Without symbols.
   b. With symbols.

7. Make a cause-and-effect diagram for the following problem: Monthly expenses exceed monthly income.
8. Make two Pareto charts for the data in the following list, one for the number of defectives and one for dollar loss. In each case, include a cumulative percentage graph.

Department	Defectives	Dollar Loss
A	20	100
B	120	60
C	80	800
D	100	500
E	50	200
F	30	90

9. The seven production lines in a company had the following number of defects last month. Make a Pareto chart and include the cumulative percentage graph.

Line	Number of Defects
1	100
2	350
3	45
4	855
5	920
6	170
7	60

10. **a.** For each process make a scatterplot for the pairs of values listed below. Put the RPM values on the horizontal axis.
    b. Classify each scatterplot:
    (1) As very good correlation
    (2) As good correlation
    (3) As poor correlation
    (4) As no correlation

Process I		Process II	
RPM	Measurement	RPM	Measurement
650	.906	620	1.420
675	.909	660	1.430
640	.905	680	1.425
660	.907	700	1.432
700	.908	600	1.424
600	.902	720	1.431

11. **a.** Make a scatterplot from the ordered pairs (T, m):

(120, 2.581) (118, 2.580) (134, 2.595)
(125, 2.592) (130, 2.594) (118, 2.578)
(110, 2.575) (123, 2.590) (132, 2.594)

b. Draw the line of best fit using a linear regression calculator.
c. What m value corresponds to $T = 115$, according to your line of best fit?
d. How would you describe the scatterplot: good correlation, poor correlation, or no correlation?

Table 4.27 consists of data from a control chart. The specifications are .290±.005 with a target measurement of .290.

TABLE 4.27

.287	.288	.283	.288	.284	.287	.289	.290	.286	.290	.291	.291
.292	.284	.285	.286	.288	.290	.288	.291	.291	.293	.292	.293
.285	.285	.286	.286	.289	.291	.289	.288	.292	.289	.294	.288
.286	.285	.290	.289	.289	.288	.291	.287	.290	.290	.289	.289
.284	.290	.295	.291	.288	.286	.291	.289				
.289	.286	.294	.293	.293	.290	.294	.291				
.290	.291	.294	.290	.291	.292	.290	.295				
.292	.290	.293	.289	.292	.293	.287	.295				

12. Organize the data in Table 4.27 in a stemplot and analyze the stemplot with respect to the center, the spread, and the shape. Use a split stemplot with two units per stem.

13. Determine the best group range and put the data from Table 4.27 in a frequency distribution.

14. **a.** Draw the histogram for the frequency distribution in Exercise 13.
b. Analyze the histogram with respect to center, spread, and shape.
c. How does the stemplot analysis in Exercise 12 compare to that of the histogram?

15. **a.** Make a split stemplot for the set of data values in Table 4.26.
b. If the LSL = 1.75 and USL = 1.95, analyze the distribution from the split stemplot (position, spread, and shape).
c. If the LSL = 1.60 and USL = 2.00, analyze the distribution from the split stemplot (position, spread, and shape).

Table 4.28 consists of data from a control chart. The measurements have a target of .60 and the specification limits are .55 to .65.

TABLE 4.28

.60	.57	.59	.58	.60	.63	.64	.66	.63	.62	.64	.63	.56	.57	.56	.58	.57
.58	.60	.60	.59	.62	.64	.62	.62	.64	.63	.65	.64	.62	.58	.55	.58	.60
.59	.60	.59	.61	.61	.61	.61	.62	.65	.65	.62	.61	.65	.64	.65	.59	.59
.58	.59	.58	.58	.60	.58	.61	.62	.65	.65	.62	.61	.65	.64	.65	.59	.58
.58	.60	.61	.60													

16. Make an ordered split stemplot for the data in Table 4.28. Use two units per stem.

17. Determine the best number of groups to use for the data in Table 4.28 and make a frequency distribution.

18. Make a histogram for the frequency distribution in Exercise 17 and analyze its position, spread, and shape.

19. You are an inspector in the calculator factory. Design a checksheet for inspecting quality features in calculators like yours.

20. You are a roving inspector for McDonald's Restaurants. Design a checksheet of quality features that you would use.

21. Make a cause-and-effect diagram for the following:
    a. The TV won't work.
    b. You are often late to your first class.
    c. Poor grades in (choose a subject).
22. a. Draw the scatterplot for the following pairs:

Temperature	Measurement
158	.625
164	.631
147	.617
152	.623
161	.629
142	.613
169	.633

    b. Draw the line of best fit using a linear regression calculator.
    c. Classify the correlation as perfect, good, poor, or no correlation.

5

THE NORMAL PROBABILITY DISTRIBUTION

OBJECTIVES

- Apply the definition of probability.
- Change a frequency distribution to a probability distribution.
- Know the central limit theorem.
- Know how the width of an $\bar{x}$ distribution compares to the population distribution of x.
- Know the relationship between the area under the normal distribution curve and the percentage of product measurements that it represents.
- Use the normal distribution tables.
- Calculate the percentage of product that is out of specification.

5.1 PROBABILITY DISTRIBUTIONS

The word *probability* is used to describe the chance that a specific event will occur. An event must be classified as either a success or a failure, so all of the possible events will be split into the two groups. The probability that a success will occur is defined as the total number of successful events divided by the total number of events.

$P(\text{success})$ means the probability that a success will occur.

$$P(\text{success}) = \frac{\#\text{successes}}{\#\text{events}}$$

EXAMPLE 5.1 If one card is drawn from a standard deck of cards, what is the probability that it is a heart?

Solution

There are 52 cards in a standard deck of cards, so the total number of possible events would be 52. There are 13 hearts in a standard deck of cards, so the total number of possible successes would be 13:

$$P(\text{heart}) = \frac{13}{52} = \frac{1}{4} = .25 = 25\%$$

The probability of getting a heart is 13 chances out of 52. This is equivalent to 1 chance out of 4 (the usual reducing fractions). Decimal fractions can be used: There is a .25 chance of getting a heart. Percentages are used most often: There is a 25% chance of getting a heart.

If the two possible extremes are investigated, then the range of probability values can be determined.

1. *The "sure thing":* If a coin is tossed, what is the probability of getting either a head or a tail? There are two possible successes and two possible events:

$$P(\text{head or tail}) = \frac{2}{2} = 1 \text{ or } 100\%$$

2. *The impossible event:* If a single die is tossed, what is the probability of getting a seven? There aren't any sevens on a single die, so 0 successes and 6 possible events:

$$P(\text{seven on a single die}) = \frac{0}{6} = 0$$

NOTE: ***All probability values will lie between 0 and 1.***

This concept will be applied to Probability Distributions. Dividing the total number of measurements in a class or group of a frequency distribution by the total number of measurements in the distribution gives the probability that a measurement from the distribution will be in that group. It also gives the percentage of measurements in that group (when the decimal fraction is multiplied by 100).

In previous discussions it was mentioned that the frequency distribution for a large sample should have approximately the same basic shape as the frequency distribution of the population. The concept is correct, but if you try to visualize the comparison realistically, you will have a population distribution with a bar height in the thousands (or more) and a sample distribution with a comparable bar height of 10 or 20. Such a comparison can be made in two ways. First (and least important), the frequency scale could be much larger for the population histogram, as illustrated in Figure 5.1 and 5.2.

The best way to show the relationship between distributions is to change them from frequency distributions to probability distributions. Change each frequency value to a probability value by dividing by the total number in the sample. The result gives the fraction or decimal fraction (or percentage if you multiply by 100) that represents the probability of each specific value. This is shown in Example 5.2.

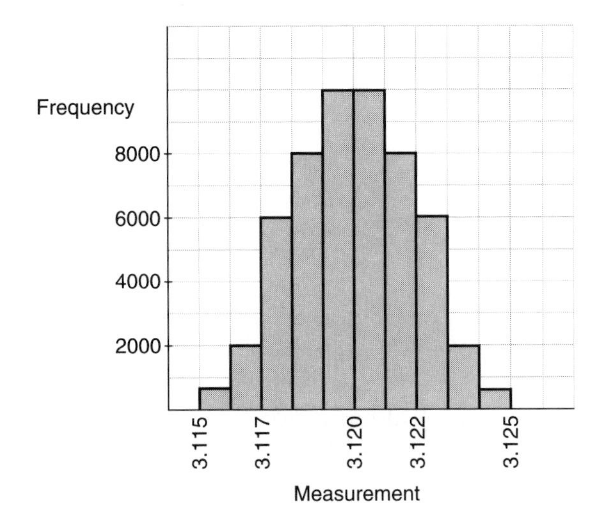

FIGURE 5.1
The histogram for the population frequency distribution.

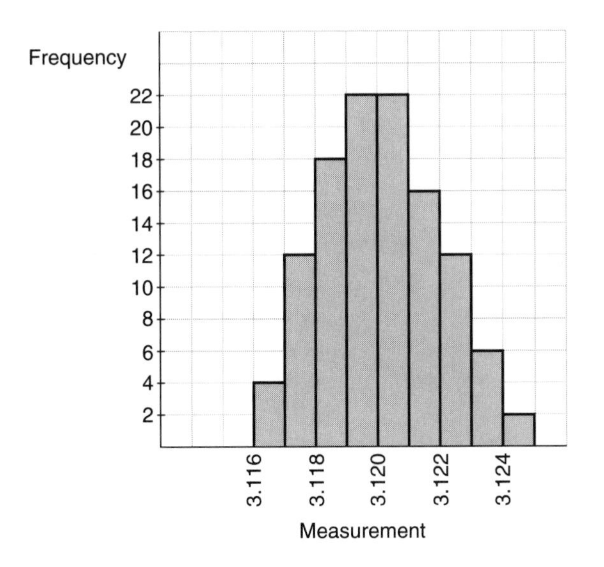

FIGURE 5.2
The histogram for a sample frequency distribution.

EXAMPLE 5.2 Change the given frequency distribution (Table 5.1) to a probability distribution (Table 5.2), show the probability histogram, and interpret the results.

TABLE 5.1
The frequency distribution

Measurement	Frequency (f)
1.847	3
1.848	12
1.849	19
1.850	9
1.851	5
1.852	2
	$N = 50$

TABLE 5.2
The probability distribution

Measurement	f/N = Probability	Percent
1.847	3/50 = .06	6
1.848	12/50 = .24	24
1.849	19/50 = .38	38
1.850	9/50 = .18	18
1.851	5/50 = .10	10
1.852	2/50 = .04	4
	1.00	100

Solution

Each frequency value is divided by $N = 50$ to change it to a probability value. The decimal value is multiplied by 100 to change it to a percentage.

Figure 5.3 shows the histogram for the frequency distribution in Table 5.1. Figure 5.4 shows the histogram for the probability distribution in Table 5.2. The two histograms have the same shape.

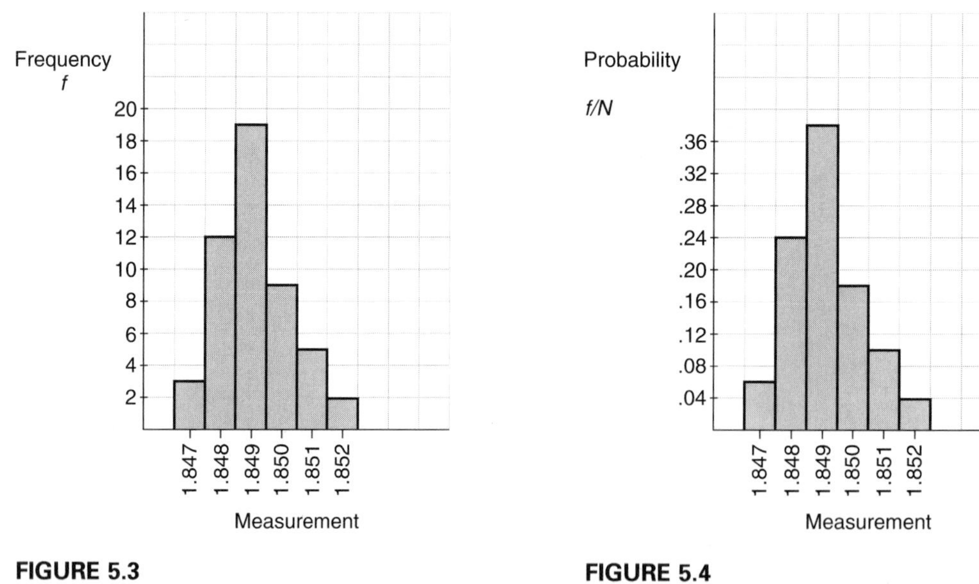

FIGURE 5.3
The frequency histogram.

FIGURE 5.4
The probability histogram.

These are all equivalent statements:

- Three measurements out of a total of 50 have a value of 1.847.
- Six percent of the measurements have a value of 1.847.
- The probability or chance that the measurement is 1.847 is 3/50, or 6%.

These are also equivalent:

- Nine of the 50 measurements have a value of 1.850.
- Eighteen percent of the measurements have a value of 1.850.
- The probability or chance of getting a measurement of 1.850 is 9 out of 50, or 18%.

One further comment about terminology: Both Table 5.1 and Figure 5.3 can be called frequency distributions. One is in table form and one is in graph form. Table 5.2 and Figure 5.4 show the table form and graph form of a probability distribution.

The concept of distributions is one of the basic building blocks of statistics. The first distribution that is usually encountered in the study of statistics is the frequency distribution. When a sample is taken, the measurements are distributed among the classes of the frequency distribution in a specific pattern with a set number in each class. Figure 4.5 illustrates how the frequency distribution takes on a specific shape as the number of data items increases. Eventually the sample size gets large enough to establish the pattern and any further data additions result in just minor fluctuations within the set pattern.

Once the frequency distribution pattern is set, it can be changed to a probability distribution. The sum of the frequencies from each column gives the total number of measurements in the table. When each class frequency is divided by the number of measurements n, the result is the probability associated with that class. This forms a probability distribution: The probabilities are distributed to each class.

EXAMPLE 5.3

1) Make a probability distribution for the frequency distribution in Table 5.3.
2) Find $P(.626 \le x < .630)$
3) Find $P(x \ge .638)$
4) Find $P(.626 \le x < .638)$

TABLE 5.3
Frequency distribution

Class	Frequency
.618 to .622	2
.622 to .626	5
.626 to .630	10
.630 to .634	16
.634 to .638	10
.638 to .642	6
.642 to .646	1
	$n = 50$

TABLE 5.4
Probability distribution

Class	f/N = Probability
.618 to .622	2/50 = .04
.622 to .626	5/50 = .10
.626 to .630	10/50 = .20
.630 to .634	16/50 = .32
.634 to .638	10/50 = .20
.638 to .642	6/50 = .12
.642 to .646	1/50 = .02

Solution

From Table 5.4, the probability that a measurement is in the .626 to .630 class is .20, or 20%. The $P(x \geq .638) = .12 + .02 = .14$, or 14%. The $P(.626 \leq x < .638) = .20 + .32 + .20 = .72$, or 72%. This probability distribution is specific to the set of data from which it was formed. The probabilities refer specifically to the sample data.

Statisticians assume that once the sample distribution stabilizes, it looks like the population distribution. This means that the population *probability* distribution must look like the sample *probability* distribution. The histogram for the sample is used to find a standard probability distribution with the same shape or pattern, and that pattern is used to calculate probabilities concerning the population. Those probabilities are then used to make predictions or decisions concerning the population. The most often used standard probability distribution is the *normal* probability distribution. This procedure is illustrated as follows.

1. The data from a sample form a sample histogram (Figure 5.5).
2. The sample histogram has a probability distribution with the same shape (Figure 5.6).
3. The probability distribution of the population has the same basic shape as that of the sample, usually slightly wider (Figure 5.7).

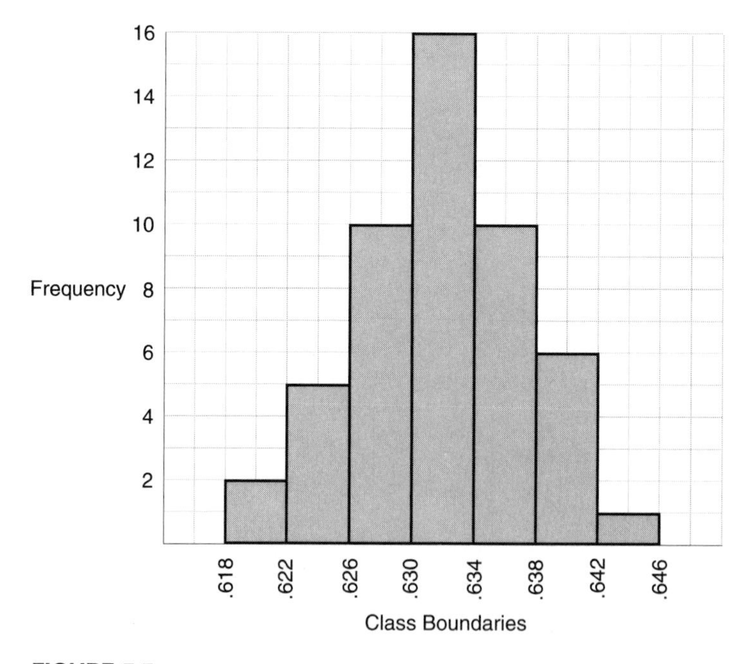

FIGURE 5.5
The frequency distribution (histogram) for the sample.

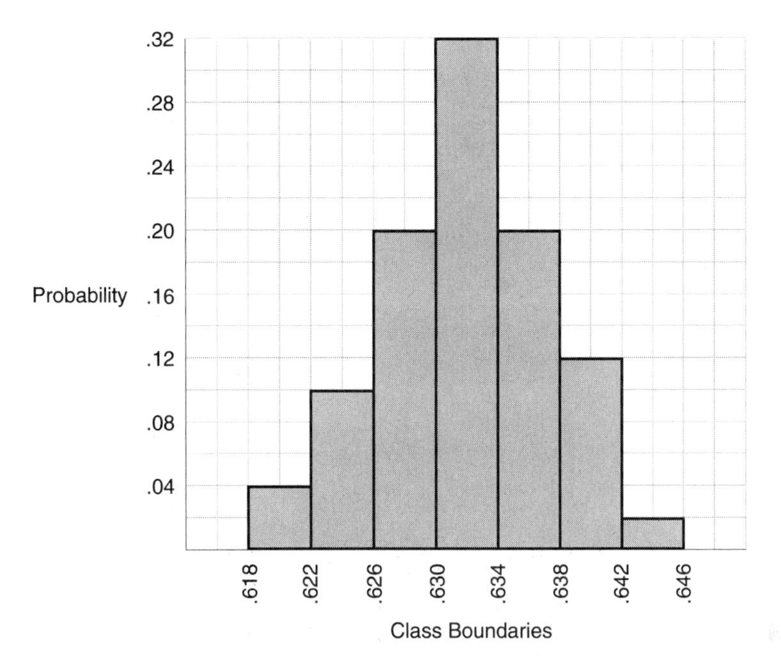

FIGURE 5.6
The probability distribution for the sample.

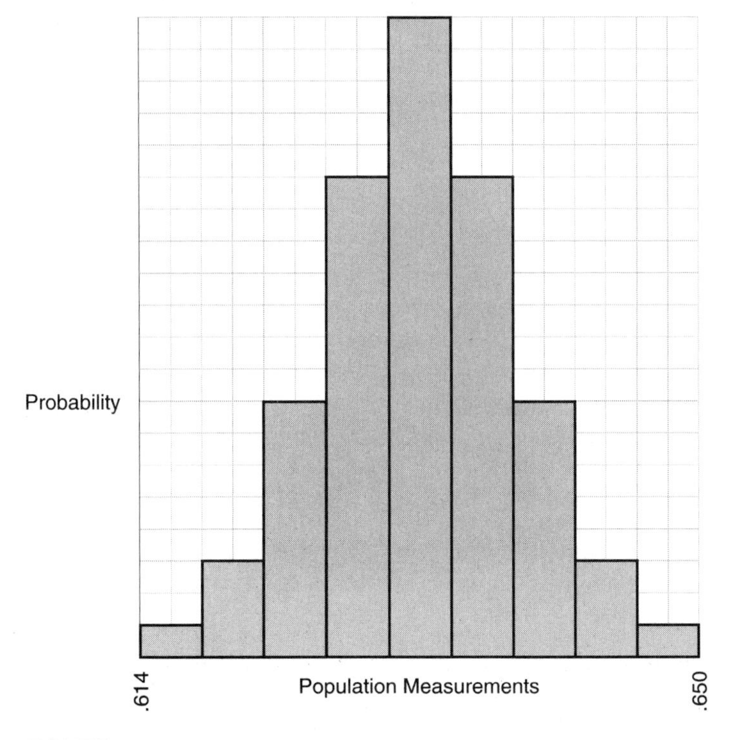

FIGURE 5.7
The population probability distribution.

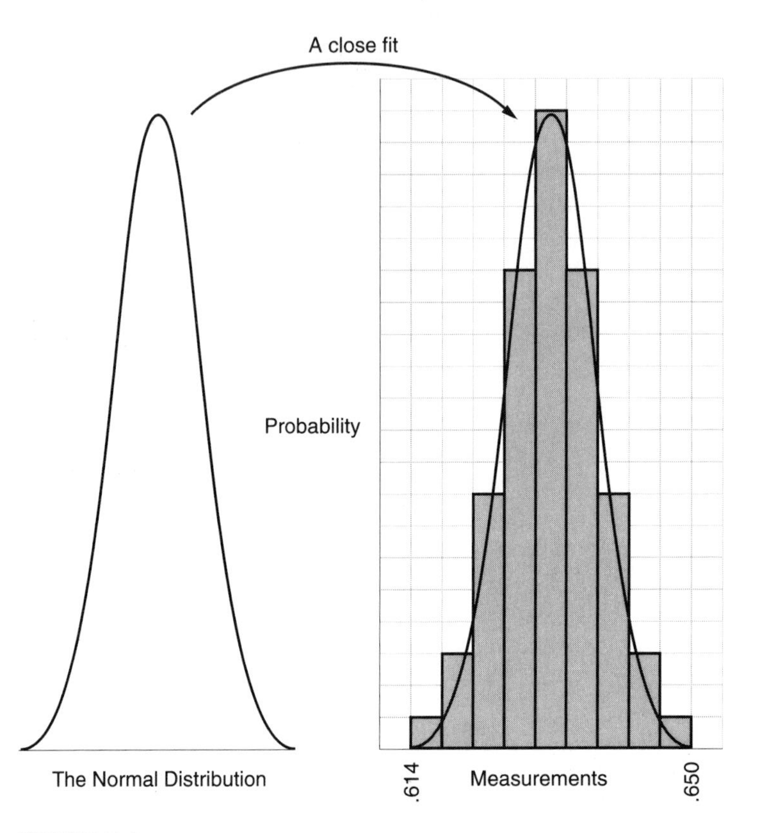

FIGURE 5.8
The normal distribution fits the measurement distribution.

4. There is a standard probability distribution that has approximately the same shape as the sample distribution and therefore the same shape as the population distribution. In this case, the *normal* probability distribution matches (Figure 5.8).
5. The normal distribution is used to calculate probabilities for the population. Those probabilities are used in decision making for action on the process.

5.2 THE NORMAL PROBABILITY DISTRIBUTION

The *normal* distribution is pictured as a symmetric, bell-shaped curve, as shown in Figure 5.8. There is no single normal curve, but rather a whole family of normal curves that have the same basic properties.

Properties of the Normal Distribution

1. The distribution curve has a single peak and is bell-shaped.
2. The curve is symmetrical about its center line: The shape on one side of the center is the mirror image of the other side.

3. The mean and median of the distribution are at the center.
4. The width of the curve is determined by the standard deviation of the distribution.
5. The tails of the curve flatten out and extend indefinitely along the horizontal axis, always approaching the axis but never touching it or crossing it.
6. The area under the curve represents probability or the percentage associated with specific sets of measurement values.
7. The total area under the curve is 1 (or 100%).
8. The normal curve is used to predict the percentage of population values in various categories.

When a sample of measurements is taken from a population, the percentage of measurements in any category reflects the distribution pattern in that specific sample. When another sample is taken from the population, a different percentage could occur in the category. For example, in Figure 5.5 the percentage of measurements that are smaller than .626 is 14%. The total frequency of the two columns to the left of .626 is 7 (5 for one column and 2 for the other). Divide that by 50, the total number of frequencies for the distribution, and multiply by 100 to get the 14%. If another sample was taken from the same population, the distribution pattern would most likely be different and the percentage of measurements smaller than .626 could be as large as 17 or 18% or as small as 9 or 10%.

The important question to be answered in this example, based on the information in the sample, is what percentage of the *population* measurements are smaller than .626? The answer is obtained using the normal probability distribution. The real stabilizing value from a large sample is the standard deviation of the sample. If several large samples were taken, their standard deviations would be very close in value and, in turn, be a very good approximation of the population standard deviation (an important application of statistical theory). The standard deviation determines the spread of the normal curve for the population and the percentage of population measurements smaller than .626 can then be determined.

The normal curve for a population is specifically determined by the mean and the standard deviation of the population. The mean locates the center of the curve, and the standard deviation determines the width of the curve. Areas under the normal curve are given in terms of z values. The z value represents the distance from the center measured in standard deviation units. The formula for calculating z is

$$z = \frac{x - \mu}{\sigma}$$

x = an individual measurement
μ = the population mean
σ = the population standard deviation
z = the number of standard deviation units from the mean

The x is the measurement of interest, μ is the population mean, and σ is the population standard deviation. Since, for any population, the mean and standard deviation are fixed values, the z formula matches the z values one-to-one with the x values. So, for each x value there is exactly one z value and for each z value there is exactly one x value. The z values are matched with specific areas under the normal curve in a normal distribution table. To find the percentage of product associated with a measurement x, find its matched z value using the z formula. The z value leads to the area under the curve (from the normal curve table), which is probability, and that probability gives the desired percentage for x. These concepts are demonstrated in the following examples.

EXAMPLE 5.4

Given the mean and standard deviation for a population, find the z-scale values that correspond to the x values shown in Figure 5.9:

$$\mu = 35 \qquad \sigma = 4$$

Solution

The z scale can be determined from the x values with the z formula. Calculate z for $x = 43$:

$z = \dfrac{x - \mu}{\sigma}$ Substitute values.

$= \dfrac{43 - 35}{4}$ Subtract first because the fraction line is a grouping symbol.

$= \dfrac{8}{4}$ Divide by 4.

$= 2$ The z value 2 corresponds to the x value 43 on the normal curve shown in Figure 5.9.

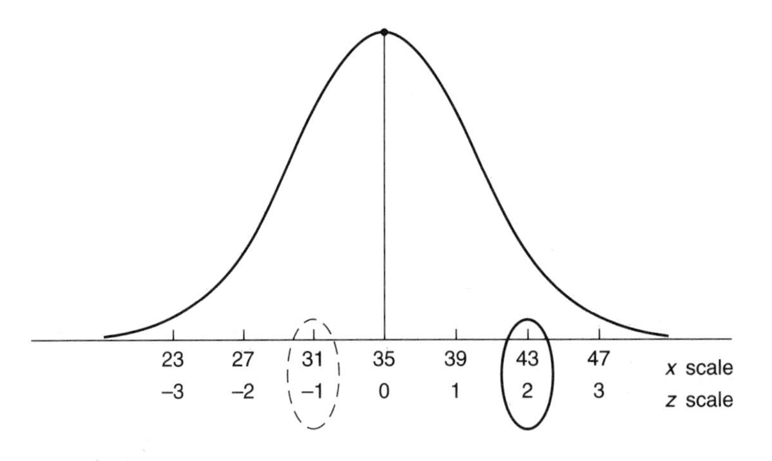

FIGURE 5.9
The normal curve with the scale of measurements (x) matched with the z scale.

If $x = 31$, substitute $x = 31$, $\mu = 35$, and $\sigma = 4$. Subtract first, then divide by 4:

$$\begin{aligned} z &= \frac{x - \mu}{\sigma} \\ &= \frac{31 - 35}{4} \\ &= \frac{-4}{4} \\ &= -1 \end{aligned}$$

Because $z = 0$ at the center, the negative z value simply indicates that z is left of center. Here a z value of -1 matches an x value of 31 (see Figure 5.9). The other z values shown in Figure 5.9 are calculated as follows:

$x = 23$	$x = 27$	$x = 35$	$x = 39$	$x = 47$
$z = \frac{23 - 35}{4}$	$z = \frac{27 - 35}{4}$	$z = \frac{35 - 35}{4}$	$z = \frac{39 - 35}{4}$	$z = \frac{47 - 35}{4}$
$z = -3$	$z = -2$	$z = 0$	$z = 1$	$z = 3$

Area Under a Normal Curve

The area under the curve indicates a proportion, or percentage, of the product or measurements; the area of various sections can be found using a statistical table. The complete table is Table B.8 in Appendix B. You should find the area values in this table with each illustration.

The normal distribution table gives the *area* in the *tail section* of the curve that corresponds to a specific z value. A positive z value indicates that the area is in the right tail section, and a negative z value indicates that the area is in the left tail section. The z values for the table range from 0.00 to 4.00. Because of symmetry, the negative z values are not needed since their corresponding positive values have the same area.

To use the table, put the z value into a three-digit format (for example, .86 is changed to 0.86 and 1.4 becomes 1.40).

1. Find the first two digits of z in the z column on the left. That designates the *row*.
2. Match the third z digit with the appropriate column number on the right.
3. The *area* value is at the intersection of the *row* and the *column*.

The following display shows the area values that correspond to given z values. They are found in the normal distribution table (Table B.8, Appendix B).

If $z = 1.52$	Row 1.5	Column 2	Area = .0643
If $z = -1.46$	Row 1.4	Column 6	Area = .0721

If $z = .7$ (0.70)	Row 0.7	Column 0	Area = .2420
If $z = -2.03$	Row 2.0	Column 3	Area = .0212
If $z = 3$ (3.00)	Row 3.0	Column 0	Area = .00135
If $z = 2.43$	Row 2.4	Column 3	Area = .0075
If $z = -1.86$	Row 1.8	Column 6	Area = .0314

In Example 5.4, the x values were given as 43 and 31, and the corresponding z values were calculated to be 2 and -1. The tail area for each is found in the normal distribution table, as demonstrated in Example 5.5.

EXAMPLE 5.5 Find the area under the curve from Example 5.4 that gives the percentage of measurements that are larger than 43 and smaller than 31.

Solution

From the previous calculations, when $x = 43$, $z = 2$, and when $x = 31$, $z = -1$. For $z = 2$ (or 2.00), row 2.0 and column 0 give a tail area of .0228. Therefore, 2.28% of the area under the curve is to the *right* of $x = 43$, and 2.28% of the product has a value larger than 43. This is illustrated in Figure 5.10.

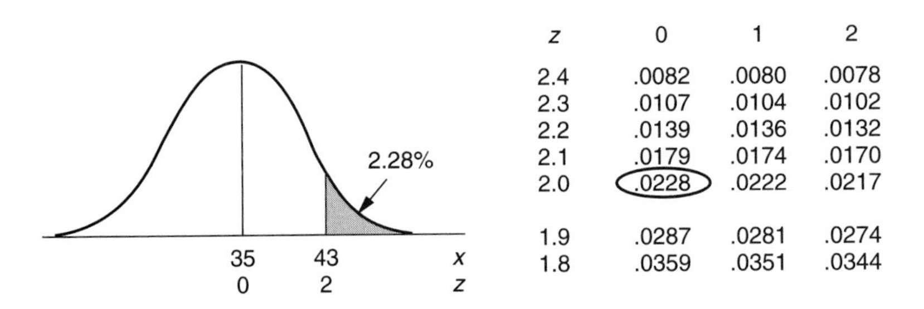

z	0	1	2
2.4	.0082	.0080	.0078
2.3	.0107	.0104	.0102
2.2	.0139	.0136	.0132
2.1	.0179	.0174	.0170
2.0	.0228	.0222	.0217
1.9	.0287	.0281	.0274
1.8	.0359	.0351	.0344

FIGURE 5.10
In Example 5.5, 2.28% of the measurements are larger than 43.

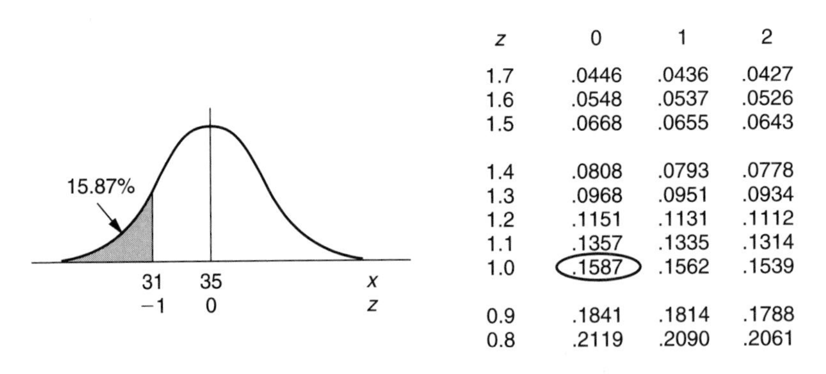

z	0	1	2
1.7	.0446	.0436	.0427
1.6	.0548	.0537	.0526
1.5	.0668	.0655	.0643
1.4	.0808	.0793	.0778
1.3	.0968	.0951	.0934
1.2	.1151	.1131	.1112
1.1	.1357	.1335	.1314
1.0	.1587	.1562	.1539
0.9	.1841	.1814	.1788
0.8	.2119	.2090	.2061

FIGURE 5.11
In Example 5.5, 15.87% of the measurements are less than 31.

For $z = -1$ (or -1.00), row 1.0 and column 0 give a tail area of .1587. This means that 15.87% of the area under the curve is to the *left* of $x = 31$ and 15.87% of the product has a value less than 31. This is illustrated in Figure 5.11.

EXAMPLE 5.6

A population has a mean μ of 44.3 and a standard deviation σ of 2.1. What percentage of the measurements is larger than 46?

Solution

1. Analyze the problem with a sketch of the normal curve.
   a. Sketch the normal curve. See Figure 5.12.
   b. Locate $\mu = 44.3$ at the center.
   c. Locate $x = 46$ at a point to the right of center.
   d. Shade the area of interest. Measurements larger than 46 correspond to the area to the right of 46.

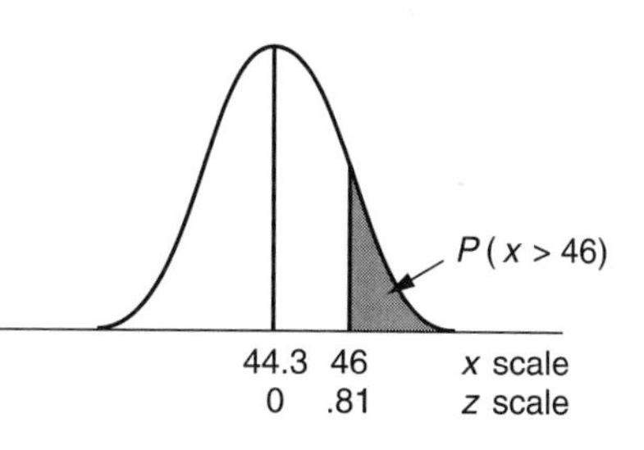

FIGURE 5.12
Analyze the problem with a sketch of the normal curve.

2. Calculate z:

$$z = \frac{x - \mu}{\sigma}$$
$$= \frac{46 - 44.3}{2.1}$$
$$= \frac{1.7}{2.1}$$
$$= .81$$

3. Find the area in the normal-curve table that corresponds to $z = 0.81$.
   a. Use Table B.8 in Appendix B.
   b. The drawing at the top of Table B.8 shows that the table gives the tail area under the curve.
   c. This is a three-digit table: The first two digits are located in the left column under $| z |$, (0.8), and the third digit corresponds to the appropriate column to the right (column marked 1). According to the table, the area is .2090.

4. Translate the table area to the desired probability. In this case, the table area equals the desired probability (both are tail areas). $P(x > 46)$ is .2090, so 20.9% of the measurements are larger than 46.

EXAMPLE 5.7 The average measurement is $\mu = .954$, and the standard deviation is $\sigma = .007$. Find the percentage of measurements less than .960. Next, find the percentage of measurements between .940 and .960.

Solution

1. Sketch the curve. This is shown in Figure 5.13.
   a. Locate $\mu = .954$ at the center.
   b. Locate $x = .960$ to the right.
   c. Shade the area of interest: All the area to the left of .960 corresponds to measurements less than .960.

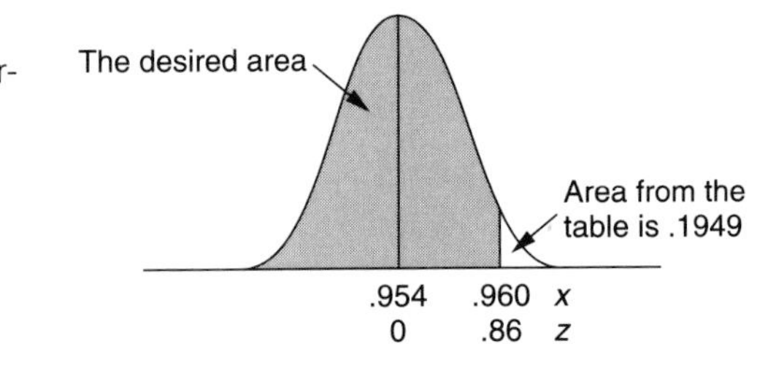

FIGURE 5.13
Analyze Example 5.7 with a sketch of the normal curve.

2. Calculate z:

$$
\begin{aligned}
z &= \frac{x - \mu}{\sigma} \\
&= \frac{.960 - .954}{.007} \\
&= \frac{.006}{.007} \\
&= 0.86
\end{aligned}
$$

3. In table B.8, Appendix B, find 0.8 in the $|z|$ column, then go across to the column marked 6. Read the area. Area = .1949.
4. The area in the table corresponds to the area in the clear section of the sketch (the right tail area). The total area under the curve is 1, so 1 – .1949 will give the desired area:

$$
\begin{aligned}
P(x < .960) &= 1 - .1949 \\
&= .8051
\end{aligned}
$$

This means that 80.51% of all the measurements in the population are less than .960.

For the second part of the question, find the percentage of measurements between .940 and .960:

1. Sketch the curve. The sketch is shown in Figure 5.14.
   a. Locate $\mu = .954$ at the center.
   b. Locate .960 to the right of center and .940 to the left of center.
   c. Shade the area between .960 and .940. This represents the probability that measurements will be between the two values.
2. Calculate the two z values (one for each measurement). Use subscripts when dealing with more than one measurement. We have calculated x_1, z_1, and area$_1$ in the first part of the example: $x_1 = .960$, $z_1 = .86$, and area$_1 = .1949$:

$$z = \frac{x - \mu}{\sigma}$$

$$z_2 = \frac{.940 - .954}{.007}$$

$$= \frac{-.014}{.007}$$

$$= -2$$

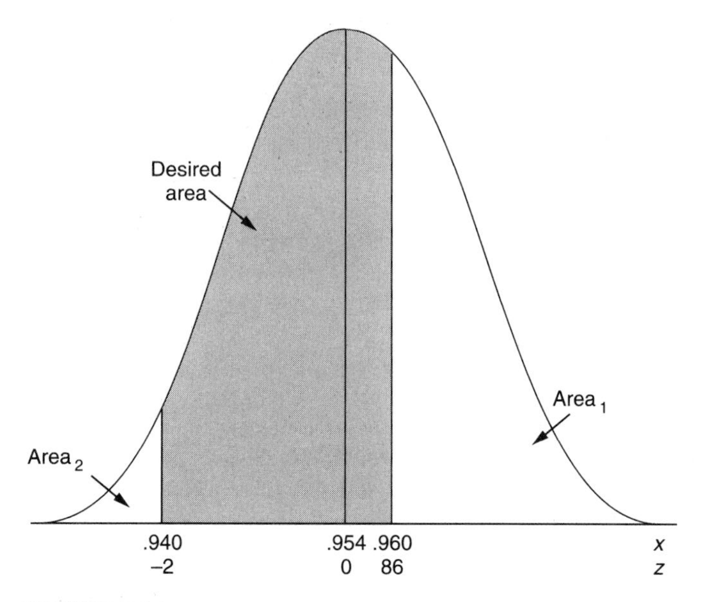

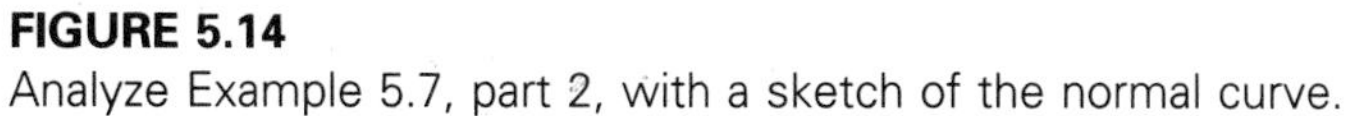

FIGURE 5.14
Analyze Example 5.7, part 2, with a sketch of the normal curve.

The negative z value just means that the x value and the area lie to the left of the center.

3. In Table B.8 Appendix B, find 2.0 in the | z | column, go across to the column marked 0, and read the area. $\text{Area}_2 = .0228$.
4. The area_1 value is the area of the clear tail section on the right of the sketch in Figure 5.14. The area_2 value is the area of the clear section on the left of the sketch. Add the two tail areas, $.1949 + .0228 = .2177$, to get the area under the curve that is *not* wanted.

The total area under the curve is 1. Subtract the area that is not wanted from 1 to get the desired area ($1.0000 - .2177 = .7823$). $P(.940 < x < .960) = .7823$, so 78.23% of the measurements in the population are between .940 and .960.

Other Normal-Curve Tables

Calculations of probability that involve area under the normal curve rely on a few basic concepts that apply to any normal-curve table. Other versions of the normal-curve table are occasionally encountered, and the areas have to be juggled around to transform the table area to the shaded area in the sketch.

Most of the work in this text uses the "tail area" table, Table B.8 in Appendix B, but there are three different versions of normal-curve tables. Each is identified by a drawing that shows the area given in the table that corresponds to the z values. With all versions, the basic concepts to keep in mind are the following:

- Areas of sections can be added or subtracted.
- The total area under the curve is 1.000.
- The midline, at the μ value, divides the curve in half so that the areas to the right and left of the middle are .500.

The version of normal-curve table used in the previous examples gave the area in the "tail."

The drawing in Table B.8 of Appendix B indicates that the table values give the area in the tail. A second version measures the area from the center to the calculated z value. The drawing in Table B.9 of Appendix B indicates that the area in the table spans from the center to the calculated z value. The third version measures the total area to the left of the calculated z value. The drawing in Table B.10 of Appendix B indicates that the total area to the left of the z value is given in the table.

An example to illustrate the use of the three tables follows.

EXAMPLE 5.8 A population of measurements has a mean and standard deviation of $\mu = 70$ and $\sigma = 6$. Find the percentage of measurements greater than $x = 75$. Then find the percentage of measurements between $x = 60$ and $x = 75$.

Solution

The first step, no matter which table is used, is to calculate the z values for each x value. There are two different x values and two different z values to work with, so use subscripts:

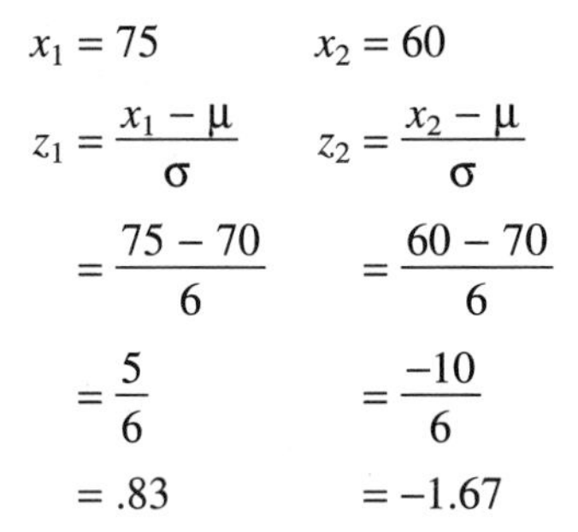

$$x_1 = 75 \qquad x_2 = 60$$

$$z_1 = \frac{x_1 - \mu}{\sigma} \qquad z_2 = \frac{x_2 - \mu}{\sigma}$$

$$= \frac{75 - 70}{6} \qquad = \frac{60 - 70}{6}$$

$$= \frac{5}{6} \qquad = \frac{-10}{6}$$

$$= .83 \qquad = -1.67$$

The sketch for the first part of this example is shown in Figure 5.15. Here $x = 75$, $z = .83$, and the table area, found in Table B.8 in Appendix B, is .2033. This means that 20.33% of the measurements are greater than 75.

From Table B.9 in Appendix B with the same values of x and z, the table area is .2967. The area spans from the center to the z value, as shown in Figure 5.16. The total area to the right of the center is .50. Notice that it is in two sections: the desired area and the area from the table. Subtraction will provide the desired area:

.5000	Area of the right half
− .2967	Table area
.2033	Tail area

The results using Table B.9 also indicate that 20.33% of the measurements are greater than 75.

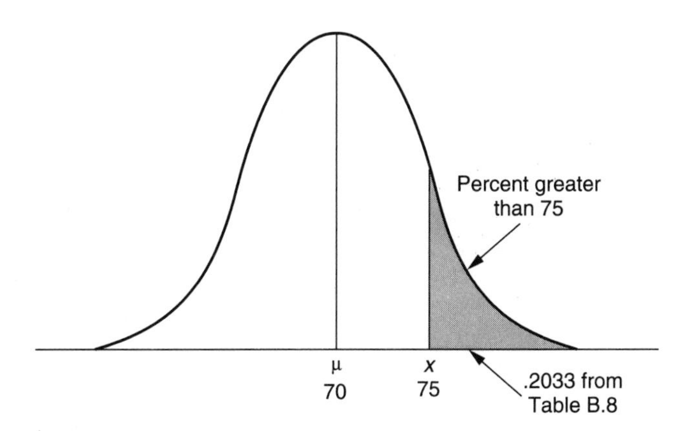

FIGURE 5.15
Tail area with Table B.8 from Appendix B.

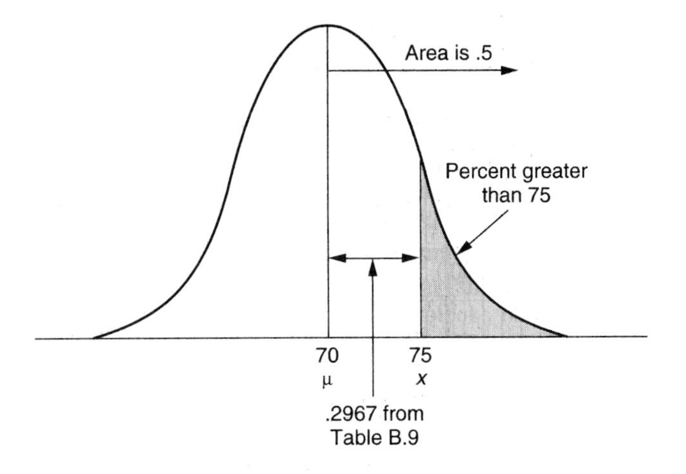

FIGURE 5.16
Tail area with Table B.9 from Appendix B.

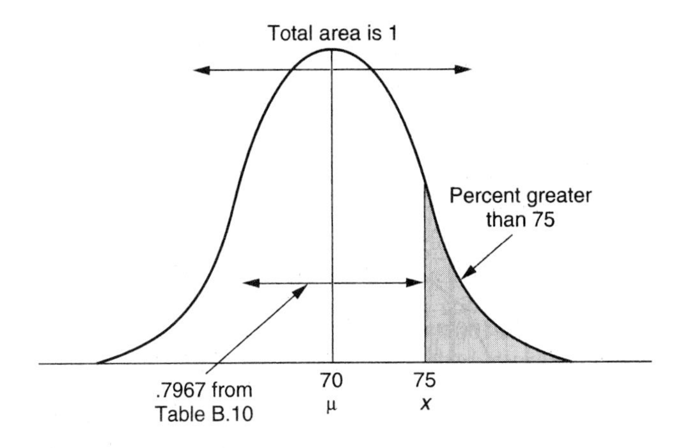

FIGURE 5.17
Tail area with Table B.10 from Appendix B.

Table B.10 in Appendix B gives a table area of .7967 for $x = 75$ and $z = .83$. The total area under the curve, illustrated in Figure 5.17 is partitioned into two sections: the table area and the tail area. Subtraction will give the tail area:

1.0000	Total area under the curve
− .7967	Table area
.2033	Tail area

Again, 20.33% of the measurements are greater than 75.

For the second part of the example, find the percentage of measurements between 60 and 75:

$$x_1 = 75 \qquad x_2 = 60$$
$$z_1 = .83 \qquad z_2 = -1.67$$

The total area under the curve in Figure 5.18 is divided into three sections, table area$_1$, table area$_2$, and the section in between, which is the desired area. From Table B.8 in Appendix B, area$_1$ = .2033 and area$_2$ = .0475. Add the two tail areas and subtract the sum from 1:

.2033	Right tail area
+ .0475	Left tail area
.2508	Tail area sum

1.0000	Total area under the curve
− .2508	Tail area sum
.7492	Area between 60 and 75

Therefore, 74.92% of the measurements are between 60 and 75.

Table B.9 in Appendix B gives a table area$_1$ of .2967 and table area$_2$ of .4525. Both areas are measured from the center: one to the right and the other to the left. The sum of the two areas will give the area between 60 and 75:

.2967	Area from the center to 75
+ .4525	Area from the center to 60
.7492	Area between 60 and 75

Again, 74.92% of the measurements are between 60 and 75. See Figure 5.19.

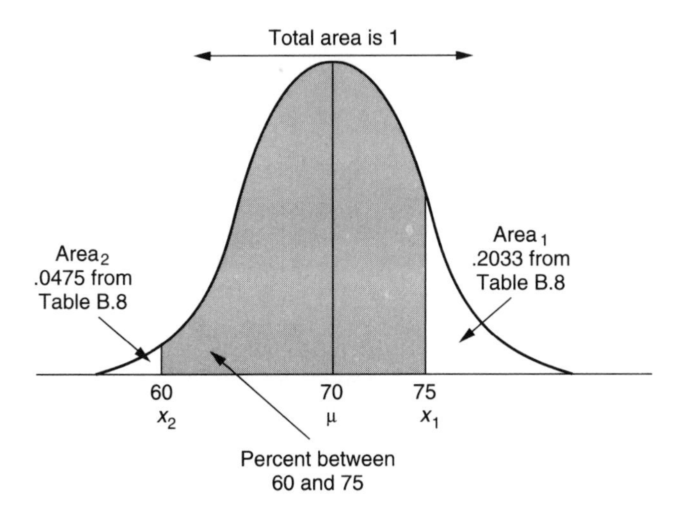

FIGURE 5.18

Area between $x_1 = 75$ and $x_2 = 60$ using Table B.8 from Appendix B.

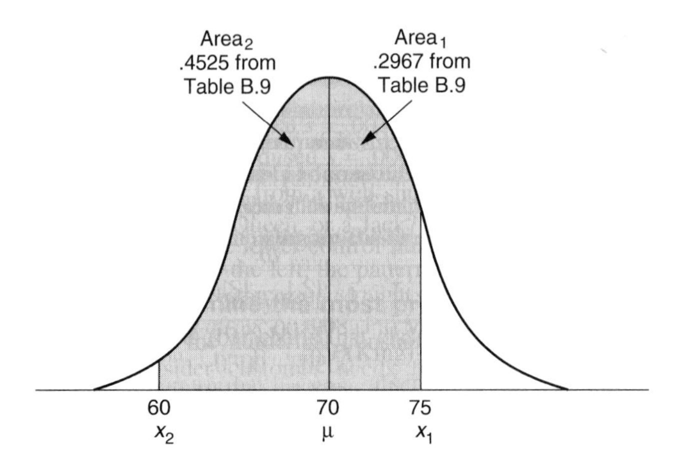

FIGURE 5.19
Area between the values using Table B.9 from Appendix B.

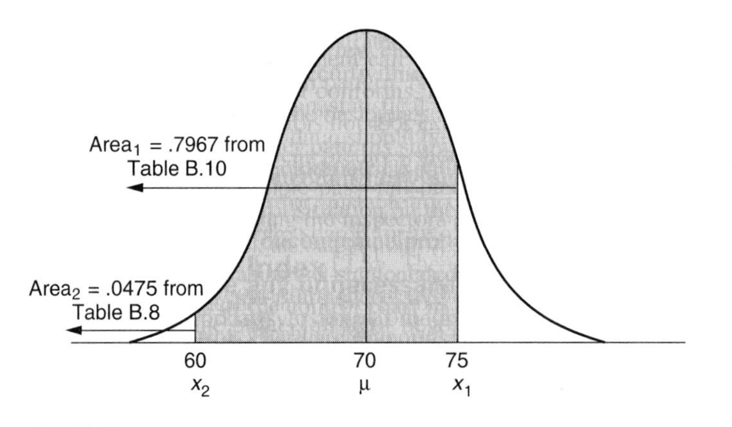

FIGURE 5.20
Area between the values of x using Table B.10 from Appendix B.

According to Table B.10 in Appendix B, table area$_1$ = .7967. Table area$_2$ = .0475 from Table B.8. The area between 60 and 75 is overlapped by area$_1$, as shown in Figure 5.20. Subtraction will give the area between these values:

.7967	Total area to the left of 75
− .0475	Area to the left of 60
.7492	Area between 60 and 75

And again, 74.92% of the measurements lie between 60 and 75.

Whenever a normal curve is used to find percentage or probabilities associated with measurements of a population, the mean and standard deviation of the population are needed for the z formula. In virtually all applications, those population measures are unknown, so estimates of those two values must be used. Statistical theory indicates that, for large samples, the sample mean and sample standard deviation are very good estimates of the population mean and standard deviation. These estimates are used in the z formula. In SPC applications, small samples are taken from a process and their average values are tracked on a control chart. The completed control chart, or a sequential set of control charts, constitutes a large set of data from the population. The mean $\bar{\bar{x}}$ and standard deviation s from that large set of data are excellent estimates of their corresponding population measures. When these values are used to find various population percentages, the z formula becomes

$$z = \frac{x - \bar{\bar{x}}}{s}$$

x = an individual measurement from a large sample
$\bar{\bar{x}}$ = the mean of a large sample
s = the standard deviation of the large sample
z = the number of standard deviation units from the mean

Both versions of the z formula are used in exactly the same way. The different symbols for the mean and standard deviation just indicate the source of the information.

EXAMPLE 5.9 A control chart contains a large sample of measurements from a process with a mean of .917 and a standard deviation of .014. Find the percentage of product that is within one standard deviation of the mean, two standard deviations from the mean, and three standard deviations from the mean.

Solution

Add and subtract .014 to and from the mean to provide measurements that are one standard deviation from the mean. Add and subtract .028, or 2 × .014, to and from the mean to obtain measurements that are two standard deviations from the mean. Adding and subtracting .042, or 3 × .014, will give measurements three standard deviations from the mean:

$$\begin{array}{cccccc} .917 & .917 & .917 & .917 & .917 & .917 \\ +.014 & -.014 & +.028 & -.028 & +.042 & -.042 \\ \hline x = .931 & x = .903 & x = .945 & x = .889 & x = .959 & x = .875 \end{array}$$

The z values for measurements $x = .931$ and $x = .903$ are $z = 1$ and $z = -1$, respectively, corresponding to one standard deviation from the mean:

$$z = \frac{x - \bar{\bar{x}}}{s} = \frac{.931 - .917}{.014} = 1 \qquad z = \frac{x - \bar{\bar{x}}}{s} = \frac{.903 - .917}{.014} = -1$$

The z values for measurements $x = .945$ and $x = .889$ are $z = 2$ and $z = -2$, respectively, corresponding to two standard deviations from the mean:

$$z = \frac{x - \bar{\bar{x}}}{s} = \frac{.945 - .917}{.014} = 2 \qquad z = \frac{x - \bar{\bar{x}}}{s} = \frac{.889 - .917}{.014} = -2$$

The z values for measurements $x = .959$ and $x = .875$ are $z = 3$ and $z = -3$, respectively, corresponding to three standard deviations from the mean:

$$z = \frac{x - \bar{\bar{x}}}{s} = \frac{.959 - .917}{.014} = 3 \qquad z = \frac{x - \bar{\bar{x}}}{s} = \frac{.875 - .917}{.014} = -3$$

Using Table B.9 in Appendix B, we find

$z = 1$ Area = .3413

$z = 2$ Area = .4773

$z = 3$ Area = .4987

The center area is given in each case. This must be translated to the area in question, which is measured from the center both ways. Using the symmetry property of the normal curve, double the center area to get the total area within one, two, or three standard deviations from the mean. The results are shown in Figures 5.21, 5.22, and 5.23.

$z = 1$ center area = .3413

$2 \times .3413 = .6826$

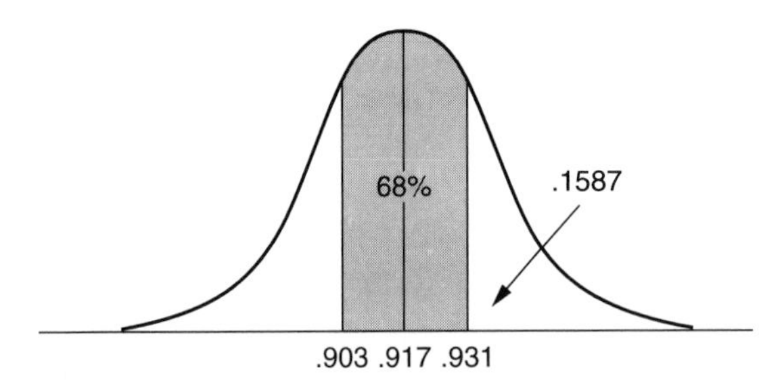

FIGURE 5.21
Sixty-eight percent (.6826) of the measurements are within one standard deviation of the mean.

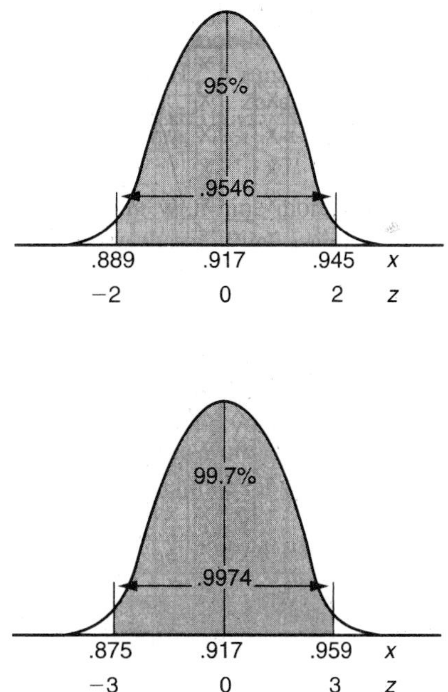

FIGURE 5.22
Ninety-five percent (.9546) of the measurements are within two standard deviations of the mean.

FIGURE 5.23
Ninety-nine point seven percent (.9974) of the measurements are within three standard deviations of the mean.

The mean, .917, is at the center. Approximately 68%, or .6826, of the product measures between .903 and .931, which are each one standard deviation from the mean. This is shown in Figure 5.21:

$$z = 2 \qquad \text{center area} = .4773$$
$$2 \times .4773 = .9546$$

Approximately 95% of the measures are within two standard deviations of the mean. This is shown in Figure 5.22. About 99.7% of the product is within three standard deviations from the mean, between .875 and .959:

$$z = 3 \qquad \text{center area} = .4987$$
$$2 \times .4987 = .9974$$

This is shown in Figure 5.23.

The results of Example 5.9 may be generalized for any application of the normal curve. Sixty-eight percent of the values will always be within one standard deviation of the mean, 95% of the values will always be within two standard deviations of the mean, and 99.7% of the values will always be within three standard deviations of the mean. This is illustrated in Figure 5.24.

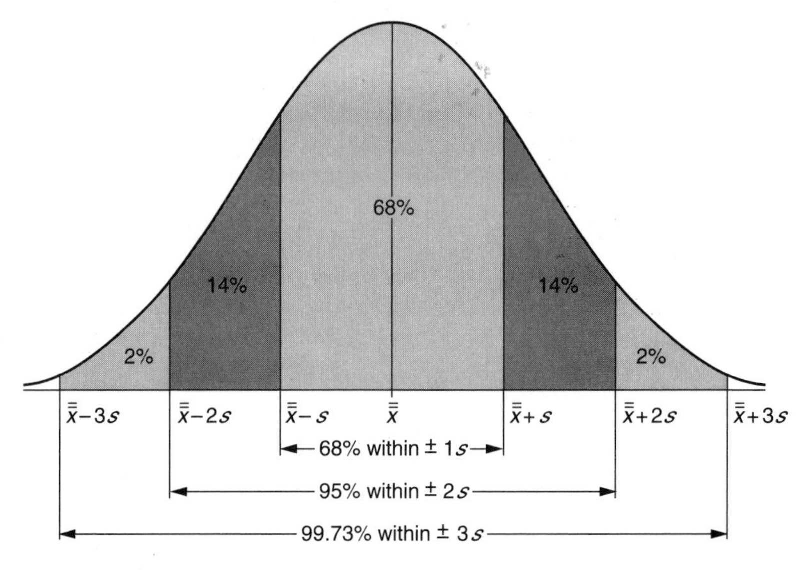

FIGURE 5.24
The standard deviation zones on the normal curve.

If the histogram of the product measurements is bell-shaped, indicating a normal distribution, the normal-curve analysis can provide much useful information about the true distribution of the measurements. More specifically, a normal-curve analysis will show how the measurements compare with the target value and with the product specifications. This is illustrated in Example 5.10.

EXAMPLE 5.10 Continuing the work from Example 5.9, in which the product mean was .917 and the standard deviation was .014, suppose the product specifications are .880 to .930. Find the percentage of the product that is out of specification.

Solution

If the data from this control chart were organized in a stemplot or a histogram with the specification limits included, we would have a visual indication that the sample crowds or overlaps the specification limits. The percentage of the sample that is out of specification could also be calculated. However, to project the sample information to the entire population, the normal curve has to be used. The area (or percentage) evaluation with the normal curve involves the formula

$$z = \frac{x - \bar{\bar{x}}}{s}$$

since μ and σ are unknown. The z gives the number of standard deviations from the center for each specific x value. Find the z value for each specification limit.

FIGURE 5.25
The total percentage of out-of-specification measurements using both USL and LSL in the z formula.

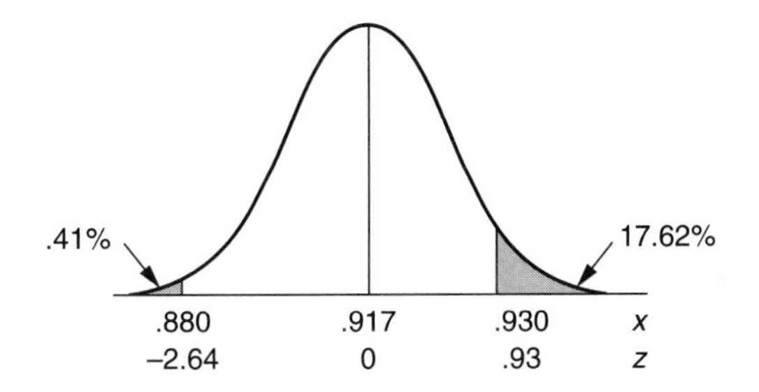

Using $x = .930$, the upper specification limit, make substitutions and calculate z:

$$z = \frac{x - \bar{\bar{x}}}{s}$$

$$= \frac{.930 - .917}{.014}$$

$$= \frac{.013}{.014}$$

$$= .93$$

$$= 0.93$$

Row 0.9 and column 3 give the tail area .1762 for $z = .93$. Next sketch the normal curve and put $x = .917$ at the center and $x = .930$ to the right. The z scale can be matched below the x scale: Place $z = 0$ at the center and $z = .93$ under the x value of .930. Shade the area to the right of $x = .930$. The shaded section illustrates the percentage of out-of-specification product on the high side. This sketch is shown in Figure 5.25. Row 0.9 and column 3 give the tail area .1762, which means that 17.62% of the product is out of specification on the high side.

Now use $x = .880$, the lower specification limit. Substitute .880 for x, .917 for $\bar{\bar{x}}$, and .014 for s and calculate z:

$$z = \frac{x - \bar{\bar{x}}}{s}$$

$$= \frac{.880 - .917}{.014}$$

$$= \frac{-.037}{.014}$$

$$= -2.64$$

The negative z value indicates that the value is to the left of the center.

Using the same sketch of the normal curve, put $x = .880$ to the left of the center and $z = -2.64$ under $x = .880$. Shade the area to the left of .880. This represents the percentage out of specification on the low side, as shown in Figure 5.25. Row 2.6 and column 4 give the tail area .0041, so .41% of the product is out of specification on the low side. The total percentage out of specification (.41% + 17.62%) is 18.03%.

Excessive variation is the major battle manufacturers have today in achieving high-quality products. When variation decreases, product quality increases.

EXAMPLE 5.11

Suppose improvement is made in the process from Example 5.10 that decreases the standard deviation, the measure of variation, from .014 to .005. Find the percentage of product that is now out of specification.

Solution

Use the upper specification limit (USL) for x, substitute $x = .930$, $\bar{\bar{x}} = .917$, and $s = .005$ into the equation, and calculate z:

$$\begin{aligned} z &= \frac{x - \bar{\bar{x}}}{s} \\ &= \frac{.930 - .917}{.005} \\ &= \frac{.013}{.005} \\ &= 2.60 \end{aligned}$$

The value $\bar{\bar{x}} = .917$ is at the center, and the x value of .930 is at the right. The corresponding z scale is shown below the x scale, 0 at the center, and 2.60 under .930. Row 2.6 and column 0 lead to a tail area of .0047, which indicates that .47% of the product is out of specification on the high side.

For the low side use the LSL for x and substitute $x = .880$, $\bar{\bar{x}} = .917$, and $s = .005$ into the equation:

$$\begin{aligned} z &= \frac{x - \bar{\bar{x}}}{s} \\ &= \frac{.880 - .917}{.005} \\ &= -7.4 \end{aligned}$$

Again, the negative z value indicates that the area is left of the center. The z value of 7.4 is not in the table, which means that the area to the left of −7.4 is 0.00000. There is area under the curve to the left of −7.4, but it is so small that this table, with accuracy to only five places, cannot measure it.

In the sketch of the normal curve, $\bar{\bar{x}} = .917$ is at the center, $x = .880$ is to the left, and $x = .930$ is to the right. The z scale can be shown below the x scale, with 0 under the center, 2.6 below .930, and − 7.4 below .880. Shade the area to the left of .880 and to the right of .930. This shaded area represents the percentage that is out of specification and is shown in Figure 5.26.

By decreasing the variation, the percentage of out-of-specification product dropped from 18.03% to .47%. If the process could be readjusted so the mean was at the mid-

FIGURE 5.26
The percentage that is out of specification in Example 5.11 when the standard deviation is reduced.

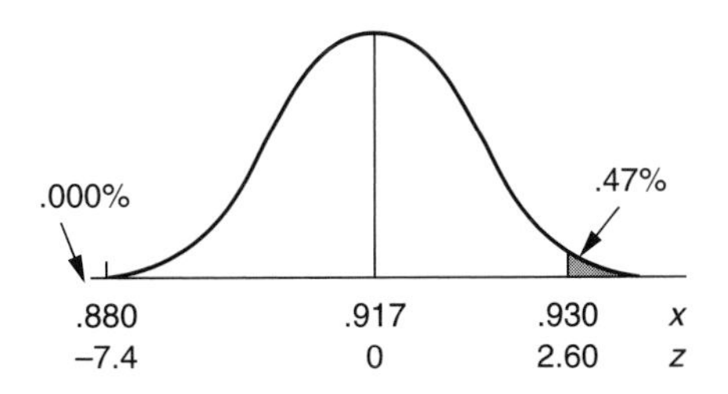

specification value of .905, virtually all of the product would be in specification. This is shown in Example 5.12.

EXAMPLE 5.12 The process from Example 5.11 has been readjusted so the average measurement is $\bar{\bar{x}} = .905$. Find the percentage of product that is out of specification.

Solution

Substitute $x = .930$, $\bar{\bar{x}} = .905$, and $s = .005$. Subtract first, then divide:

$$z = \frac{x - \bar{\bar{x}}}{s}$$
$$= \frac{.930 - .905}{.005}$$
$$= 5.0$$

Sketch the curve, put $\bar{\bar{x}} = .905$ at the center and $x = .930$ to the right, and match the z scale below it. Shade the area to the right, the out-of-specification section. The z value of 5.0 is not in the table, which means the tail area to the right is .00000 (not measurable with this table). The result is .000% out-of-specification product on the high end. Next, check the low side.

Substitute $x = .880$, $\bar{\bar{x}} = .905$, and $s = .005$:

$$z = \frac{x - \bar{\bar{x}}}{s}$$
$$= \frac{.880 - .905}{.005}$$
$$= \frac{-.025}{.005}$$
$$= -5.0$$

On the same sketch of the normal curve, mark $x = .880$ to the left and match the z scale below it. Shade the area to the left of .880 to represent the percentage out of specification,

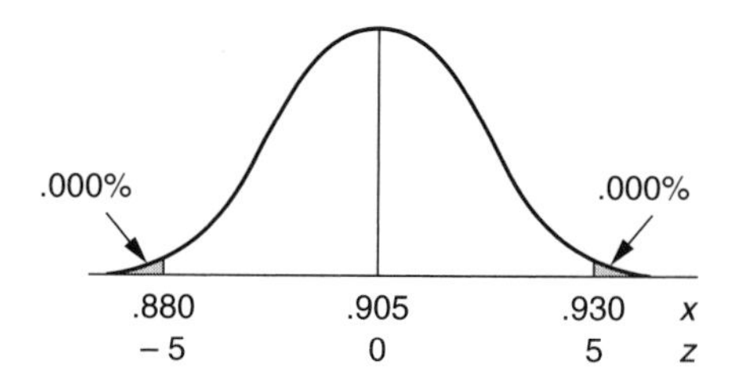

FIGURE 5.27
USL and LSL analysis of Example 5.12: Tail area is .000%.

as shown in Figure 5.27. As before, $z = 5.0$ is not in the table, so the area to the left is .00000. We therefore have .000% out of specification on the low end, too.

The following is the general procedure for finding the area or percentage in a section using the normal curve:

1. The measurements must be normally distributed; the histogram of the measurements should be approximately bell-shaped.
2. Determine the mean and standard deviation for the measurements.
3. Sketch the normal curve, locate the $\bar{\bar{x}}$ value at the center, label the x value(s) accordingly, and shade the area of interest.
4. Find the z score for the x values that border the area of interest using $z = (x - \bar{\bar{x}})/s$ and show the z scale.
5. Convert the z scores to area values using a normal-curve table.
6. Transform the table area to the area of interest.
7. Convert area to percentage by multiplying by 100.

EXAMPLE 5.13 The process mean is $\bar{\bar{x}} = 1.747$, and the standard deviation is $s = .006$. If the specification limits are 1.740 to 1.760, what percentage of the product is out of specification?

Solution

Sketch the distribution. Draw a normal curve and locate $\bar{\bar{x}}$ at the center. Set point 1.760 to the right and 1.740 to the left. Shade the tail areas to represent the percentage of out-of-specification product, as shown in Figure 5.28. Calculate the z value for each specification limit:

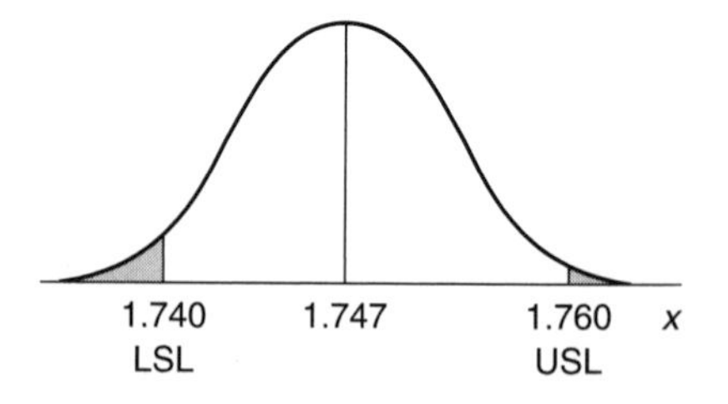

FIGURE 5.28
The curve, LSL, USL, and $\bar{\bar{x}}$ for Example 5.13.

$$z_1 = \frac{x_1 - \bar{\bar{x}}}{s} \qquad z_2 = \frac{x_2 - \bar{\bar{x}}}{s}$$

$$= \frac{1.740 - 1.747}{.006} \qquad = \frac{1.760 - 1.747}{.006}$$

$$= \frac{-.007}{.066} \qquad = \frac{.013}{.066}$$

$$= -1.17 \qquad = 2.17$$

$\text{Area}_1 = .1210$ $\qquad$ $\text{Area}_2 = .0150$

12.1% out (low) $\qquad$ 1.5% out (high)

The total product out of specification (12.1% + 1.5%) is 13.6%.

EXAMPLE 5.14 Suppose an improvement in the preceding process reduced the standard deviation from .006 to .003. What percentage of the product is out of specification now?

Solution

The drawing will be the same as the one in Figure 5.28. Calculate the new z values:

$$z_1 = \frac{x_1 - \bar{\bar{x}}}{s} \qquad z_2 = \frac{x_2 - \bar{\bar{x}}}{s}$$

$$= \frac{1.740 - 1.747}{.003} \qquad = \frac{1.760 - 1.747}{.003}$$

$$= \frac{-.007}{.003} \qquad = \frac{.013}{.003}$$

$$= -2.33 \qquad = 4.33$$

$\text{Area}_1 = .0099$ $\qquad$ $\text{Area}_2 = .00000$

.99% out (low) $\qquad$ .000% out (high)

If the process mean could be adjusted to the midspecification position of 1.750, both z calculations would be

$$z = \frac{.010}{.003}$$

$$= 3.33$$

$$\text{Area} = .00043$$

Both the high and the low side would have .043% out of specification for a total of .086% out of specification.

The distributions that have been considered up to this point have been the distributions of individual measurements x. Large samples of measurements were analyzed visually by comparing their distributions with specification limits. This was illustrated by putting the specification limits on stemplots and on histograms to see if the spread of individual values crowded or overlapped them. The sample-to-sample variability in the large samples was eliminated in that capability check by using the average and standard deviation of the sample with a normal-curve analysis. That information was then used to project the capability of the population with respect to the specification limits by finding the percentage of measurements that were out of specification. However, more information is needed about the process that is producing the measurements.

It was mentioned in Chapter 3 that there are two sources of variation: common-cause variation, which is imbedded in the process, and special-cause variation, which can be totally eliminated from the process. The pattern of the distribution can indicate the presence of special-cause variation when the pattern is nonnormal, but with such a large sample the source of the special-cause variation sometimes cannot be found. One alternative would be to look at the piece-to-piece variation as the pieces are produced. This has been done effectively in some situations with what is called a run chart and is demonstrated in Chapter 6. The one problem that will be noted about run charts is that it is easy to overreact to the piece-to-piece variation that occurs.

5.3 APPLICATION OF THE CENTRAL LIMIT THEOREM

If small samples with $n = 4, 5,$ or 6 are taken and their average values are charted sequentially, the $\bar{x}$ values will have less variation than a chart of individual measurements. This can readily be seen in Table 5.5, which has seven samples of size $n = 5$ and their corresponding $\bar{x}$ values. The individual measurements in the table vary from a low of 6.0 to a high of 7.9. The average values $\bar{x}$ represent middle numbers and they vary from a low of 7.0 to a high of 7.4. This decrease in variability is also noted in the *central limit theorem.*

TABLE 5.5

	7.6	7.9	7.2	6.8	7.5	6.5	7.9
	6.4	7.8	6.9	7.2	7.5	6.9	7.4
	7.3	6.9	7.6	7.4	6.0	7.3	7.0
	7.1	7.2	6.9	7.1	6.8	7.1	6.7
	7.1	7.2	7.4	7.0	7.2	7.2	7.5
$\bar{x}$	7.1	7.4	7.2	7.1	7.0	7.0	7.3

Central Limit Theorem: The distribution of sample means $\bar{x}$ taken from any large population approaches that of the normal distribution as the sample size n increases:

- The mean of the sample means $\bar{\bar{x}}$ will be equal to the population mean μ.
- The standard deviation of the distribution of sample means will be $\sigma/\sqrt{n}$.
- The symbol for the standard deviation of the sample means is $s_{\bar{x}}$, read s sub x-bar, so $s_{\bar{x}} = \frac{\sigma}{\sqrt{n}}$

FIGURE 5.29
The population distribution.

FIGURE 5.30
The distribution of $\bar{x}$ for $n = 3$.

FIGURE 5.31
The distribution of $\bar{x}$ for $n = 12$.

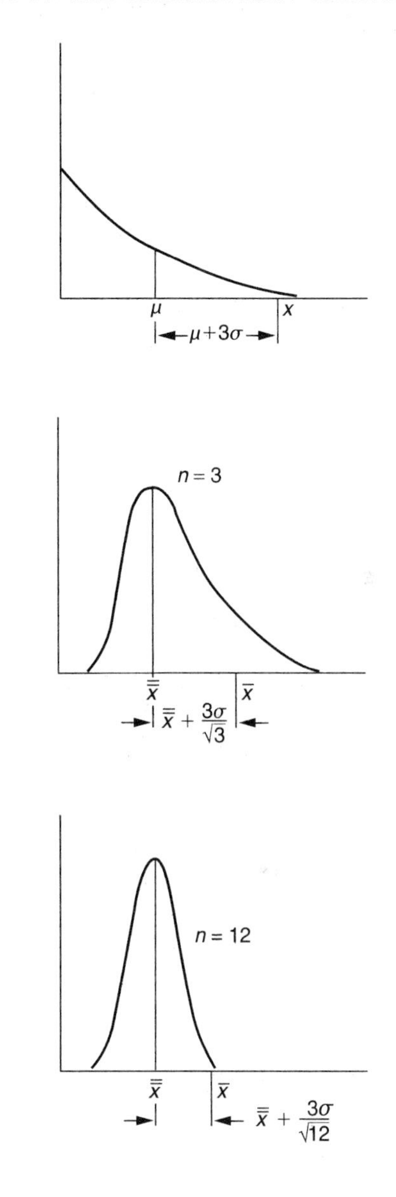

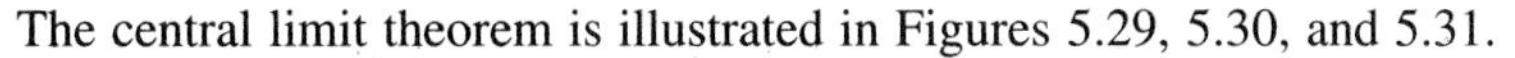

The central limit theorem is illustrated in Figures 5.29, 5.30, and 5.31.

Figure 5.29 could have any shape. It represents any nonnormal distribution of measurements x. Figure 5.30 is the distribution of sample averages $\bar{x}$ when the sample size is $n = 3$. The shape of the curve is becoming bell-shaped and the width of the distribution has decreased. It was noted earlier that virtually all measurements in a distribution will lie within $\pm\, 3\sigma$ of the mean. Exceptions to this observation occurred when the distributions were highly skewed (as in Figure 5.29) or when the distribution contained outliers.

Figure 5.30 shows that the 3 standard deviation width is almost half of the original 3σ ($\frac{3\sigma}{\sqrt{3}} \approx$ 1.7σ). Figure 5.31 shows a curve that is very close to a normal pattern. The 3 standard deviation width in Figure 5.31 is less than a third of the original 3σ ($\frac{3\sigma}{\sqrt{12}} \approx .9\sigma$). The size of the sample size n that will give a normal distribution depends on how abnormal the original population is.

The distinction between the *distribution of individual data values* and the *distribution of sample means* is important in the analysis of control charts. The relationship was demonstrated numerically in Table 5.5 and can also be seen in the comparison of distributions in Figure 5.32. The distribution of sample means is always narrower than the distribution of the data values from which the samples are taken:

$$\text{Width of the sample mean distribution} \approx \frac{6\sigma}{\sqrt{n}}$$

$$\text{Width of the population distribution} \approx 6\sigma$$

The central limit theorem provided part of the statistical theory that Walter Shewhart used to develop the control chart in the 1920s at Bell Labs. First, the distribution of $\bar{x}$ for samples taken from a *normal* population will have a distribution that is also normal for *any* sample size n. When small samples of size n = 4, 5, or 6 are taken from a normal population (a stable process with no special-cause variation), the sample-to-sample variation is relatively small (as shown in Table 5.5) and their pattern will be random when they are charted sequentially on a control chart. (Control charts are discussed in Chapter 6.) Virtually all the $\bar{x}$ values will lie within three standard deviations ($3\sigma/\sqrt{n}$) of their mean $\bar{\bar{x}}$, and about two-thirds of them (68%) will lie within one standard deviation ($\frac{\sigma}{\sqrt{n}}$) of the mean. When a special-cause variation occurs in the process, the distribution of measurements becomes nonnormal and the corresponding distribution of $\bar{x}$ will also be nonnormal. The n value is small enough for this to occur.

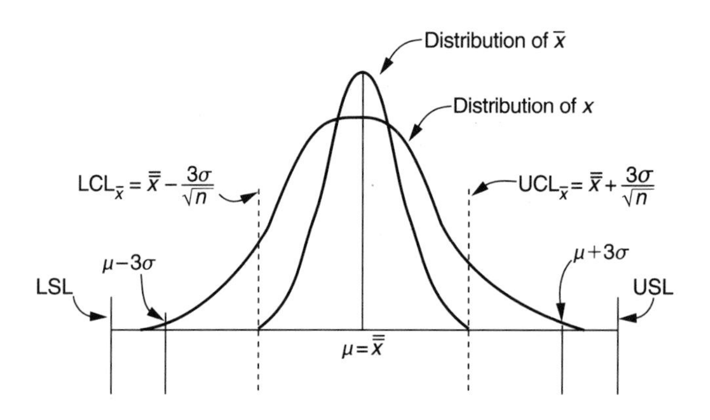

FIGURE 5.32
A distribution of sample means compared to a distribution of measurements.

The central limit theorem states that the $\bar{x}$ distribution will stay normal only if n is large. When n is small and special cause variation is present, patterns show up on the control chart that provide clues to the source of the special cause variation (discussed in Chapters 6–10). The patterns can be noted in a timely manner (within one to several samples). When that source has been identified, the special-cause variation can be eliminated. The control chart will indicate when all the sources of special-cause variation have been eliminated because it returns to a random pattern with about two thirds of the data values within one standard deviation of $\bar{\bar{x}}$ and no data values beyond three standard deviations of $\bar{\bar{x}}$ ($UCL_{\bar{x}}$ and $LCL_{\bar{x}}$).

NOTE: The control charts are the primary statistical tool for the identification and elimination of special-cause variation in a process.

When all the special-cause variation has been eliminated, the remaining variation is called common-cause variation. Unlike the special-cause variation, common-cause variation cannot be eliminated; it can only be reduced by making improvements in the process. Common-cause variation is measured by comparing the amount of variation with the specification limits. One way of doing this is to calculate the percentage of the measurements that are out of specification. This procedure was demonstrated in Examples 5.10–5.14.

The language of the previous discussion implies that control charts apply only to manufacturing systems, but they can be applied to any source of data that is produced sequentially (over time). The use of small sample means eliminates the unnecessary reaction to seemingly excessive variation between successive data values that can occur when a run chart is used. It was previously pointed out that overreaction to the process will result in an increase in overall variation.

EXAMPLE 5.15 A population of measurements has a mean $\mu = 1.750$ inches and a standard deviation $\sigma = .055$. Samples of size $n = 6$ are taken from the population.

1. What is the mean of all the sample means $\bar{\bar{x}}$?
2. What is the standard deviation of all the sample means $s_{\bar{x}}$?
3. What percentage of the measurements lie within $\pm\,\sigma$ of 1.750?
4. What percentage of the $\bar{x}$ values lie within $\pm\,2s_{\bar{x}}$ of 1.750?
5. Find the three standard deviation spread of measurements about 1.750.
6. Find the three standard deviation spread of $\bar{x}$ values about 1.750.

Solution

1. By the Central Limit Theorem, $\bar{\bar{x}} = \mu$, so $\bar{\bar{x}} = 1.750$.
2. By the Central Limit Theorem, $s_{\bar{x}} = \sigma/\sqrt{n}$:

$$s_{\bar{x}} = \frac{.055}{\sqrt{6}} = .0225$$

3. From Figure 5.24, approximately 68% (or 68.26% with four-place accuracy).
4. From Figure 5.24, approximately 95% (or 95.44% with four-place accuracy).
5. The standard deviation of the measurements is $\sigma = 055$. $1.750 \pm 3 \times .055$ is from $1.750 - .165$ to $1.750 + .165$, which is 1.585 to 1.915.

6. The standard deviation for $\bar{x}$ is $s_{\bar{x}} = .0225$ from part 2: $1.750 \pm 3 \times .0225$ is from $1.750 + .0675$ to $1.750 - .0675$, which is 1.6825 to 1.8175.

Notice that the numerical results in Example 5.15 fit the picture in Figure 5.32. The mean for the distribution of measurements, μ, is the same as the mean of the distribution of the sample means $\bar{\bar{x}}$. The six standard deviation width of the distribution of sample means, $6s_{\bar{x}}$, (from 1.6825 to 1.8175) is much smaller than the six standard deviation width of the distribution of measurements, 6σ (from 1.585 to 1.915). $65_{\bar{x}} = .0675$ is less than half of $6\sigma = .165$.

EXERCISES

---

1. As parts are made they are put into trays which will hold 30 parts. If a tray contains 6 defective parts and you sample one part from the tray, what is the probability that the part will be a good one?

2. Suppose a tray of 30 parts contains 4 defective parts. If you randomly choose one, what is the probability that it will be defective?

3. If a secretary entered 280 product inquiries last month and 20 of them contained errors that led to lost customers, what is the probability that the secretary will lose the next customer due to errors?

4. Use the normal distribution table, Table B.8 in Appendix B, to find the tail area associated with the following z values:

a. $z = 2.07$
b. $z = -1.53$
c. $z = -.64$
d. $z = 4.73$
e. $z = .5$

5. Use the normal distribution table, Table B9 in Appendix B, to find the tail area associated with the following z values. Check your results using Table B8.

a. $z = 2.18$ Right tail
b. $z = -.92$ Left tail
c. $z = -4.51$ Left tail

6. Calculate z, given the following sets of values. In each case, find the percentage of product larger than x.

a. $x = 9.2$, $\bar{\bar{x}} = 8.3$, $s = .48$
b. $x = 5.3$, $\bar{\bar{x}} = 7.9$, $s = 1.41$
c. $x = 3.475$, $\bar{\bar{x}} = 3.494$, $s = .0051$
d. $x = 6.924$, $\bar{\bar{x}} = 6.910$, $s = .0092$

7. Calculate z given the following sets of values. In each case, find the percentage of product smaller than x:

a. $x = 8.43$, $\bar{\bar{x}} = 9.05$, $s = .317$
b. $x = 4.52$, $\bar{\bar{x}} = 3.98$, $s = .715$
c. $x = .786$, $\bar{\bar{x}} = .799$, $s = .0069$

8. If $\bar{\bar{x}} = 1.832$ and $s = .009$, find the percentage of measurements that are:

a. Larger than 1.850
b. Smaller than 1.81
c. Between 1.81 and 1.82
d. Smaller than 1.825
e. Between 1.825 and 1.845

9. If $\bar{\bar{x}} = 3.51$ and $s = .0074$, find the percentage of measurements that are:
a. Smaller than 3.50.
b. Larger than 3.525.
c. Between 3.495 and 3.525.
d. Between 3.52 and 3.525.

10. If the specification limits are 2.230 to 2.250, $\bar{\bar{x}} = 2.238$, and $s = .0065$, what percentage of the measurements would be out of specification?

11. If the standard deviation in Exercise 10 is decreased to .0025, what percentage of the product is out of specification?

12. If the process in exercise 10 is adjusted so that the mean is at the midspecification value and the standard deviation is .0020, what percentage of the product is out of specification?

13. A population of measurements has a mean of 1.450 inches and a standard deviation of .0175 inches. If a large number of samples of size $n = 5$ are taken from the population,
a. What is the mean of all the $\bar{x}$ values?
b. What is the standard deviation of all the $\bar{x}$ values?
c. Find the values of $\bar{\bar{x}} \pm 3s_{\bar{x}}$ and compare them with $\mu \pm 3\sigma$.

14. If a population of measurements has a mean of .875 inches and a standard deviation of .019 inches and samples of size 4 are taken:
a. What is the standard deviation of $\bar{x}$?
b. What is the mean of all the $\bar{x}$ values?
c. What percentage of all the $\bar{x}$ values lie within two standard deviations of $\bar{\bar{x}}$?

15. A population of measurements has a mean of .760 inches and a standard deviation of .023 inches.
a. Between what two values will 99.73% of all measurements fall?
b. If a large number of samples of size $n = 4$ are taken from the population, between what two values will 99.73% of all the sample means, $\bar{x}$, fall?

16. If a population of measurements has a mean of 2.45 inches and a standard deviation of .15 inches:
a. Find the $\mu \pm 3\sigma$ spread of measurements.
b. What percentage of the measurements will lie within the values found in part (a)?
c. If samples of size $n = 5$ are taken from the population, between what values will 99.73% of the sample means fall?

17. A control chart is a distribution of $\bar{x}$ values over time and the upper control limit is three standard deviations of $\bar{x}$ above the average value $\bar{\bar{x}}$. If the standard deviation of the individual measurements is .025 inches and the average measurement is 3.50 inches (so $\bar{\bar{x}} = 3.50$), find the upper control limit of $\bar{x}$ when samples of size $n = 5$ are taken.

18. Change the given frequency distributions to probability distributions and draw the histograms for both the given frequency distribution and the resulting probability distribution:

a. Measurement	**Frequency**	**b. Group Boundaries**	**Frequency**
2.317	2	.581 to .584	4
2.318	8	.584 to .587	8
2.319	14	.587 to .590	16
2.320	16	.590 to .593	10
2.321	6	.593 to .596	2
2.322	4		

19. Change the frequency distribution to a probability distribution and draw the histogram of both.

Group Boundaries	Frequency
1.84 to 1.90	3
1.90 to 1.96	8
1.96 to 2.02	12
2.02 to 2.08	14
2.08 to 2.14	6
2.14 to 2.20	5
2.20 to 2.26	2

20. In Exercise 8, if the process mean is adjusted to 1.830 and the standard deviation is reduced to .006, find the percentage of measurements in each category.

21. A population of measurements has a mean of 2.350 and a standard deviation of .016.

a. Virtually all measurements in the population should lie between what two values?

b. Samples of size $n = 5$ are drawn from the population to form a distribution of $\bar{x}$ values.

(i) What is the mean of the $\bar{x}$ distribution?

(ii) What is the standard deviation of the $\bar{x}$ distribution?

(iii) Virtually all the $\bar{x}$ values should lie between what two values?

c. If the specification limits are 2.320 to 2.380, what percentage of the measurements are out of specification?

22. A process improvement decreased the standard deviation in Exercise 21 to a value of .009. Answer questions (a) to (c).

23. Define the central limit theorem:

a. In words

b. In pictures

24. The average measurement on a control chart is $\bar{\bar{x}} = 3.514$, and the standard deviation of the measurements is $s = .009$.

a. Find the percentage of measurements that are larger than 3.525.

b. Find the percentage of measurements that are less than 3.500.

c. What percentage of the measurements are between 3.510 and 3.520?

25. The population mean for measurements x is $\mu = 2.250$ and the standard deviation is $\sigma = .042$. Samples of size $n = 5$ are taken from the population.

a. What is the mean of all possible sample means $\bar{\bar{x}}$?

b. What is the standard deviation of all sample means $s_{\bar{x}}$?

c. What percentage of all measurements x lie within $\pm\ 2\sigma$ of 2.250?

d. What percentage of $\bar{x}$ values lie within $3s_{\bar{x}}$ of 2.250?

e. What two measurements x are exactly two standard deviations from 2.250?

f. Find the three standard deviation spread of $\bar{x}$ values from 2.250 (i.e., find $\bar{\bar{x}} \pm 3s_{\bar{x}}$).

6

INTRODUCTION TO CONTROL CHARTS

OBJECTIVES

- Set the chart scales for both a first chart and a continuation chart.
- Make an $\bar{x}$ and R chart.
- Recognize basic out-of-control situations such as points beyond the control limits, shifts, runs, and trends.
- Calculate and interpret process capability.

6.1 THE CONTROL CHART CONCEPT

Control charts are the tools that are used in statistical process control to indicate when special-cause variation is present in a process. Their use can provide more information about a process than workers with years of experience. A powerful combination is formed when experienced workers use the control charts, because the charts indicate when a source of variation occurs, the chart patterns provide hints to the cause of the variation, and the workers can use the hints along with their experience to identify the source of the variation problem and eliminate it.

The normal curve concepts that were discussed in Chapter 5 apply to manufacturing and service processes when results can be measured. The measurements will usually follow a normal distribution pattern when the only variation in the process is common cause variation. Quality is directly related to variation because the closer the process result adheres to a set sandard, the better the quality. When variation increases, the quality decreases. Every process has its own standard and level of quality that is achieved when the process is in statistical control, i.e. when only common cause variation is present. If that standard isn't good enough the process has to be improved by making appropriate changes to the process. When the process standard is acceptable, the process must be controlled so that other sources of variation don't creep in, specifically, special cause variation. The shape of the distribution will change to some non-normal pattern when special cause variation is present, but sometimes hundreds (or more) of measurements are needed to detect a true non-normal pattern. Whenever data is gathered over time and the measurements are plotted on a histogram, the shape gradually evolves. This concept was illustrated in Figure 4.5. The histogram may develop in spurts and its growth pattern is sensitive to the samples taken. Different sample plans will result in different intermediate patterns on the histogram. When can a developing histogram be interpreted? Does a spurt of measurements in one section indicate a bimodal pattern or will the pattern become more normal looking later? Is special cause variation creeping into the process? Clearly a more sensitive interpretation of the data is needed to answer these questions within a reasonable amount of time. Walter Shewhart developed a method to do this and he presented his findings in his book *Economic Control of Quality of Manufactured Product* in 1931. Shewhart's methods evolved from the concept that more information can be obtained from data in a timely manner by using rational subgrouping. Small samples of consecutive pieces with sizes n = 4, 5, or 6, taken over time, will provide the three basic keys of data interpretation: position, spread (or variation), and shape. The average value of the small samples will show the position of the measurements relative to the target value and the use of an average measurement decreases the possibility of overreaction to individual measurements straying from the target in normal variation. The range of the sample will give an immediate measure of the spread and, as more samples are taken, any changes in the sample variation may be noted as well as changes in position. When enough samples have been taken, the individual measurements can be plotted on a histogram and the shape of the distribution can be determined.

The "rational" part of the subgrouping is achieved by sampling consecutive pieces or occurences in the small samples to measure the piece-to-piece variation and position. Then randomly chosen small samples will lead to a true picture of the population of measurements. The rational subgroup information, position and spread, are plotted on a control chart over time. The pattern of points on the chart can be interpreted so that the advent of special cause variation can be noted in sometimes one point or up to several points on the chart. The chart is said to be out of statistical control when this happens and the pattern of the points will provide hints to the source of the special cause variation. When the process is "fixed" and the special cause variation eliminated, the control chart will return to a random pattern.

After all of the special-cause variation problems have been eliminated, the process is said to be in statistical control. The only variation remaining is common-cause variation. The control chart information is then used to measure the extent of common-cause variation and that translates directly to a measure of quality.

Variables and Attributes

There are two basic types of control charts: variables and attributes. *Variables* control charts use actual measurements for charting. The types of variables control charts include the following:

1. *Average and range charts ($\bar{x}$ and R):* The average and range chart, commonly called the x-bar and R chart, consists of two separate charts on the same sheet of chart paper. One graph tracks the sample mean $\bar{x}$ and the other tracks the sample range R. Small samples of consecutive pieces are taken. The sample size must be the same for all samples and usually consists of three to seven pieces. The dimension of interest is measured, and the measurements are recorded on the chart for each sample. The mean and range for each sample are calculated, recorded, and charted. The chart is analyzed as it develops for indications of special-cause variation, and after about 25 samples, it is analyzed again to determine the location, spread, and shape of the distribution of measurements. The x-bar and R chart is the most commonly used control chart for variables.
2. *Median and range charts ($\tilde{x}$ and R):* The median and range chart is exactly like the x-bar and R chart, with one exception: The median is calculated and charted instead of the mean. The $\tilde{x}$ and R chart analysis also shows indications of special-cause variation and the location, spread, and shape of the distribution of measurements. Some companies prefer the $\tilde{x}$ and R charts because the calculation of the median is easier and faster than the calculation of the mean. The Saginaw Division of General Motors has used the $\tilde{x}$ and R charts extensively. One reason that the x-bar and R chart is usually preferred over the $\tilde{x}$ and R chart is that the x-bar and R chart is more sensitive to variation analysis.
3. *Average and standard deviation charts ($\bar{x}$ and s):* The $\bar{x}$ and s chart tracks x-bar on one part of the chart and the sample standard deviation s on the other. The $\bar{x}$ and s chart does everything that the x-bar and R chart does and is used in the same way. Although it is preferred to the R chart because the standard deviation is a better measure of variation than the range, $\bar{x}$ and s charts are not used as much as the other variables charts because the standard deviation is more difficult to calculate than the range. When production workers are provided with statistical calculators or when the sample measurements are taken automatically and computerized, however, the sample standard deviation is calculated automatically and found at the touch of a button. For this reason, the $\bar{x}$ and s charts will be used more extensively in the future and may eventually become the standard control chart in use.
4. *Individual and moving range charts (x and MR):* This chart tracks consecutive individual measurements x on one chart and the piece-to-piece variability MR on the other. The x and MR chart is less reliable than the $\bar{x}$ and R chart because it is based on fewer data values. It is susceptible to overreaction because it does not have the damping effect of averaging. The x and MR chart is generally used when just a small number of data are available, such as for small or short production runs or for batch processes.
5. *Run charts:* The run chart is the simplest measurement-type control chart. It tracks individual measurements against a target measurement. There is no measure of variation and the run chart is very susceptible to over-adjustment. The run chart is often misused as a control chart with samples of size one by periodically checking the measurement of just one piece.

Attributes control charts use pass–fail information for charting. An item passes inspection when it conforms to the standards; a nonconforming item fails inspection. The types of attributes control charts include the following:

1. *p charts:* The *p* chart is a single chart that tracks the percentage of nonconforming items in each sample. The sample sizes are large, usually 100 pieces or more. The sample sizes may vary on a single chart when time is the controlling factor because the sample will consist of all the pieces produced in some time period, such as an hour or a day.
2. *np charts:* The *np* chart tracks the number of nonconforming items in each sample. It is easier to use than the *p* chart because the percentage of defective items does not have to be calculated, but it has one restriction: All the samples must be the same size.
3. *c charts:* The *c* chart graphs the total number of nonconformances found in each piece or unit that is inspected.
4. *u charts:* Samples are taken and the total number of nonconformances in the sample is determined. The *u* chart then tracks the average number of nonconformances per unit.

Control charts for variables are more useful than charts for attributes, but attributes charts work best in some situations, such as when tracking paint flaws or surface smoothness. Most processes have measurable characteristics, however, and the measurement contains more information than a pass–fail or go–no go judgment. Small samples of three to seven are used in variables charts, whereas large samples of 100 or more are used with attribute data. Also, although measured data are more expensive (time consuming) on individual pieces than go–no go gauges, fewer pieces need to be checked when using variables charts, so the time lapse for corrective action is less for variables charts than for attributes charts. Ongoing process analysis is also better with variables charts because the location (the middle value) is charted along with the piece-to-piece variability (the range); the shape of the distribution of measurements can be determined as well.

Control charts are the basic tools of SPC, and variables control charts provide the most information. They can quickly indicate if a process is in statistical control at specific process points, and their analysis can suggest causes of any out-of-control occurrence. They provide the basis for a capability study by indicating when a study can be made and by providing the data needed for the study. Incoming inspection is giving way to demands for proof of quality from a supplier, and variables control charts can be used to supply that proof.

The current industry standard is the $\bar{x}$ and R chart. It is favored over the median chart because the average is more sensitive to an individual measurement change. The mean is also the more traditional measure of central tendency. The R chart is currently favored over the s chart as a measure of piece-to-piece variability because it can be calculated without a statistical calculator. For smaller samples both R and s give adequate measures of variability. When the sample size is larger than 7, however, s gives the best measure of variability.

6.2 PREPARATION FOR CONTROL CHARTING

Management Direction

Management must establish an environment suitable for action and be totally committed to the use of control charts as an integral part of the move toward higher quality and productivity. People in every phase of the process have to be trained in basic statistical skills and be convinced of their importance. The application of statistics will direct the move toward better quality, higher productivity, and lower costs. Everyone's contribution is needed, so everyone must be involved. It is important to realize that SPC charts by themselves do not improve the process. The charts will show when problems exist; they are trouble indicators. When the charts show trouble, the operator or quality team must find the source of the problem and eliminate it. Management support is necessary to do this because the search for the source of special cause variation may interrupt the process flow.

Planning

The total process, from start to finish, should be outlined. Everything that can affect quality should be detailed: the machines, the materials, the methods, and the labor resources involved. The process can be charted in various ways. A process flowchart is a good start because it can evolve into a more complete storyboard that includes details to be considered for process improvement.

A process control team approach should be used. The size and complexity of the process determine whether one team will work for the entire process or subgroups should be formed for different phases.

The points in the process that are most promising for quality improvement must be determined by the process control team. Pareto charts and customer needs assessment can be important here. Where are the obvious trouble areas in the process? Look for evidence such as scrap, rework, and excessive overtime in specific areas. Talk to the operators. They often know of specific problem areas. Chapter 11 gives more details on the planning process.

Minimize any unnecessary variation and obvious trouble before the control charting begins. Let the charts concentrate on the harder, more subtle problems; do not complicate the process by allowing known trouble spots to remain until they are "officially" uncovered by the control chart indicators.

Chart Commentary

Operators should keep a process log. They should note all relevant events that may somehow affect the process, such as tool changes, adjustments, new material lots, breaks in the work schedule, and power downs. Each time a sample is taken, the date, time, and other details must be noted on the chart, the log, or both.

Gauge Capability

Before measurements for charting are taken, the gauges must undergo a gauge capability study to determine the extent of gauge variation. If the study shows too much variability in the gauge, the gauge should be repaired or replaced with a better one; excessive gauge variation can "muddy the water" and confuse interpretation of the control charts. The control charts should primarily reflect process variation, not gauge variation. Gauge capability procedures are introduced in Chapter 12.

Sampling Plan Preparation

A sampling plan must be carefully devised. The samples should fairly represent the population of measurements from which they are taken, so every effort must be made to ensure that *random samples* are attained. Decisions regarding the process at each charting situation rely on information gained from the samples. Each sample consists of a set of consecutive pieces that measure piece-to-piece variation. The number of pieces per sample is kept constant for each charting situation. These randomly chosen samples measure variation over time. At least 25 samples should be taken before the process can be analyzed for statistical control and process capability.

The sample size that is chosen is always a compromise between cost and reliability. In general, larger samples give more reliable results but cost more in worker time. Samples of three, four, or five are most common, but the situation may dictate the optimum sample size. For example, a sample of six is best from a machine with six spindles. In that case, if the samples are kept in order so that the first piece in the sample is always from the first spindle, the second piece is always from the second spindle, and so forth, individual spindle problems may be spotted as well as variation within the general control chart for the specific dimension being charted.

Sampling frequency also varies with the specific situation. The time it takes to produce one piece is a factor as well as the length of the production period. When control charts are first used at a specific point in a process, frequent samples will help to bring it into statistical control. When statistical control is attained, sampling frequency can be decreased. If the process is in statistical control but has too much variability so that it is not capable, more frequent samples are needed to assess process changes. There are several ways to organize a random sample, as the following examples illustrate.

EXAMPLE 6.1 Prepare a plan for 10 samples drawn during a shift that begins at 8:00 A.M.

Solution

Number a set of cards 0 through 7 and put the cards in a box marked "hour." Be sure that all the cards are the same size and shape. Mark a second set of cards 00 through 59 and put the cards in a box marked "minute." Again, be sure they are all the same size and shape.

Decide on the number of samples to be taken on each shift; for this example we will use 10. Draw and replace a card from the "hour" box 10 times. Be sure to mix the cards thoroughly before each draw. Record the numbers as shift hours. Simultaneously, draw and replace a card from the "minute" box and record the number with the corresponding "hour" draw. Add the hour and minute combinations to the starting time for the shift. If the shift begins at 8:00 A.M., the draw proceeds as follows:

Hour	Minute	Sample Time	Ordered Sample Time
0	22	8:22 A.M.	8:22 A.M.
3	37	11:37 A.M.	9:45 A.M.
4	36	12:36 P.M.	10:39 A.M.
5	22	1:22 P.M.	11:37 A.M.
1	45	9:45 A.M.	12:36 P.M.
4	42	12:42 P.M.	12:42 P.M.
5	53	1:53 P.M.	1:22 P.M.
6	32	2:32 P.M.	1:53 P.M.
6	21	2:21 P.M.	2:21 P.M.
2	39	10:39 A.M.	2.32 P.M.

If any sample times overlap, take two consecutive samples beginning with the first of the overlapping times.

EXAMPLE 6.2 The procedure shown in Example 6.1 can also be done using a random-number table or a random-number generator on a statistical calculator. Using three digits, decide which digit will correspond to the hour and which two will correspond to the minutes. If any random-number value is not in the appropriate domain (0 through 7 for the hour number and 00 through 59 for the minute number), skip it and go on to the next random number.

Solution

To illustrate, the three-digit random number on a statistical calculator will be used. The first digit will be the hour and the second two will be the minutes. The shift begins at 7:00 A.M. Random numbers that do not apply are crossed out:

Random Number(s)	Procedure	Sample Time
~~969~~, 617	7:00 A.M. + 6 hours, 17 minutes	1:17 P.M.
~~863~~, 430	7:00 A.M. + 4 hours, 30 minutes	11:30 A.M.
~~476~~, 758	7:00 A.M. + 7 hours, 58 minutes	2:58 P.M.
509	7:00 A.M. + 5 hours, 9 minutes	12:09 P.M.
151	7:00 A.M. + 1 hours, 51 minutes	8:51 A.M.
~~939~~, ~~075~~, ~~398~~, ~~592~~, 204	7:00 A.M. + 2 hours, 4 minutes	9:04 A.M.
~~292~~, ~~489~~, ~~932~~, 541	7:00 A.M. + 5 hours, 41 minutes	12:41 P.M.
735	7:00 A.M. + 7 hours, 35 minutes	2:35 P.M.
~~780~~, ~~864~~, 555	7:00 A.M. + 5 hours, 55 minutes	12:55 P.M.
143	7:00 A.M. + 1 hours, 43 minutes	8:43 A.M.

If one sample time runs into or overlaps another, depending on how long it takes to gather a sample, then consecutive samples are taken. Be sure to order the sample times.

6.3 CONTROL CHARTS AND RUN CHARTS

Making control charts and run charts involves interpreting a numerical scale and graphing a point. Figure 6.1 features a vertical scale marked off in five-thousandths from .700 to .715. Four data values are given in their production sequence and are marked at the bottom of the chart. Graph the four points and connect them with a broken-line graph.

To graph three of the four numbers in Figure 6.1, the position of the point must be estimated. The value .713 lies between .710 and .715. Imagine that the space between each line is divided into five equal spaces (four imaginary dividing lines) to estimate the position of the point more easily. Connect the points with line segments.

Visualizing four units between the lines in Figure 6.2 will make the estimation of point position easier. Imagine that the space between the lines is divided in half and that the two smaller spaces are divided in half again. Then estimate the position of the measurements and connect the points with straight-line segments.

The Run Chart

A run chart can be made for tracking the measurements of consecutively produced pieces. A target line is shown on the chart, the relationship between the charted points and the target line, as well as the relationship between the points themselves, is continually analyzed as the pieces are being produced. The two initial considerations are the closeness of the points to the target line and the extent of the variation in the measurements. The run chart can be used for process diagnosis and also for process improvement, in which case it is used in a cause-and-effect fashion. The effect of each process change is analyzed on the run chart.

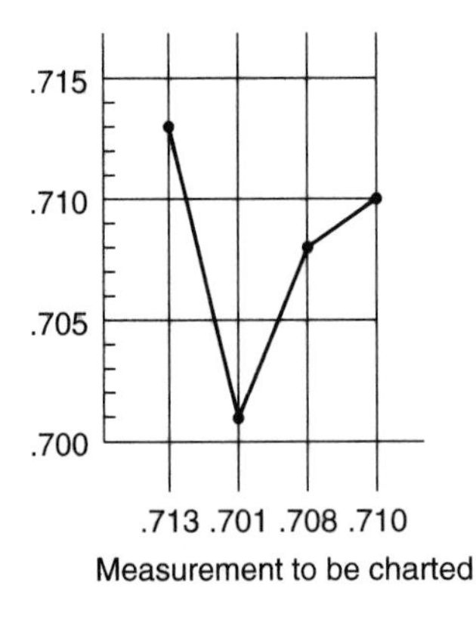

FIGURE 6.1
Charting points on a control chart or run chart.

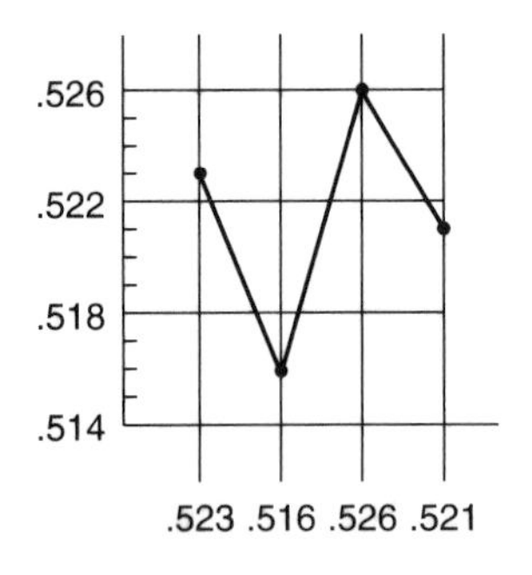

FIGURE 6.2
Charting points on a control chart or run chart.

In some situations, a high and low warning line is included on the run chart. The relation of points to the warning line is then watched as well as the previously mentioned relationships. The setup and charting procedure will follow that of the x and MR charts discussed in Chapter 8. The run chart does not include the moving range analysis.

EXAMPLE 6.3 Use the data from Table 6.1 to make a run chart. Assume that the order of production is by column and the target measurement is .910.

TABLE 6.1

.912	.910	.911	.904	.905	.910	.910
.914	.912	.914	.910	.913	.908	.909
.907	.909	.913	.913	.912	.909	
.902	.906	.908	.909	.907	.906	
.915	.909	.909	.910	.911	.912	

Solution

The data range is .013, so a scale of .001 units per line will be used (Figure 6.3). It is important not to make the scale too large. The height of the chart should be between 1 and 2 inches for easier interpretation.

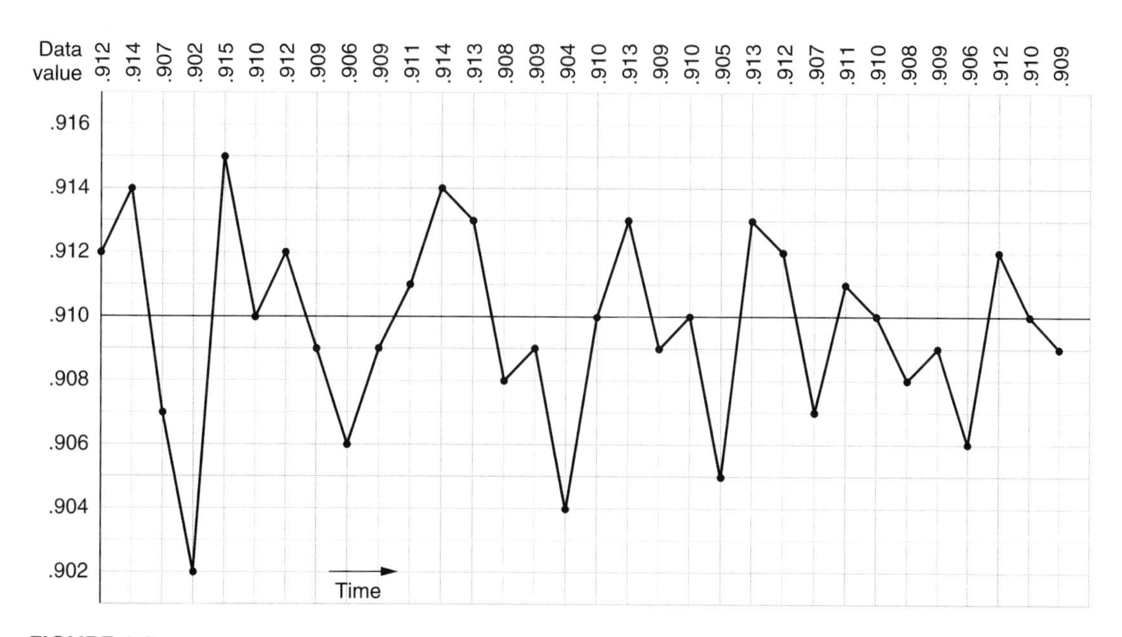

FIGURE 6.3
The run chart for Table 6.1 (target = .910).

6.4 THE BASIC $\bar{x}$ AND R CHART

The following step-by-step procedure is designed for student participation. The top of a control chart is partially shown in Figure 6.4. It contains 10 samples and each sample consists of the measurements of five consecutively produced pieces. The sample size is $n = 5$. Beneath each sample of five measurements the sample average (the sum of the five numbers divided by five) and sample range (the largest measurement in the sample minus the smallest) is shown. Figure 6.5 shows the first 10 samples on an $\bar{x}$ and R chart. The charting section contains two separate charts, one for the $\bar{x}$ (average) values and one for the R (range) values. As each sample average is calculated, its value is charted and a straight line is drawn from the new point to the previous point. The same is done with the range chart. After tracking the procedure on Figure 6.5, students will be directed to complete the control chart in Figure 6.6 and then check their work with the chart in Figure 6.7. As the control limit calculations are completed, students can then compare their chart with Figure 6.8.

EXAMPLE 6.4 Make an $\bar{x}$ and R chart.

Solution

Step 1. As each sample is taken the sample measurements are recorded on the control chart. This is shown in Figure 6.4. Write any comments on the process log sheet (usually located on the back of the control chart).

Step 2. Calculate the average and the range for each sample and record them in the space provided on the control chart. In Figure 6.4, the average and range value is shown for each sample column.

Step 3. Directly in line with the column of sample data, plot a point for the average and the range on their respective charts and connect them to the previous point with a straight line. This produces two *broken-line* graphs on the control chart, as illustrated in Figure 6.5. At this point you should be able to make the control chart in Figure 6.6. It should look like Figure 6.7 when you finish.

	Sample Number	1	2	3	4	5	6	7	8	9	10
Step 1	Record Sample Measurements	933	911	889	882	903	890	892	908	895	916
		897	898	915	913	930	940	912	920	920	890
		885	900	905	930	890	895	895	896	922	891
		900	905	902	900	890	909	896	894	928	920
		879	862	873	871	900	915	902	906	926	915
Step 2 Calculate	Average	899	895	897	899	903	910	899	905	918	906
	Range	54	49	42	59	40	50	20	26	33	30

FIGURE 6.4
Measurements, averages, and ranges on the $\bar{x}$ and R chart.

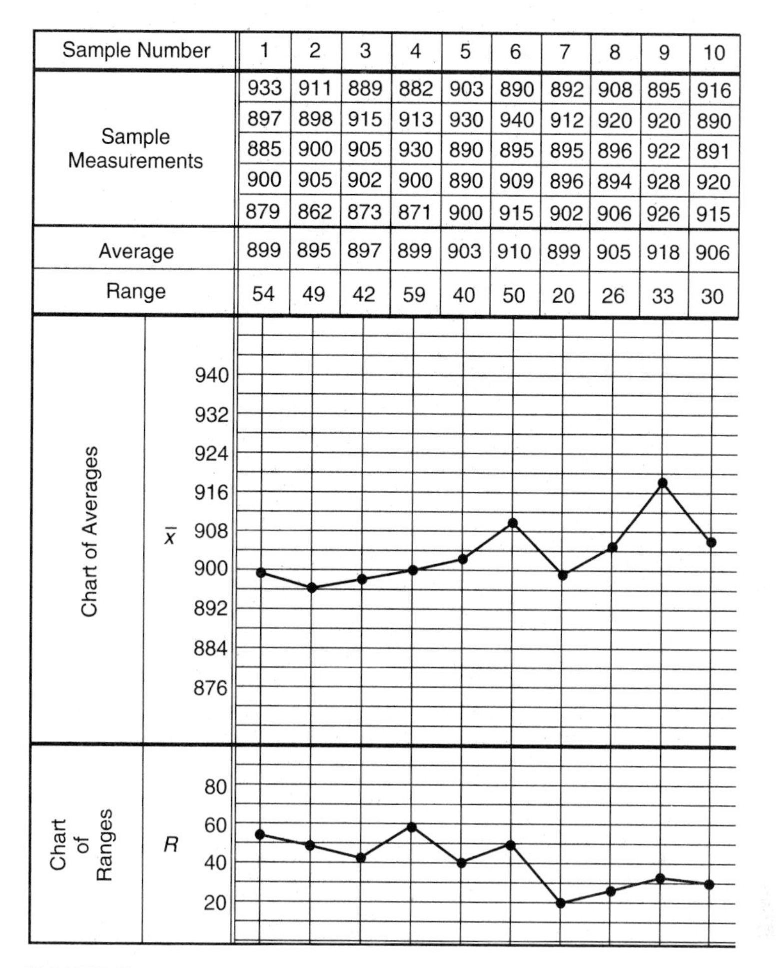

Sample Number	1	2	3	4	5	6	7	8	9	10
Sample Measurements	933	911	889	882	903	890	892	908	895	916
	897	898	915	913	930	940	912	920	920	890
	885	900	905	930	890	895	895	896	922	891
	900	905	902	900	890	909	896	894	928	920
	879	862	873	871	900	915	902	906	926	915
Average	899	895	897	899	903	910	899	905	918	906
Range	54	49	42	59	40	50	20	26	33	30

FIGURE 6.5
Plotting on the $\bar{x}$ and R chart.

Step 4. Calculate the averages and control limit values using the appropriate statistical formulas. Average the sample means. The symbol for the average of the sample means is $\bar{\bar{x}}$:

$$\bar{\bar{x}} = \frac{\Sigma \bar{x}}{k}$$

$$\bar{\bar{x}} = \frac{899 + 895 + 897 + \cdots + 900}{25}$$

$$= \frac{22{,}675}{25}$$

$$= 907$$

Specification: 850 to 950

Sample Number	1	2	3	4	5	6	7	8	9	10	11	12	13	14	15	16	17	18	19	20	21	22	23	24	25
Sample Measurements	933	911	889	882	903	890	892	908	895	916	901	908	909	895	893	909	885	897	912	882	896	912	926	917	884
	897	898	915	913	930	940	912	920	920	890	892	895	904	902	906	907	892	904	896	894	912	909	903	917	889
	885	900	905	930	890	895	895	896	922	891	892	896	906	902	917	904	942	916	932	941	907	913	908	918	912
	900	905	902	900	890	909	896	894	928	920	895	925	872	932	910	923	911	912	936	934	928	915	910	914	919
	879	862	873	871	900	915	902	906	926	915	898	933	927	932	925	888	916	920	913	917	926	928	885	925	898
Average	899	895	897	899																					
Range	54	49	42	59																					

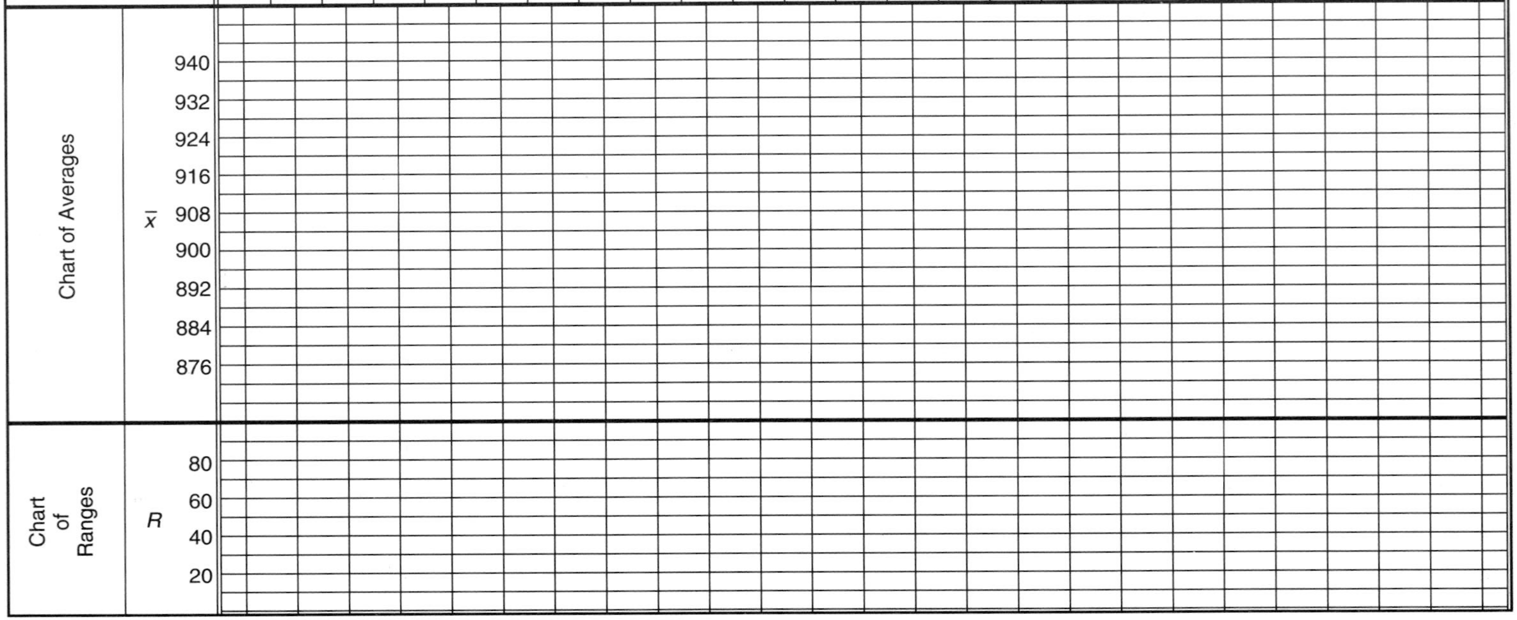

Chart of Averages — $\bar{x}$: 940, 932, 924, 916, 908, 900, 892, 884, 876

Chart of Ranges — R: 80, 60, 40, 20

FIGURE 6.6
Calculate $\bar{x}$ and R and make the $\bar{x}$ and R chart.

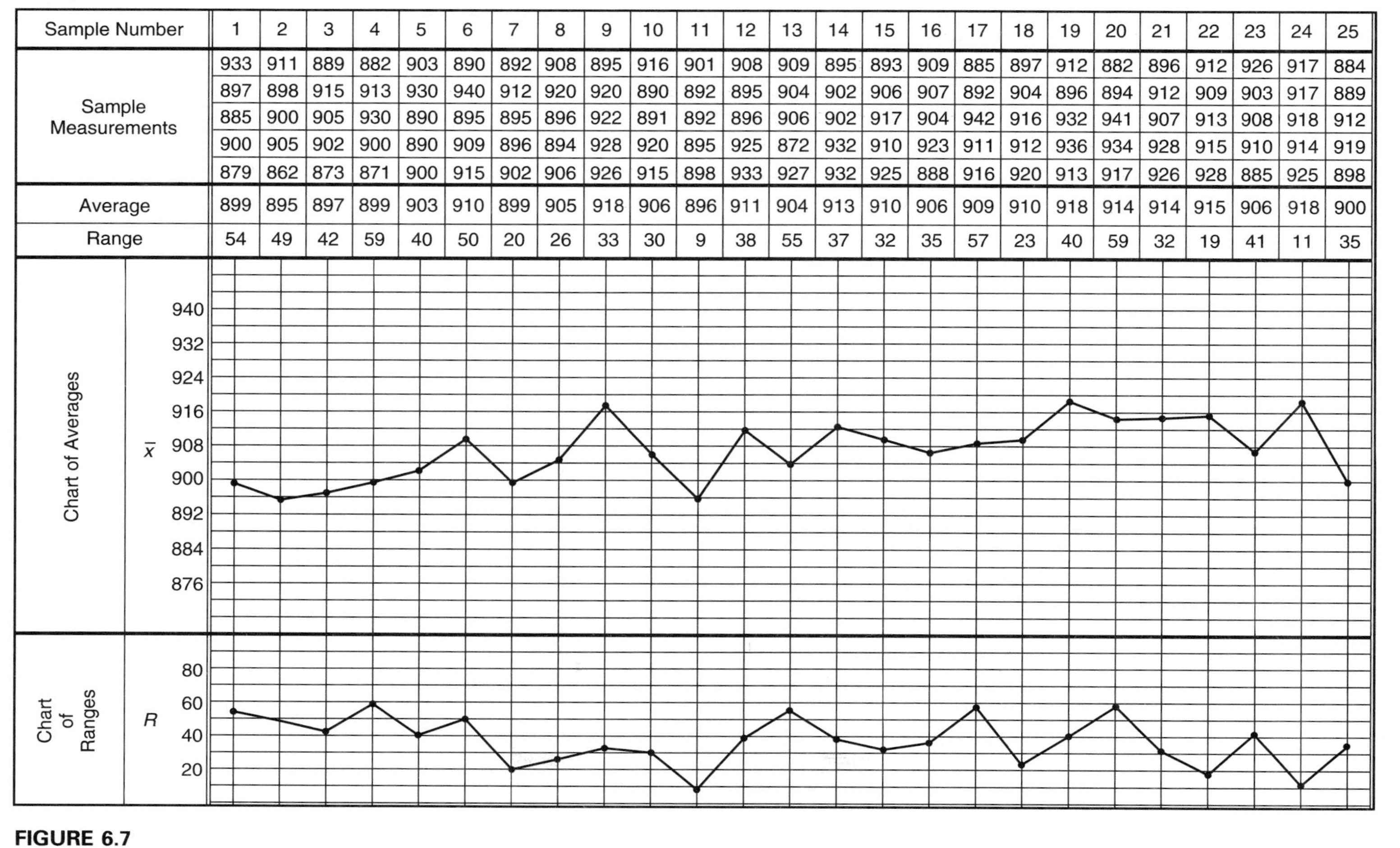

Specification: 850 to 950

Sample Number	1	2	3	4	5	6	7	8	9	10	11	12	13	14	15	16	17	18	19	20	21	22	23	24	25
Sample Measurements	933	911	889	882	903	890	892	908	895	916	901	908	909	895	893	909	885	897	912	882	896	912	926	917	884
	897	898	915	913	930	940	912	920	920	890	892	895	904	902	906	907	892	904	896	894	912	909	903	917	889
	885	900	905	930	890	895	895	896	922	891	892	896	906	902	917	904	942	916	932	941	907	913	908	918	912
	900	905	902	900	890	909	896	894	928	920	895	925	872	932	910	923	911	912	936	934	928	915	910	914	919
	879	862	873	871	900	915	902	906	926	915	898	933	927	932	925	888	916	920	913	917	926	928	885	925	898
Average	899	895	897	899	903	910	899	905	918	906	896	911	904	913	910	906	909	910	918	914	914	915	906	918	900
Range	54	49	42	59	40	50	20	26	33	30	9	38	55	37	32	35	57	23	40	59	32	19	41	11	35

FIGURE 6.7
The sample mean $\bar{x}$ and sample range R values are charted and connected with straight-line segments.

Specification: 850 to 950

Sample Number	1	2	3	4	5	6	7	8	9	10	11	12	13	14	15	16	17	18	19	20	21	22	23	24	25
Sample Measurements	933	911	889	882	903	890	892	908	895	916	901	908	909	895	893	909	885	897	912	882	896	912	926	917	884
	897	898	915	913	930	940	912	920	920	890	892	895	904	902	906	907	892	904	896	894	912	909	903	917	889
	885	900	905	930	890	895	895	896	922	891	892	896	906	902	917	904	942	916	932	941	907	913	908	918	912
	900	905	902	900	890	909	896	894	928	920	895	925	872	932	910	923	911	912	936	934	928	915	910	914	919
	879	862	873	871	900	915	902	906	926	915	898	933	927	932	925	888	916	920	913	917	926	928	885	925	898
Average	899	895	897	899	903	910	899	905	918	906	896	911	904	913	910	906	909	910	918	914	914	915	906	918	900
Range	54	49	42	59	40	50	20	26	33	30	9	38	55	37	32	35	57	23	40	59	32	19	41	11	35

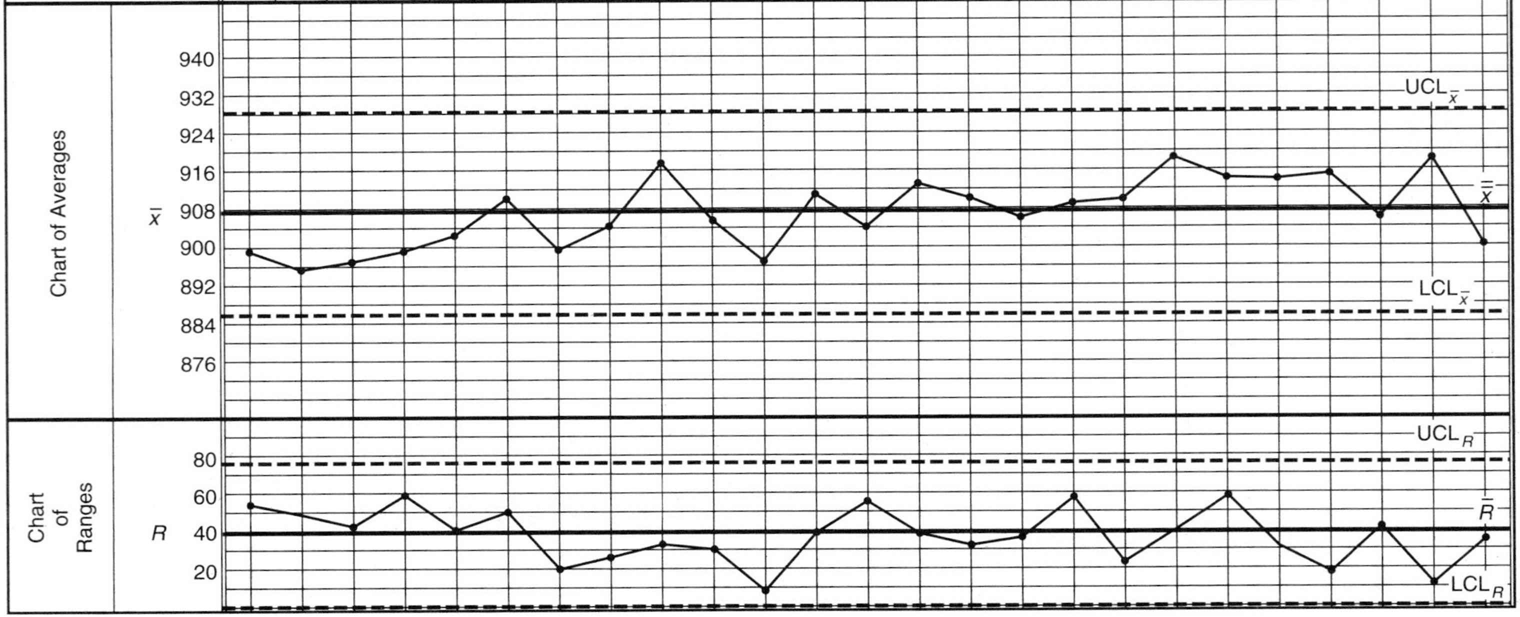

FIGURE 6.8
The $\overline{x}$ and R chart with center lines and control limits calculated with the Shewhart formulas.

Next average the range values from the control chart:

$$\overline{R} = \frac{\Sigma R}{K}$$

$$\overline{R} = \frac{54 + 49 + 42 + \cdots + 35}{25}$$

$$= \frac{926}{25}$$

$$= 37.04$$

Control Limit Formulas

$UCL_R = D_4\overline{R}$	UCL_R, LCL_R = upper and lower control limits for the range
$LCL_R = D_3\overline{R}$	$UCL_{\bar{x}}$, $LCL_{\bar{x}}$ = control limits for $\bar{x}$
$UCL_{\bar{x}} = \bar{\bar{x}} + A_2\overline{R}$	D_3, D_4, A_2 = Shewhart constants (Appendix B1)
$LCL_{\bar{x}} = \bar{\bar{x}} - A_2\overline{R}$	

Calculate the upper control limit for the $\bar{x}$ chart ($UCL_{\bar{x}}$):

$$UCL_{\bar{x}} = \bar{\bar{x}} + (A_2 \cdot \overline{R})$$

A_2 is a numerical constant that corresponds to the sample size. The table of constants is given in Appendix B, Table B.1. The sample size is given in column n. Find 5 in the n column, go straight across to the A_2 column, and read the value: A_2 = .577.

$$\begin{aligned} UCL_{\bar{x}} &= \bar{\bar{x}} + (A_2 \cdot \overline{R}) \\ &= 907 + (.577 \times 37.04) \\ &= 907 + 21 \\ &= 928 \end{aligned}$$

Calculate the lower control limit for the $\bar{x}$ chart ($LCL_{\bar{x}}$):

$$LCL_{\bar{x}} = \bar{\bar{x}} - (A_2 \cdot \overline{R})$$

The product in the parentheses is the same as that used in the $UCL_{\bar{x}}$ calculation.

$$\begin{aligned} LCL_{\bar{x}} &= \bar{\bar{x}} - (A_2 \cdot \overline{R}) \\ &= 907 - 21 \\ &= 886 \end{aligned}$$

Calculate the upper control limit for the R chart (UCL_R):

$$UCL_R = D_4 \cdot \overline{R}$$

The value for the numerical constant D_4 is found in Table B.1 of Appendix B. Locate the n value, 5, and go straight across to the D_4 column: $D_4 = 2.114$.

$$\begin{aligned} \text{UCL}_R &= D_4\overline{R} \\ &= 2.114 \times 37.04 \\ &= 78.3 \end{aligned}$$

Calculate the lower control limit for the R chart (LCL_R):

$$\text{LCL}_R = D_3 \cdot \overline{R}$$

The value of the numerical constant D_3 is 0 for sample sizes less than 7. Therefore, $\text{LCL}_R = 0$.

Draw a solid line at the average values $\overline{\overline{x}}$ and $\overline{R}$ and a dashed line at the control limits $\text{UCL}_{\overline{x}}$, $\text{LCL}_{\overline{x}}$, UCL_R, and LCL_R. When you draw these lines on your chart in Figure 6.6, your completed chart should look like the chart in Figure 6.8.

Step 5. Interpret the pattern on the charts formed by the points and their broken-line graph for signs of special-cause variation. If there is no obvious pattern, the points will be randomly scattered, with about two-thirds of them close to the average line and the other third spread out with a few closer to the control limit lines. Figure 6.9 illustrates a control chart with a random pattern. Nonrandom patterns will be discussed thoroughly in Chapter 10, but a few of the basic ones are presented here in Figures 6.10 to 6.12.

When these or other patterns show up, the worker in charge of the process control team analyzes the chart commentary and other process information to determine what caused the out-of-control situation on the chart. When the cause is found, it is eliminated from the process and the out-of-control data are removed from the charts. The average and control limits values are recalculated with the remaining set of "good" data.

An analysis of the control chart in Figure 6.8 reveals that the R chart appears to be in statistical control. There are no points beyond the control limits, and the pattern of the points within the control limits does not match any of the trouble patterns that were illustrated. The $\overline{x}$ chart does not have any points beyond the control limits, but a trend is evident. A trend is a gradual increase or decrease of the points on a control chart. There may be fluctuations within the gradual change, so there is no specified number of points to

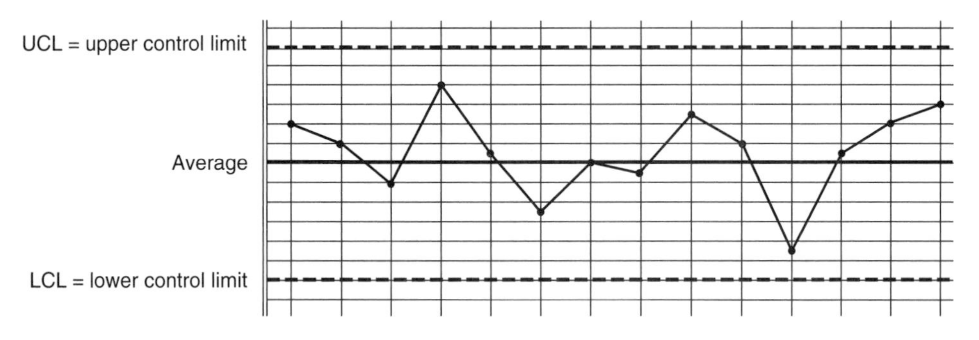

FIGURE 6.9
A random pattern of points.

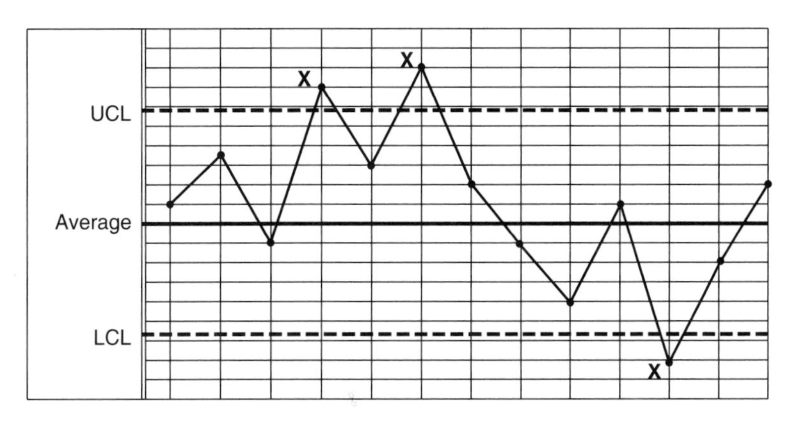

FIGURE 6.10
Points beyond the control limits, marked X, indicate special-cause variation.

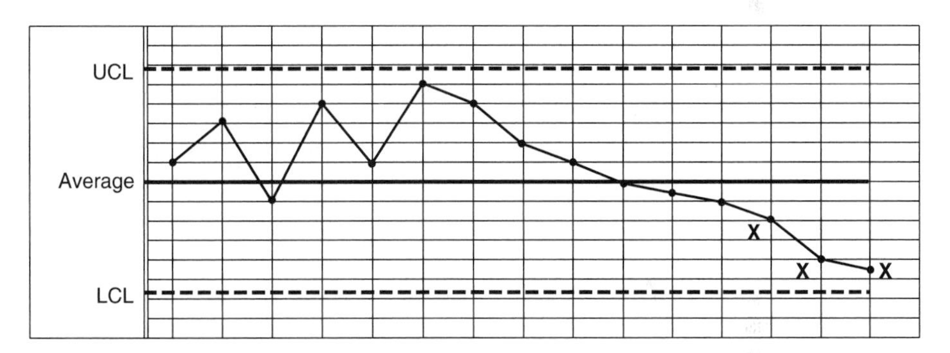

FIGURE 6.11
A run: seven consecutive decreasing points or seven consecutive increasing points.

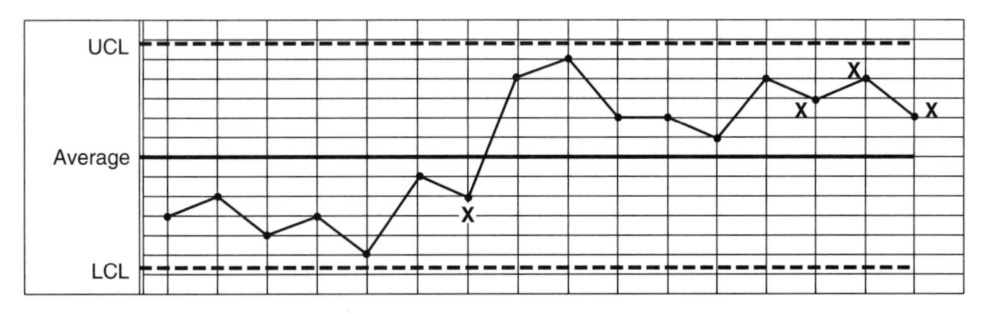

FIGURE 6.12
A shift: seven consecutive points on the same side of the average line.

indicate the trend. The process is not in statistical control, and the cause for the out-of-control situation should be determined.

A tally histogram of the data values can be formed in conjunction with step 1: Take the measurements, write them on the control chart, and mark the X's on the tally histogram. The shape of the histogram, when it differs from the expected shape, is a trouble indicator and may provide hints regarding the source of the out-of-control situation. The setup procedure for the tally histogram is as follows:

1. Make an L-shaped coordinate system with horizontal and vertical axes.
2. Assign the measurements to one axis (usually the horizontal axis).
3. Assign the frequency numbers to the other axis.
4. Put an X on the chart for each measurement. Start at the bottom (the squares next to the horizontal axis) and stack up the X's as repeated values of the different measurements are encountered.
5. The height of each column shows the number of times that measurement occurred in the set of data.

In Figure 6.13, the measurements are grouped instead of arranged in the simpler single measurement per column. (The grouping procedure was discussed in Chapter 4.) The two high columns show a bimodal distribution, which could indicate that the data are coming from two sources (two different operators, two different machines, etc.). The machines or operator procedures should be checked and corrected as needed to eliminate the problem.

This analysis procedure continues until all the special-cause problems have been eliminated. The process is then said to be in statistical control. Statistical control indicates that the process is currently working as well as it possibly can without process changes. The next question is, "Is that good enough?"

Step 6. Calculate the process capability. When statistical control is achieved, the distribution of the individual measurements may be analyzed statistically using the concepts in Chapter 5. The application of statistics will indicate what percentage of the product is out of specification or, if the output is entirely within specification, how much leeway or margin for error exists. The histogram also gives a good indication of the process capability. The histogram in Figure 6.13 shows that the measurements in that large sample crowd the specification limits. This crowding implies that the population of measurements may contain some that are out of specification.

If the process is not capable, that is, if it has too much variation embedded in the process, management action is necessary to make any improvements. The process must be changed somehow, and management must direct that change.

One example of how the pattern on a control chart can lead to improved quality and decreased costs involves Figures 6.11 and 6.12. Companies that do not use SPC deal with tool sharpening and adjustments in three basic ways:

1. *Sharpen or adjust tools when product measurements are beyond the specification limits.* This rule is costly because some pieces of unacceptable quality must be made to indicate the need for tool adjustment.

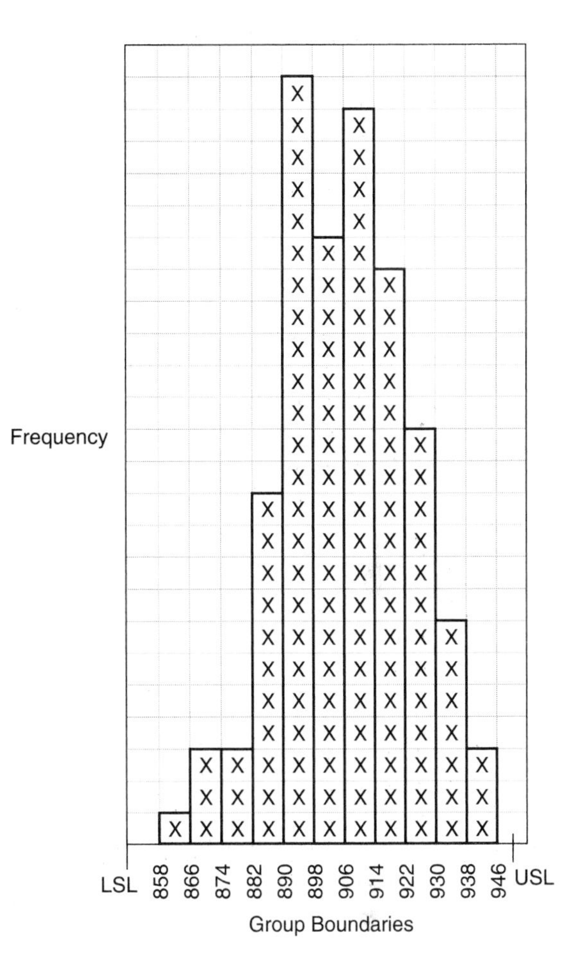

FIGURE 6.13
The tally histogram for the control chart data with the specification limits.

2. *Sharpen or adjust tools according to some specified time period.* This rule leads to unnecessary downtime when the time period is too short, and poor or unacceptable quality when the time period is too long.
3. *Rely on the operator's experience.* This rule can lead to either overadjustment or underadjustment. The overadjustment not only increases downtime but also decreases quality because "chasing" the target measurement increases overall variability. Every time the operator shuts down and adjusts the machine, a new distribution results. If this is done excessively, the overall distribution of measurements flattens out with several modes and a wider pattern. Underadjustment results in the production of some pieces of unacceptable quality.

When SPC is used, a *trend* or *run* on a control chart can show a need for sharpening or adjusting before poor-quality items are made. Shifts can also indicate the same problem.

The $\bar{x}$ and R chart consists of two separate, broken-line graphs. One tracks the averages $\bar{x}$ and the other tracks the range R of small samples of consecutive pieces. The sample size is usually three to seven, most commonly with samples of four or five. Consecutive pieces are used in each sample so that a measure of piece-to-piece variability can be attained as well as the overall variability that is determined by using all the data on the chart. The goals for using the chart were illustrated in Figures 3.11 to 3.14. *First,* the chart shows the presence of any out-of-control situations. The elimination of those problems will bring the process into statistical control. *Second,* the extent of embedded process problems can be measured in a process capability study. The measure of quality can be seen in a comparison of the $\bar{\bar{x}} \pm 3s$ measurement spread with the design specifications. *Third,* if the process is not capable (if it has too much product variation and poor product quality), the chart is used to measure the amount of improvement that results when the process is changed. *Fourth,* the $\bar{x}$ and R chart provides the evidence when the process is both in control and capable. Within that goal structure there are four main functions of the chart.

First, the extent of piece-to-piece variation may be measured using the range values for the samples taken. The individual sample ranges may then be compared with the average of all the range values and the values of the control limits. Product variability is in statistical control when no R values exceed the upper control limit and no trouble (nonrandom) patterns crop up between the control limits.

Second, the overall process variation may be measured by statistical calculations with the sample range values. The process variation will be used to determine the process capability with respect to the tolerance or specification limits once statistical control is achieved.

Third, the average measurement from the $\bar{x}$ chart may be found for comparison with the midspecification or target value. The sample means are also analyzed with respect to the expected, normal spread of the $\bar{x}$ values indicated by the control limits. The product measurements are in statistical control when all the $\bar{x}$ values are between the control limits and no trouble patterns are apparent.

Finally, a histogram made from the measurements on the chart may determine if the measurements are normally distributed. Other process problems may show up on the histogram as well; histogram analysis was discussed in Chapter 4 using Figures 4.7 to 4.12.

Two types of variation show up on the control charts:

- *Special-cause variation* shows up as out-of-control points or point patterns and can be eliminated by either the operator in charge or the process control team.
- *Common-cause variation,* the inherent process variation, shows up on the control charts as the variation in the points after the process is in statistical control and can be decreased only by management-directed action on the process.

In the 1920s when Walter Shewhart developed the control chart concept, he realized that the most effective use of the charts would be right on the shop floor where the workers could react to the chart information about the process. Two big obstacles for doing this was the lack of statistical training in the work force and the extensive calculations necessary to make the charts. The initial calculations that you had to do for the control chart in

Example 6.4 involved finding averages and ranges. However, if you had to use the basic statistical formulas to calculate the control limits it would have involved the following steps:

1. Find the standard deviation of the range values s_R:
   (a) Without a calculator, $s_R = \sqrt{\dfrac{\Sigma(R - \bar{R})^2}{k - 1}}$, where k = the number of samples.
   (b) With a calculator, enter all the R values and press s (for s_R) and $\bar{x}$ (for the range average $\bar{R}$).
2. Draw the control limit lines:

$$\text{UCL}_R = \bar{R} + 3s_R$$

$$\text{LCL}_R = \bar{R} - 3s_R \qquad \text{Use the default value 0 if this is negative.}$$

3. Calculate the standard deviation for the $\bar{x}$ values:
   (a) Without a calculator, $s_{\bar{x}} = \sqrt{\dfrac{\Sigma(\bar{x} - \bar{\bar{x}})^2}{k - 1}}$.
   (b) With a calculator, enter all the $\bar{x}$ values and press s (for $s_{\bar{x}}$) and $\bar{x}$ (for the average of the sample means $\bar{\bar{x}}$).
4. Draw the control limit lines:

$$\text{UCL}_{\bar{x}} = \bar{\bar{x}} + 3s_{\bar{x}}$$

$$\text{LCL}_{\bar{x}} = \bar{\bar{x}} - 3s_{\bar{x}}$$

5. To calculate the process capability, the standard deviation of the individual measurements is needed.
   (a) Without a calculator, $s = \sqrt{\dfrac{\Sigma(x - \bar{\bar{x}})^2}{N - 1}}$, where N = the number of mea-sure-ments or data values.
   (b) With a calculator, enter all the data values in the calculator and press s.

When the standard deviation formula was introduced in Chapter 3, a small sample of data values was used to demonstrate its use. Imagine following that procedure three times without a calculator for the three different standard deviations needed for a control chart, first for the 25 range values, then for the 25 $\bar{x}$ values, and finally for all 125 data values. There was one other consideration that was thoroughly investigated by Shewhart during his development of the control chart concept. For every process there is an ideal distribution that reflects the effects of common cause variation. Given that this level of variation is acceptable, it represents a set of "good" data from the process. Now suppose that a source of special cause variation crops up. A set of "bad" data is produced that contains more variability and, as a result, substandard products. The bad data could inflate $\bar{R}$ and the control limits for R so the bad data may be hidden on the R chart. The effect of bad data on $\bar{\bar{x}}$ and the control limits for $\bar{x}$ can be masked by the averaging process. A sample with little variation such as 901, 900, and 899 could have the same $\bar{x}$ value as a sample with larger variation such as 890, 900, and 910. With no variation factor involved, the upper and lower control limits on the $\bar{x}$ chart would be unaffected and the increased variation may be undetected there too. Another, more important problem that occurs, is that most initial applications of SPC are on processes that are out-of-control and the samples that are taken contain bad data. How can you tell if $\bar{R}$ is inflated in this situation? How can you spot an out-of-control occurrence? Shewhart's response was first, use small rational subgroups so out-of-control situations can be identified as soon as possible. Since the rational subgroups represent short, consistent conditions, any changes in the conditions will be quickly noted. Second, incorporate the variability and the subgroup size when calculating control limits. The Shewhart control limits have the double effect of making the control charts easier to calculate and making them approximate what the control limits would be if only good data were used. Thus, the charts are more sensitive to identifying the bad data when it is included in the calculations. The Shewhart formulas and constants were used in Example 6.4.

Many industries are using computers to generate the various SPC charts. As the measurements are taken, they are entered into a computer and the computer program will offer a choice between control limits from the statistical formulas and control limits from the Shewhart formulas. The results based on the Shewhart formulas are the better ones to use.

The only logical time that the basic statistical formulas could be used on control charts would be when only "good" data is used, i.e., when you know that all the data comes from a process that is in statistical control. Even in that situation, the formulas should reflect the subgroup size and the control limits for $\bar{x}$ should contain a subgroup variability factor.

6.5 THE $\bar{x}$ AND R CHART PROCEDURE

The following procedure will provide more detail than the introductory steps given in Section 6.4.

Step 1. Select a process measurement. Choose a critical process measurement to chart. There may be obvious measurements for which process control is important, but in some cases, process analysis (as discussed in Chapter 11) may be necessary to determine the critical measurements. For initial efforts measurements should have good potential for process improvement or should be important in the self-certification plan. Self-certification includes the use of control charts and histograms as proof of good quality for the customer. Considerations could include the following:

- There *may* be a trouble spot in the process that the chart can detect.
- There *is* a trouble spot determined by evidence from scrap analysis or customer (or downstream) complaints that the chart will help resolve.
- The process capability must be known at this point in the process for the self-certification quality plan.

Step 2. Decrease variability. Eliminate any obvious sources of variation before starting the chart. Chart interpretation should concentrate on the less conspicuous problems.

Step 3. Check the gauges. Be sure the gauges are working properly and that gauge variation is at an acceptable minimum. Variation that shows up on the chart should primarily reflect process variation. Excessive gauge variation makes the interpretation of process variability more difficult and sometimes impossible. Gauge capability methods are discussed in Chapter 12. The *rule of 10* for gauges states that the gauge should be *10 times* more accurate than the measurement accuracy. For example, if a measurement is needed to the nearest thousandth, the gauge should measure to the nearest ten-thousandth. The rule of 10 is absolutely necessary to measure and control variability realistically.

Step 4. Make a sample plan. Devise a sampling plan that consists of two parts. First, choose the sample size. Larger samples of six or seven can lead to more reliable estimates of variation and average value but are more costly. The extra time involved to take the larger sample of measurements could be a factor. The cost of samples of four or five may be more reasonable. Small samples of two or three are appropriate for monitoring a process that is in statistical control, but if the product output is small, small samples used with an $\bar{x}$ and R chart can give more reliable process information than an individual and moving range chart (discussed in Chapter 8). Second, determine the sampling frequency. The number of samples per shift depends on the product output, operator time available for measuring and charting, and the time pressure for statistical control.

There are two basic concepts involved in determining sample times. First, the sample must be a random sample. This is important in determining the process capability. A random sample taken when the process is in statistical control (no special-cause variation) will give the most reliable information for determining the capability of the process. Second, sampling must be done often enough to indicate any changes that may occur in the process. A change will usually show either the presence of special-cause variation or the elimination of special-cause variation. Both of these sampling concepts can be utilized by setting the basic sampling plan using the random sample method illustrated in Examples 6.1 or 6.2. In addition, the worker in charge is given the leeway of taking additional samples whenever a potential process change is suspected.

Step 5. Set up the charts and process log. Choose the scale for the $\bar{x}$ chart and for the R chart. For a continuous charting situation, the chart scales will have been established on the previous chart. For a new chart, use an educated guess to set up the scales. When establishing scales, avoid the two extremes if possible: Keep scales from being too large or too small.

If the scale (or units per line) is too large, the points end up close together, vertically, and the broken-line graph looks almost straight. The middle line and the control limit lines will be close together and the chart will be hard to interpret. On the other hand, if the scale (or units per line) is too small, points end up off the chart and the broken-line graph looks like a very steep zig zag. The control limit lines will end up off the chart or at the very top and bottom of the chart. Interpretation becomes too confusing.

Several educated guess routines are built around the fact that the chart needs some space above and below the control limits in case there are out-of-control points that fall beyond the control limits. Two methods for setting the scales on the initial control chart are shown as follows.

Method 1. Using the tolerance (quick & easy): For the range chart, divide the tolerance by the number of spaces on the chart and round off to an easy charting number. Let 0 be the bottom line on the range chart and set the scale from there. For the $\bar{x}$ chart, divide three times the tolerance by the sample size and then divide by the number of spaces on the chart. Round off to an easy charting number. Put the midspecification value (target measurement) at the middle of the $\bar{x}$ chart and scale in both directions from there.

EXAMPLE 6.5 Use method 1 to set the scales for the control chart in Example 6.4.

Solution

The range chart. Subtract the specification limits at the top of the chart to find the tolerance:

$$\text{Tolerance} = 950 - 850 = 100$$

$$\frac{\text{Tolerance}}{\#\text{spaces}} = \frac{100}{10} = 10 \text{ units per space}$$

This is the scale that was used on the range chart.

The $\bar{x}$ chart. Let T = tolerance and n = sample size:

$$\frac{3T}{n \times \#\text{spaces}} = \frac{3 \times 100}{5 \times 22} = \frac{300}{110} = 2.7$$

Round off to an easy charting number: three units per space. The $\bar{x}$ chart in Example 6.4 used four units per space, but three units per space would have worked too. The chart and control limits would be slightly wider, as shown in Figure 6.14. The three-unit scale is easy to chart because the in-between numbers are either one-third above a line or one-third below a line. The first average value on the control chart, 899, is one-third below the 900 line and the second average value, 895, is one-third above the 894 line.

Method 2. Using several initial samples. Take several samples and find the range of ranges R_R:

$$R_{\text{largest}} - 0 = R_R$$

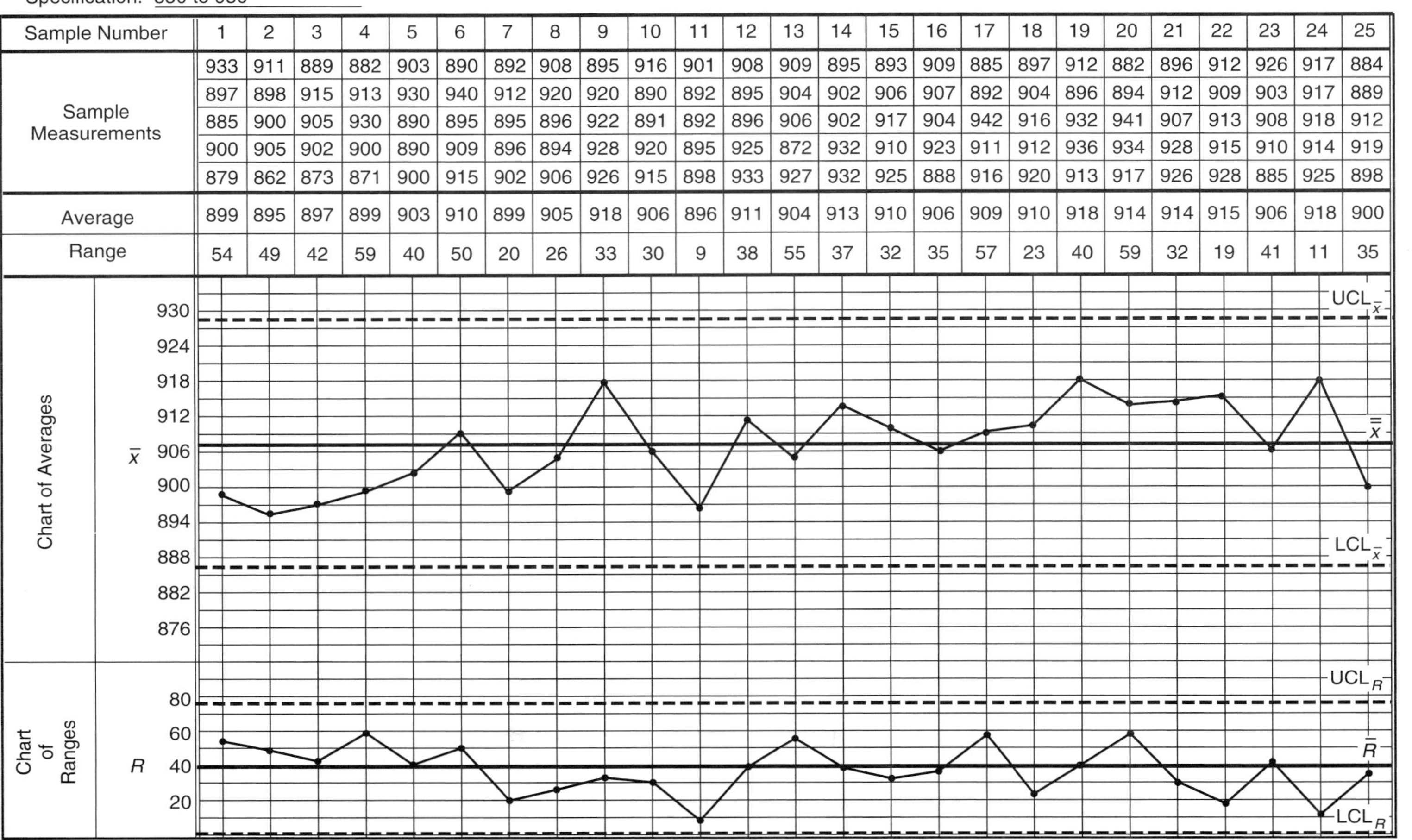

Specification: 850 to 950

Sample Number	1	2	3	4	5	6	7	8	9	10	11	12	13	14	15	16	17	18	19	20	21	22	23	24	25
Sample Measurements	933	911	889	882	903	890	892	908	895	916	901	908	909	895	893	909	885	897	912	882	896	912	926	917	884
	897	898	915	913	930	940	912	920	920	890	892	895	904	902	906	907	892	904	896	894	912	909	903	917	889
	885	900	905	930	890	895	895	896	922	891	892	896	906	902	917	904	942	916	932	941	907	913	908	918	912
	900	905	902	900	890	909	896	894	928	920	895	925	872	932	910	923	911	912	936	934	928	915	910	914	919
	879	862	873	871	900	915	902	906	926	915	898	933	927	932	925	888	916	920	913	917	926	928	885	925	898
Average	899	895	897	899	903	910	899	905	918	906	896	911	904	913	910	906	909	910	918	914	914	915	906	918	900
Range	54	49	42	59	40	50	20	26	33	30	9	38	55	37	32	35	57	23	40	59	32	19	41	11	35

FIGURE 6.14
The control chart with 3 units per space on the $\bar{x}$ scale and Shewhart control limits.

Since the R chart starts at 0, use 0 for R_{smallest}. Next find the range of $\bar{x}$'s, $R_{\bar{x}}$:

$$\bar{x}_{\text{largest}} - \bar{x}_{\text{smallest}} = R_{\bar{x}}$$

A conservative estimate of the standard deviation of a statistic (a sample measurement such as $\bar{x}$ or R) is found by dividing the range of that statistic by 4. The symbol "≈" means "approximately equal to." The standard deviation for ranges is

$$\frac{R_R}{4} \approx \sigma_R$$

The standard deviation for $\bar{x}$ is

$$\frac{R_{\bar{x}}}{4} \approx \sigma_{\bar{x}}$$

Three standard deviations, plus a little more, is approximately four standard deviations:

$$4\sigma_R \approx \frac{4R_R}{4} = R_R$$

$$4\sigma_{\bar{x}} \approx \frac{4R_{\bar{x}}}{4} = R_{\bar{x}}$$

The range scale should go from 0 to $\bar{R} + R_R$ or from 0 to $2R_R$; use the second choice, which is simpler. The $\bar{x}$ scale should go from $\bar{\bar{x}} - R_{\bar{x}}$ to $\bar{\bar{x}} + R_{\bar{x}}$ or from the midspecification $-R_{\bar{x}}$ to midspecification $+R_{\bar{x}}$. Again, use the second choice.

Mark the scale as follows. Divide the number of lines in the R chart section into $2R_R$ and round off to an easy number for setting up the chart (see Example 6.6, which follows). Start the R scale from 0 on the *bottom* line. Next, divide the number of lines in the $\bar{x}$ chart section into $2R_{\bar{x}}$. Round off to an easy number for charting. Start at the *middle* of the chart and label the middle line with the midspecification value. Mark the $\bar{x}$ scale up and down from the midspecification center line.

Keep a process log during the control charting, and in it, be sure to note the time and make a comment about any occurrence that may have some effect on the process (either good or bad). The date and time should accompany each sample. The process log may be kept on the control chart or may be a separate comment sheet attached to the chart. When variation problems occur, the combination of a process log and control chart can be very beneficial to the operator or process control team as they attempt to isolate and eliminate the problems.

Step 6. Set up the tally histogram. Prepare the tally histogram using the specification limits to set the measurement scale. Aim for about 10 sections or groups depending on the number of units between the specification limits. Mark the target measurement and the specification limits. Some control charts have a built-in histogram positioned at the scale end of the $\bar{x}$ chart. This is a nice feature because it emphasizes the difference between the distribution of measurements, which are compared with the specification limits, and the distribution of sample means, which are compared with the control limits. This also demonstrates the effect of the central limit theorem, since the width of the $\bar{x}$ distribution will be narrower than the distribution of measurements.

$$\text{Width of the distribution of measurements} \approx 6s.$$

$$\text{Width of the distribution of } \bar{x} \approx \frac{6s}{\sqrt{n}}.$$

Some instructors and supervisors are firmly against this idea because if the people using the control chart don't understand the central limit theorem they confuse *control limits* with *specification limits.* They think that as long as the charted $\bar{x}$ values are in specification everything is OK. But, if they are directed to tally the individual measurements as they are taken, then an *individual measurement* beyond the *specification limit* is also an out-of-control situation.

Step 7. Take the samples and chart the points. As the samples are taken, write the measurements on the control chart and put the X's on the tally histogram. Calculate the sample average and range, chart their points, and draw the next leg of the broken-line graph. If this is a first chart, any analysis usually must wait until the chart is completed. If this is a continuation chart, the scale, average lines, and control limit lines are transferred from the previous chart and chart analysis is an ongoing process as the chart develops.

When continuation charts are used (which is most of the time) the operator or worker in charge must know the various trouble patterns on control charts and histograms to be able to recognize them as soon as they appear so that immediate action can be taken. Furthermore, they should learn to anticipate problems. For example, shifts and runs are slow to develop (they need seven samples in their patterns). When four points show up on the chart that could be interpreted as the start of one of these patterns, the worker in charge should start thinking about what possible process change could be occurring. It may be a false alarm, but if not, the problem will be identified sooner. It was mentioned in step 4 that sampling times should be frequent enough to reflect any process changes, so if a change is suspected, additional samples should be taken. Take action when problems show up. It is much easier to diagnose problems when they occur than to analyze the problem when it shows up on a control chart analysis a few days later.

Step 8. Calculate averages and control limits. Use the Shewhart formulas:

Average range	$\bar{R} = \dfrac{\Sigma R}{k}$	For k samples
Control limits	$UCL_R = D_4\bar{R}$	Values for D_3, D_4, and A_2 from
	$LCL_R = D_3\bar{R}$	Appendix B, Table B.1

Average $\bar{x}$	$\bar{\bar{x}} = \dfrac{\Sigma\bar{x}}{k}$
Control limits	$UCL_{\bar{x}} = \bar{\bar{x}} + A_2\bar{R}$
	$LCL_{\bar{x}} = \bar{\bar{x}} - A_2\bar{R}$

It was mentioned in Chapter 5 that the control limits corresponded to lines that are three standard deviations from the center. The simpler UCL_R formula does not show the separate parts, but the other formula does:

$$UCL_{\bar{x}} = \bar{\bar{x}} + 3s_{\bar{x}} \qquad UCL_{\bar{x}} = \bar{\bar{x}} + A_2\bar{R}$$

This shows that the $A_2\bar{R}$ product represents the three-standard-deviation value. For the smaller sample sizes, where $D_3 = 0$, the lower control limit, $LCL_R = \bar{R} - 3s_R$, would extend below zero, and since a negative range value is impossible, $LCL_R = 0$ becomes the default line for the lower control limit. This is why the control limits can be unsymmetric with respect to $\bar{R}$, whereas the control limits on the $\bar{x}$ chart will always be symmetric about $\bar{\bar{x}}$.

If this is a continuation chart, good record keeping saves calculation time. The previous chart should have the continuing sums and continuing number of samples:

$$\bar{R} = \frac{\Sigma R_{\text{previous}} + \Sigma R}{k_{\text{previous}} + k} \qquad \bar{\bar{x}} = \frac{\Sigma \bar{x}_{\text{previous}} + \Sigma \bar{x}}{k_{\text{previous}} + k}$$

Recalculate the control limits, adjust them on the chart, and reanalyze if necessary.

The Shewhart formulas for the control limits are all functions of the average range $\bar{R}$. Whenever there is an apparent change in $\bar{R}$, the control limits for the continuation charts must be recalculated. A good-news example of this occurring would be a situation where SPC efforts resulted in a decrease in variation. A shift below the $\bar{R}$ line would be noted on the control chart. A bad-news example could occur when cost cutting leads to the use of inferior material in the process and a shift above the $\bar{R}$ line would be noted.

Action Procedures

There are three different out-of-control situations: Something could visibly change or break down, one of the pieces in a sample could be out of specification, or a trouble pattern could show up on a statistical chart. Companies have various action plans for out-of-control problems. The best ones call for immediate investigation and problem elimination. This may involve just the operator, the operator and supervisor, or a quality-control team. In addition, they institute 100% inspection that extends back to the last point when the process was in control and continues until the process is in control again.

Step 9. Calculate the capability. When the process is in statistical control, determine the capability. There are various ways this can be done. The first step, in any case, is to calculate the standard deviation with the Shewhart formula:

$$s = \frac{\bar{R}}{d_2}$$

Appendix B, Table B.1 contains the d_2 value.

- The percentage out of specification can be calculated using the normal curve as shown in Chapter 5.

- A capability ratio can show what percentage of the tolerance is being used.
- A capability index can be used as a capability indicator.

The capability ratio and index are discussed in Section 6.7.

Step 10. Monitor the process. When the process is in control and capable, a monitoring process should be used. Continuous control charts with one or two samples per shift work well. Another monitoring method would use the precontrol procedure that is discussed in Chapter 8.

Step 11. Continuous improvement. Quality improvement is a continuous process. The operator should stay alert to occurrences that lead to errors or are related to measurement variability. A sequence of small improvements over a period of a year or two can result in a substantial improvement in quality.

CASE STUDY 6.1

Quality manager James Smith and machine shop supervisor Michael Clark,[1] at Ingersoll–Rand's Von Duprin Incorporated, report that finished product rejections were reduced by 77% over a 5-month period when SPC techniques were used.

After employees were trained in its use, SPC was implemented on a plantwide basis to assess each department's quality performance, to determine causes for defects, and to target the corrective action needed. A quick analysis indicated that the defect rate was higher than anticipated and the capability of the machining operations was better than their performance. A major problem identified was a lack of feedback to operators regarding their individual performance. Further analysis showed a problem with the setup operation.

Corrective action was taken. A better setup was instituted, and operator control charts were started. The charts indicated that the defective pieces came from just 5 of the 70 operators. Over a 90-day period the first-piece defect level dropped from 42.9% to 14.7%. It has since dropped further to 4.8%, and the entire machine shop's reject rate has decreased from 16.3% to 3.7%. Chart visibility turned out to be a strong motivating factor for the operators; they became more alert and took pride in their performance. Responsibility for quality was transferred from the inspectors to the shop floor. Operators began to consider themselves to be team members participating in problem solving.

The emphasis on control of a process through the use of control charts is an important feature of SPC. When changes have to be made within a process, such as tools sharpened and dies rebuilt, the control chart indicates when the change should be made. The control chart optimizes the use of anything that is subject to wear. A change will not be needed as long as the chart is in control, and as soon as the chart indicates

[1]James Smith and Michael Clark, "SPC Sharply Drops Rejects" (*Modern Machine Shop,* May 1986), 66–69.

that the wear is leading to a problem, the change can be made before any defective items are made. An article in the *Quality Control Supervisor's Bulletin*[2] stresses that the use of control charts on the line leads to logical decisions regarding the process. Problems such as tool wear are discovered and acted on right away before defectives are made.

EXAMPLE 6.6 A measurement has a target value of .025 and specification limits .0200 to .0300. Make a control chart and analyze it for statistical control.

- Use Table 6.2 to set the chart scales. This is a first chart, and the scales are set from the first eight samples.
- Complete the chart in Figure 6.15.
- Calculate average lines and control limits and draw them on the chart.
- Make the tally histogram.
- Analyze the chart and the histogram for out-of-control problems.
- Eliminate the out-of-control data, recalculate average lines and control limits, and analyze again.

Solution

The sample size is 4, so each $\bar{x}$ is found by adding the measurements and then dividing by 4. Use a calculator to check the $\bar{x}$ results and the R values.

The range of R for the first eight samples is

$$R_R = .0062 - 0 = .0062$$
$$2R_R = .0124$$

TABLE 6.2
Specification: .0200 to .0300

Sample values	.0218	.0247	.0244	.0254	.0265	.0259	.0266	.0259
	.0243	.0255	.0252	.0238	.0232	.0274	.0245	.0282
	.0232	.0282	.0265	.0249	.0294	.0228	.0231	.0264
	.0256	.0261	.0267	.0275	.0281	.0254	.0280	.0234
$\bar{x}$	.0237	.0261	.0257	.0254	.0268	.0254	.0256	.0260
R	.0038	.0035	.0023	.0037	.0062	.0046	.0049	.0048

[2] "Statistical Methods Slam the Door on Defects" (*Quality Control Supervisor's Bulletin*, 24 Rope Ferry Road, Waterford, CT 06386).

$\bar{R} = \frac{\Sigma R}{k}$ = ______ = ______

$\bar{\bar{x}} = \frac{\Sigma \bar{x}}{k}$ = ______ = ______

Specification: .0200 to .0300

$UCL_R = D_4 \times \bar{R}$ = ____ × ____ = ____

$A_2 \times \bar{R}$ = ____ × ____ = ☐

$UCL_{\bar{x}} = \bar{\bar{x}} + A_2\bar{R}$ = ____ + ☐ = ____

$LCL_{\bar{x}} = \bar{\bar{x}} - A_2\bar{R}$ = ____ − ☐ = ____

Constants

n	A_2	$\tilde{A}_2$	D_4	d_2
3	1.023	1.19	2.574	1.693
4	0.729	—	2.282	2.059
5	0.577	.69	2.114	2.326

Chart of Averages — $\bar{x}$

Chart of Ranges — R

Sample Number		1	2	3	4	5	6	7	8	9	10	11	12	13	14	15	16	17	18	19	20	21	22	23	24	25
Sample Measurements	1	.0218	.0247	.0244	.0254	.0265	.0259	.0266	.0259	.0302	.0221	.0219	.0205	.0264	.0251	.0259	.0254	.0263	.0258	.0273	.0213	.0223	.0252	.0248	.0253	.0260
	2	.0243	.0255	.0252	.0238	.0232	.0274	.0245	.0282	.0271	.0247	.0242	.0253	.0267	.0274	.0221	.0257	.0267	.0243	.0266	.0256	.0251	.0224	.0232	.0262	.0229
	3	.0232	.0282	.0265	.0249	.0294	.0228	.0231	.0264	.0250	.0256	.0223	.0241	.0253	.0236	.0245	.0263	.0254	.0249	.0264	.0294	.0248	.0263	.0250	.0221	.0232
	4	.0256	.0261	.0267	.0275	.0281	.0254	.0280	.0234	.0254	.0246	.0258	.0236	.0278	.0265	.0255	.0230	.0240	.0230	.0232	.0275	.0241	.0247	.0208	.0243	.0251
	5																									
Sum																										
Average, $\bar{x}$		.0237	.0261	.0257	.0254	.0268	.0254	.0256	.0260																	
Range, R		.0038	.0035	.0023	.0037	.0062	.0046	.0049	.0048																	

FIGURE 6.15
Calculate $\bar{x}$ and R and set the scales for the $\bar{x}$ and R charts. Plot and connect the points. Check your results with Figure 6.16.

There are nine lines on the range chart in Figure 6.15. Divide 9 into $2R_R$:

$$\begin{array}{r} .0013+ \\ 9\overline{).0124} \end{array}$$

Round to .001 units per line. Start at 0 at the bottom of the R chart and count up .001 units per line as shown:

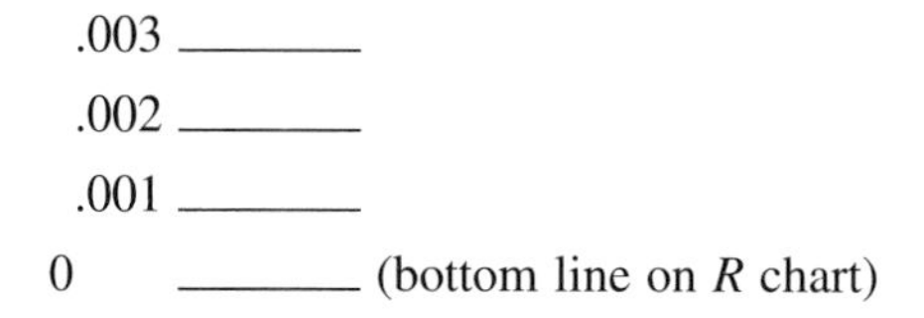

Note: If the above quotient was rounded up to .002 units per line the width of the R chart would be smaller.

The midspecification value is (.0200 + .0300)/2 = .0250. Start the $\bar{x}$ chart with .0250 at the center of the vertical scale. Find the range of $\bar{x}$ for the first eight samples:

$$R_{\bar{x}} = .0268 - .0237 \qquad \begin{array}{r} .00029+ \\ 21\overline{).0062} \end{array}$$

$$R_{\bar{x}} = .0031$$

$$2R_{\bar{x}} = .0062$$

There are 21 lines on the chart for $\bar{x}$ in Figure 6.15. Divide 21 into the $2R_{\bar{x}}$ value and round up to .0005 units per line. Start at the midspecification value in the center and mark segments in increments of .0005 units per line in both directions as shown:

.0260 _______

.0255 _______

.0250 _______ (middle of $\bar{x}$ chart)

.0245 _______

.0240 _______

Note: If the above quotient was rounded up to .0004 units per line the width of the $\bar{x}$ chart would be wider. A round up to .0003 units per line would be even wider and probably crowd the chart boundaries.

Chart the points. Three charts are presented for this $\bar{x}$ and R illustration. The first, Figure 6.15, has the measurements listed for the 25 samples. Calculate the missing $\bar{x}$ and R values. Set the $\bar{x}$ scale and the R scale as directed, plot the points, and draw the connecting lines from point to point on Figure 6.15 for two separate broken-line graphs (one for $\bar{x}$ and one for R). Check your results with the second chart, Figure 6.16.

Calculate the averages and the control limits. To calculate $\bar{R}$, add all the range values and then divide by the number of values k:

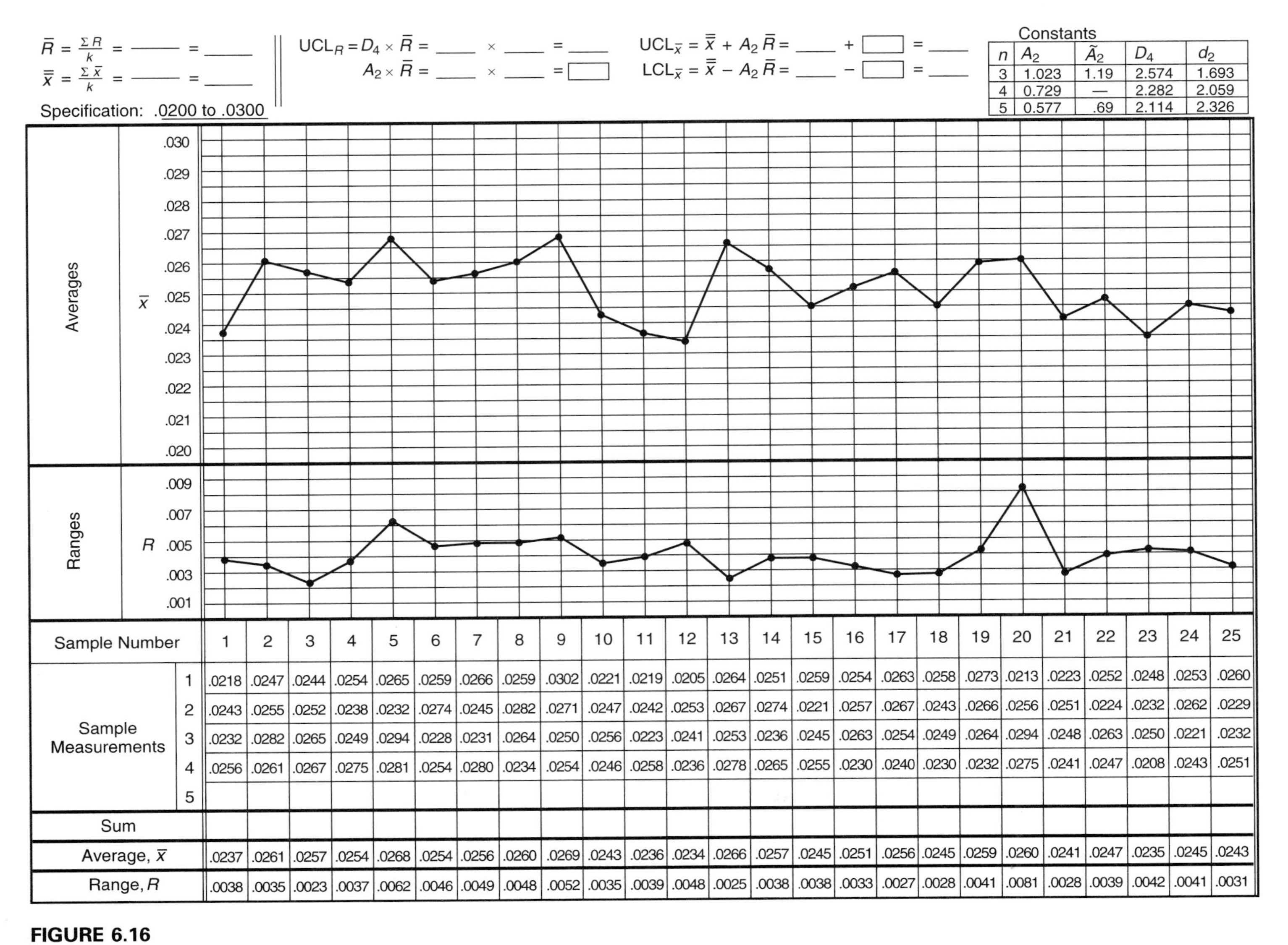

Constants

n	A_2	$\tilde{A}_2$	D_4	d_2
3	1.023	1.19	2.574	1.693
4	0.729	—	2.282	2.059
5	0.577	.69	2.114	2.326

Sample Number		1	2	3	4	5	6	7	8	9	10	11	12	13	14	15	16	17	18	19	20	21	22	23	24	25
Sample Measurements	1	.0218	.0247	.0244	.0254	.0265	.0259	.0266	.0259	.0302	.0221	.0219	.0205	.0264	.0251	.0259	.0254	.0263	.0258	.0273	.0213	.0223	.0252	.0248	.0253	.0260
	2	.0243	.0255	.0252	.0238	.0232	.0274	.0245	.0282	.0271	.0247	.0242	.0253	.0267	.0274	.0221	.0257	.0267	.0243	.0266	.0256	.0251	.0224	.0232	.0262	.0229
	3	.0232	.0282	.0265	.0249	.0294	.0228	.0231	.0264	.0250	.0256	.0223	.0241	.0253	.0236	.0245	.0263	.0254	.0249	.0264	.0294	.0248	.0263	.0250	.0221	.0232
	4	.0256	.0261	.0267	.0275	.0281	.0254	.0280	.0234	.0254	.0246	.0258	.0236	.0278	.0265	.0255	.0230	.0240	.0230	.0232	.0275	.0241	.0247	.0208	.0243	.0251
	5																									
Sum																										
Average, $\bar{x}$		.0237	.0261	.0257	.0254	.0268	.0254	.0256	.0260	.0269	.0243	.0236	.0234	.0266	.0257	.0245	.0251	.0256	.0245	.0259	.0260	.0241	.0247	.0235	.0245	.0243
Range, R		.0038	.0035	.0023	.0037	.0062	.0046	.0049	.0048	.0052	.0035	.0039	.0048	.0025	.0038	.0038	.0033	.0027	.0028	.0041	.0081	.0028	.0039	.0042	.0041	.0031

FIGURE 6.16
The completed broken-line graphs.

$$\overline{R} = \frac{\Sigma R}{k}$$
$$= \frac{.1004}{25}$$
$$= .00402$$

The calculations will be shown here in a step-by-step format. The same calculations are shown in concise form in Figure 6.17.

Calculate UCL_R and LCL_R. The subscript R indicates that the control limits are for ranges. Here UCL_R is the upper control limit for ranges. The Shewhart formulas for the control limits are

$$UCL_R = D_4\overline{R} \qquad LCL_R = D_3\overline{R}$$

The values for the numerical constants in the formulas, D_4 and D_3, are found in Appendix B, Table B.1. For a sample of four ($n = 4$), $D_4 = 2.282$ and $D_3 = 0$:

$$UCL_R = D_4 \times \overline{R} \qquad LCL_R = D_3 \times \overline{R}$$
$$= 2.282 \times .00402 \qquad = 0 \times .00402$$
$$= .0092 \qquad = 0$$

Draw a solid line for $\overline{R}$ and a dashed line for UCL_R and LCL_R. Check for out-of-control patterns on the R chart, patterns characterized by points above the UCL_R, shifts (seven or more consecutive points above $\overline{R}$ or below $\overline{R}$), or runs (seven or more consecutive points going up or going down). Because there are no apparent out-of-control patterns in this case (see Figure 6.17), proceed with the $\overline{x}$ chart. Had we discovered any points out of control, we would have marked them with an X, found and eliminated the cause, removed the sample(s) from the data set, and recalculated $\overline{R}$. When a point on a chart is hard to read, relative to the average or control limit lines, check the calculated values. Sample 19 on Figure 6.17 has a R value of .0041. That is larger than $\overline{R}$ = .00402, so the point is above the $\overline{R}$ line and the potential shift pattern is broken.

The range chart must be in statistical control before any interpretation can be made on the $\overline{x}$ chart. Excessive variation can cause the $\overline{x}$ chart to appear out of control, but the underlying problem is then the variation, not necessarily a needed adjustment toward the target measurement.

In the next step, we will calculate $\overline{\overline{x}}$ and the control limits. To find $\overline{\overline{x}}$, the average of the $\overline{x}$ values, add all the $\overline{x}$ values and divide by the number of samples k:

$$\overline{\overline{x}} = \frac{\Sigma\overline{x}}{k}$$
$$= \frac{.6279}{25}$$
$$= .0251$$

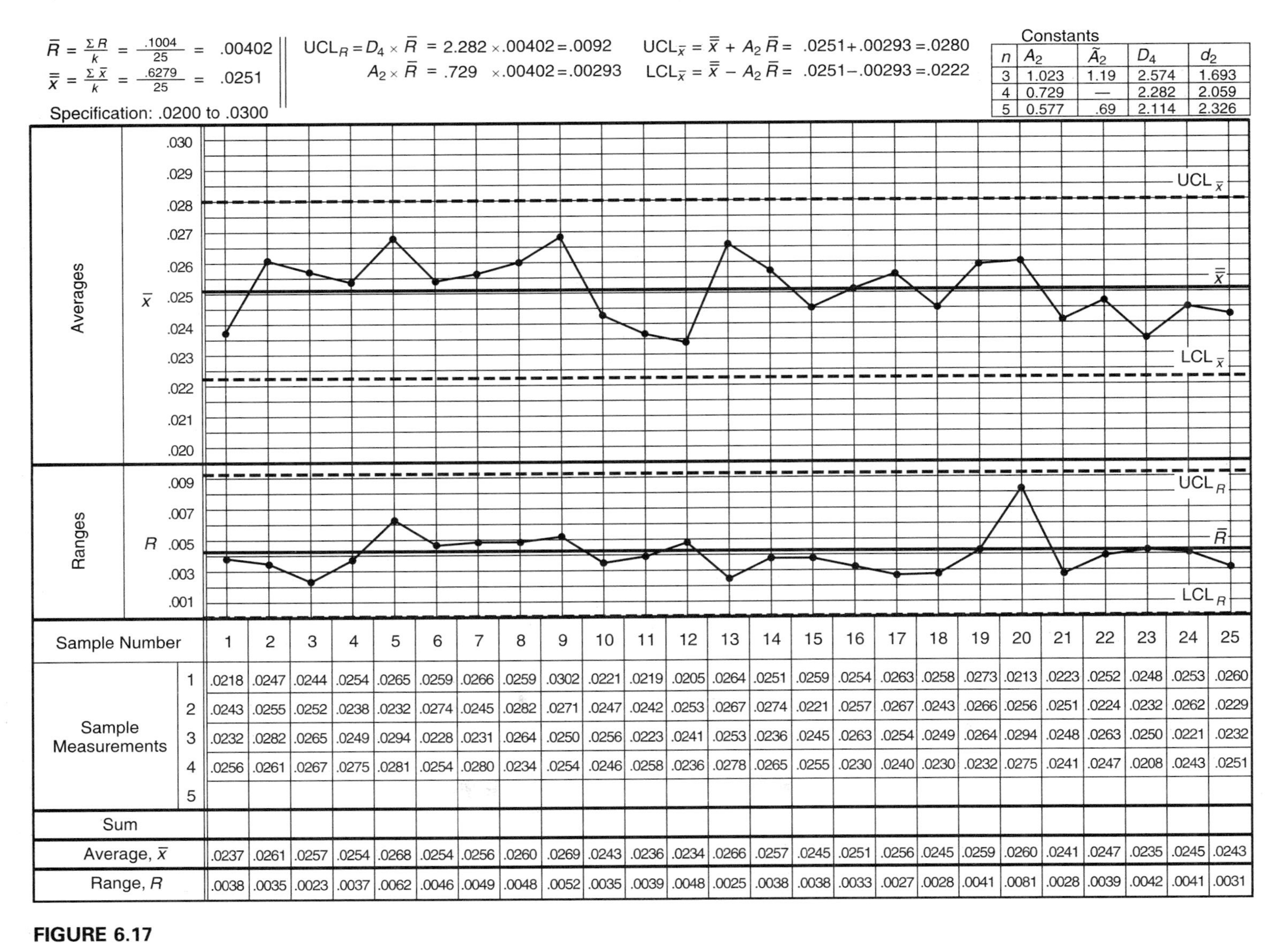

n	A_2	$\tilde{A}_2$	D_4	d_2
3	1.023	1.19	2.574	1.693
4	0.729	—	2.282	2.059
5	0.577	.69	2.114	2.326

Sample Number		1	2	3	4	5	6	7	8	9	10	11	12	13	14	15	16	17	18	19	20	21	22	23	24	25
Sample Measurements	1	.0218	.0247	.0244	.0254	.0265	.0259	.0266	.0259	.0302	.0221	.0219	.0205	.0264	.0251	.0259	.0254	.0263	.0258	.0273	.0213	.0223	.0252	.0248	.0253	.0260
	2	.0243	.0255	.0252	.0238	.0232	.0274	.0245	.0282	.0271	.0247	.0242	.0253	.0267	.0274	.0221	.0257	.0267	.0243	.0266	.0256	.0251	.0224	.0232	.0262	.0229
	3	.0232	.0282	.0265	.0249	.0294	.0228	.0231	.0264	.0250	.0256	.0223	.0241	.0253	.0236	.0245	.0263	.0254	.0249	.0264	.0294	.0248	.0263	.0250	.0221	.0232
	4	.0256	.0261	.0267	.0275	.0281	.0254	.0280	.0234	.0254	.0246	.0258	.0236	.0278	.0265	.0255	.0230	.0240	.0230	.0232	.0275	.0241	.0247	.0208	.0243	.0251
	5																									
Sum																										
Average, $\bar{x}$		.0237	.0261	.0257	.0254	.0268	.0254	.0256	.0260	.0269	.0243	.0236	.0234	.0266	.0257	.0245	.0251	.0256	.0245	.0259	.0260	.0241	.0247	.0235	.0245	.0243
Range, R		.0038	.0035	.0023	.0037	.0062	.0046	.0049	.0048	.0052	.0035	.0039	.0048	.0025	.0038	.0038	.0033	.0027	.0028	.0041	.0081	.0028	.0039	.0042	.0041	.0031

FIGURE 6.17
Mark all points that are of control. Check your results with Figure 6.18.

Calculate the control limits, $UCL_{\bar{x}}$ and $LCL_{\bar{x}}$. The $\bar{x}$ subscript indicates that the control limits are for sample averages. The Shewhart formulas for the control limits are

$$UCL_{\bar{x}} = \bar{\bar{x}} + (A_2\bar{R}) \qquad LCL_{\bar{x}} = \bar{\bar{x}} - (A_2\bar{R})$$

Notice that the same product, $A_2\bar{R}$, is added to the mean for the $UCL_{\bar{x}}$ and subtracted from the mean for the $LCL_{\bar{x}}$. The value for the numerical constant A_2 is found in Appendix B, Table B.1; $A_2 = .729$ for $n = 4$:

$$\begin{aligned} UCL_{\bar{x}} &= \bar{\bar{x}} + (A_2\bar{R}) \\ &= .0251 + (.729 \times .00402) \\ &= .0251 + .00293 \\ &= .0280 \end{aligned} \qquad \begin{aligned} LCL\bar{x} &= \bar{\bar{x}} - (A_2\bar{R}) \\ &= .0251 - .00293 \\ &= .0222 \end{aligned}$$

Draw a solid line for $\bar{\bar{x}}$ and dashed lines for the control limits $UCL_{\bar{x}}$ and $LCL_{\bar{x}}$. At this point, your control chart in Figure 6.15 should look like the completed control chart in Figure 6.17. Check for out-of-control patterns of the $\bar{x}$ chart: points above $UCL_{\bar{x}}$ or below $LCL_{\bar{x}}$, shifts (seven or more consecutive points on one side of $\bar{\bar{x}}$), or runs (seven or more consecutive points going up or going down).

The second through the ninth points on the chart are all above $\bar{\bar{x}}$, and that indicates an out-of-control situation. Mark the points for samples 8 and 9 in the shift pattern with an X to show the points that are out of control. When an out-of-control situation such as this occurs, the operator, the process control team, or both analyze the chart and the process log to determine the cause of the problem and then work to eliminate the problem. Work procedures are amended to prevent the problem from recurring.

At this point, the control chart is used in two ways: to find out-of-control situations, as illustrated in this case, and to measure the process for capability. That measuring takes place after all the out-of-control situations have been resolved and the process is in statistical control. Statistical control on the chart means that all the points on the chart are *in control* and that the process at this point has no special-cause variation. The quickest way to determine the process capability is to eliminate all of the out-of-control points from the control chart (after the out-of-control problem has been resolved). Then recalculate the averages and control limits and reanalyze the chart to see if the process is still in control. The alternative would be to wait until a chart has no out-of-control situations on it and calculate the process capability from the information on that chart.

When the out-of-control pattern is a run or shift, it takes several points before the situation can be recognized. If the initial out-of-control point can be determined on the control chart, remove all the out-of-control points back to that initial point. Otherwise, just remove the ones marked with an X. Connect the points on either side of the removed points by a dotted line, eliminate the data values, and recalculate $\bar{\bar{x}}$ and $\bar{R}$. Figure 6.18 shows this adjustment.

Eliminate $\bar{x} = .0260$ and $\bar{x} = .0269$ from $\Sigma\bar{x}$:

$$\begin{array}{r} .0260 \\ +.0269 \\ \hline .0529 \end{array} \quad \text{Add them.}$$

$\bar{R} = \frac{\Sigma R}{k} = \frac{.0904}{23} = .00393$

$\bar{\bar{x}} = \frac{\Sigma \bar{x}}{k} = \frac{.5750}{23} = .0250$

Specification: .0200 to .0300

$UCL_R = D_4 \times \bar{R} = 2.282 \times .00393 = .0090$

$A_2 \times \bar{R} = .729 \times .00393 = .00286$

$UCL_{\bar{x}} = \bar{\bar{x}} + A_2 \bar{R} = .0250 + .00286 = .0279$

$LCL_{\bar{x}} = \bar{\bar{x}} - A_2 \bar{R} = .0250 - .00286 = .0221$

Constants

n	A_2	$\tilde{A}_2$	D_4	d_2
3	1.023	1.19	2.574	1.693
4	0.729	—	2.282	2.059
5	0.577	.69	2.114	2.326

Averages: $\bar{x}$ .030, .029, .028, .027, .026, .025, .024, .023, .022, .021, .020; $UCL_{\bar{x}}$, $\bar{\bar{x}}$, $LCL_{\bar{x}}$

Ranges: R .009, .007, .005, .003, .001; UCL_R, $\bar{R}$, LCL_R

Sample Number		1	2	3	4	5	6	7	8	9	10	11	12	13	14	15	16	17	18	19	20	21	22	23	24	25
Sample Measurements	1	.0218	.0247	.0244	.0254	.0265	.0259	.0266	.0259	.0302	.0221	.0219	.0205	.0264	.0251	.0259	.0254	.0263	.0258	.0273	.0213	.0223	.0252	.0248	.0253	.0260
	2	.0243	.0255	.0252	.0238	.0232	.0274	.0245	.0282	.0271	.0247	.0242	.0253	.0267	.0274	.0221	.0257	.0267	.0243	.0266	.0256	.0251	.0224	.0232	.0262	.0229
	3	.0232	.0282	.0265	.0249	.0294	.0228	.0231	.0264	.0250	.0256	.0223	.0241	.0253	.0236	.0245	.0263	.0254	.0249	.0264	.0294	.0248	.0263	.0250	.0221	.0232
	4	.0256	.0261	.0267	.0275	.0281	.0254	.0280	.0284	.0254	.0246	.0258	.0236	.0278	.0265	.0255	.0230	.0240	.0230	.0232	.0275	.0241	.0247	.0208	.0243	.0251
	5																									
Sum																										
Average, $\bar{x}$		.0237	.0261	.0257	.0254	.0268	.0254	.0256	.0260	.0269	.0243	.0236	.0234	.0266	.0257	.0245	.0251	.0256	.0245	.0259	.0260	.0241	.0247	.0235	.0245	.0243
Range, R		.0038	.0035	.0023	.0037	.0062	.0046	.0049	.0048	.0052	.0035	.0039	.0048	.0025	.0038	.0038	.0033	.0027	.0028	.0041	.0081	.0028	.0039	.0042	.0041	.0031

FIGURE 6.18
Out-of-control data are eliminated from the control chart *after* the cause has been determined and eliminated. Recalculate.

.6279	Previous $\Sigma\bar{x}$
$-.0529$	Subtract their sum from $\Sigma\bar{x}$.
.5750	The new $\Sigma\bar{x}$ value

$$\bar{\bar{x}} = \frac{\Sigma\bar{x}}{k}$$ Recalculate $\bar{\bar{x}}$.

$$= \frac{.5750}{23}$$ The new value for $\Sigma\bar{x}$ is .5750. The new value for k is 23.

$$= .0250$$ The new $\bar{\bar{x}}$ value

Eliminate $R = .0048$ and $R = .0052$ from ΣR. Follow the procedure shown for eliminating $\bar{x}$ data:

.0048	Add them.
$+.0052$	
.0100	

.1004	Previous ΣR
$-.0100$	Subtract their sum from ΣR.
.0904	The new ΣR value

$$\bar{R} = \frac{\Sigma R}{k}$$ Recalculate $\bar{R}$.

$$= \frac{.0904}{23}$$

$$= .00393$$ The new $\bar{R}$ value

Recalculate the control limits:

$$\text{UCL}R = D_4\bar{R}$$
$$= 2.282 \times .00393$$
$$= .0090$$ The new UCL_R value

$$\text{UCL}\bar{x} = \bar{\bar{x}} + (A_2\bar{R})$$
$$= .0250 + (.729 \times .00393)$$
$$= .0250 + .00286$$
$$= .0279$$ The new $\text{UCL}_{\bar{x}}$ value

$$\text{LCL}\bar{x} = \bar{\bar{x}} - A_2\bar{R}$$
$$= .0250 - .00286$$
$$= .0221$$ The new $\text{LCL}_{\bar{x}}$ value

There is no substantial change in the average and control limit lines, and no other out-of-control point patterns are evident. If all the causes of the out-of-control points have been determined and eliminated, the process is now in statistical control and the capability analysis can commence. The elimination of the out-of-control data points may seem useless because the control limit lines and averages did not really change; however, the change in $\bar{R}$ is quite important for the capability study because the standard deviation of all the measurements is determined from the $\bar{R}$ value.

The recalculation of the averages and the control limits was made easier because the initial sums, $\Sigma\bar{x}$ and ΣR, were noted on the worksheet. It was much easier to subtract the deleted data values from the original sums than to have to recalculate the sums from the remaining 23 data values. Whenever you do control chart calculations, be sure to write down the formulas, show your numerical substitutions, and keep your calculated results. This procedure allows easy checking and reference.

Check the histogram for further evidence of out-of-control problems. The shape should be normal. If it is not, analysis as discussed in Chapter 4 may uncover the problem. If the specification limits are marked on the histogram's measurement scale, it provides an ongoing capability check as the data are gathered. Remember, the range of measurements in a sample will be smaller than that of the population. If the sample measurements crowd the specification limits, it may be an indication that the population of measurements is overlapping them. The histogram in Figure 6.19 has a normal pattern, so the population should have a normal distribution. There is a variation problem because

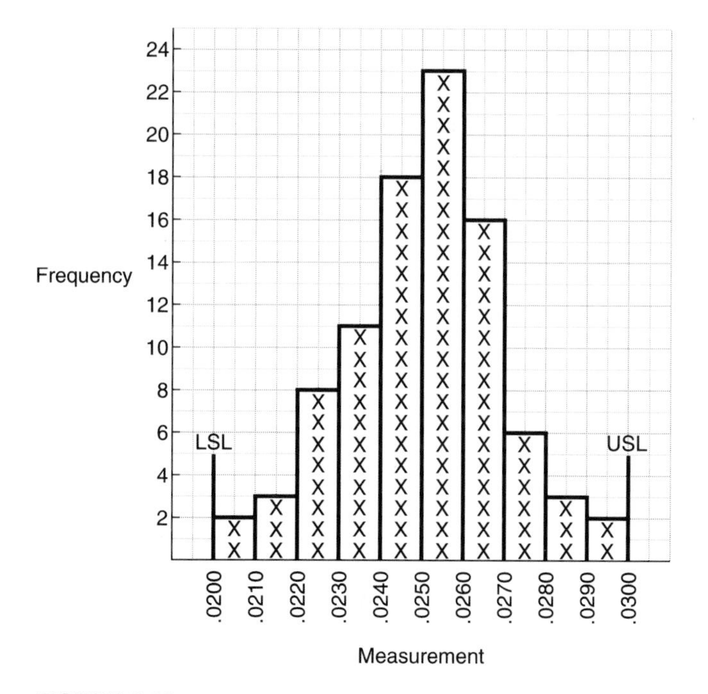

FIGURE 6.19
The histogram of all the control chart values. Measurements that crowd the specification limits indicate poor capability.

the histogram crowds the specification limits, and the population, with its wider range, is sure to overlap them.

6.6 THE CONTINUATION CONTROL CHART

A continuation control chart is one that continues from a previous chart. All of the averages and control limits are carried over from the previous chart so the process can be analyzed for statistical control as the data are gathered and plotted. Continuation charts are used most often; the first chart and charts with recalculated control limits are the exceptions. The continuation chart is a better tool than a first chart because the worker is immediately aware that the process is out of control when a point is plotted beyond a control limit and also the worker can anticipate an out-of-control situation when a shift or run appears to be developing.

When continuation charts are used, a recalculation is occasionally needed.

Recalculate control limits under the following circumstances:

1. After a sequence of several in-control charts to fine tune the average and control limit lines. This also gives a better data base for a capability evaluation.
2. When there has been an apparent change in variation. The measure of variation is the standard deviation and the control limits are three standard deviations from the average value. So, any variation change will result in a control limit change and the new control limits have to be calculated. The Shewhart formulas contain a $\overline{R}$ factor, so any change in $\overline{R}$ will change the control limits.

EXAMPLE 6.7 Make a continuation control chart from Figure 6.18. Use the data in Table 6.3, in order by columns, to simulate an on-the-job situation and indicate any action taken. Assume that you are using a list of random sample times with 8 samples per shift and that your machine is running from the previous shift (no warm-up time is necessary).

TABLE 6.3

.0255	.0255	.0238	.0245	.0236	.0231	.0232	.0251	
.0238	.0261	.0239	.0241	.0239	.0242	.0237	.0246	
.0231	.0246	.0237	.0252	.0258	.0255	.0257	.0259	
.0236	.0259	.0254	.0258	.0244	.0257	.0236	.0255	
.0253	.0251	.0262	.0255	.0248	.0254	.0247	.0252	
.0219	.0238	.0256	.0254	.0249	.0253	.0238	.0251	
.0226	.0254	.0241	.0236	.0241	.0251	.0247	.0256	
.0262	.0264	.0266	.0245	.0245	.0252	.0245	.0253	
.0241	.0259	.0237	.0258	.0252	.0255	.0249	.0252	.0260
.0239	.0236	.0246	.0256	.0255	.0256	.0251	.0247	.0261
.0244	.0246	.0239	.0244	.0251	.0249	.0238	.0258	.0257
.0231	.0256	.0241	.0245	.0257	.0256	.0262	.0255	.0257

Solution

The blank continuation chart is shown in Figure 6.20. The control limits and average lines are drawn. The data values are entered on Figure 6.21. To simulate an on-the-job application, cover the data values with a sheet of paper and slide it down uncovering one sample at a time.

The thought and action sequence for the continuation chart would proceed as follows:

- The previous chart (Figure 6.18) ended with 5 consecutive points below the $\bar{\bar{x}}$ line. *I wonder if a shift has occurred.*
- The first sample is also below the line. *Perhaps I should take the next sample early, as soon as I get a chance to do it.*
- The second sample is below the line, too. This makes 7 consecutive points below the $\bar{\bar{x}}$ line, which is a shift. Action: The company action plan indicates that the machine must be shut down and the source of the special-cause variation determined. Also, all the pieces made since the start of the control chart shift have to be 100% inspected. A cutting tool was sharpened and adjusted and a loose fixture was tightened. This action was noted on the process log. The machine is restarted and another early sample taken. *Everything seems OK.*
- The next few samples show a possible shift on the range chart.
- The ninth sample confirms the shift; seven consecutive points below the $\bar{R}$ line. *Since the shift is down on the range chart, this is good news; it indicates that the variation has decreased. No action has to be taken at this time, but I should try to determine why the process has improved. The only change in addition to the sharpening and adjustment was the loose fixture I found. That must have contributed to the higher variation numbers on the previous chart. The rest of the chart shows the range values staying below their previous level. This means that I'll have to recalculate the control limits. I'll use the data beginning with sample 3 since that was when the change took place.*

Figure 6.22 shows the chart with the new calculations at the top and the new control limits. Normally, the new control limits would be drawn on the current chart (Figure 6.21), but since a new scale was warranted the chart analysis would be easier with the expanded scale. The new control limits show that the chart is in control at the new level.

6.7 THE CAPABILITY ANALYSIS

The main function of a capability analysis is to determine how good the measurements are when compared to the specifications. The introductory work in Chapter 5 emphasized using the normal curve and calculating the percentage of product that is out of specification. In section 6.5, step 9 in the control chart procedure alluded to capability ratios and indexes. In either case we start with the calculation of the population standard deviation.

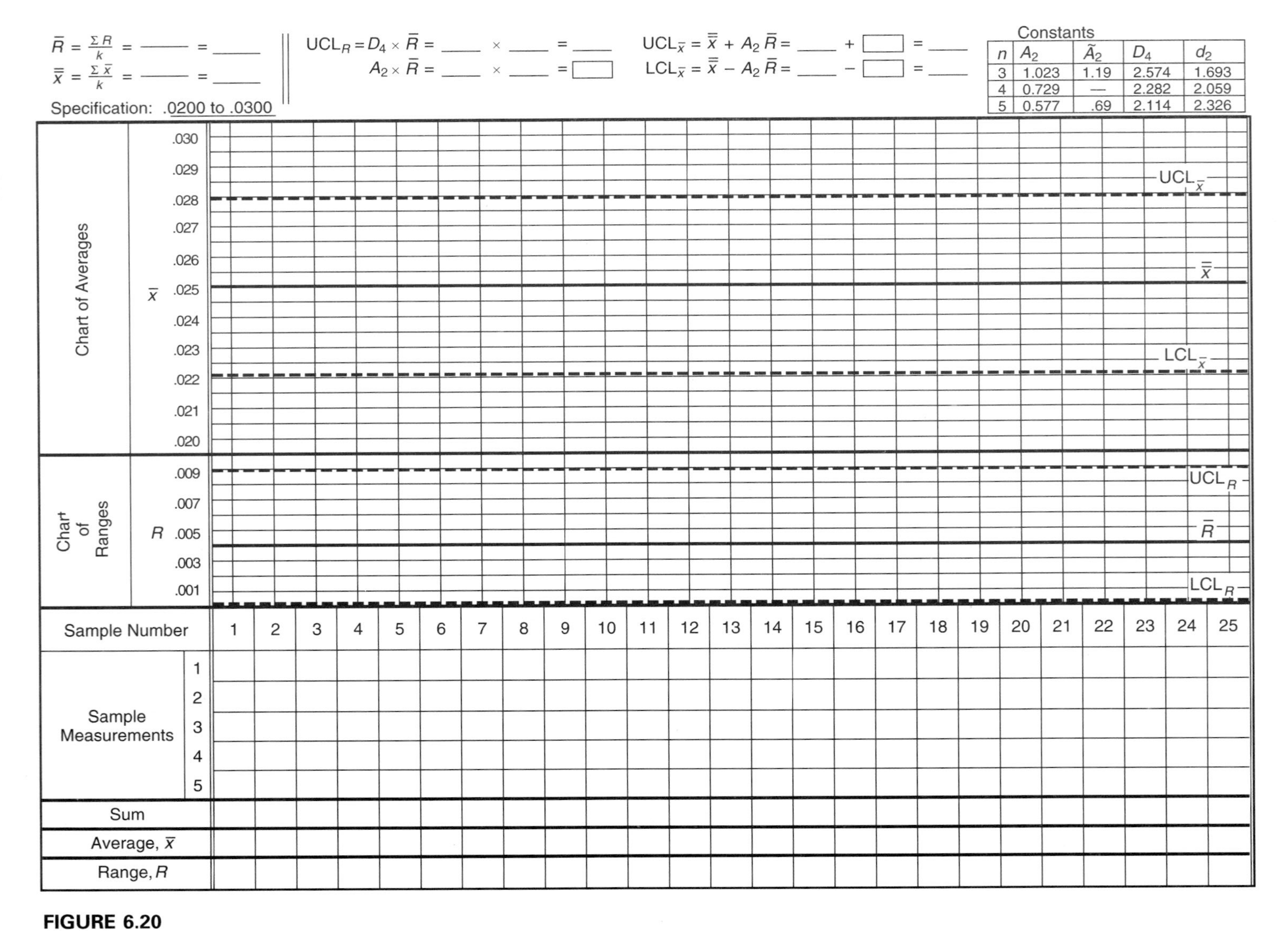

FIGURE 6.20
Enter the data from Table 6.3 and analyze the continuation chart as it develops.

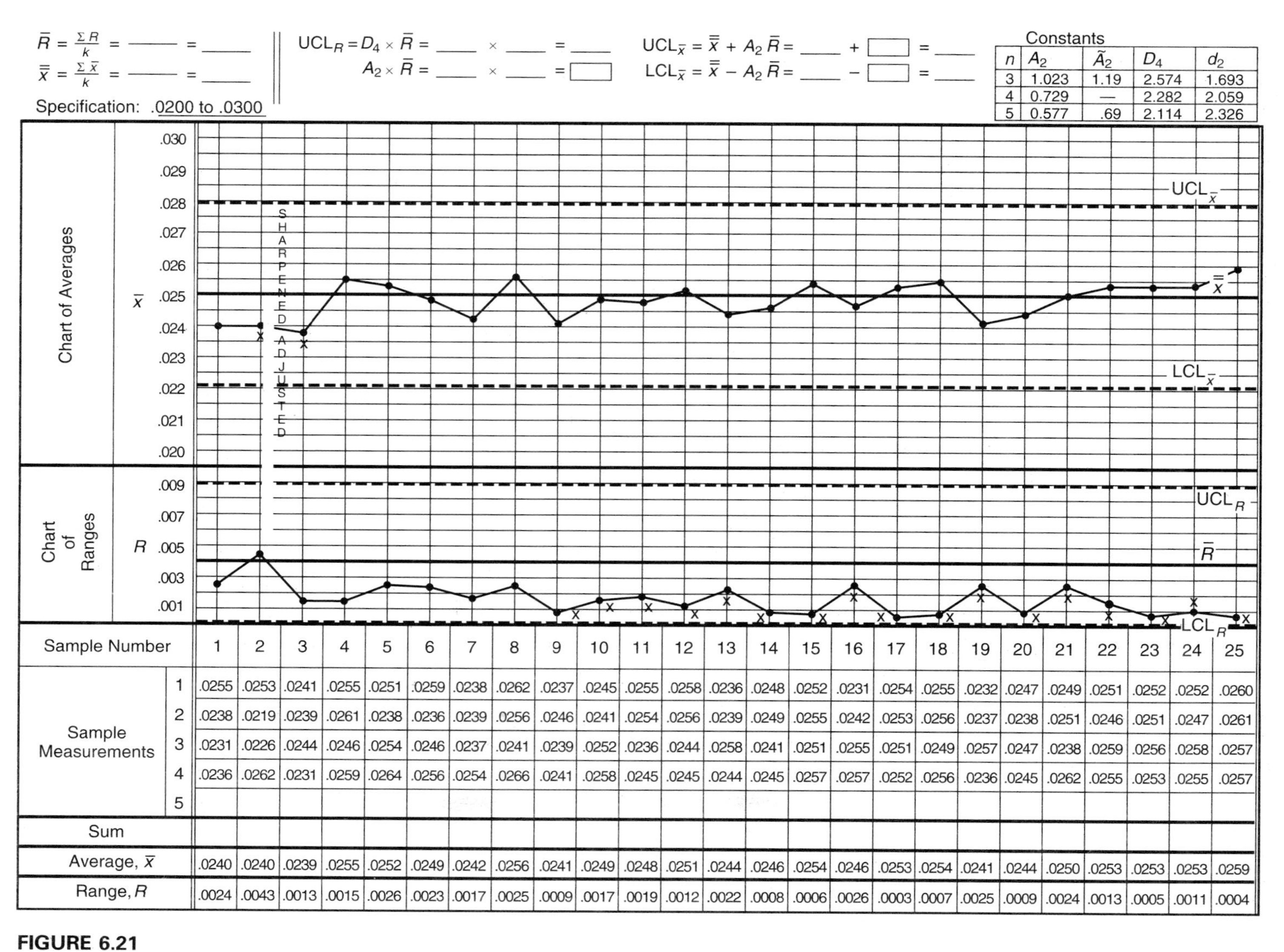

Constants

n	A_2	$\tilde{A}_2$	D_4	d_2
3	1.023	1.19	2.574	1.693
4	0.729	—	2.282	2.059
5	0.577	.69	2.114	2.326

Sample Number		1	2	3	4	5	6	7	8	9	10	11	12	13	14	15	16	17	18	19	20	21	22	23	24	25
Sample Measurements	1	.0255	.0253	.0241	.0255	.0251	.0259	.0238	.0262	.0237	.0245	.0255	.0258	.0236	.0248	.0252	.0231	.0254	.0255	.0232	.0247	.0249	.0251	.0252	.0252	.0260
	2	.0238	.0219	.0239	.0261	.0238	.0236	.0239	.0256	.0246	.0241	.0254	.0256	.0239	.0249	.0255	.0242	.0253	.0256	.0237	.0238	.0251	.0246	.0251	.0247	.0261
	3	.0231	.0226	.0244	.0246	.0254	.0246	.0237	.0241	.0239	.0252	.0236	.0244	.0258	.0241	.0251	.0255	.0251	.0249	.0257	.0247	.0238	.0259	.0256	.0258	.0257
	4	.0236	.0262	.0231	.0259	.0264	.0256	.0254	.0266	.0241	.0258	.0245	.0245	.0244	.0245	.0257	.0257	.0252	.0256	.0236	.0245	.0262	.0255	.0253	.0255	.0257
	5																									
Sum																										
Average, $\bar{x}$		.0240	.0240	.0239	.0255	.0252	.0249	.0242	.0256	.0241	.0249	.0248	.0251	.0244	.0246	.0254	.0246	.0253	.0254	.0241	.0244	.0250	.0253	.0253	.0253	.0259
Range, R		.0024	.0043	.0013	.0015	.0026	.0023	.0017	.0025	.0009	.0017	.0019	.0012	.0022	.0008	.0006	.0026	.0003	.0007	.0025	.0009	.0024	.0013	.0005	.0011	.0004

FIGURE 6.21
A shift on the R chart signals a need for recalculating averages and control limits.

$\bar{R} = \frac{\Sigma R}{k} = \frac{.00339}{23} = .001474$

$\bar{\bar{x}} = \frac{\Sigma \bar{x}}{k} = \frac{.5732}{23} = .0249$

Specification: .0200 to .0300

$UCL_R = D_4 \times \bar{R} = 2.282 \times .001474 = .0034$

$A_2 \times \bar{R} = .729 \times .001474 = .0011$

$UCL_{\bar{x}} = \bar{\bar{x}} + A_2\bar{R} = .0249 + .0011 = .0260$

$LCL_{\bar{x}} = \bar{\bar{x}} - A_2\bar{R} = .0249 - .0011 = .0238$

Constants

n	A_2	$\tilde{A}_2$	D_4	d_2
3	1.023	1.19	2.574	1.693
4	0.729	—	2.282	2.059
5	0.577	.69	2.114	2.326

Chart of Averages ($\bar{x}$): .0262, .0258, .0254, .0250, .0246, .0242, .0238, .0234, .0230 — SHARPENED, ADJUSTED, $UCL_{\bar{x}}$, $\bar{\bar{x}}$, $LCL_{\bar{x}}$

Chart of Ranges (R): .004, .003, .002, .001 — x, UCL_R, $\bar{R}$, LCL_R

Sample Number		1	2	3	4	5	6	7	8	9	10	11	12	13	14	15	16	17	18	19	20	21	22	23	24	25
Sample Measurements	1	.0255	.0253	.0241	.0255	.0251	.0259	.0238	.0262	.0237	.0245	.0255	.0258	.0236	.0248	.0252	.0231	.0254	.0255	.0232	.0247	.0249	.0251	.0252	.0252	.0260
	2	.0238	.0219	.0239	.0261	.0238	.0236	.0239	.0256	.0246	.0241	.0254	.0256	.0239	.0249	.0255	.0242	.0253	.0256	.0237	.0238	.0251	.0246	.0251	.0247	.0261
	3	.0231	.0226	.0244	.0246	.0254	.0246	.0237	.0241	.0239	.0252	.0236	.0244	.0258	.0241	.0251	.0255	.0251	.0249	.0257	.0247	.0238	.0259	.0256	.0258	.0257
	4	.0236	.0262	.0231	.0259	.0264	.0256	.0254	.0266	.0241	.0258	.0245	.0245	.0244	.0245	.0257	.0257	.0252	.0256	.0236	.0245	.0262	.0255	.0253	.0255	.0257
	5																									
Sum																										
Average, $\bar{x}$		.0240	.0240	.0239	.0255	.0252	.0249	.0242	.0256	.0241	.0249	.0248	.0251	.0244	.0246	.0254	.0246	.0253	.0254	.0241	.0244	.0250	.0253	.0253	.0253	.0259
Range, R		.0024	.0043	.0013	.0015	.0026	.0023	.0017	.0025	.0009	.0017	.0019	.0012	.0022	.0008	.0006	.0026	.0003	.0007	.0025	.0009	.0024	.0013	.0005	.0011	.0004

FIGURE 6.22
The control chart redrawn with a new scale and new control limits.

Actually, four different standard deviations have entered into the discussion so far. Every distribution has its own mean and standard deviation. The measure that is frequently used with the various distributions is the value three standard deviations from the mean:

- The R chart is a distribution of R values and has a standard deviation s_R. The upper control limit is three standard deviations above the mean. We don't have to calculate s_R directly because it was incorporated in the Shewhart formula. However, if all the R values from in-control data were entered into a statistical calculator, s_R could be calculated directly.
- The $\bar{x}$ chart is a distribution of $\bar{x}$ values and has a standard deviation $s_{\bar{x}}$. The control limits are three standard deviations from the mean. From the Shewhart formula $A_2\bar{R}$ corresponds to $3s_{\bar{x}}$ with additional consideration to subgroup variability. The standard deviation $s_{\bar{x}}$ can be calculated directly by entering all the in control $\bar{x}$ values into a statistical calculator.
- All the pieces that have been made at a process point would form the population of measurements. If it were possible (and it generally is not) to enter all the measurements from those pieces for a specific dimension into a statistical calculator, the standard deviation σ could be determined. Virtually all the measurements would fall between boundaries that extend from three standard deviations below the mean to three standard deviations above the mean, $\mu \pm 3\sigma$. This is the measurement spread that is compared to the specifications in any capability study.
- A large sample has a standard deviation s that is approximately equal to σ. The usual way to calculate s is to use the Shewhart formula, $\bar{R}/d_2$. Also, s could be determined by entering all the in control sample measurements into a statistical calculator.

EXAMPLE 6.8 For the in-control data from Example 6.6, calculate

1. s
2. Control limits for measurement x
3. Percentage of product out of specification

Solution

1. The standard deviation of the individual measurements is approximated by

$$s = \frac{\bar{R}}{d_2} = \frac{.00393}{2.059} = .00191$$

The value for d_2 is found in Appendix B, Table B.1.

2. The six-standard-deviation spread of the individual measurements x is found by adding and subtracting $3s$ or $3\bar{R}/d_2$ to the average value $\bar{\bar{x}}$.

Again, according to the normal-curve analysis, 99.7% of all the individual measurements will lie between the $\bar{\bar{x}} \pm (3\bar{R})/d_2$ values. This spread of values can be compared with the specification limits to determine the process capability:

$$\begin{aligned} UCLx &= \bar{\bar{x}} + 3s \\ &= .0250 + 3 \times .00191 \\ &= .0250 + .00573 \\ &= .0307 \end{aligned} \qquad \begin{aligned} LCLx &= \bar{\bar{x}} - 3s \\ &= .0250 - 3 \times .00191 \\ &= .0250 - .00573 \\ &= .0193 \end{aligned}$$

As before, 99.7% of all process measurements should be between .0193 and .0307 and be normally distributed. When the control limits extend beyond the specification limits, some of the process output is out of specification.

3. If the specification limits are put on a sketch of the normal curve for the individual measurements, as shown in Figure 6.23, the percentage of product that is out of specification can be calculated. Use the z formula for analysis of the sections under the normal curve. Substitute the lower specification value for x and the values for $\bar{\bar{x}}$ and s. Subtract first, then divide:

$$\begin{aligned} z &= \frac{x - \bar{\bar{x}}}{s} \\ &= \frac{.0200 - .0250}{.00191} \\ &= \frac{-.0050}{.00191} \\ &= -2.62 \end{aligned}$$

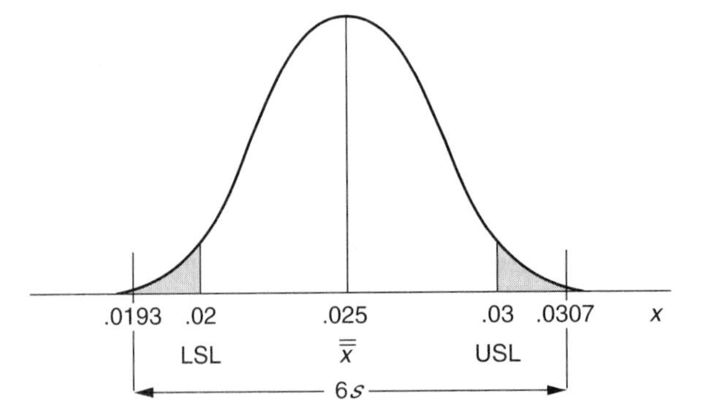

FIGURE 6.23
Specification limits on a normal curve.

The area from Table B.8 in Appendix B is .0044, so .44% is out of specification on the low side.

To calculate the z value for the high side, substitute the upper specification limit for x and the values for $\bar{\bar{x}}$ and s:

$$z = \frac{x - \bar{\bar{x}}}{s}$$

$$= \frac{.0300 - .0250}{.00191}$$

$$= 2.62$$

The area is again .0044, meaning that .44% is out of specification on the high side. The total percentage of product out of specification is

$$.44\% + .44\% = .88\%$$

This process analysis shows that the average measurement .0250 is equal to the midspecification value, so no adjustment is needed there. However, with almost 1% of the product out of specification, process improvements are needed to decrease the amount of product variability. The three-standard-deviation spread of the measurements should be well within the specification limits. The ultimate goal is to have the ± three-standard-deviation spread of the measurements use less than 50% of the tolerance (the specification spread). If the standard deviation could be decreased to .00083 in Example 6.6, this would occur.

Calculating the Capability Ratio

The capability ratio compares the six-standard-deviation spread of measurements to the tolerance. Two notations for the capability ratio are PCR and CR and are shown in two equivalent formulas:

Capability Ratio

$$PCR = \frac{6s}{\text{tolerance}} \qquad CR = \frac{6s}{USL - LSL}$$

LSL = Lower Specification Limit
USL = Upper Specification Limit

The capability ratio gives the fraction of the tolerance that is used by the distribution of measurements. If you multiply by 100, you get the percentage of tolerance used by the distribution.

EXAMPLE 6.9 Find the capability ratio, PCR, for the population of Example 6.6:

1. Using the initial $s = .00191$.
2. Using the reduced $s = .00083$.

Solution

$$1.\ \text{PCR} = \frac{6s}{\text{USL} - \text{LSL}} = \frac{6 \times .00191}{.0300 - .0200} = 1.146 \quad \text{or} \quad 114.6\%$$

This indicates that 114.6% of the tolerance is used by the distribution. An overlapping picture results, which is shown in Figure 6.23. Part of the product is out of specification:

$$2.\ \text{PCR} = \frac{6s}{\text{tolerance}} = \frac{6 \times .00083}{.01} = .498 \quad \text{or} \quad 49.8\%$$

This shows that less than 50% of the tolerance is used.

The Capability Index

There are two working versions of the capability index, Cp and Cpk. The Cp is just the reciprocal of the PCR:

Capability Index, Cp

$$\text{Cp} = \frac{\text{tolerance}}{6s} \quad \text{or} \quad \text{Cp} = \frac{\text{USL} - \text{LSL}}{6s}$$

The reciprocal relationships are

$$\text{Cp} = \frac{1}{\text{PCR}} \qquad \text{PCR} = \frac{1}{\text{Cp}}$$

The index numbers are being used more than the PCR. When an index number is used, it is either compared to some required value or goal or it can be changed to a percentage of tolerance by the reciprocal formula. A Cp value less than 1 means that the tolerance is smaller than the $6s$ measurement spread and there are some out-of-specification pieces in the population.

EXAMPLE 6.10

1. If CP = 1.2, what percentage of the tolerance is being used by the distribution?
2. Calculate Cp for the distribution in Example 6.6.

Solution

1. $$\text{PCR} = \frac{1}{\text{Cp}} \times 100 = \frac{1}{1.2} \times 100$$
$$= 83\%$$

The multiplication by 100 changed the ratio to a percentage.

2. $s = .00191$

$$\text{Cp} = \frac{\text{tolerance}}{6s} = \frac{.01}{6 \times .00191}$$
$$= .87$$

A second capability index that is used in conjunction with Cp is the Cpk. The Cp value compares the whole tolerance with $6s$ and indicates how good the process could be. The Cpk compares the worst half of the distribution with $3s$. It shows the worst-case scenario when the distribution is off center: In Figure 6.24 the distribution overlaps the USL and the shaded section represents the percentage of the measurements that are out of specification. This is the worst side of the distribution because it has the most product out of specification. Notice that $\bar{\bar{x}}$ is closer to the USL than it is to the LSL. In Figure 6.25, the left side of the distribution is the worst side because it has the largest amount of product that is out of specification. In this case, $\bar{\bar{x}}$ is closer to the LSL:

Capability Index, C_{pk}

$$C_{pk} = \text{the minimum of } \frac{USL - \bar{\bar{x}}}{3s} \text{ or } \frac{\bar{\bar{x}} - LSL}{3s}$$

The minimum occurs with the specification limit that is closest to $\bar{\bar{x}}$

In Figure 6.24, the USL value would be used in the Cpk formula. In Figure 6.25, the LSL value would be used in the Cpk formula.

FIGURE 6.24
Use the USL in the Cpk formula.

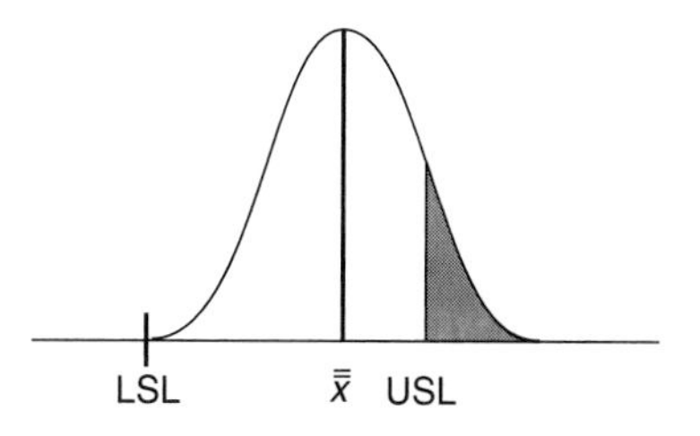

FIGURE 6.25
Use the LSL in the Cpk formula.

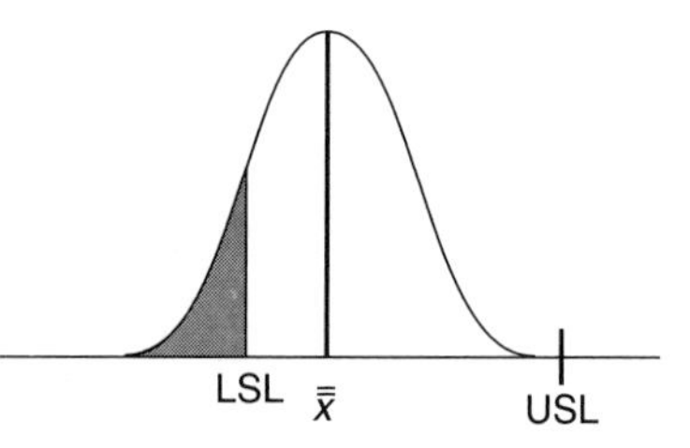

It is possible to have a negative value for C_{pk}. This would indicate excessive poor quality because both specification limits would be on the same side of $\bar{\bar{x}}$; either both above, or both below.

EXAMPLE 6.11

1. Calculate the Cpk for Example 6.6 and compare with the Cp from Example 6.10.
2. Calculate the Cp and Cpk for Example 6.4.

Solution

1. USL $- \bar{\bar{x}} = .0300 - .0250 = .005$
   $\bar{\bar{x}} -$ LSL $= .0250 - .0200 = .005$ Both differences will be the same when the distribution is centered.
   $s = .00191$

$$\text{Cpk} = \frac{USL - \bar{\bar{x}}}{3s}$$

$$\text{Cpk} = \frac{.005}{3 \times .00191}$$

$$= .87$$

From Example 6.10, Cp = .87. When the distribution is centered, Cp = Cpk.

2. From Example 6.4: Specifications are 850 to 950; $\bar{\bar{x}} = 907$; $\bar{R} = 37.04$ and $n = 5$. Calculate s.

$$s = \frac{\bar{R}}{d_2}$$

$$= \frac{37.04}{2.326}$$

$$= 15.92$$

The USL is closer to $\bar{\bar{x}}$, so use

$$\text{Cpk} = \frac{\text{USL} - \bar{\bar{x}}}{3s}$$

$$= \frac{950 - 907}{3 \times 15.92}$$

$$= \frac{43}{47.76}$$
$$= .90$$
$$Cp = \frac{USL - LSL}{6s}$$
$$= \frac{950 - 850}{6 \times 15.92}$$
$$= 1.05$$

The reciprocals show the percentage of tolerance used:

$$\frac{1}{Cp} \times 100 = \frac{1}{1.05} \times 100 = 95\%$$

This shows that if the distribution were centered, it would use slightly less than all the tolerance.

$$\frac{1}{Cpk} \times 100 = \frac{1}{.90} \times 100 = 111\%$$

This shows that on the worst side, the distribution overlaps the specification limit.

The three different capability measures all give a different view of capability. The normal-curve application generally concentrates on the percentage of product that is out of specification. There are limitations because most normal-curve tables do not go beyond $z = 4$, and special tables have to be used for highly capable processes. The PCR is a very visual indicator of capability. When the ratio is less than 1 or the percentage is less than 100%, it promotes an image of a distribution totally within specification, as in Figure 6.26. When the percentage is more than 100%, an image of the distribution overlapping the specification limits, as in Figure 6.23, comes to mind. The combination of Cp and Cpk is used extensively because the combination shows both the potential capability in Cp and any off-center problem when the Cpk is different.

The index numbers are not as visual as the ratios but can easily be changed to ratios (or percentages) using the reciprocal formulas:

$\frac{1}{Cp} \times 100$ Gives the percentage of tolerance used if the distribution is centered.

$\frac{1}{Cpk} \times 100$ Gives the percentage of tolerance used on the worst side.

Many companies are using a company standard of Cpk = 1.33, and some have also set a goal of Cp = 2. So, the Cpk interpretation is that the company requirement is Cpk = 1.33 or more. If it is less than 1.33, then every effort should be made to bring it into compliance. If Cpk is less than 1, then 100% inspection has to be instituted as well because there will be some out-of-specification product to deal with.

TABLE 6.4
Cp and parts per million defective

Cp	z	Tail Area	ppm
.8	2.4	.0082	16,400
1.1	3.3	.00048	960
1.33	3.99	.00003	60
1.5	4.50	.0000034	7
2	6	.0000000009	.0018

Another interpretation of Cp is to match it to the number of defective parts it would correspond to, assuming that the distribution is normal. Comparing the formulas for z and Cpk, we can see that $z = 3 \times \text{Cpk}$. Using the USL as the closest specification limit,

$$z = \frac{\text{USL} - \bar{\bar{x}}}{s} \qquad \text{Cpk} = \frac{\text{USL} - \bar{\bar{x}}}{3s}$$

$$3\text{Cpk} = \frac{\text{USL} - \bar{\bar{x}}}{s}$$

$$z = 3\text{Cpk}$$

when the distribution is on target, $C_p = C_{pk}$ and the first three entries in Table 6.4 can be determined by using Table B.8, Appendix B:

Multiply Cp by 3 to get the z value.

Find the tail area in Table B.8.

Double it (for the area in both tails) and multiply by 1,000,000 for defective parts per million.

If C_{pk} is used instead of C_p, Table 6.4 would give a maximum number of defective parts per million. Since C_{pk} represents the side of the distribution where $\bar{\bar{x}}$ is closest to the specification limit, the tail area on that side is larger than the other side. The doubling process would then overstate the number of defective parts per million.

The normal-curve pattern contains virtually all the measurements (99.7% of them) in a $6s$ spread from $3s$ units below the mean to $3s$ units above. Process 1, shown in Figure 6.26 is a *good* process. The distribution is centered, and the $6s$ spread of the measurements is well within the specification limits.

EXAMPLE 6.12

1. Determine $6s$, $3s$, and the tolerance from the sketch of the distribution in Figure 6.26.
2. Calculate the PCR, Cp, and Cpk for the distribution in Figure 6.26.

Solution

1.
$$\begin{array}{r} 2.317 \\ -2.313 \\ \hline .004 = 6s \\ .002 = 3s \end{array} \qquad \begin{array}{r} 2.320 \\ -2.310 \\ \hline .010 = \text{tolerance} \end{array}$$

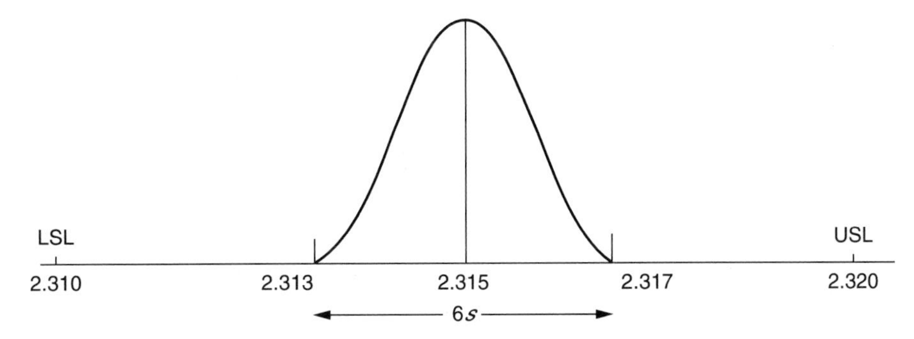

FIGURE 6.26
Process 1: a good process. Distribution is well within the specification limits.

2. Now calculate the PCR and Cpk: the process is on target, so either USL or LSL could be used for C_{pk}

$$\text{Cp} = \frac{\text{tolerance}}{6s} \qquad \text{PCR} = \frac{6s}{\text{tolerance}} \qquad \text{Cpk} = \frac{\text{USL} - \bar{\bar{x}}}{3s}$$

$$\text{Cp} = \frac{.010}{.004} \qquad = \frac{.004}{.010} \qquad = \frac{2.320 - 2.315}{.002}$$

$$= 2.5 \qquad = .40 \qquad = 2.5$$

The PCR in Example 6.12 indicates that the process uses .4, or 40%, of the tolerance. The Cp and Cpk also indicate that the process uses 40% of the tolerance (1/Cpk = 1/2.5 = .4). The calculations for Cp and Cpk are the same because the process is centered. This is classified as an A process: The PCR is less than 50%, and the Cpk is greater than 2.0.

This process classification system, the A, B, C, and D classification, is used by the Saginaw Division of General Motors and is illustrated in Table B.11 in Appendix B.

Process 2, shown in Figure 6.27, is a *poor* process because it crowds the LSL and overlaps the USL with its $6s$ spread.

EXAMPLE 6.13 Using the information in Figure 6.27, find the PCR, Cp, and Cpk.

Solution

Using the sketch of the distribution, we can determine $6s$, $3s$, and the tolerance:

$$\begin{array}{r} 5.1210 \\ -5.1020 \\ \hline .0190 = 6s \\ .0095 = 3s \end{array} \qquad \begin{array}{r} 5.120 \\ -5.100 \\ \hline .020 = \text{tolerance} \end{array}$$

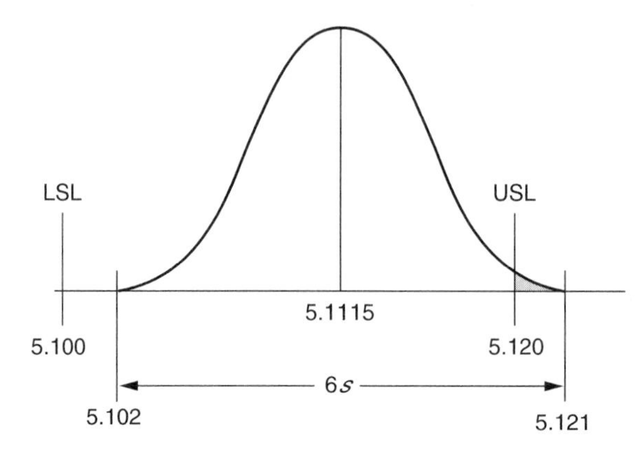

FIGURE 6.27
Process 2: a poor process. The distribution overlaps and crowds the specification limits.

Calculate the PCR and Cp:

$$\text{PCR} = \frac{6s}{\text{tolerance}} = \frac{.019}{.020} = .95 \qquad \text{Cp} = \frac{1}{\text{PCR}} = \frac{1}{.95} = 1.05$$

Next, calculate the Cpk. The USL is the closest specification limit to $\bar{\bar{x}}$, so

$$\text{Cpk} = \frac{\text{USL} - \bar{\bar{x}}}{3s} = \frac{5.120 - 5.1115}{.0095} = \frac{.0085}{.0095} = .895$$

The PCR in Example 6.13 indicates that the process uses .95, or 95%, of the tolerance; because the distribution is off center, however, the Cpk indicates that the process uses 1.12, or 112%, of the tolerance (1/Cpk = 1/.895) on the "worst" side. The PCR shows that if the process were centered, it would use 95% of the tolerance. The Cpk shows the result of the existing, noncentered distribution: 112% of the tolerance is used (the distribution overlaps the specifications). This is a D process because the process capability is more than 91% and Cpk is less than 1.10.

Process 3, illustrated in Figure 6.28, shows the discrepancy between the PCR and the Cpk values.

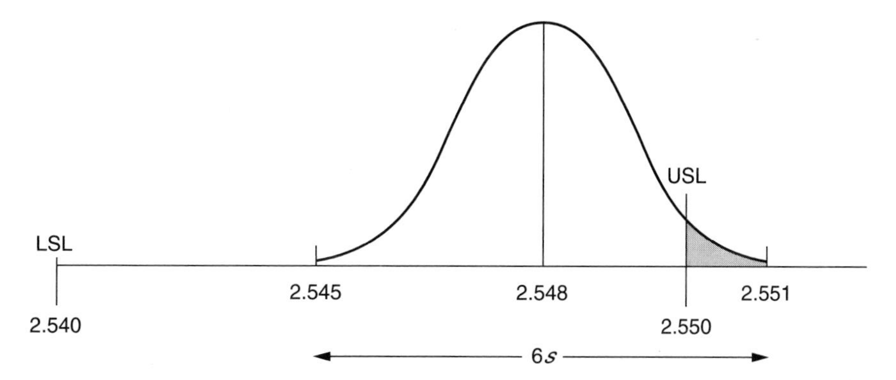

FIGURE 6.28
Process 3: a fair process, but off center.

EXAMPLE 6.14 Determine the PCR, Cp, and Cpk for Figure 6.28.

Solution

Determine 6*s*, 3*s*, and the tolerance from the sketch of the distribution:

$$\begin{array}{r} 2.551 \\ -2.545 \\ \hline .006 = 6s \\ .003 = 3s \end{array} \qquad \begin{array}{r} 2.550 \\ -2.540 \\ \hline .010 = \text{tolerance} \end{array}$$

Calculate the PCR and Cp:

$$\text{PCR} = \frac{6s}{\text{tolerance}} = \frac{.006}{.010} = .60 \qquad \text{Cp} = \frac{1}{\text{PCR}} = \frac{1}{.60} = 1.67$$

Sixty percent of the tolerance is used by the distribution of product measurements. To calculate the Cpk, notice that $\bar{\bar{x}}$ is closest to the USL.

$$\text{Cpk} = \frac{\text{USL} - \bar{\bar{x}}}{3s} = \frac{2.550 - 2.548}{.003} = \frac{.002}{.003} = .67$$

This is classified as a B process because the PCR indicates a capability between 51% and 70% of the tolerance. However, it may also be classified a D process because its Cpk is less than 1.10 and because 150% of its tolerance [(1/Cpk) × 100 = 150%] is used by the distribution on the worst side.

From the three illustrations we can conclude that a good process has a process capability of 50% or less, in which the 6*s* spread of variation is less than half the tolerance. A poor process has a process capability of 91% or more, which means that the data we calculated on the $\bar{x}$ and *R* chart in Figure 6.18 came from a process with a poor capability.

In general, the PCR shows what fraction of the tolerance is used by the product distribution. Because the product variability should be totally within tolerance, the numerator of the PCR formula should be less than the tolerance in the denominator and, ideally, less than half of it.

The Cpk formula uses the worst half of the product distribution: the one with $\bar{\bar{x}}$ closest to the specification limit. The formula reverses the tolerance and measurement variation in the fraction, so the reciprocal of the Cpk value gives the fraction or percentage of the tolerance used by the distribution on that worst side. The Cp and Cpk are often given together to signify the process potential, Cp, and the actual, Cpk, due to being off target.

A nice feature of using the PCR, Cp, and Cpk as indicators of the process capability is that they are basically independent of the shape of the distribution of measurements. Any distribution of measurements will have virtually all the measurements within three standard deviations of the mean, so all three will give a realistic comparison of the true measurement spread and the product tolerance. In contrast, the use of the *z* formula and the normal-curve analysis requires that the distribution of measurements have a normal pattern.

One additional comment on Cp and Cpk. The A, B, C, and D classification system is used at the Saginaw Division of General Motors and perhaps at other divisions as well. Every company will set its own standards for Cp and Cpk. The most common standard is Cpk = 1.33. Many quality consultants, including Juran, consider that adequate for the job. The Cp standard for a company often depends on how it translates to trouble. For example, a Cp = 1.2 uses 83% of the tolerance and the percentage defective is .03%. When that is translated to 318 parts per million, the 318 defective units may be considered excessive.

SIX-SIGMA QUALITY

There are some companies, such as Motorola, who are striving for $C_p = 2$. When that value is achieved for a process it is referred to as a six-sigma process or that it has six-sigma quality. Table 6.4 indicates that when $C_p = 2$, the *z* value for the specification limits is 6 and represents six standard deviations, (6σ), from the mean. The distribution of measurements has a spread of $\pm 3\sigma$ and the tolerance has a spread of $\pm 6\sigma$, so the distribution uses half of the tolerance. Given that the centers of many distributions gradually change as tools wear, a six-sigma process allows this change to occur in a controlled fashion. Suppose that the measurements in a process gradually decrease as the tools wear. When tools are sharpened or replaced, the set-up can be at a point where $\bar{\bar{x}}$ is larger than the target value. The specific point would be governed by the C_{pk} value. If 3ppm defective is acceptable, the $\bar{\bar{x}}$ value would be set so that $C_{pk} = 1.5$. Here the worst half of the distribution would have about 3 ppm defective and the other half would have 0 ppm defective. If less than 1 ppm defective is desired, use a setting that gives an initial C_{pk} = 1.63. As the tools wear, the measurements will gradually decrease to the target (mid-specification) value and the C_{pk} will increase to 2.0. As the measurements continue to decrease and $\bar{\bar{x}}$ becomes smaller than the target value, the C_{pk} will decrease too. When the C_{pk} again matches the set-up C_{pk}, the process is stopped, tools sharpened or replaced, and it is re-set at the previous set-up point. If the process has measurements that increase as tools wear, the set-up is at a point smaller than the target and the stopping point will be above the target, again governed by the desired C_{pk} value.

EXAMPLE 6.15 Twenty-five samples with $n = 5$ are shown in Figure 6.29. Create a control chart and analyze the capability of the process by completing the following steps:

1. Assume that this is the first chart for the process and set the scale using the tolerance (method 1).
2. Chart the points and draw the graphs.
3. Calculate the center lines and the control limit lines on the calculation sheet and draw the lines.
4. Interpret the charts.
5. Make the histogram for all the data.
6. Calculate the process capability: the PCR, the Cpk, and the percentage out of specification.
7. Check the results with Figures 6.30 to 6.33.

Solution

First, set the scale:

$$\text{Tolerance} = .755 - .745 = .01$$

There are 15 lines available:

$$15\overline{).0100} \quad .0006+$$

Use .001 units per line for the range chart:

$$\frac{3T}{n \times \#\text{spaces}} = \frac{3(.01)}{5(21)}$$
$$= .00028$$

Use .0005 units per line on the $\bar{x}$ chart. (.0003 or .0004 units per line would also work.)

Second, make the control chart, plot the points, and draw the broken-line graphs on Figure 6.29. Check your results with Figure 6.30. Third, calculate the center lines and the control limit lines using the formulas on Figure 6.30. Compare your results with Figure 6.31.

$\bar{R} = \frac{\Sigma R}{k} =$ ——— = _____

$\bar{\bar{x}} = \frac{\Sigma \bar{x}}{k} =$ ——— = _____

Specification: .745 to .755

$UCL_R = D_4 \times \bar{R} =$ ____ × ____ = ____

$A_2 \times \bar{R} =$ ____ × ____ = ☐

$UCL_{\bar{x}} = \bar{\bar{x}} + A_2 \bar{R} =$ ____ + ☐ = ____

$LCL_{\bar{x}} = \bar{\bar{x}} - A_2 \bar{R} =$ ____ − ☐ = ____

Constants

n	A_2	$\tilde{A}_2$	D_4	d_2
3	1.023	1.19	2.574	1.693
4	0.729	—	2.282	2.059
5	0.577	.69	2.114	2.326

Chart of Averages ($\bar{x}$)

Chart of Ranges (R)

Sample Number		1	2	3	4	5	6	7	8	9	10	11	12	13
Sample Measurements	1	.751	.750	.749	.748	.753	.755	.754	.748	.751	.751	.752	.750	.748
	2	.747	.748	.749	.749	.749	.752	.751	.753	.750	.753	.752	.749	.749
	3	.752	.749	.752	.749	.752	.753	.752	.749	.751	.751	.751	.749	.751
	4	.750	.750	.750	.751	.751	.744	.750	.748	.752	.752	.751	.748	.747
	5	.751	.752	.748	.748	.751	.749	.750	.748	.751	.751	.751	.751	.750
Sum														
Average, $\bar{x}$		.7502	.7498	.7496	.7490	.7512	.7506	.7514	.7492	.7510	.7516	.7514	.7494	.7490
Range, R		.005	.004	.004	.003	.004	.011	.004	.005	.002	.002	.001	.003	.004

Sample Number		14	15	16	17	18	19	20	21	22	23	24	25
Sample Measurements	1	.749	.752	.752	.756	.749	.752	.754	.750	.750	.750	.754	.749
	2	.750	.751	.750	.754	.749	.750	.753	.750	.752	.751	.745	.749
	3	.750	.751	.750	.752	.750	.753	.750	.750	.751	.749	.747	.749
	4	.751	.752	.748	.744	.751	.750	.750	.750	.753	.748	.746	.749
	5	.751	.751	.750	.747	.749	.754	.751	.748	.750	.748	.755	.750
Sum													
Average, $\bar{x}$		.7502	.7514	.7500	.7506	.7496	.7518	.7516	.7496	.7512	.7492	.7494	.7492
Range, R		.002	.001	.004	.012	.002	.004	.004	.002	.003	.003	.010	.001

FIGURE 6.29
Set the scales, plot the points, and draw the graphs. Calculate averages and control limits and check your results with Figure 6.30.

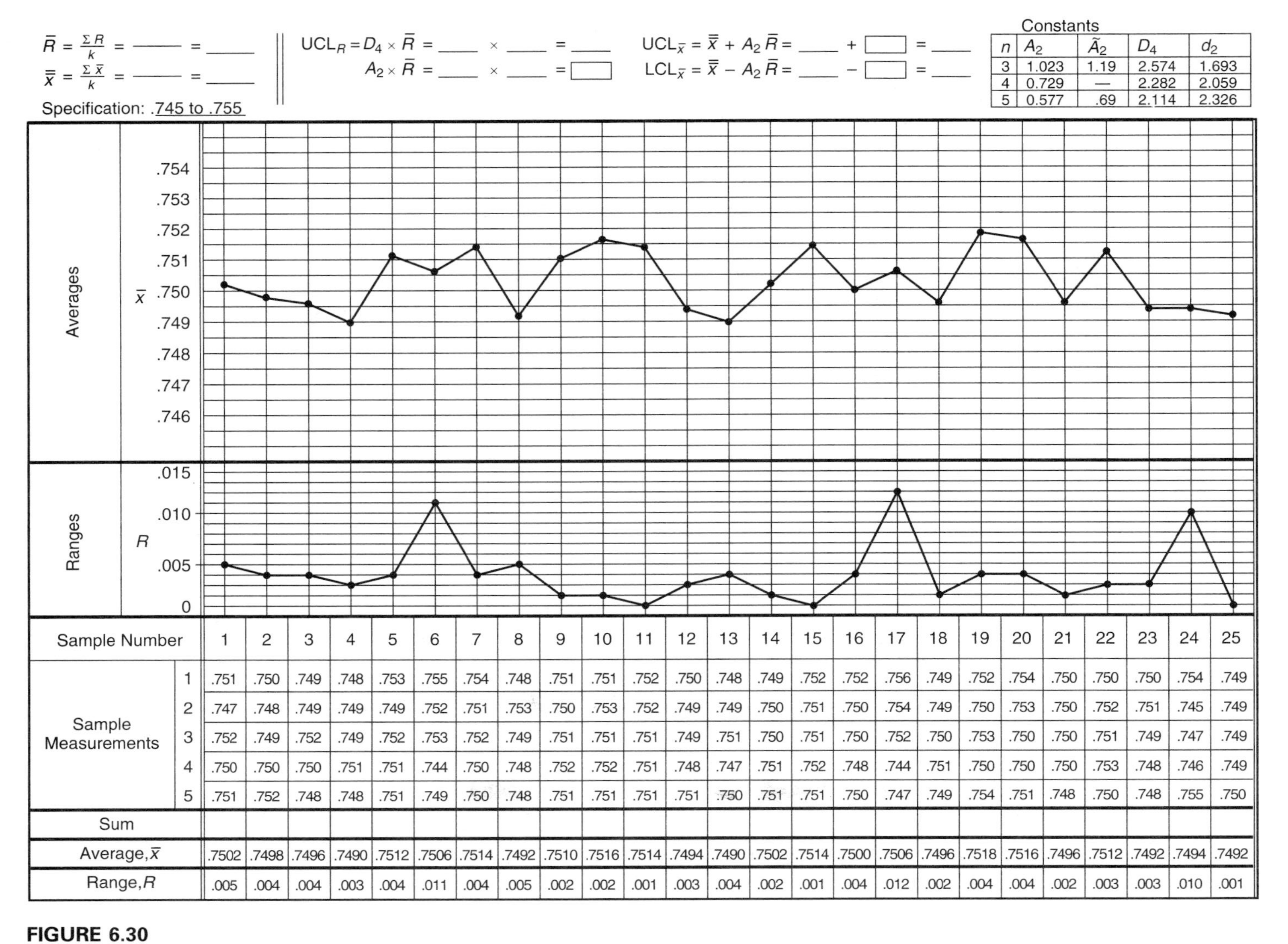

n	A_2	$\tilde{A}_2$	D_4	d_2
3	1.023	1.19	2.574	1.693
4	0.729	—	2.282	2.059
5	0.577	.69	2.114	2.326

Sample Number		1	2	3	4	5	6	7	8	9	10	11	12	13	14	15	16	17	18	19	20	21	22	23	24	25
Sample Measurements	1	.751	.750	.749	.748	.753	.755	.754	.748	.751	.751	.752	.750	.748	.749	.752	.752	.756	.749	.752	.754	.750	.750	.750	.754	.749
	2	.747	.748	.749	.749	.749	.752	.751	.753	.750	.753	.752	.749	.749	.750	.751	.750	.754	.749	.750	.753	.750	.752	.751	.745	.749
	3	.752	.749	.752	.749	.752	.753	.752	.749	.751	.751	.751	.749	.751	.750	.751	.750	.752	.750	.753	.750	.750	.751	.749	.747	.749
	4	.750	.750	.750	.751	.751	.744	.750	.748	.752	.752	.751	.748	.747	.751	.752	.748	.744	.751	.750	.750	.750	.753	.748	.746	.749
	5	.751	.752	.748	.748	.751	.749	.750	.748	.751	.751	.751	.751	.750	.751	.751	.750	.747	.749	.754	.751	.748	.750	.748	.755	.750
Sum																										
Average, $\bar{x}$		.7502	.7498	.7496	.7490	.7512	.7506	.7514	.7492	.7510	.7516	.7514	.7494	.7490	.7502	.7514	.7500	.7506	.7496	.7518	.7516	.7496	.7512	.7492	.7494	.7492
Range, R		.005	.004	.004	.003	.004	.011	.004	.005	.002	.002	.001	.003	.004	.002	.001	.004	.012	.002	.004	.004	.002	.003	.003	.010	.001

FIGURE 6.30
Calculate middle lines and control limits.

$\bar{R} = \frac{\Sigma R}{k} = \frac{.100}{25} = .004$

$\bar{\bar{x}} = \frac{\Sigma \bar{x}}{k} = \frac{18.7572}{25} = .7503$

Specification: .745 to .755

$UCL_R = D_4 \times \bar{R} = 2.114 \times .004 = .0085$

$A_2 \times \bar{R} = .577 \times .004 = .0023$

$UCL_{\bar{x}} = \bar{\bar{x}} + A_2 \bar{R} = .7503 + .0023 = .7526$

$LCL_{\bar{x}} = \bar{\bar{x}} - A_2 \bar{R} = .7503 - .0023 = .7480$

Constants

n	A_2	$\tilde{A}_2$	D_4	d_2
3	1.023	1.19	2.574	1.693
4	0.729	—	2.282	2.059
5	0.577	.69	2.114	2.326

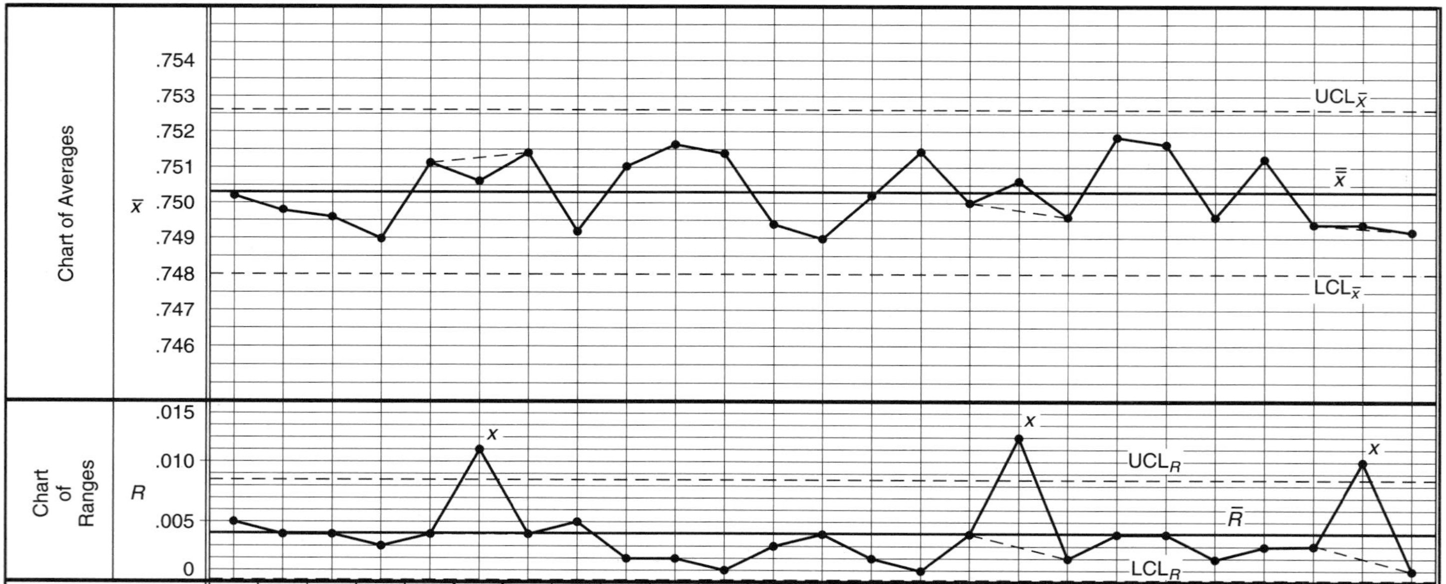

Sample Number		1	2	3	4	5	6	7	8	9	10	11	12	13	14	15	16	17	18	19	20	21	22	23	24	25
Sample Measurements	1	.751	.750	.749	.748	.753	.755	.754	.748	.751	.751	.752	.750	.748	.749	.752	.752	.756	.749	.752	.754	.750	.750	.750	.754	.749
	2	.747	.748	.749	.749	.749	.752	.751	.753	.750	.753	.752	.749	.749	.750	.751	.750	.754	.749	.750	.753	.750	.752	.751	.745	.749
	3	.752	.749	.752	.749	.752	.753	.752	.749	.751	.751	.751	.749	.751	.750	.751	.750	.752	.750	.753	.750	.750	.751	.749	.747	.749
	4	.750	.750	.750	.751	.751	.744	.750	.748	.752	.752	.751	.748	.747	.751	.752	.748	.744	.751	.750	.750	.750	.753	.748	.746	.749
	5	.751	.752	.748	.748	.751	.749	.750	.748	.751	.751	.751	.751	.750	.751	.751	.750	.747	.749	.754	.751	.748	.750	.748	.755	.750
Sum																										
Average, $\bar{x}$		.7502	.7498	.7496	.7490	.7512	.7506	.7514	.7492	.7510	.7516	.7514	.7494	.7490	.7502	.7514	.7500	.7506	.7496	.7518	.7516	.7496	.7512	.7492	.7494	.7492
Range, R		.005	.004	.004	.003	.004	.011	.004	.005	.002	.002	.001	.003	.004	.002	.001	.004	.012	.002	.004	.004	.002	.003	.003	.010	.001

FIGURE 6.31
Three samples are eliminated from the set of data. Recalculate averages and control limits.

Fourth, interpret the control chart for statistical control. Mark any points that are out of control with an X.

The ranges of three samples are out of control. Assume that the out-of-control problems have been corrected, eliminate that data from the data set, and recalculate the center lines and the control limits. For the follow-up process capability, the process must be in statistical control, so only in-control points should be used. This is illustrated on Figure 6.32.

To calculate the new value of $\overline{R}$, deduct .033 (.011 + .012 + .010) from the previous ΣR:

.100	Previous ΣR
−.033	Excluded R values
.067	The new ΣR

After eliminating the out-of-control points, $k = 22$:

$$\text{New } \overline{R} = \frac{.067}{22}$$

$$\overline{R} = .003$$

Calculate the new $\overline{\overline{x}}$ value. Deduct 2.2506 (.7506 + .7506 + .7494) from 18.7572:

18.7572	Previous $\Sigma\overline{x}$
−2.2506	Excluded $\overline{x}$ values
16.5066	The new $\Sigma\overline{x}$

As above, $k = 22$:

$$\text{New } \overline{\overline{x}} = \frac{15.6066}{22}$$

$$\overline{\overline{x}} = .7503$$

Calculate the new control limits:

$$\begin{aligned} \text{UCL}_R &= D_4\overline{R} \\ &= 2.114 \times .003 \\ &= .0063 \end{aligned}$$

$$\begin{aligned} \text{UCL}_{\overline{x}} &= \overline{\overline{x}} + A_2\overline{R} \\ &= .7503 + (.577 \times .003) \\ &= .7503 + .0017 \\ &= .7520 \end{aligned}$$

$$\begin{aligned} \text{LCL}_{\overline{x}} &= \overline{\overline{x}} - A_2\overline{R} \\ &= .7503 - .0017 \\ &= .7486 \end{aligned}$$

Recheck for control on the $\overline{x}$ and R chart.

$\bar{R} = \frac{\Sigma R}{k} = \frac{.067}{22} = .003$

$\bar{\bar{x}} = \frac{\Sigma \bar{x}}{k} = \frac{16.5066}{22} = .7503$

Specification: .745 to .755

$UCL_R = D_4 \times \bar{R} = 2.114 \times .003 = .0063$

$A_2 \times \bar{R} = .577 \times .003 = .0017$

$UCL_{\bar{x}} = \bar{\bar{x}} + A_2 \bar{R} = .7503 + .0017 = .7520$

$LCL_{\bar{x}} = \bar{\bar{x}} - A_2 \bar{R} = .7503 - .0017 = .7486$

Constants

n	A_2	$\tilde{A}_2$	D_4	d_2
3	1.023	1.19	2.574	1.693
4	0.729	—	2.282	2.059
5	0.577	.69	2.114	2.326

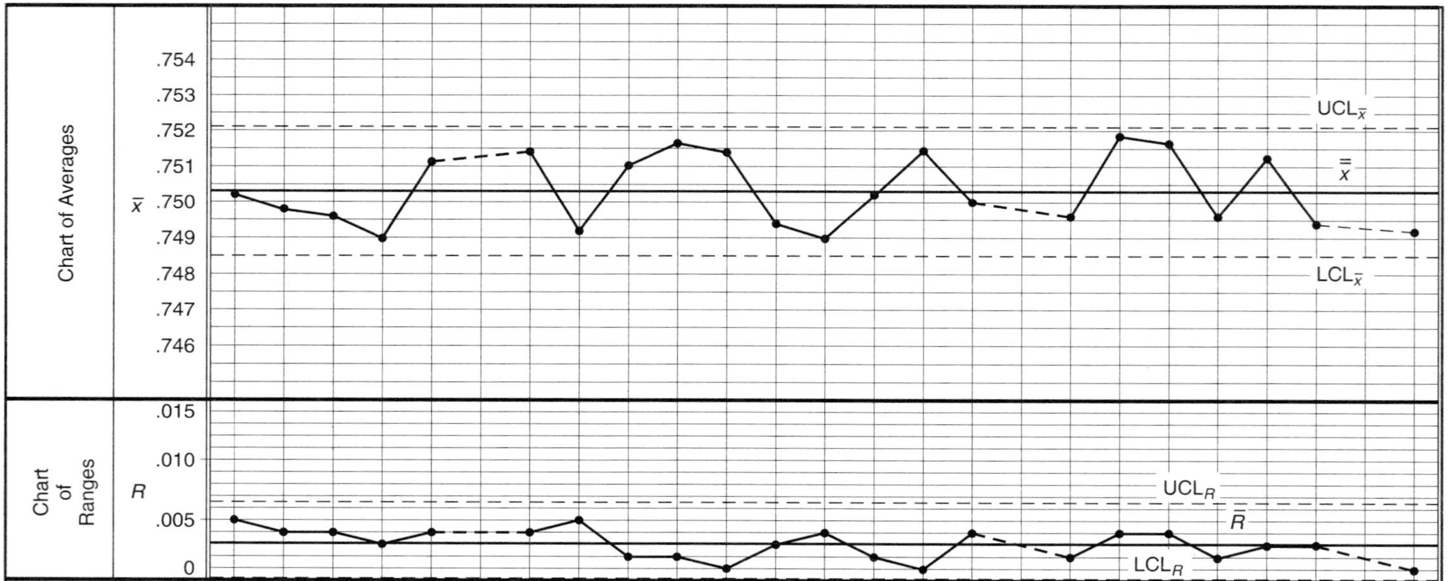

Chart of Averages: $\bar{x}$ axis .754, .753, .752, .751, .750, .749, .748, .747, .746; lines $UCL_{\bar{x}}$, $\bar{\bar{x}}$, $LCL_{\bar{x}}$

Chart of Ranges: R axis .015, .010, .005, 0; lines UCL_R, $\bar{R}$, LCL_R

Sample Number	1	2	3	4	5	6	7	8	9	10	11	12	13	14	15	16	17	18	19	20	21	22	23	24	25
Sample Measurements 1	.751	.750	.749	.748	.753	.755	.754	.748	.751	.751	.752	.750	.748	.749	.752	.752	.756	.749	.752	.754	.750	.750	.750	.754	.749
2	.747	.748	.749	.749	.749	.752	.751	.753	.750	.753	.752	.749	.749	.750	.751	.750	.754	.749	.750	.753	.750	.752	.751	.745	.749
3	.752	.749	.752	.749	.752	.753	.752	.749	.751	.751	.751	.749	.751	.750	.751	.750	.752	.750	.753	.750	.750	.751	.749	.747	.749
4	.750	.750	.750	.751	.751	.744	.750	.748	.752	.752	.751	.748	.747	.751	.752	.748	.744	.751	.750	.750	.750	.753	.748	.746	.749
5	.751	.752	.748	.748	.751	.749	.750	.748	.751	.751	.751	.751	.750	.751	.751	.750	.747	.749	.754	.751	.748	.750	.748	.755	.750
Sum																									
Average, $\bar{x}$	.7502	.7498	.7496	.7490	.7512	.7506	.7514	.7492	.7510	.7516	.7514	.7494	.7490	.7502	.7514	.7500	.7506	.7496	.7518	.7516	.7496	.7512	.7492	.7494	.7492
Range, R	.005	.004	.004	.003	.004	.011	.004	.005	.002	.002	.001	.003	.004	.002	.001	.004	.012	.002	.004	.004	.002	.003	.003	.010	.001

FIGURE 6.32
The control chart with new control limits.

Everything seems fine: There are no points beyond the control limits and no trouble patterns for the points between the control limits on either chart. The process is in statistical control.

The fifth step instructs us to draw the histogram for all the in-control measurements to check for a normal distribution of data. Check the results with Figure 6.33. There appears to be a slight skewness to the right, but the pattern is close enough to normal for a capability check with the normal-curve analysis.

Sixth, determine the process capability:

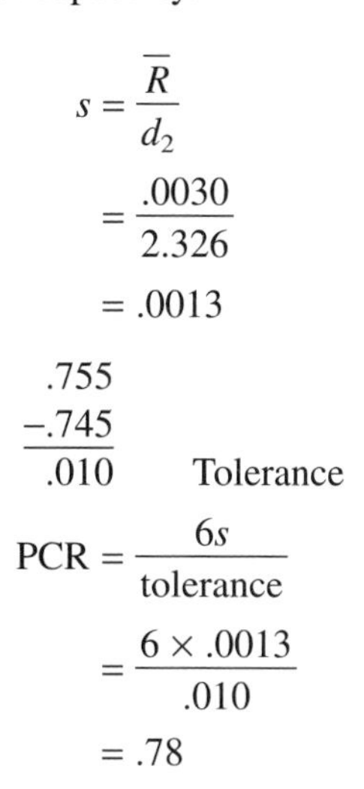

$$s = \frac{\overline{R}}{d_2} = \frac{.0030}{2.326} = .0013$$

$$\begin{array}{r} .755 \\ -.745 \\ \hline .010 \end{array} \quad \text{Tolerance}$$

$$\text{PCR} = \frac{6s}{\text{tolerance}} = \frac{6 \times .0013}{.010} = .78$$

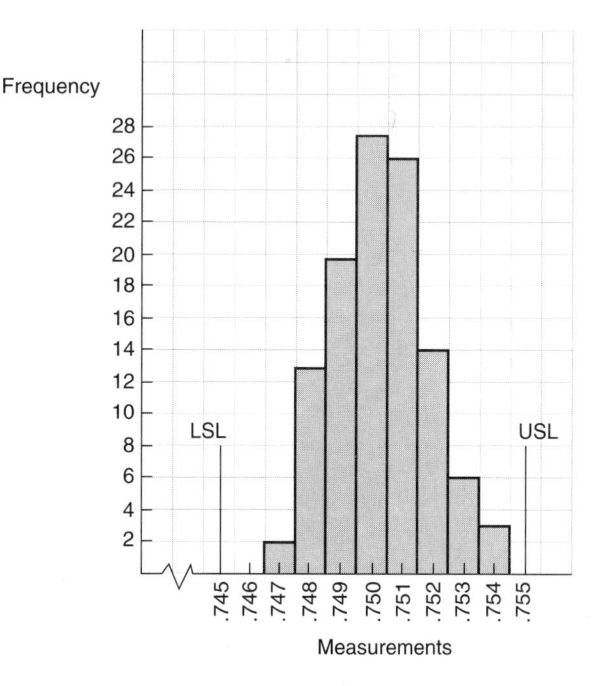

FIGURE 6.33
The tally histogram of the in-control data.

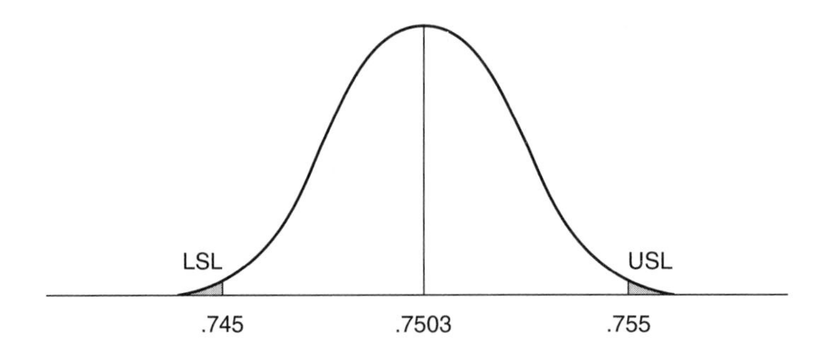

FIGURE 6.34
Find the percentage of product beyond the specification limits.

This shows the process using 78% of the tolerance.

$\bar{\bar{x}}$ is closer to the USL, so

$$\text{Cpk} = \frac{\text{USL} - \bar{\bar{x}}}{3s}$$
$$= \frac{.755 - .7503}{3 \times .0013}$$
$$= \frac{.0047}{.0039}$$
$$= 1.21$$
$$\frac{1}{\text{Cpk}} = \frac{1}{1.21}$$
$$= .83$$

Eighty-three percent of the tolerance (.83) is used by the process on the "short" side.

Both the PCR and the Cpk values classify this as a C process according to the classification chart in Appendix B, Table B.11.

What percentage of the process is out of specification? Substitute the specification limits for x in the z formula, as suggested by the sketch of the normal curve in Figure 6.34. Use Table B.8 in Appendix B to find the area:

$$z = \frac{x - \bar{\bar{x}}}{s} \qquad z = \frac{x - \bar{\bar{x}}}{s}$$
$$= \frac{\text{LSL} - \bar{\bar{x}}}{s} \qquad = \frac{\text{USL} - \bar{\bar{x}}}{s}$$
$$= \frac{.745 - .7503}{.0013} \qquad = \frac{.755 - .7503}{.0013}$$
$$z = -4.08 \qquad z = 3.62$$
$$\text{Area} = .00000 \qquad \text{Area} = .00015$$

A total of .00015, or .015%, is out of specification.

EXERCISES

1. What two types of variation show up on a variables control chart such as the $\overline{x}$ and R chart?
2. What does it mean when a process is said to be in statistical control?
3. What 11 steps are involved in charting a process?
4. The specification limits for a process are given as 2.745 to 2.765. The first seven samples are given in Table 6.5:

TABLE 6.5

Sample	1	2	3	4	5	6	7
	2.754	2.741	2.758	2.743	2.748	2.762	2.749
	2.759	2.753	2.753	2.747	2.743	2.761	2.753
	2.748	2.747	2.747	2.752	2.749	2.758	2.768
	2.759	2.751	2.750	2.749	2.751	2.755	2.763
	2.752	2.754	2.751	2.747	2.745	2.757	2.762

   a. Set up the scales that could be used on the $\overline{x}$ and R chart when there are 15 lines on the R chart and 25 lines on the $\overline{x}$ chart.
   b. What action should be taken due to samples 4, 5, and 7?
5. Given that $\overline{\overline{x}}$ is 4.725, $\overline{R} = .0096$, and $n = 4$, calculate:
   a. s
   b. UCL_R, LCL_R
   c. $UCL_{\overline{x}}$, $LCL_{\overline{x}}$
6. Given that USL = 6.450, $\overline{R} = .042$, $n = 5$, LSL = 6.350, and $\overline{\overline{x}} = 6.391$, find:
   a. PCR, Cp, and the percentage of tolerance used
   b. Cpk and the percentage of tolerance used
   c. The percentage of product out of specification
7. When a control chart is out of control, what type of variation is it revealing?
8. What action procedures should be taken when a control chart shows an out-of-control pattern?
9. Why is it important to keep a process log?
10. When should 100% inspection be instituted in a process that is being charted and when can it be stopped?
11. In a first chart situation, why do we set the chart scales from the first several samples instead of waiting for all 25?
12. When analyzing a control chart, why is it helpful to make note of the final $\Sigma\,\overline{x}$, $\Sigma\,R$, and k?
13. When analyzing a control chart, why is it necessary to analyze the R chart first?
14. Given the set of data on Figure 6.35, make the $\overline{x}$ and R control chart. Include center lines and all control limit lines.
15. For the chart completed in exercise 14, eliminate any out-of-control points and calculate process capability:
   a. Cpk and the percentage of tolerance used
   b. PCR, Cp, and the percentage of tolerance used
   c. The percentage of product out of specification

$\bar{R} = \frac{\Sigma R}{k} = \_\_\_\_ = \_\_\_\_$

$\bar{\bar{x}} = \frac{\Sigma \bar{x}}{k} = \_\_\_\_ = \_\_\_\_$

Specification: .7450 to .7470

$UCL_R = D_4 \times \bar{R} = \_\_\_\_ \times \_\_\_\_ = \_\_\_\_$

$A_2 \times \bar{R} = \_\_\_\_ \times \_\_\_\_ = \square$

$UCL_{\bar{x}} = \bar{\bar{x}} + A_2 \bar{R} = \_\_\_\_ + \square = \_\_\_\_$

$LCL_{\bar{x}} = \bar{\bar{x}} - A_2 \bar{R} = \_\_\_\_ - \square = \_\_\_\_$

Constants

n	A_2	$\tilde{A}_2$	D_4	d_2
3	1.023	1.19	2.574	1.693
4	0.729	—	2.282	2.059
5	0.577	.69	2.114	2.326

Chart of Averages — $\bar{x}$

Chart of Ranges — R

Sample Number	1	2	3	4	5	6	7	8	9	10	11	12	13	14	15	16	17	18	19	20	21	22	23	24	25
Sample Measurements 1	.7465	.7460	.7455	.7465	.7460	.7455	.7455	.7460	.7465	.7455	.7445	.7455	.7450	.7455	.7460	.7470	.7460	.7470	.7460	.7465	.7450	.7460	.7450	.7460	.7460
2	.7450	.7465	.7450	.7470	.7465	.7440	.7465	.7470	.7460	.7455	.7465	.7465	.7460	.7470	.7465	.7460	.7445	.7455	.7450	.7460	.7460	.7465	.7455	.7465	.7465
3	.7455	.7445	.7450	.7460	.7465	.7460	.7465	.7455	.7460	.7455	.7460	.7450	.7450	.7460	.7465	.7475	.7445	.7455	.7465	.7460	.7455	.7455	.7455	.7460	.7475
4																									
5																									
Sum																									
Average, $\bar{x}$																									
Range, R																									

FIGURE 6.35
Make an $\bar{x}$ and R chart.

16. Set up the scales for an $\bar{x}$ and R chart for a process that has specification limits of 1.345 inches to 1.355 inches and the chart has 25 spaces on the $\bar{x}$ section and 15 spaces on the R section.
17. Use copies of the blank control chart in Figure 6.36 and the data in Table 6.6 to make the $\bar{x}$ and R chart with $n = 4$. The data are in order by column (column 1 contains the first three samples):

TABLE 6.6

.302	.304	.300	.310	.321	.299	.312	.299	.289
.306	.306	.312	.320	.310	.280	.304	.310	.286
.299	.297	.288	.310	.310	.290	.288	.290	.299
.301	.309	.292	.312	.323	.299	.300	.285	.314
.306	.312	.289	.318	.288	.298	.280	.289	
.310	.312	.296	.300	.320	.289	.289	.280	
.308	.308	.294	.314	.313	.289	.289	.307	
.304	.304	.313	.300	.307	.280	.302	.292	
.312	.305	.308	.310	.286	.280	.287	.299	
.314	.319	.308	.310	.289	.289	.289	.289	
.298	.310	.300	.322	.309	.289	.288	.280	
.296	.302	.312	.322	.288	.286	.308	.312	

18. What measure on a control chart indicates the position of the data distribution?
19. Do a complete capability analysis for the chart in exercise 17 (PCR, Cp, Cpk, and percentage out of specification). Be sure to eliminate any out-of-control data first. Specification limits are .275 to .325.
20. What are the three basic pieces of information needed for the interpretation of a data distribution?
21. The control chart in Figure 6.37 contains average and control limit lines from Figure 6.8. For each sample calculate $\bar{x}$ and R, plot the points, and draw the broken-line graphs. Analyze the chart point by point as they're charted by indicating.
    a. First indication of trouble at sample # ________.
    b. Action recommended.
    c. Out-of-control at sample # ________.
    d. Action recommended.
22. Name four types of variables control charts.
23. Name four types of attributes control charts.

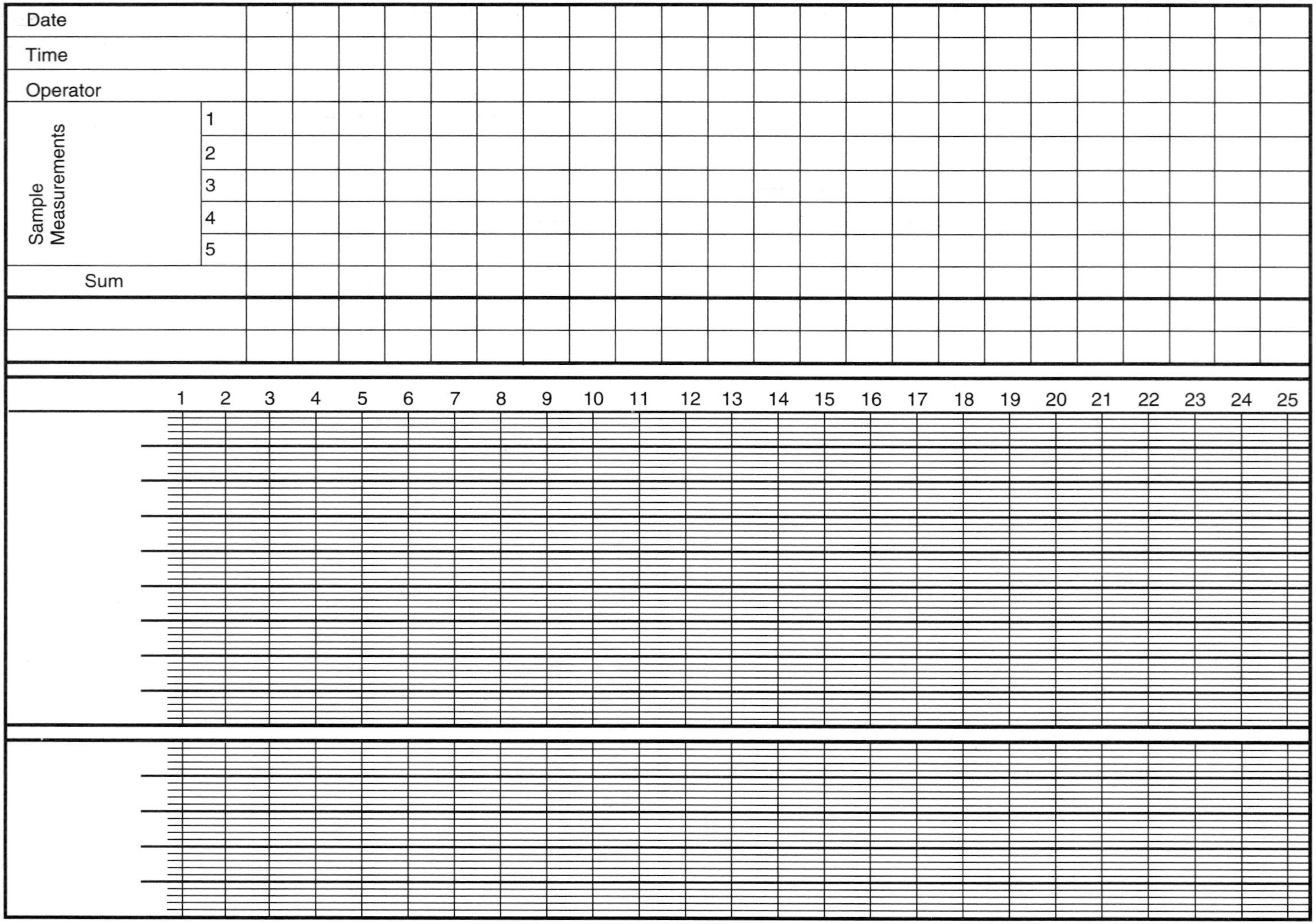

Date																										
Time																										
Operator																										
Sample Measurements	1																									
	2																									
	3																									
	4																									
	5																									
Sum																										

1 2 3 4 5 6 7 8 9 10 11 12 13 14 15 16 17 18 19 20 21 22 23 24 25

FIGURE 6.36
A blank control chart.

Specification: 850 to 950

Sample Number	1	2	3	4	5	6	7	8	9	10	11	12	13	14	15	16	17	18	19	20	21	22	23	24	25
Sample Measurements	883	901	889	882	903	890	892	908	895	906	901	908	909	895	893	909	895	897	912	898	896	912	920	917	884
	897	898	905	913	880	890	902	870	870	890	892	895	904	902	906	907	892	904	896	899	912	909	903	917	889
	885	900	905	880	890	895	895	896	872	891	892	896	906	902	912	904	902	916	902	921	907	913	908	908	912
	900	905	902	900	890	909	896	894	898	900	895	905	892	902	910	903	911	912	906	904	908	915	910	904	919
	899	882	893	871	900	905	902	906	886	905	898	903	907	903	905	898	916	900	913	917	920	908	899	905	898
Average																									
Range																									

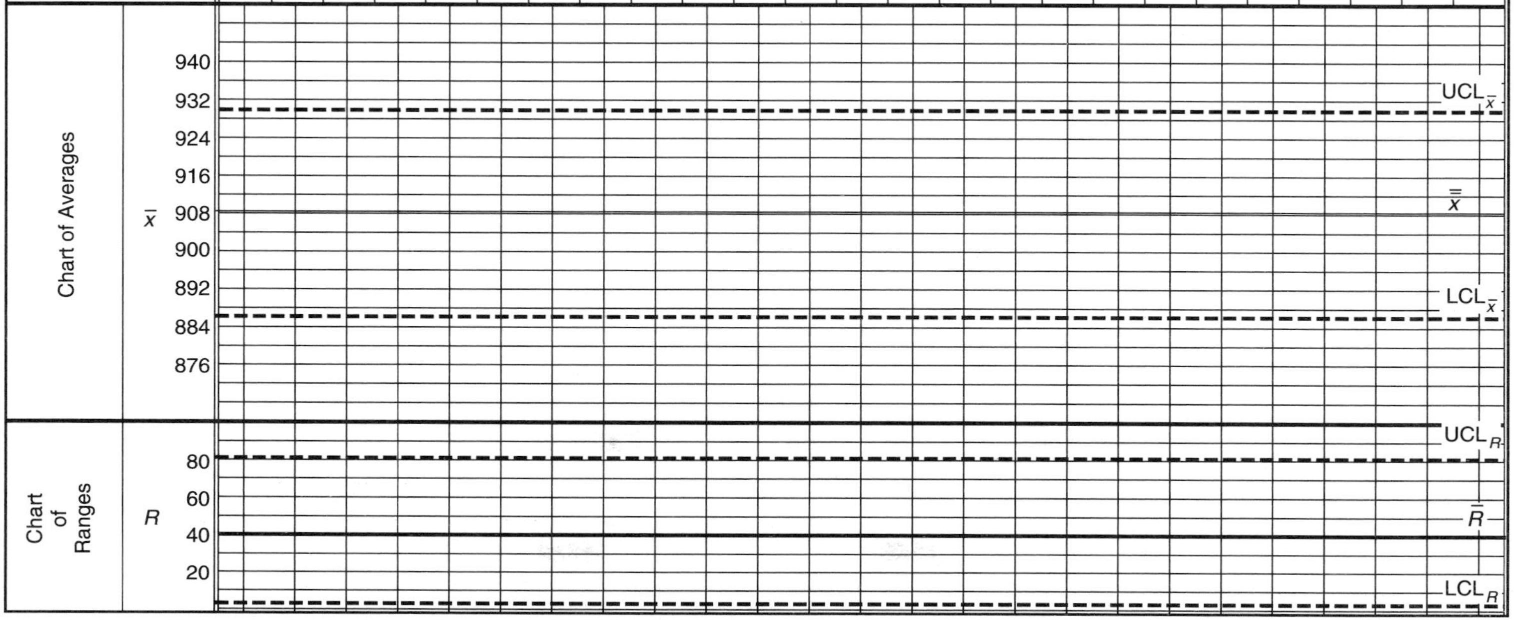

Chart of Averages

$\bar{x}$: 940, 932, 924, 916, 908, 900, 892, 884, 876

$UCL_{\bar{x}}$, $\bar{\bar{x}}$, $LCL_{\bar{x}}$

Chart of Ranges

R: 80, 60, 40, 20

UCL_R, $\bar{R}$, LCL_R

FIGURE 6.37
A continuation chart from Figure 6.8

24. Prepare a random sampling plan like Example 6.2 that tells the specific times that each sample should be taken. Ten samples are to be taken in an 8-hour shift that begins at 3:00 P.M. Each sample consists of five pieces that take 15 minutes to produce. Use the sequence of random numbers (in order columnwise) generated by a statistical calculator given in Table 6.7:

TABLE 6.7

838	676	816	853	305
918	203	358	745	508
470	533	694	961	941
422	410	928	657	104
007	641	645	828	090
287	733	753	058	754
956	357	699	022	163
143	404	079	190	636

25. Which of the charts in Figures 6.38 to 6.40 appears out of control and why?

26. Why is it necessary to eliminate the data from points that are out of control before doing a capability study? Why is s recalculated?

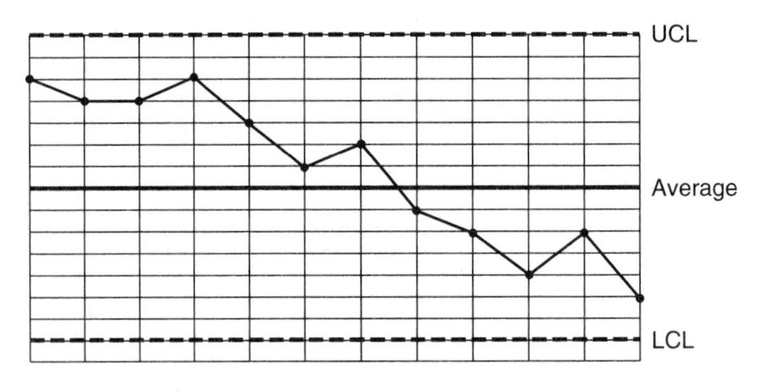

FIGURE 6.38
Exercise 25(**a**).

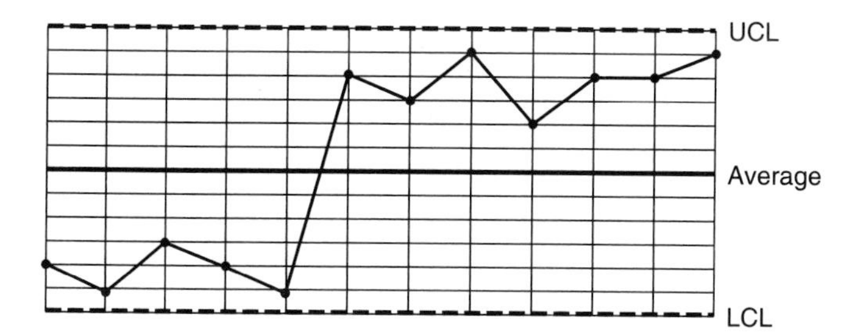

FIGURE 6.39
Exercise 25(**b**).

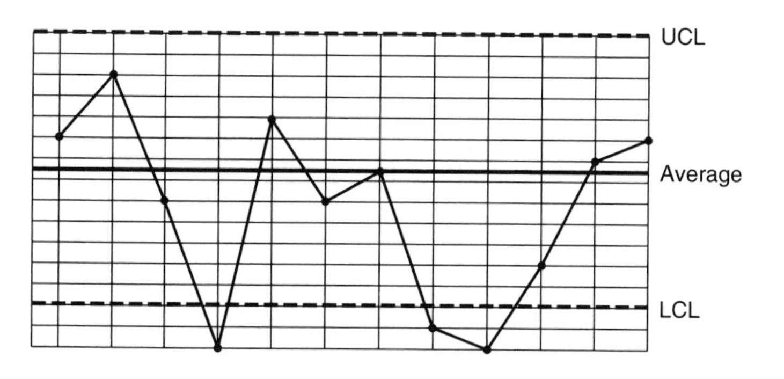

FIGURE 6.40
Exercise 25(**c**).

27. Complete and analyze the control chart in Figure 6.41:

a. Make a tally histogram for the data in the given control chart and analyze its shape. Use a separate column for each measurement.

b. Calculate the average and range for each sample in the control chart.

c. Chart the points and draw the broken-line graphs.

d. Calculate the averages and control limits with the Shewhart formulas.

e. Analyze the shape of the control chart for any out-of-control points. Circle them.

f. Calculate PCR, Cp, and Cpk and interpret them if USL = 3.5 and LSL = 2.5.

g. Calculate the percentage that is out of specification.

28. Use a random-number generator on a calculator to plan random-sample times for 10 samples that are to be taken during an 8-hour shift that begins at 7:00 A.M.

29. Describe three basic methods for determining the spread of a distribution of data from the information on a $\bar{x}$ and R control chart.

30. What are the two basic types of control charts and how do they differ?

Variables Control Chart $\overline{x}$ & R Averages & Ranges

Date																					
Time																					
Operator																					
Sample Measurements	1	2.8	3.2	3.3	3.0	3.5	3.5	3.5	3.4	3.0	3.3	3.1	3.0	3.0	2.9	2.8	2.9	2.6	2.7	2.8	2.8
	2	2.9	3.1	3.6	3.4	3.5	3.4	3.1	3.0	3.4	3.2	3.3	3.4	2.9	2.9	2.8	2.8	2.9	2.7	2.9	2.8
	3	3.1	3.4	3.3	3.6	3.5	3.6	3.3	3.4	3.4	3.0	3.2	3.4	3.1	2.9	3.0	3.0	2.9	3.1	2.6	2.6
	4	2.8	3.1	3.4	3.6	3.5	3.5	3.3	3.4	3.0	3.3	2.8	3.0	3.0	2.9	3.0	2.9	2.8	2.7	2.5	3.0
	5																				
Sum																					
Average $\overline{x}$																					
Range R																					

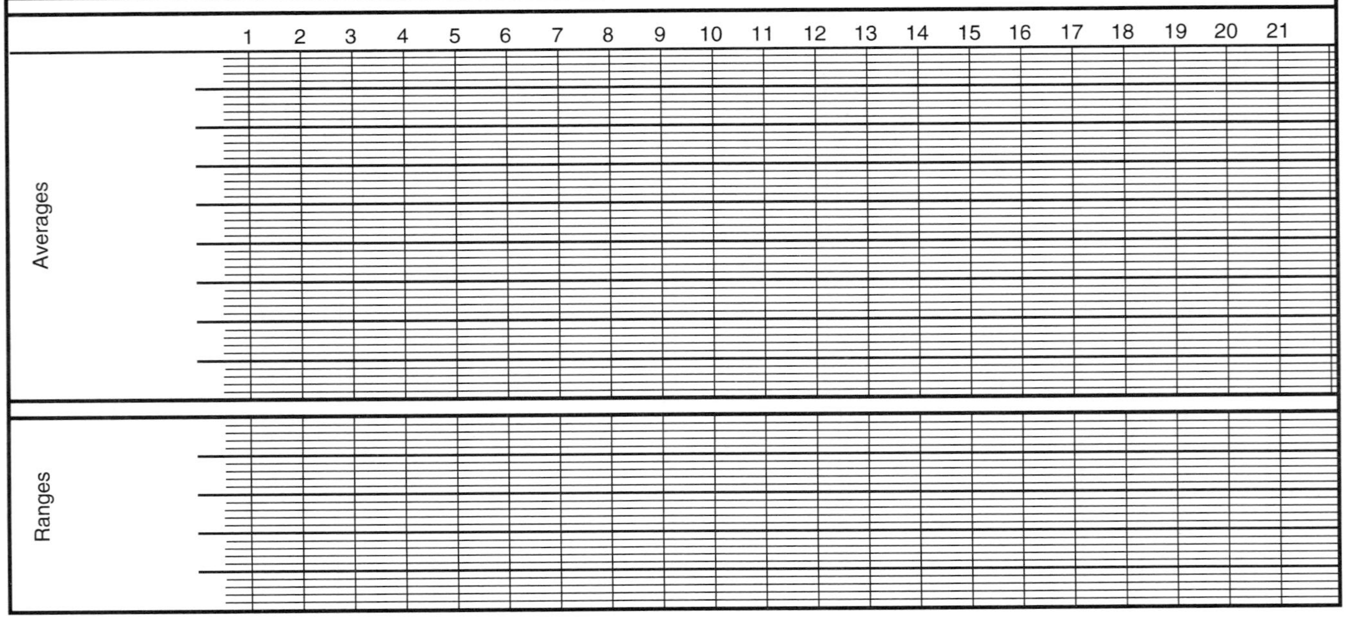

Averages

1 2 3 4 5 6 7 8 9 10 11 12 13 14 15 16 17 18 19 20 21

Ranges

FIGURE 6.41
Complete the $\overline{x}$ and R chart and analyze the results. Exercise 24.

7

ADDITIONAL CONTROL CHARTS FOR VARIABLES

OBJECTIVES

- Make an $\tilde{x}$ and R chart.
- Make an $\bar{x}$ and s chart.
- Chart and interpret coded data.
- Apply $\bar{x}$ and R charts to short runs and small runs.
- Learn the modified $\bar{x}$ and R chart method.
- Know the nominal $\bar{x}$ and R method and the criteria for its use.
- Learn the transformation $\bar{x}$ and R chart method and the criteria for its use.

7.1 THE MEDIAN AND RANGE CHART ($\tilde{x}$ AND R)

When the distribution of measurements is normal, the mean and the median will both be at the center. If there are no extreme data values in the samples (if the range is in control), the median values can be charted as an alternative to the average values. The median chart is not quite as sensitive for interpretation as the averages, but it is preferred in some companies because it is simpler to use.

The $\tilde{x}$ and R Chart Procedure

The procedure for using $\tilde{x}$ and R charts is basically the same as the one for the $\bar{x}$ and R charts:

1. Select a process measurement to track.
2. Eliminate obvious sources of variability before charting.

3. Check the gauges. The gauges must be capable before the charting begins.
4. Devise a sample plan. The sample plan should be a random sample plan similar to the ones shown in Examples 6.1 and 6.2. The one additional consideration for medians is the requirement that for easy figuring, the sample size should be an odd number such as 3, 5, or 7. The median is then the middle value, not the average of the two in the middle.
5. Set up the charts and process log using the format that was introduced for the $\bar{x}$ and R charts.
6. Prepare the tally histogram.
7. Take the samples and chart the points.
8. Calculate the center lines and the control limits. A slight change in the formulas should be noted:

Control Chart Formulas	
$UCL_R = D_4\bar{R}$	$\bar{\tilde{x}} = \frac{\Sigma\tilde{x}}{k}$
$LCL_R = D_3\bar{R}$	$\tilde{A}_2$, D_3, and D_4 are Shewhart
$UCL_{\tilde{x}} = \bar{\tilde{x}} + \tilde{A}_2\bar{R}$	constants from Table B2
$LCL_{\tilde{x}} = \bar{\tilde{x}} - \tilde{A}_2\bar{R}$	For sample sizes $n<7$, $D_3 = 0$

The R chart in the $\tilde{x}$ and R chart is the same as the R chart in the $\bar{x}$ and R chart. The chart interpretation is the same as before. Out-of-control patterns include points beyond the control limits, trends, shifts, and runs.

9. Process capability is calculated as before, but with $\bar{\tilde{x}}$ instead of $\bar{\bar{x}}$.
10. Monitor the process.
11. Continuous improvement.

We will explain this procedure by applying it to an example.

EXAMPLE 7.1 Twenty-five samples are recorded on the control chart in Figure 7.1. Set the scales, chart the points, and calculate the center lines and control limits. Then interpret the charts, and draw the histogram for all *in-control* data. Determine the process capability with the in-control data.

Solution

Calculate the median and range for all the samples in Figure 7.1. Check the results with Figure 7.2. Use the first 10 samples (again, simulating a first-run charting situation) to set the $\tilde{x}$ and R scales:

$$K_R = 59 - 0$$
$$= 59$$

$\bar{R} = \frac{\Sigma R}{k} = \frac{\_\_\_\_}{} = \_\_\_\_$

$\bar{\tilde{x}} = \frac{\Sigma \tilde{x}}{k} = \frac{\_\_\_\_}{25} = \_\_\_\_$

Specification: 850 to 950

$UCL_R = D_4 \times \bar{R} = \_\_\_\_ \times \_\_\_\_ = \_\_\_\_$

$\tilde{A}_2 \times \bar{R} = \_\_\_\_ \times \_\_\_\_ = \square$

$UCL_{\tilde{x}} = \bar{\tilde{x}} + \tilde{A}_2 \bar{R} = \_\_\_\_ + \square = \_\_\_\_$

$LCL_{\tilde{x}} = \bar{\tilde{x}} - \tilde{A}_2 \bar{R} = \_\_\_\_ - \square = \_\_\_\_$

Constants

n	A_2	$\tilde{A}_2$	D_4	d_2
3	1.023	1.19	2.574	1.693
4	0.729	—	2.282	2.059
5	0.577	.691	2.114	2.326

Sample Numbers	1	2	3	4	5	6	7	8	9	10	11	12	13	14	15	16	17	18	19	20	21	22	23	24	25
Sample Measurements	915	900	905	871	940	912	897	920	928	890	895	901	906	932	906	907	885	904	913	882	907	913	910	918	889
	891	905	889	930	890	895	933	908	926	900	925	895	872	902	893	909	942	897	936	941	912	909	885	917	919
	920	911	902	882	909	892	879	894	920	890	908	892	927	932	917	904	916	912	932	917	896	912	908	914	898
	890	862	873	913	915	902	885	896	895	930	933	898	909	895	925	923	892	916	912	934	926	928	903	917	912
	916	898	915	900	895	896	900	906	922	903	896	892	904	902	910	888	911	920	896	894	928	915	926	925	884
Median $\tilde{x}$																									
Range R																									

Chart of Medians $\tilde{x}$

Chart of Ranges R

Figure 7.1
Calculate $\tilde{x}$ and R for each sample. Set the scales for each chart, plot the $\tilde{x}$ and R values, and draw the lines. Calculate center lines and control limits.

$\bar{R} = \frac{\Sigma R}{k} = \frac{926}{25} = 37.04$

$\bar{\tilde{x}} = \frac{\Sigma \tilde{x}}{k} = \frac{22676}{25} = 907.04$

Specification: 850 to 950

$UCL_R = D_4 \times \bar{R} = 2.114 \times 37.04 = 78.3$

$\tilde{A}_2 \times \bar{R} = .691 \times 37.04 = \boxed{25.6}$

$UCL_{\tilde{x}} = \bar{\tilde{x}} + \tilde{A}_2 \bar{R} = 907.04 + \boxed{25.6} = 932.6$

$LCL_{\tilde{x}} = \bar{\tilde{x}} - \tilde{A}_2 \bar{R} = 907.04 - \boxed{25.6} = 881.4$

Constants

n	A_2	$\tilde{A}_2$	D_4	d_2
3	1.023	1.19	2.574	1.693
4	0.729	—	2.282	2.059
5	0.577	.691	2.114	2.326

Sample Numbers	1	2	3	4	5	6	7	8	9	10	11	12	13	14	15	16	17	18	19	20	21	22	23	24	25
Sample Measurements	915	900	905	871	940	912	897	920	928	890	895	901	906	932	906	907	885	904	913	882	907	913	910	918	889
	891	905	889	930	890	895	933	908	926	900	925	895	872	902	893	909	942	897	936	941	912	909	885	917	919
	920	911	902	882	909	892	879	894	920	890	908	892	927	932	917	904	916	912	932	917	896	912	908	914	898
	890	862	873	913	915	902	885	896	895	930	933	898	909	895	925	923	892	916	912	934	926	928	903	917	912
	916	898	915	900	895	896	900	906	922	903	896	892	904	902	910	888	911	920	896	894	928	915	926	925	884
$\tilde{x}$	915	900	902	900	909	896	897	906	922	900	908	895	906	902	910	907	911	912	913	917	912	913	908	917	898
R	30	49	42	59	50	20	54	26	33	40	38	9	55	37	32	35	57	23	40	59	32	19	41	11	35

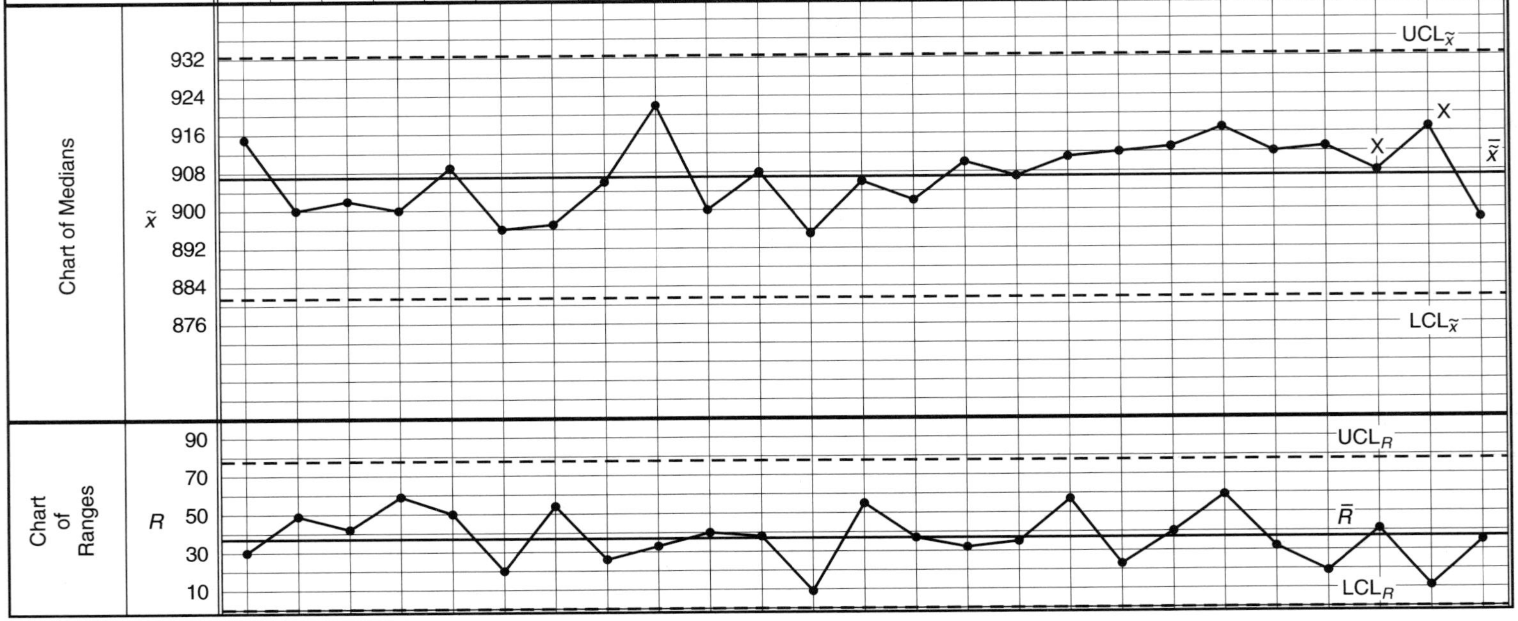

Figure 7.2
Interpret the charts.

Scale from 0 to $2R_R$ (0 to 118). There are nine lines:

$$9\overline{)118}\quad 13+$$

Round to 10. Use 10 units per line:

$$R_{\tilde{x}} = 922 - 896$$
$$= 26$$

Start at the midspecification value of 900 and scale 26 units above and below. There are 21 lines:

$$21\overline{)52}\quad 2+$$

Round up to 4 units per line. (A choice of 3 units per line would make the chart about 5 lines wider and 5 units per line would make the chart about 2 lines narrower.)

Chart the points on Figure 7.1 and check the results with Figure 7.2. When all the data are plotted, calculate the averages. There are 25 samples to average, so $k = 25$. The average of the medians:

$$\bar{\tilde{x}} = \frac{\Sigma\tilde{x}}{k}$$
$$= \frac{\tilde{x}_1 + \tilde{x}_2 + \tilde{x}_3 + \cdots + \tilde{x}_{25}}{25}$$
$$= \frac{915 + 900 + \cdots + 898}{25}$$
$$= \frac{22{,}676}{25}$$
$$= 907.04$$

The average of the ranges:

$$\bar{R} = \frac{\Sigma R}{k}$$
$$= \frac{R_1 + R_2 + R_3 + \cdots + R_{25}}{25}$$
$$= \frac{30 + 49 + \cdots + 35}{25}$$
$$= \frac{926}{25}$$
$$= 37.04$$

Draw a solid line on each chart for $\bar{\tilde{x}}$ and $\bar{R}$.

To continue, calculate the upper and lower control limits for the ranges and medians. Use the calculation sheet on Figure 7.1. It has the same format as the one used with the $\bar{x}$ and R chart. Check your results with the following calculations.

The numerical constants for the control chart are found in Table B.2, Appendix B. The sample size is $n = 5$:

$$D_4 = 2.114$$
$$D_3 = 0 \qquad \tilde{A}_2 \text{ is the Shewhart}$$
$$\tilde{A}_2 = .691 \qquad \text{constant for medians.}$$

The upper control limit for the range is 77.8:

$$\begin{aligned} \text{UCL}_R &= D_4\bar{R} \\ &= 2.114 \times 37.04 \\ &= 78.3 \end{aligned}$$

The lower control limit for the range is 0:

$$\begin{aligned} \text{LCL}_R &= D_3\bar{R} \\ &= 0 \times 37.04 \\ &= 0 \end{aligned}$$

The upper and lower control limits for the medians have the same $\tilde{A}_2\bar{R}$ product in the formulas:

$$\begin{aligned} \text{UCL}_{\tilde{x}} &= \bar{\tilde{x}} + \tilde{A}_2\bar{R} \\ &= 907.04 + (.691 \times 37.04) \\ &= 907.04 + 25.6 \\ &= 932.6 \end{aligned} \qquad \begin{aligned} \text{LCL}_{\tilde{x}} &= \bar{\tilde{x}} - \tilde{A}_2\bar{R} \\ &= 907.04 - 25.6 \\ &= 881.4 \end{aligned}$$

Draw a dashed line on the charts for each control limit. Check your results with Figure 7.2.

Now interpret the charts on Figure 7.2, analyzing the pattern of data points with respect to the center line and control limit lines. Look for points that fall outside the control limits and nonrandom patterns such as runs of seven or more points going up or going down, all above the middle line or all below the middle line. These would be considered out of control. Check your findings with Figure 7.3. The cause of the out-of-control situations must be found and prevented from happening again. Reference to the control chart comments and observations will be helpful in determining the cause(s) for the out-of-control situations.

Always do the R chart first. The R chart must be in statistical control before the median chart can be realistically analyzed.

When all of the out-of-control occurrences have been eliminated, delete their data points from the set of data and recalculate the control limits. Eliminate the out-of-control data from the control chart by marking an X by each point and connecting the good points on either side by a dotted line.

$\bar{R} = \frac{\Sigma R}{k} = \frac{874}{23} = 38$

$\bar{\tilde{x}} = \frac{\Sigma \tilde{x}}{k} = \frac{20851}{23} = 906.6$

Specification: 850 to 950

$UCL_R = D_4 \times \bar{R} = 2.114 \times 38 = 80.3$

$\tilde{A}_2 \times \bar{R} = .691 \times 38 = 26.3$

$UCL_{\tilde{x}} = \bar{\tilde{x}} + \tilde{A}_2 \bar{R} = 906.6 + 26.3 = 932.9$

$LCL_{\tilde{x}} = \bar{\tilde{x}} - \tilde{A}_2 \bar{R} = 906.6 - 26.3 = 880.3$

Constants

n	A_2	$\tilde{A}_2$	D_4	d_2
3	1.023	1.19	2.574	1.693
4	0.729	—	2.282	2.059
5	0.577	.691	2.114	2.326

Sample Numbers	1	2	3	4	5	6	7	8	9	10	11	12	13	14	15	16	17	18	19	20	21	22	23	24	25
Sample Measurements	915	900	905	871	940	912	897	920	928	890	895	901	906	932	906	907	885	904	913	882	907	913	910	918	889
	891	905	889	930	890	895	933	908	926	900	925	895	872	902	893	909	942	897	936	941	912	909	885	917	919
	920	911	902	882	909	892	879	894	920	890	908	892	927	932	917	904	916	912	932	917	896	912	908	914	898
	890	862	873	913	915	902	885	896	895	930	933	898	909	895	925	923	892	916	912	934	926	928	903	917	912
	916	898	915	900	895	896	900	906	922	903	896	892	904	902	910	888	911	920	896	894	928	915	926	925	884
$\tilde{x}$	915	900	902	900	909	896	897	906	922	900	908	895	906	902	910	907	911	912	913	917	912	913	908	917	898
R	30	49	42	59	50	20	54	26	33	40	38	9	55	37	32	35	57	23	40	59	32	19	41	11	35

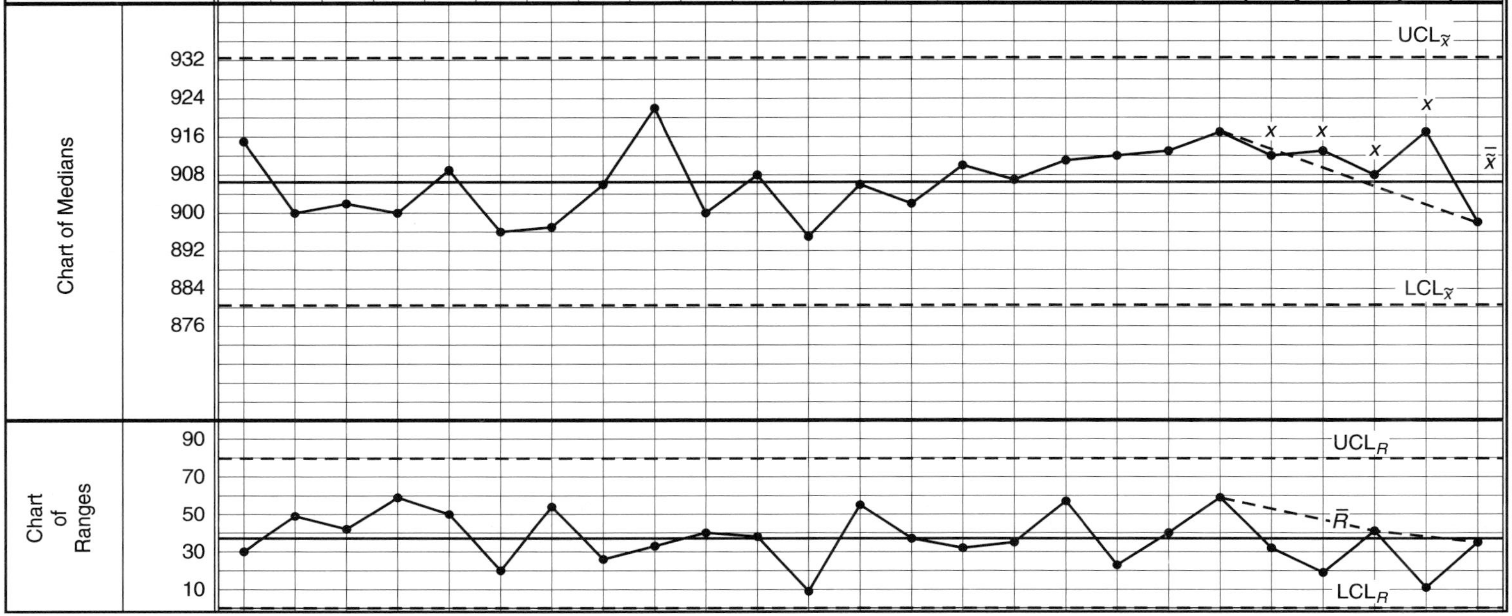

Figure 7.3
The $\tilde{x}$ and R chart is now in statistical control.

When all the data points are between the control limits in a random pattern, the process is in statistical control.

The R chart in Figure 7.2 is in statistical control. The $\tilde{x}$ chart, however, shows the pattern of a shift: seven points in a row on one side of $\bar{\tilde{x}}$. Sample points 23 and 24 are classified out of control and marked with an X, and the points on either side are connected with a dashed line. The data values for those samples are eliminated as well.

To recalculate, subtract the sum of the medians from the eliminated samples from the previous $\Sigma\tilde{x}$:

$$\begin{array}{rl} 908 & \text{Medians of eliminated samples} \\ \underline{+917} & \\ 1{,}825 & \\ 22{,}676 & \text{The original } \Sigma\tilde{x} \\ \underline{-1{,}825} & \\ 20{,}851 & \text{The new } \Sigma\tilde{x} \end{array}$$

$$\text{The new } \bar{\tilde{x}} = \frac{\Sigma\tilde{x}}{k} = \frac{20{,}851}{23} = 906.6$$

Subtract the sum of the eliminated R values from ΣR:

$$\begin{array}{rl} 41 & \\ \underline{+11} & \\ 52 & \text{Eliminated } R \text{ values} \\ 926 & \\ \underline{-52} & \\ 874 & \text{The new } \Sigma R \end{array}$$

$$\text{The new } \overline{R} = \frac{\Sigma R}{k} = \frac{874}{23} = 38.0$$

In both calculations, two samples were eliminated from the data set, so the new k value is 23.

Calculate the new control limits:

$$\text{UCL}_R = D_4\overline{R} = 2.114 \times 38.0 = 80.3$$

$$\begin{aligned} \text{UCL}_{\tilde{x}} &= \bar{\tilde{x}} + \tilde{A}_2\overline{R} \\ &= 906.6 + (.691 \times 38) \\ &= 906.6 + 26.3 \\ &= 932.9 \end{aligned}$$

$$
\begin{aligned}
\text{LCL}_{\tilde{x}} &= \bar{\tilde{x}} - \tilde{A}_2\overline{R} \\
&= 906.6 - 26.3 \\
&= 880.3
\end{aligned}
$$

The new value for $\bar{\tilde{x}}$ lowers the center line just enough to create another shift pattern for samples 15 through 22. This double recalculation could be avoided by analyzing the initial shift pattern carefully to determine when the process changed to cause the shift and eliminating all the data values back to that time of change. The shift may have started at sample 15 or an increasing trend may have started at sample 12, 13, or 14. For the completion of this example we will eliminate samples 21 and 22.

$$
\begin{array}{r}
20851 \\
\underline{-1825} \\
19026
\end{array}
\quad \text{The new } \Sigma\tilde{x}
\qquad\qquad
\begin{array}{r}
874 \\
\underline{-51} \\
823
\end{array}
\quad \text{The new } \Sigma \text{R}
$$

$$
\bar{\tilde{x}} = \frac{19026}{21} \qquad\qquad \overline{R} = \frac{823}{21}
$$

$$
= 906 \text{ The new } \bar{\tilde{x}} \qquad\qquad = 39.2 \text{ The new } \overline{R}
$$

The changes to the control limits would be minimal and the chart is now in statistical control.

Make a histogram with the remaining in-control data values. The data range is eighty (942 to 862). For 105 data values, the G chart in Table B.7, Appendix B, indicates that 10 groups or classes should be used. Eighty divided by 10 is 8, so the group range should be greater than 8 if just 10 classes are wanted. A class range of 9 will be used. The frequency distribution is shown in Table 7.1. The histogram is shown in Figure 7.4.

The next step is to find out if the product being produced is good enough, which is determined by a process capability analysis. The shape of the histogram shows that the data is normally distributed, so the normal-curve formulas can be used for the capability analysis as well as the PCR and Cpk. The histogram crowds the specification

TABLE 7.1
Group boundaries for the data in Figure 6.38

859–867
868–876
877–885
886–894
895–903
904–912
913–921
922–930
931–939
940–948

Figure 7.4
The data appear to be normally distributed.

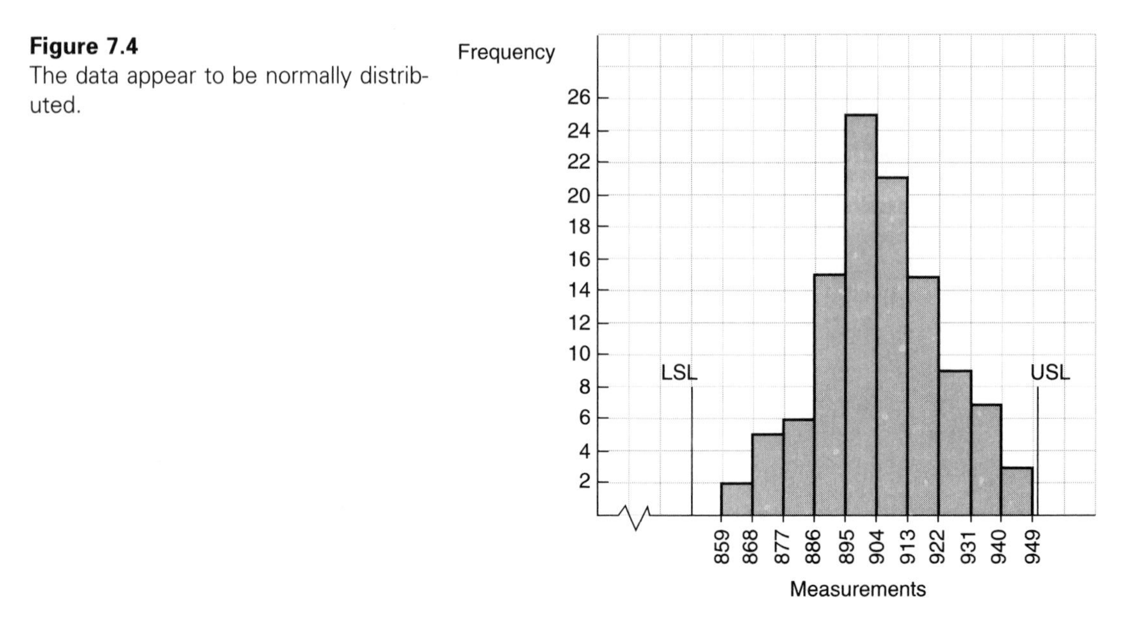

limits, so there is probably some out-of-specification product. An effort should be made to reduce variability.

Calculate the standard deviation for all the measurements. The d_2 value from Table B.2 in Appendix B is 2.326 for a sample size of 5. Divide the average range value by 2.326:

$$s = \frac{\overline{R}}{d_2}$$
$$= \frac{39.2}{2.326}$$
$$= 16.853$$
$$3s = 50.6$$
$$6s = 101.1$$

The tolerance is

$$\begin{array}{r} 950 \\ -850 \\ \hline 100 \end{array}$$

Calculate the PCR:

$$\text{PCR} = \frac{6s}{\text{tolerance}}$$
$$= \frac{101.1}{100}$$
$$= 1.01$$

The process uses 101% of the tolerance.

To calculate the Cpk, notice that $\bar{\tilde{x}}$ is closer to the USL.

$$\text{Cpk} = \frac{\text{USL} - \bar{\tilde{x}}}{3s}$$

$$= \frac{950 - 906}{50.6}$$

$$= .870$$

$$\frac{1}{\text{Cpk}} = \frac{1}{.870}$$

$$= 1.15$$

The process uses 115% of the tolerance on the "short" side.

To find the percentage of product that is out of specification, substitute the specification limits into the z formula:

$$z = \frac{x - \bar{\tilde{x}}}{s} \qquad z = \frac{x - \bar{\tilde{x}}}{s}$$

$$z_1 = \frac{\text{LSL} - \bar{\tilde{x}}}{s} \qquad z_2 = \frac{\text{USL} - \bar{\tilde{x}}}{s}$$

$$= \frac{850 - 906}{16.853} \qquad = \frac{950 - 906}{16.853}$$

$$= -3.32 \qquad = 2.61$$

Find the tail area values in the normal-curve table in Table B.8, Appendix B:

Area 1 = .00045 Area 2 = .0045

The total amount of product that is out of specification, .00045 + .0045, is .00495, or .5%.

The charting procedure, analysis, and interpretation of the $\tilde{x}$ and R chart and the $\bar{x}$ and R chart are primarily the same. The 11 steps outlined for the $\bar{x}$ and R chart follow the same process as the 11 steps outlined for the $\tilde{x}$ and R chart with these minor differences:

1. An *odd*-number sample size should be used when making the $\tilde{x}$ and R chart. The sample size n can be odd *or* even for the $\bar{x}$ and R chart.
2. The numerical constants for the control limits are different: A_2 for the $\bar{x}$ chart and $\tilde{A}_2$ for the $\tilde{x}$ chart.
3. The average values used are slightly different, but they represent the same quantity. The average of the sample means $\bar{\bar{x}}$ and the average of the sample medians $\bar{\tilde{x}}$ both represent the average of all the individual measurements.
4. The sample median $\tilde{x}$ will not necessarily show a "freak" measurement (one measurement that differs drastically from the others) in the sample. The sample mean $\bar{x}$ will show its existence. The R chart shows its existence in both cases.

EXAMPLE 7.2

The target measurement for shaft diameters is .500 inches, and the tolerance is .495 to .505 inches. Twenty-five samples are recorded on the control chart in Figure 7.5. Make the $\tilde{x}$ and R chart and interpret the results. Check your results on steps 1 to 4 with the completed chart in Figure 7.6. The process capability analysis is completed after Figure 7.7 has been assessed; check your results from step 5:

1. Calculate the median and range for each sample.
2. Set the $\tilde{x}$ and R scales and make the charts. Assume that this is the first chart; use the first eight samples.
3. The calculation format is at the top of the control chart. Do the calculations and draw the averages and control limit lines.
4. Analyze the charts (R chart first) and mark any out-of-control points with an X.
5. Remove any out-of-control data from the chart and calculate the process capability with the PCR, Cpk, and the percentage out of specification. Remember to check the data histogram to see if the calculation of the out-of-specification percentage is credible.

Solution

To set the scales, calculate the range of the $\tilde{x}$'s and the R's from the first eight samples:

$$R_{\tilde{x}} = .502 - .499$$
$$= .003$$
$$2R_{\tilde{x}} = 2 \times .003$$
$$= .006$$

Count the number of lines in the $\tilde{x}$ section of the control chart. Divide the number of lines into the .006:

$$21\overline{).006}\quad = .0002+$$

Round up to .0005 units per line. It is easiest to use two lines per thousandth. Start at the midspecification value of .500 at the center of the chart and use two lines for each .001.

Calculate the range of the ranges and set the scale:

$$R_R = .006 - 0$$
$$= .006$$
$$2R_R = 2 \times .006$$
$$= .012$$

Count the number of lines on the R section of the control chart. Divide the number of lines into the .012:

$$10\overline{).012}\quad = .0012$$

Round to .001 units per line. Start at 0 on the bottom line and use one line for each .001. Make the $\tilde{x}$ and R chart on Figure 7.5.

$\bar{R} = \frac{\Sigma R}{k} = \_\_\_\_\_ = \_\_\_\_\_$

$\bar{\tilde{x}} = \frac{\Sigma \tilde{x}}{k} = \_\_\_\_\_ = \_\_\_\_\_$

Specification: .495 to .505

$UCL_R = D_4 \times \bar{R} = \_\_\_\_ \times \_\_\_\_ = \_\_\_\_$

$\tilde{A}_2 \times \bar{R} = \_\_\_\_ \times \_\_\_\_ = \square$

$UCL_{\tilde{x}} = \bar{\tilde{x}} + \tilde{A}_2 \bar{R} = \_\_\_\_ + \square = \_\_\_\_$

$LCL_{\tilde{x}} = \bar{\tilde{x}} - \tilde{A}_2 \bar{R} = \_\_\_\_ - \square = \_\_\_\_$

Constants

n	A_2	$\tilde{A}_2$	D_4	d_2
3	1.023	1.19	2.574	1.693
4	0.729	—	2.282	2.059
5	0.577	.691	2.114	2.326

Sample Numbers	1	2	3	4	5	6	7	8	9	10	11	12	13	14	15	16	17	18	19	20	21	22	23	24	25
Sample Measurements	.502	.500	.496	.501	.501	.499	.502	.500	.500	.501	.498	.501	.502	.494	.498	.500	.499	.501	.504	.499	.497	.502	.501	.502	.502
	.503	.501	.500	.500	.501	.501	.500	.499	.499	.500	.500	.502	.502	.499	.496	.498	.502	.500	.500	.499	.499	.500	.505	.502	.496
	.499	.497	.502	.500	.500	.502	.504	.501	.495	.500	.499	.500	.503	.503	.499	.502	.504	.501	.499	.495	.501	.501	.502	.501	.499
	.498	.499	.499	.500	.501	.504	.503	.497	.498	.500	.501	.501	.501	.496	.500	.500	.502	.501	.500	.498	.500	.500	.501	.500	.498
	.503	.500	.499	.502	.498	.501	.500	.501	.501	.497	.502	.501	.503	.505	.499	.501	.500	.503	.502	.498	.501	.501	.500	.503	.499
$\tilde{x}$																									
R																									

Chart of Medians — $\tilde{x}$

Chart of Ranges — R

Figure 7.5
Make the $\tilde{x}$ and R chart.

$\bar{R} = \frac{\Sigma R}{k} = \frac{.107}{25} = .00428$

$\bar{\tilde{x}} = \frac{\Sigma \tilde{x}}{k} = \frac{12.01}{24} = .50042$

Specification: .495 to .505

$UCL_R = D_4 \times \bar{R} = 2.114 \times .00428 = .00905$

$\tilde{A}_2 \times \bar{R} = .691 \times .004 = .00276$

Recalculate

$\bar{R} = \frac{.096}{24} = .004 \qquad UCL_R = 2.114 \times .004 = .0085$

$UCL_{\tilde{x}} = \bar{\tilde{x}} + \tilde{A}_2 \bar{R} = .50042 + .00276 = .5032$

$LCL_{\tilde{x}} = \bar{\tilde{x}} - \tilde{A}_2 \bar{R} = .50042 - .00276 = .4977$

Constants

n	A_2	$\tilde{A}_2$	D_4	d_2
3	1.023	1.19	2.574	1.693
4	0.729	—	2.282	2.059
5	0.577	.691	2.114	2.326

Sample Numbers	1	2	3	4	5	6	7	8	9	10	11	12	13	14	15	16	17	18	19	20	21	22	23	24	25
Sample Measurements	.502	.500	.496	.501	.501	.499	.502	.500	.500	.501	.498	.501	.502	.494	.498	.500	.499	.501	.504	.499	.497	.502	.501	.502	.502
	.503	.501	.500	.500	.501	.501	.500	.499	.499	.500	.500	.502	.502	.499	.496	.498	.502	.500	.500	.499	.499	.500	.505	.502	.496
	.499	.497	.502	.500	.500	.502	.504	.501	.495	.500	.499	.500	.503	.503	.499	.502	.504	.501	.499	.495	.501	.501	.502	.501	.499
	.498	.499	.499	.500	.501	.504	.503	.497	.498	.500	.501	.501	.501	.496	.500	.500	.502	.501	.500	.498	.500	.500	.501	.500	.498
	.503	.500	.499	.502	.498	.501	.500	.501	.501	.497	.502	.501	.503	.505	.499	.501	.500	.503	.502	.498	.501	.501	.500	.503	.499
$\tilde{x}$	.502	.500	.499	.500	.501	.501	.502	.500	.499	.500	.500	.501	.502	.499	.499	.500	.502	.501	.500	.498	.500	.501	.501	.502	.499
R	.005	.004	.006	.002	.003	.005	.004	.004	.006	.004	.004	.002	.002	.011	.004	.004	.005	.003	.005	.004	.004	.002	.005	.003	.006

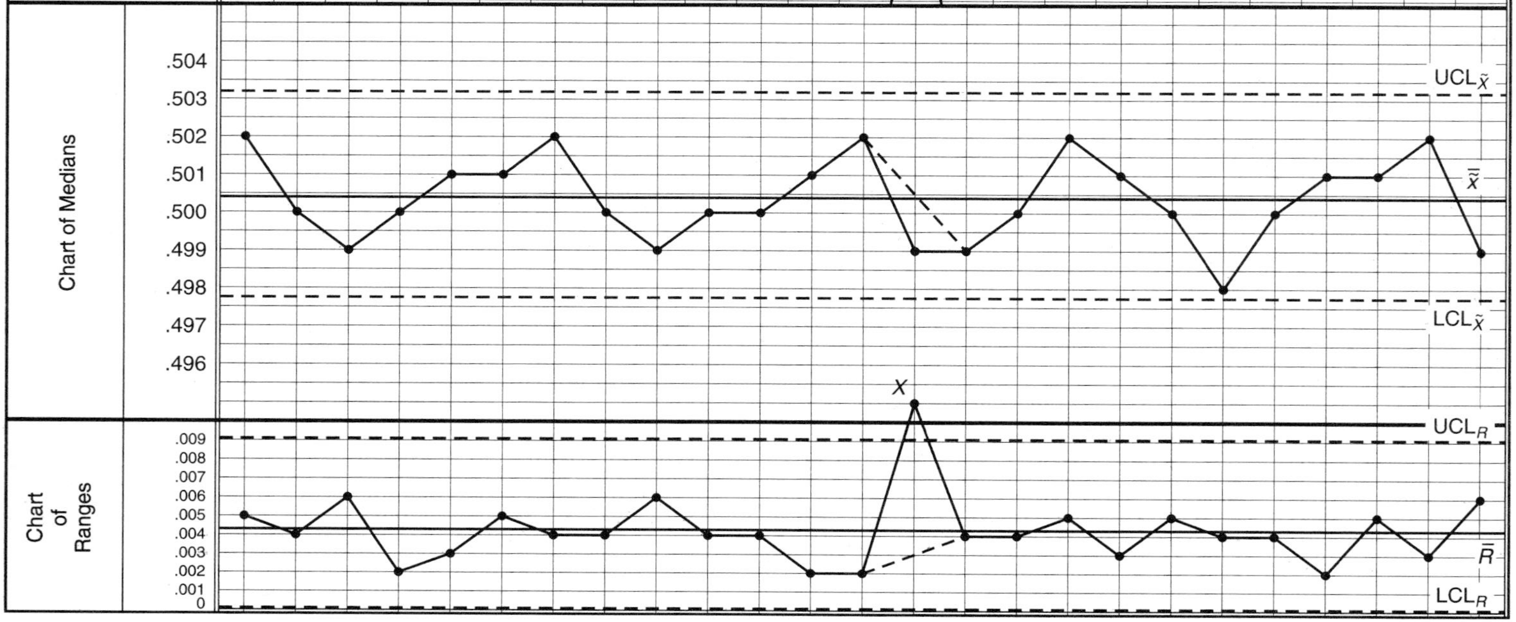

Chart of Medians: .504, .503, .502, .501, .500, .499, .498, .497, .496; $UCL_{\tilde{x}}$, $\bar{\tilde{x}}$, $LCL_{\tilde{x}}$

Chart of Ranges: .009, .008, .007, .006, .005, .004, .003, .002, .001, 0; X; UCL_R, $\bar{R}$, LCL_R

Figure 7.6
Recalculate. The chart is now in statistical control.

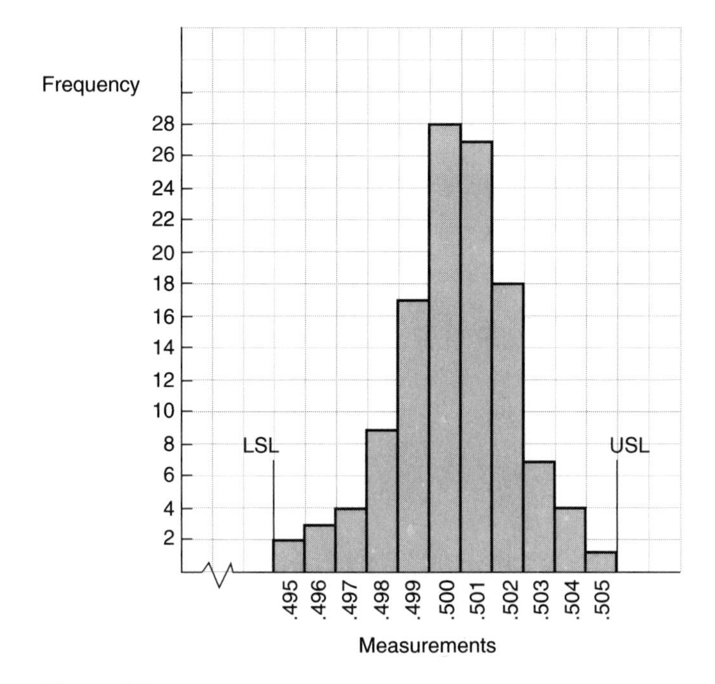

Figure 7.7
The tally histogram shows that the distribution is normal and crowds the specification limits.

When the chart is completed, compare your results with the chart in Figure 7.6. Do the calculations for the average and control limits of the range chart first. Analyze the range chart and eliminate any out-of-control points before proceeding with the $\tilde{x}$ chart. Assume that the out-of-control problem in the process has been solved and eliminated. Bringing the R chart into control this way before starting calculations on the $\tilde{x}$ chart saves time. Otherwise, the calculations on the $\tilde{x}$ will be done twice: before and after out-of-control data are removed as a result of the R chart analysis. The median chart calculations, shown on Figure 7.6, indicate that one data point has been removed from the chart. Further chart analysis reveals no other out-of-control situations.

For the process capability, the new value of R is as follows:

.107	Original ΣR
−.011	The out-of-control R value
.096	The new ΣR

$$\bar{R} = \frac{\Sigma R}{k} \qquad UCL_R = 2.114 \times .004$$

$$UCL_R = .0085$$

$$= \frac{.096}{24}$$

$$= .004$$

There is no substantial difference on the chart.

$$s = \frac{\overline{R}}{d_2}$$
$$= \frac{.004}{2.326}$$
$$= .00172$$
$$3s = .00516$$
$$6s = .01032$$

.505	USL
−.495	LSL
.010	Tolerance

$$\text{PCR} = \frac{6s}{\text{tolerance}}$$
$$= \frac{.01032}{.010}$$
$$= 1.032$$

The PCR indicates that this is a D process that uses 103% of the tolerance. To calculate the Cpk, notice that $\overline{\overline{x}}$ is closer to the USL.

$$\text{Cpk} = \frac{\text{USL} - \overline{\overline{x}}}{3s}$$
$$= \frac{.505 - .50042}{.00516}$$
$$= \frac{.00458}{.00516}$$
$$= .8876$$

The Cpk indicates that this is a D process that uses 1/Cpk = 1/.8876 = 1.13, or 113%, of the tolerance on the "worst" side of the mean, where the mean is closest to the specification limit.

The histogram in Figure 7.7 shows that the distribution of the data has a normal shape, so the calculation of the percentage of product that is out of specification should give credible results. Substitute the specification limits for x in the z formula:

$$z = \frac{x - \overline{\overline{x}}}{s}$$
$$= \frac{\text{USL} - \overline{\overline{x}}}{s}$$
$$= \frac{.505 - .50042}{.00172}$$
$$= 2.66$$

The area is .0039 according to Table B.8 in Appendix B, which indicates that .39% of the product is out of specification on the high side:

$$\begin{aligned} z &= \frac{x - \bar{\bar{x}}}{s} \\ &= \frac{\text{LSL} - \bar{\bar{x}}}{s} \\ &= \frac{.495 - .50042}{.00172} \\ &= -3.15 \end{aligned}$$

The area is .00082, which indicates that .082% of the product is out of specification on the low side. The total percentage of product that is out of specification is

$$.39\% + .082\% = .472\%$$

7.2 $\bar{x}$ AND s CHARTS

The $\bar{x}$ and s charts follow the same step-by-step procedure that was outlined for the $\bar{x}$ and R charts, and the $\bar{x}$ chart section is exactly the same for both charts. The difference between the two charts is the measurement of the variability: The $\bar{x}$ and s chart uses the sample standard deviation s instead of the sample range R. Also, use of the $\bar{x}$ and s charts requires that the person doing the charting use a statistical calculator or an on-station computer. Our discussion will focus on the use of a statistical calculator.

EXAMPLE 7.3 For this problem, we will use the data from the previous shaft diameter problem (Example 7.2). Make the $\bar{x}$ and s chart and analyze the process capability after achieving statistical control.

Solution

The data are presented in Figure 7.8. Calculate $\bar{x}$ and s for each sample using a statistical calculator. Be sure to clear the statistical registers before starting each new sample. One way to see if the registers are clear is to press the $\bar{x}$ key. If a numerical value appears, the calculator has not been cleared. If either a zero or an error message comes up, the statistical registers are clear. Enter the five sample values with the data entry key. Press the $\bar{x}$ key and record the sample mean. Press the s or σ_{n-1} key and record the sample standard deviation. Be careful, because the calculator will probably have two different standard deviation keys. Check your results with Figure 7.9.

Set the scales from the first eight samples and chart the points. Find the range of $\bar{x}$ and double it to set the scale:

$$\begin{aligned} R_{\bar{x}} &= .5018 - .4992 \\ &= .0026 \\ 2R_{\bar{x}} &= 2 \times .0026 \\ &= .0052 \end{aligned}$$

$\bar{s} = \frac{\Sigma s}{k} =$ ______ = ______

$\bar{\bar{x}} = \frac{\Sigma \bar{x}}{k} =$ ______ = ______

Specification: .495 to .505

$UCL_S = B_4 \times \bar{s} =$ ____ × ____ = ____

$A_3 \times \bar{s} =$ ____ × ____ = ☐

$UCL_{\bar{x}} = \bar{\bar{x}} + A_3\,\bar{s} =$ ____ + ☐ = ____

$LCL_{\bar{x}} = \bar{\bar{x}} - A_3\,\bar{s} =$ ____ − ☐ = ____

Constants

n	A_3	B_3	B_4	C_4
3	1.954	0	2.568	.8862
4	1.628	0	2.266	.9213
5	1.427	0	2.089	.9400

Sample Numbers	1	2	3	4	5	6	7	8	9	10	11	12	13
Sample Measurements	.502	.500	.496	.501	.501	.499	.502	.500	.500	.501	.498	.501	.502
	.503	.501	.500	.500	.501	.501	.500	.499	.499	.500	.500	.502	.502
	.499	.497	.502	.500	.500	.502	.504	.501	.495	.500	.499	.500	.503
	.498	.499	.499	.500	.501	.504	.503	.497	.498	.500	.501	.501	.501
	.503	.500	.499	.502	.498	.501	.500	.501	.501	.497	.502	.501	.503
$\bar{x}$													
s													

Sample Numbers	14	15	16	17	18	19	20	21	22	23	24	25
Sample Measurements	.494	.498	.500	.499	.501	.504	.499	.497	.502	.501	.502	.502
	.499	.496	.498	.502	.500	.500	.499	.499	.500	.505	.502	.496
	.503	.499	.502	.504	.501	.499	.495	.501	.501	.502	.501	.499
	.496	.500	.500	.502	.501	.500	.498	.500	.500	.501	.500	.498
	.505	.499	.501	.500	.503	.502	.498	.501	.501	.500	.503	.499
$\bar{x}$												
s												

Chart of Averages — $\bar{x}$

Chart of Standard Deviations — s

Figure 7.8
Make the $\bar{x}$ and s chart.

Divide by the number of lines on the chart. Round up to .0005 units/line:

$$21\overline{).0052} = .0002+$$

Next find the range of s and double it. The s value could be 0; subtract 0 as we did with the range values:

$$\begin{aligned} R_s &= .0023 - 0 \\ &= .0023 \\ 2R_s &= 2 \times .0023 \\ &= .0046 \end{aligned}$$

Divide by the number of lines on the chart. Round up to .0005 units per line:

$$10\overline{).0046} = .00046$$

Start the $\bar{x}$ chart in the center at the midspecification value of .500. Start the s chart on the bottom line at 0. Complete the control chart in Figure 7.8 for $\bar{x}$ and s. Check your results with Figure 7.9.

Control Chart Formulas	
$UCL_s = B_4\bar{s}$	B_3, B_4, and A_3 are Shewhart constants from Table B3
$LCL_s = B_3\bar{s}$	For sample sizes, $n < 6$, $B_3 = 0$
$UCL_{\bar{x}} = \bar{\bar{x}} + A_3\bar{s}$	$\bar{s} = \dfrac{\Sigma s}{k}$
$LCL_{\bar{x}} = \bar{\bar{x}} - A_3\bar{s}$	

The formulas for calculating the values for the averages and the control limits are given at the top of the control chart in Figure 7.8. Do the s chart first and bring it into statistical control by eliminating any out-of-control points. Assume that the cause for the out-of-control situation has been found and eliminated.

Do your calculations, draw the average and control limit lines, and analyze the chart for points out of control. Check your results with Figure 7.9.

The calculation section has been completed on Figure 7.9. The $\bar{s}$ calculation shows $\bar{s}$ calculated from all 25 s values. The UCL_s is .0035. When the average and control limit lines are drawn on the s chart, the point for sample 14 is beyond the control limit. Assuming that the cause for the out-of-control point was eliminated from the process, delete the sample 14 data. The average $\bar{\bar{x}}$ is determined from the remaining 24 sample values. Also, $\bar{s}$ is recalculated by subtracting .0046 (from sample 14) from the previous sum and dividing by 24:

.0423	Original Σs
−.0046	Sample 14
.0377	New Σs

First $\bar{s} = \frac{.0423}{25} = .00169$

$\bar{s} = \frac{\Sigma s}{k} = \frac{.0377}{24} = .00157$

$\bar{\bar{x}} = \frac{\Sigma \bar{x}}{k} = \frac{12.0072}{24} = .5003$

Specification: .495 to .505

$UCL_s = B_4 \times \bar{s} = 2.089 \times .00169 = .0035$

$A_3 \times \bar{s} = 1.427 \times .00157 = .00224$

$UCL_{\bar{x}} = \bar{\bar{x}} + A_3 \bar{s} = .5003 + .00224 = .50254$

$LCL_{\bar{x}} = \bar{\bar{x}} - A_3 \bar{s} = .5003 - .00224 = .49806$

Final calculations

$\bar{s} = \frac{.0361}{23} = .0016$ $\quad \bar{\bar{x}} = \frac{11.5094}{23} = .5004$

$UCL_s = .0033$

$UCL_{\bar{x}} = .5027$ $\quad LCL_{\bar{x}} = .4981$

Constants

n	A_3	B_3	B_4	C_4
3	1.954	0	2.568	.8862
4	1.628	0	2.266	.9213
5	1.427	0	2.089	.9400

Sample Numbers	1	2	3	4	5	6	7	8	9	10	11	12	13	14	15	16	17	18	19	20	21	22	23	24	25
Sample Measurements	.502	.500	.496	.501	.501	.499	.502	.500	.500	.501	.498	.501	.502	.494	.498	.500	.499	.501	.504	.499	.497	.502	.501	.502	.502
	.503	.501	.500	.500	.501	.501	.500	.499	.499	.500	.500	.502	.502	.499	.496	.498	.502	.500	.500	.499	.499	.500	.505	.502	.496
	.499	.497	.502	.500	.500	.502	.504	.501	.495	.500	.499	.500	.503	.503	.499	.502	.504	.501	.499	.495	.501	.501	.502	.501	.499
	.498	.499	.499	.500	.501	.504	.503	.497	.498	.500	.501	.501	.501	.496	.500	.500	.502	.501	.500	.498	.500	.500	.501	.500	.498
	.503	.500	.499	.502	.498	.501	.500	.501	.501	.497	.502	.501	.503	.505	.499	.501	.500	.503	.502	.498	.501	.501	.500	.503	.499
$\bar{x}$	.5010	.4994	.4992	.5006	.5002	.5014	.5018	.4996	.4986	.4996	.5000	.5010	.5022	.4994	.4984	.5002	.5014	.5012	.5010	.4978	.4996	.5008	.5018	.5016	.4988
s	.0023	.0015	.0022	.0009	.0013	.0018	.0018	.0017	.0023	.0015	.0016	.0007	.0008	.0046	.0015	.0015	.0019	.0011	.0020	.0016	.0017	.0008	.0019	.0011	.0022

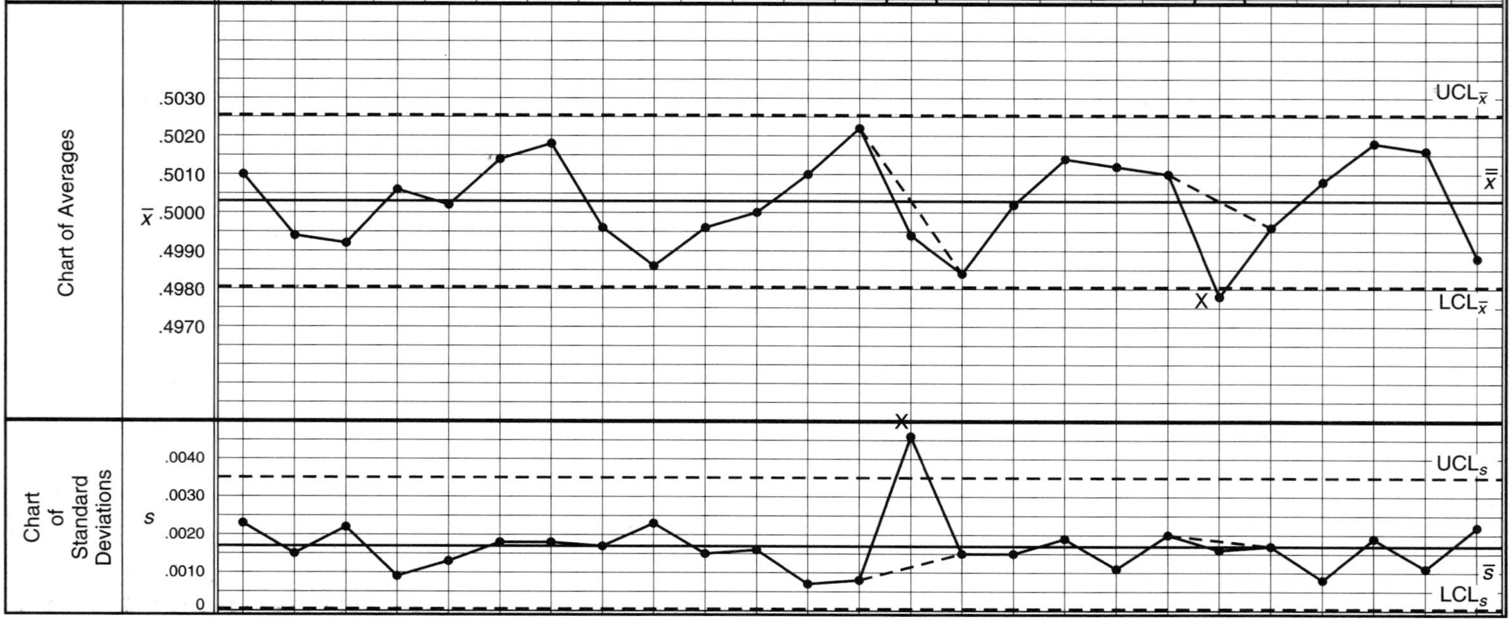

Figure 7.9
The $\bar{x}$ and s chart is now in statistical control.

$$\text{The new } \bar{s} = \frac{.0377}{24} = .00157$$

The new $\bar{s}$ is then used to calculate $A_3\bar{s}$ for the $\bar{x}$ chart control limits.

Analyze the $\bar{x}$ chart for points that are out of control; look for points beyond the control limits and runs or shifts of seven or more consecutive points. A more complete chart analysis will be discussed in Chapter 10.

One additional point is out of control, as shown in Figure 7.9. With that data point eliminated, the process is in statistical control. The final calculations are based on the remaining 23 samples. The adjusted sums are $\Sigma s = .0361$ and $\Sigma \bar{x} = 11.5094$:

$$\bar{s} = \frac{\Sigma s}{k} = \frac{.0361}{23} = .0016 \qquad \bar{\bar{x}} = \frac{\Sigma \bar{x}}{x} = \frac{11.5094}{23} = .5004$$

$$\begin{aligned} \text{UCL}_s &= B_4\bar{s} \\ &= 2.089 \times .0016 \\ &= .0033 \end{aligned}$$

$$\begin{aligned} \text{UCL}_{\bar{x}} &= \bar{\bar{x}} + A_3\bar{s} \\ &= .5004 + 1.427 \times .0016 \\ &= .5004 + .0023 \\ &= .5027 \\ \text{LCL}_{\bar{x}} &= .5004 - .0023 \\ &= .4981 \end{aligned}$$

The first step in the capability analysis is to determine the standard deviation of the *individual* measurements. The s values calculated for the control chart were *sample* standard deviations. Fortunately, a simple conversion is available, similar to the one used with the R charts. In the previous capability studies, the symbol s was used for the estimate of the process standard deviation σ. However, when working with $\bar{x}$ and s charts the symbol s represents the sample standard deviation. To avoid confusion with different meanings for the same symbol, $\hat{\sigma}$ will be used for the Shewhart estimate of the process standard deviation:

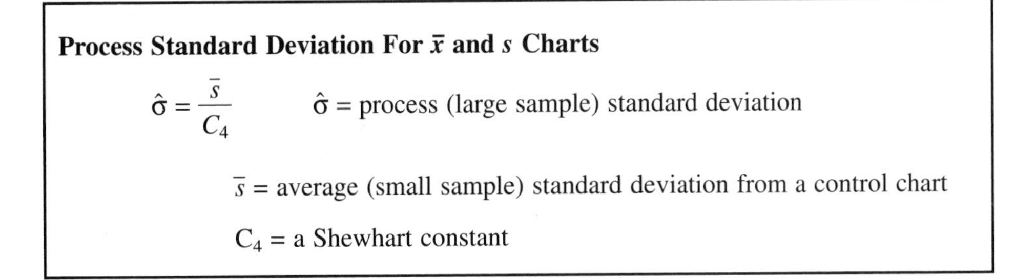

Process Standard Deviation For $\bar{x}$ and s Charts

$$\hat{\sigma} = \frac{\bar{s}}{C_4}$$

$\hat{\sigma}$ = process (large sample) standard deviation

$\bar{s}$ = average (small sample) standard deviation from a control chart

C_4 = a Shewhart constant

The numerical constant C_4 depends on the sample size and is shown in a table in the calculation section of the control chart in Figure 7.8. A more complete table of constants is shown in Table B.3 in Appendix B:

$$\hat{\sigma} = \frac{.0016}{.94}$$
$$= .0017$$
$$3\hat{\sigma} = .0051$$
$$6\hat{\sigma} = .0102$$

$$\begin{array}{rl} .505 & \text{USL} \\ -.495 & \text{LSL} \\ \hline .010 & \text{Tolerance} \end{array}$$

$$\text{PCR} = \frac{6\hat{\sigma}}{\text{tolerance}}$$
$$= \frac{.0102}{.01}$$
$$= 1.02$$

This indicates a D process that uses 102% of the tolerance:
Since $\bar{x}$ is closer to the USL,

$$\text{Cpk} = \frac{\text{USL} - \bar{\bar{x}}}{3\hat{\sigma}}$$
$$= \frac{.505 - .5004}{.0051}$$
$$= \frac{.0046}{.0051}$$
$$= .90$$
$$\frac{1}{\text{Cpk}} = \frac{1}{.90}$$
$$= 1.11$$

This calculation shows that the process uses 111% of the tolerance on the worst side.

The histogram for these data was shown in Figure 7.7 to have a normal shape. The normal-curve analysis can be used to find the percentage of product that is out of specification:

$$z = \frac{\text{USL} - \bar{\bar{x}}}{\hat{\sigma}}$$
$$= \frac{.505 - .5004}{.0017}$$
$$= 2.71$$
$$\text{Area} = .0034$$

This analysis indicates that .34% of the product is out of specification on the high side.

$$z = \frac{\text{LSL} - \bar{\bar{x}}}{\hat{\sigma}}$$
$$= \frac{.495 - .5004}{.0017}$$
$$= -3.18$$
$$\text{Area} = .00074$$

On the low side, .074% is out of specification. The total is .414% out of specification.

This same set of data was analyzed twice: first with the $\tilde{x}$ and R chart in Example 7.2 and then here with the $\bar{x}$ and s chart. The results are comparable. The slight difference is due to the varying estimations of the standard deviation of the individual measurements: One is calculated from R and the other from s:

	$\bar{x}$ and s	$\tilde{x}$ and R
PCR	102%	103%
Cpk	111%	113%
Percent out	.414%	.472%

The $\bar{x}$ chart seems a little more sensitive than the $\tilde{x}$ chart. The $\bar{x}$ chart showed one point out of specification that the median chart missed. It also seemed more obvious with the $\bar{x}$ chart that the process was still not in statistical control. Two point patterns suggest control problems: (1) too many points near the control limits and (2) rhythmic ups and downs. These out-of-control patterns are discussed thoroughly in Chapter 10.

EXAMPLE 7.4 The run of shaft diameters (from Example 7.3) continues, as does the effort to bring the process into statistical control. Continue the charting and look for potential out-of-control situations as they occur:

1. Set up the continuation chart in Figure 7.10 and check your results with Figure 7.11.
2. In reaction to the cycle pattern on the previous control chart (Chapter 10 deals with this and other control chart trouble patterns), a loose fixture was found and repaired. The new samples are shown in Table 7.2. Chart the first six samples on Figure 7.10. Why was an adjustment made at this point?
3. Chart the rest of the samples on Figure 7.10. Indicate any suspicion of a trouble pattern developing as you chart the points. Check your results with Figure 7.11.
4. Eliminate out-of-control data, recalculate, and reevaluate.
5. Check the histogram of new data as well as a combined histogram.

$\bar{s} = \frac{\Sigma s}{k} = \_\_\_\_ = \_\_\_\_$

$\bar{\bar{x}} = \frac{\Sigma \bar{x}}{k} = \_\_\_\_ = \_\_\_\_$

$UCL_s = B_4 \times \bar{s} = \_\_\_\_ \times \_\_\_\_ = \_\_\_\_$

$A_3 \times \bar{s} = \_\_\_\_ \times \_\_\_\_ = \square$

$UCL_{\bar{x}} = \bar{\bar{x}} + A_3 \bar{s} = \_\_\_\_ + \square = \_\_\_\_$

$LCL_{\bar{x}} = \bar{\bar{x}} - A_3 \bar{s} = \_\_\_\_ - \square = \_\_\_\_$

Specification: .495 to .505

Constants

n	A_3	B_3	B_4	C_4
3	1.954	0	2.568	.8862
4	1.628	0	2.266	.9213
5	1.427	0	2.089	.9400

Sample Numbers	1	2	3	4	5	6	7	8	9	10	11	12	13	14	15	16	17	18	19	20	21	22	23	24	25
$\bar{x}$																									
s																									
Chart of Averages $\bar{x}$																									
Chart of Standard Deviations s																									

Figure 7.10
The continuation chart from Example 7.3. Transfer scales, averages and control limits.

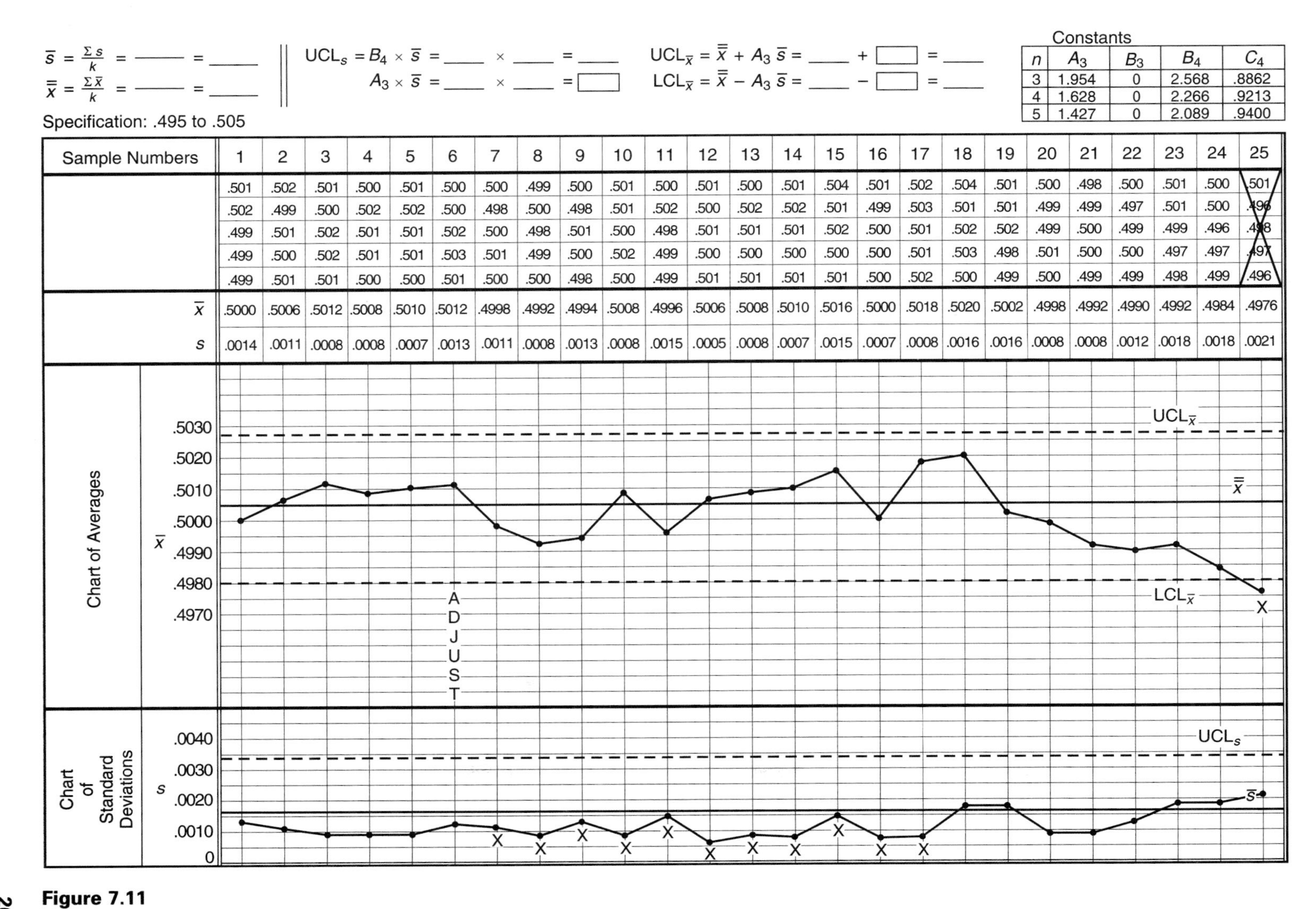

Constants

n	A_3	B_3	B_4	C_4
3	1.954	0	2.568	.8862
4	1.628	0	2.266	.9213
5	1.427	0	2.089	.9400

Sample Numbers	1	2	3	4	5	6	7	8	9	10	11	12	13	14	15	16	17	18	19	20	21	22	23	24	25
	.501	.502	.501	.500	.501	.500	.500	.499	.500	.501	.500	.501	.500	.501	.504	.501	.502	.504	.501	.500	.498	.500	.501	.500	.501
	.502	.499	.500	.502	.502	.500	.498	.500	.498	.501	.502	.500	.502	.502	.501	.499	.503	.501	.501	.499	.499	.497	.501	.500	.496
	.499	.501	.502	.501	.501	.502	.500	.498	.501	.500	.498	.501	.501	.501	.502	.500	.501	.502	.502	.499	.500	.499	.499	.496	.498
	.499	.500	.502	.501	.501	.503	.501	.499	.500	.502	.499	.500	.500	.500	.500	.500	.501	.503	.498	.501	.500	.500	.497	.497	.497
	.499	.501	.501	.500	.500	.501	.500	.500	.498	.500	.499	.501	.501	.501	.501	.500	.502	.500	.499	.500	.499	.499	.498	.499	.496
$\bar{x}$	.5000	.5006	.5012	.5008	.5010	.5012	.4998	.4992	.4994	.5008	.4996	.5006	.5008	.5010	.5016	.5000	.5018	.5020	.5002	.4998	.4992	.4990	.4992	.4984	.4976
s	.0014	.0011	.0008	.0008	.0007	.0013	.0011	.0008	.0013	.0008	.0015	.0005	.0008	.0007	.0015	.0007	.0008	.0016	.0016	.0008	.0008	.0012	.0018	.0018	.0021

Figure 7.11
The completed chart for Example 7.4.

TABLE 7.2
Sample measurements for Example 7.4

Sample		1	2	3	4	5	6	Adjustment Made	
		.501	.502	.501	.500	.501	.500		
		.502	.499	.500	.502	.502	.500		
		.499	.501	.502	.501	.501	.502		
		.499	.500	.502	.501	.501	.503		
		.499	.501	.501	.500	.500	.501		
7	8	9	10	11	12	13	14	15	16
.500	.499	.500	.501	.500	.501	.500	.501	.504	.501
.498	.500	.498	.501	.502	.500	.502	.502	.501	.499
.500	.498	.501	.500	.498	.501	.501	.501	.502	.500
.501	.499	.500	.502	.499	.500	.500	.500	.500	.500
.500	.500	.498	.500	.499	.501	.501	.501	.501	.500
17	18	19	20	21	22	23	24	25	
.502	.504	.501	.500	.498	.500	.501	.500	.501	
.503	.501	.501	.499	.499	.497	.501	.500	.496	
.501	.502	.502	.499	.500	.499	.499	.496	.498	
.501	.503	.498	.501	.500	.500	.497	.497	.497	
.502	.500	.499	.500	.499	.499	.498	.499	.496	

Solution

1. The carryover figures from Example 7.3 are

$$\Sigma s = .0361 \qquad \bar{\bar{x}} = .5004 \qquad \bar{s} = .0016$$
$$\Sigma \bar{x} = 11.5094 \qquad UCL_{\bar{x}} = .5027 \qquad UCL_s = .0033$$
$$k = 23 \qquad LCL_{\bar{x}} = .4981 \qquad LCL_s = 0$$

   The averages and control limits are drawn on the control chart.

2. The adjustment was made because there were five consecutive points above the mean. Normally, the indicator for a shift is seven consecutive points on one side of the mean. In this case, however, a repair was made (the loose fixture) and the reset may have been off. Another (and better) strategy to use when there is a suspicion of trouble is to increase the sampling frequency so the trouble pattern, if there is one, will show up sooner. At the same time, be more alert to potential sources of trouble.
3. The standard deviation chart has a shift pattern from sample 7 to sample 17. Since this is a shift down on a variation chart, it indicates good news: a decrease in variation. This is most likely the result of the fixture repair.

   There is a gradual increasing trend on the $\bar{x}$ chart from sample 7 to sample 18. You may have noticed the increase at about sample 12 or sample 13. There is no "official" out-of-control pattern, but as an operator or supervisor, that should alert you to watch for possible causes as the process continues to run.

Trouble is occurring toward the end of the chart. The last point is beyond the control limit, but that point is also part of a shift pattern and the shift pattern seems to be imbedded within a downward trend. Each pattern gives a set of hints for the problem cause. If just the last point was the trouble indicator, we would look for some mistake that would affect that one sample and maybe some other pieces back to the previous sample. If the shift was the trouble indicator, we would look for one thing that had changed that could affect the process: a different operator, new material introduced, a sudden change in a machine setting, a chipped or broken tool, etc. If the downward trend was the main trouble pattern, we would look for something that is changing gradually, such as tool wear or a fixture slowly loosening. Since the standard deviation chart is showing an increase in variation along with the trend, the loose fixture problem may be recurring. In any case, the operation should be stopped and the source of the problem found and corrected. If the process is not stopped, 100% inspection must be used while the search for the source of trouble is under way. Also, all the pieces produced since sample 18 should be inspected because that could be the starting point of the problem.

4. The only "official" out-of-control point is point 25. That was both the seventh point in a shift pattern and a point beyond the control limit. We eliminate the data of point 25, combine $\Sigma\bar{x}$ and Σs with the previous data, and recalculate. If a third chart continued the pattern of lower sample standard deviations, the third calculation would involve just the second and third charts. This is because the added variation due to the loose fixture has been eliminated and the lower $\bar{s}$ will show that the process has a better capability.

   Recalculations for charts 1 and 2 yield

Chart 1	$\Sigma s = .0361$	$\Sigma\bar{x} = 11.5094$	$k = 23$
Chart 2	$\Sigma s = .0262$	$\Sigma\bar{x} = 12.0072$	$k = 24$
Totals	$\Sigma s = .0623$	$\Sigma\bar{x} = 23.5166$	$k = 47$

$$\bar{s} = \frac{\Sigma s}{k} = \frac{.0623}{47} = .0013 \qquad \bar{\bar{x}} = \frac{\Sigma\bar{x}}{k} = \frac{23.5166}{47} = .5004$$

$$\begin{aligned} UCL_s &= B_4\bar{s} \\ &= 2.089 \times .0013 \\ &= .0027 \end{aligned} \qquad \begin{aligned} UCL_{\bar{x}} &= \bar{\bar{x}} + A_3\bar{s} \\ &= .5004 + 1.427 \times .0013 \\ &= .5023 \end{aligned}$$

$$\begin{aligned} LCL_{\bar{x}} &= .5004 - 1.427 \times .0013 \\ &= .4985 \end{aligned}$$

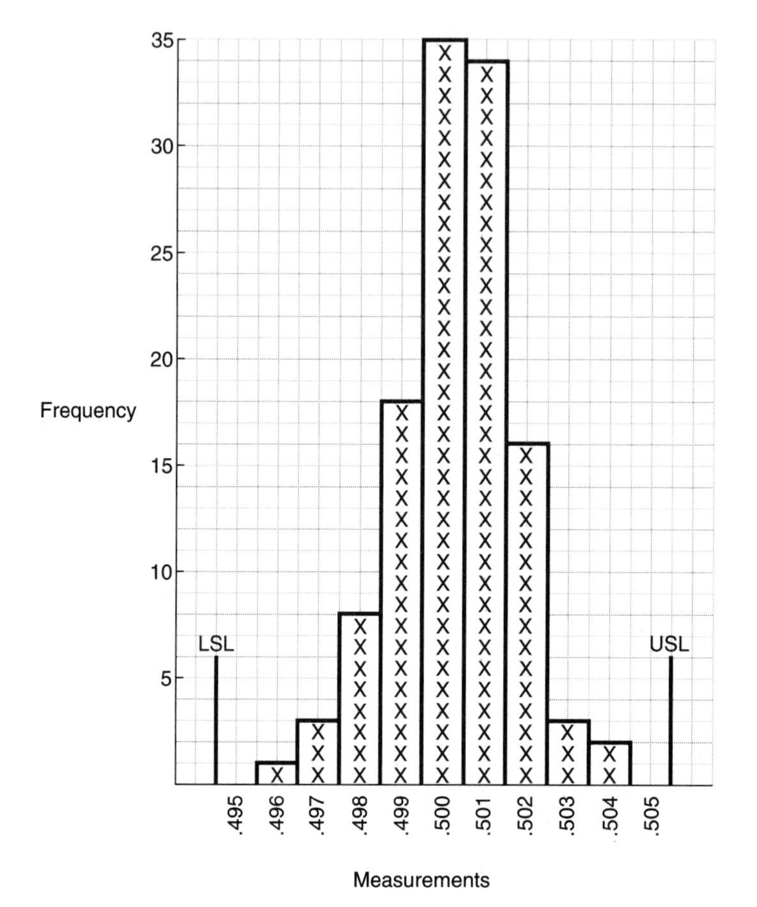

Figure 7.12
The tally histogram for the new data.

The capability analysis yields

$$\hat{\sigma} = \frac{\bar{s}}{C_4}$$

$$\hat{\sigma} = \frac{.0013}{.94}$$

$\bar{\bar{x}}$ is the closest to the USL, so use the USL in the C_{pk} calculation.

$$= .0014$$

$$\text{PCR} = \frac{6\hat{\sigma}}{\text{tolerance}} \qquad \text{Cp} = \frac{1}{\text{PCR}} \qquad \text{Cpk} = \frac{\text{USL} - \bar{\bar{x}}}{3\hat{\sigma}}$$

$$\text{Cpk} = \frac{6 \times .0014}{.01} \qquad = \frac{1}{.84} \qquad = \frac{.0046}{3 \times .0014}$$

$$= .84 \qquad = 1.19 \qquad = 1.10$$

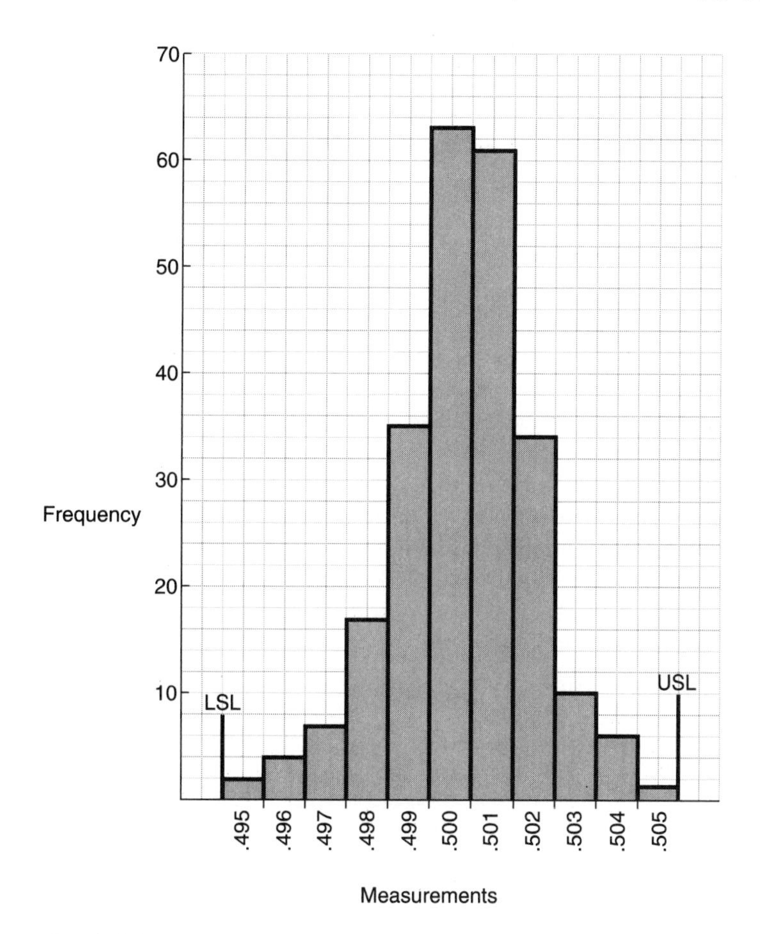

Figure 7.13
The histogram for the combined data.

$$\frac{1}{\text{Cpk}} \times 100 = 91\%$$

Eighty-four percent of the tolerance is used by the whole distribution.

Ninety-one percent of the tolerance is used on the high side.

If we drew the new averages and control limits on the chart, it would result in one additional point out of control: 24. That trouble point is already included in the analysis of the shift and trend patterns. Recalculating with the 46 points, $\bar{s}$ remains substantially the same, so there is no resulting change in capability.

5. The histogram of new data in Figure 7.12 shows that the data since the fixture repair is still normally distributed and the variation has decreased. In the combined data histogram shown in Figure 7.13, the total distribution is normal and there is no other hint of trouble.

7.3 CODING DATA

Coding measurements is a method used to simplify the arithmetic associated with averaging and finding the ranges and standard deviations for measurements. Instead of dealing with measurements, coding uses a base value and translates each measurement to a coded number that reflects its positive or negative distance from the base value.

Coding Rule 1: The measurement unit is the base factor and the coded value for a measurement is the number of measurement units from the base value.

If the measurements differ by thousandths, then the coded numbers represent thousandths and the base factor is .001. If the measurements differ by hundredths, then the coded numbers represent hundredths and the base factor is .01. Often the target value is used for the base value. This keeps the coded numbers small in absolute value, but the arithmetic for signed numbers must be used. Another choice for a base value is a number smaller than all the measurements. The coded numbers are then larger, but they are all positive and the more familiar arithmetic of whole numbers applies.

In the following illustration, the base value 2.315 is used. The coded numbers represent thousandths, and the base factor is .001. See Table 7.3.

TABLE 7.3

Measurement	Coded Value	Base = 2.315
2.314	−1	1 coded unit below base
2.321	6	6 coded units above base
2.316	1	1 coded unit above base
2.310	−5	5 coded units below base
2.315	0	the base codes as 0
2.319	4	4 coded units above base
2.309	−6	6 coded units below base
2.313	−2	2 coded units below base

- The coded value 1 corresponds to .001 units above the base, so 2.316 codes to 1.
- The coded value 2 corresponds to .002 units above the base, so 2.317 codes to 2.
- The coded value −4 corresponds to .004 units below the base, so 2.311 codes to −4.
- The coded value .23 corresponds to .00023 units above the base, so 2.31523 codes to .23.

Coding Rule 2: To change the coded number back to measurement units, multiply by the base factor.

The concepts regarding the transformation from measurement to coded value, the calculations with the coded values, and the transformation back to measurement values can best be illustrated with an example.

If the preceding measurements are averaged directly, their average value will be the sum divided by 8. The sum of the eight measurements is 18.517:

$$\frac{18.517}{8} = 2.314625$$

The average of the coded values is the sum divided by 8. The coded sum is –3:

$$\frac{-3}{8} = -.375$$

The coded average gives the number of coded units from the base measurement to the average measurement. When the coded average is multiplied by the base factor, the result is the measurement change from the base measurement to the average measurement:

$$-.375 \times .001 = -.000375 \qquad \text{(measurement change)}$$

Coding Rule 3: When the measurement change is added to the measurement base, the measurement value results.

To continue the example, add the measurement change and the measurement base:

$$\begin{aligned} &- .000375 + 2.315 \\ &= 2.314625 \end{aligned}$$

Subtract their absolute values and keep the positive sign. This is the same as the average calculated directly from the measurements.

To simulate the range of samples and the calculation of the standard deviation from a control chart, pair the values from Table 7.3 for samples of size 2. Calculate the sample range by subtracting the smaller value from the larger, and then calculate the average-range value. For the illustration, this is done for both the measurements and the coded values. See Table 7.4.

The average range for the measurements is the sum, .021, divided by 4:

$$\frac{.021}{4} = .00525$$

The average range for coded values is again the sum divided by 4:

$$\frac{21}{4} = 5.25$$

Notice that the product of the coded range and the base factor is the measurement range.

The standard deviation for a set of measurements is determined by dividing the average range by a numerical constant that is determined from the sample size (constant d_2, found in Appendix B, Table B.1). The constant for a sample of size 2 is 1.128. The standard deviation for the measurements is:

$$\hat{\sigma} = \frac{\bar{R}}{d_2}$$

$$= \frac{.00525}{1.128} = .004654$$

TABLE 7.4

Base = 2.315				
Range	Measurement	Sample Number	Coded Value	Range
	2.314		−1	
.007	2.321	1	6	7
	2.316		1	
.006	2.310	2	−5	6
	2.315		0	
.004	2.319	3	4	4
	2.309		−6	
.004	2.313	4	−2	4

The coded standard deviation is

$$\text{Coded } \hat{\sigma} = \frac{5.25}{1.128} = 4\ .654$$

To change the coded value back to the measurement value, apply rule 2 and multiply by the base factor:

$$\hat{\sigma} = 4.654 \times .001 = .004654$$

This is the same as the direct calculation using the measurements.

EXAMPLE 7.5 Six samples of size 3 are given in Table 7.5. Code the measurements and calculate the average and the standard deviation. For samples of this size, divide the average range by 1.693 (d_2 for samples of size 3) to calculate the standard deviation. A table containing these numerical constants is shown in Appendix B, Table B.1.

Solution

To get only positive coded numbers this time, a base value of 23.50 will be used (a number smaller than all the measurements). The measurements differ by hundredths, so the base factor is .01.

For 18 coded values,

$$\begin{aligned}\text{Average coded value} &= \frac{\text{sum}}{18}\\ &= \frac{144}{18}\\ &= 8\end{aligned}$$

TABLE 7.5

Base = 23.50											
Sample 1	Code	Sample 2	Code	Sample 3	Code	Sample 4	Code	Sample 5	Code	Sample 6	Code
23.59	9	23.57	7	23.58	8	23.55	5	23.58	8	23.54	4
23.62	12	23.54	4	23.51	1	23.54	4	23.62	12	23.57	7
23.61	11	23.63	13	23.62	12	23.59	9	23.57	7	23.61	11
Range	3		9		11		5		5		7

$$\begin{aligned}\text{Measurement change} &= \text{coded average} \times \text{base factor}\\ &= 8 \times .01\\ &= .08\end{aligned}$$

$$\begin{aligned}\text{Average} &= \text{base} + \text{measurement change}\\ &= 23.50 + .08\\ &= 23.58\end{aligned}$$

For six sample ranges,

$$\begin{aligned}\text{Average coded value} &= \frac{\text{sum}}{6}\\ &= \frac{40}{6}\\ &= 6.66667\end{aligned}$$

$$\hat{\sigma} = \frac{\overline{R}}{d_2} \qquad d_2 = 1.693$$

$$\begin{aligned}\text{Coded } \hat{\sigma} &= \frac{6.6667}{1.693}\\ &= 3.938\end{aligned}$$

$$\begin{aligned}\text{Measurement standard deviation} &= \text{coded } \hat{\sigma} \times \text{base factor}\\ &= 3.938 \times .01\\ \hat{\sigma} &= .03938\end{aligned}$$

CASE STUDY 7.1

The ABC company had downsized several times and was in competition with a Brazilian company for a contract to rebuild some diesel engines. One of the contract stipulations was that the company had to use SPC. A consultant was called in, the work force received SPC training, and the contract was won. During the training, the consultant took the class out on the shop floor for on-the-job applications. Cam roller bushings were being run at that time and were chosen for a control chart analysis.

$\bar{s} = \frac{\Sigma s}{k} = \_\_\_\_ = \_\_\_\_$ || $UCL_s = B_4 \times \bar{s} = \_\_\_\_ \times \_\_\_\_ = \_\_\_\_$ | $UCL_{\bar{x}} = \bar{\bar{x}} + A_3\,\bar{s} = \_\_\_\_ + \square = \_\_\_\_$

$\bar{\bar{x}} = \frac{\Sigma \bar{x}}{k} = \_\_\_\_ = \_\_\_\_$ || $A_3 \times \bar{s} = \_\_\_\_ \times \_\_\_\_ = \square$ | $LCL_{\bar{x}} = \bar{\bar{x}} - A_3\,\bar{s} = \_\_\_\_ - \square = \_\_\_\_$

Constants

n	A_3	B_3	B_4	C_4
3	1.954	0	2.568	.8862
4	1.628	0	2.266	.9213
5	1.427	0	2.089	.9400

Specification: 1.3280 to 1.3340 (–20 to 40 coded)

Sample Numbers	1	2	3	4	5	6	7	8	9	10	11	12	13	14	15	16	17	18	19	20	21	22	23	24	25
Coded 0 = 1.3300	29	29	28	28	22	33	33	33	27	28	38	36	25	26	20	16	29	17	21	29	24	27	20	28	08
	30	19	27	27	26	36	27	20	21	21	29	27	25	35	31	19	29	04	28	16	10	25	21	20	30
	28	25	14	14	08	32	20	21	31	17	25	24	11	20	19	23	05	27	16	19	30	21	18	29	16
	37	21	18	18	27	27	33	27	30	34	26	26	27	25	30	16	19	26	17	33	16	27	26	27	09
	29	35	34	34	24	28	21	33	27	25	27	21	36	24	35	25	21	16	21	15	21	30	31	22	00
Average $\bar{x}$																									
Standard Deviation s																									

Chart of Averages $\bar{x}$

Chart of Standard Deviations s

Figure 7.14
Calculate $\bar{x}$ and s for each sample and make an $\bar{x}$ and s chart. Calculate averages and control limits and determine process capability.

The data values presented in Figure 7.14 are lengths of cam roller bushings that have a tolerance of 1.328 to 1.334 inches. The measurements are to the nearest ten-thousandth and *coded* measurements are used. The base measurement is 1.3300, and the coded values represent the last two digits. For example, 1.3324 codes to 24. Measurements smaller than 1.3300 will have a negative coded value, so 1.3286, for example, codes to −14 (14 below 00). Coded averages are multiplied by .0001 and added to the base, 1.3300, for the true mean value: The true value for coded $\bar{x} = 13$ is $\bar{x} = 1.3313$. Coded standard deviations are simply multiplied by .0001: A coded $s = 3.6$ is really $s = .00036$. Make an $\bar{x}$ and s chart and interpret the results.

Solution

The calculations on the first few samples are illustrated in Table 7.6. The entire set of data is in coded form on the control chart in Figure 7.14. Complete the calculations for $\bar{x}$ and s for the remaining samples on Figure 7.14 and check your results with Figure 7.15. Enter the coded measurements into your statistical calculator with the data entry key. Then press $\bar{x}$ and s (or σ_{n-1}) and record the values.

Continue the rest of the $\bar{x}$ and s calculations with the coded data given in Figure 7.14. Set the scales in coded form and make the $\bar{x}$ an s chart. The formulas, constants, and calculation format are shown at the top of the control chart in Figure 7.14. Do the calculations, draw the center lines and control limit lines, and interpret the results. Check your results with the completed chart in Figure 7.15.

TABLE 7.6
Coding sample data

Sample 1	Coded Form	Coded $\bar{x}$	Coded s
1.3329	29		
1.3330	30		
1.3328	28		
1.3337	37		
1.3329	29	30.6	3.65
Sample 2			
1.3329	29		
1.3319	19		
1.3325	25		
1.3321	21		
1.3335	35	25.8	6.42
Sample 3			
1.3328	28		
1.3327	27		
1.3314	14		
1.3318	18		
1.3334	34	24.2	8.07

$\overline{s} = \frac{\Sigma s}{k} = \frac{165.07}{25} = 6.60$

$\overline{\overline{X}} = \frac{\Sigma \overline{X}}{k} = \frac{604.6}{25} = 24.2$

$UCL_s = B_4 \times \overline{s} = 2.089 \times 6.60 = 13.8$

$A_3 \times \overline{s} = 1.427 \times 6.60 = \boxed{9.4}$

$UCL_{\overline{X}} = \overline{\overline{X}} + A_3\overline{s} = 24.2 + \boxed{9.4} = 33.6$

$LCL_{\overline{X}} = \overline{\overline{X}} - A_3\overline{s} = 24.2 - \boxed{9.4} = 14.8$

Constants

n	A_3	B_3	B_4	C_4
3	1.954	0	2.568	.8862
4	1.628	0	2.266	.9213
5	1.427	0	2.089	.9400

Specification: 1.3280 to 1.3340 (–20 to 40 coded)

Sample Numbers	1	2	3	4	5	6	7	8	9	10	11	12	13	14	15	16	17	18	19	20	21	22	23	24	25
Coded 0 = 1.3300	29	29	28	28	22	33	33	33	27	28	38	36	25	26	20	16	29	17	21	29	24	27	20	28	08
	30	19	27	27	26	36	27	20	21	21	29	27	25	35	31	19	29	04	04	16	10	25	21	20	30
	28	25	14	14	08	32	20	21	31	17	25	24	11	20	19	23	05	27	27	19	30	21	18	29	16
	37	21	18	18	27	27	33	27	30	34	26	26	27	25	30	16	19	26	26	33	16	27	26	27	09
	29	35	34	34	24	28	21	33	27	25	27	21	36	24	35	25	21	16	21	15	21	30	31	22	00
$\overline{X}$	30.6	25.8	24.2	24.2	21.4	31.2	26.8	26.8	27.2	25.0	29.0	26.8	24.8	26.0	27.0	19.8	20.6	18.0	19.8	22.4	20.2	26.0	23.2	25.2	12.6
s	3.65	6.42	8.07	8.07	7.73	3.70	6.26	6.26	3.90	6.52	5.24	5.63	8.96	5.52	7.11	4.09	9.84	9.30	9.26	8.11	7.63	3.32	5.26	3.96	11.26

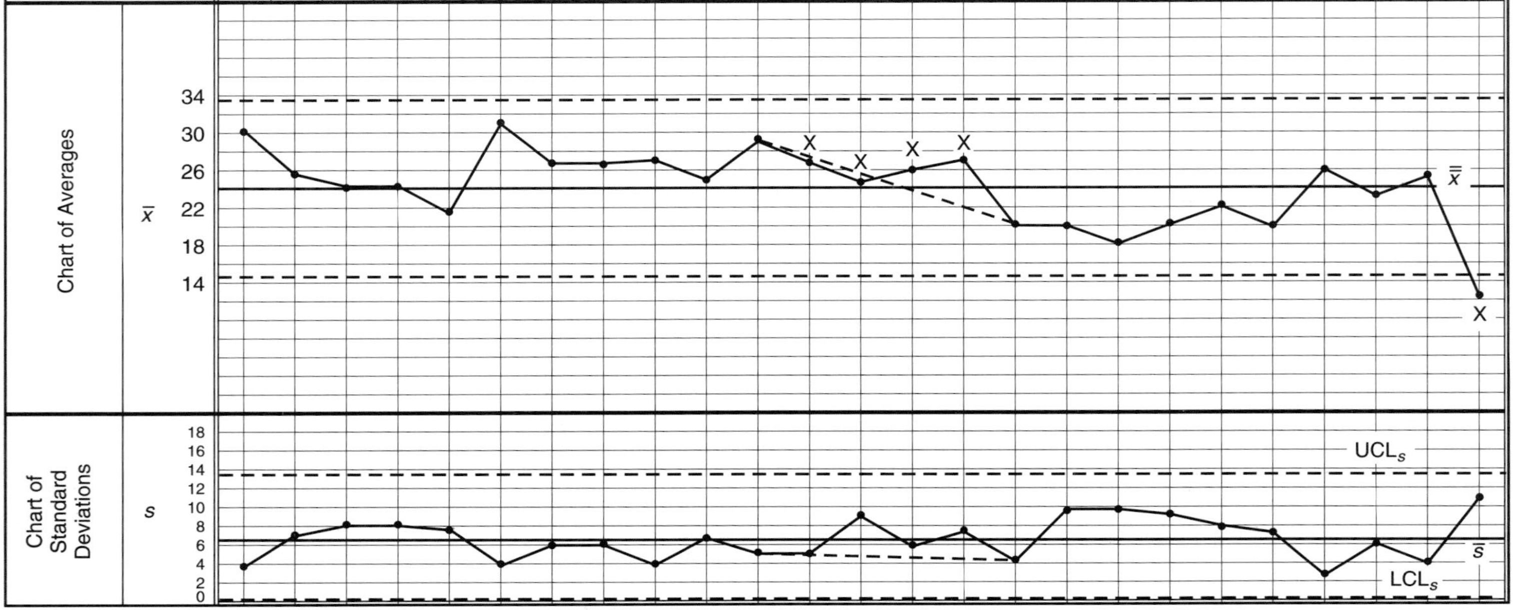

Figure 7.15
Recalculate averages and control limits and calculate PCR and Cpk. Make a histogram to check for a normal pattern; determine the percentage out of specification.

Set the scale from the first eight values, assuming a first chart situation:

$$R_s = 8.07 - 0 = 8.07$$
$$2R_s = 16.14$$

$$10\overline{)16.14}\quad 1.6+$$

Round off to two units per line.

$$R_{\bar{x}} = 30.6 - 21.4$$
$$2R_{\bar{x}} = 18.4$$

$$22\overline{)18.4}\quad .8+$$

Round off to 2 units per line. The average $\bar{s}$ for the first eight samples is 6.3 and $A_3\bar{S}$ is about 9. The 18 units between control limits is too wide at the more obvious 1 unit per line.

A scan of the data indicates that the average value appears to be above the midspecification value of 10. For the first eight samples $\bar{\bar{x}} = 26$. Use that value for the center of the chart and mark the scale from there.

Calculation for $\bar{s}$, $\bar{\bar{x}}$, and control limits:

$$\bar{s} = \frac{\Sigma s}{k} = \frac{165.07}{25} = 6.60 \qquad \bar{\bar{x}} = \frac{\Sigma \bar{x}}{k} = \frac{604.6}{25} = 24.2$$

$$\text{UCL}_s = B_4\bar{s} = 2.089 \times 6.60 = 13.8 \qquad \text{UCL}_{\bar{x}} = \bar{\bar{x}} + A_3\bar{s} = 24.2 + 1.427 \times 6.60 = 33.6$$

$$\text{LCL}_{\bar{x}} = \bar{\bar{x}} - A_3\bar{s} = 14.8$$

Chart Interpretation On the s chart, $\bar{s} = 6.60$ (coded). The base is ten-thousandths, so multiplication by the base factor .0001 will change the coded value to the actual value:

$$\bar{s} = 6.60 \times .0001 = .000660 \quad \text{(actual)}$$

There are no apparent out-of-control patterns on the s chart. Proceed with the $\bar{x}$ chart.

On the $\bar{x}$ chart, $\bar{\bar{x}} = 24.2$ (coded). Multiply the coded value by the base factor .0001 and add the product to the base value of 1.3300 to change the coded value back to the actual value:

$$\bar{\bar{x}} = (24.2 \times .0001) + 1.3300 = 1.33242 \quad \text{(actual)}$$

One source of trouble that is apparent in the $\bar{x}$ chart is that the cut for bushing length is set too high. This is a common problem with machining operators when SPC is not used. They have a tendency to hedge on the large side, thinking that any excess can be reworked off it necessary. The average value $\bar{\bar{x}}$ should be near the midspecification value of 1.3310 (the coded $\bar{\bar{x}} = 10$).

A second trouble indicator is the shift pattern that shows up in samples 6 through 15. Seven consecutive points on one side of the average line indicate a shift, and the seventh point and any successive points in the shift pattern are marked with an X.

A third trouble indicator occurs at sample 25, where the point falls beyond the control limits.

Each trouble indicator must be investigated. If the out-of-control situation is determined and resolved, the out-of-control data values can be removed from the chart and the remaining in-control data can be used to calculate the process capability. Assume that the trouble has been eradicated, cross out the out-of-control data values, and calculate the process capability.

Process Capability Remove the out-of-control data values:

Excluded $\bar{x}$'s	Excluded s's
26.8	5.63
24.8	8.96
26.0	5.52
27.0	7.11
12.6	11.26
117.2	38.48

604.6	Previous $\Sigma\bar{x}$	165.07	Previous Σs
−117.2	Excluded $\bar{x}$'s	−38.48	Excluded s's
487.4	New $\Sigma\bar{x}$	126.59	New Σs

$$\bar{\bar{x}} = \frac{\Sigma\bar{x}}{k} = \frac{487.4}{20} = 24.4 \quad \text{New } \bar{\bar{x}}$$

$$\bar{s} = \frac{\Sigma s}{k} = \frac{126.59}{20} = 6.33 \quad \text{New } \bar{s}$$

$$UCL_s = 13.2$$
$$UCL_{\bar{x}} = 33.2$$
$$LCL_{\bar{x}} = 15.2$$

There is no appreciable change in the averages or control limits, so further analysis for out-of-control point patterns is unnecessary.

The standard deviation for the individual measurements is found with the formula $\hat{\sigma} = \bar{s}/C_4$:

$$\hat{\sigma} = \frac{\bar{s}}{C_4}$$
$$= \frac{6.33}{.9400}$$
$$= 6.73 \qquad \text{(coded)}$$

Apply the base factor:

$$\hat{\sigma} = .0001 \times 6.73$$
$$= .000673\text{`}\ \text{(actual)}$$

$$\begin{array}{rl} 1.3340 & \text{USL} \\ -1.3280 & \text{LSL} \\ \hline .0060 & \text{Tolerance} \end{array}$$

$$\text{PCR} = \frac{6\hat{\sigma}}{\text{tolerance}}$$
$$= \frac{6 \times .000673}{.0060}$$
$$= .673$$

This classifies the process as a B process (see Table B.11 in Appendix B), which uses 67% of the tolerance.

$\bar{\bar{x}}$ is closer to the USL, so

$$\text{Cpk} = \frac{\text{USL} - \bar{\bar{x}}}{3\hat{\sigma}}$$
$$= \frac{1.3340 - 1.3324}{3 \times .000673}$$
$$= \frac{.0016}{.002019}$$
$$= .79$$

$$\frac{1}{\text{Cpk}} = \frac{1}{.79}$$
$$= 1.27$$

The Cpk classifies this as a D process, which uses 127% of the tolerance on the worst side. The Cpk value, along with the PCR measurement, verifies the previous conclusion that the cut on the bushings is set too high. The process is good enough to center the distribution at the midspecification value: The PCR shows that 65% of the tolerance is used by the distribution of measurements. The Cpk indicates that the present setting is producing some bushings that are out of specification on the high side: The $3\hat{\sigma}$ spread of measurements above the mean overlaps the upper specification limit. The present setting is costly because the out-of-specification bushings must be found (inspection costs) and reworked to bring them into specification (processing costs).

Although the histogram of the in-control data, shown in Figure 7.16, indicates a slight skewness to the left, the pattern is close enough to the normal distribution to apply the normal-curve analysis:

$$
\begin{aligned}
z &= \frac{\text{USL} - \bar{\bar{x}}}{\hat{\sigma}} \\
&= \frac{1.3340 - 1.3324}{.000673} \\
&= 2.38
\end{aligned}
$$

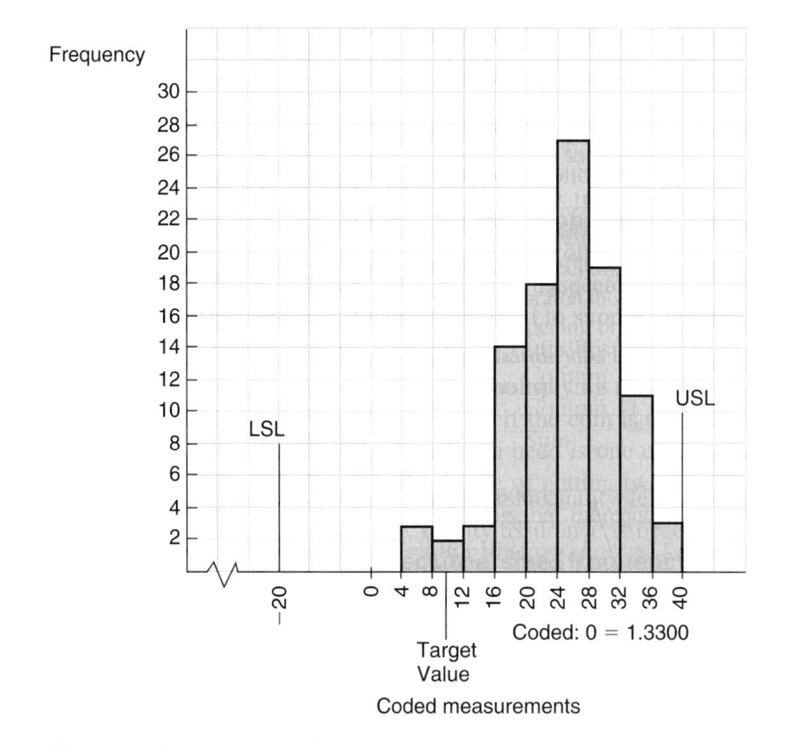

Figure 7.16
The histogram of the in-control data.

The area, according to Appendix B, Table B.8, is .0087. This indicates that .87% of the bushings must be found and reworked to bring them into specification:

$$z = \frac{\text{LSL} - \bar{\bar{x}}}{\hat{\sigma}}$$

$$= \frac{1.3280 - 1.3324}{.000673}$$

$$= -6.54$$

$$\text{Area} = .00000$$

The z value is off the chart, so the measurable tail area is .000%. This large negative value for z also shows that there is room to shift the process toward the midspecification value, 1.3310. With the process centered, there should be no problem with out-of-specification bushings. The consultant made the recommendation that the run should be stopped and the machine adjusted to the target measurement of 1.3310. Unfortunately, the SPC lessons were considered just a necessary "window dressing" for the contract battle. The decision was made that the bushings were "close enough" since all the ones checked were in specification.

The bushings were used in a group of diesel engines that were subsequently sold to the U.S. Navy. The Navy put the engines in storage, took them out two years later, and was faced with several engine failures due to oversized cam roller bushings. The ABC Company had to deal with expensive warranty costs because of a "close-enough" attitude.

Control charts, along with accompanying histograms of the data they contain, serve two basic purposes. First, they indicate trouble, and the patterns that occur offer hints of the source of trouble. In this situation, the control charts demand action. If there is no operator or management response to trouble patterns, charting is a waste of time and effort. SPC is used as "window dressing," and no quality improvement results. Second, control charts and histograms present proof of good quality. Purchasing departments should demand this proof and be able to interpret it. Sales and marketing departments should understand it and use it convincingly.

7.4 A MODIFIED $\bar{x}$ AND R CHART FOR SMALL SETS OF DATA

Two situations that can result in a limited small amount of data are the small run and the short run. The small run, where just a small number of pieces are produced, is analyzed in Section 8.4 with an x and MR chart. This section will illustrate an alternative method. The short run refers to short with respect to time. A sampling procedure may produce just several samples, but there may have been hundreds or thousands of pieces made over a short sample period.

Short and small runs have been a major problem for smaller companies. Many have not used SPC because it does not seem to apply to their limited data situation. Now, the small companies are being asked to supply proof that they are producing quality products. Large companies are also seeing more instances of small and short runs because of just-in-time (JIT) manufacturing. They are reducing their parts inventories to save on overhead costs and to improve quality by making trouble situations more traceable. In the past many sources of poor quality were hidden in large inventories. Even when JIT manufacturing is not followed strictly, a modified version using low inventories is becoming more popular. Consequently, both large and small companies have to use SPC for limited data situations to improve and maintain quality levels.

One major requirement when applying SPC for small and short runs is to put more emphasis on prerun planning. There is often a history of previous runs and information that can be used to improve the process for the coming run. Lists of possible sources of variation and problems can be analyzed to create a potentially better process.

When just a small amount of data are available, there is generally less variation than for a large set. The average range $\bar{R}$ will consequently be smaller than the true $\bar{R}$ of the process. Control limits, based on a smaller $\bar{R}$, will be narrower than they should be and the control charts will produce false out-of-control patterns. To compensate for the smaller $\bar{R}$, inflated D_4 and A_2 values can be used to establish control limits that are better estimates of the process control limits. When more data are gathered, updated values of D_4 and A_2 are used to recalculate the control limits. Eventually the standard D_4 and A_2 values are used when the number of samples increases to 25 or more. There are seven steps in the procedure:

1. Check the setup:
   a. Run 3 to 10 pieces without adjustment.
   b. Calculate $T = |\text{average} - \text{target}|/\text{range}$.
   c. Compare the calculated T value with the critical table value found in Appendix B, Table B.5. If T is less than the table value, accept the setup and begin the run. If T is greater than or equal to the table value, adjust the setup and start over.
2. Run the initial number of subgroups g.
3. Calculate $\bar{R}$ and UCL_R using D_{4F} (F for first). The values for D_{4F} and the other constants are found in Appendix B, Table B.6.
4. Eliminate any subgroups whose R value is larger than UCL_R and recalculate. This procedure continues until all remaining R values are less than the current UCL_R.
5. Calculate $UCL_{\bar{x}}$ and $LCL_{\bar{x}}$ using the final information from step 4 and A_{2F}. Again, eliminate any further samples that have $\bar{x}$ values beyond the control limits and recalculate. This also continues until all remaining $\bar{x}$ values lie between the control limits.
6. Calculate the control limits for the rest of the run using the final $\bar{\bar{x}}$ and $\bar{R}$ from step 5 with D_{4S} and A_{2S} (S for second). These are the control limits that approximate the true process control limits.
7. If the run continues past 25 subgroups, recalculate the control limits using the standard D_4 and A_2 values from Appendix B, Table B.1.

EXAMPLE 7.6

The BCD Company produces automatic-entry systems for automobiles. The specification limits for the transmitter frequency are 314.69 and 315.31 MHz. A run of 25 transmitters is going to be made:

1. Check the setup with six pieces.
2. If the setup is OK, use those initial pieces along with nine more in groups of three to establish control limits for the run using the modified $\bar{x}$ and R chart.
3. Analyze the remainder of the run.
4. A second run is made two days later:
   a. Check the setup with six pieces.
   b. If OK, use the previous control limits and averages to check for statistical control.

Solution

1. Table 7.7 contains the first 15 measurements.

Average of first six	$\bar{x} = 315.05$
Target value	315
Range of first six	$R = .37$

$$T = \frac{|\text{average} - \text{target}|}{\text{range}} = \frac{315.05 - 315}{.37} = .135$$

From Table B.5, $T_{\text{critical}} = .312$. Since T is less than T_{critical}, the setup is OK.

2. For the five samples

$$\bar{R} = \frac{\Sigma R}{5} = \frac{.82}{5} = .164$$

$$\text{UCL}_R = D_{4F} \times \bar{R} = 2.4 \times .164 = .39$$

All five R values are less than .39. Sample elimination and recalculation is not necessary.

$$\bar{\bar{x}} = \frac{\Sigma \bar{x}}{5} = \frac{1575.27}{5} = 315.054$$

TABLE 7.7
Transmitter frequencies (in order by column)

	315.09	315.12	315.05	315.15	314.89
	315.09	315.15	314.94	315.15	315.00
	315.06	314.78	315.10	315.12	315.12
$\bar{x}$	315.08	315.02	315.03	315.14	315.00
R	.03	.37	.16	.03	.23

$$\begin{aligned} UCL_{\bar{x}} &= \bar{\bar{x}} + A_{2F} \times \overline{R} \\ &= 315.054 + 1.2 \times .164 \\ &= 315.25 \\ LCL_{\bar{x}} &= \bar{\bar{x}} - A_{2F} \times \overline{R} \\ &= 315.054 - 1.2 \times .164 \\ &= 314.86 \end{aligned}$$

There are no samples whose $\bar{x}$ values are beyond the first control limits, so calculate the second control limits:

$$\begin{aligned} UCL_R &= D_{4s} \times \overline{R} \\ &= 3.4 \times .164 \\ &= .56 \end{aligned}$$

$$\begin{aligned} UCL_{\bar{x}} &= \bar{\bar{x}} + A_{2s} \times \overline{R} \\ &= 315.054 + (1.5)(.164) \\ &= 315.30 \end{aligned} \qquad \begin{aligned} LCL_{\bar{x}} &= \bar{\bar{x}} - A_{2s} \times \overline{R} \\ &= 315.054 - .246 \\ &= 314.81 \end{aligned}$$

3. The control chart with the final control limits and the remaining samples is shown in Figure 7.17. Both the R chart and the $\bar{x}$ chart are in statistical control. The average is slightly above the target.
4. The second run is shown in Figure 7.17 commencing at sample 14. For the first six measurements, $\bar{x} = 315.01$ and $R = .21$:

$$T = \frac{|\bar{x} - \text{target}|}{R} = \frac{315.01 - 315}{.21} = .05$$

T is less than the critical table value so the run can continue.

The average and control limit lines are drawn (from the first run) and the values plotted. The R chart is in control, but there may be an upward shift developing in the last five samples. The $\bar{x}$ chart shows a downward shift. Before the next run, possible causes of increased variation and lower frequencies should be investigated. Also, all the data from the two runs can be used to recalculate the control limits in preparation for the next run.

7.5 THE NOMINAL $\bar{x}$ AND R CHART

In small- and short-run situations, the operator may produce several different parts using the same machine with different setups and the same material. This consistency often results in the same standard deviation (or approximately the same) on all the parts produced. When this occurs, all the parts can be tracked on one control chart, the nominal $\bar{x}$ and R chart.

The criteria for using a nominal $\bar{x}$ and R chart are as follows:

- $6s$ is approximately the same for all measurements being tracked.
- The sample size n is the same for all parts.
- A target $\bar{x}$ may be substituted when there is no specified nominal.

Specification: 314.69 to 315.31 — First Run (samples 1–8) — Second Run (samples 14–21)

Sample Number	1	2	3	4	5	6	7	8	9	10	11	12	13	14	15	16	17	18	19	20	21	22	23	24	25
Coded 0 = 315.00	9	12	5	15	−11	−2	12	7						5	−3	−15	−8	−21	−12	−13	10				
	9	15	−6	15	0	5	−9	3						−3	−9	−13	7	18	14	9	15				
	6	−22	10	12	12	1	−5	1						4	12	−1	−19	−6	6	−1	−7				
Average $\bar{x}$	8	2	3	14	0	1	−1	4						2	0	−10	−7	−3	3	−2	6				
Range R	3	37	16	3	23	7	23	6						8	21	14	26	39	26	22	22				

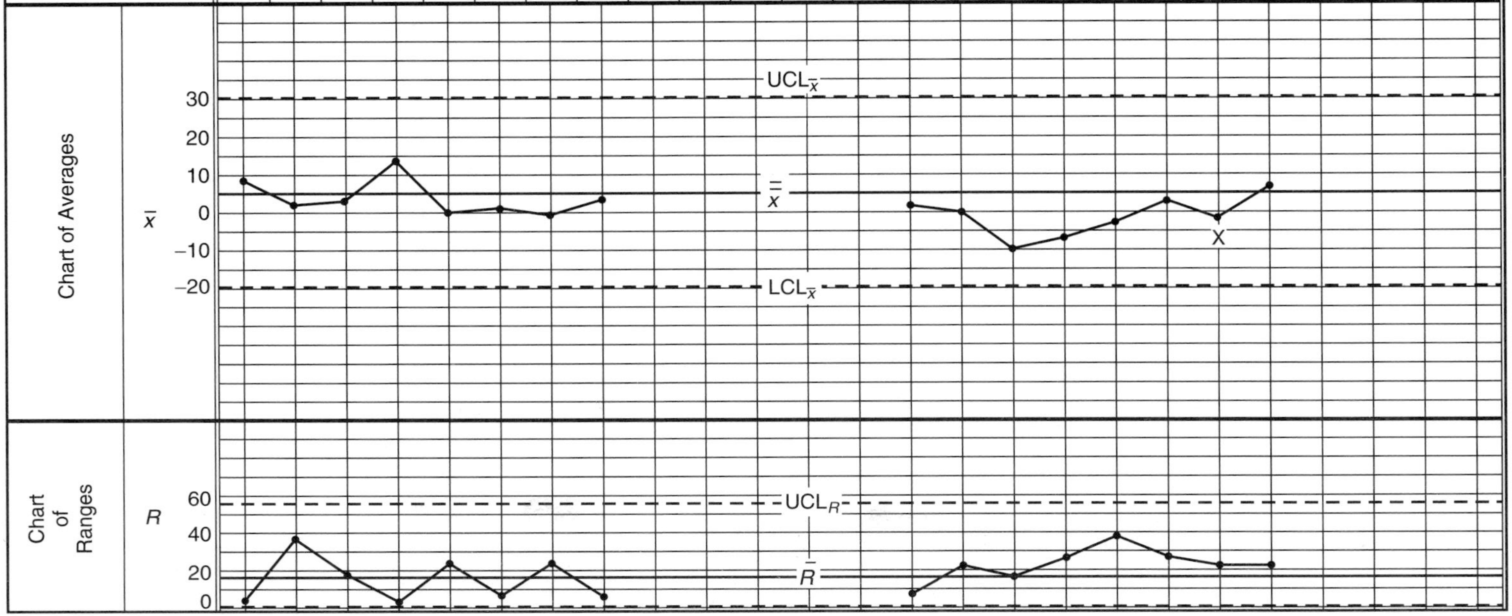

Figure 7.17
The modified $\bar{x}$ and R chart for Example 7.6.

The nominal $\bar{x}$ and R procedure.

1. Code each measurement by subtracting the nominal, x – nominal.
2. Chart the coded measurements on the $\bar{x}$ chart and separate the different part runs with vertical lines.
3. Chart the range values on the R chart as usual.
4. Calculate the averages and control limits, draw the lines, and analyze the chart in the usual manner. If one of the parts shows a shift on the range chart, it indicates that the assumption of equal standard deviations was wrong; the one part has a different standard deviation and should be tracked on its own control chart.

$$\text{Coded } x = x - \text{nominal}$$

$$\text{Coded } \bar{x} = \frac{\Sigma\,(x - \text{nominal})}{n}$$ For each sample (just the usual calculation of average values)

$$\bar{\bar{x}}_c = \frac{\Sigma\,(\text{coded } \bar{x})}{k}$$ Average coded $\bar{x}$ (k is number of samples)

$$\text{UCL}_{\bar{x}} = \bar{\bar{x}}_c + A_2\bar{R}$$ The rest of the formulas are the same as the regular $\bar{x}$ and R charts.

$$\text{LCL}_{\bar{x}} = \bar{\bar{x}}_c - A_2\bar{R}$$

$$\bar{R} = \frac{\Sigma R}{k}$$

$$\text{UCL}_R = D_4 \times R$$
$$\text{LCL}_R = D_3 \times R$$

EXAMPLE 7.7 Table 7.8 contains the data for three parts that were made from the same material on the same machine by the same operator. Make a nominal $\bar{x}$ and R chart and analyze the results. Use all the data to set the scales.

Solution

The coded x values, x – nominal, are entered on the control chart, and the average and range are calculated. All coded values represent thousandths.

The range of $\bar{x}$ is 2.5 – (–3.5) = 6 coded units.

Two spaces per coded unit will use 12 spaces.

The range of R is 4 – 0 = 4. Use one space per unit.

$$\bar{\bar{x}}_c = \frac{\Sigma \bar{x}_c}{k} = \frac{2.75}{22} = .125$$

$$\text{UCL}_{\bar{x}} = \bar{\bar{x}}_c + A_2 \times \bar{R}_c = .125 + .729 \times 2.23 = 1.75$$

$$\bar{R}_c = \frac{\Sigma R_c}{k} = \frac{49}{22} = 2.23$$

$$\text{UCL}_R = D_4 \times \bar{R}_c = 2.282 \times 2.23 = 5.1$$

TABLE 7.8
Samples with $n = 4$ of parts A, B, and C

Part A. Specification: 2.395 to 2.405

Sample	1	2	3	4	5	6	7	
	2.398	2.403	2.402	2.400	2.397	2.399	2.399	
	2.402	2.399	2.401	2.398	2.399	2.400	2.400	
	2.399	2.399	2.401	2.397	2.399	2.402	2.400	
	2.400	2.401	2.402	2.400	2.397	2.400	2.400	

Part B. Specification: 1.245 to 1.255

Sample	8	9	10	11	12	13	14	15
	1.251	1.250	1.252	1.253	1.247	1.252	1.253	1.251
	1.251	1.249	1.248	1.249	1.250	1.251	1.252	1.251
	1.252	1.251	1.251	1.250	1.251	1.250	1.253	1.251
	1.252	1.250	1.249	1.252	1.248	1.250	1.252	1.251

Part C. Specification: .695 to .705

Sample	16	17	18	19	20	21	22	
	.696	.698	.699	.700	.701	.703	.703	
	.697	.697	.698	.700	.700	.701	.702	
	.697	.699	.700	.702	.702	.701	.702	
	.696	.697	.700	.699	.700	.702	.702	

$$\begin{aligned} LCL_{\bar{x}} &= \bar{\bar{x}}_c - A_2 \times \bar{R}_c \\ &= .125 - .729 \times 2.23 \\ &= -1.50 \end{aligned} \qquad \begin{aligned} LCL_R &= D_3 \times \bar{R}_c \\ &= 0 \times 2.23 \\ &= 0 \end{aligned}$$

All the calculated values are in coded form. If the capability had to be determined, the $\bar{R}_c$ value would have to be multiplied by .001 to change it to a true range average. There are not enough "good" data for a capability analysis: Too many points are out of control. The R chart is in control in all three sections. There is a problem with the $\bar{x}$ chart: Five points are beyond the specification limits and a run pattern is showing on part C (technically seven increasing points are needed for a run).

The chart is presented as Figure 7.18.

7.6 THE TRANSFORMATION $\bar{x}$ AND R CHART

The transformation $\bar{x}$ and R chart can be used to track several measurements on one chart with no restrictions other than that they all have to have the same sample size. The transformation formulas make the charted points independent of the units used and the standard deviations of the different measurements. The calculations for the transformation are com-

Specification: 2.395 to 2.405 A; 1.245 to 1.255 B; .695 to .705 C

Sample Number	1	2	3	4	5	6	7	8	9	10	11	12	13	14	15	16	17	18	19	20	21	22	23	24	25
Coded Target = Nominal R_x .001	−2	3	2	0	−3	−1	−1	1	0	2	3	−3	2	3	1	−4	−2	−1	0	1	3	3			
	2	−1	1	−2	−1	0	0	1	−1	−2	−1	0	1	2	1	−3	−3	−2	0	0	1	2			
	−1	−1	1	−3	−1	2	0	2	1	1	0	1	0	3	1	−3	−1	0	2	2	1	2			
	0	1	2	0	−3	0	0	2	0	−1	2	−2	0	2	1	−4	−3	0	−1	0	2	2			
Average $\bar{x}$	−.25	.5	1.5	−1.25	−2	.25	−.25	1.5	0	0	1	−1	.75	2.5	1	−3.5	−2.25	−.75	.25	.75	1.75	2.25			
Range R	4	4	1	3	2	3	1	1	2	4	4	4	2	1	0	1	2	2	3	2	2	1			

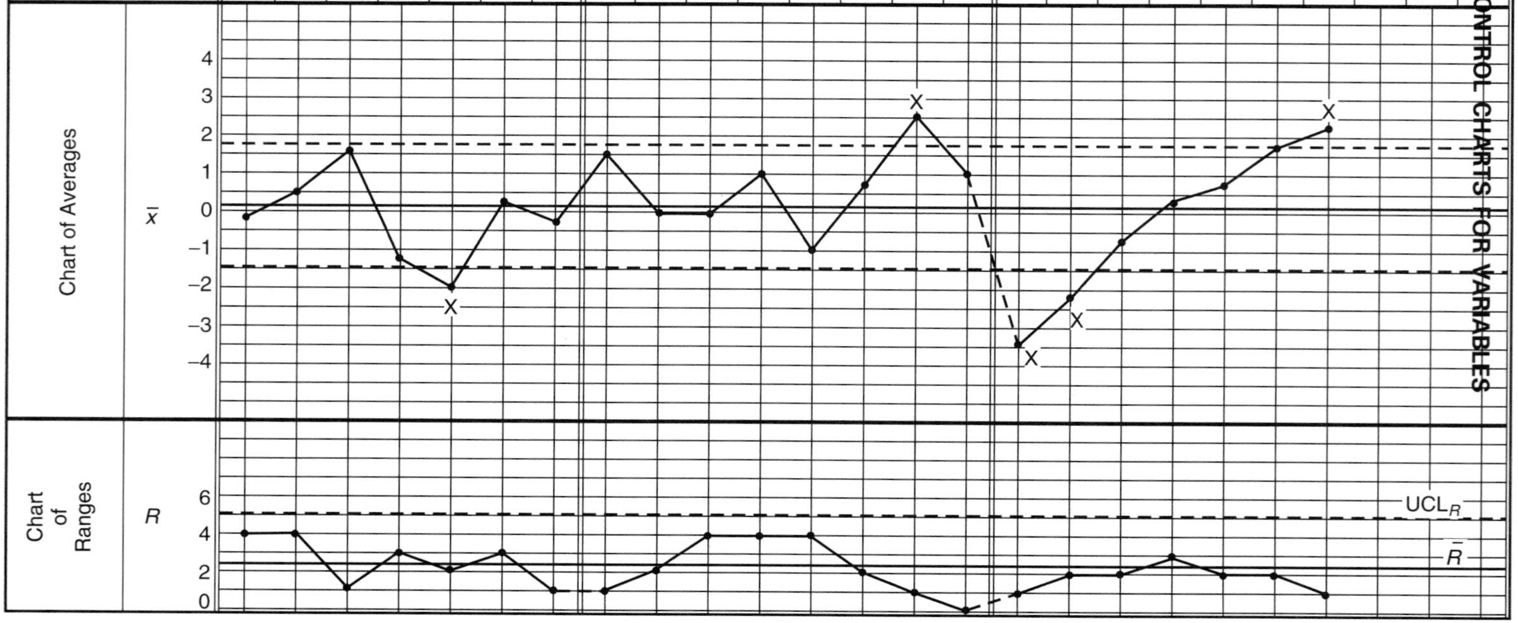

Figure 7.18
The nominal $\bar{x}$ and R chart for Example 7.7.

plicated, but it does allow $\bar{x}$ and R charting for several small or short runs together. The transformation $\bar{x}$ and R chart is an alternative to using the x and MR chart for small and short runs. It would not be used in place of the nominal $\bar{x}$ and R chart because when the nominal $\bar{x}$ and R chart applies to a situation, it is easier to use than the transformation chart.

The Transformation Formulas

The target R value, $\overline{R}_T$, is determined from the capability:

$$\text{Since Cp} = \frac{\text{tolerance}}{6s} \quad \text{then} \quad s = \frac{\text{tolerance}}{6\text{Cp}} \quad \text{(Solve for } s\text{)}$$

$$s = \frac{\overline{R}_T}{d_2} \quad \text{The Shewhart formula with the target range } \overline{R}_T$$

$$\frac{\overline{R}_T}{d_2} = \frac{\text{tolerance}}{6\text{Cp}} \quad \text{Substitute.}$$

$$\overline{R}_T = \frac{d_2 \times \text{tolerance}}{6\text{Cp}} \quad \text{Solve for } \overline{R}_T.$$

The transformation from R to R_{point}:

$$D_3\overline{R}_T < R < D_4\overline{R}_T \quad \text{The } R \text{ values lie between the control limits.}$$

$$\frac{D_3\overline{R}_T}{\overline{R}_T} < \frac{R}{\overline{R}_T} < \frac{D_4\overline{R}_T}{\overline{R}_T} \quad \text{Divide by } \overline{R}_T.$$

$$D_3 < \frac{R}{\overline{R}_T} < D_4 \quad \text{The quotient lies between } D_3 \text{ and } D_4.$$

$$R_{\text{point}} = \frac{R}{\overline{R}_T}$$

Here R_{point} is the value that is charted and the control limits will be D_3 and D_4. The average line will always be 1.

The transformation from $\bar{x}$ to $\bar{x}_{\text{point}}$:

$$\bar{\bar{x}} - A_2\overline{R}_T < \bar{x} < \bar{\bar{x}} + A_2\overline{R}_T \quad \bar{x} \text{ lies between the control limits.}$$

$$-A_2\overline{R}_T < \bar{x} - \bar{\bar{x}} < A_2\overline{R}_T \quad \text{Subtract } \bar{\bar{x}}.$$

$$\frac{-A_2\overline{R}_T}{\overline{R}_T} < \frac{\bar{x} - \bar{\bar{x}}}{\overline{R}_T} < \frac{A_2\overline{R}_T}{\overline{R}_T} \quad \text{Divide by } \overline{R}_T.$$

$$-A_2 < \frac{\bar{x} - \bar{\bar{x}}}{\overline{R}_T} < A_2$$

$$\bar{x}_{\text{point}} = \frac{\bar{x} - \bar{\bar{x}}}{\overline{R}_T}$$

Here $\bar{x}_{\text{point}}$ is the value that is charted. The average line will always be 0 and the control limits will always be $-A_2$ and A_2.

The Transformation $\bar{x}$ and R Procedure

1. Calculate $\overline{R}_T$:

$$\overline{R}_T = \frac{d_2 \times \text{tolerance}}{6\text{Cp}} \qquad \text{Use the target Cp or the historical Cp.}$$

2. Transform each R to R_{point} for plotting:

$$R_{\text{point}} = \frac{R}{\overline{R}_T}$$

3. Transform each $\bar{x}$ to $\bar{x}_{\text{point}}$ for plotting:

$$\bar{x}_{\text{point}} = \frac{\bar{x} - \bar{\bar{x}}}{\overline{R}_T}$$

The transformation provides $\bar{x}$ and R capabilities to small data situations by blending them together on one chart for analysis. The chart and the analysis are independent of both the units and standard deviations of the individual measurements used.

EXAMPLE 7.8 An audit is performed on three- and four-button transmitters for automatic-entry systems. The power and the frequency are checked on each transmitter. Five samples of size $n = 5$ are measured and the data are analyzed on a transformation $\bar{x}$ and R chart. The specifications are

Power: 65 to 75

Frequency: 4.69 to 5.31

Cp = 1.33 (the target Cp)

Solution

The samples are recorded on the transformation $\bar{x}$ and R chart Figure 7.19. Do the following work on Figure 7.19 and check your results with Figure 7.20.

Calculate all sample averages and ranges. Enter them on the chart.

Calculate $\overline{R}_T$ for both frequency and power:

$$\overline{R}_T = \frac{d_2 \times \text{tolerance}}{6 \times \text{Cp}} = \frac{2.326 \times .62}{6 \times 1.33} = .1807 \quad \text{for frequency}$$

$$\overline{R}_T = \frac{d_2 \times \text{tolerance}}{6 \times \text{Cp}} = \frac{2.326 \times 10}{6 \times 1.33} = 2.9148 \quad \text{for power}$$

Change each R value to R_{point} for plotting with $R_{\text{point}} = \dfrac{R}{\overline{R}_T}$

Calculate $\bar{\bar{x}}$ for each different set of data.

Change each $\bar{x}$ to $\bar{x}_{\text{point}}$ for plotting with $\bar{x}_{\text{point}} = (\bar{x} - \bar{\bar{x}})/\overline{R}_T$.

Specification:

		3 - Button Frequency						3 - Button Power						4 - Button Frequency						4 - Button Power						
Date							$\bar{\bar{x}}$						$\bar{\bar{x}}$						$\bar{\bar{x}}$						$\bar{\bar{x}}$	
$\bar{x}$ point																										
R point																										
Sample Measurements	1	5.09	4.78	5.15	4.98	4.95		72.43	66.23	68.22	67.10	67.60		5.05	5.12	5.07	4.88	4.99		68.09	65.72	69.47	65.52	72.23		
	2	5.09	5.05	5.12	5.05	5.07		70.27	69.99	71.76	71.77	71.36		4.97	4.85	4.81	5.14	5.10		67.31	66.74	71.77	68.73	52.73		
	3	5.06	4.94	4.89	5.01	5.03		71.84	72.09	70.58	68.27	71.64		5.04	4.87	4.79	5.06	5.15		66.31	71.91	71.94	69.19	70.11		
	4	5.12	5.10	5.00	5.12	5.01		71.09	69.75	64.71	70.65	68.43		4.97	4.99	5.18	4.87	4.93		71.32	63.28	71.41	54.07	63.32		
	5	5.15	5.15	5.12	4.91	4.75		66.93	72.48	69.71	69.92	65.82		4.91	4.92	4.94	5.09	4.84		72.05	65.42	64.46	71.88	66.33		
Sum																										
Average $\bar{x}$																										
Range R																										

	1	2	3	4	5	6	7	8	9	10	11	12	13	14	15	16	17	18	19	20	21	22	23	24	25
Averages																									
Ranges																									

Figure 7.19
Complete the transformation $\bar{x}$ and R chart for Example 7.8.

Specification:		4.69 to 5.31 3 - Button Frequency						65 to 75 3 - Button Power						4.69 to 5.31 4 - Button Frequency						65 to 75 4 - Button Power						
Date							$\bar{\bar{x}}$						$\bar{\bar{x}}$						$\bar{\bar{x}}$						$\bar{\bar{x}}$	
$\bar{x}$ point		.39	−.17	.17	−.11	−.39		.30	.16	−.22	−.03	−.23		.06	−.17	−.11	.17	.11		.61	−.22	.88	−.47	−.79		
R point		.498	2.048	1.439	1.162	1.771		1.887	2.010	2.419	1.602	1.997		.775	1.494	2.158	1.494	1.716		1.969	2.961	2.566	6.110	6.690		
Sample Measurements	1	5.09	4.78	5.15	4.98	4.95		72.43	66.23	68.22	67.10	67.60		5.05	5.12	5.07	4.88	4.99		68.09	65.72	69.47	65.52	72.23		
	2	5.09	5.05	5.12	5.05	5.07		70.27	69.99	71.76	71.77	71.36		4.97	4.85	4.81	5.14	5.10		67.31	66.74	71.77	68.73	52.73		
	3	5.06	4.94	4.89	5.01	5.03		71.84	72.09	70.58	68.27	71.64		5.04	4.87	4.79	5.06	5.15		66.31	71.91	71.94	69.19	70.11		
	4	5.12	5.10	5.00	5.12	5.01		71.09	69.75	64.71	70.65	68.43		4.97	4.99	5.18	4.87	4.93		71.32	63.28	71.41	54.07	63.32		
	5	5.15	5.15	5.12	4.91	4.75		66.93	72.48	69.71	69.92	65.82		4.91	4.92	4.94	5.09	4.84		72.05	65.42	64.46	71.88	66.33		
Sum																										
Average $\bar{x}$		5.10	5.00	5.06	5.01	4.96	5.03	70.51	70.11	69.00	69.54	68.97	69.63	4.99	4.95	4.96	5.01	5.00	4.98	69.02	66.61	69.81	65.88	64.94	67.25	
Range R		.09	.37	.26	.21	.32		5.50	5.86	7.05	4.67	5.82		.14	.27	.39	.27	.31		5.74	8.63	7.48	17.81	19.50		

Figure 7.20
The completed transformation $\bar{x}$ and R chart for Example 7.8.

First-sample calculations:

$$R_{\text{point}} = \frac{R}{\overline{R}_T} = \frac{.09}{.1807} = .498$$

$$\overline{\overline{x}} = \frac{\Sigma \overline{x}}{5} = \frac{25.13}{5} = 5.03$$

$$\overline{x}_{\text{point}} = \frac{\overline{x} - \overline{\overline{x}}}{\overline{R}_T} = \frac{5.10 - 5.03}{.1807} = .39$$

Set the scale on the R chart from $D_3 = 0$ to $D_4 = 2.115$. There are 25 lines, so use .1 unit per line.

Chart the R_{point} values.

Draw the average line at 1.0 and the upper control limit line at $D_4 = 2.115$.

Set the scale on the $\overline{x}$ chart. The center line is 0, the upper control limit is $A_2 = .577$, and the lower control limit is $-A_2 = -.577$. Let each line be .05 units.

Chart Analysis

In the R chart, all but two of the points are above the average, $\overline{R} = 1$, line. This is a shift up on a variation chart that indicates that there is excessive variation for both power and frequency on the transmitters. The Cp = 1.33 requirement is not being satisfied. The frequency sections have just one range value beyond the control limit and the $\overline{x}$ sections are both in control. The three-button power range has one point beyond the control limit that corresponds to one individual power value below specification. The four-button power ranges have four of the five points beyond the control limit (and off the chart), and they correspond to several individual power values below specification.

7.7 CONTROL CHART SELECTION

The first decision in choosing the type of control chart is to determine the type of data that are to be collected. If the data consist of counts, then one of the attributes charts should be selected. The second decision for attributes charts is whether the number of nonconforming units in a sample is counted or the total number of nonconformances in a sample is counted. The third decision is whether the sample size will remain constant or whether it will vary from sample to sample. The attributes charts, *p, np, u,* and *c charts,* are discussed in Chapter 9. Figure 7.21 illustrates the basic choices among the attributes charts.

If the data are measured data, then the control charts that were discussed in this chapter and the charts that will be discussed in Chapter 8 will apply. The second decision for measured data depends on what sample sizes make sense for the situation. The most common choices for the measured data would be between $\overline{x}$ and R, $\overline{x}$ and s, and $\tilde{x}$ and R charts. If the mathematical abilities of the work force are limited, the $\tilde{x}$ and R charts would be the best choice, since it is easier to find a middle number in a small sample than

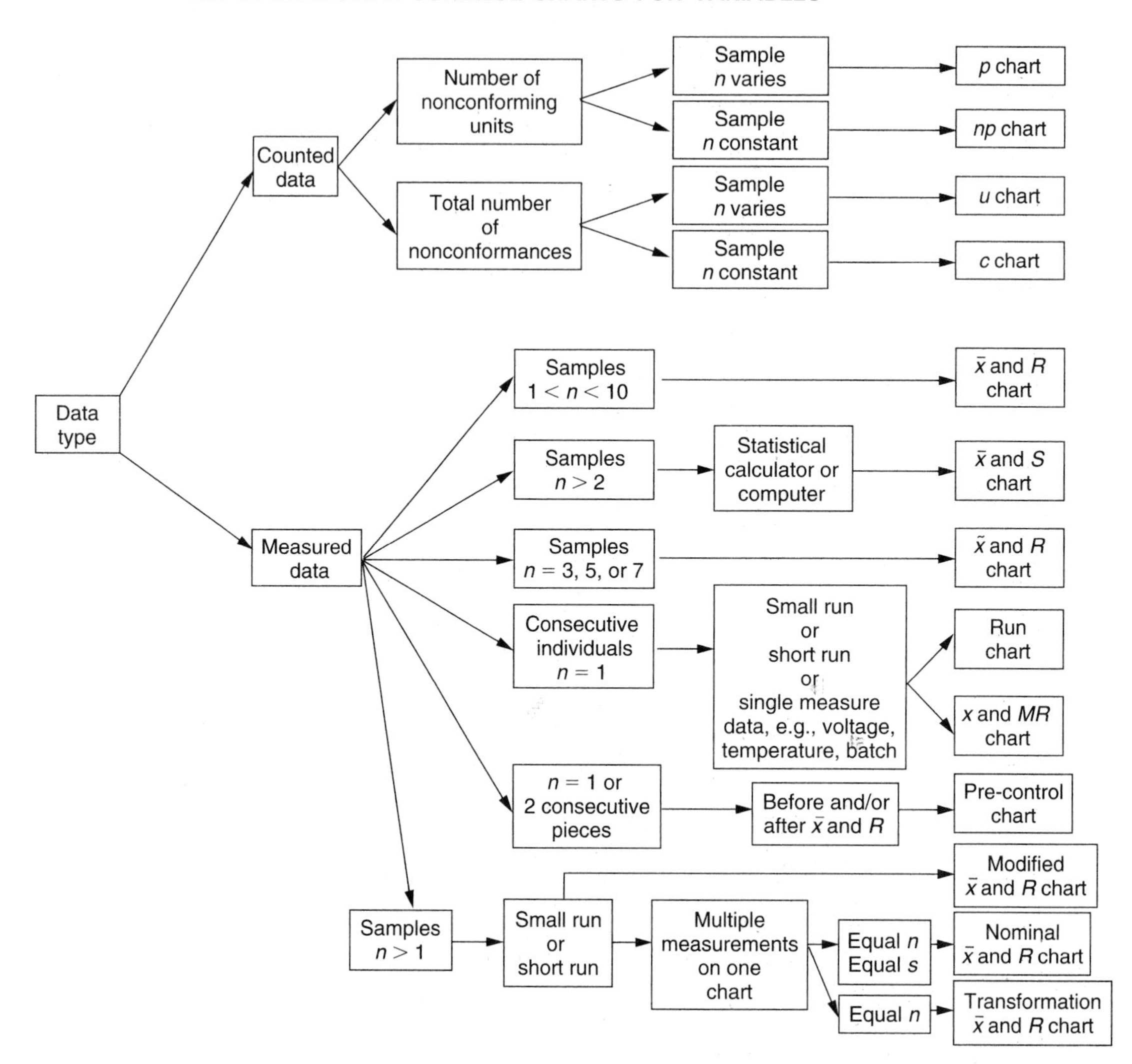

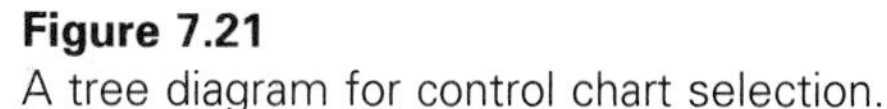

Figure 7.21
A tree diagram for control chart selection.

to calculate the average. If the work force is supplied with statistical calculators or computers, $\bar{x}$ and s charts would be the most efficient. Once the data from a sample are entered in a statistical calculator it is quicker to find s with the press of a button than to inspect the sample, find the largest and smallest, and subtract for the R value. The usual choice, however, would be the $\bar{x}$ and R chart that is currently the industry standard and is most widely understood. The other measurement control charts apply for special situations. Small runs with limited data or short runs with limited time could utilize the *modified, nominal,* or *transformation* $\bar{x}$ and R charts. The charts that are discussed in Chapter 8 are also special-use charts. The precontrol charts are used before $\bar{x}$ and R charts to determine if the charts are needed and also after the $\bar{x}$ and R charts to verify that the process is staying in control. The *run* chart and the x and MR charts apply for both limited data and batch data. The tree diagram in Figure 7.21 displays the control chart decision process.

EXERCISES

1. The specification limits for a process are given as 2.740 to 2.770. The first seven samples are given in Table 7.9:
   a. Set the scale for a median chart with 25 lines.
   b. Set the scale for a standard deviation chart with 10 lines.

TABLE 7.9

Sample	1	2	3	4	5	6	7
	2.754	2.741	2.758	2.743	2.748	2.762	2.749
	2.759	2.753	2.753	2.747	2.743	2.761	2.753
	2.748	2.747	2.747	2.752	2.749	2.758	2.768
	2.759	2.751	2.750	2.749	2.751	2.755	2.763
	2.752	2.754	2.751	2.747	2.745	2.757	2.762

2. Make an $\bar{x}$ and s chart for the data in Figure 7.22. Analyze it and compare the results with Example 6.7.
3. Make a range chart to match with the $\bar{x}$ chart in Figure 7.14. How does the $\bar{x}$ and R analysis compare with the $\bar{x}$ and s analysis? Do a capability analysis using $\bar{R}$.
4. Table 7.10 contains the measurements of the first pieces produced after setup. Use the T test and the first eight (by column) to determine if the run can commence. The target value is 70.

TABLE 7.10
The first measurements after setup

72.43	66.93	69.75
70.27	66.23	72.48
71.84	69.99	68.22
71.09	72.09	71.76

5. Use the three samples ($n = 4$) in Table 7.10 to calculate the control limits for a process run with a modified $\bar{x}$ and R chart.
6. Figure 7.23 contains short-run data in samples of size $n = 4$. Make a modified $\bar{x}$ and R chart using the averages and control limits from Exercise 5. Analyze for statistical control.
7. Acme produces a small run of 21 shell blanks containing three critical dimensions: hole diameter, body diameter, and hole depth. Since all three were done on one machine by the same operator, make a nominal $\bar{x}$ and R chart with $n = 3$ and analyze the results. The sample

$\overline{S} = \frac{\Sigma S}{k} = $ ______ = ______

$\overline{\overline{x}} = \frac{\Sigma \overline{x}}{k} = $ ______ = ______

$UCL_S = B_4 \times \overline{S} = $ ____ × ____ = ____

$A_3 \times \overline{S} = $ ____ × ____ = ☐

$UCL_{\overline{x}} = \overline{\overline{x}} + A_3\,\overline{S} = $ ____ + ☐ = ____

$LCL_{\overline{x}} = \overline{\overline{x}} - A_3\,\overline{S} = $ ____ − ☐ = ____

Specification: .0200 to .0300

Chart of Averages — $\overline{x}$

Chart of Standard Deviation — S

Sample Number		1	2	3	4	5	6	7	8	9	10	11	12	13
Sample Measurements	1	.0218	.0247	.0244	.0254	.0265	.0259	.0266	.0259	.0302	.0221	.0219	.0205	.0264
	2	.0243	.0255	.0252	.0238	.0232	.0274	.0245	.0282	.0271	.0247	.0242	.0253	.0267
	3	.0232	.0282	.0265	.0249	.0294	.0228	.0231	.0264	.0250	.0256	.0223	.0241	.0253
	4	.0256	.0261	.0267	.0275	.0281	.0254	.0280	.0234	.0254	.0246	.0258	.0236	.0278
	5													
Sum														
Average, $\overline{x}$		.0237	.0261	.0257	.0254	.0268	.0254	.0256	.0260					
Standard Deviation s		.0016	.0015	.0011	.0016	.0027	.0019	.0022	.0020					

Sample Number		14	15	16	17	18	19	20	21	22	23	24	25
Sample Measurements	1	.0251	.0259	.0254	.0263	.0258	.0273	.0213	.0223	.0252	.0248	.0253	.0260
	2	.0274	.0221	.0257	.0267	.0243	.0266	.0256	.0251	.0224	.0232	.0262	.0229
	3	.0236	.0245	.0263	.0254	.0249	.0264	.0294	.0248	.0263	.0250	.0221	.0232
	4	.0265	.0255	.0230	.0240	.0230	.0232	.0275	.0241	.0247	.0208	.0243	.0251
	5												
Sum													
Average, $\overline{x}$													
Standard Deviation s													

Figure 7.22
Calculate $\overline{x}$ and s and set the scales for the $\overline{x}$ and S charts. Plot and connect the points.

Sample Number	1	2	3	4	5	6	7	8	9	10	11	12	13	14	15	16
	72.43	66.93	69.75	70.58	71.77	67.60	68.09	72.05	63.28	71.94	68.73	72.23				
	70.27	66.23	72.48	64.71	68.27	71.36	67.31	65.72	65.42	71.41	69.19	52.73				
	71.84	69.99	68.22	69.71	70.65	71.64	66.31	66.74	69.47	64.46	54.07	70.11				
	71.09	72.09	71.76	67.10	69.92	68.43	71.32	71.91	71.77	65.52	71.88	63.32				
Average $\bar{x}$																
Range R																
Chart of Averages $\bar{x}$																
Chart of Ranges R																

Figure 7.23
Make a modified $\bar{x}$ and R chart for exercise 6.

measurements are given in Table 7.11. Code the data and chart it on Figure 7.24. Pieces 1, 2, and 3 (by row) will form the first sample, pieces 4, 5, and 6 the second, etc. The nominal values are:

Hole diameter .770

Body diameter .871

Hole depth .900

8. In a production line, a part is worked on by three different operators. Chart the small run using a transformation $\bar{x}$ and R chart. The sample data are entered on the control chart in Figure 7.25. Complete the chart and analyze it.

TABLE 7.11

Data for exercise 7

Piece	1	2	3	4	5	6	7	8
Hole diameter	.7695	.7695	.7690	.7700	.7700	.7720	.7710	.7715
Body diameter	.8655	.8710	.8706	.8705	.8710	.8720	.8710	.8715
Hole depth	.8985	.8980	.8980	.8990	.8990	.8975	.8985	.8990
Piece	9	10	11	12	13	14	15	16
Hole diameter	.7715	.7715	.7715	.7670	.7715	.7715	.7705	.7695
Body diameter	.8730	.8715	.8715	.8715	.8720	.8715	.8720	.8720
Hole depth	.8980	.8990	.9000	.9000	.8985	.8985	.8975	.8980
Piece	17	18	19	20	21			
Hole diameter	.7695	.7695	.7695	.7715	.7715			
Body diameter	.8715	.8715	.8720	.8710	.8720			
Hole depth	.9000	.8980	.8990	.8990	.8960			

9. Make an s chart to accompany the $\bar{x}$ chart in Figure 6.6.
   a. Is there any difference in the control analysis based on $\bar{s}$?
   b. Compare the capability analysis based on $\bar{s}$ with that of $\bar{R}$.

10. Use copies of the blank control chart in Figure 7.26 and the data in Table 7.12. The data are in order by column.
   a. Make an $\tilde{x}$ and R chart with $n = 5$.
   b. Make an $\bar{x}$ and s chart with $n = 4$. Compare with Exercise 17 in Ch 6.

11. Do a complete capability analysis for the two charts in Exercise 10. Specification limits are .275 to .325. Be sure to eliminate any out-of-control data first. (PCR, C_p, C_{pk}, percent out of specification). Compare the results of 11b with Exercise 19 in Chapter 6.

12. Given the measurements .812, .824, .817, .820, .815, .813, and .817.
   a. Code the measurements using the base .810.
   b. Find the average measurement using the coded values.
   c. Check your answer to part (b) by averaging the measurements.

TABLE 7.12

.302	.304	.300	.316	.321	.297	.291	.293	.284
.305	.306	.312	.323	.314	.285	.284	.284	.286
.299	.297	.286	.310	.317	.290	.286	.290	.286
.301	.306	.290	.315	.323	.292	.285	.286	.280
.306	.312	.285	.318	.288	.292	.287	.283	
.310	.312	.296	.307	.328	.284	.282	.280	
.308	.308	.294	.314	.314	.289	.285	.281	
.306	.304	.312	.303	.307	.285	.284	.292	
.312	.304	.308	.317	.286	.282	.287	.293	
.314	.315	.307	.319	.284	.287	.287	.289	
.298	.310	.302	.322	.301	.288	.288	.288	
.299	.302	.312	.325	.289	.283	.291	.291	

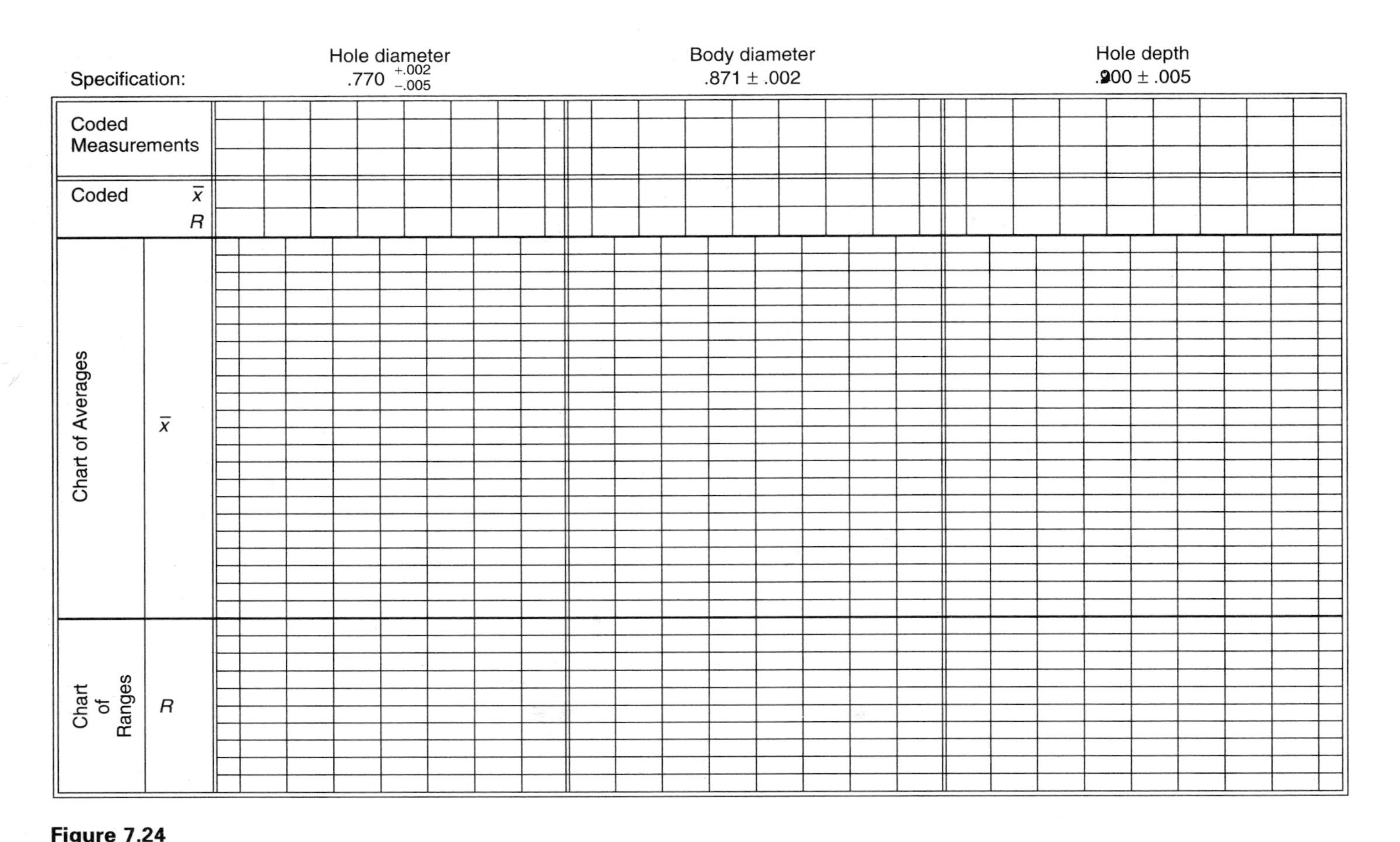

Figure 7.24
Make the nominal $\bar{x}$ and R chart for exercise 7.

Specification: Body length $2.500 \pm .003$ | Depth of groove $.150 \pm .002$ | Hole diameter $1.25 \pm .05$

Specification:		Body length 2.500 ± .003								Depth of groove .150 ± .002								Hole diameter 1.25 ± .05					
Date																							
$\bar{x}$ point																							
R point																							
Sample Measurements	1	2.504	2.502	2.498	2.499	2.500	2.500	2.502		.1504	.1515	.1509	.1482	.1505	.1510	.1493		1.23	1.21	1.25	1.24	1.29	1.24	1.27
	2	2.500	2.500	2.497	2.498	2.502	2.501	2.503		.1508	.1508	.1500	.1495	.1510	.1512	.1500		1.22	1.26	1.27	1.28	1.27	1.22	1.23
	3	2.499	2.501	2.499	2.501	2.499	2.500	2.501		.1499	.1520	.1493	.1500	.1497	.1503	.1497		1.26	1.25	1.26	1.22	1.25	1.26	1.25
	4	2.501	2.500	2.501	2.500	2.499	2.499	2.501		.1490	.1510	.1488	.1492	.1496	.1495	.1510		1.25	1.24	1.28	1.25	1.27	1.23	1.26
	5																						
Sum																							
Average $\bar{x}$																							
Range R																							

	1	2	3	4	5	6	7	8	9	10	11	12	13	14	15	16	17	18	19	20	21	22	23
Averages																							
Ranges																							

Figure 7.25
Complete the transformation $\bar{x}$ and R chart for exercise 8.

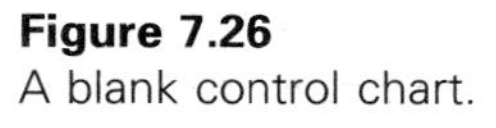

Figure 7.26
A blank control chart.

13. Samples of size $n = 4$ are given:

.314	.312	.316	.317	.313
.318	.315	.315	.314	.315
.312	.314	.315	.316	.315
.315	.314	.316	.314	.312

a. Code the measurements using the base .315.
b. Find the average measurement using the coded values.
c. Check the answer by averaging the measurements.
d. Find the five range values, R, from the coded values and then from the measurements.
e. Using

$$\hat{\sigma} = \frac{\overline{R}}{2.059}$$

find $\hat{\sigma}$ from the coded values and then from the measurements.
f. Code the measurements using the base .310 and repeat (**b**), (**d**), and (**e**).

14. Given the samples in Figure 7.27.
a. Complete the median and range chart.
b. Draw the average and control limit lines and analyze for statistical control.
c. Eliminate any out-of-control data and do a complete capability study:
1) PCR, C_p, C_{pk}, and the percentage tolerance used.
2) Calculate the percentage of product that is out of specification.

15. Given the samples in Figure 7.28.
a. Complete the average and standard deviation chart.
b. Draw the average and control limit lines and analyze for statistical control.
c. Eliminate any out-of-control data and do a complete capability study.
1) PCR, C_p, C_{pk}, and the percentage of tolerance used.
2) Calculate the percentage of product that is out of specification.

$\bar{R} = \frac{\Sigma R}{k} = \_\_\_\_ = \_\_\_\_$

$\bar{\tilde{x}} = \frac{\Sigma \tilde{x}}{k} = \_\_\_\_ = \_\_\_\_$

Specification: .7450 to .7470

$UCL_R = D_4 \times \bar{R} = \_\_\_\_ \times \_\_\_\_ = \_\_\_\_$

$\tilde{A}_2 \times \bar{R} = \_\_\_\_ \times \_\_\_\_ = \square$

$UCL_{\tilde{x}} = \bar{\tilde{x}} + \tilde{A}_2 \bar{R} = \_\_\_\_ + \square = \_\_\_\_$

$LCL_{\tilde{x}} = \bar{\tilde{x}} - \tilde{A}_2 \bar{R} = \_\_\_\_ - \square = \_\_\_\_$

Constants

n	A_2	$\tilde{A}_2$	D_4	d_2
3	1.023	1.19	2.574	1.693
4	0.729	–	2.282	2.059
5	0.577	.69	2.114	2.326

	.7465	.7460	.7455	.7465	.7460	.7455	.7455	.7460	.7465	.7455	.7445	.7455	.7450	.7455	.7460	.7470	.7460	.7470	.7460	.7465	.7450	.7460	.7450	.7460	.7460
	.7450	.7465	.7450	.7470	.7465	.7440	.7465	.7470	.7460	.7455	.7465	.7465	.7460	.7470	.7465	.7460	.7445	.7455	.7450	.7460	.7460	.7465	.7455	.7465	.7465
	.7455	.7445	.7450	.7460	.7465	.7460	.7465	.7455	.7460	.7455	.7460	.7450	.7450	.7460	.7465	.7475	.7445	.7455	.7465	.7460	.7455	.7455	.7455	.7460	.7475
$\tilde{x}$																									
R																									

Chart of Medians — $\tilde{x}$

Chart of Ranges — R

Figure 7.27
Make a median and range chart.

$\bar{s} = \frac{\Sigma s}{k} =$ ______ = ______

$\bar{\bar{x}} = \frac{\Sigma x}{k} =$ ______ = ______

Specification: .7450 to .7470

$UCL_s = B_4 \times \bar{s} =$ ____ × ____ = ____

$A_3 \times \bar{s} =$ ____ × ____ = ☐

$UCL_{\bar{x}} = \bar{\bar{x}} + A_3\,\bar{s} =$ ____ + ☐ = ____

$LCL_{\bar{x}} = \bar{\bar{x}} - A_3\,\bar{s} =$ ____ − ☐ = ____

Constants

n	A_3	B_3	B_4	C_4
3	1.954	0	2.568	.8862
4	1.628	0	2.266	.9213
5	1.427	0	2.089	.9400

	1	2	3	4	5	6	7	8	9	10	11	12	13	14	15	16	17	18	19	20	21	22	23	24	25
	.7465	.7460	.7455	.7465	.7460	.7455	.7455	.7460	.7465	.7455	.7445	.7455	.7450	.7455	.7460	.7470	.7460	.7470	.7460	.7465	.7450	.7460	.7450	.7460	.7460
	.7450	.7465	.7450	.7470	.7465	.7440	.7465	.7470	.7460	.7455	.7465	.7465	.7460	.7470	.7465	.7460	.7445	.7455	.7450	.7460	.7460	.7465	.7455	.7465	.7465
	.7455	.7445	.7450	.7460	.7465	.7460	.7465	.7455	.7460	.7455	.7460	.7450	.7450	.7460	.7465	.7475	.7445	.7455	.7465	.7460	.7455	.7455	.7455	.7460	.7475
$\bar{x}$																									
s																									

Chart of Averages — $\bar{x}$

Chart of Standard Deviations — s

Figure 7.28
Make an $\bar{x}$ and s chart.

8

VARIABLES CHARTS FOR LIMITED DATA

OBJECTIVES

- Know the standard precontrol method.
- Learn the "and" and "or" rules for compound probability.
- Apply the modified precontrol method for tight control.
- Learn the x and MR charting method.
- Understand the limitations of both precontrol and x and MR charts.

The precontrol method and the x and MR charts are similar to the charting techniques discussed in Chapter 6 because, like the $\bar{x}$ and R, and the $\tilde{x}$ and R, and the $\bar{x}$ and s charts, these two charts use the actual measurements and are classified as variables control charts. However, their use is somewhat different. The $\bar{x}$ and R chart is usually the first choice among control charts for variables because it provides the most information for analyzing, improving, and controlling a process. The precontrol method and the x and MR chart, on the other hand, work best under different circumstances.

The precontrol method is best used for monitoring the process after it has been brought under control and improved to an acceptable quality level. The x and MR chart can be used when there is a limited amount of data or when the data consist of only single values. In a small-run manufacturing situation in which a small number of pieces are made, the x and MR chart can give valuable analysis for both control and improvements during the run. Over a longer period of time, after several small runs are made, the data may be grouped into samples and reanalyzed using an $\bar{x}$ and R chart. Situations that produce single data values include accounting figures for specified times and process measurements such as temperature, voltage, and pressure. The x and MR chart works well for these situations, too.

8.1 PRECONTROL[1] OR RAINBOW CHARTS

Precontrol charting is an excellent monitoring method for a process that is in statistical control and has been judged capable. When a process has been charted with the $\bar{x}$ and R or the $\tilde{x}$ and R chart, brought under control, and improved to the point at which it is considered best to make improvements elsewhere, the switch to a precontrol chart for monitoring is a welcome change because it is so much easier to use.

Precontrol is also a good screening method that can pinpoint where $\bar{x}$ and R charts should be applied in a process. If instability shows up on a precontrol chart, it signals the presence of special-cause variation. A switch to $\bar{x}$ and R charts will provide the information needed to find the source of the problem.

The Precontrol Procedure

Step 1. Be sure the process capability is safely less than the specification: Six standard deviations must be less than .88 times the tolerance, or even less if the process is subject to drifts.

Step 2. Divide the tolerance by 4 to set up four equal zones. Figure 8.1 illustrates how the zones can be color coded for easier analysis: Sections 2 and 3 make up the green (go) zone, and sections 1 and 4 are designated the yellow (caution) zone. The areas to the left of 1 and to the right of 4 are the red (stop) zones. When the charts are printed with the three colors, they are often called rainbow charts.

Precontrol Formulas

$$\frac{USL-LSL}{4} = w \qquad w = \text{the width factor for the precontrol zones}$$

Precontrol boundaries:

LSL	Boundary between the lower Red zone and the lower Yellow zone
LSL + w	Boundary between the lower Yellow zone and the Green zone
LSL + $3w$	Boundary between the Green zone and the upper Yellow zone
USL	Boundary between the upper Yellow zone and the upper Red zone

[1]Robert W. Traver, "Pre-Control. A Good Alternative to $\bar{x}$ and R Charts" (*Quality Progress*, September 1985): 11–13. Reprinted with permission of the American Society for Quality Control.

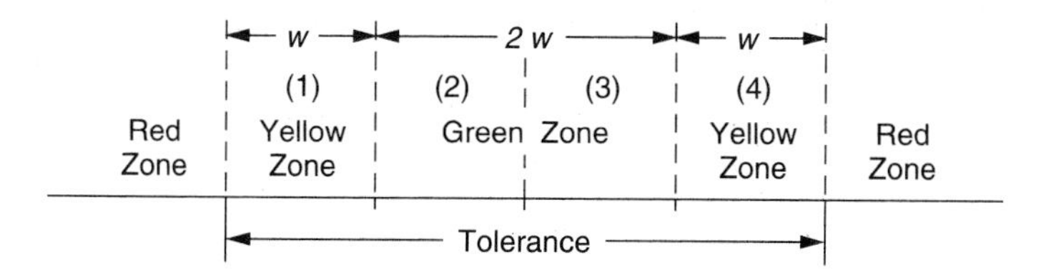

FIGURE 8.1
Precontrol zones.

Step 3. Remember the following precontrol rules:

1. A setup job is safe to run when five consecutive pieces have measurements in the target (green) zone.
2. When the process is running, sample one or two consecutive pieces:
   a. If the first piece measures in the green zone, let the process run. Do not bother to check the second piece.
   b. If the first piece measures in the yellow zone, check the second. If the second is in the green zone, let the process run, but if it is not in the green zone, adjust the process and return to the setup procedure.
   c. If the first piece is in the red zone, adjust the process and return to the setup procedure.

These rules for checking a running process are capsulized in the table that follows, where A is the first measurement and B is the second:

Measurement	Zone	Decision
A	Green	Run (B not needed)
A B	Yellow Green	Check the next piece Run
A B	Yellow Yellow	Check the next piece Adjust or correct: A and B in same yellow zone, $\bar{x}$ is out of control. A and B in opposite yellow zones, the range, R, is out of control and variation has increased.
A	Red	Adjust or correct

Figure 8.2 compares the normal curve and the four zones.

Calculating Zones

Suppose the 6σ spread of the normal distribution of the measurements is within the tolerance. The precontrol rules specify that 6σ should be less than $.88 \times$ tolerance, so for this

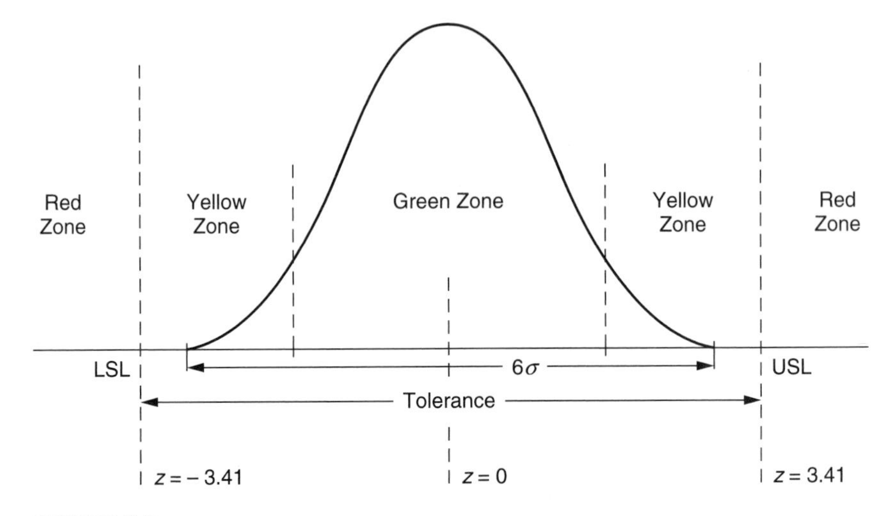

FIGURE 8.2
The normal distribution and the precontrol zones ($6\sigma = .88T$).

explanation, the maximum spread of measurements will be used: $6\sigma = .88 \times$ tolerance. The laws of probability can be used to explain the precontrol rules.

To set up the zones, we must use algebra to solve the equation in the preceding paragraph. The following equation has the same form as the one we want to solve:

$$3x = 15$$
$$3(5) = 15$$
$$x = 5$$

The solution in this case is obvious: A 5 makes the statement true when it replaces the variable *x,* so the value of *x* must be 5.

To solve more complicated equations in which the answer is not obvious, algebraic rules are formed to change the equation from its complicated form to a simple form, or answer statement. The answer statement isolates the variable on one side of the "=" and the answer on the other side. The equation $3x = 15$ can be solved by eliminating the number 3, which can be accomplished by using the opposite operation (in this case, division). The equation states that the quantities on each side are exactly the same amount, so any mathematical operation that changes one side must be performed on the other side as well:

$3x = 15$	To solve for *x,* eliminate the 3. Because 3 is *multiplied* by *x,* use the opposite operation, division, to eliminate it.
$\frac{3x}{3} = \frac{15}{3}$	Divide both sides of the equation by 3.
$x = 5$	The solution

This algebra technique can be used to solve the equation $6\sigma = .88 \times$ tolerance. Let the variable T represent the tolerance:

$.88T = 6\sigma$	To solve for T, eliminate the number .88 by division.
$\frac{.88T}{.88} = \frac{6\sigma}{.88}$	Divide both sides of the equation by .88.
$T = 6.82\sigma$	The right side of the equation was simplified by dividing the .88 into 6.

The resulting equation, $T = 6.82\sigma$, means that the tolerance T is 6.82 standard deviation units. As we learned in our study of the normal curve in Chapter 5, standard deviation units are the z values that are used in the normal-curve table. So, with $z = 0$ at the center of the curve and the 6.82 split half to each side, the z values at the specification units are 3.41 at the USL and −3.41 at the LSL. This is illustrated in Figure 8.2. Four equal-width zones are needed, so divide the z values by 2 to get the green-yellow boundaries:

$\frac{-3.41}{2} = -1.70$	$\frac{3.41}{2} = 1.70$
Left green boundary	Right green boundary

According to the normal-curve table in Appendix B (Table B.8), the tail area for $z = 1.70$ is .0446. That gives the total area for both the yellow and red zones on one side. We can calculate the area in the red zone and subtract it from .0446 to find the area in one yellow zone. This is shown in Figure 8.3:

$z = 3.41$	The red-zone boundary
Tail area = .00032	The red-zone area
$z = 1.70$	The yellow- and green-zone boundary
Tail area = .0446	
.0446	Area in yellow and red zones
−.00032	Area in red zone
.04428	Area in yellow zone (one side)

The area in each yellow zone is therefore .04428, or about 4%.

The total area under the normal curve is 1, and the center line divides it in half, so the area under the right half of the curve is .5. Figure 8.3 shows that the tail area to the right of the green zone is .0446. By subtracting that number from .5, we can determine the area in the green zone on one side:

.5000	
−.0446	
.4554	Area in green zone (right half)

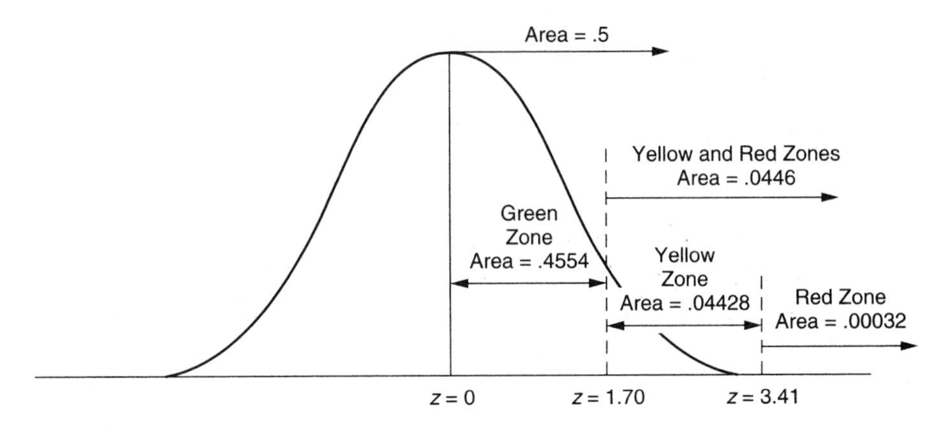

FIGURE 8.3
The area under the right side of the normal curve in the precontrol zones.

To find the total area in the green zone, we simply need to multiply the area in the right half by 2:

$$2 \times .4554 = .9108 \qquad \text{Total area in green zone (both halves)}$$

8.2 COMPOUND PROBABILITY

The laws of probability can be used to explain the precontrol rules. What is the chance that a single piece will be in the green zone? The preceding calculations show a 91% chance, making it highly likely. This is what we expect to happen when the process is running well.

What is the chance that a single piece will be in the red zone? The LSL and USL correspond to $z = 3.41$, which we found in the normal-curve table to be .00032, or .032%. There is a red zone at both tails, so the total chance that a measurement will fall in the red zone is .064%. This is classified as a rare event and is not very likely to occur if the process is running right. When it does occur, we can conclude that the process changed somehow and needs to be adjusted. The red zone is a trouble indicator; the process should be stopped and the trouble eliminated.

The concept of probability that has been used up to this point refers directly to the normal curve. The area under specific sections corresponds to the percentage of data that falls in those sections, which is the same as the percent chance that an individual measurement will fall in those specific sections. To illustrate, in the foregoing work, the area of the yellow zone on the right side is .04428. Because the total area under the curve is 1, the fraction of the total area that is in the right yellow zone is .04428, or 4.428%, of the area. The curve represents the distribution of product produced, so 4.428% of all the product will have measurements that are in the right yellow zone. If one specific piece is randomly chosen from production, the probability, or chance, that its measurement will be in the right yellow zone will be 4.428%.

To answer the probability questions that lead to the precontrol rules, a few more probability concepts must be developed. By definition, the *probability,* or chance, that some event will happen is equal to the number of ways the event can occur divided by the total number of events that can occur. For example, if a coin is tossed, the chance of getting a tail is 50%. There is one way a tail can occur (the numerator = 1), and there are two events: a head or a tail (the denominator = 2). Therefore, the probability of getting a tail is ½, or 50%.

EXAMPLE 8.1 A bag of marbles contains 4 red, 5 white, and 3 blue marbles. If one marble is randomly drawn from the bag, what is the probability of getting a red one?

Solution

There are four ways a red marble could be drawn (the numerator = 4) and 12 events that can occur because each of the 12 different marbles could be drawn (the denominator = 12). The probability of getting a red marble is $\frac{4}{12}$, or $\frac{1}{3}$.

It does not take too much imagination to create a complicated event in which the probability will be difficult to calculate. If four marbles are drawn, what is the chance of not getting a red one? Or, what is the chance that the third one will be blue? Questions such as these require compound probability laws to resolve them. For immediate purposes, we will introduce two laws of compound probability. For convenient notation let $P(A)$ represent the probability of event A.

Rule 1: If event A and event B cannot occur simultaneously,

$$P(A \text{ or } B) = P(A) + P(B)$$

We will use several examples to explain this rule.

EXAMPLE 8.2 If one marble is drawn from the bag of 4 red, 5 white, and 3 blue marbles, what is the probability that it will be either a red or a blue marble?

Solution

$$P(\text{Red or Blue}) = P(\text{Red}) + P(\text{Blue})$$

$$P(\text{Red}) = \frac{4}{12}$$

$$P(\text{Blue}) = \frac{3}{12}$$

$$P(\text{Red or Blue}) = \frac{4}{12} + \frac{3}{12}$$

$$= \frac{7}{12}$$

This could also be found by using the definition. There are seven possible successful events (draw any of the 4 red and 3 blue marbles) and 12 possible events all together.

EXAMPLE 8.3 If one card is drawn from a well-shuffled standard deck of cards, what is the probability that it is a King, a Queen, or a Jack?

Solution

$$P(\text{King}) = \frac{4}{52} \qquad \text{Four Kings in deck of 52 cards}$$

$$P(\text{Queen}) = \frac{4}{52} \qquad \text{Four Queens in deck}$$

$$P(\text{Jack}) = \frac{4}{52} \qquad \text{Four Jacks in deck}$$

$$\begin{aligned} P(\text{King or Queen or Jack}) &= P(\text{King}) + P(\text{Queen}) + P(\text{Jack}) \\ &= \frac{4}{52} + \frac{4}{52} + \frac{4}{52} \\ &= \frac{12}{52}, \text{ or } \frac{3}{13} \end{aligned}$$

The second law of compound probability follows.

Rule 2: If A and B are successive events and $P(B)$ is unaffected by event A,

$$P(A \text{ and then } B) = P(A) \cdot P(B)$$

When two pieces are checked, the probability of an event, such as getting two measurements in the yellow zone, is found by multiplying their individual probabilities. This idea is illustrated with a coin toss example. If the coin is tossed once, the chance of getting a head is 50%. The chance of getting a head is one chance out of two possibilities, or ½. If the coin is tossed twice, the chance of getting two heads is $\frac{1}{2} \cdot \frac{1}{2} = \frac{1}{4}$ or 25%. The ¼ chance can also be seen by observing that there is just one favorable outcome in the four possibilities:

Head and head
Head and tail
Tail and head
Tail and tail

Head and head is one chance out of four possibilities.

Now back to the zones. What is the chance of getting the first measurement in the yellow zone and the second in the green zone? Because the yellow zone could be in the left strip or the right, the total probability of being in the yellow first, by rule 1, is

$$\begin{aligned} P(Y_{\text{left}} \text{ or } Y_{\text{right}}) &= P(Y_{\text{left}}) + P(Y_{\text{right}}) \\ &= .0443 + .0443 \\ &= .0886 \end{aligned}$$

The chance of being in the yellow zone first and the green zone second is found by multiplying their probabilities (rule 2):

$$\begin{aligned} P(Y \text{ and then } G) &= P(Y) \cdot P(G) \\ &= .0886 \cdot .9108 \\ &= .0807 \end{aligned}$$

It is not highly likely to occur, at approximately 8%, but 8 times out of 100 is often enough to expect it to happen occasionally. This probability is not an indication of trouble with the process.

What is the chance of getting two successive pieces in the yellow zones? This can happen four different ways:

$$\begin{array}{ccc} Y_{\text{right}} & \text{and} & Y_{\text{right}} \\ Y_{\text{right}} & \text{and} & Y_{\text{left}} \\ Y_{\text{left}} & \text{and} & Y_{\text{right}} \\ Y_{\text{left}} & \text{and} & Y_{\text{left}} \end{array}$$

In each case we can use rule 2 and substitute the area values we found earlier for the yellow zone:

$$\begin{aligned} P(Y \text{ and then } Y) &= P(Y) \cdot P(Y) \\ &= .0443 \cdot .0443 \end{aligned}$$

Counting all four ways, the probability of two yellows, then, is $.0443 \times .0443 \times 4 = .0078$, or approximately .8%. This means that with the process running as it should, .8% of the time a sample of two consecutive pieces will be in the yellow zones. This is considered a small enough chance to be classified as a trouble indicator: Getting two in the yellow zones is a rare enough event to believe that a problem with the process caused it to happen. Further, the type of trouble is usually indicated by a pattern: Two in the same yellow zone may indicate process drift, and two in opposite yellow zones suggest increased variability.

8.3 MODIFIED PRECONTROL FOR TIGHT CONTROL

When a process has been classified as an A or B process (as defined in Appendix B, Table B.11) *and* when tight control is desired, the $6s$ value should be used to set up the zones instead of the tolerance. Otherwise, the process may regress to a C or D classification, and the precontrol monitoring will not necessarily indicate the change. This is shown in Figures 8.4 through 8.6.

Each figure shows product distributions that come from A processes (processes in which the $6s$ spread is less than half the tolerance). In order to maintain the A process classification, any trouble with the process should be noted as soon as possible. In all three illustrations, the zones in the (*a*) diagrams are figured from the tolerance and the zones in the (*b*) diagrams are figured from the $6s$ product distribution spread. In each case the A and B indicators show trouble on the (*b*) diagram, but the same measurements (in the same position on the tolerance scale) are considered acceptable on the (*a*) diagram. The zones calculated from the tolerance neglected to indicate the trouble. The pictures show that if the goal is to stay within the tolerance, the (*a*) diagram, with the zones calculated from the tolerance, will suffice. If the goal is to maintain a tight distribution well within the tolerance limits, the (*b*) diagram, with the zones calculated from $6s$, is needed.

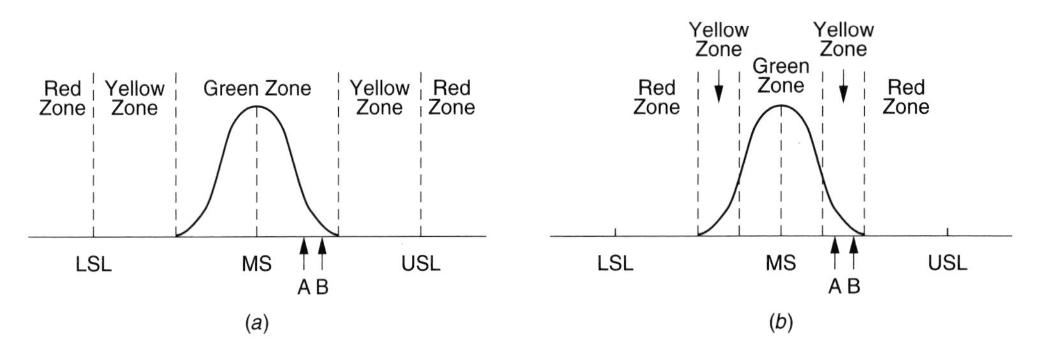

FIGURE 8.4
(*a*) Zones calculated from the specification limits. Trouble is not indicated; both measurements are in the green zone. (*b*) Zones calculated from the midspecification (MS) point $\pm$ $3s$. Trouble is indicated; both measurements are in the same yellow zone.

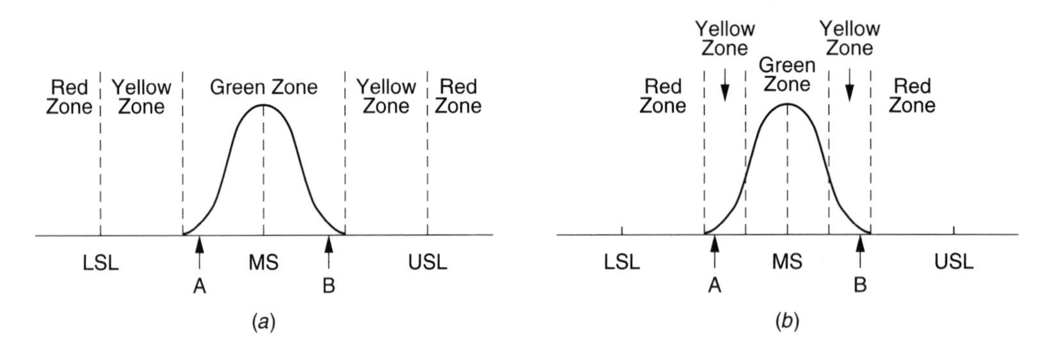

FIGURE 8.5
(*a*) Zones calculated from the specification limits. No trouble is indicated; both measurements are in the green zone. (*b*) Zones calculated from MS $\pm$ $3s$. Measurements in opposite yellow zones indicate trouble.

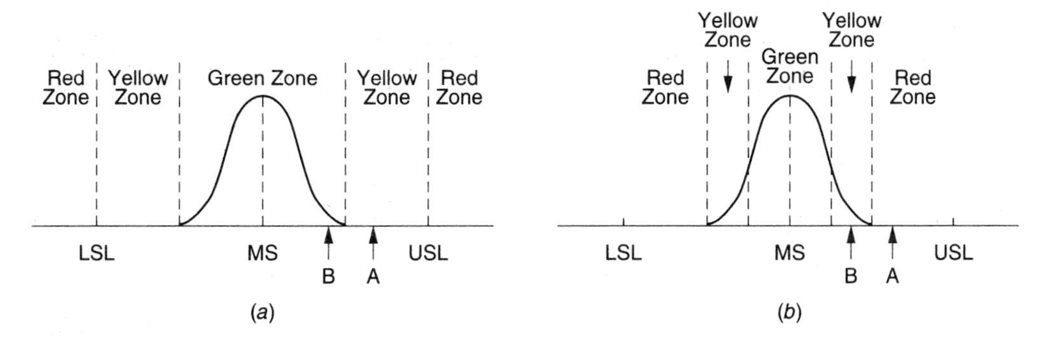

FIGURE 8.6
(*a*) Zones calculated from specification limits. Trouble is not indicated; A is in the yellow and B in the green. (*b*) Zones calculated from MS ± $3s$. Point A is in the red zone, signaling trouble.

Modified Precontrol

Use $w = \frac{6s}{4}$ to set up the precontrol zones.

$MS = \frac{LSL + USL}{2}$ = mid-specification value

MS − $2w$ = Boundary between the lower red zone and the lower yellow zone

MS − w = Boundary between the lower yellow zone and the green zone

MS + w = Boundary between the green zone and the upper yellow zone

MS + $2w$ = Boundary between the upper yellow zone and the upper red zone

---

EXAMPLE 8.4

Apply precontrol to a run of top adapters. The top adapters are being turned out on an automatic screw machine. The specification limits are .212 to .218, and the goal is to stay within those limits. Measurements are sampled hourly.

Solution

Set up the zones from the tolerance. First divide the tolerance by 4:

$$w = \frac{.218 - .212}{4} = \frac{.006}{4}$$

$$= .0015$$

Start at the lower specification limit and add .0015 for the boundary between the first yellow zone and the green zone. Then add $3w$ (.0045) to the LSL for the boundary between the green zone and the upper yellow zone. The LSL and the USL are the boundary values between the red zones and the yellow zones. See Figure 8.7.

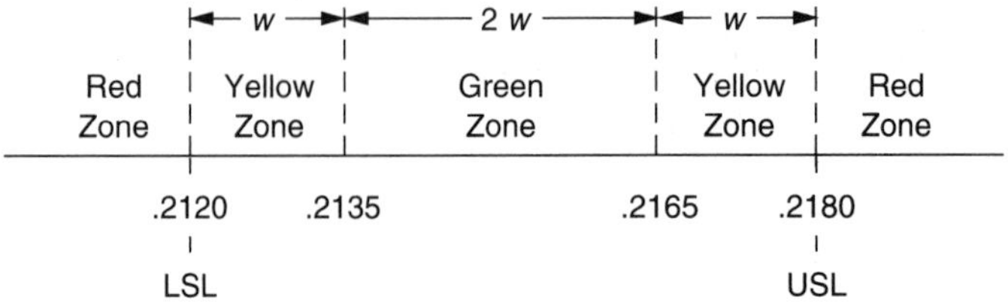

	w	2 w	w	
Red Zone	Yellow Zone	Green Zone	Yellow Zone	Red Zone

.2120 (LSL) .2135 .2165 .2180 (USL)

FIGURE 8.7
Precontrol zones based on the tolerance.

Samples 2 through 6 had both A and B in the same yellow zone, which indicated a process shift. The process shift was the result of a dull tool. After sharpening the tool, consecutive pieces were measured until five consecutive pieces measured in the green zone, which indicated that the process shifted back into the target (green) zone. The chart shows that for at least four hours, the measurements crowded the lower specification limit. The past discussion on product distributions indicates that some of the product was most likely out of specification during that period. The decision to shut down and sharpen should have been made sooner.

Sample	A Value	Zone	B Value	Zone	Decision

Five consecutive pieces are needed in the green zone for the start.

Measurement:	.2145	.2140	.2145	.2150	.2140	OK
Zone:	(G)	(G)	(G)	(G)	(G)	to run

Sample	A Value	Zone	B Value	Zone	Decision
1	.214	Green	Not needed		OK
2	.2125	Yellow	.2125	Yellow	Reported trouble
3	.2120	Yellow	.2120	Yellow	Reported trouble
4	.2125	Yellow	.2120	Yellow	Reported trouble
5	.2123	Yellow	.2124	Yellow	Reported trouble
6	.2125	Yellow	.2125	Yellow	Given OK to sharpen

Five consecutive pieces are needed in the green zone for the restart.

Measurement:	.2140	.2136	.2134	.2139	.2138	.2138	.2139	.2135	OK
Zone:	(G)	(G)	(Y)	(G)	(G)	(G)	(G)	(G)	to run

Sample	A Value	Zone	B Value	Zone	Decision
7	.2136	Green	Not needed		OK
8	.2138	Green	Not needed		OK
9	.2135	Green	Not needed		OK
10	.2130	Yellow	.2139	Green	OK
11	.2137	Green	Not needed		OK
12	.2138	Green	Not needed		OK
13	.2133	Yellow	.2135	Green	OK
14	.2135	Green	Not needed		OK
15	.2140	Green	Not needed		OK

EXAMPLE 8.5 Use the modified precontrol to maintain tight control of an A process. The process has specification limits of .008 to .010 and a standard deviation $s = .000158$.

Solution

The process capability is

$$\begin{aligned} \text{PCR} &= \frac{6s}{\text{tolerance}} \\ &= \frac{6 \times .000158}{.002} \\ &= \frac{.00948}{.002} \\ &= .474 \end{aligned}$$

The process uses 47.4% of the tolerance and is classified as an A process. The goal is to maintain the A process classification, so the $6s$ spread is used to set up the precontrol zones, (as illustrated in Figure 8.8):

$$\begin{aligned} w = \frac{6s}{4} &= \frac{6 \times .000158}{4} \\ &= \frac{.00948}{4} \\ &= .000237 \end{aligned}$$

$$\begin{aligned} \text{MS} &= \text{midspecification point} \\ &= \frac{.008 + .010}{2} \\ &= .009 \end{aligned}$$

$$\begin{aligned} \text{Green zone} &= \text{MS} \pm w \\ &= .009 \pm .00024 \\ &= .00876 \text{ to } .00924 \end{aligned}$$

$$\begin{aligned} \text{Red - Yellow boundary (left)} &= \text{MS} - 2w \\ &= .009 - .00047 \\ &= .00853 \end{aligned}$$

$$\begin{aligned} \text{Red - Yellow boundary (right)} &= \text{MS} + 2w \\ &= .009 + .00047 \\ &= .00947 \end{aligned}$$

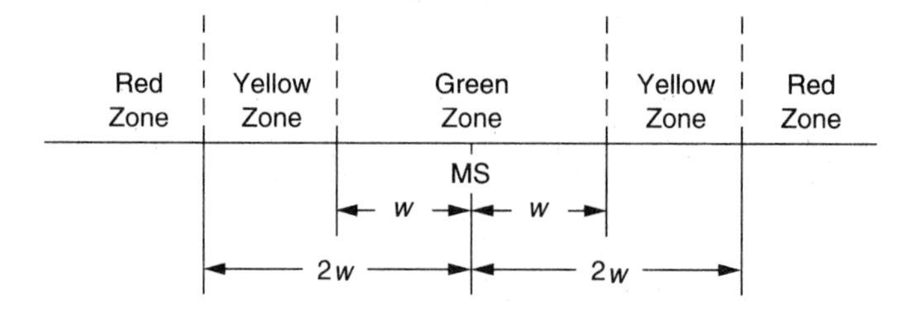

FIGURE 8.8
Precontrol zones based on MS ± 3*s*.

The red zones are to the left of .00853 and to the right of .00947. Use Figure 8.9 to zone the A and/or B measurement and indicate the decision for samples 1 to 15 (see the table that follows).

The process should be shut down at the first indication of trouble. Two measurements occur in opposite yellow zones at sample 7, which indicates that something has changed to increase the variation. Examination of the process indicates a loose fixture, which will be tightened. The required five-consecutive-piece startup procedure follows.

If the tolerance had been used to set up the zones in this case, the loose fixture would not have been found until the variation had increased enough to show the problem with the wider zones. In the meantime the process would have deteriorated to a C or D classification and quality would have suffered.

Sample	A Value	Zone	B Value	Zone	Decision
.0089	.0090	.0089	.0090	.0091	OK
1	.0093	Yellow	.0090	Green	OK
2	.0091	Green	Not needed		OK
3	.0089	Green	Not needed		OK
4	.0088	Green	Not needed		OK
5	.0090	Green	Not needed		OK
6	.0089	Green	Not needed		OK
7	.0084	Yellow	.0093	Yellow	Tightened loose fixture
.0090	.0091	.0090	.0089	.0091	OK
8	.0090	Green	Not needed		OK
9	.0092	Green	Not needed		OK
10	.0093	Yellow	.0090	Green	OK
11	.0089	Green	Not needed		OK
12	.0090	Green	Not needed		OK
13	.0091	Green	Not needed		OK
14	.0087	Yellow	.0091	Green	OK
15	.0090	Green	Not needed		OK

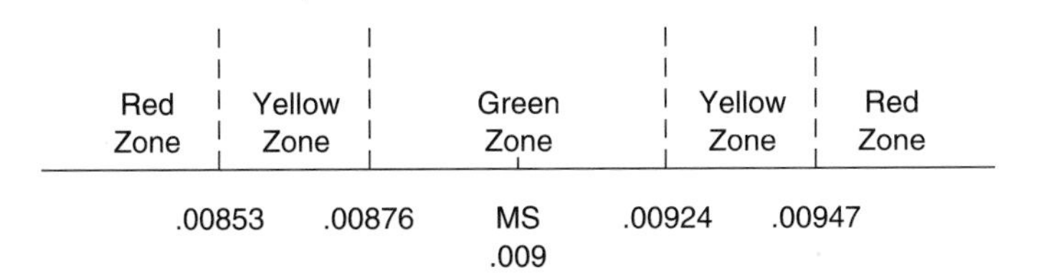

FIGURE 8.9
Precontrol zones based on MS ± 3s.

In some situations precontrol may be the first choice for use as a control chart. An automatic screw machine, for example, may be considered in control with only occasional shutdowns for tool sharpening or adjustment. The precontrol chart will tell when a shutdown is needed and if the initial in-control assumption is correct. Successive readings in different zones indicate excessive variability. Drifts or shifts of the measurements from the desired midspecification value eventually show up as two successive measurements in the same yellow zone or in a yellow and red zone on the same side. If the in-control assumption is wrong, a switch to an $\bar{x}$ and R chart may provide better information for bringing the process under control.

8.4 CHARTS FOR INDIVIDUAL MEASUREMENTS

The Run Chart

The run chart is the crudest control chart. A target line and a scale of measurements are drawn on a chart. Then, individual pieces are periodically measured, charted, and checked against the target value. Run charts have been used as a first charting experience by companies and as a preliminary investigation for troubled areas.

The information on a run chart is limited to a comparison with the target measurement over time. One of the main problems that stem from using run charts is that they can lead to overadjustment. As a first experience chart, the operator often does not understand the concept of variation in measurement: the bell curve of normal variation. Consequently, as the measurement varies from the target line, the temptation is to adjust it back

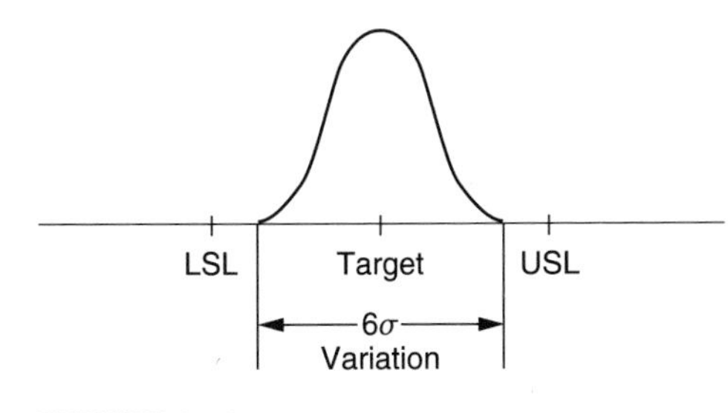

FIGURE 8.10
The single-setting variation.

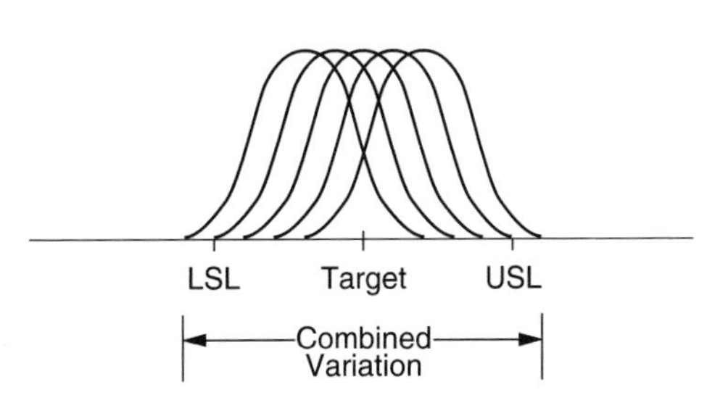

FIGURE 8.11
Overadjustment variation.

on target. Each setting of the machine has its own bell curve of variation with its 6σ spread of measurements. Overadjustment has the effect of using several bell curves simultaneously (one for each adjustment). The resulting variation is much worse than the basic product variation from a single setting. This is illustrated in Figures 8.10 and 8.11.

CASE STUDY 8.1

Richard Stanula,[2] statistical process engineer at Mercury Marine, Brunswick Company, and Robert Hart of the University of Wisconsin report on the use of the run chart to bring a process into control.

A run chart is like a control chart, but without the controls; it simply specifies a target area. The chart is used to provide communication from the process to the operator. In this case, it was used in a cause-and-effect relationship.

Two close-tolerance dowel holes and a water passage hole were being machined in a cast drive-shaft plate. Problems with the operation were caused by out-of-tolerance dowel holes and excessive downtime for tool changes.

The initial chart showed a large percentage of oversized holes that were being produced intermittently. The first step was the change from hand feed to automatic feed. This brought about some decrease in variability on the chart, but the holes were still too large and too variable. A second change featured the installation of an undersized drill reamer. The holes were nearer the target size, but variability was still excessive. Two changes were tried next: New bushings were installed on the drill reamers, and the precision hole boring operation was separated from the less critical water passage hole boring. Standard tooling was reinstalled. The charts showed that variations were reduced again but that the dowel holes were still too large. A change to undersized tooling brought the measurements into the target area, and the variation was more reasonable.

At each change the effect was noted on the run chart. The process was brought under control so that assignable causes of variation could be detected. The step-by-step improvements were the result of cooperation between the operator, inspector, engineer, and control statistician, with the run chart providing the main communication link to the process.

The authors cited the following benefits from the experience:

- Increased operator involvement
- Worker satisfaction and pride of workmanship
- Elimination of scrap and rework
- Lower costs
- Higher productivity

The *x* and *MR* Chart

The x and MR chart is a refinement on the run chart. Individual measurements are tracked as with the run chart, but the variation in measurements MR is tracked as well. By tracking variation, too, it allows the use of control limits for process analysis similar to the $\bar{x}$ and R

---

[2]Richard Stanula and Robert Hart, "A Process Improvement Tool" (*Quality*, September 1987), 59.

charts. The *MR* value represents the moving range, a measure of piece-to-piece variability. Usually *MR* is the positive difference between consecutive measurements, $MR = |x_2 - x_1|$.

The best charts for analyzing process variation are the $\bar{x}$ and *R*, $\tilde{x}$ and *R*, and $\bar{x}$ and *s* charts. They are more sensitive to the process because samples are used. One of Walter Shewhart's important discoveries during his development of the control chart was that small samples of size $n = 3$, 4, or 5 will have a normal distribution *as long as the population distribution stays the same.* Any change in the population distribution will have a corresponding change in the sample distribution, and that is on what we base our control chart analysis. Any nonnormal pattern on the control chart or histogram can signal a change in the population of measurements.

One of the attempts to make the *x* and *MR* charts more sensitive to process changes is to spread the variation measure over three, four, or five consecutive pieces. For example, using $n = 4$, MR_1 is the range for measurements x_1 to x_4. MR_2 is the range for measurements x_2 to x_5, and MR_3 is the range for measurements x_3 to x_6. Sensitivity may be improved, but it is not as good as the three sample-based charts. This is why a small sample control chart is best when there are enough data.

The most common use of *x* and *MR* charts is for the limited-data situation. Its use is effective when the number of items produced is too small to form the 25 or more samples needed for an $\bar{x}$ and *R* chart. Consecutive pieces are measured and charted in this case. The *x* and *MR* chart can be used for a preliminary analysis at a process point and then followed with an $\bar{x}$ and *R* chart if there is some evidence of special-cause variation or poor capability. When data values naturally occur as single values such as temperature, humidity, batch measures, or voltage, the *x* and *MR* chart can be effectively used. Individual charts can also be used for the final inspection of assembled units.

CASE STUDY 8.2

A manufacturer of automotive transmissions runs a final unit test on 44 key characteristics. The transmissions are organized in blocks of 50 for tracking purposes. Each characteristic measurement is charted on an *x* and *MR* chart, so there are 44 different charts. The charts have the average line, $\bar{x}$; the control limits, $\bar{x} \pm 3s$; and the specification limits, USL and LSL. The measurements are automatically fed into a computer so the final unit inspection consists of calling up the forty-four *x* and *MR* charts on the computer screen, noting any chart discrepancies and targeting any faulty transmissions. These transmissions would then be carefully inspected for specific problems. The problems would be fixed and categorized, and an upstream search would track the source of the problem. The source of the problem would then be eliminated.

Procedure for the Individuals and Moving Range Chart

Step 1. Take measurements of consecutive pieces. More than 10 measurements are required for a chart, but approximately 20 or more measurements are commonly used.

Step 2. Record the measurements *x* and the moving range *MR* on the chart. The *MR* values will be offset one position with the *first* position blank when

$n = 2$. The moving range MR is the difference between consecutive measurements without regard to sign (the absolute value of the differences):

$$MR_1 = |x_2 - x_1|$$
$$MR_2 = |x_3 - x_2|$$
$$MR_3 = |x_4 - x_3|$$

The values are entered on the chart, as shown in Figure 8.12 for the first few values.

x	x_1	x_2	x_3	x_4	
MR	———	MR_1	MR_2	MR_3	

FIGURE 8.12
Selected data values on an x and MR chart with $n = 2$.

The value n represents the number of consecutive pieces used to find the moving range. Example 8.6 illustrates the chart entries for different n values. The most commonly used x and MR chart, however, is the basic piece-to-piece analysis with $n = 2$.

EXAMPLE 8.6 Given the data values in order of production, enter the values on an individual and moving range chart. The measurements are 4.213, 4.219, 4.209, 4.207, 4.215, 4.215, 4.210, and 4.205.

1. Use $n = 2$.
2. Use $n = 4$.

Solution

1. $MR_1 = 4.219 - 4.213 = .006$ $MR_2 = 4.219 - 4.209 = .010$ $MR_3 = 4.209 - 4.207 = .002$ $MR_4 = 4.215 - 4.207 = .008$ . . .

In each case the smaller of the two consecutive values is subtracted from the larger to get the absolute value. Figure 8.13 shows the chart entries.

x	4.213	4.219	4.209	4.207	4.215	4.215	4.210	4.205
MR	———	.006	.010	.002	.008	.000	.005	.005

FIGURE 8.13
Entries for the x and MR chart with $n = 2$.

2. MR_1 = Range of x_1 to x_4 = 4.219 – 4.207 = .012
   MR_2 = Range of x_2 to x_5 = 4.219 – 4.207 = .012
   MR_3 = Range of x_3 to x_6 = 4.215 – 4.207 = .008
   MR_4 = Range of x_4 to x_7 = 4.215 – 4.207 = .008
   MR_5 = Range of x_5 to x_8 = 4.215 – 4.205 = .010

Figure 8.14 shows the chart entries.

x	4.213	4.219	4.209	4.207	4.215	4.215	4.210	4.205
MR	——	——	——	.012	.012	.008	.008	.010

FIGURE 8.14
Entries for the *x* and *MR* chart with $n = 4$.

Step 3. Set the scale for the *x* chart. The easiest method is to use the specification limits.

Do not scale it too wide; use about 10 divisions between the specification limits. First locate the middle of the chart so it has a value about halfway between the specification limits. This can be done by averaging the two:

$$\text{Middle chart value} \approx \frac{\text{LSL} + \text{USL}}{2}$$

The scale is approximately

$$\frac{\text{USL} - \text{LSL}}{10}$$

units per line. Round up or down to find an easy number for charting. Since individual measurements are being charted, their position relative to the specifications can be seen directly.

After several *MR* values have been calculated, set the *MR* scale to accommodate twice the range of the initial values.

EXAMPLE 8.7 A measurement has specification limits of .749 to .751. Set up an *x* and *MR* chart in preparation for tracking the measurement.

Solution

$$\begin{aligned} \text{MS} &= \frac{\text{USL} + \text{LSL}}{2} \\ &= \frac{.751 + .749}{2} \\ &= .750 \end{aligned}$$

$$\text{Scale} = \frac{\text{USL} - \text{LSL}}{10}$$
$$= \frac{.751 - .749}{10}$$
$$= \frac{.002}{10}$$
$$= .0002 \text{ units per line}$$

Start at the MS value of .750 and count .0002 units per line in both directions. The *MR* scale cannot be determined until several *MR* values have been calculated.

Step 4. Plot the *x* measurements in the upper section of the chart and the *MR* values in the lower section.

Step 5. Calculate the averages and draw a solid line for each.

$$\bar{x} = \frac{\Sigma x}{N} \qquad \overline{MR} = \frac{\Sigma MR}{(N - n) + 1}$$

N is the number of measurements x.
n is the number of consecutive measurements used to calculate *MR*.
$(N - n) + 1$ in the *MR* formula represents the number of *MR* values on the chart.

Step 6. Calculate the control limits and draw a dashed line for each.

Control Limit Formulas

$$UCL_{MR} = D_4 \times \overline{MR} \qquad UCL_x = \bar{x} + E_2 \times \overline{MR}$$
$$LCL_{MR} = D_3 \times \overline{MR} \qquad LCL_x = \bar{x} - E_2 \times \overline{MR}$$

The constants D_3, D_4, and E_2 are found in Table B.4, Appendix B.

Step 7. Interpret the chart.

1. Look for runs of seven or more consecutive points going up or going down. Look for shifts of seven or more consecutive points above or below the average. The shifts or runs may indicate that an adjustment is necessary.
2. Look for points beyond the control limits.
3. Check to see if the pattern of points avoids one of the control limits. This may indicate a skewed distribution.
4. Look for cycles and grouping, or bunching, of points. Either pattern can be an indicator of process problems.
5. Look for an absence of points near $\bar{x}$. That may mean that there are two or more specific groups, each with its own distribution, blended at or before the point at which the measurements are taken.

6. Mark the specification limits on the control chart. The relationship of the individual values to the specification limits shows up directly on the chart when the product specification limits are included.

Step 8. Make the tally histogram of the measurements. Caution must be taken with interpretation because the control limits are less precise and less sensitive than for $\bar{x}$ and R charts. The shape of the distribution is also more critical. Remember, the big difference between this chart and an $\bar{x}$ chart is that the distribution of $\bar{x}$ will be normal for a stable population. The distribution of an individual measurement can be all over the place before there are enough data values to imply the shape of the population distribution. Usually 100 or more data values are needed for any shape stability. When possible, check the interpretations with a follow-up $\bar{x}$ and R chart.

Step 9. Do a preliminary capability analysis. The more reliable capability analysis with the $\bar{x}$ and R chart is done in step 10.

Any capability study from an x and MR chart must be considered preliminary information. The chart provides a hint of how the process is behaving, but nothing conclusive. Bad news is cause for immediate action, but good news should be received with a wait-and-see attitude until enough data have been gathered.

A normal-curve analysis for the percentage of product that is out of specification can be applied to the data from an x and MR chart when there are enough data to verify that the measurements follow a normal distribution. The PCR and Cpk *can* be used to assess the capability, however, because they compare the tolerance with the $6s$ spread of the data. The PCR uses the entire $6s$ value and the Cpk uses half of it, assessing only the side of the mean that is closest to the specification limit. Virtually all of the data of any distribution, normal or otherwise, will be within $3s$ of the mean, so the PCR and Cpk values can be treated as realistic capability measures. It must be stressed, however, that even those measures must be considered temporary when a small amount of data are used in the calculations.

The control limit lines on all of the charts represent a six-standard-deviation spread of the variable being charted. In this case, with the individual values being charted, $\text{UCL}_x - \text{LCL}_x = 6s$:

$$\text{PCR} = \frac{6s}{\text{tolerance}}$$

$$\text{Cpk} = \text{minimum of } \frac{\text{USL} - \bar{x}}{3s} \text{ or } \frac{\bar{x} - \text{LSL}}{3s}$$

Step 10. Follow up with an $\bar{x}$ and R chart when sufficient data are available. Assign the individual measurements to the samples. If a sample of five is used, the first five measurements taken are sample 1, the second five measurements are sample 2, and so on. Note any breaks in the production schedule on the chart. When 25 or more samples have been formed, calculate $\bar{x}$, R, and the control limits and analyze the results.

When time is a factor, process information can be determined sooner with the x and MR chart. Some process problems, indicated by trends, cycles, and mixtures, show up better (or sooner) on the x and MR chart. However, unnatural pattern analysis is more reliable on the $\bar{x}$ and R chart. (Mixtures, cycles, and other pattern analyses are discussed in Chapter 10.)

EXAMPLE 8.8 Demonstrate the x and MR chart for the first 36 data values from Figure 6.16 using $n = 2$.

Solution

Follow the 10 steps outlined:

1. Record the consecutive data values on the x and MR chart. The data are presented in Figure 8.15.
2. Calculate the MR values and enter them on the chart. Remember to offset the MR values one place to the right.

$$\begin{aligned} x_2 - x_1 &= MR_1 & .0243 - .0218 &= .0025 \\ x_2 - x_3 &= MR_2 & .0243 - .0232 &= .0011 \\ x_4 - x_3 &= MR_3 & .0256 - .0232 &= .0024 \\ x_4 - x_5 &= MR_4 & .0256 - .0247 &= .0009 \end{aligned}$$

   In each case, take consecutive measurements and subtract the smaller from the larger. Continue the calculations on Figure 8.15 and check your results with Figure 8.16.

3. Set the scales: USL = .0300 and LSL = .0200. The average of these two measurements will give the approximate middle of the chart. Assuming a first-chart situation, the first eight MR values have a high of .0027. Double it and divide by the 10 lines:

$$10\overline{).0054}\quad .0005+ \qquad \text{Round up to .001.}$$

$$\frac{\text{USL} + \text{LSL}}{2} = \frac{.0300 + .0200}{2} = \frac{.0500}{2} = .0250 \quad \text{Middle of chart}$$

   Aiming for a chart of about 10 lines, divide the difference of the largest and smallest by 10:

$$\frac{.0300 + .0200}{10} = \frac{.0100}{10} = .001 \text{ units per line}$$

   This chart will be centered at .0250 and will use .001 units per line. Set up the scale on Figure 8.15 and check the results on Figure 8.16.

$\overline{MR} = \frac{\Sigma MR}{k-1} = \text{\_\_\_\_\_\_} = \text{\_\_\_\_\_\_}$

$\bar{x} = \frac{\Sigma x}{k} = \text{\_\_\_\_\_\_} = \text{\_\_\_\_\_\_}$

Specification: .0200 to .0300

$UCL_{MR} = 3.267 \times \overline{MR} = \text{\_\_\_\_\_\_}$

$2.659 \times \overline{MR} = 2.659 \times \text{\_\_\_\_\_\_} = \square$

$UCL_x = \bar{x} + 2.659\overline{MR} = \text{\_\_\_\_\_\_} + \square = \text{\_\_\_\_\_\_}$

$LCL_x = \bar{x} - 2.659\overline{MR} = \text{\_\_\_\_\_\_} - \square = \text{\_\_\_\_\_\_}$

Sample Numbers	1	2	3	4	5	6	7	8	9	10	11	12	13	14	15	16	17	18
Time																		
x	.0218	.0243	.0232	.0256	.0247	.0255	.0282	.0261	.0244	.0252	.0265	.0267	.0254	.0238	.0249	.0275	.0265	.0232
MR	X	.0025	.0011	.0024	.0009													

Sample Numbers	19	20	21	22	23	24	25	26	27	28	29	30	31	32	33	34	35	36
Time																		
x	.0294	.0281	.0259	.0274	.0228	.0254	.0266	.0245	.0231	.0280	.0259	.0282	.0264	.0234	.0302	.0271	.0250	.0254
MR																		

x

MR

FIGURE 8.15
Set the scales and chart the x values. Calculate and chart the MR values, $\bar{x}$, $\overline{MR}$, and control limits. Check your results with Figure 8.16.

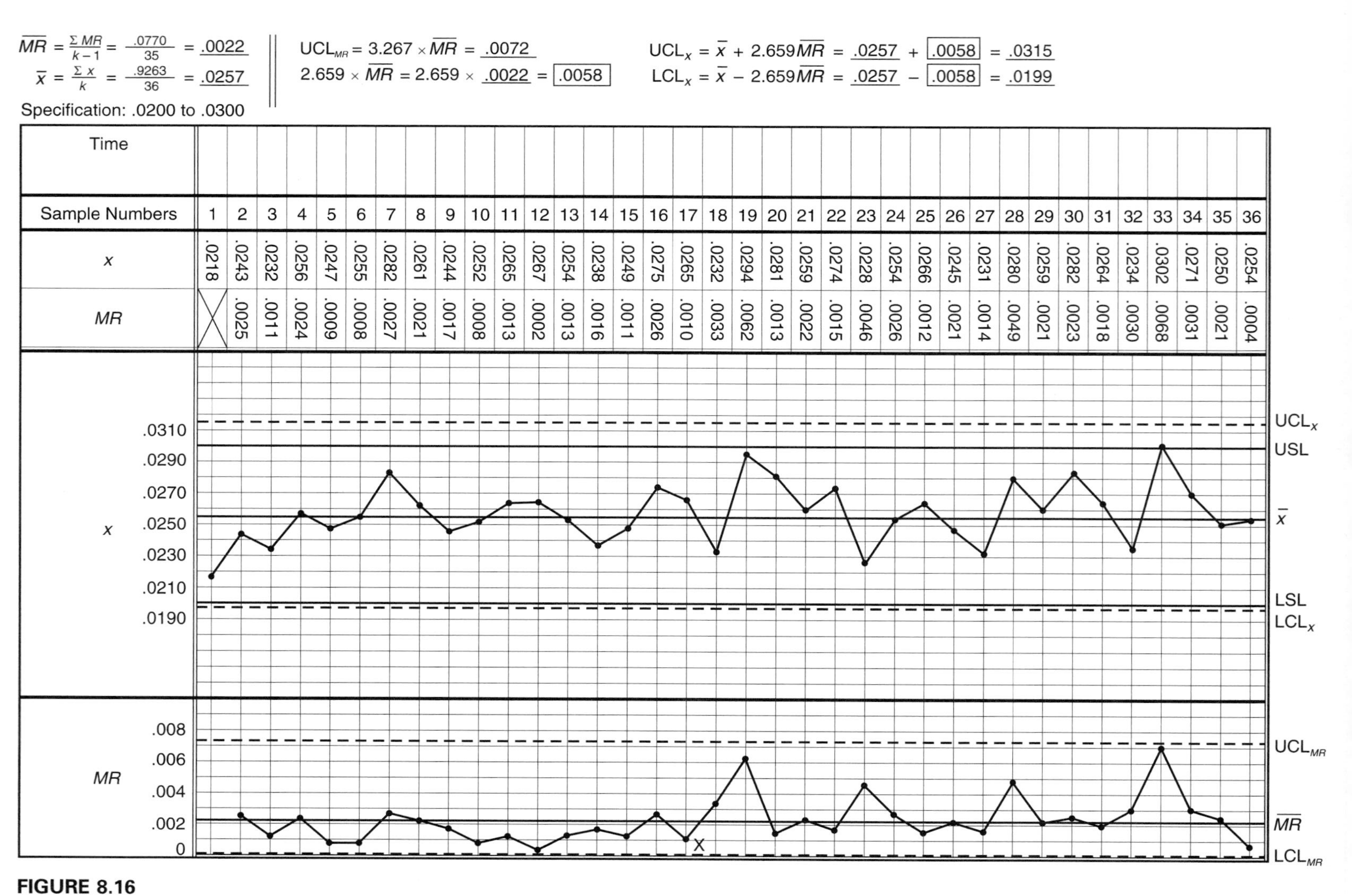

$\overline{MR} = \frac{\Sigma MR}{k-1} = \frac{.0770}{35} = .0022$

$\bar{x} = \frac{\Sigma x}{k} = \frac{.9263}{36} = .0257$

Specification: .0200 to .0300

$UCL_{MR} = 3.267 \times \overline{MR} = .0072$

$2.659 \times \overline{MR} = 2.659 \times .0022 = \boxed{.0058}$

$UCL_x = \bar{x} + 2.659\overline{MR} = .0257 + \boxed{.0058} = .0315$

$LCL_x = \bar{x} - 2.659\overline{MR} = .0257 - \boxed{.0058} = .0199$

Sample Numbers	1	2	3	4	5	6	7	8	9	10	11	12
Time												
x	.0218	.0243	.0232	.0256	.0247	.0255	.0282	.0261	.0244	.0252	.0265	.0267
MR		.0025	.0011	.0024	.0009	.0008	.0027	.0021	.0017	.0008	.0013	.0002

Sample Numbers	13	14	15	16	17	18	19	20	21	22	23	24
Time												
x	.0254	.0238	.0249	.0275	.0265	.0232	.0294	.0281	.0259	.0274	.0228	.0254
MR	.0013	.0016	.0011	.0026	.0010	.0033	.0062	.0013	.0022	.0015	.0046	.0026

Sample Numbers	25	26	27	28	29	30	31	32	33	34	35	36
Time												
x	.0266	.0245	.0231	.0280	.0259	.0282	.0264	.0234	.0302	.0271	.0250	.0254
MR	.0012	.0021	.0014	.0049	.0021	.0023	.0018	.0030	.0068	.0031	.0021	.0004

FIGURE 8.16
The completed x and MR chart.

4. Graph the x and MR values and connect the points with a broken-line graph.
5. Find $\bar{x}$ and draw the solid line:

$$\bar{x} = \frac{\Sigma x}{K} = \frac{.9263}{36} = .0257$$

6. Calculate $\overline{MR}$ and draw the solid line:

$$\overline{MR} = \frac{\Sigma MR}{K-1} = \frac{.0770}{35} = .0022$$

7. Calculate the control limits and draw the dashed lines. The constants E_2, D_3, and D_4 are found in Appendix B, Table B.4.

$$UCL_{MR} = D_4 \times \overline{MR} = 3.267 \times .0022 = .0072$$

$$UCL_x = \bar{x} + E_2 \times \overline{MR} = \bar{x} + 2.659 \times \overline{MR} = .0257 + 2.659 \times .0022 = .0257 + .0058 = .0315$$

$$LCL_{MR} = D_3 \times \overline{MR} = 0 \times .0022 = 0$$

$$LCL_x = \bar{x} - E_2 \times \overline{MR} = \bar{x} - 2.659 \times \overline{MR} = .0257 - 2.659 \times .0022 = .0257 - .0058 = .0199$$

8. Check for statistical control.

The MR chart has a shift pattern from 9 to 15. This is a downward shift on a variation chart, so it is a good-news pattern and not a trouble pattern. It should be investigated, however, to try to determine why the variation decreased. If the cause is found, it may lead to a decrease in overall variation (improved capability). After measurement 15 there seems to be an increasing variation trend with several larger variation values. Something may have happened to the process at that point.

The x chart does not have any runs, shifts, trends, or points beyond the control limits. There is an indication of excessive variation though. There is an instability pattern (discussed in Chapter 10) that points to that as well as to the fact that the control limits (for individual measurements) overlap the specification limits. Also one measurement, sample 33, is out of specification. This indicates that the process is not capable.

9. A look at the data histogram in Figure 8.17 shows that the distribution of the points seems to be following a normal pattern. It also shows that the distribution crowds one specification limit and overlaps the other. This is another indication of poor capability.

   The process shift that showed up on the $\bar{x}$ and R chart on the same data (shown in Figure 6.19) failed to show up on this chart. The shift on the $\bar{x}$ chart was accompanied by an increase in variability on the R chart. Analyzed together, the x and MR chart and the $\bar{x}$ and R chart show the basic problem to be one of excessive variation.

   This example did not follow the general x and MR recommendation in that the pieces measured were not all consecutively produced. There was a break after every fourth piece because of the sampling in Example 6.7. When pieces are randomly or haphazardly chosen for an x and MR chart, a multitude of chart patterns and their process implications can result. This will be illustrated in exercise 8 where every third data point from Example 6.7 is used for an x and MR chart.

10. To determine the process capability, subtract the control limits for the $6s$ value:

$$.0315 - .0199 = .0116$$
$$6s = .0116$$

The specification limits for the data were given as .0200 to .0300, so the tolerance will be .0300 − .0200 = .0100:

$$\text{PCR} = \frac{6s}{\text{tolerance}} = \frac{.0116}{.01} = 1.16$$

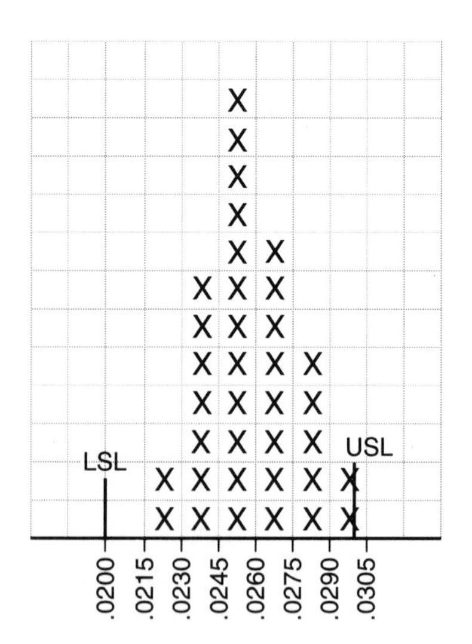

FIGURE 8.17
The histogram for Example 8.8.

The PCR is 116%. According to the classification chart (Appendix B, Table B.11), this classifies as a D process. Notice that this value is close to the PCR of 114.6% that was calculated from the data on the $\bar{x}$ and R chart in Examples 6.7 and 6.10. The Cpk calculation follows:

$$3s = \frac{6s}{2} = \frac{.0116}{2}$$
$$= .0058$$

Calculate the difference between $\bar{x}$ and the nearest specification limit:

$$\textbf{USL} - \bar{\boldsymbol{x}}$$
$$.0300 - .0257$$
$$.0043$$

$$\text{Cpk} = \frac{\text{USL} - \bar{x}}{3s}$$
$$= \frac{.0043}{.0058}$$
$$= .74$$

The chart in Appendix B shows that the Cpk analysis also classifies this as a D process. This Cpk value compares with the value of .87 that was calculated using all the data in Examples 6.7 and 6.11.

EXAMPLE 8.9 The first 30 measurements are taken from a process for an x and MR analysis. Specification limits are .745 to .755. The measurements are entered in order on Figure 8.18. Complete each of the following steps and compare your results with the calculations below and Figure 8.19. Use $n = 3$.

1. Calculate the MR values.
2. Set the x scale using the specification limits. Set the MR scale at piece 10.
3. Chart the values and draw the broken-line graphs.
4. Calculate the averages and draw the solid lines. Draw solid lines at the specification limits too.
5. Calculate the control limits and draw the dashed lines.
6. Interpret the results.
7. Draw the histogram.
8. Calculate the preliminary process capability.

Solution

Calculations

$$\text{MS} = \frac{.745 + .755}{2}$$
$$= .750 \quad \text{(center of chart)}$$

$\overline{MR} = \frac{\Sigma MR}{k-2} = \_\_\_\_ = \_\_\_\_$ ‖ $UCL_{MR} = 2.574 \times \overline{MR} = \_\_\_\_$ ‖ $UCL_x = \bar{x} + 1.772\overline{MR} = \_\_\_\_ + \square = \_\_\_\_$

$\bar{x} = \frac{\Sigma x}{k} = \_\_\_\_ = \_\_\_\_$ ‖ $1.772 \times \overline{MR} = 1.772 \times \_\_\_\_ = \square$ ‖ $LCL_x = \bar{x} - 1.772\overline{MR} = \_\_\_\_ - \square = \_\_\_\_$

Specification: .745 to .755

Sample Numbers	1	2	3	4	5	6	7	8	9	10	11	12	13	14	15
Time															
x	.751	.747	.752	.750	.751	.750	.748	.749	.750	.752	.749	.749	.752	.750	.748
MR	X	X													
x															
MR															

Sample Numbers	16	17	18	19	20	21	22	23	24	25	26	27	28	29	30
Time															
x	.748	.749	.749	.751	.748	.753	.749	.752	.751	.751	.755	.752	.753	.744	.749
MR															
x															
MR															

FIGURE 8.18
Make the x and MR chart with $n = 3$.

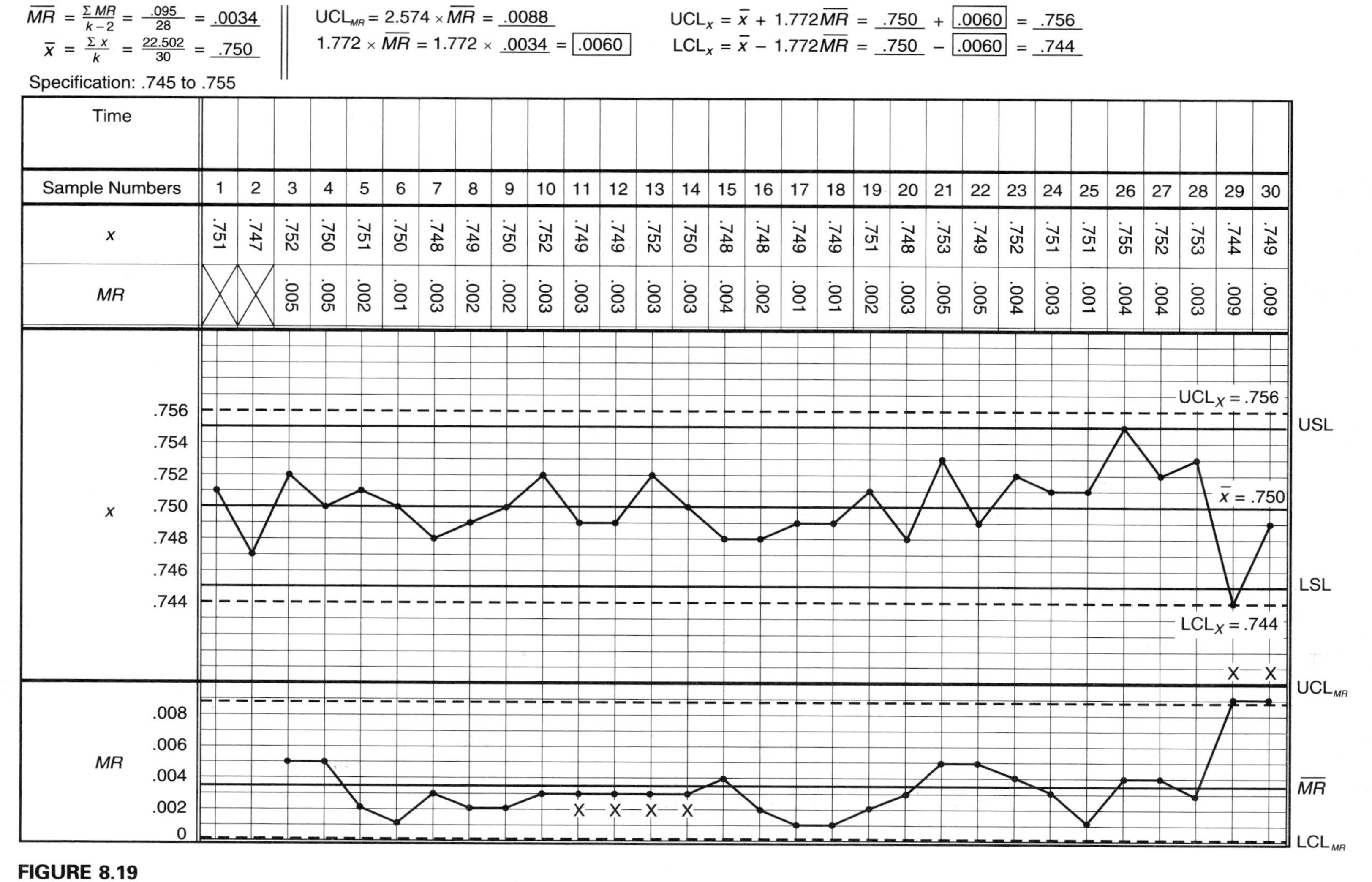

Sample Numbers	1	2	3	4	5	6	7	8	9	10	11	12	13	14	15	16	17	18	19	20	21	22	23	24	25	26	27	28	29	30
x	.751	.747	.752	.750	.751	.750	.748	.749	.750	.752	.749	.749	.752	.750	.748	.748	.749	.749	.751	.748	.753	.749	.752	.751	.751	.755	.752	.753	.744	.749
MR			.005	.005	.002	.001	.003	.002	.002	.003	.003	.003	.003	.003	.004	.002	.001	.001	.002	.003	.005	.005	.004	.003	.001	.004	.004	.003	.009	.009

FIGURE 8.19
The x and MR chart for Example 8.9.

$$\frac{.755 - .745}{10} = \frac{.010}{10}$$
$$= .001 \quad \text{units per line}$$

$$\bar{x} = \frac{\Sigma x}{K}$$
$$= \frac{.751 + .747 + \cdots + .749}{30}$$
$$= \frac{22.502}{30}$$
$$= .750$$

$$\overline{MR} = \frac{\Sigma MR}{K - 2} = \frac{.005 + .005 + \cdots + .009}{28}$$
$$= \frac{.095}{28}$$
$$= .0034$$

$$\begin{aligned} UCL_{MR} &= D_4 \times \overline{MR} \\ &= 2.574 \times .0034 \\ &= .0088 \\ LCL_{MR} &= D_3 \times \overline{MR} \\ &= 0 \times .0034 \\ &= 0 \end{aligned}$$

$$\begin{aligned} UCL_x &= \bar{x} + E_2 \times \overline{MR} \\ &= .750 + 1.772 \times .0034 \\ &= .750 + .006 \\ UCL_x &= .756 \\ LCL_x &= \bar{x} - E_2 \times \overline{MR} \\ &= .750 - .006 \\ &= .744 \end{aligned}$$

The values for D_3, D_4, and E_2 are found in Appendix B, Table B.4.

Interpretation The *MR* chart shows a shift pattern from piece 5 to 14. Since this is a downward shift on a variation chart, it does not suggest trouble and could be "good news." At the end of the *MR* chart there are two points beyond the control limits that are caused by the one measurement that is out of specification, sample 29. If the source of trouble for that one part can be determined, the data can be eliminated and the resulting *MR* calculation will be below .003. That will remove the earlier shift pattern, so there is really no good-news situation to investigate. The main trouble that shows up on the *x* chart is that there is too much variation in the measurements, resulting in poor capability. This can be seen in the relationship of the control limits to the specification limits. For an individual measurement, the control limits correspond to the 6*s* spread of measurement values. For a capable process, that should be well inside the specification limits. The chart shows the control limits beyond the specific limits. Even with piece 29 removed, the control limits would still be very close to the specification limits.

There are not enough data for the histogram to show a normal distribution pattern (see Figure 8.20). What it does show is that there are two measurements that are contributing significantly to the poor capability, samples 26 and 29. Without those two pieces, the distribution is well within specification. If there is a specific problem associated with the two pieces and that problem is eliminated, then the capability of the process should improve dramatically.

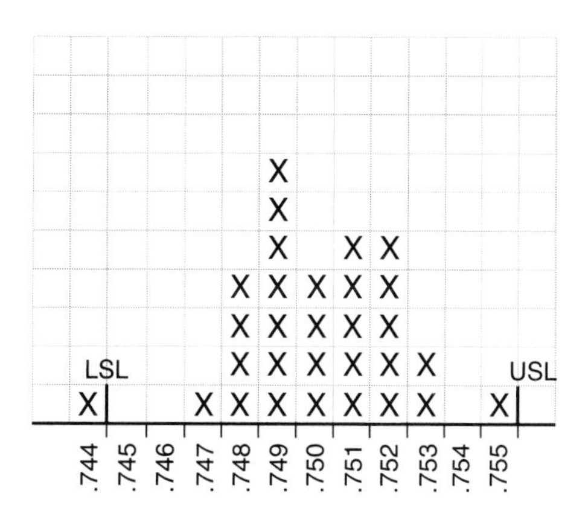

FIGURE 8.20
The histogram for Example 8.9.

This is another good illustration of what SPC is all about. The interpretation of statistical charts can pinpoint where problems exist. The charts, by themselves, do nothing to improve quality. Improvement can occur only when action is taken and the cause of the problem is eliminated.

Process Capability The process capability can also be seen in the PCR and Cpk calculations because they are somewhat independent of the shape of the product distribution. The control limits are for the individual measurements, so they coincide with $\bar{x} + 3s$ and $\bar{x} - 3s$. Their difference is $6s$:

$$\text{PCR} = \frac{6s}{\text{USL} - \text{LSL}} = \frac{.012}{.010} = 1.2$$

.756	UCL	.755	USL
−.744	LCL	−.745	LSL
.012	$6s$	.010	Tolerance

The process uses 120% of the tolerance. This is a rough estimate because of the small amount of data.

Because $\bar{x}$ is the same as the midspecification value, the process is on target and the Cpk will yield the same information as the PCR.

The chart in Table B.11 of Appendix B labels this a D process. This is a trouble indicator because the D process can overlap the specification limits and more out-of-specification products may be made.

EXERCISES

1. The specification limits for a process are .925 to .945. In Table 8.1, 15 samples of two are given in order by column for precontrol analysis. The goal is to stay within the specification limits. Set up the precontrol zones using the tolerance and analyze each sample pair.
   a. Indicate the zone for each measurement used.
   b. Indicate when the B measurement is not needed.
   c. Indicate your decision after each measurement or pair to let the process run or to report trouble.

TABLE 8.1

Measurement	A	B	A	B	A	B
	.932	.940	.937	.939	.938	.935
	.935	.931	.936	.925	.929	.942
	.942	.935	.918	.934	.931	.940
	.938	.936	.937	.932	.943	.944
	.927	.933	.927	.929	.944	.949

2. A process has an A classification and is in statistical control. The decision is made to switch from $\bar{x}$ and R charting to the precontrol chart, but the A classification must be maintained. Specification limits are .520 to .540, and the analysis of the $\bar{x}$ and R chart gives $\bar{\bar{x}} = .530$ and $s = .0015$. Twenty pairs of measurements are given in order by column in Table 8.2. Set up the precontrol chart from $\bar{\bar{x}} \pm 3s$ (because the A classification must be maintained) and analyze each sample pair.
   a. Indicate the zone for each measurement used.
   b. Indicate when the B measurement is not needed.
   c. Indicate your decision after each measurement or pair to let the process run or to report trouble.

TABLE 8.2

Measurement	A	B	A	B	A	B
	.531	.532	.528	.530	.531	.530
	.530	.532	.527	.527	.536	.532
	.529	.528	.529	.530	.532	.531
	.532	.531	.531	.530	.533	.526
	.530	.531	.530	.529	.529	.528
	.527	.529	.530	.529	.529	.530
	.531	.531	.534	.533		

3. If both the A and B measurements are in the same yellow zone on a precontrol chart, what type of trouble is indicated?
4. What type of trouble is indicated by A and B measurements in opposite yellow zones on a precontrol chart?
5. On an x and MR chart, what trouble is indicated when the control limits are beyond the specification limits?
6. For the data in Table 8.3, ordered by column, make and interpret an x and MR chart using Figure 8.21. Use the specification limits .849 to .851 to set the chart scale. Also, calculate process capability (PCR) after removing any out-of-control data (assume that any problem has been resolved). Use the following measurements:

TABLE 8.3

.8503	.8500	.8508	.8502	.8507
.8498	.8506	.8498	.8503	.8508
.8497	.8497	.8502	.8504	.8507
.8502	.8508	.8498	.8504	.8508
.8499	.8493	.8500	.8505	.8506
.8501	.8492	.8502	.8507	.8507

a. Use $n = 2$.
b. Use $n = 4$.

7. Make an x and MR chart on Figure 8.22 for the data in Table 8.4. The specification limits are .021 to .027. The following measurements are for consecutive pieces and are listed in order by column:

TABLE 8.4

.0258	.0213	.0252	.0253
.0243	.0256	.0224	.0262
.0249	.0274	.0263	.0221
.0230	.0275	.0247	.0243
.0273	.0223	.0248	.0260
.0266	.0251	.0232	.0229
.0264	.0248	.0250	
.0232	.0241	.0228	

Use the given specification limits to set the x chart scale. Interpret the chart. If any points are out of control, remove them from the data set (assume that the out-of-control problem has been resolved) and calculate process capability.

a. Use $n = 2$.
b. Use $n = 3$.

8. Use data point 1 and every third point thereafter in Figure 6.16 to make an x and MR chart with $n = 2$. Compare the results with Example 6.7. (Data values will be .0218, .0256, .0282, . . .)

9. For the data in Table 8.1, estimate the capability of the process from which it was taken. Should precontrol be used for this process?

10. For the data in Table 8.2,

a. Set up a precontrol chart based on the specification limits.
b. Indicate the zone for each measurement and when the B measurement is not needed.
c. Indicate your decision after each measurement.
d. Estimate the capability of the process. It is really an A process?

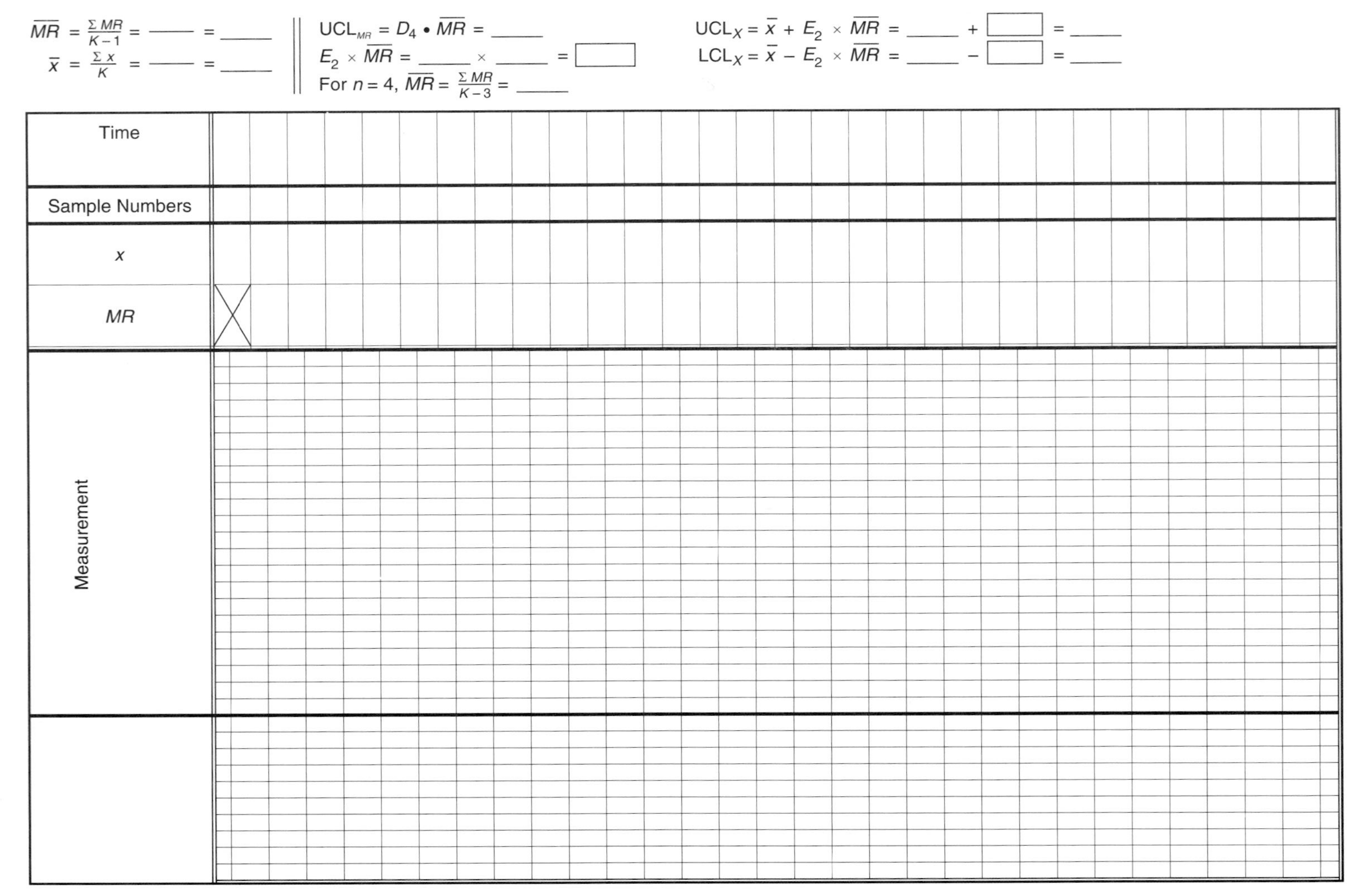

FIGURE 8.21
Make an x and MR chart for exercise 6.

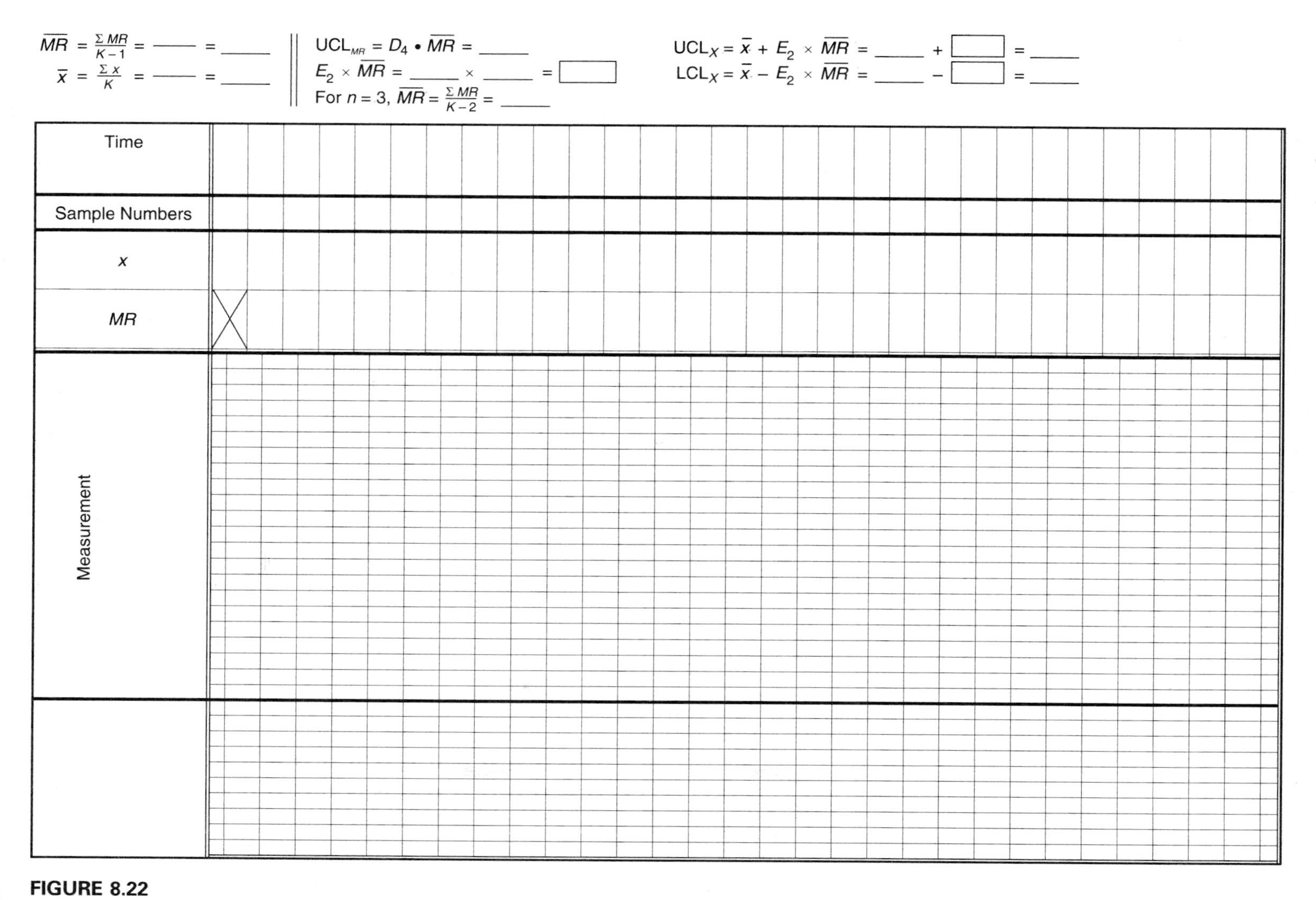

FIGURE 8.22
Make an x and MR chart for exercise 7.

11. Explain why specification limits should *not* be shown on an $\overline{x}$ and R chart and why they *should* be shown on an x and MR chart.
12. Set up a precontrol chart for the data in Table 8.3 based on the specification limits .849 to .851. Use the data, as needed by column, classify the measurements as start-up, A, or B measurements, and indicate your decisions.
13. Set up a modified precontrol chart for the data in Table 8.4. The specification limits are .015 to .035. Use the data as needed by column, classify the measurements as start-up, A, or B measurements, and indicate your decisions.

9

ATTRIBUTES CONTROL CHARTS

OBJECTIVES

- Know the four types of attributes charts.
- Make a p chart.
- Make an np chart.
- Make a c chart.
- Make a u chart.
- Make a q or nq learning chart.

The four types of attributes control charts are discussed in this chapter. Attributes charts, as compared to the previously discussed variables charts, do not use the actual measurements in their construction. Measurements are simply classified as acceptable or not acceptable (pass–fail, or go–no go). Then either the number of failures or the fraction, or percent, of failures is charted.

Attributes control charts are used when measurements are too difficult to take, when measurements do not apply to the situation (such as visual checks for mars or flaws), or when measurements are too costly to take because of time lost. Because measurements are not used in the charting process, attributes charts are not as sensitive to analysis as the variables control charts.

A *defect* is a single failure to specification. It differs from a defective item in that a defect is one specification discrepancy and a defective item contains one or more defects. An item sampled for a p or an np chart is classified as acceptable if it contains no defects but is counted as a single item if it contains one or more defects.

9.1 THE FOUR TYPES OF ATTRIBUTES CHARTS

The *p* Chart

The *p* chart is the most commonly used attributes chart. The value *p* is the fraction, or percentage, of the number of items checked that are defective (or unacceptable). Large samples of 50 or more are needed. When the *p* value is very small, the sample size must be increased so that primarily nonzero values of *p* are charted. For example, if $\bar{p} = .0002$, the sample size necessary for charting is in the range of $n = 20{,}000$. The sample size may vary in a *p* chart. Sampling in time blocks naturally leads to different sample sizes: In time block sampling, the sample consists of all the items produced in a specific time period.

The *np* Chart

The *np* chart is sometimes used instead of the *p* chart because it is easier; *np* is simply the number, rather than the fraction, of defective items in the sample. When a sample is taken and the number of defective items is counted, it is easier to chart that number than to divide it by the sample size to change it into *p*, the fraction defective. The *p* and the *np* charts differ by that constant divisor, so they do the same job with respect to control, process analysis, assessing priority, and so forth. The one detail necessary for an *np* chart is the requirement that all samples be the *same size*. If the sample sizes differ, the *p* chart must be used.

The *c* Chart

The *c* chart tracks the number of defects in constant size units. There may be a single type of defect or different types, but the *c* chart tracks the total number of defects in each unit. The unit may be a single item or a specified section of an item.

The *u* Chart

When samples of different size are taken, *u* is the *average* number of defects per unit. The *u* chart is quite similar to the *c* chart in function.

The attributes charts fall into two categories. The first category contains the *p* and *np* charts, which keep track of the number of defective items or the number of nonconforming items. The *np* chart tracks the *number* of nonconforming items in constant size samples and the *p* chart tracks the *fraction* of nonconforming items in the samples. The second category contains the *c* and *u* charts, which keep track of the total number of defects in the samples (one nonconforming item may contain several defects). The *c* chart tracks the *number* of defects in constant size samples and the *u* chart tracks the *average number* of defects per unit when the number of units in the sample varies.

Applications of the Charts

The following examples list several situations in which the various attributes charts apply. In the first case, a printed circuit board is produced in high volume. The fraction defective has been reduced to approximately .007, and control is to be maintained at that level. Samples of 700 circuit boards are tested. The test criterion is pass–fail: The circuit either works or is defective. Here a p chart may be used to track the *proportion* of defective circuits in each sample or an np chart may track the *number* of defective circuits in each sample.

In the second situation, there has been a problem with incorrect invoices. To isolate the problem, the invoices are separated according to vendor. Samples of 50 invoices are selected for each vendor. Invoices are classified as correct or incorrect. Again, a p chart may be formed for each vendor to track the *proportion* of incorrect invoices in each sample, or an np chart may be formed for each vendor to track the *number* of incorrect invoices in each sample.

A new automatic frame welder has been installed in the third case, and an initial performance standard must be established for the operation of the welder. According to the test criterion, a weld is either acceptable or faulty. Several samples of 100 welds are checked: One has two faulty welds and three have one faulty weld. The sample size is increased to 500. Here a p chart may track the *fraction* of faulty welds in each sample, or an np chart may track the *number* of faulty welds in each sample.

In the final example of attributes chart applications, a textile machine produces bolts of fabric. A quality check for flaws is to be made. A random square yard is checked in each bolt, and the number of flaws in that square yard is counted. The various types of flaws are defined by picture example. In this case a c chart is initially formed to track the *number* of flaws per square yard. The c chart is found to be too inconsistent for a quality analysis in this case, so a more intensive sampling plan is devised. Four random square yards are checked in each bolt, and five consecutive bolts complete the sample for a total of 20 square yards checked in each sample. Dividing the total number of flaws in each sample by 20 gives the value u, the number of flaws per square yard in the sample. A u chart is formed to chart the *average* number of flaws per square yard.

9.2 THE *p* CHART

The p chart tracks the fraction of nonconforming items in a sample. Samples may consist either of n consecutive items taken at specified times during the day according to a random-sample plan or of 100% of the production for a specified time period (hour, day, week, and so forth).

The conforming/nonconforming decision may be based on one characteristic or on several, but a defective item is counted as defective only once, even though it may contain several defects.

Preparation for the *p* Chart

The preparatory steps are similar to those of the variables charts.

Step 1. Establish an environment suitable for action. The appropriate people must be trained in charting techniques and interpretation and made aware of their importance in the quality improvement process.

Step 2. Define the process. Brainstorming with the quality control team can be effective for developing flow charts, cause-and-effect diagrams, or storyboards for process analysis. These are discussed in Chapter 11.

Step 3. Determine the most promising characteristics to be managed. Use of Pareto charts (bar charts that classify defects) can be helpful in setting priorities. Be sure to consider customer needs for both the final customer and the downstream customers (the people using the item in subsequent process steps) along with current and potential problem areas. Quality control personnel trained in statistics can check for correlation between characteristics. When high correlation (mathematical interdependence) exists, charting one characteristic may be sufficient.

Step 4. Define the measurement system. This may be something simple, such as a go–no-go gauge at a specific measurement point, or something quite difficult for consistency, such as a judgment call on pass–fail. When individual judgment is used for deciding whether a product conforms, inspectors must have total agreement as to exactly what is defective and what is not. For example, if a visual check is made for surface flaws of the item produced, clear pictures should be available to illustrate what is acceptable and what is not, or sample pass–fail pieces should be used for comparison. Occasional consistency checks among the inspectors can keep the measurement system fine-tuned.

Step 5. Minimize any unnecessary variation before the charting begins. Do not wait for the charts to point out obvious problems when those problems can be eliminated beforehand. What is obvious to one inspector may not be obvious to another, however, and in that case, the chart can be used to supply the proof that trouble exists so that action may be authorized.

The *p* Chart Procedure

There are six basic steps in the p chart procedure.

Step 1. Gather the data. Select the size, frequency, and number of samples. Large samples of 50, 100, or more are common. The sample size should be large enough to ensure that most of the samples will have a nonzero number of defectives. The chart will be easier to interpret if the average number of defectives in the sample is 5 or more. If an estimate of $\overline{p}$, the average proportion of defective items, can be made, the necessary sample size n can be determined by choosing an n value large enough to make the product $n\overline{p}$ approximately 5. The calculations and interpretation of the chart are easier when the sample size is kept constant. When constant sample times are used and the sample size varies, a single set of control limits can be used as long as the individual sample sizes do not differ too much. If any individual samples or groups of samples differ in size by more than 25% of the average sample size $\bar{n}$, separate control limit calculations will be needed. Chart interpretation is trickier because it is more difficult to relate the point pattern to fluctuating control limits.

Sampling frequency should make sense in terms of production schedules. The chart should give as accurate a picture as possible of the process at the specific point in question. To do this, every item produced during the charting program must have an equal chance of being chosen in a sample. A random sampling plan similar to the plans discussed in Example 3.3 should therefore be used.

The number of samples on a p chart should be 20 or more, but if time is a factor, a preliminary analysis may be made after 10 samples have been charted. The data collection period should be long enough to track all possible sources of variation.

Step 2. Calculate *p* for each sample. Record the sample size n and the number of defective items np on the chart. Calculate and record p:

$$p = \frac{np}{n}$$

Step 3. Set the scale for the control chart. After several p values have been calculated, make the p scale from 0 to approximately twice the largest p value. For easier interpretation, do not make the scale too wide.

Step 4. Plot the *p* values and connect them with a broken-line graph.

Step 5. Calculate $\bar{p}$, the average *p* value, and the control limits after about 20 points have been charted. The number of samples is k. Draw a solid line at $\bar{p}$ and dashed lines at the control limits:

Control Limit Formulas

$$\bar{p} = \frac{\Sigma np}{\Sigma n}$$

$$= \frac{np_1 + np_2 + np_3 + \cdots + np_k}{n_1 + n_2 + \cdots + n_k}$$

$$\text{UCL}_p = \bar{p} + 3\sqrt{\frac{\bar{p}(1-\bar{p}}{\bar{n}}}, \bar{n} = \frac{\Sigma n}{\text{k}}$$

$$\text{LCL}_p = \bar{p} - 3\sqrt{\frac{\bar{p}(1-\bar{p})}{\bar{n}}}$$

If the sample size is constant, the denominator under the radical in the preceding control limit formulas is n. When a constant sample size is *not* used, the average sample size is used in the control limit calculations. For k samples,

$$\overline{n} = \frac{\Sigma n}{k}$$

When the sample size variation is minimal, the one set of control limits is adequate. However, any individual sample sizes that vary by more than 25% from $\overline{n}$ require separate control limits. This is not the best type of p chart because it is harder to interpret than a p chart with constant sample size, but unequal sample sizes can be used when necessary.

Step 6. Interpret the chart.

Shifts Because p represents the defective proportion, a shift of seven or more points to a higher or lower level could mean that something has affected the proportion defective or that the definition of what is defective has somehow changed, either intentionally or unintentionally.

Runs A run of seven or more points up or down indicates that something is causing the proportion that is defective to change in a gradual manner.

An erratic pattern indicates that something irregular is affecting the process. There are several possible trouble spots: Poorly trained operators may be responsible for widely fluctuating values of p, different inspection criteria may be used by some inspectors, or an assembly point may be using poorly controlled parts. The erratic pattern usually signals a need for either separate p charts at various upstream positions in the process or a switch to a variables control chart for further analysis.

Interpretation of p charts demands more caution than interpretation of the variables control charts. Do not assume too quickly that the p chart is in control. Sometimes a situation that appears to be in control is just a balance of different characteristics that are individually out of control. It is important to stay aware of the number of different characteristics that are involved in a specific pass–fail decision. Decision standards should be checked periodically to ensure that they have not changed.

Often, several p charts are dealt with simultaneously. Do not concentrate on just the charts with high $\overline{p}$ values: It may be more beneficial to work with a chart that has a low $\overline{p}$ value but shows a lack of control. A similar concept was discussed with the variables control charts. The first step toward achieving control on a variables chart is control of the variation on the R chart. A fluctuating p chart signals a variation problem. Solving the variation problem can have a ripple effect and possibly reduce some of the $\overline{p}$ values on other charts.

EXAMPLE 9.1 Figure 9.1 contains data that show samples of 200, the number of defective items, and the first few p values. Calculate the rest of the p values and check your results with Figure 9.2.

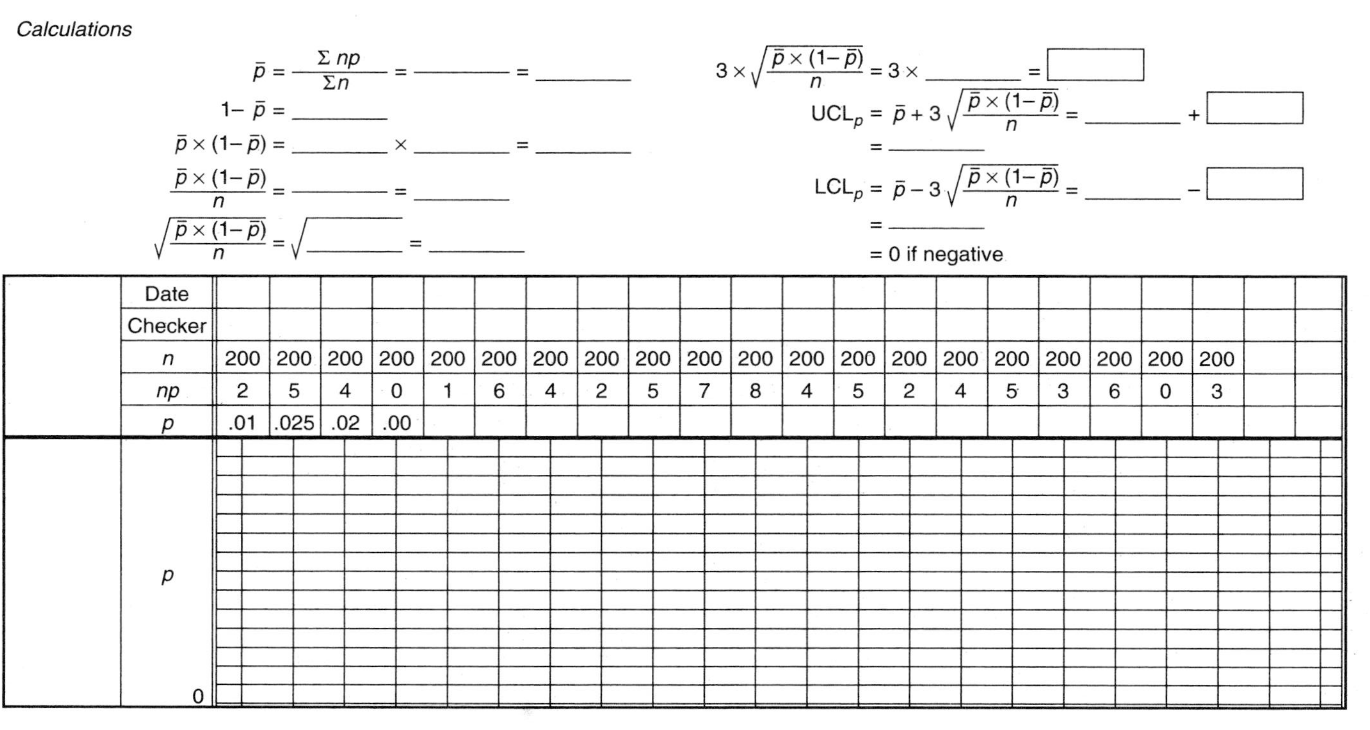

Calculations

$\bar{p} = \dfrac{\Sigma np}{\Sigma n} = \_\_\_\_\_ = \_\_\_\_\_$

$1 - \bar{p} = \_\_\_\_\_$

$\bar{p} \times (1-\bar{p}) = \_\_\_\_\_ \times \_\_\_\_\_ = \_\_\_\_\_$

$\dfrac{\bar{p} \times (1-\bar{p})}{n} = \_\_\_\_\_ = \_\_\_\_\_$

$\sqrt{\dfrac{\bar{p} \times (1-\bar{p})}{n}} = \sqrt{\_\_\_\_\_} = \_\_\_\_\_$

$3 \times \sqrt{\dfrac{\bar{p} \times (1-\bar{p})}{n}} = 3 \times \_\_\_\_\_ = \boxed{}$

$UCL_p = \bar{p} + 3\sqrt{\dfrac{\bar{p} \times (1-\bar{p})}{n}} = \_\_\_\_\_ + \boxed{}$

$= \_\_\_\_\_$

$LCL_p = \bar{p} - 3\sqrt{\dfrac{\bar{p} \times (1-\bar{p})}{n}} = \_\_\_\_\_ - \boxed{}$

$= \_\_\_\_\_$

= 0 if negative

Date																					
Checker																					
n	200	200	200	200	200	200	200	200	200	200	200	200	200	200	200	200	200	200	200	200		
np	2	5	4	0	1	6	4	2	5	7	8	4	5	2	4	5	3	6	0	3		
p	.01	.025	.02	.00																		

Figure 9.1
Follow the procedure for Example 9.1 and check the results with Figure 9.2.

Calculations

$$\bar{p} = \frac{\Sigma\, np}{\Sigma n} = \frac{76}{4000} = .019$$

$$1 - \bar{p} = .981$$

$$\bar{p} \times (1 - \bar{p}) = .019 \times .981 = .018639$$

$$\frac{\bar{p} \times (1 - \bar{p})}{n} = \frac{.018639}{200} = .000093195$$

$$\sqrt{\frac{\bar{p} \times (1 - \bar{p})}{n}} = \sqrt{.000093195} = .0096538$$

$$3\sqrt{\frac{\bar{p} \times (1 - \bar{p})}{n}} = 3 \times .0096538 = .029$$

$$UCL_p = \bar{p} + 3\sqrt{\frac{\bar{p} \times (1 - \bar{p})}{n}}$$

$$= .019 + .029 = .048$$

$$LCL_p = .019 - .029$$

$$= -.01$$

$$= 0 \text{ (because of negative value)}$$

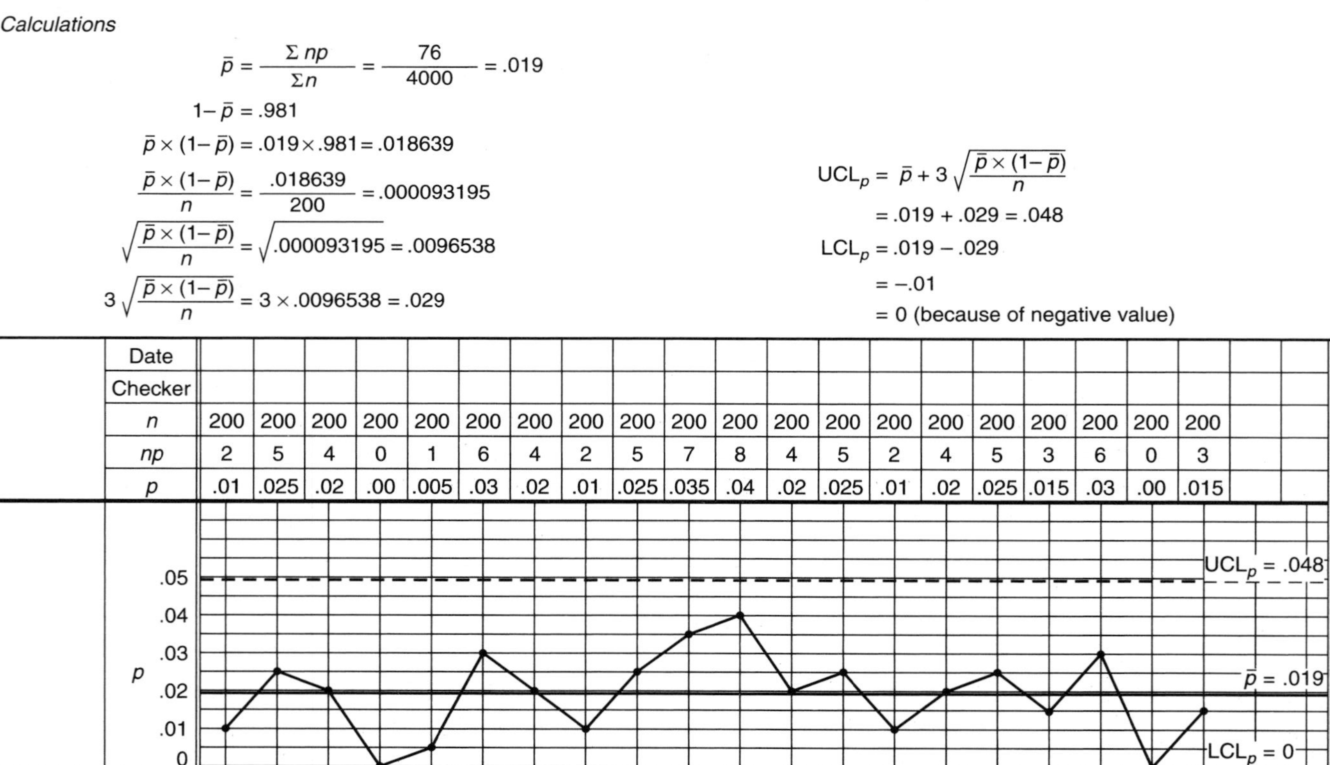

Date																						
Checker																						
n	200	200	200	200	200	200	200	200	200	200	200	200	200	200	200	200	200	200	200	200		
np	2	5	4	0	1	6	4	2	5	7	8	4	5	2	4	5	3	6	0	3		
p	.01	.025	.02	.00	.005	.03	.02	.01	.025	.035	.04	.02	.025	.01	.02	.025	.015	.03	.00	.015		

Figure 9.2
The p chart for Example 9.1.

Solution

$$p_1 = \frac{np_1}{n} \qquad p_2 = \frac{np_2}{n}$$
$$= \frac{2}{200} \qquad = \frac{5}{200}$$
$$= .01 \qquad = .025$$
$$p_3 = \frac{np_3}{n} \qquad p_4 = \frac{np_4}{n}$$
$$= \frac{4}{200} \qquad = \frac{0}{200}$$
$$= .02 \qquad = .000$$

The proportion p varies from 0 to .04. Aiming for approximately 10 lines on the chart, divide 10 into the range of p and round up to .005 units per line:

$$10\overline{).04}\quad = .004$$

This should allow enough room for control limits without making the chart too wide. Put the scale on the chart in Figure 9.1 and graph the p values. Connect the points with a broken-line graph. Check your results with Figure 9.2.

Calculate $\overline{p}$ and draw the solid line:

$$\overline{p} = \frac{\Sigma np}{\Sigma n}$$
$$= \frac{2 + 5 + 4 + 0 + 1 + \cdots + 3}{20 \times 200}$$
$$= \frac{76}{4000}$$
$$= .019$$

The denominator, 200 + 200 + 200 + · · ·, is calculated by multiplying 200 by 20.

Next, calculate the control limits and draw the dashed lines:

$$\text{UCL}_p = \overline{p} + 3\sqrt{\frac{\overline{p}(1 - \overline{p})}{n}} \qquad \text{LCL}_p = \overline{p} - 3\sqrt{\frac{\overline{p}(1 - \overline{p})}{n}}$$

The more complicated second term will be calculated first, then added to and subtracted from $\overline{p}$:

1. Find $1 - \overline{p}$:

$$1 - \overline{p} = 1 - .019$$
$$= .981$$

2. Multiply by $\overline{p}$:

$$\overline{p}(1 - \overline{p}) = .019 \times .981$$
$$= .018639$$

3. Divide by n:

$$\frac{\overline{p}(1 - \overline{p})}{n} = \frac{.018639}{200}$$
$$= .000093195$$

4. Calculate the square root:

$$\sqrt{\frac{\overline{p}(1 - \overline{p})}{n}} = \sqrt{.000093195}$$
$$= .0096538$$

With the .000093195 in the calculator display, press the radical button on the calculator, "$\sqrt{\ }$."

5. Multiply by 3:

$$3\sqrt{\frac{\overline{p}(1 - \overline{p})}{n}} = 3 \times .0096538$$
$$= .029$$

6. Add .029 to $\overline{p}$ for UCL_p:

$$\text{UCL}_p = .019 + .029$$
$$= .048$$

7. Subtract .029 from $\overline{p}$ for LCL_p. When the LCL comes out negative, use LCL = 0:

$$\text{LCL}_p = .019 - .029$$
$$= -.01$$
$$= 0$$

Check your results with Figure 9.2.

Interpretation is the final step. There are no obvious out-of-control patterns such as points beyond the control limits, seven points in a row above or below the mean, or seven points steadily increasing or decreasing. There may be a downward trend beginning with the eleventh sample, but the chart will have to be extended for several more samples to see if a trend really is occurring.

CASE STUDY 9.1

The final assembly point for remote lock controls at the ABC Corporation must final test each unit. Workers are responsible for testing only the locks they have assembled on their shift. Consequently the number in each sample varies widely with the production schedule. The workers are directed to keep a p chart. They were able to follow directions for calculating p and charting it, but they could not interpret the results. A consultant was called in to teach them how.

The data are given on the chart in Figure 9.3. Calculate the p values as the workers were directed to do, set the scale, and plot the points. Check your results with the completed chart in Figure 9.4.

Solution

Calculate $p = np/n$ for each sample. After all the p values have been calculated, determine the scale so about half the chart is used.

Other than the obvious out-of-control value, $p = .111$, the largest p value is .041. Use the scale of .01 for each long line and let the $p = .111$ value run off the chart. Using a scale that will accommodate the one large value too (such as .02 for each line) will shrink the chart by half and make it hard to read. Chart the p values.

The charts that the workers had been using had one set of control limits on them. Out-of-control situations were either overestimated or underestimated depending on where the limits ended up. The consultant mentioned that different control limits are recommended when sample sizes differ by 25%. If this were done following the sampling order, almost every sample would need its own control limits. The sample sizes may be grouped so that each extreme is within 25% of the middle n value.

For the first set, if $n = 50$, then all the smaller samples are within 25% of 50. Sixty is approximately 25% larger $[50 + (.25 \times 50)]$. To find the next middle n value so that 60 is the lower boundary (25% below n), solve the equation

$$\begin{aligned} n - .25n &= 60 \\ .75n &= 60 \\ n &= 80 \end{aligned}$$

To find the upper boundary for the middle value, $n = 80$, calculate the number that is 25% larger:

$$80 + .25 \times 80 = 100$$

The value 100 will be the upper bound of the group with a middle value of 80, and it will be the lower bound of the next higher group. To find the middle value of the next higher group, solve the equation

$$\begin{aligned} n - .25n &= 100 \\ .75n &= 100 \\ n &= 133 \end{aligned}$$

The rest of the middle n values are calculated in a similar manner. Table 9.1 shows the results. The group endpoints have been changed by 1 unit to make the group selection for each n value easier. The table also contains the 3σ calculation and the upper control limit for each middle n value. The lower control limit in each case is the default value $p = 0$ ($\bar{p} - 3\sigma_p$ would be negative).

Sample Size (n)	234	190	173	587	51	148	212	229	96	74	110	112	179	49	272	149	744	540	387	54	325	61	45	365	206
Discrepancies: Number (np)	7	2	0	5	0	0	0	0	0	3	1	1	7	0	6	6	3	1	0	1	5	1	5	3	2
Discrepancies: Proportion (p)																									
Date	7/7 11:00P	7/17	7/20	7/23	7/24	7/24	7/24	7/25	8/4	8/8	8/13	8/18	8/18	8/25	8/26	9/2	9/10	9/11	9/14	9/16	9/16	9/18	9/18	9/18	9/18
	1	2	3	4	5	6	7	8	9	10	11	12	13	14	15	16	17	18	19	20	21	22	23	24	25
p																									

Figure 9.3
Complete the *p* chart for Case Study 9.1.

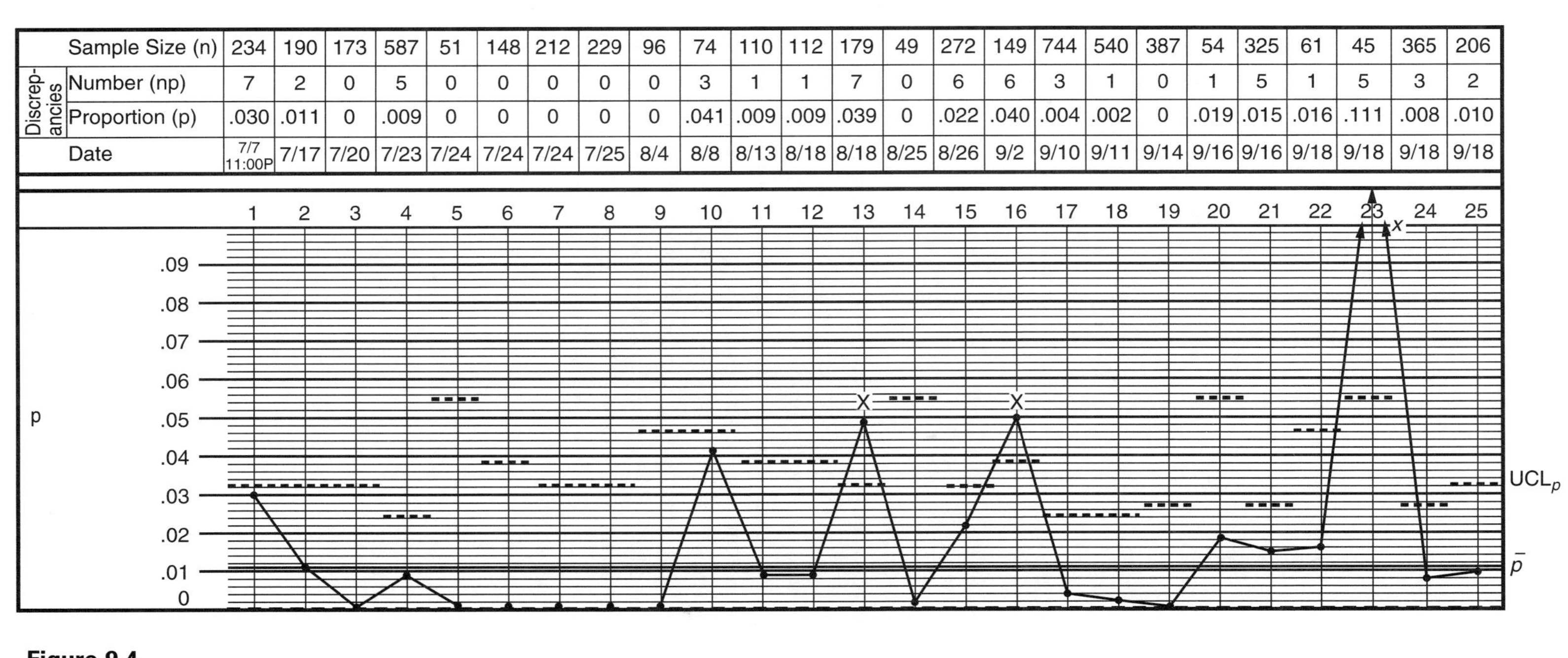

Figure 9.4
Completed *p* chart for Case Study 9.1.

Calculate $\bar{p}$ and draw the line:

$$\bar{p} = \frac{\Sigma np}{\Sigma n} = \frac{59}{5592}$$
$$= .011$$

Calculate $3\sigma_p$ for each middle n value. The results are shown in Table 9.1:

For $n = 50$ $\quad 3\sqrt{\frac{\bar{p}(1-\bar{p})}{n}} = 3\sqrt{\frac{.011(.989)}{50}} = .044$

For $n = 80$ $\quad 3\sqrt{\frac{\bar{p}(1-\bar{p})}{n}} = 3\sqrt{\frac{.011(.989)}{80}} = .035$

Calculate the upper control limit by adding .011 to each $3\sigma_p$ value, $(\bar{p} + 3\sigma_p)$. These are also shown in Table 9.1.

Draw the control limits on the p chart. The lower control limit, $\bar{p} - 3\sigma_p$, is less than zero in each case, so $LCL_p = 0$.

Chart Interpretation

There are three points that are beyond the control limits. The staggered control limits make it difficult to spot any of the other trouble patterns. The 1.1% average defect rate is excessive and must cut into the profit margin.

With a 1.1% defect rate, the sample size should be about 400 (so that $.011 \times n \approx 5$). The consultant recommended samples of 400, pooling the work as needed. Each worker would still be responsible for checking his or her own work. Any pooled samples could contain a note (chart comment) giving the pooled sample breakdown, indicating the number produced and the number of defects for each worker contributing to the sample. The resulting p chart could be much simpler for everyone to use and interpret. With the constant sample size they could even use the easier np chart and just track the number of defectives in each sample.

TABLE 9.1
Middle n values, their 25% bounds, and $3\sigma_p$ and upper control limits

Middle n	$n - .25n$ to $n + .25n$	$3\sqrt{\frac{\bar{p}(1-\bar{p})}{n}}$	UCL_p
50	38–60	.044	.055
80	61–100	.035	.046
133	101–166	.027	.038
221	167–276	.021	.032
368	277–460	.016	.027
610	461–760	.013	.024

9.3 THE *np* CHART

The *np* chart can be used in place of the *p* chart whenever the sample size is constant. The *np* chart is the preferred chart because it is easier to construct. An easy conversion is available for changing the charting results to a proportion format. That conversion will be demonstrated at the end of Example 9.2.

Control Chart Formulas for the np Chart

np = Number of defective pieces / sample

$$\overline{np} = \frac{\Sigma np}{k} \quad \bar{p} = \frac{\overline{np}}{n}$$

$$\text{UCL}_{np} = \overline{np} + 3\sqrt{\overline{np}(1-\bar{p})}$$

$$\text{LCL}_{np} = \overline{np} + 3\sqrt{\overline{np}(1-\bar{p})}$$

EXAMPLE 9.2

Make an *np* chart for the data in Figure 9.5, the data have been copied from Figure 9.1. Follow the directions below and complete the chart; check your results with Figure 9.6.

Solution

The *np* values range from 0 to 8, so scale the *np* chart in one unit values from 0 to 10. Chart the *np* values and draw the broken-line graph. Find the value of the center line $\overline{np}$ and draw a solid line:

$$\overline{np} = \frac{\Sigma np}{k}$$

$$= \frac{2+5+4+\cdots+3}{20}$$

$$= \frac{76}{20}$$

$$= 3.8$$

Calculate the control limits and draw the dashed lines:

$$\text{UCL}_{np} = \overline{np} + 3\sqrt{\overline{np}\,(1-\bar{p})} \quad \text{LCL}_{np} = \overline{np} - 3\sqrt{\overline{np}\,(1-\bar{p})}$$

1. Calculate $\bar{p}$:

$$\bar{p} = \frac{\overline{np}}{n}$$

$$= \frac{3.8}{200}$$

$$= .019$$

2. Subtract $\bar{p}$ from 1:

$$1 - \bar{p} = 1 - .019$$

$$= .981$$

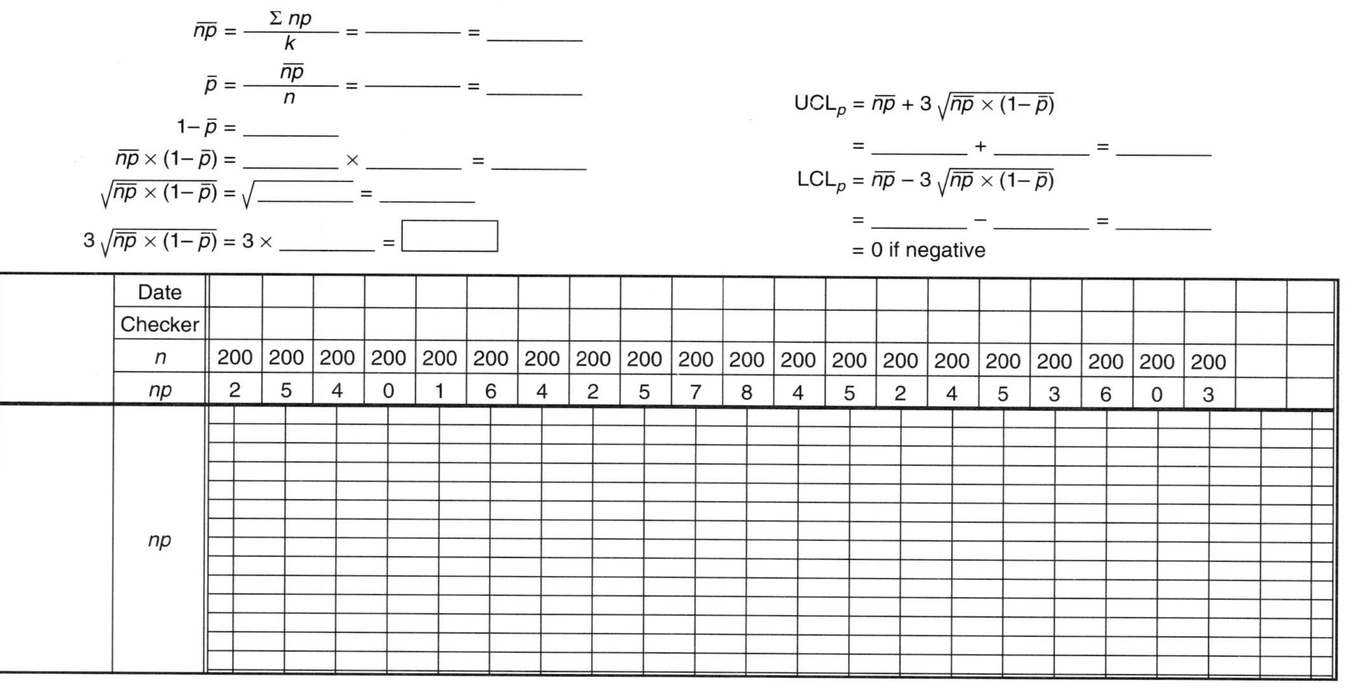

Calculations

$\overline{np} = \frac{\Sigma\, np}{k} = \_\_\_\_\_\_ = \_\_\_\_\_\_$

$\bar{p} = \frac{\overline{np}}{n} = \_\_\_\_\_\_ = \_\_\_\_\_\_$

$1 - \bar{p} = \_\_\_\_\_\_$

$\overline{np} \times (1 - \bar{p}) = \_\_\_\_\_\_ \times \_\_\_\_\_\_ = \_\_\_\_\_\_$

$\sqrt{\overline{np} \times (1 - \bar{p})} = \sqrt{\_\_\_\_\_\_} = \_\_\_\_\_\_$

$3\sqrt{\overline{np} \times (1 - \bar{p})} = 3 \times \_\_\_\_\_\_ = \square$

$UCL_p = \overline{np} + 3\sqrt{\overline{np} \times (1 - \bar{p})}$

$= \_\_\_\_\_\_ + \_\_\_\_\_\_ = \_\_\_\_\_\_$

$LCL_p = \overline{np} - 3\sqrt{\overline{np} \times (1 - \bar{p})}$

$= \_\_\_\_\_\_ - \_\_\_\_\_\_ = \_\_\_\_\_\_$

= 0 if negative

Date																					
Checker																					
n	200	200	200	200	200	200	200	200	200	200	200	200	200	200	200	200	200	200	200	200		
np	2	5	4	0	1	6	4	2	5	7	8	4	5	2	4	5	3	6	0	3		

Figure 9.5
Follow the directions given in Example 9.2. Check your results with Figure 9.6.

Calculations

$$\overline{np} = \frac{\Sigma np}{k} = \frac{76}{20} = 3.8$$

$$\bar{p} = \frac{\overline{np}}{n} = \frac{3.8}{200} = .019$$

$$1-\bar{p} = .981$$

$$\overline{np} \times (1-\bar{p}) = 3.8 \times .981 = 3.7278$$

$$\sqrt{\overline{np} \times (1-\bar{p})} = \sqrt{3.7278} = 1.931$$

$$3\sqrt{\overline{np} \times (1-\bar{p})} = 3 \times 1.931 = 5.793$$

$$UCL_p = \overline{np} + 3\sqrt{\overline{np} \times (1-\bar{p})}$$

$$= 3.8 + 5.8 = 9.6$$

$$LCL_p = \overline{np} - 3\sqrt{\overline{np} \times (1-\bar{p})}$$

$$= 3.8 - 5.8 = -2$$

$$= 0 \text{ if negative}$$

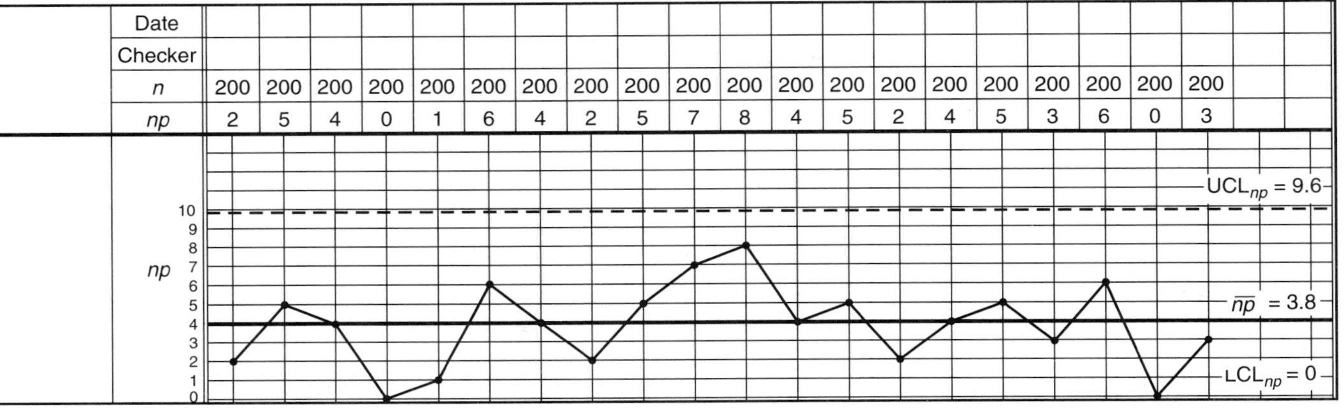

Figure 9.6
The *np* chart for Example 9.2.

3. Multiply by $\overline{np}$:

$$\overline{np}\,(1 - \overline{p}) = 3.8 \times .981$$
$$= 3.7278$$

4. Calculate the square root:

$$\sqrt{\overline{np}\,(1 - \overline{p})} = \sqrt{3.7278}$$
$$= 1.931$$

5. Multiply by 3:

$$3\sqrt{\overline{np}\,(1 - \overline{p})} = 3 \times 1.931$$
$$= 5.793$$

6. Add to $\overline{np}$ for the UCL_{np}:

$$UCL_{np} = 3.8 + 5.8$$
$$= 9.6$$

7. Subtract from $\overline{np}$ for the LCL_{np}:

$$LCL_{np} = 3.8 - 5.8$$
$$= -2$$
$$= 0$$

Just as before, if the LCL comes out negative, use LCL = 0.

The *np* chart interpretation is the same as that for the *p* chart. There are no apparent out-of-control situations, but if the chart is continued for several more samples, a downward trend may show up.

Process capability is a measure of how well the process is doing with respect to some standard. With variables control charts, the standard is always the specification limits. With attributes control charts, both the *p* and *np* charts have a standard of zero defects. The measure of process capability is the number of defects produced per sample on the average, $\overline{np}$, or the fraction (or percentage) of defects produced on the average, $\overline{p}$.

If you compare the *p* chart and the *np* chart carefully, you will notice that the scales match exactly and the broken-line graphs look the same. Further, the averages and control limits match. The information from an *np* chart can be translated to the corresponding *p* chart information by dividing by the sample size *n*. Also, *p* chart information can be translated to *np* values by multiplying by the sample size *n:*

$$\overline{np} = 3.8$$
$$\frac{\overline{np}}{n} = \frac{3.8}{200}$$
$$= .019$$
$$= \overline{p}$$
$$UCL_{np} = 9.6$$

$$\frac{\text{UCL}_{np}}{n} = \frac{9.6}{n}$$
$$= .048$$
$$= \text{UCL}_p$$

EXAMPLE 9.3 For another example of the connection between p values and np values, suppose that a p chart has $\overline{p} = .032$ and the UCL_p is .0654 for constant samples of $n = 250$. What is the average number of defectives per sample and the number of defectives three standard deviations above the mean?

Solution

The average number of defectives per sample is $\overline{np}$:

$$\overline{np} = n \times \overline{p}$$
$$= 250 \times .032$$
$$= 8$$

The upper control limit gives the number of defectives three standard deviations above the mean:

$$\text{UCL}_{np} = n \times \text{UCL}_p$$
$$= 250 \times .0654$$
$$= 16.35$$

9.4 THE c CHART

The c chart tracks the number of defects in some unit. In the following example, the unit used is one shirt.

Control Chart Formulas for the c Chart

$$\overline{c} = \frac{\Sigma c}{k}$$

c = Number of defects / units

k = Number of units

$$\text{UCL}_c = \overline{c} + 3\sqrt{\overline{c}}$$

$$\text{LCL}_c = \overline{c} - 3\sqrt{\overline{c}}$$

EXAMPLE 9.4 A shirt manufacturer is trying to upgrade the quality of the dress shirts it produces in order to introduce a new line in men's stores. After the cloth quality is improved, the management decides that with the introduction of the new cloth into the process, the number of defects should be tracked. The inspectors are provided with a list of items to check on each shirt and are shown samples of acceptable quality. Ten consecutive shirts are checked several times a day according to a random-sample plan, and the data in Table 9.2

TABLE 9.2
Number of flaws per shirt

c	c	c	c	c	c	c	c	c	c
0	2	0	3	0	1	2	0	0	2
5	0	1	2	0	0	0	4	4	2
0	1	0	0	1	3	4	0	0	2
0	1	1	5	3	2	2	0	1	3
0	4	1	2	0	1	0	0	2	3
3	0	0	1	5	3	2	4	0	1
0	0	1	4	4	3	0	2	1	1
0	2	1	3	2	0	0	2	2	1
0	0	0	4	6	2	0	3	0	1
2	0	5	6	2	0	1	2	4	1

are collected. With these data, construct a c chart. The data are in order by column. The number of defects on each shirt is c.

Solution

The c values range from 0 to 6, so set the scale in units from 0 to 7. A quick scan of the data shows that $\bar{c}$ should be relatively small and that the upper control limit should be less than 7. There is enough room on the chart to put three c charts for shirts 1 to 30, 31 to 65, and 66 to 100. Set three different scales from 0 to 7 on the chart in Figure 9.7, plot the c values, and draw the broken-line graph. Check your results with Figure 9.8.

Calculate c and draw the solid line on each of the three c charts (one $\bar{c}$ for all the data). For k units,

$$\bar{c} = \frac{\Sigma c}{k} = \frac{157}{100} = 1.57$$

Calculate the control limits and draw the dashed lines:

$$\begin{aligned} \text{UCL}_c &= \bar{c} + 3\sqrt{\bar{c}} \\ &= 1.57 + 3\sqrt{1.57} \\ &= 1.57 + (3 \times 1.253) \\ &= 1.57 + 3.76 \\ &= 5.33 \end{aligned} \qquad \begin{aligned} \text{LCL}_c &= \bar{c} - 3\sqrt{\bar{c}} \\ &= 1.57 - 3\sqrt{1.57} \\ &= 1.57 - (3 \times 1.253) \\ &= 1.57 - 3.76 \\ &= 0 \end{aligned}$$

A negative LCL is calculated, so LCL = 0.

A few shirts fall in the out-of-control category because their number of flaws exceeds the control limit value of 5.33. The process capability is the average number of defects, $\bar{c}$, and a $\bar{c}$ of 1.57 defects per shirt must be improved before the men's stores market can be attempted.

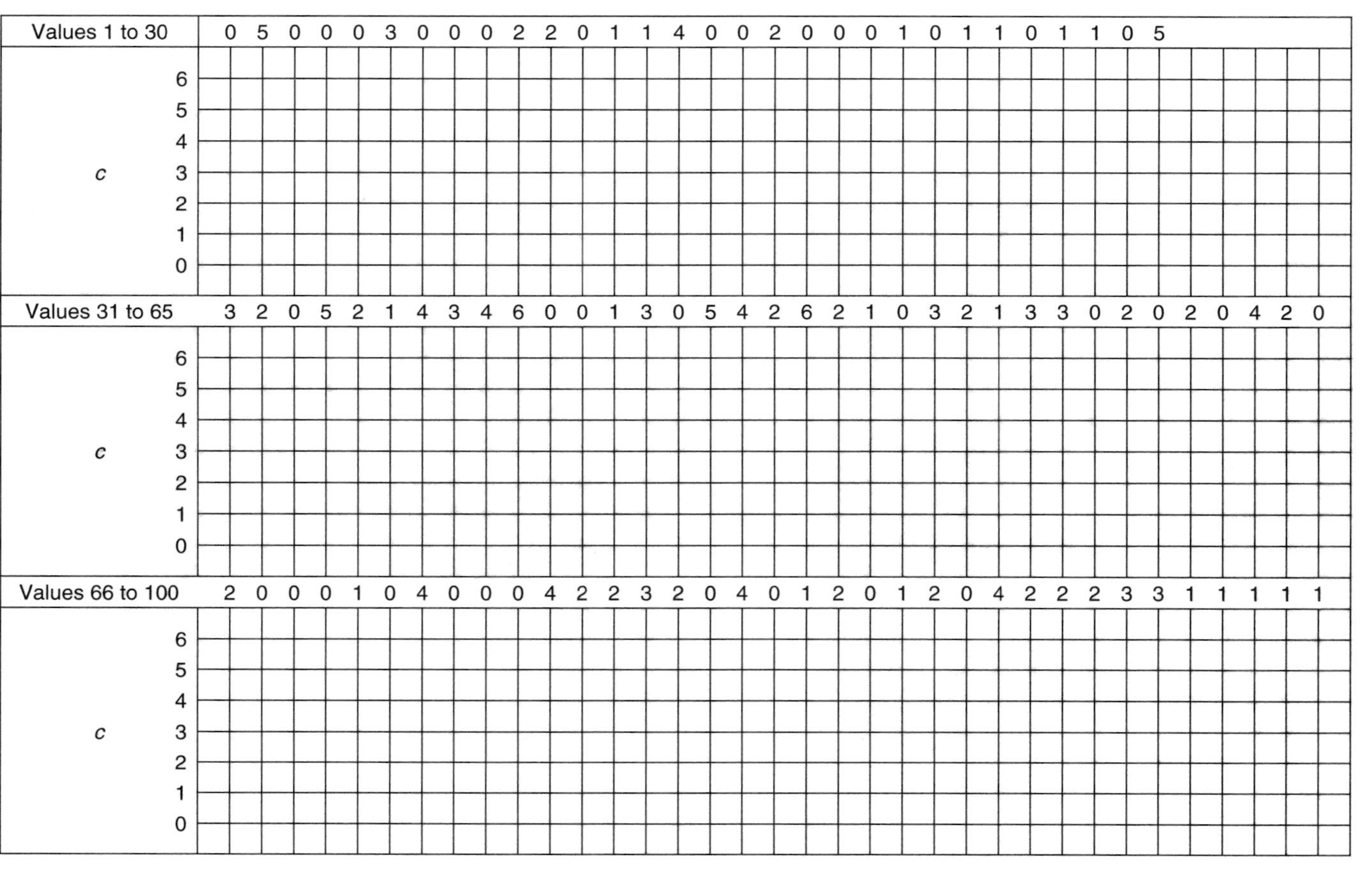

Figure 9.7
Construct the *c* chart, calculate the average and control limits, and draw the lines.
Check your results with Figure 9.8.

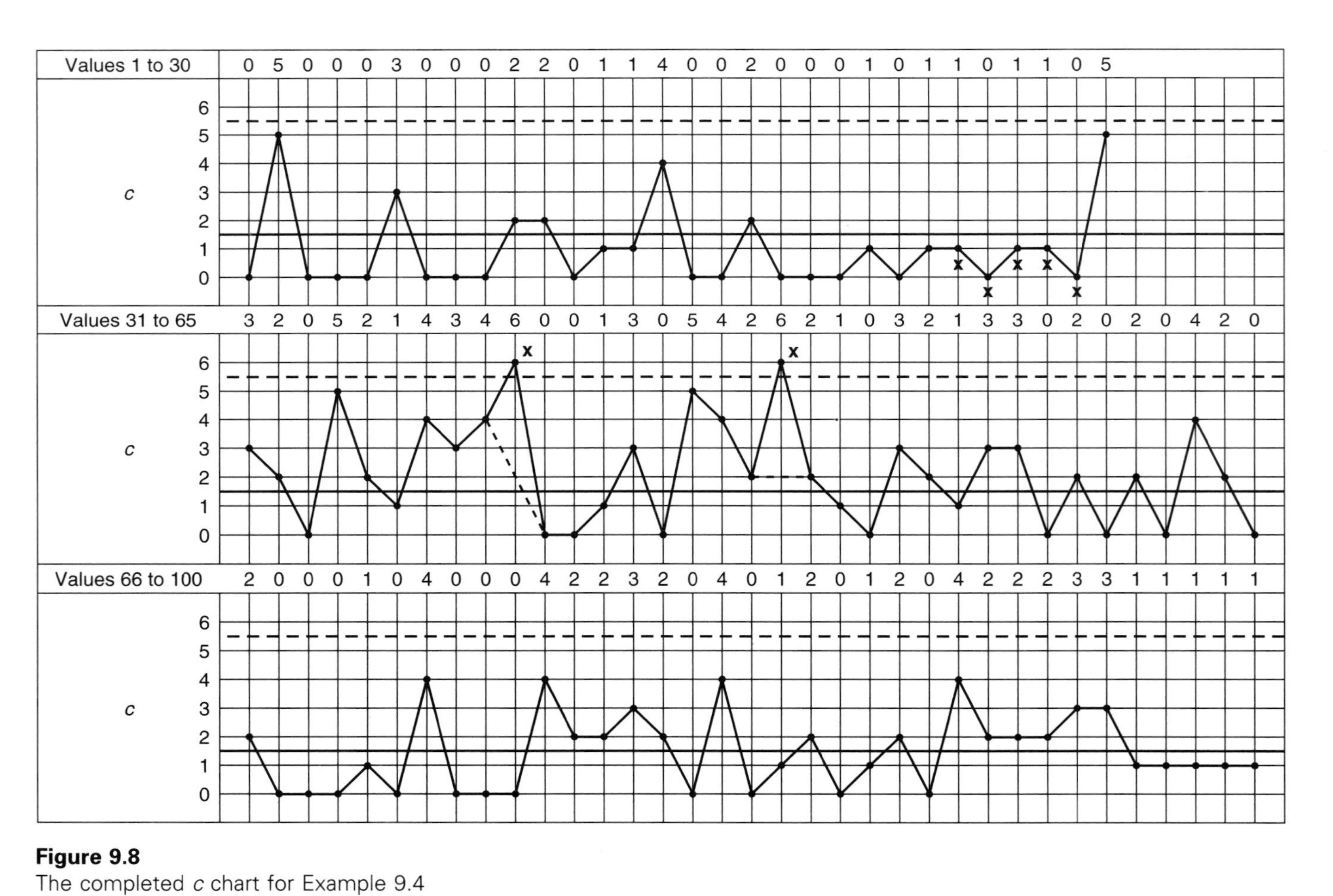

Figure 9.8
The completed c chart for Example 9.4

9.5 THE *u* CHART

The data from Example 9.4 can also be analyzed with a u chart by grouping the shirts into samples. In this example the data are grouped into samples of $n = 10$ shirts. The previous table from the c chart is reorganized into samples by taking 10 consecutive shirts per sample. The first 10 samples in the following table are taken from Example 9.4 and an additional 10 samples have been added.

Control Chart Formulas for the *u* Chart

u = average number of defects / sample

$$u = \frac{\Sigma c}{n}$$

$$\bar{u} = \frac{\Sigma c}{\Sigma n} \qquad n = \text{samplesize}$$

$$UCL_u = \bar{u} + 3\sqrt{\frac{\bar{u}}{n}}$$

$$LCL_u = \bar{u} - 3\sqrt{\frac{\bar{u}}{n}}$$

EXAMPLE 9.5

Make a u chart for the data in Table 9.3. For each u value in the table, the total number of defects in the sample is divided by the sample size, 10. The average number of defects per shirt in each sample is u.

Solution

The data have been entered on the chart in Figure 9.9. Set the scale on the chart. The maximum u value in the table is 3.0, so a scale of .2 units per line is convenient. Make the scale from 0 to 3.0. Chart the points and check your results with the completed chart in Figure 9.10.

Calculate $\bar{u}$ and the control limits:

$$\Sigma n = 10 \times 20$$
$$= 200$$
$$\bar{u} = \frac{\Sigma c}{\Sigma n}$$
$$= \frac{266}{200}$$
$$= 1.33$$

$$UCL_u = \bar{u} + 3\sqrt{\frac{\bar{u}}{n}}$$

TABLE 8.3
Samples of grouped data

Sample	Σc	u	Sample	Σc	u
1	10	1.0	11	9	.9
2	10	1.0	12	8	.8
3	10	1.0	13	11	1.1
4	30	3.0	14	7	.7
5	23	2.3	15	12	1.2
6	15	1.5	16	10	1.0
7	11	1.1	17	9	.9
8	17	1.7	18	15	1.5
9	14	1.4	19	14	1.4
10	17	1.7	20	14	1.4

$$= 1.33 + 3\sqrt{\frac{1.33}{10}}$$

$$= 1.33 + 3\sqrt{.133}$$
$$= 1.33 + (3 \times .36)$$
$$= 1.33 + 1.08$$
$$= 2.41$$

$$\text{LCL}_u = \bar{u} - 3\sqrt{\frac{\bar{u}}{n}}$$
$$= 1.33 - 1.08$$
$$= .25$$

Draw a solid line at $\bar{u}$ and dashed lines at the control limits. Your completed chart should look like Figure 9.10.

The process capability is $\bar{u}$, which is 1.33 flaws per shirt. Because the capability is considered too large, a follow-up study is planned that will chart individual flaws as a first step toward process improvement.

9.6 SPC APPLIED TO THE LEARNING PROCESS

An attribute chart that was not discussed previously is the q chart: q is the opposite of p, $q = 1 - p$. The value p represents the proportion defective, so the value q represents the proportion acceptable or not defective. In like manner, if np represented the number of defectives, then nq would represent the number of good pieces in the sample. In a process the p value is usually small and it is generally considered easier to deal with numbers like $p = .005$, $p = .01$, and $p = .008$ than to use the corresponding

Calculations

$\bar{u} = \frac{\Sigma c}{\Sigma n} = \_\_\_ = \_\_\_$

$3\sqrt{\frac{\bar{u}}{n}} = \_\_\_$

$\frac{\bar{u}}{n} = \_\_\_$

$UCL_u = \bar{u} + 3\sqrt{\frac{\bar{u}}{n}} = \_\_\_ + \_\_\_ = \_\_\_$

$\sqrt{\frac{\bar{u}}{n}} = \_\_\_$

$UCL_u = \bar{u} - 3\sqrt{\frac{\bar{u}}{n}} = \_\_\_ - \_\_\_ = \_\_\_$

Sample	1	2	3	4	5	6	7	8	9	10	11	12	13	14	15	16	17	18	19	20					
n	10	10	10	10	10	10	10	10	10	10	10	10	10	10	10	10	10	10	10	10					
c	10	10	10	30	23	15	11	17	14	17	9	8	11	7	12	10	9	15	14	14					
u	1	1	1	3	2.3	1.5	1.1	1.7	1.4	1.7	.9	.8	1.1	.7	1.2	1	.9	1.5	1.4	1.4					

u

Figure 9.9
Complete the u chart as instructed in Example 9.5

Calculations

$\bar{u} = \frac{\Sigma c}{\Sigma n} = \frac{266}{200} = 1.33$

$\frac{\bar{u}}{n} = \frac{1.33}{10} = .133$

$\sqrt{\frac{\bar{u}}{n}} = \sqrt{.133} = .36$

$3\sqrt{\frac{\bar{u}}{n}} = 3 \times .36 = 1.08$

$UCL_u = \bar{u} + 3\sqrt{\frac{\bar{u}}{n}} = 1.33 + 1.08 = 2.41$

$LCL_u = \bar{u} - 3\sqrt{\frac{\bar{u}}{n}} \quad 1.33 - 1.08 = .25$

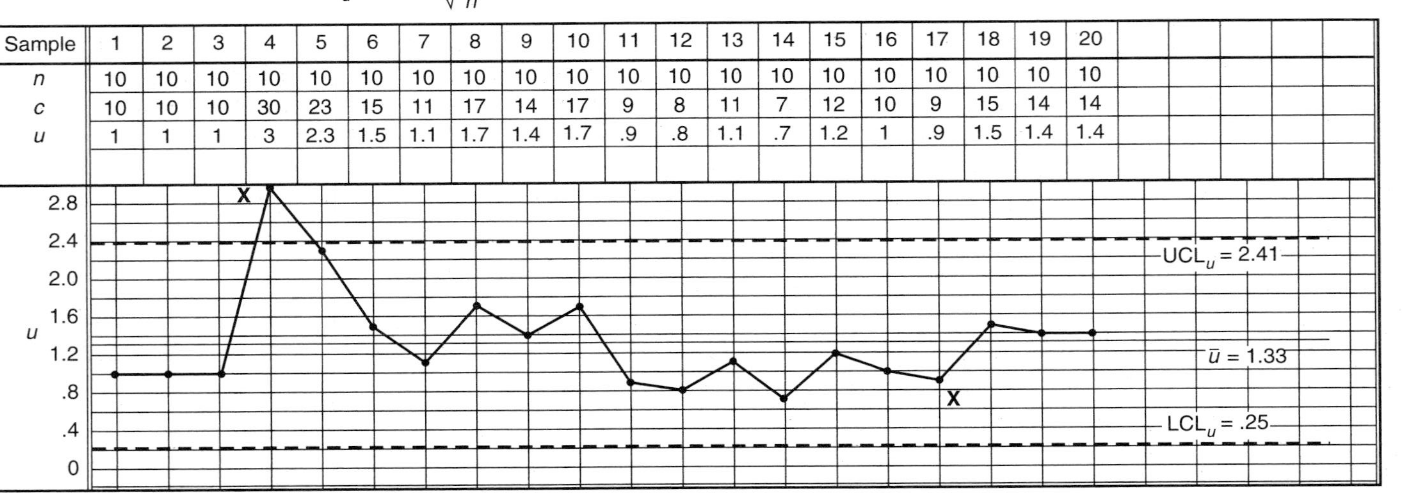

Sample	1	2	3	4	5	6	7	8	9	10	11	12	13	14	15	16	17	18	19	20
n	10	10	10	10	10	10	10	10	10	10	10	10	10	10	10	10	10	10	10	10
c	10	10	10	30	23	15	11	17	14	17	9	8	11	7	12	10	9	15	14	14
u	1	1	1	3	2.3	1.5	1.1	1.7	1.4	1.7	.9	.8	1.1	.7	1.2	1	.9	1.5	1.4	1.4

Figure 9.10
The completed chart for Example 9.5.

q values of q = .995, q = .99, and q = .992. Consequently, a q chart is seldom used in industry. The following example, however, does illustrate a good application of an nq chart.

Statistical process control can be applied in any work environment as long as pertinent data are available. In the school learning process, two potential sources of data can be used to measure learning: daily homework and daily quizzes (one or the other). When homework is used as the source of data, a small sample of $n = 4$ should be sufficient for an nq control chart. The samples can be chosen in two ways. Either a random sample is chosen (e.g., using a random-number generator on a statistical calculator) or a planned sample that includes questions on all the basic concepts or ideas in the assignment (called a stratified sample). The student grades each sample (with grading hints from the instructor) and charts the total score as the nq value. Various scoring methods may be used such as 0 to 4 or 0 to 1, as long as just one is consistently used. The 0-to-4 scoring method would be used with an nq chart. The 0 to 1 would be used with a q chart. In either case the emphasis is on either the number correct or the proportion correct.

Try to visualize the nq chart described above. If things are going well, most of the homework problems will be correct and the chart should be varying between 3 and 4 using the 0-to-4 scoring method (a maximum of 1 point for each homework problem). This is illustrated in Figure 9.11. When you analyze the chart, the question in mind should be, "What happened on the ninth assignment?" Can you see that more information is needed? What were the topics or concepts on that assignment? What was going on when that assignment was done? If a test is scheduled soon, have the topics or concepts from the ninth assignment been learned?

One of the important parts of a control chart is the accompanying log or comment sheet. By using a log with the learning chart, it will become more useful and can be used to improve your learning process. The log entries should include some assignment information such as the page number or the main topic or concept. Coded log entries can be used to describe the circumstances accompanying each assignment. There are different sets of log entries that can be used. The important thing is that most of the circumstances that can affect your learning process should be included. Table 9.4 shows a basic list of coded log entries. Any other events that occur that may have some effect on your learning process should be written in.

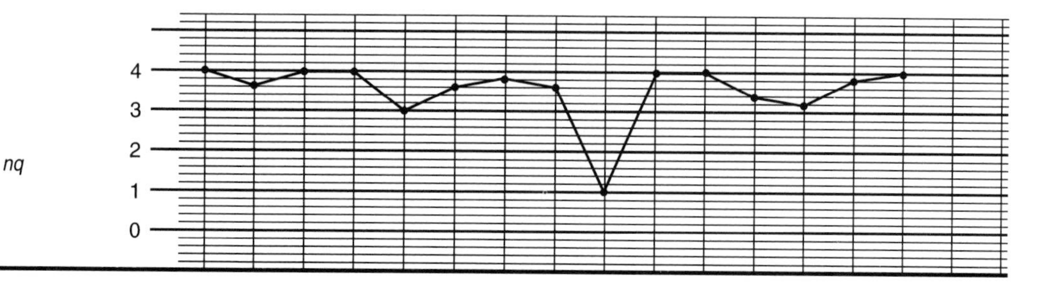

Figure 9.11
A learning chart showing the number correct in each sample.

TABLE 9.4
Log entries for a learning chart

Log Entry	Code
Missed the class that pertained to the homework	M
Out of school Sick	S
Assignment was completed Late	L
At the end of class did you have Questions?	Q
Do you need Help?	H
Did you Get help?	G
Did you Redo the assignment?	R
Did you read the text and Work the illustrated problems?	W
Did you Just do the homework without reading the text?	J
When you did the assignment, was it Noisy?	N
Did you find the assignment Difficult?	D

EXAMPLE 9.6 Analyze the nq learning chart in Figure 9.12.

Solution

The major dips in the nq chart are associated with missed classes, M. There is also a carryover effect from missed classes because it takes a few more classes to get caught up with the work. There are also more Q's after a missed class because some of the work covered was based on the material covered in the missed classes. The crossed-out lines and dashed lines indicate that poorly done assignments were redone, R, and the improved grade was recorded. There is also a noticeable dip in the homework grade when the student did not read the book and work the illustrated problems, J. Working in a noisy environment, N, did not seem to affect the homework score, although it may correlate to missed problems on tests. The average score, nq, should be calculated with the redone scores instead of the originals because that should give a better estimate of the capability of the student. The $\overline{nq}$ should give a decent estimate of the corresponding test scores:

$$\frac{\overline{nq}}{4} = \frac{3.4}{4} = .85$$

Test scores on the assigned material should be near 85%.

The above example does show some cause and effect between homework scores and events. A further analysis that can improve the student's learning process is to compare the test results with the nq chart. Missed test problems are matched to the homework assignment that contained them. Clues may be found that will lead to learning process improvements. One common situation that occurs is when the student has good homework scores but poor test scores. The difficult, D, codes could indicate that the student struggled getting correct answers. A sequence of "try it, check the answer, try again, . . . , got it!" generally needs at least one redo, R, at a later time to ensure the concept was learned. Otherwise, students take the test believing that they know more than they actually do. The result of that is a poor test score.

Log Entry	*Code*
Missed the class that pertained to the homework	M
Out of school Sick	S
Assignment was completed Late	L
At the end of class did you have Questions?	Q
Do you need Help?	H
Did you Get help?	G

Did you Redo the assignment?	R
Did you read the text and Work the illustrated problems?	W
Did you Just do the homework?	J
When you did the assignment was it quiet or Noisy?	N
Did you find the assignment Difficult?	D

Assignment Page	6	13	22	31	39	45	52	59	69	78	90	96
Score	4	3	3.7	3.5	1	.5	2.5	3.2	4	2	2.8	3.6
Comment Codes	HG W	QH J	W	N,W	M,W,H GR 4	WHLQ GR 3.5	HG JQ	GQ NW	W	QH NJ	HN J	GW

Assignment Page	99	107	115	125	132	140	148	155	165	170	177	187	198
Score	4	0	0	0	3	2.7	2	3.5	3.5	4	3.9	4	3.6
Comment Codes	W	MS LH G3.2 RW	MS LH G3.6 RW	LH Q G4 RW	LH QW	LJ QD	JH QD	NW	W	W	NW	W	W

Figure 9.12
Analyze the completed *nq* learning chart.

9.7 TECHNOLOGY IN SPC

One of the things you should have noticed by this time is that making control charts is a rather time-consuming process. Even in an on-the-job situation, when only a few to several samples are taken during a shift, charting takes time. Many companies have been slow to use SPC because the management does not want the workers "wasting" time with paperwork.

Happiness is having on-station computer access with automatic measurement transmission. The operator takes a sample piece, adjusts the measuring device, and presses a button that electronically sends the measurement into the computer. The control chart can be called up on the screen at any time for analysis. The computer does all the computation and charting for you. At the press of a button you can fill the screen with all kinds of statistical information. Computers can be tremendous time-savers in keeping control charts. Their only drawback (other than the expense of getting one) is that they do so much that it can be intimidating and confusing. It is important to be able to "zero-in" on the pertinent information and to use them effectively.

The standard analysis of a control chart is to look for trouble patterns. With a chart at your finger tips it can also be used to anticipate developing patterns and watch for specific causes. One of the advantages of many computer packages is that they provide an extra analysis with using the coded comments. If you suspect that comment D contributes to instability, for example, you can look at the control chart with all the D data removed to see if there is an improvement. With the computer this can be done with a few key strokes, but by hand it can be a tedious project. This also shows the importance of including the comments when the data are entered.

EXERCISES

1. The number of defectives np and the sample size n are given on the chart in Figure 9.13.
  a. Set the scale and make a p chart.
  b. Calculate $\overline{p}$ and the control limits. Draw the lines.
  c. Mark any out-of-control situations.
  d. What is the process capability?

2. The data from exercise 1 are given in Figure 9.14.
  a. Make an np chart.
  b. Calculate $\overline{np}$ and the control limits. Draw the lines.
  c. Mark any out-of-control situations.
  d. What is the process capability?
  e. Make sure that the graphs on Figures 9.13 and 9.14 look the same and that $\overline{np}$ and UCL_{np} convert to $\overline{p}$ and UCL_p by dividing by n.

3. The number of defects per item is shown on Figure 9.15.
  a. Make a c chart.
  b. Calculate $\overline{c}$ and the control limits. Draw the lines.
  c. Interpret for statistical control.

Calculations

$\bar{p} = \frac{\Sigma np}{\Sigma n} = \frac{\_\_\_}{\_\_\_} = \_\_\_$

$1 - \bar{p} = \_\_\_$

$\bar{p} \times (1 - \bar{p}) = \_\_\_ \times \_\_\_ = \_\_\_$

$\frac{\bar{p} \times (1 - \bar{p})}{n} = \frac{\_\_\_}{\_\_\_} = \_\_\_$

$\sqrt{\frac{\bar{p} \times (1 - \bar{p})}{n}} = \sqrt{\_\_\_} = \_\_\_$

$3\sqrt{\frac{\bar{p} \times (1 - \bar{p})}{n}} = 3 \times \_\_\_ = \square$

$UCL_p = \bar{p} + 3\sqrt{\frac{\bar{p} \times (1 - \bar{p})}{n}} = \_\_\_ + \square$

$UCL_p = \_\_\_$

$LCL_p = \bar{p} - 3\sqrt{\frac{\bar{p} \times (1 - \bar{p})}{n}} = \_\_\_ - \square$

$LCL_p = \_\_\_$

np	12	10	8	14	21	16	18	15	12	19	16	10	22	17	14	20	21	12	8	10	16	12	18	14	15
n	400	400	400	400	400	400	400	400	400	400	400	400	400	400	400	400	400	400	400	400	400	400	400	400	400
p																									

Figure 9.13
Make the *p* chart for exercise 1.

Calculations

$\overline{np} = \dfrac{\Sigma\, np}{n} = \text{\_\_\_\_} = \text{\_\_\_\_}$

$\bar{p} = \dfrac{\overline{np}}{n} = \text{\_\_\_\_}$

$1 - \bar{p} = \text{\_\_\_\_}$

$\overline{np}\,(1 - \bar{p}) = \text{\_\_\_\_} \times \text{\_\_\_\_} = \text{\_\_\_\_}$

$\sqrt{\overline{np}\,(1 - \bar{p}} = \sqrt{\text{\_\_\_\_}} = \text{\_\_\_\_}$

$3\sqrt{\overline{np}\,(1 - \bar{p})} = 3 \times \text{\_\_\_\_} = \square$

$UCL_p = \overline{np} + 3\sqrt{\overline{np}\,(1 - \bar{p})}$

$UCL_p = \text{\_\_\_\_} + \text{\_\_\_\_} = \text{\_\_\_\_}$

$LCL_p = \overline{np} - 3\sqrt{\overline{np}\,(1 - \bar{p})}$

$LCL_p = \text{\_\_\_\_} - \text{\_\_\_\_} = \text{\_\_\_\_}$ = 0 if negative

n = 400

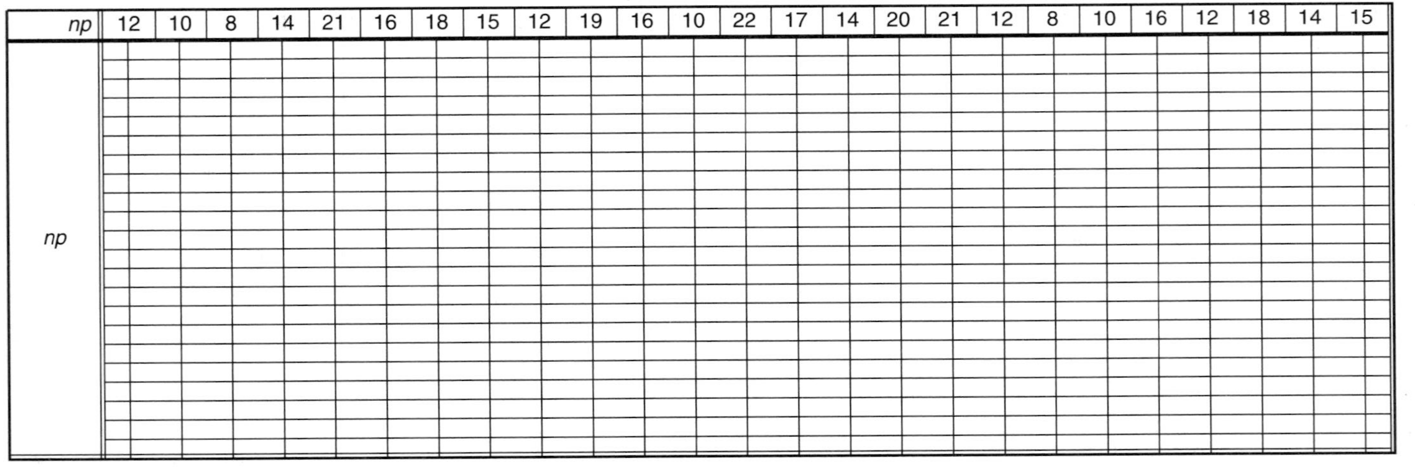

np	12	10	8	14	21	16	18	15	12	19	16	10	22	17	14	20	21	12	8	10	16	12	18	14	15

Figure 9.14
Make the *np* chart for exercise 2.

Calculations

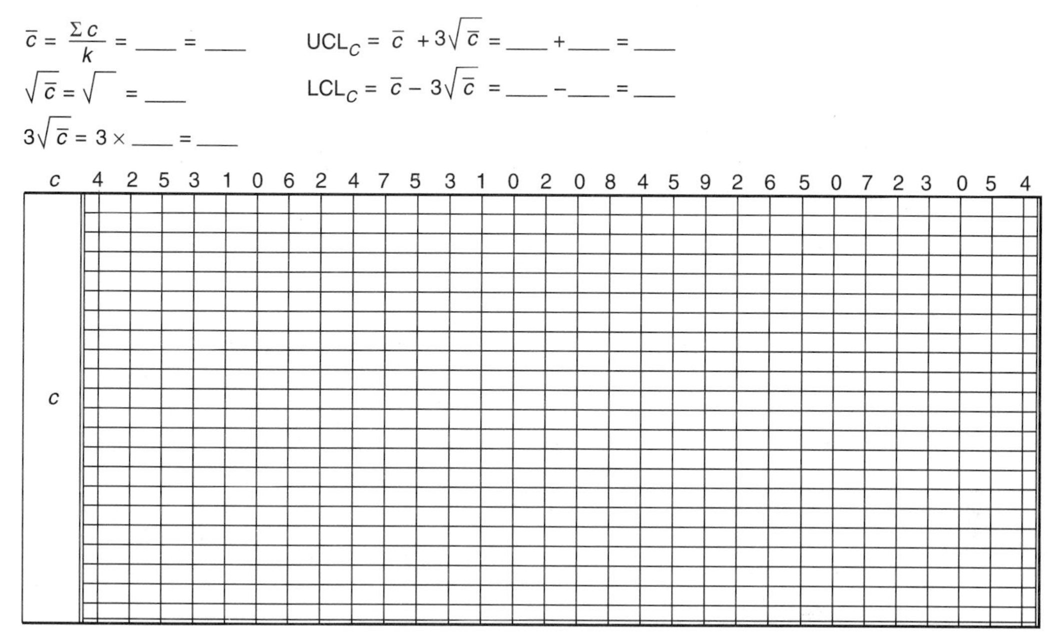

Figure 9.15
Make the c chart for exercise 3.

TABLE 9.5
Number of defects per item

	0	2	1	6	4	5	5	0	1	3	2	4	0	1	2	5	8	6	2	0
	2	1	2	4	2	1	5	0	5	2	0	3	6	1	4	2	4	6	1	4
	6	2	0	2	0	4	2	2	2	2	0	2	2	0	8	1	6	2	4	2
	2	4	5	1	2	8	1	6	3	5	2	1	4	1	6	4	2	0	4	3
	4	3	0	3	4	6	6	0	5	6	4	5	0	3	4	0	1	2	2	0
	0	8	7	5	4	3	4	2	1	4	4	6	4	4	5	1	2	7	1	0
	3	4	3	5	4	3	0	2	1	4	2	4	4	7	3	2	4	4	0	2
	0	1	3	0	1	0	0	3	2	0	1	1	2	3	1	3	0	0	2	1
Σc																				
u																				

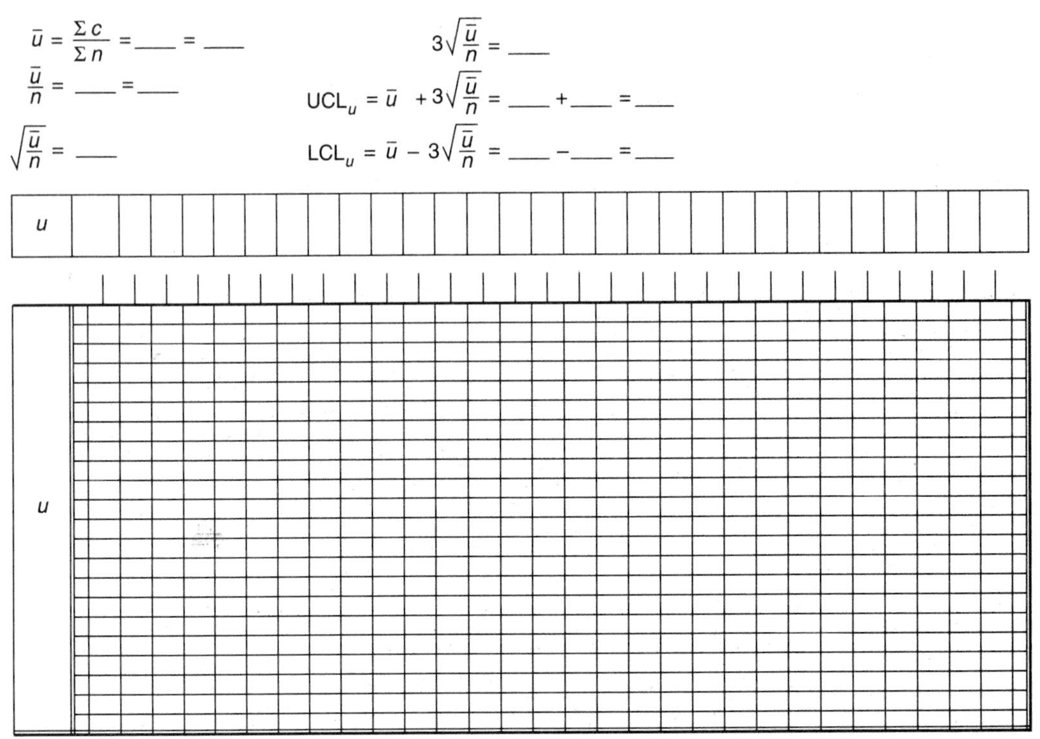

Figure 9.16
Make the u chart for exercise 4.

d. What is the process capability?

4. Samples of size 8 are shown in Table 9.5 with the number of defects in each item listed. Each column is a sample.
   a. Make a u chart for the 20 samples using Figure 9.16.
   b. Calculate $\bar{u}$ and the control limits.
   c. Interpret for statistical control.
   d. What is the process capability?
5. Make a learning chart (nq or q) for one or more of your classes.
6. The housekeeping supervisor in a major hotel checks n items in each room and records the number that are nonconforming (not adequately cleaned or arranged). The n value varies with the quality (price) of the room.
   a. Make a p chart for the data in Table 9.6.
   b. Mark any out-of-control situations.

TABLE 9.6

n	25	25	30	30	35	25	25	30	30	25	40	35	25	25
Nonconforming	1	0	2	3	0	6	2	3	8	0	0	1	2	3

TABLE 9.7

Date	2/3	2/5	2/10	2/13	2/18	2/23	2/25	2/28	3/5	3/6	3/11	3/15	3/20
Nonconforming	3	4	0	2	8	2	1	0	4	3	1	12	2

7. A national fast-food chain provides each manager with a quality item list that should be checked at random times twice a week. The list contains 40 items. Table 9.7 records the number of nonconforming items in each quality check.

a. Make an *np* chart for the number of nonconforming quality items.

b. Mark any out-of-control patterns.

c. Make a *p* chart for the data in Table 9.7 ($n = 40$) and analyze it for control and capability.

10

INTERPRETING CONTROL CHARTS

OBJECTIVES

- Distinguish between a random pattern of points on a control chart and a trouble-indicating pattern.
- Classify the various trouble-indicating patterns.
- Understand how probability concepts are used to classify the trouble-indicating patterns.
- Identify the basic possible causes of an out-of-control pattern of points on a control chart.

When a process is in statistical control the points on the control chart should follow a completely random pattern and the measurements will have a normal distribution. The visible connection between the control chart and the normal distribution is that the chart pattern will be random but the distribution of the points on the chart will have about 68% of them within one standard deviation of the center line, 28% between one and two standard deviations from the mean, and 4% between the two standard deviation mark and the control limits. This is illustrated in Figure 10.1. As soon as the control limits are drawn on a control chart, the space between the center line and each control limit can be divided into three sections of approximately equal width. A quick count of points in each section should be somewhat close to the predicted normal pattern. When the population of measurements has some other distribution pattern, one of the trouble-indicating patterns discussed in this chapter will show it on the control chart.

When a process is out of control, the pattern of points on the control chart will be affected in some way. Freaks, shifts, trends, and cycles are some of the recognizable control chart patterns. Along with other patterns, these four signify that a change has occurred in the process. Each of these trouble patterns must be spotted as nonrandom as soon as possible, without misinterpreting a truly random pattern. Rules concerning the minimum number of points that are needed to indicate a trouble pattern will be discussed in the following sections along with the mathematics of chance that were used to form the rules.

Remember one of the basic rules in chart interpretation: *Always get the range chart in statistical control first.* The accompanying averages or median chart cannot be analyzed if the range chart is out of control. Excessive variation or changes in variation affect the $\bar{x}$ or $\tilde{x}$ chart by changing the control limits. This occurs because $\bar{R}$ is used in the formulas for control limits: $CL = \bar{\bar{x}} \pm A_2\bar{R}$ and $CL = \bar{\tilde{x}} \pm \tilde{A}_2\bar{R}$. A range chart that is out of control can make the $\bar{x}$ chart look out of control as well. It can also hide an out-of-control situation because $\bar{R}$ may be based on bad data. Efforts to control the average measurement size when variability is out of control can be a complete waste of time. Solving the variation problem will often eliminate what was originally thought to be an accompanying problem with average measurement. When standard deviations were discussed relative to control charts, the control limits were always three standard deviations away from the average line. However, there were different standard deviations that were used. The $A_2\bar{R}$ product for an $\bar{x}$ chart represented three standard deviations of $\bar{x}$, $3\sigma_{\bar{x}}$. The $\tilde{A}_2\bar{R}$ product for a median chart represented three standard deviations of $\tilde{x}$, $3\sigma_{\tilde{x}}$. The value $3\sqrt{\bar{p}(1-\bar{p})/n}$ represented three standard deviations of p, $3\sigma_p$. In this chapter there will often be reference to measurements from the average value in standard deviation units. Instead of using a subscript for a specific standard deviation, as shown above, we will just use σ to keep the notation simple and general. If a specific chart is referred to, σ will correspond to the specific σ for that chart. For example, the average + σ would mean $\bar{\bar{x}} + \sigma_{\bar{x}}$ on an $\bar{x}$ chart and $\bar{p} + \sigma_p$ on a p chart.

10.1 THE RANDOM DISTRIBUTION OF POINTS

Figure 10.1 illustrates a random distribution of points on a control chart. The expected percentages in the different sections are shown on the left side. Some charts include upper warning lines (UWL) and lower warning lines (LWL) at two standard deviations from the average to help recognize some of the trouble patterns. The characteristics of a normally distributed random pattern of points include the following:

- The sequence of points is not predictable (random).
- The majority of the points (about 68%) should be within one standard deviation of the center line.
- About 28% should be between 1σ and 2σ units from the center line, with about 14% on each side.
- About 2% should be near each of the control limit lines.
- None of the points should be beyond the control limits.

When control charts are analyzed, points classified as out of control should be marked with an X. A consistent marking pattern should be used. The X should be placed outside the broken-line graph, away from the center line. If a single point is out of control, just that point is marked. If several points are out of control within a classification

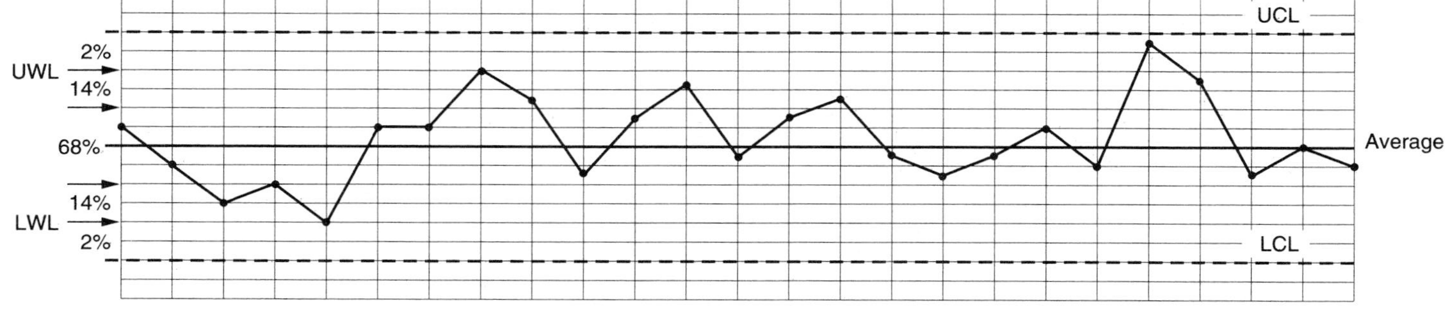

FIGURE 10.1
A random distribution of points.

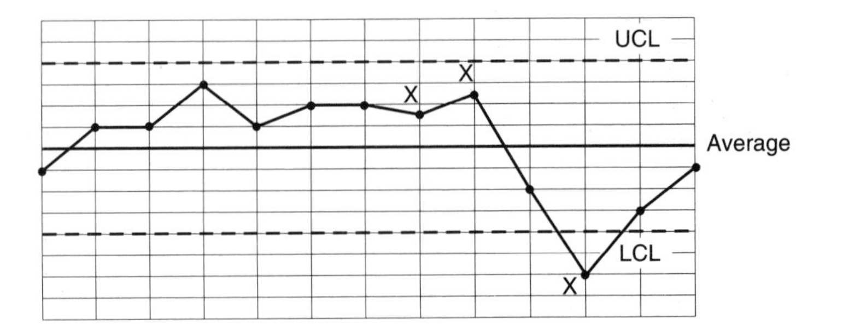

FIGURE 10.2
Out-of-control points marked with X's.

such as a shift or trend, the first point that satisfies the rule and all points after it are marked out of control by the same classification. In Figure 10.2, for example, the first out-of-control classification is a shift: seven or more consecutive points on one side of the center line. The seventh and eighth points in the shift pattern are marked for that classification. The second out-of-control classification is a freak: a single point beyond the control limit. Just that one point is marked. Sometimes the out-of-control patterns are circled, either instead of or in conjunction with the X marks.

When a control chart indicates that a process is out of statistical control, the ideal action is to stop the process and eliminate the cause. However, this is not always possible. First, it may not be easy to determine the cause. Each different out-of-control pattern has its own "laundry-list" of possible causes, and several out-of-control occurrences may be needed before enough information is gathered to pinpoint the trouble. Second, some out-of-control situations can be temporarily tolerated because the measurements are not critical with respect to the quality of the final product. Third, the relative position of the process within the quality program is important. If SPC is just being introduced to the process, there may be several problems contributing to poor quality at one of the process points. If the process was running with some degree of satisfaction before the application of SPC, it may make sense to continue the process while the sources of trouble are being determined and eliminated. A common rule at many companies is to 100% inspect when an out-of-control process is allowed to run.

10.2 FREAKS

A *freak* is a single point that is beyond a control limit. It signifies that something changed dramatically in the process for a short time or that a mistake was made.

The most common process cause of freaks is a sudden change in material. If the raw material is subject to occasional inconsistencies or if the control of the mix of material is affected by either human error or flow gauge irregularity, freaks will result. These process problems indicated by freaks should be tracked because they can seriously affect the quality of the product.

A freak can also falsely signal process trouble when a mistake is made in measurement, recording, arithmetic, or plotting. When a freak occurs, always check for errors of this type first. Figure 10.3 shows two separate occasions in which freaks occur.

The most common causes of freaks in three types of charts are as follows:

***R* chart:**

- A sudden change in material
- A mistake in measurement
- An error in recording
- An error in arithmetic
- An error in plotting
- An incomplete or omitted operation
- Damage in handling

$\bar{x}$ chart with *R* chart in control:

- Same as the *R* chart list

***p* chart:**

- Variation in sample size
- Sample taken from a different population
- An occasional very good or very poor lot

The identification of freaks is based on the mathematics associated with the normal distribution. The control limits for $\bar{x}$ and R are three standard deviations from the center line. The work with the normal curve revealed that 99.7% of all the values lie within three standard deviations of the center. Therefore, the chance that a point lies beyond the control limits is .3%, a very small chance of occurrence. When a freak occurs, the question arises, "Did a rare event occur or did something happen to cause the freak to occur?" The second reason is assumed, and the cause for the freak is investigated. If no cause is found and the freak does not happen again, then the first reason, that rare events do happen, is accepted.

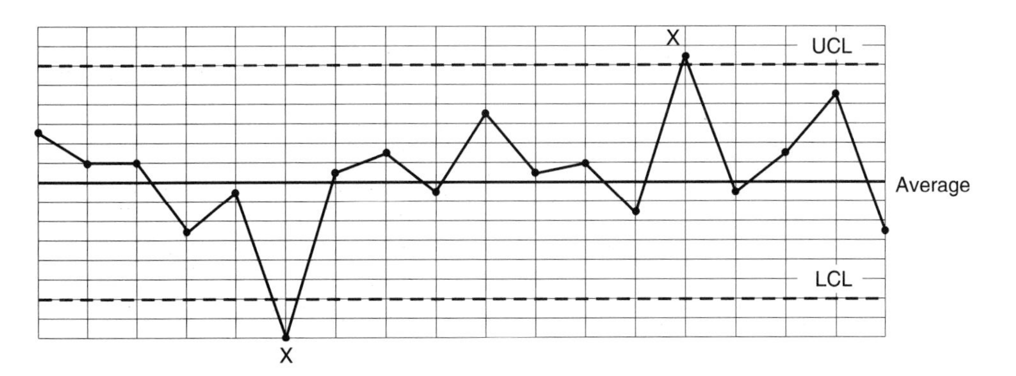

FIGURE 10.3
Freaks.

EXAMPLE 10.1 Indicate the most likely cause for the freaks in Figures 10.4 to 10.7 from the following list:

1. Change in material
2. Mistake in measurement
3. Recording error
4. Arithmetic error
5. Plotting error

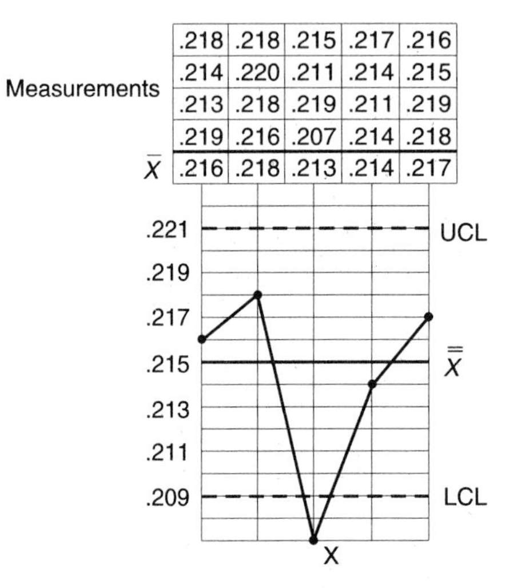

FIGURE 10.4
Suggest a likely cause of the freak.

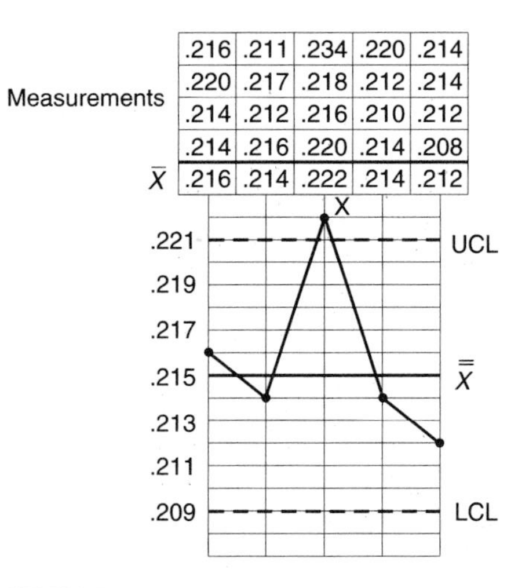

FIGURE 10.5
Suggest a likely cause of the freak.

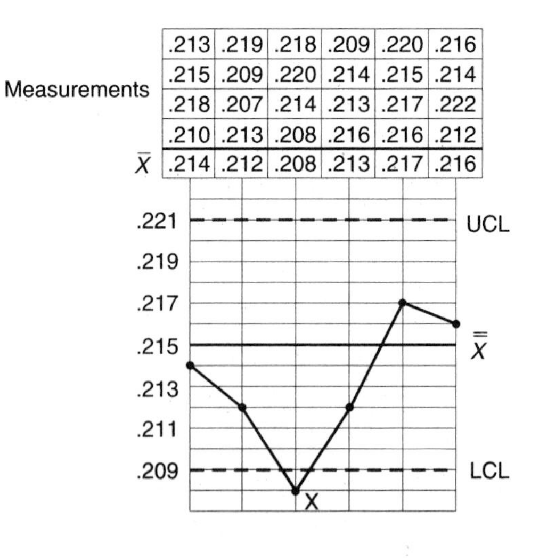

FIGURE 10.6
Suggest a likely cause of the freak.

FIGURE 10.7
Suggest a likely cause of the freak.

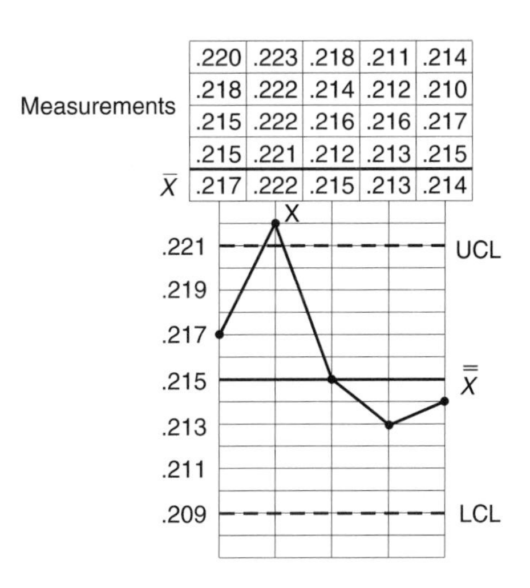

Solution

In Figure 10.4 the average value of .213 is incorrectly plotted. Perhaps the last sample value is plotted instead of $\bar{x}$.

The most likely cause for the freak in Figure 10.5 is either a measuring error or a recording error. The other measurements in that sample are within limits.

The sample measurements at the freak in Figure 10.6 are incorrectly averaged.

All four measurements at the freak in Figure 10.7 are high. This most likely points to a temporary change in material.

10.3 BINOMIAL DISTRIBUTION APPLICATIONS

A binomial experiment is one in which there are only two outcomes. Tossing a coin is a binomial experiment: The two outcomes are a head and a tail. Any situation in which the outcomes can be categorized as successes or failures with no other possibilities can be classified as binomial. If the experiment is repeated several times and the individual probabilities of success and failure are known, probabilities of specific events can be determined. Binomial distribution tables are available for cases in which the individual probabilities range from .1 to .9, and algebraic structures can be used when probabilities are in hundredths or smaller units. For example, if a coin is tossed eight times, a binomial table may be used to find the probability of events such as getting five heads and three tails. Our present needs can be better served, however, by relying on a few basic rules of chance and applying them to any binomial situations that occur. The "and" and "or" rules of compound probability discussed in Section 8.2 can be applied to binomial probabilities.

The easiest illustration of a binomial problem is the coin-tossing problem. Suppose a coin is tossed five times. What is the chance of getting five heads? This can be answered by using the "and" rule for probability:

$$P(A \text{ and } B) = P(A) \times P(B)$$

When successive events occur, such as tossing a coin several times, and when the probability of any one of the events is unaffected by the preceding events, the probability of that successive event is found by multiplying the individual probabilities. If H is a head,

$$\begin{aligned} P(5H) &= P(H \text{ and } H \text{ and } H \text{ and } H \text{ and } H) \\ &= P(H) \times P(H) \times P(H) \times P(H) \times P(H) \\ P(H) &= .5 \\ P(5H) &= .5 \times .5 \times .5 \times .5 \times .5 \\ &= .01325 \end{aligned}$$

The problem becomes more complicated when both heads and tails are involved in the outcome. Suppose the coin is tossed three times. Find the probability of getting two tails (T) and one head. There are three ways that this event can occur:

$$\begin{array}{ccccc} H & \text{and} & T & \text{and} & T \\ T & \text{and} & H & \text{and} & T \\ T & \text{and} & T & \text{and} & H \end{array}$$

This combines the "or" rule, in which the probabilities are added,

$$P(A \text{ or } B) = P(A) + P(B)$$

with the "and" rule:

$$\begin{aligned} P(2T \text{ and } 1H) &= P(H \cdot T \cdot T) + P(T \cdot H \cdot T) + P(T \cdot T \cdot H) \\ &= (.5 \times .5 \times .5) + (.5 \times .5 \times .5) + (.5 \times .5 \times .5) \\ &= 3 \times .125 \\ &= .375 \end{aligned}$$

In the last equation, the 3 is the number of ways of getting two tails and one head. The probability, .125, is the same for each way.

What is the chance of getting two heads and three tails when the coin is tossed five times? This can happen in 10 ways:

$$\begin{array}{ccccccccccc} H & H & T & T & T & \quad & T & T & H & T & H \\ H & T & H & T & T & & T & T & T & H & H \\ H & T & T & H & T & & T & T & H & H & T \\ H & T & T & T & H & & T & H & H & T & T \\ T & H & T & T & H & & T & H & T & H & T \end{array}$$

The probability of *each* of the 10 events is found by applying the "and" rule:

$$.5 \times .5 \times .5 \times .5 \times .5 = .03125$$

The probability that one of the 10 will occur is then

$$P(2H \text{ and } 3T) = 10 \times .03125$$
$$= .3125$$

The 10 represents the number of possible ways of getting two heads and three tails. The probability, .03125, is the same for each way.

Multiplying by the number of ways that the event can happen is a shortcut for the "or" rule. The repeated addition of the "or" probability can be done by multiplying by the number of products that are being added, because they are all the same product.

In general, the binomial distribution is a probability distribution that applies to a two-outcome situation with repeated trials. The outcomes may be labeled specifically, such as heads and tails, or generally, such as success and failure. The repeated trials may be tossing a coin several times or considering the position of several consecutive points on a control chart. Of course, it is the latter application that pertains to SPC.

In a normal distribution most points should be close to the mean; 68% should be within one standard deviation of the center. How many points should be beyond the 1σ boundaries before trouble is suspected to have changed the process measurements? The percent chance from both the normal curve and the binomial distribution can be used to answer the question. The binomial situation works because the position of a point on a control chart can be categorized in one of two ways: It is either inside the $\pm 1\sigma$ boundary or outside.

Several points on the control chart that fall beyond the 1σ limit may indicate a problem with the process because only 16% of the points are expected to be in that area on each side of the mean. One classification for this situation is called *freak patterns.* Testing for freak patterns involves one side of the center line. The control chart is informally divided into sections that are 1σ in width on each side of the mean. Because each control limit is three standard deviations from the center line, there are six sections on an $\bar{x}$ chart and usually five complete sections on an R chart. Remember, the lower control limit on the R chart cannot be below zero, so the section below $\bar{R}$ is usually narrower than the 3σ width above.

The 3σ widths above and below the center line are divided into thirds, as shown in Figure 10.8. Each section is now 1σ wide. The 2σ lines are sometimes referred to as the

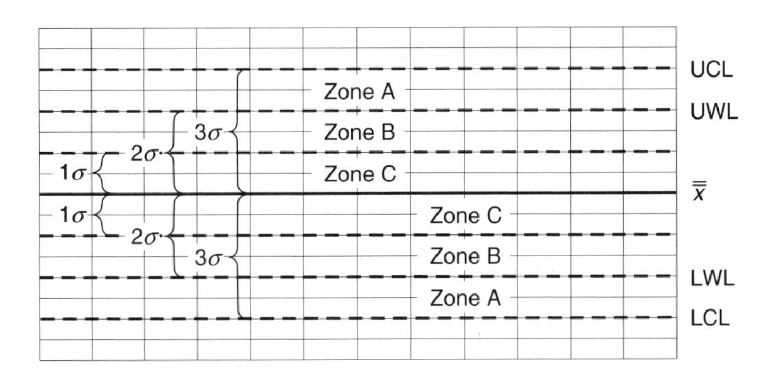

FIGURE 10.8
Control chart zones.

upper and lower warning lines and marked UWL and LWL. Each of the 1σ widths are called zones and are lettered C, B, and A from the center line out.

The probability of getting four out of five points in zone B or beyond can be calculated using the binomial distribution application and information from the normal curve. The area beyond 1σ can be found with the normal-curve table in Appendix B. The z value is the number of standard deviations from the center, so $z = 1$. The area is .16 and the probability that a measurement falls in that tail area is .16. The total area under the curve is 1. The probability that a measurement is not in that area can be found by subtraction:

1.00	Total area under the curve
− .16	Area in zone B and beyond (on one side of $\bar{x}$)
.84	Area outside of zone B and beyond

The shaded area under the curve in Figure 10.9 corresponds to the probability that a measure is in zone B and beyond on one side of the mean, as defined in Figure 10.8.

Let I represent a measurement inside zone B or beyond.

Let O represent a measurement outside that zone in the direction of the center.

This sets up the binomial situation: A point is in zone B or beyond with a probability of .16 or a point is not in that section with a probability of .84.

There are five ways in which four measurements may be inside and one measurement outside:

$$\begin{array}{ccccc} I & I & I & I & O \\ I & I & I & O & I \\ I & I & O & I & I \\ I & O & I & I & I \\ O & I & I & I & I \end{array}$$

Each of the five ways represents an "and" probability such as

$$\begin{aligned} P(I \text{ and } I \text{ and } I \text{ and } I \text{ and } O) &= .16 \times .16 \times .16 \times .16 \times .84 \\ P(I \cdot I \cdot I \cdot I \cdot O) &= .00055 \\ P(4 \text{ out of 5 in B or beyond}) &= 5 \times .00055 \\ &= .0028 \end{aligned}$$

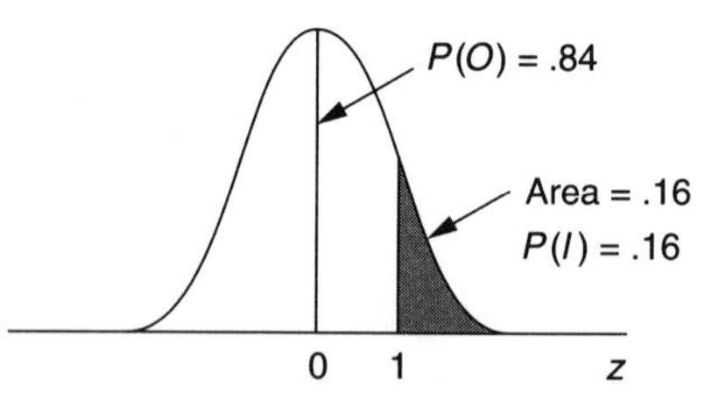

FIGURE 10.9
Area under the normal curve beyond 1σ.

The 5 represents the number of ways of getting four inside and one outside. The probability, .00055, is the same for each way.

The calculation shows that the chance of getting four out of five points in zone B or beyond in a random distribution is about .3%, a slim chance. When that occurs on a control chart, the question again arises, "Did a rare event occur, or did the process change somehow to cause that pattern?" The four-of-five pattern is considered a trouble indicator, and the chart should subsequently be marked as out of control and an investigation started to determine the cause.

Points that crowd the control limit line can also indicate trouble. Find the probability that two out of three points will be in the A zone or beyond on one side of the mean. The probability behind the two-of-three rule can be calculated in the same way as the previous four-of-five rule.

A z value of 2 corresponds to 2σ units from the center to zone A or beyond, as defined in Figure 10.8. According to the normal-curve table, when $z = 2$, the tail area is .023. That is the probability that a measurement is inside zone A or beyond. Subtraction from 1 gives the probability that a measurement is outside of zone A on the center line side. This is illustrated in Figure 10.10.

1.000	Total area under the normal curve
− .023	Area in zone A or beyond on one side of mean
.977	Area outside of zone A on the center line side

$$P(I) = .023 \qquad P(O) = .977$$

There are three ways in which there may be two measurements inside the zone and one measurement outside:

$$\begin{matrix} I & I & O \\ I & O & I \\ O & I & I \end{matrix}$$

$$P(I \text{ and } I \text{ and } O) = .023 \times .023 \times .997$$
$$= .00053$$
$$P(2 \text{ of } 3 \text{ in A or beyond}) = 3 \times .00053$$
$$= .0016$$

The probability of getting two of three points outside the warning line, or 2σ line, in a random distribution of points is .16%. This is used as a trouble indicator because the probability is so small that it is assumed that something in the process caused that distribution of points.

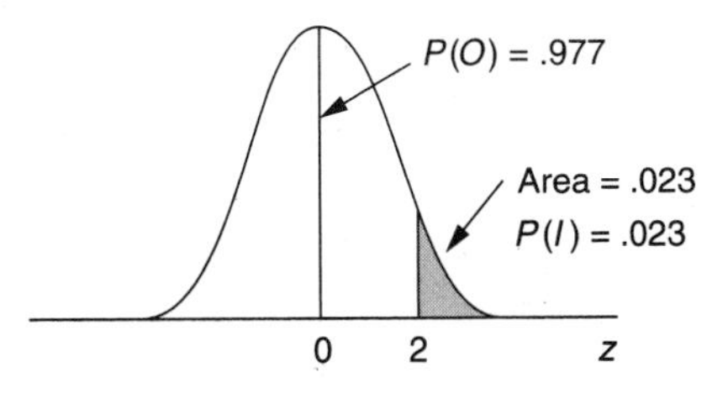

FIGURE 10.10
Area under the normal curve beyond 2σ.

10.4 FREAK PATTERNS

The preceding work with the binomial distribution leads to two out-of-control tests that are labeled *freak patterns:*

Test 1: If four out of five consecutive points are in zone B or beyond on the same side of the center line, a freak pattern is formed.

Test 2: A freak pattern is formed if two out of three consecutive points fall in zone A or beyond on the same side of the center line.

The charts shown for Examples 10.2 through 10.8 (Figures 10.11 through 10.17) are repeated three times in each figure. Analyze (*a*) to see if you can spot the out-of-control situations. The out-of-control sections are circled in (*b*), and the out-of-control points are correctly marked with X's in (*c*).

EXAMPLE 10.2 Mark the points that are out of control in Figure 10.11.

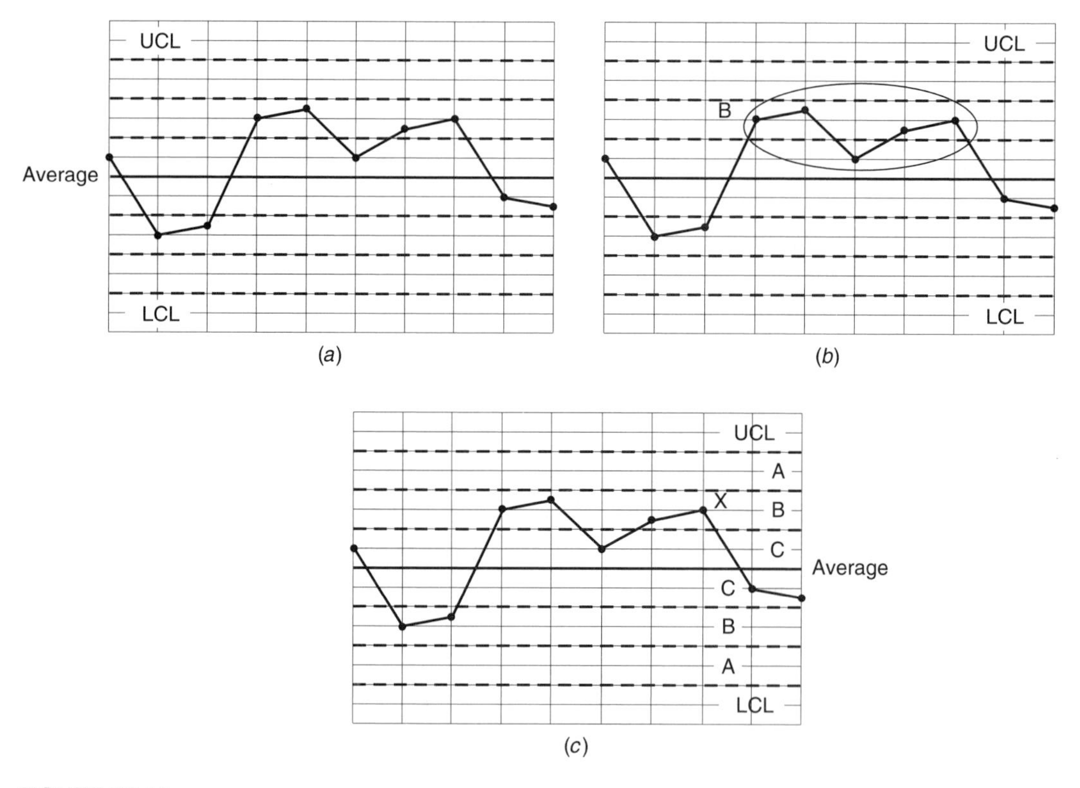

FIGURE 10.11
The chart for Example 10.2.

Solution

According to test 1, any sequence of five consecutive points that has four of the points in zone B or beyond on the same side of the center line forms a four-of-five freak pattern. The fourth through eighth points on the chart satisfy that description. The second and third points in Figure 10.11 cannot be counted with the circled points because the four points in zone B or beyond have to be on the same side of the center line. Only one of the five points can be on the center line side of the B boundary. Try to interpret (*a*) as it is pictured in (*c*). The fifth point in the pattern is marked with an X because it is the fourth point in the B zone. Cover (*b*) and (*c*) and see if the freak pattern in (*a*) is obvious.

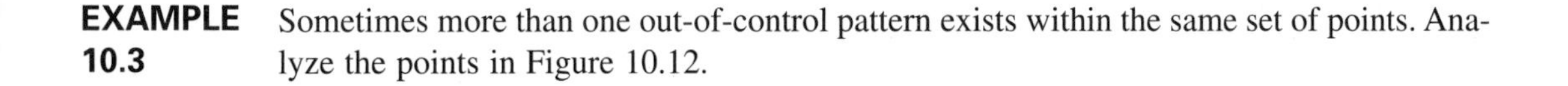

EXAMPLE 10.3 Sometimes more than one out-of-control pattern exists within the same set of points. Analyze the points in Figure 10.12.

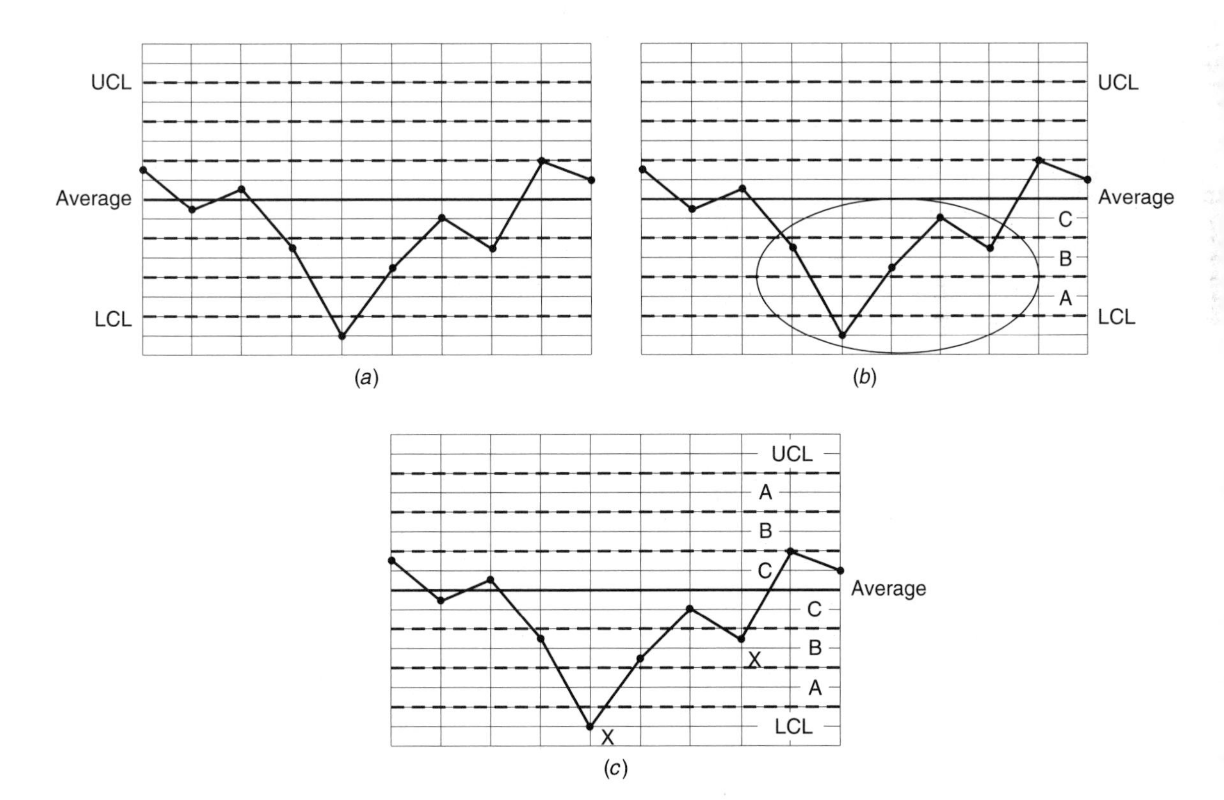

FIGURE 10.12
The chart for Example 10.3.

Solution

The first point marked in Figure 10.12*c* is classified as a freak because it is a single point beyond the control limit. The second point marked satisfies the four-of-five rule for a freak pattern. The first point marked is also part of the four-of-five freak pattern and is counted as one of the four points beyond the 1σ line.

EXAMPLE 10.4 Identify the points that are out of control in Figure 10.13.

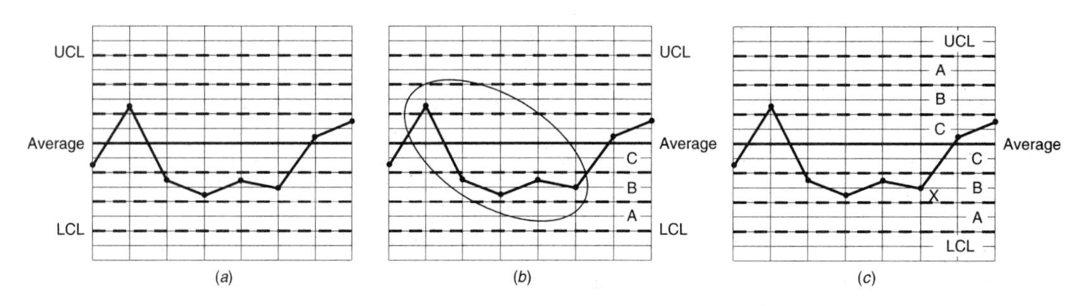

FIGURE 10.13
The chart for Example 10.4.

Solution

The four consecutive points in zone B in Figure 10.13 satisfy the four-of-five rule because the second point on the chart can be counted as the first of the five.

EXAMPLE 10.5 Mark the points that are out of control in Figure 10.14.

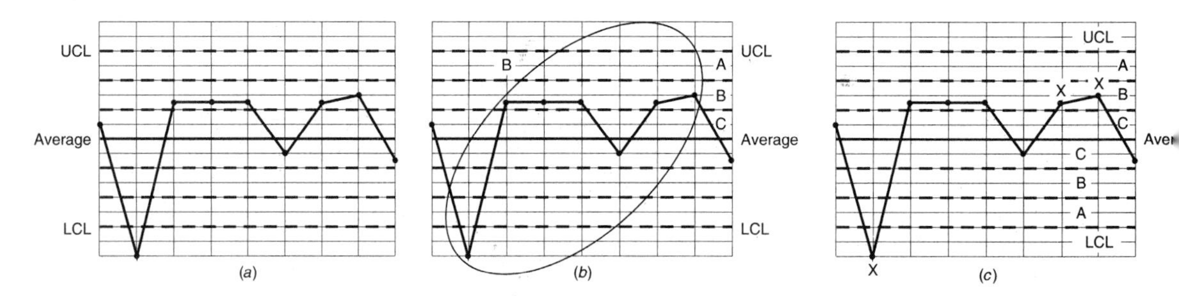

FIGURE 10.14
The chart for Example 10.5.

Solution

The first X in Figure 10.14 marks a freak. The second and third X's are part of a freak pattern satisfying the four-of-five rule. When the out-of-control pattern continues, the additional points that are out of control are marked too.

EXAMPLE 10.6 Identify the points that are out of control in Figure 10.15.

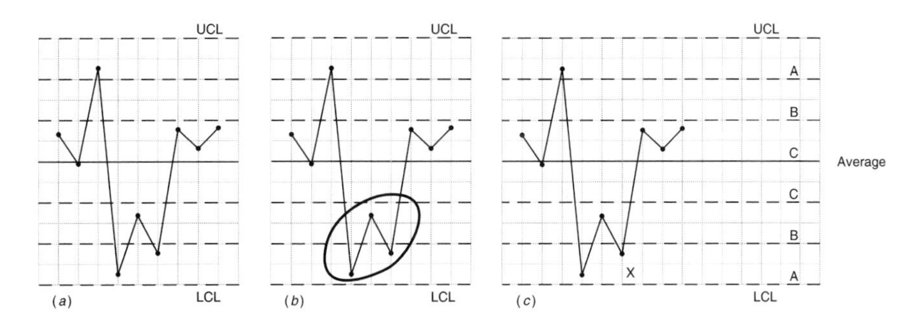

FIGURE 10.15
The chart for Example 10.6.

Solution
In Figure 10.15*b* the circled section shows that a freak pattern is formed by the two-of-three rule. Two of the three points circled are in zone A. The third point in that pattern is marked with an X to signal the out-of-control situation.

EXAMPLE 10.7 Identify the points that are out of control in Figure 10.16.

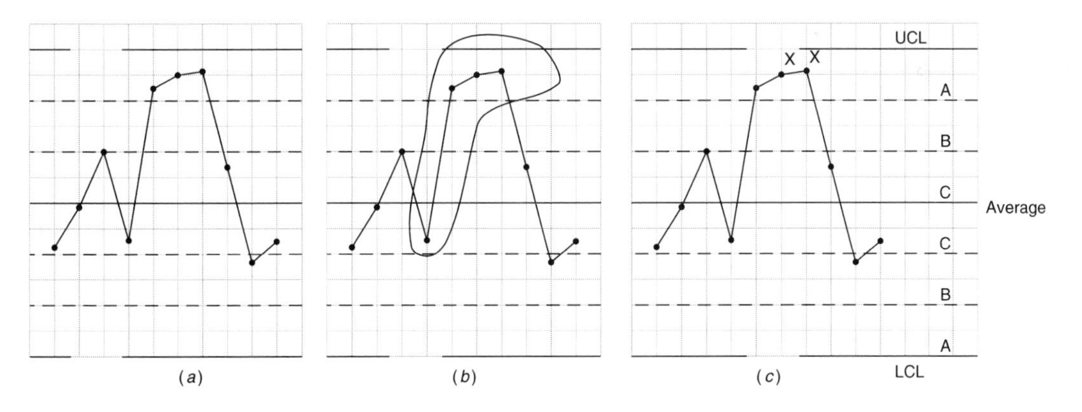

FIGURE 10.16
The chart for Example 10.7.

Solution
In Figure 10.16 a freak pattern is formed according to the two-of-three rule. The first three points circled satisfy the two-of-three rule. The fourth point is part of the same out-of-control pattern and is marked as well.

EXAMPLE 10.8 Mark the points that are out of control in Figure 10.17.

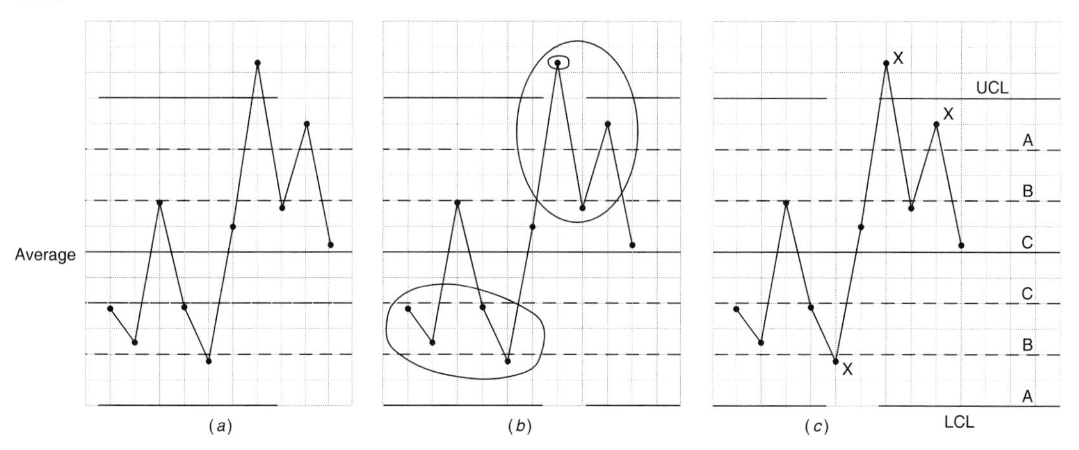

FIGURE 10.17
The chart for Example 10.8.

Solution

There are three different out-of-control situations in Figure 10.17. The first group circled in (*b*) is a freak pattern by the four-of-five rule: only the fifth point in that group is marked. The second group circled has a freak embedded within a two-of-three freak pattern. The freak is marked along with the third point in the freak pattern.

Each out-of-control pattern has its own list of common causes. Some process problems may produce more than one out-of-control pattern, and consequently, that problem will be on more than one list. The following is a list of the most common causes of freak patterns:

***R* chart:**

- Operator errors
- Poor technique owing to improper training or instruction
- Excessive vibration and increased variability caused by inadequate fixtures
- Inconsistent materials or materials from different suppliers
- Defective parts used in assembly
- Measurement or gauge problems
- "Chasing" the target measurement and excessive variability caused by over-adjustment of the machine

$\bar{x}$ chart with the *R* chart in control:

- The preceding possible causes
- Planned rework resulting from deliberate crowding of one side of the specifications as a hedge against producing scrap

***p* chart:**

- Poor maintenance of the machine
- Defective parts in an assembly
- Variations in sample size
- Nonrandom sampling
- Sampling different distributions

10.5 SHIFTS

Shifts, sets of seven or more consecutive points that are all on one side of the center line, indicate that the center of the distribution has changed. Something was introduced to the process that changed the whole process. Shifts are usually temporary. The points on the control chart change to a different level for a while, then change back again. When diagnosing a shift, look for something that affects the whole process. Any special cause of the shift in the process is generally in one of six major categories:

Operator	Method	Machine
Material	Tooling	Environment

A change in one of those categories can affect the entire process and shift the points on the control chart.

Shifts can indicate improvements as well as problems in the process. A shift away from the target line on the $\bar{x}$ chart signifies that the parts being measured and charted are now too large or too small, depending on the direction of the shift. A shift *up* on the R chart is a trouble indicator because product variation has increased. However, a shift *down* on the R chart indicates process improvement: Variation has decreased. A shift on a p chart may be an improvement or worsening of the defective proportion, but quick judgments are not advisable because a change in the inspection criteria, for example, has an accompanying shift on the p chart without any real change in the product.

The seventh point and any following points in the same shift pattern should be marked as out of control. This is illustrated in Figure 10.18.

The mathematical justification for the seven-in-a-row rule is based on a random pattern of points about a center line. Each point has a 50% chance of being on one side of the center line. Points that appear to fall on the line are not really an exception to this because both the measurement and the calculation of the average can theoretically be carried to enough places (such as thousandths or ten-thousandths) to show whether the point is above or below the line. The chance of getting seven consecutive points above the line can be calculated with the "and" rule:

$$
\begin{aligned}
P(\text{7 points above}) &= P(\text{point 1 is above } \textit{and} \text{ point 2 is above } \textit{and} \ldots) \\
&= .5 \times .5 \times .5 \times .5 \times .5 \times .5 \times .5 = (.5)^7 \\
&= .008 \\
&= .8\%
\end{aligned}
$$

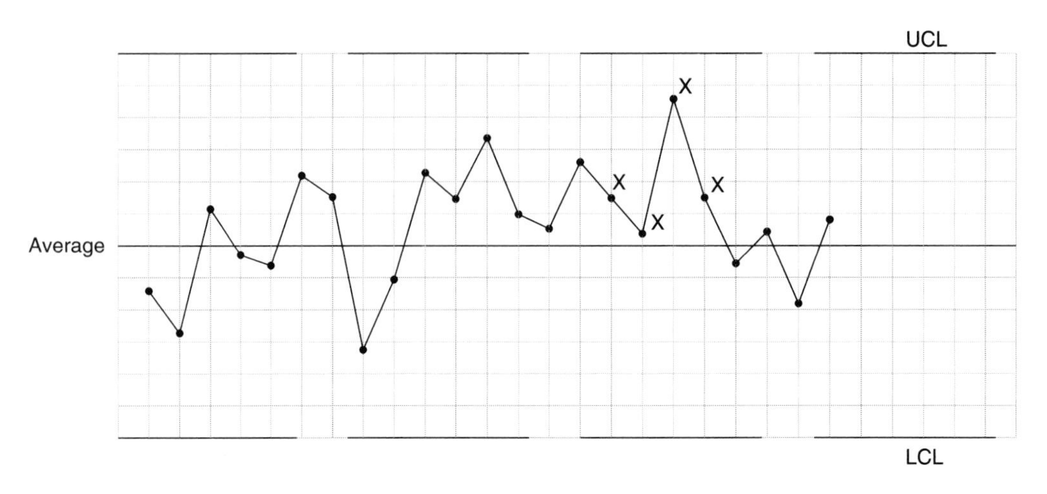

FIGURE 10.18
A shift.

The probability is small enough to classify the event of getting seven consecutive points on one side of the center line as a rare event. As before, with the other patterns, this is classified as a trouble indicator. Something has changed the process and the cause of the change must be found.

The following are possible causes of shifts:

***R* chart:**

- A careless or poorly trained operator
- Maintenance problems
- Change in material: poorer material will shift up; better material will shift down; a mixture of material will shift up
- Fixtures not holding the work in place
- Downshifts caused by better labor quality
- Downshifts owing to improved process quality

$\bar{x}$ chart when *R* chart is in control:

- A change in the machine setting or speed
- A new operator or inspector
- A change in method
- A new lot of material
- A new setup
- Use of a different gauge

***p* chart:**

- Any changes listed for the $\bar{x}$ chart
- A change in standards

The start of a shift is not obvious. Although the seventh point in the pattern is the official designator of the shift, the actual process shift may have begun at any of the preceding points in the shift pattern. It is not always at the first point in the pattern because the first few points may have been part of the normal random fluctuations. Be sure to check the process log at each point in the shift pattern.

EXAMPLE 10.9 Identify the points that are out of control in Figure 10.19. Cover (*b*) and analyze (*a*) first.

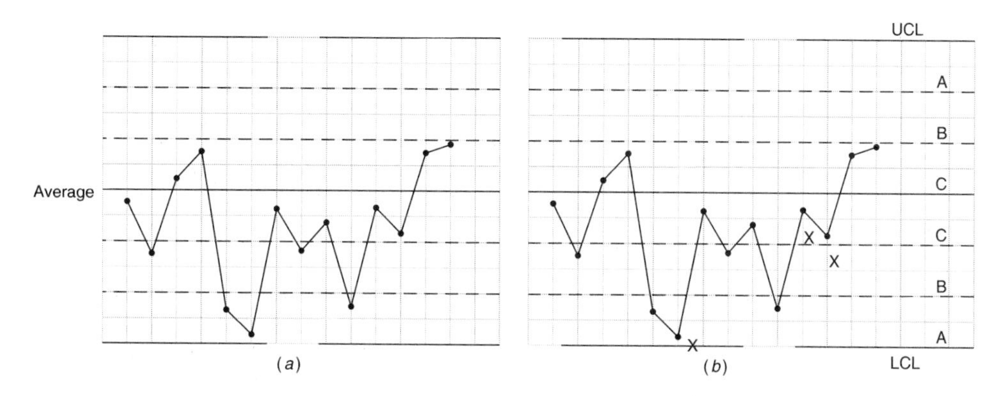

FIGURE 10.19
The chart for Example 10.9.

Solution

The first point marked in Figure 10.19*b* is for a freak pattern using the two-of-three rule. That freak pattern is embedded in a shift, and the second and third X's mark the shift.

EXAMPLE 10.10 Identify the points that are out of control in Figure 10.20. Cover (*b*) and analyze (*a*) first.

Solution

There are three out-of-control patterns in Figure 10.20. The first point marked is a freak, and the second point marked indicates a freak pattern by the four-of-five rule. The tenth, eleventh, and twelvth points, the next three marked points, denote a shift pattern. Figure 10.20 also illustrates that freaks and freak patterns can be forerunners to a shift pattern and can call attention to the trouble before the shift may be recognized. It is also important to realize that shift patterns are functions of time. For example, if samples are taken hourly, the shift pattern may not be recognized until several hours after the process shift has occurred. If a shift is suspected after three or four points have started the pattern, a change to more frequent sampling will verify the shift sooner.

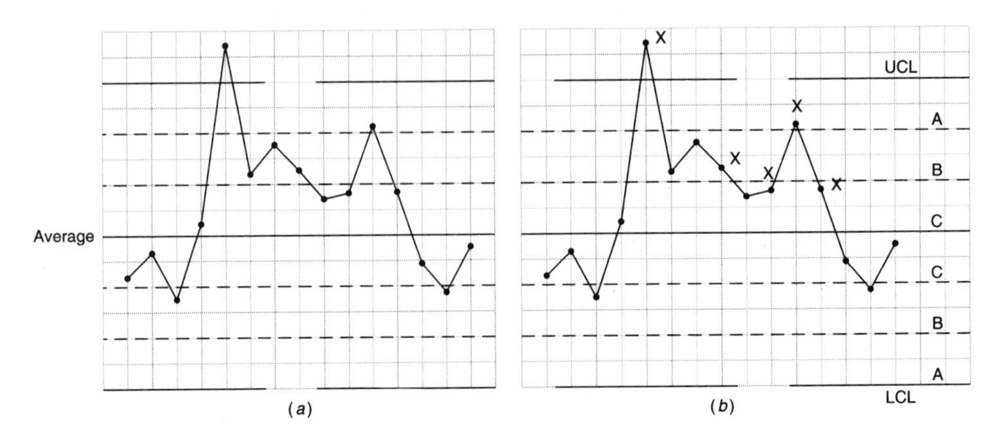

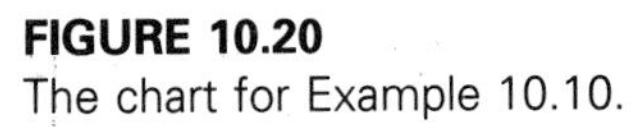
FIGURE 10.20
The chart for Example 10.10.

EXAMPLE 10.11 Using the list below, identify the most likely cause for the points that are out of control in Figures 10.21 through 10.24:

1. Machine adjustment
2. Arithmetic error
3. Improvement in process capability
4. New operator

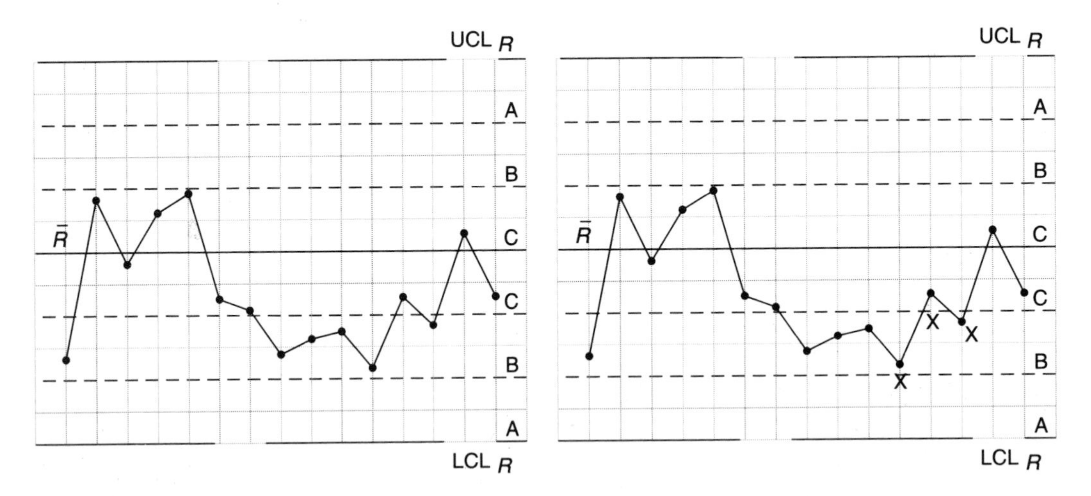

FIGURE 10.21
A chart for Example 10.11.

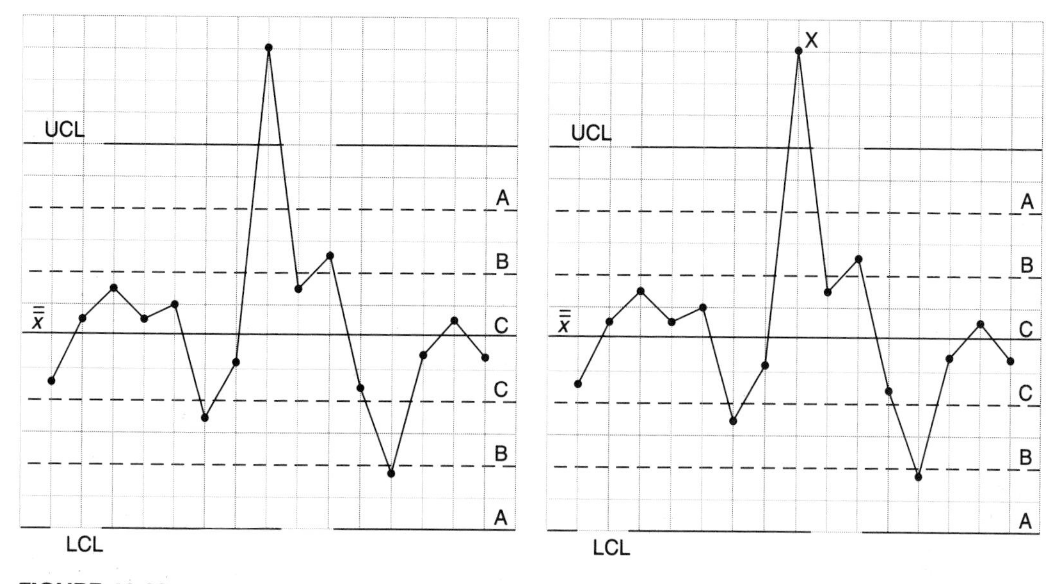

FIGURE 10.22
A chart for Example 10.11.

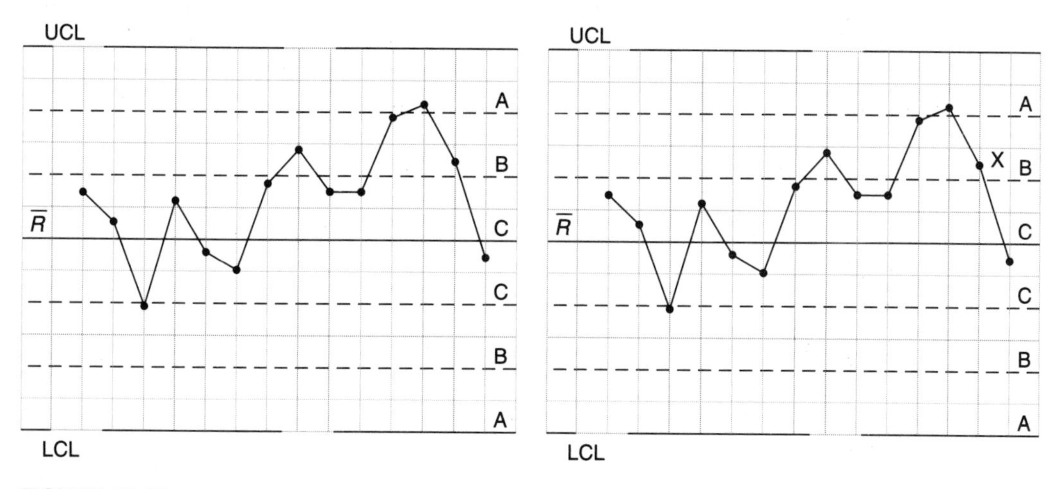

FIGURE 10.23
A chart for Example 10.11.

Solution

Figure 10.21, an R chart, has a freak pattern by the four-of-five rule that is part of a shift. The basic pattern is the shift pattern. Because the shift is down in an R chart, variability has decreased. Look for something that has improved the process. Choice 3 from the list of causes is the correct one.

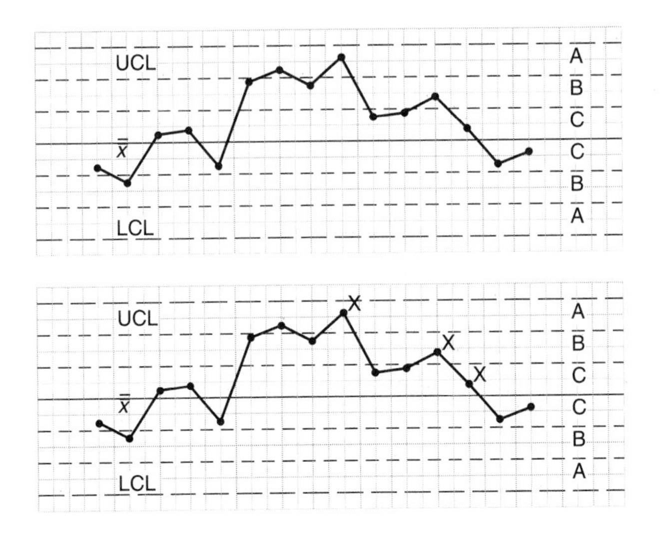

FIGURE 10.24
A chart for Example 10.11.

There is one freak on the $\bar{x}$ chart in Figure 10.22. The other points seem randomly well behaved, so the most likely cause is 2, the arithmetic error.

Figure 10.23 shows a shift upward on a range chart. This is an indication that variation has increased. The most likely cause from the list is a new operator, 4.

The $\bar{x}$ chart in Figure 10.24 has a shift as the main trouble pattern. The first X marks a freak pattern by the four-of-five rule. This could be caused by either a machine adjustment, 1, or a new operator, 4.

10.6 RUNS AND TRENDS

A chart that shows a pattern of points that are steadily climbing or steadily falling is called a *run.* Another pattern that is closely associated with a run is a *trend.* An *increasing trend* would be a sequence of points that are gradually climbing on the control chart with some decreases mixed in, but an overall pattern showing that the measurements are increasing. Likewise, a *decreasing trend* shows the sequence of points gradually falling with some increases mixed in. Both runs and trends indicate a gradual increase or a gradual decrease in the measurements, but the run is quicker to spot. A situation in which seven consecutive points are climbing or seven points are steadily falling is classified as a run. Also, if 10 out of 11 points are climbing or 10 out of 11 are falling on the chart, that is classified as a run as well. Trends are usually more gradual with more fluctuations and often indicate a slower process change. It may take half of the points on a control chart to show a trend or even several control charts laid end to end on the floor to show a gradual long-term trend.

Figure 10.25 shows a run: Seven or more points are steadily increasing. The seventh increasing point is marked along with the next two points, which are part of the same

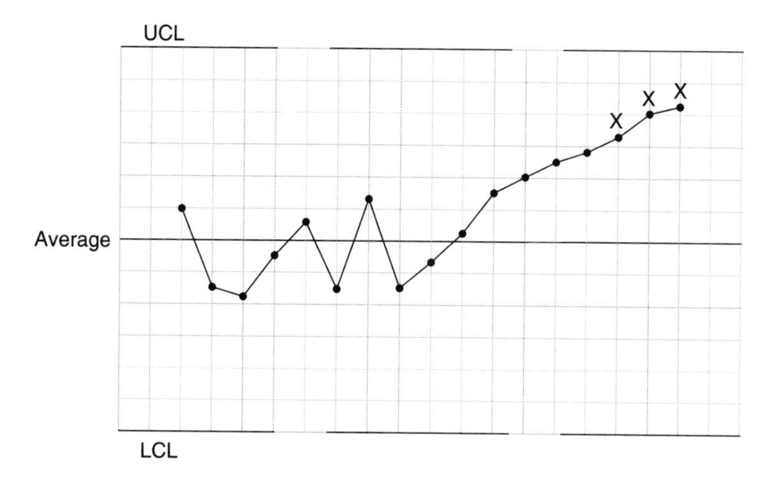

FIGURE 10.25
A run.

run pattern. The base point of the run, the eighth point from the left on the chart, is not counted as part of the run.

The general pattern in Figure 10.26 is a trend. The random fluctuations that occur within it, however, make it fail both the 7-in-a-row run test and the 10-of-11 run test. Its "official" out-of-control pattern is a shift, with seven or more consecutive points above the center line. The distinction between the trend or run pattern and the shift pattern can be important in analyzing the cause for the out-of-control situation. The shift classification implies that something in the process suddenly changed. The trend or run classifica-

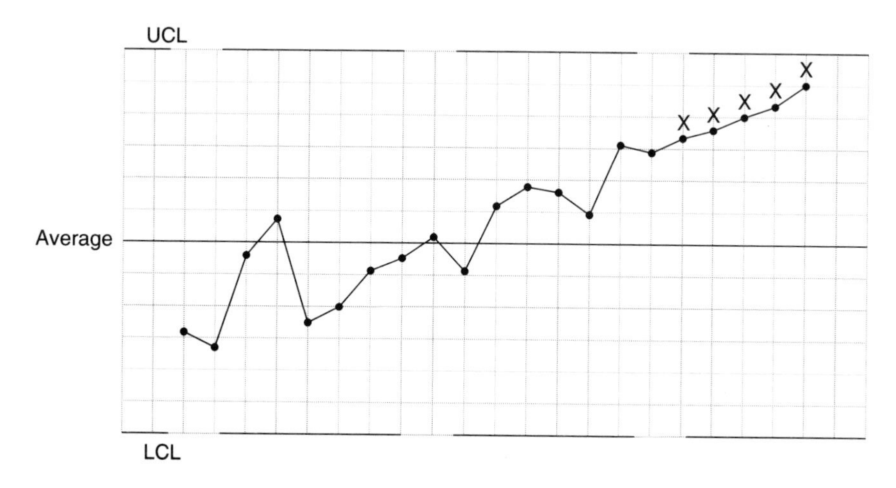

FIGURE 10.26
The general pattern is a trend, but the official pattern is a shift.

tion implies that there is a gradual change occurring. The two different clues can lead in different directions when the source of the trouble is unknown.

A run suggests that something in the process is changing gradually. If the run is on an $\bar{x}$ chart and the range chart is in control, the problem may be tool wear. If it is a boring tool such as a drill, gradual wear will make the holes smaller and the run will be in a downward direction. If it is a cutting tool, gradual wear will make the pieces larger and the run will move upward.

A run or trend can point out good news as well as bad. On an R chart, a downward trend indicates that variation is decreasing in the product. This may signal process improvements owing to use of SPC, better operator training, or improved maintenance. An upward run or trend on an R chart signals that things are getting worse: Variation is increasing. Fixtures may be loosening, operator fatigue may be setting in, or tool wear may be causing more variability.

The probability calculation for seven steadily increasing (or decreasing) points uses the same concept that is used with the shift: The "and" rule applies. The symbol ">" represents "larger than."

$$P(7\text{ increasing}) = P\ (\text{first point} > \text{base point } \textit{and} \text{ second} > \text{first } \textit{and} \text{ third} > \text{second } \textit{and} \ldots \textit{and} \text{ seventh} > \text{sixth})$$

By the "and" rule, the individual probabilities are multiplied:

$$P(7\text{ increasing}) = P\ (\text{first} > \text{base point}) \times P(\text{second} > \text{first}) \times \cdots \times P\ (\text{seventh} > \text{sixth})$$

The individual probabilities are impossible to determine using the normal distribution because the probability that a second point is larger than the first depends on the position of the first. This is illustrated in Figure 10.27.

Figure 10.27 shows two different positions of the first measurement on the measurement scale. Drawing (*a*) shows the first measurement value less than the average value, and (*b*) shows the first measurement larger than the average value. In each diagram the shaded area represents the probability that the second measurement is larger than the first. There are so many possible positions for the first variable that it is impossible to tell exactly what the probability will be. The same situation occurs when considering the probability that the third measurement is larger than the second or the fourth is larger than the third. A different approach to the problem is needed.

A run implies a gradual change in measurement from some base level. If there is some temporary working average at that base level, in case the process is not at $\bar{\bar{x}}$, then diagram (*b*) in Figure 10.27 will represent the probability that a second measurement is larger than the first in an increasing run. Each new point will be further above that initial, average value, and the probability that it is larger than the previous point will therefore be less than .5. Using .5 for each probability value will ensure that the true probability will be less than the calculated value. The symbol for "less than" is "<":

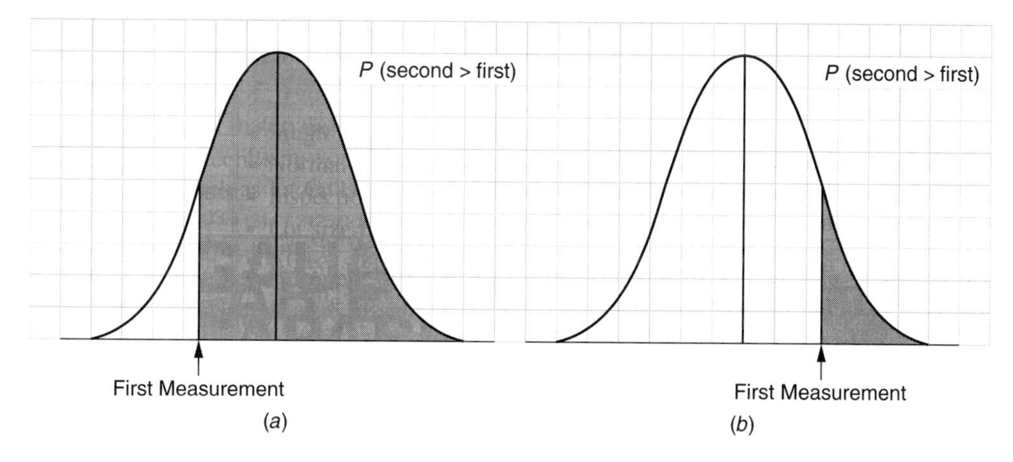

FIGURE 10.27
The shaded areas show the probability that the second measurement is larger than the first for two different first measurements.

$$
\begin{aligned}
&P(\text{1st point} > \text{base point}) < .5\\
&P(\text{2nd point} > \text{1st point}) < .5\\
&P(\text{3rd point} > \text{2nd point}) < .5\\
&P(\text{7 increasing points}) = P(\text{1st} > \text{base value}) \times P(\text{2nd} > \text{1st})\\
&\qquad\qquad \times P(\text{3rd} > \text{2nd}) \times \cdots \times P(\text{7th} > \text{6th})\\
&P(\text{7 increasing points}) < .5 \times .5 \times \cdots \times .5 = (.5)^7\\
&\qquad\qquad < .008
\end{aligned}
$$

The probability of getting seven increasing points or seven decreasing points in a row is less than .8%, a rare event. Therefore, this pattern is an indicator that the process is changing.

The probability for the 10-of-11 rule combines the "and" rule for the 11 consecutive points with the "or" rule, which takes into account the fact that the one point that is not increasing (or decreasing) with the others could be first *or* second *or* third or any of the 11 points. If I represents increasing points and D represents decreasing, the sequence of points will be, for example,

$$
\begin{array}{ll}
 & I\ I\ I\ I\ I\ I\ I\ I\ I\ I\ D\\
\text{or} & I\ I\ I\ I\ I\ I\ I\ I\ I\ D\ I\\
\text{or} & I\ I\ I\ I\ I\ I\ I\ I\ D\ I\ I\\
\text{or} & I\ I\ I\ I\ I\ I\ I\ D\ I\ I\ I
\end{array}
$$

These are 4 of 11 possibilities:

$$
\begin{aligned}
P(\text{10 of 11 increasing}) &< 11 \times (.5)^{11}\\
&< .0054
\end{aligned}
$$

Eleven is the number of possibilities, and each way has a probability less than .5. The probability of getting 10 out of 11 points that are steadily increasing (or decreasing) is therefore less than .5%.

The following list summarizes the specifications for runs and trends:

1. Seven consecutive points that steadily increase or steadily decrease indicate a run.
2. Ten out of 11 points that steadily increase or steadily decrease form a run.
3. A long fluctuating pattern that is steadily increasing or steadily decreasing is a trend.

Runs and trends may stem from a variety of causes:

***R* chart, *upward* trend (process deteriorating):**

- Gradually increasing material variability
- Loosening fixtures
- Operator fatigue (check day-to-day time patterns)
- Machine wear
- Gauge wear

***R* chart, *downward* trend (process improving):**

- Improvements owing to SPC and the quality program
- Better training of operators
- Better maintenance program
- Gradual introduction of better material

$\bar{x}$ chart with the *R* chart in control:

- Gradual introduction of new material
- Tool wear
- Machine due for adjustment

***p* chart:**

- Changing defective proportion
- Changing requirements or standards

10.7 TIME AND CONTROL CHART PATTERNS

When a sampling plan is formed for keeping a control chart, the time interval between samples may vary extensively. When control charts are first applied at a point in a process, samples may be taken frequently to establish whether or not the process is in statistical control. Once control has been achieved, the time between samples is often increased. The ideal situation is to sample often enough to note any changes in the process. But what is often enough? Experienced operators may be aware of occurrences that affect the process and would be sure to sample accordingly. But without that insight, an in-control process may go two or more hours between samples. A shift occurring in that situation could go undetected for a couple of days. When a hint of an out-of-control pattern occurs, the operator should have the freedom to sample more frequently so the recognition of the out-of-control situation occurs sooner and corrective action can be applied.

Different out-of-control patterns are affected by the time interval between samples. Samples taken every fifteen minutes that show an increasing trend would show a run if sampled hourly because the hourly increase would be enough to override any small decreases that occur on the trend line. A freak pattern that develops when sampling every two hours could very well be a shift if the sampling were every half hour. All the number rules that were applied to the various point patterns were based on probability so that processes would not be stopped on false alarms. All of the probabilities were less than 1%, which means that false alarms (stopping the process when it is really in-control) would occur less than 1% of the time. The increased sampling frequency on suspicion of trouble would not change that figure; it would just establish the out-of-control pattern sooner.

One of the early indicators of possible trouble would be two consecutive points above the C zone (or below the C zone). The probability shown in Figure 10.9 and the "and" rule show that the chance of that happening is $(.16) \times (.16) = .0256$, a rather small chance of occurrence. Two more samples taken sooner than originally planned would confirm the freak pattern and corrective action could be taken sooner, or the pattern would be broken, indicating that the process wasn't really out-of-control. Another hint of trouble would be a point in the A zone. That has about a 2.5% chance of occurring as well. Another early sample could confirm that a freak pattern has occurred or the pattern would be broken and the process would continue to run.

10.8 CYCLES

Cycles on a control chart are patterns that repeat on a regular basis. They signal that something is systematically affecting the process. The key to finding the problem that is causing the cycles is concentration on the factors that change the process periodically. Power fluctuations, varying speeds on conveyor belts, operator fatigue patterns, and temperature changes are likely causes.

There is no number rule for cycles; it involves a recognition of repeating patterns. Figure 10.28 illustrates the cycle pattern. The figure also shows several out-of-control points by the two-of-three, the four-of-five, and the shift patterns. Cycles may be short term in duration and clearly show up on a control chart such as in Figure 10.28. It is also possible to have long-range cycles in which several or many charts must be considered simultaneously. To check on long-term cycles such as seasonal effects, it may be necessary to condense data from many charts to one. This can be done by charting daily, weekly, or monthly averages.

In some cases, trouble patterns of one type exist within another. For example, runs or trends may be the trouble pattern diagnosed on a chart, but they may be just one sec-

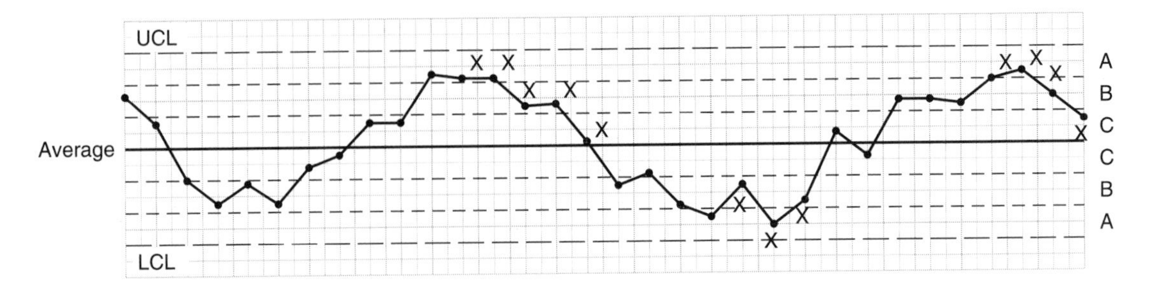

FIGURE 10.28
A cycle.

tion of a cycle pattern. Spotting the overall cycle pattern may add that extra clue needed to diagnose the process trouble. Looking at the data from a different perspective provides information that can be used to solve the problem.

Repeating patterns are usually quite obvious when they occur on a single chart. When several charts are compared simultaneously, they should be checked for repeating patterns.

There are various causes of cycles:

***R* chart:**

- Operator fatigue and shift changes
- Measurement gauge rotation
- Eccentric tooling wear, periodic changes
- Maintenance schedules
- Periodic speed changes

$\bar{x}$ chart with the *R* chart in control:

- The listed *R* chart causes
- Seasonal or environmental changes
- Worn threads and locking devices
- Power fluctuations
- Reliance on different suppliers

***p* chart:**

- Material variations
- Different suppliers
- Change in sampling methods

A manufacturing process is often likened to a river or a stream with the product becoming more complete as it moves downstream. The workers who are positioned downstream from a specific process point are the internal customers. A good-quality program emphasizes a quality product for both the downstream customer and the eventual purchaser of the finished product. When an out-of-control pattern shows up on a control chart, the problem will be at the process point being charted or at some point(s) upstream. When all the possible solutions at the process point have been eliminated and the problem still exists, an upstream search for the problem source must be undertaken.

10.9 GROUPING

This is another case in which one trouble classification may be embedded in another. *Grouping,* or *bunching,* occurs when the points on a chart occur in clusters. Large fluctuations between clusters may qualify as one of the instability patterns: freaks or freak patterns. However, the overall pattern of grouping may be the pattern that provides the necessary information for solving the problem.

Grouping indicates that several different distributions are present. They may be strictly upstream problems that are converging at a point in the process or may signify inconsistent workmanship or materials.

Figure 10.29 illustrates bunching. The circles are included for emphasis. The only out-of-control point marks a freak pattern within the bunching pattern.

The causes of bunching are the following:

***R* chart:**

- Differences in work quality
- Inconsistent materials

$\bar{x}$ chart with the *R* chart in control:

- Differences in work quality
- Inconsistent materials
- Shifting fixtures
- Inconsistencies in method
- Several upstream problems

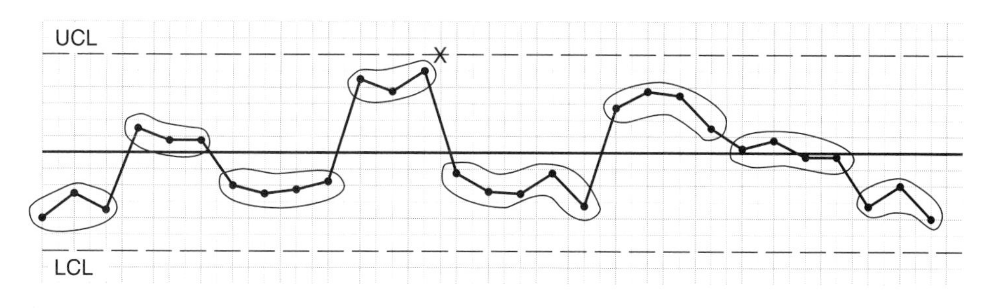

FIGURE 10.29
Bunching.

EXAMPLE 10.12 In Figure 10.30, cover (*b*) while you analyze (*a*).

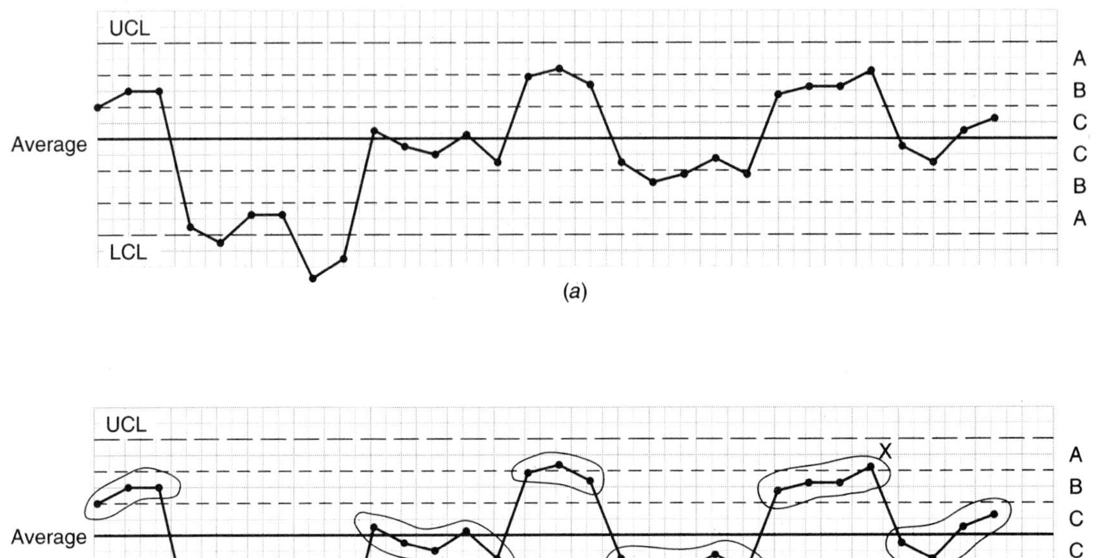

FIGURE 10.30
The chart for Example 10.12.

Solution

The main problem structure is the bunching that has been emphasized with the circles in Figure 10.30. The first X marks a freak. The second and third X's are part of a two-of-three freak pattern. The fourth and fifth X's are freaks as well as a continuation of the two-of-three freak pattern. The last X marks a freak pattern by the four-of-five rule. The bunches seem to follow a cyclic pattern as well.

EXAMPLE 10.13 In Figure 10.31 cover (*b*) while you analyze (*a*).

Solution

In Figure 10.31, the main pattern is a run. The first three X's mark the run. The next three X's simultaneously show a shift and a freak pattern by the four-of-five rule. The shift is

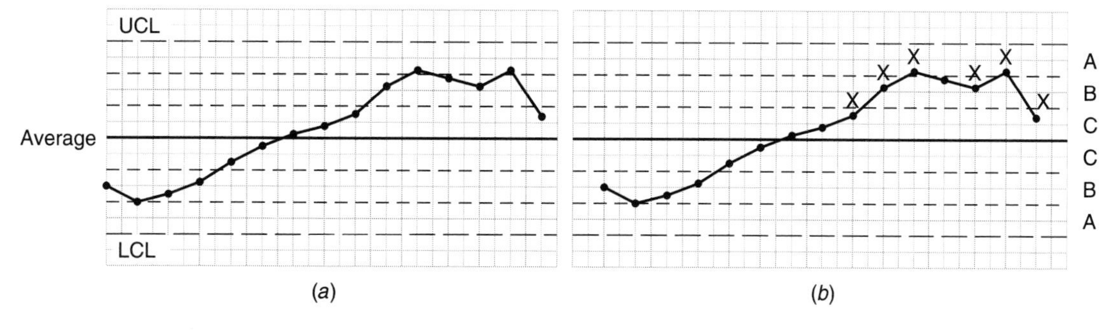

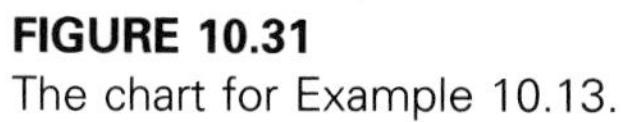

FIGURE 10.31
The chart for Example 10.13.

the pattern of secondary importance because the process held at that new level after the gradual run brought it there.

EXAMPLE 10.14 In Figure 10.32, cover (*b*) and analyze (*a*).

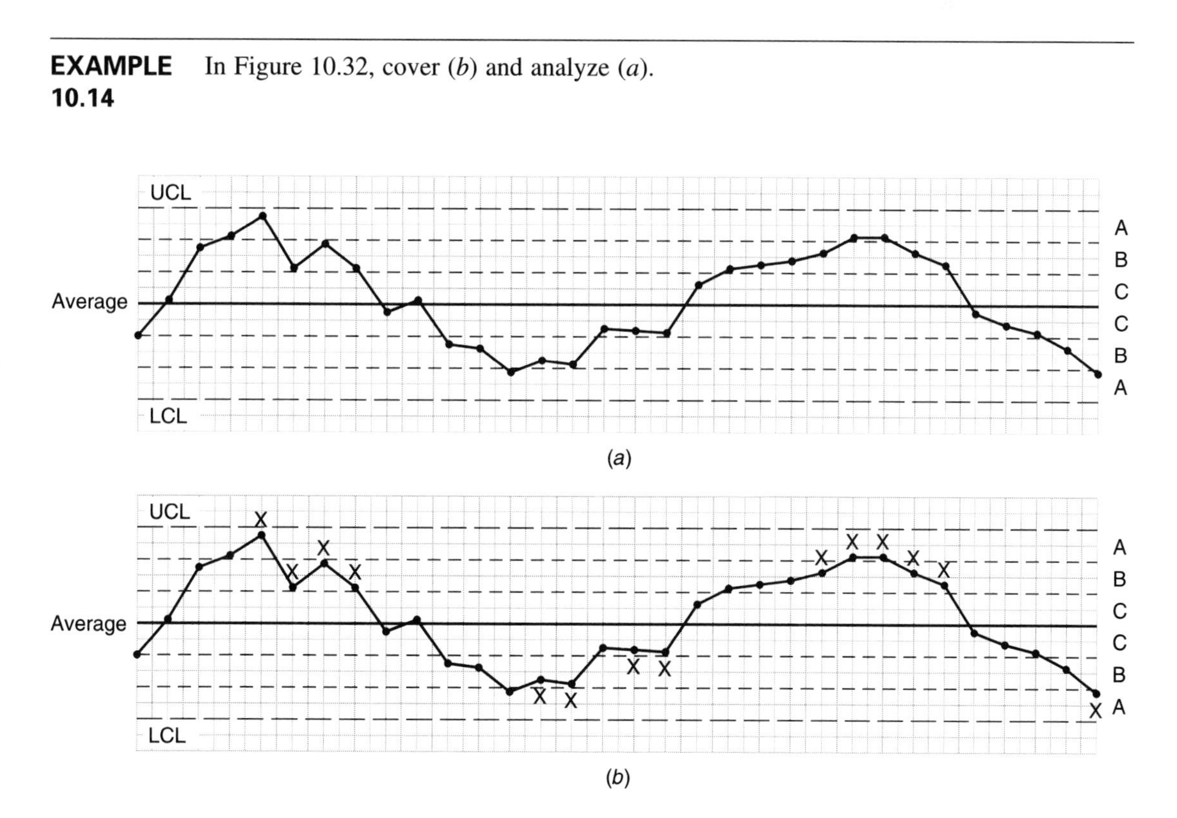

FIGURE 10.32
The chart for Example 10.14.

Solution

The main pattern in Figure 10.32 is a cycle. Within that pattern, the first X marks a two-of-three freak pattern and the second through fourth X's mark a freak pattern by the four-of-five rule. The next pair of X's also marks out-of-control points by the four-of-five rule; these points are followed by two X's marking a shift pattern. The cluster of five X's starts as a freak pattern by the four-of-five rule with the latter three points simultaneously fitting the shift pattern. The last X signifies a run.

10.10 INSTABILITY

An erratic pattern that has large fluctuations on a control chart is classified as *instability*, or *unstable mixture*. Freaks or freak patterns may exist within the erratic pattern, but the main message comes from the extended and frequent ups and downs. Instability is generally hard to track because it usually has several causes. A pattern of instability often indicates that the trouble lies upstream from the chart that indicates the trouble. For example, there may be several product distributions at an assembly point, all capable of changing with respect to each other. The trouble may be in one or several of the separate pieces. Unstable mixture, illustrated in Figure 10.33, has a good, descriptive name. The two X's mark freaks that occur within the unstable mixture pattern.

When a process is in statistical control, 68% of the points on the control chart, roughly two-thirds of the data, should be within one standard deviation of the mean. Also, only 4% of the points should lie in the two 1σ bands that are adjacent to the control limit lines (A zones). To test for unstable mixture, look for violations in that normal pattern. This process consists of a search for two characteristics:

1. More than one-third of the points lie outside the *center* $\pm\ 1\sigma$ band *or* more than 4% of the points fall in or beyond the *outer* 1σ bands.
2. The chart has a steep, zigzag pattern.

Look for several causes of instability, often upstream. The following are possible causes:

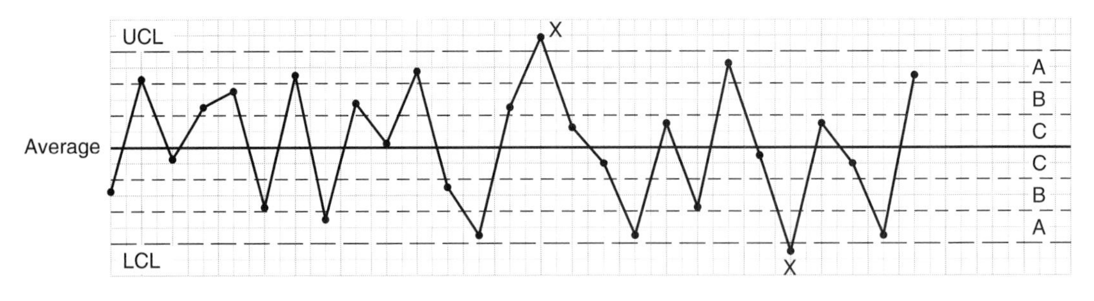

FIGURE 10.33
An unstable mixture.

***R* charts, $\bar{x}$ charts, *p* charts:**

- Frequent breakdowns and start-ups (R and $\bar{x}$)
- Overadjustment ($\bar{x}$)
- A mixture of materials or parts from different machines or spindles (R and $\bar{x}$ and p)
- Loose fixtures (R and $\bar{x}$)
- Poor sampling procedures ($\bar{x}$)
- Inconsistent materials (R and $\bar{x}$)
- A faulty gauge (R and $\bar{x}$)
- Inconsistent inspection standards (p)

10.11 STABLE MIXTURES

A mixture pattern that has erratic ups and downs similar to the instability pattern but has very few points in the middle of the chart is a *stable mixture.* The points crowd or overlap the control limits. This pattern usually indicates a mixing of two different stable distributions: one for the upper set of points and one for the lower set. Five or more consecutive points outside the C zones (beyond the 1σ lines) on either side of the center line signal a stable mixture pattern, as Figure 10.34 illustrates. The first X in Figure 10.34 signals a two-of-three freak pattern, and the remaining X's indicate a stable mixture pattern.

The probability of getting five consecutive points beyond the C zones can be calculated using the "and" rule:

$$P\text{ (5 points beyond the C zones)} = P\text{ (1st and 2nd and 3rd and 4th and 5th are beyond)}$$

Using the normal distribution, the probability of one point falling beyond the C zones is the chance that it will fall in either of the shaded sections indicated in Figure 10.35. The boundaries of the C zones are 1σ from the average value, which correspond to z values of ± 1. The number of standard deviations from the center is z.

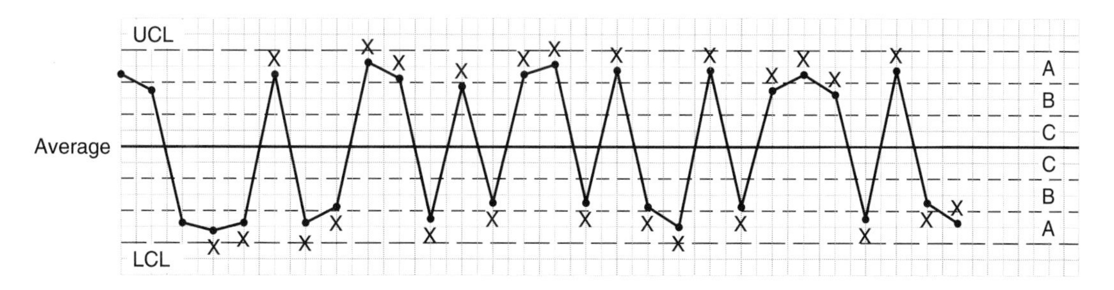

FIGURE 10.34
A stable mixture.

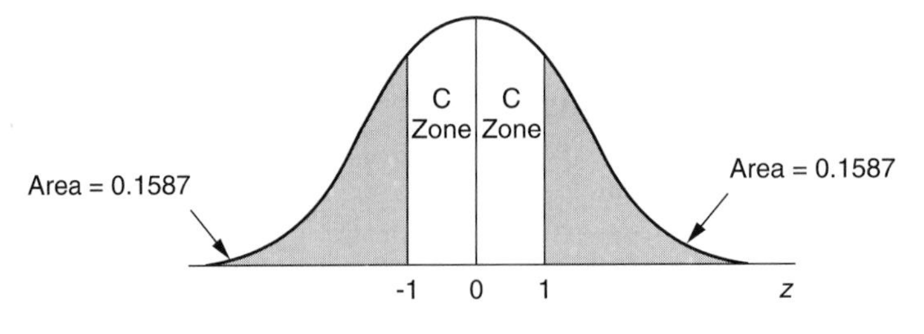

FIGURE 10.35
The probability of a point falling beyond $\pm 1\sigma$.

When $z = 1$, the shaded tail area on one side is .1587. Double this for the probability that a point is in either shaded area:

$$\begin{aligned} .1587 \times 2 &= .3174 \\ P\text{ (5 points beyond C)} &= .3174 \times .3174 \times .3174 \times .3174 \times .3174 \\ &= (.3174)^5 = .00322 \\ &= .32\% \end{aligned}$$

The small probability classifies this as a rare event and an indicator that something is affecting the process.

Test for a stable mixture by looking for two characteristics:

1. Five or more consecutive points that are outside the center $\pm 1\sigma$ band
2. Steep zigzag lines on the control chart with very few points in the middle of the chart

A stable mixture pattern on an R, $\bar{x}$, or p chart may be caused by mixing two stable distributions from the following list:

- Different suppliers
- Different inspectors
- Different operators
- Different lines
- Different lots
- Different gauges or inspection standards

10.12 STRATIFICATION

In a *stratification* pattern, the points hug the averages line on a control chart. There is a notable absence of points near the control limits. The rule for spotting this pattern is to look for 14 or more consecutive points in the two C zones, within $\pm 1\sigma$ of the average value. Figure 10.36 illustrates a stratification pattern on a control chart.

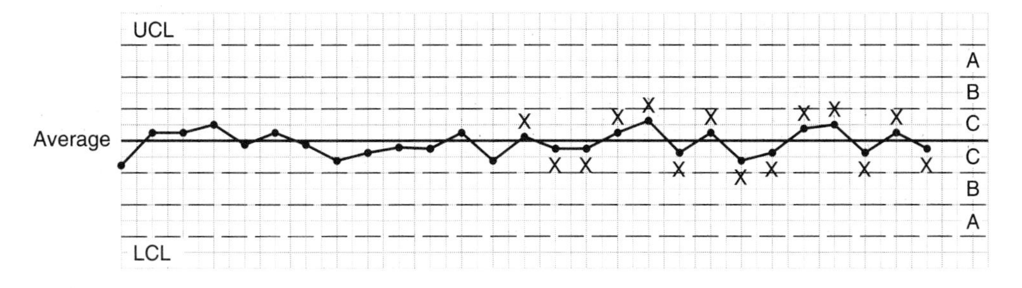

FIGURE 10.36
Stratification.

To the untrained eye, a stratification pattern seems to identify a smoothly running process. If the process has really improved, however, a downward trend or shift on the range chart will accompany the stratification pattern on the $\bar{x}$ chart. The follow-up step calls for new calculations for $\bar{\bar{x}}$ and $\bar{R}$ at the new level of control. A recalculation of the control limits will make both the $\bar{x}$ and R charts look normal again with two-thirds of the points in the $\pm 1\sigma$ center strip and the other one-third beyond. Of course, the new charts will reflect the tighter control of the improved process because the control limits will be numerically closer to the average line for both $\bar{x}$ and R.

Stratification on the $\bar{x}$ chart without the downward move on the range chart denotes trouble because it is a nonnormal pattern. There may be a measuring problem: The measuring scale may be too crude for the job. Two general rules of measurement must be followed. The first, called the *rule of 10,* states that the measuring instrument should measure to one-tenth of the tolerance. The second states that there should be at least 10 units of measurement between $\bar{x} - 3\hat{\sigma}$ and $\bar{x} + 3\hat{\sigma}$. Initially, if a process has not been charted, there should be at least 10 units of measurement within the tolerance. Then as the process capability improves, the rule of 10 units or more between $\bar{x} \pm 3\hat{\sigma}$ should be used.

EXAMPLE 10.15 Given the tolerance .215 ± .001, what unit of measurement should be on the gauges?

Solution

$$\frac{\text{tolerance}}{10} = \frac{.002}{10}$$
$$= .0002$$

The gauges should be marked in two ten-thousandths or less.

EXAMPLE 10.16 Suppose the tolerance is .002 and the process capability used 65% of the tolerance (PCR = .65, a B process according to Table B.11 in Appendix B). What should the unit of measurement be?

Solution

$$\frac{\text{tolerance} \times .65}{10} = \frac{.002 \times .65}{10}$$

$$= \frac{.0013}{10}$$

$$= .00013$$

The gauges should be marked in ten-thousandths.

Another source of trouble that may result in stratification is dishonest recording. An employee may arbitrarily record values close to the average to try to look good or may rework pieces that are not good enough and record the better, reworked measurement. People working in purchasing or incoming inspection should take extra care to spot doctored control charts.

A third possible cause for stratified $\bar{x}$ charts is nonrandom sampling. This sometimes happens inadvertently when a sampler tries to make sure that each of several different distributions are represented in every sample. For example, in a multiple-spindle operation, if each sample has one piece from each spindle, the range may be large owing to large differences in the individual spindle averages. If the variability on each spindle is low, the average from each sample will be close to $\bar{\bar{x}}$, and the control limits, calculated with $\bar{R}$, will be far from the center line. This may result in a stratified pattern. A stratified control chart may therefore show that several different distributions are being mixed in a nonrandom, representative way.

To summarize, possible causes of stratification are the following:

***R* charts, $\bar{x}$ charts:**

- Falsification of data
- Gauge scale too crude for the job
- Nonrandom, representative sampling
- Process improvement when $\bar{x}$ stratifies and R values decrease

The probability calculation for the 14-point rule uses the probabilities from the normal curve and the "and" rule. If $z = 1$, the tail area in the normal curve is .1587. Doubled, .3174, it is the combined tail area for both sides. The total area under the curve is 1, so the area in the middle $\pm 1\sigma$ section is

$$\begin{array}{r} 1.0000 \\ -\ .3174 \\ \hline .6826 \end{array}$$

The chance of getting a single measurement in the center $\pm 1\sigma$ section is .6826.

The probability of getting 14 consecutive measurements in that center section is calculated using the "and" rule:

$$\begin{aligned} P\text{ (14 in the C zones)} &= P\text{ (first in } and \text{ second in } and \text{ third in } and \ldots and \text{ fourteenth in)} \\ &= P\text{ (first in)} \times P\text{ (second in)} \times \ldots \times P\text{ (fourteenth in)} \\ &= (.6826)^{14} \\ &= .0048 \end{aligned}$$

The probability, about .5%, is small enough to classify this as a rare event and an indicator of either trouble or process improvement.

EXAMPLE 10.17 For each of the charts in Figures 10.37 through 10.41, cover (*b*) and analyze (*a*).

Solution

Figure 10.37 shows a stable mixture. After the eighth point on the chart, there is an absence of points in the C zones. The first and last X's mark a freak pattern by the two-of-three rule. The other X's show the stable mixture when five consecutive points are beyond the C zones. This chart may also be interpreted as bunching within a cycle pattern. The analyses lead in different directions in the search for a solution. The mixture analysis will attempt to identify two separate distributions that may have worked their way into the process. The cyclic bunching approach suggests a time analysis, an operator check, and possibly a gauge check. Both approaches may lead to the conclusion that material from two different lines or suppliers was being systematically mixed.

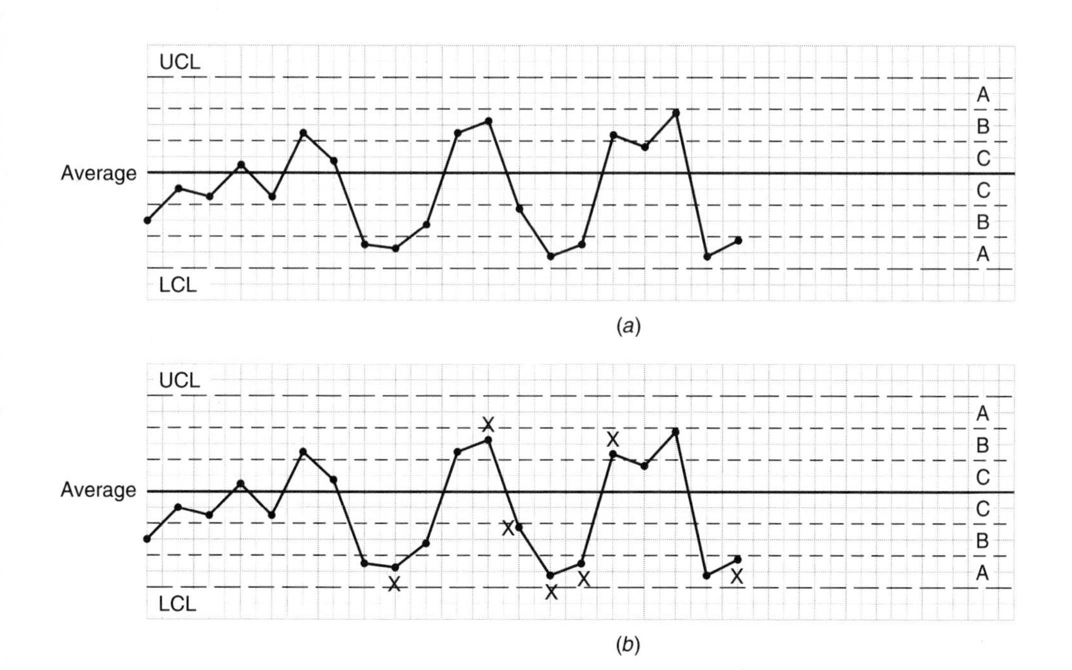

FIGURE 10.37
A chart for Example 10.17.

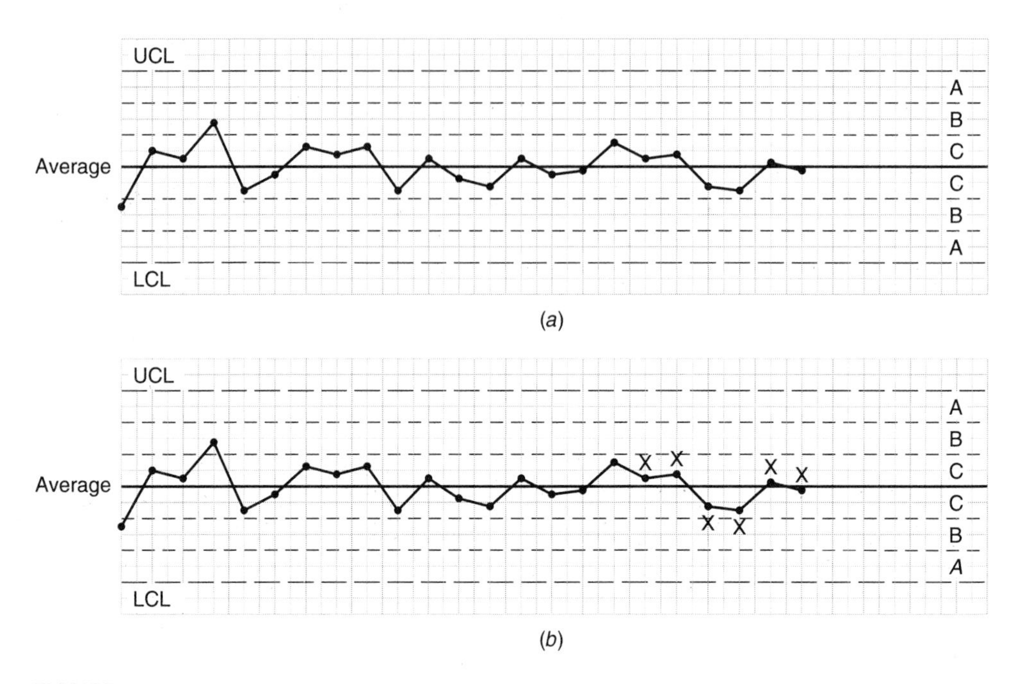

FIGURE 10.38
A chart for Example 10.17.

The X's in Figure 10.38 mark the out-of-control points by the 14-point stratification rule.

In Figure 10.39 the main problem structure seems to be bunching. Within that overall bunching format, however, are shifts, freaks, and freak patterns. The X's are numbered for easy reference. The first two X's show a freak pattern by the two-of-three rule. The third X marks a freak pattern by the four-of-five rule. The fourth, fifth, and sixth X's may be part of a freak pattern by the four-of-five rule, and the fifth and sixth indicate a shift. The seventh and eighth X's signal a stable mixture. For the group of X's on the right, X number 9 marks a freak pattern by the two-of-three rule, and number 10 indicates a stable mixture. The next three are all freaks, and the last two point out a shift.

The main problem in Figure 10.40 is instability. The erratic ups and downs and the fact that 16 of the 22 points, or 73% of them, are beyond the $\pm 1\sigma$ center strip signal a strong possibility that several factors are affecting the process. Within the instability pattern the first X marks a freak and the other two mark freak patterns by the two-of-three rule.

The X's in Figure 10.41 are numbered for easy reference. A cycle is the basic pattern, within which shifts and freak patterns exist. The first four X's show freak patterns by the two-of-three rule and the four-of-five rule, and the fourth and fifth are part of a shift. The first two X's in the middle group, numbers 6 and 7, mark a run; the next two, a shift. Numbers 10 and 11 show a run, number 12 is part of a freak pattern by the two-of-three rule, and the last marks both a freak pattern by the four-of-five rule and a shift.

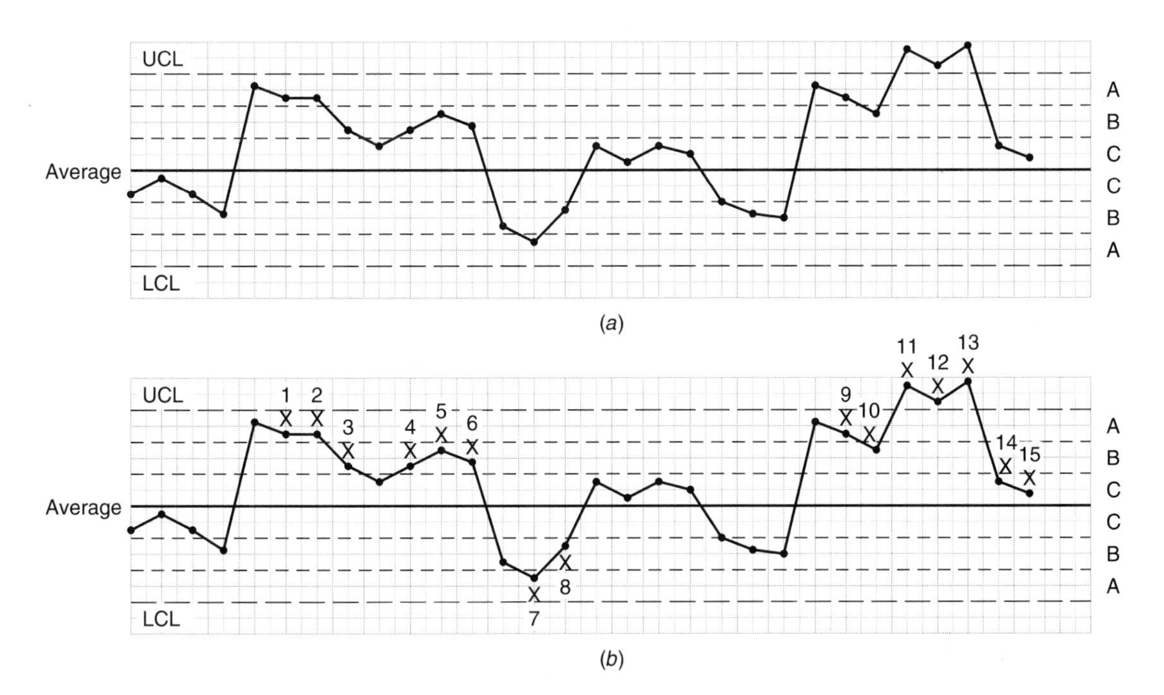

FIGURE 10.39
A chart for Example 10.17.

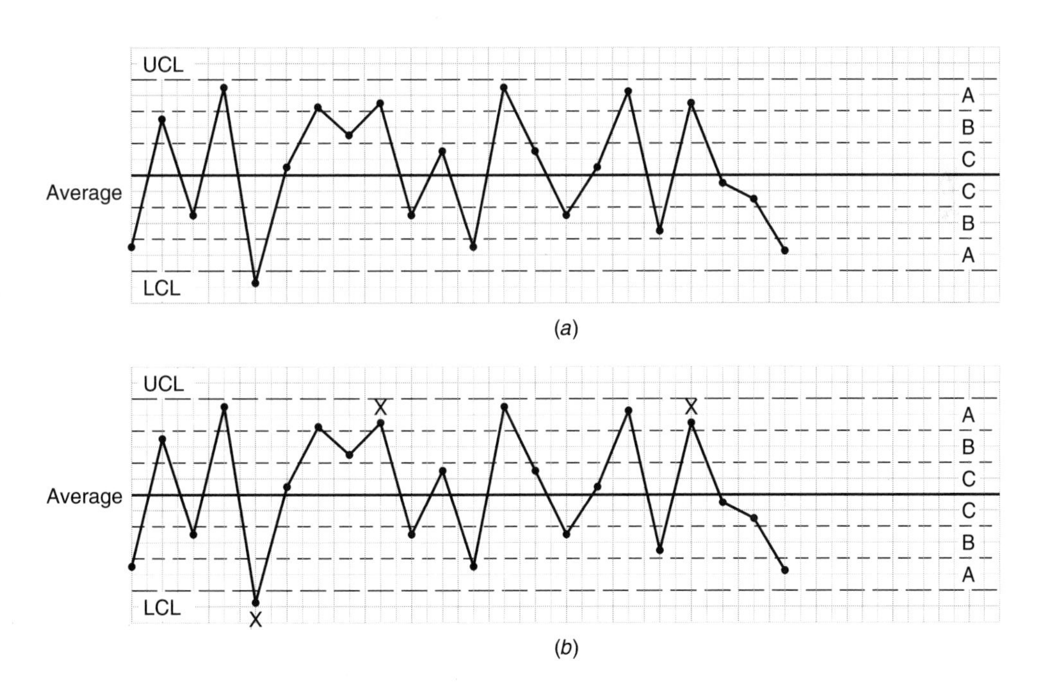

FIGURE 10.40
A chart for Example 10.17.

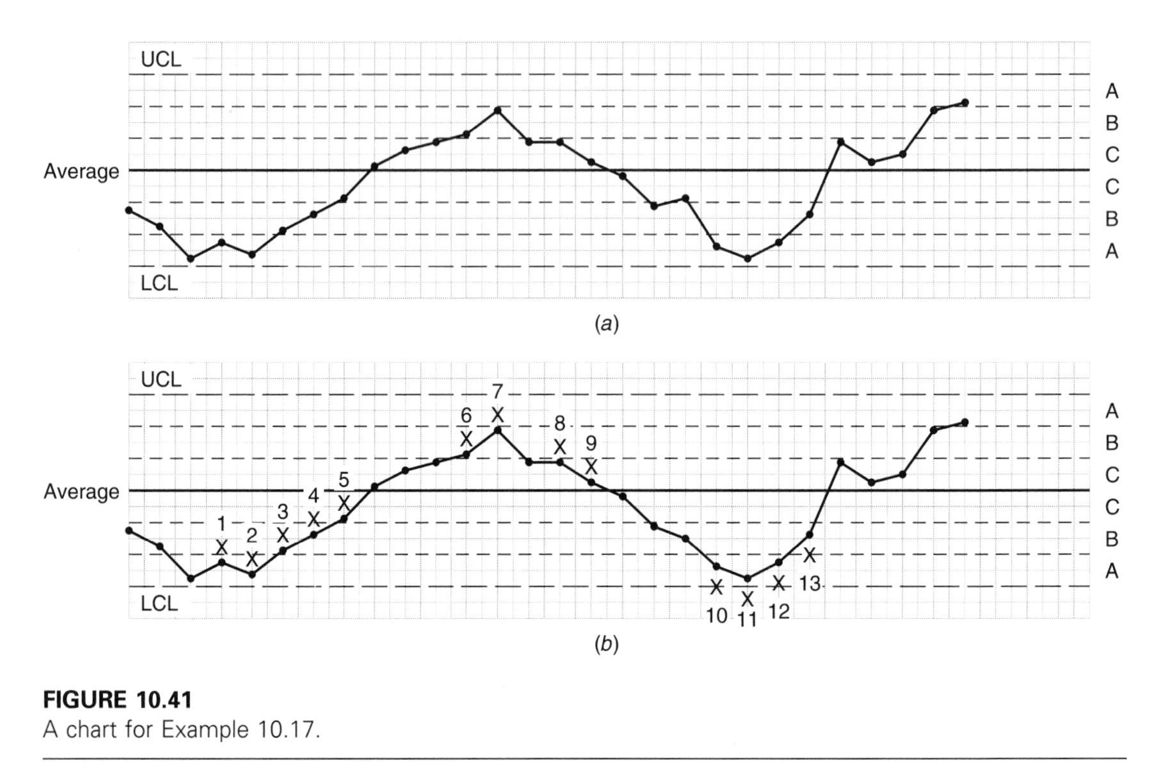

FIGURE 10.41
A chart for Example 10.17.

10.13 USING CONTROL CHART PATTERNS IN PROBLEM SOLVING

These basic control chart patterns that have been identified signal process changes or process troubles. For most patterns, a rule based on the laws of probability dictates the minimum number of points needed to declare an out-of-control situation. Each probability calculation shows that if the process is behaving normally, the chance of getting that specific point pattern is less than 1%, a rare event. The probability calculation leads to two possible conclusions:

1. The process is still behaving normally and a rare event occurred when the point pattern was formed.
2. Something affected the process and caused it to change. It is no longer in statistical control.

The second conclusion is always assumed to be correct, and the process is investigated to find the cause for the out-of-control point pattern. Investigators should be aware, however, that occasionally (in fewer than 1% of the cases) the first conclusion is the correct one. The probability calculation for each point pattern gives the chance of getting a false signal from the control charts.

There are variations in the pattern rules at different companies and sometimes within the same company. If an operation has a history of shifts or runs, the number of necessary points may be decreased to 5 or 6. Another possibility would be to have the operator increase the frequency of the samples if a shift or run of four or five points should occur. Then the existence of a problem would be verified sooner without increasing the possibility of a false alarm due to fewer pattern points.

Sometimes the necessary number of pattern points is linked to the process capability. If the Cpk is close to 2, larger pattern numbers could be used to lessen the chance of a false alarm. The product distribution is far enough from the specification limits so that process changes will not result in out-of-specification pieces. If the Cpk is close to 1, a process change is more likely to result in pieces out of specification. In that case smaller pattern numbers may be used.

This chapter emphasizes individual charts and their interpretations. It is important to realize, however, that on-the-job application of these concepts is usually more complex than simplified textbook examples. Chart analysis provides clues that can be effectively used to solve problems. Each out-of-control pattern has primary causes, but when a direct, easy solution is not apparent, tunnel vision will only frustrate efforts. The problem-solving team should keep the broad picture of the process in mind. Each phase of the process is interconnected with several others, so problem solving often becomes a systematic, upstream search.

When digging for the cause of a tough problem, remember that troubles with the process will always fall into one or more basic categories:

Operators	Machines
Methods	Materials
Tooling	Environment

At each step of the investigation, starting at the point at which the charts indicate trouble, use the chart clues and the six basic categories. Decide which categories apply most directly at that point in the process and whether the out-of-control pattern may indicate trouble in those categories.

Many of the topics in the following chapter come into play during the search for the root cause of problems. Critical at this point is a thorough understanding of the entire process. Teamwork, combined with the various problem-solving procedures at each process point, is needed for an effective search.

EXERCISES

Analyze the control chart in exercises 1–17 and mark the points that are out of control. List three possible causes for each out-of-control chart.

1. Figure 10.42

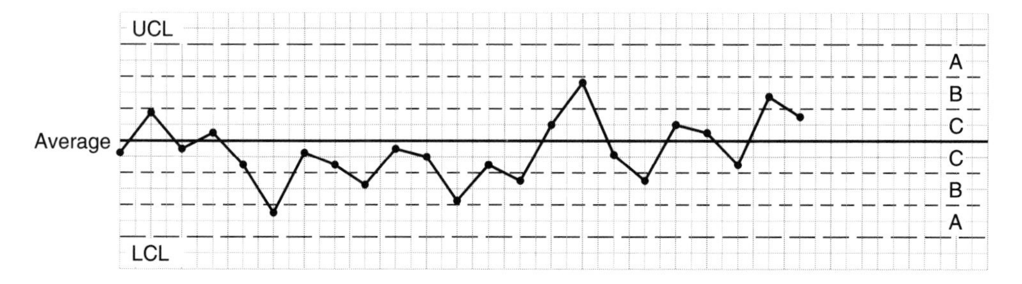

2. Figure 10.43

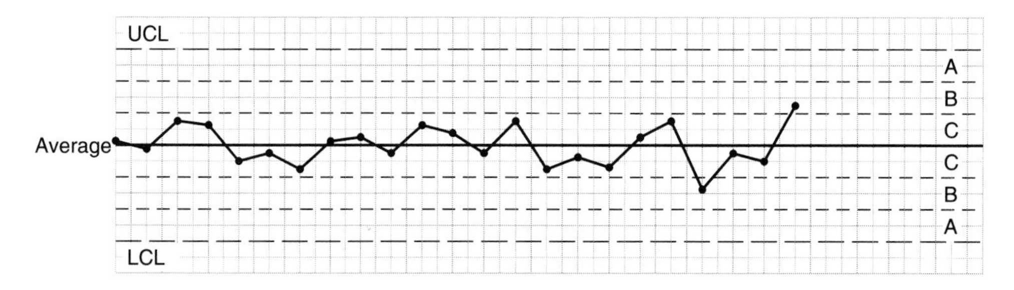

3. Figure 10.44

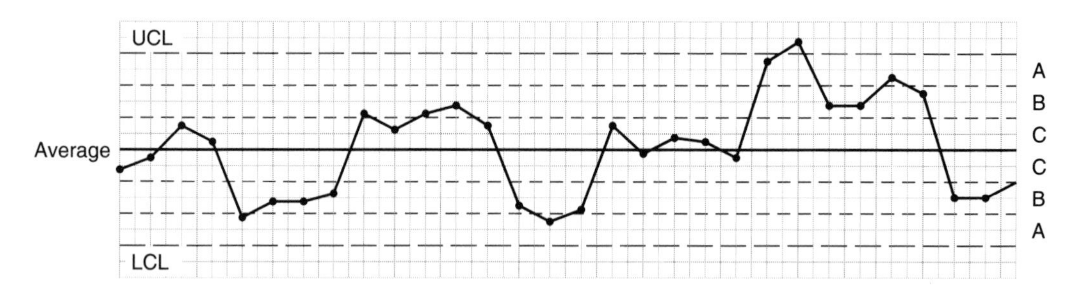

4. Figure 10.45

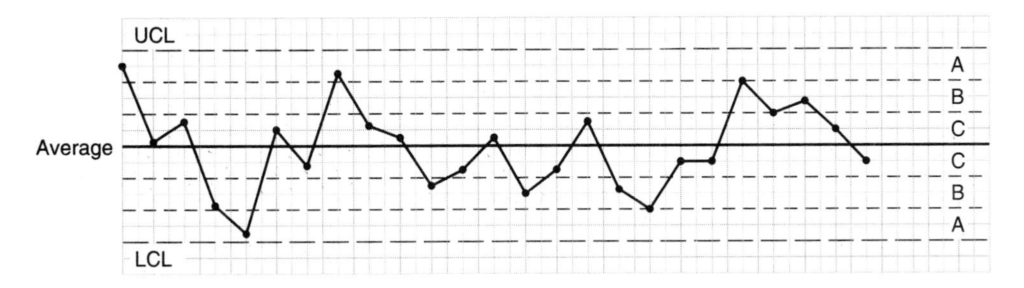

5. Figure 10.46

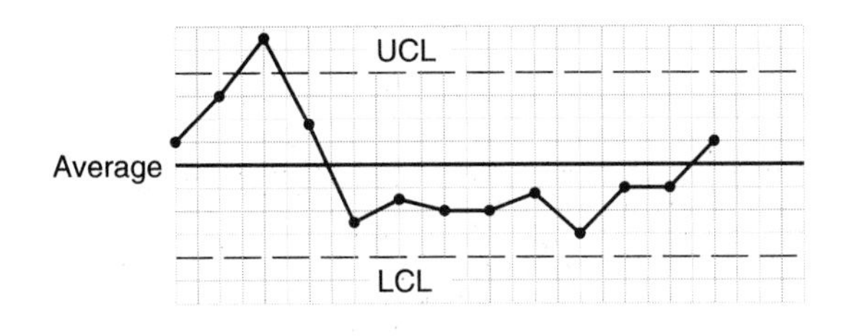

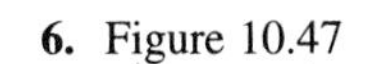
6. Figure 10.47

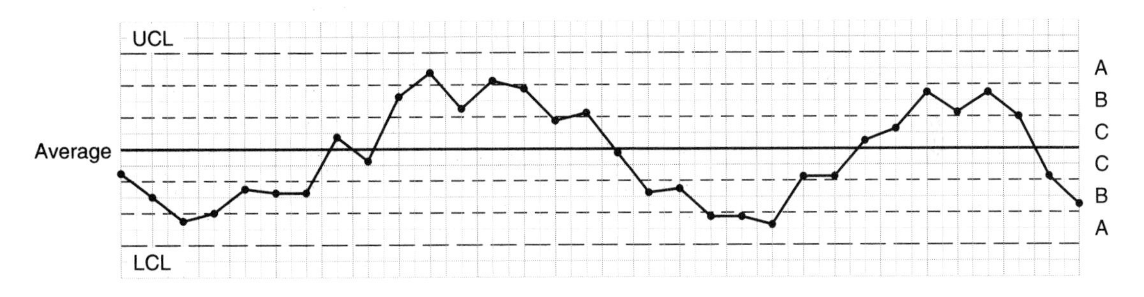

7. Figure 10.48

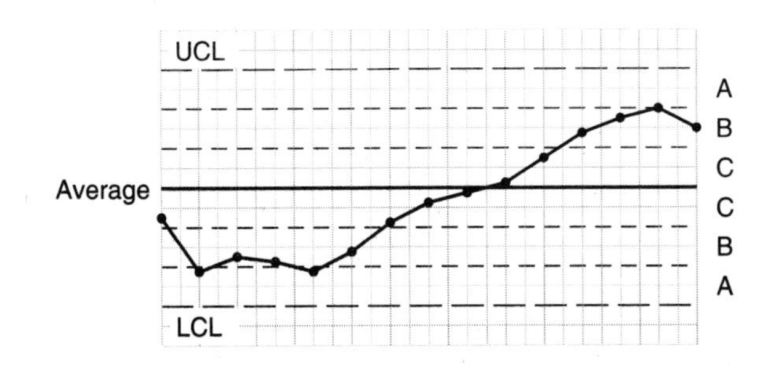

8. Figure 10.49

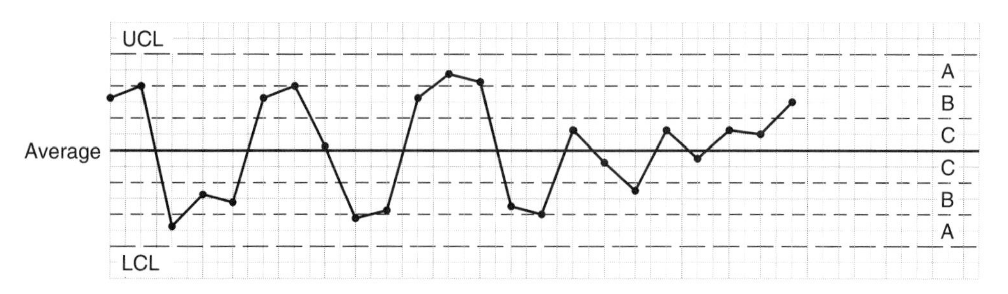

9. Figure 10.50

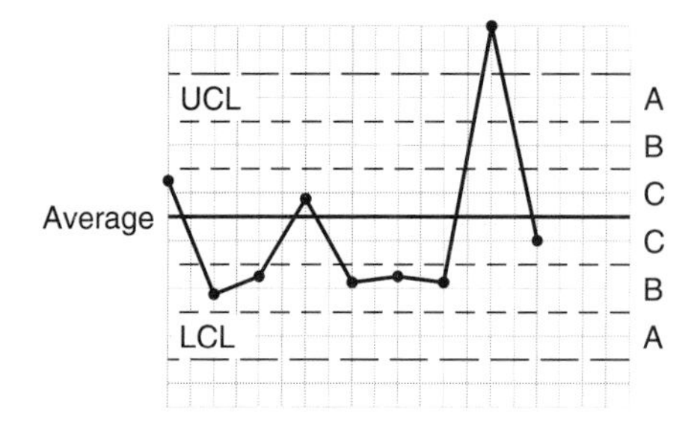

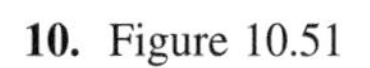

10. Figure 10.51

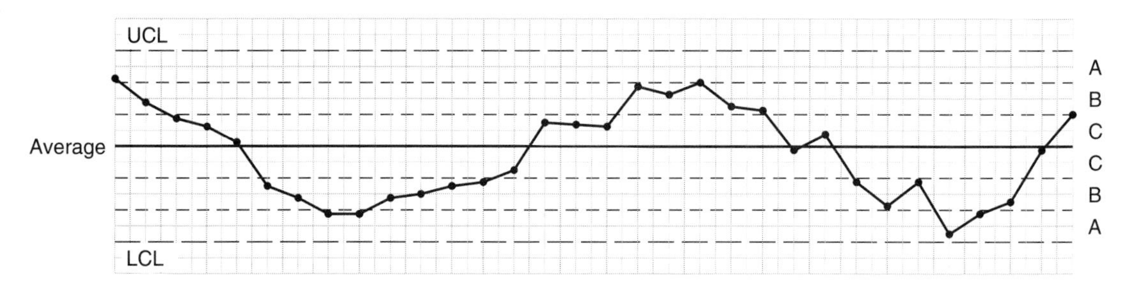

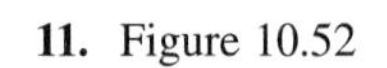

11. Figure 10.52

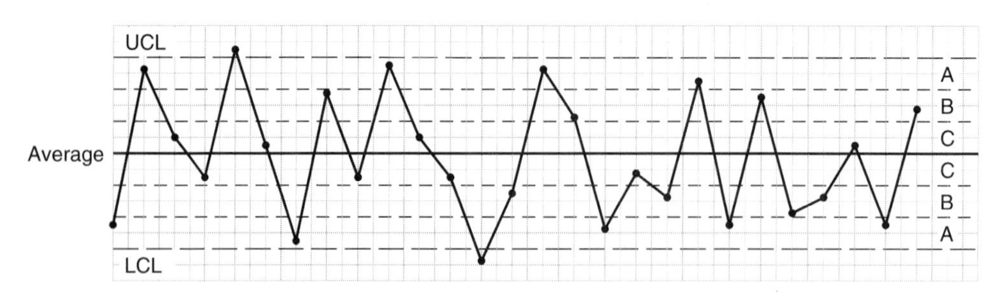

12. Figure 10.53

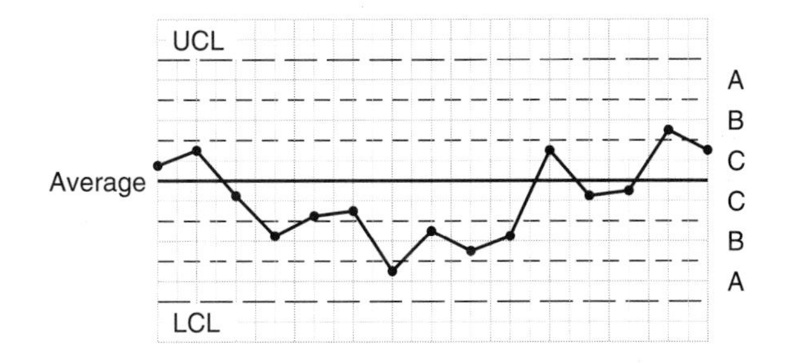

13. Figure 10.54

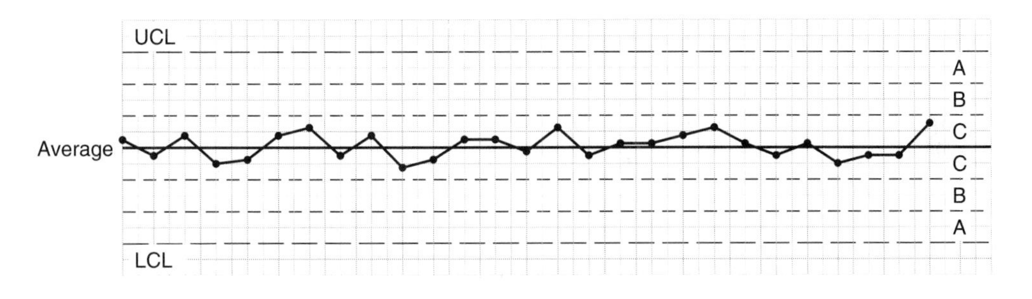

14. Figure 10.55

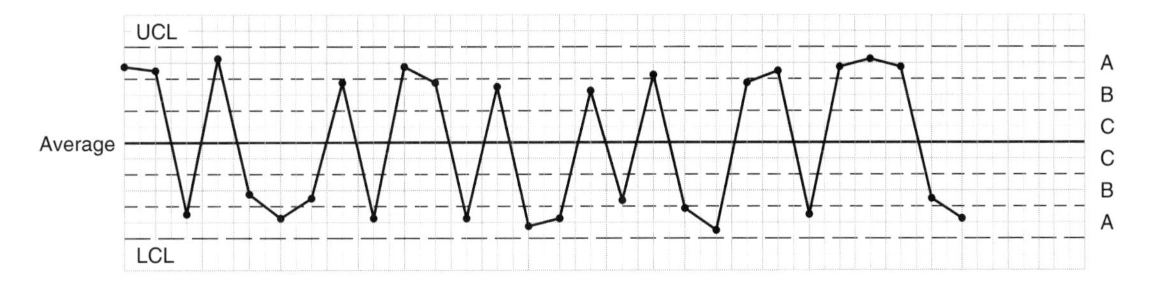

15. Figure 10.56

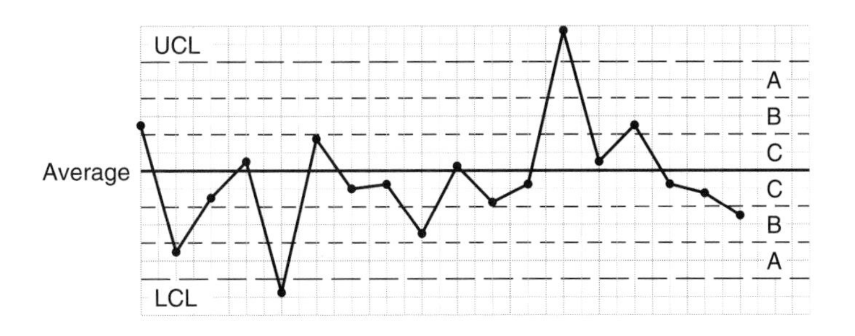

16. Figure 10.57

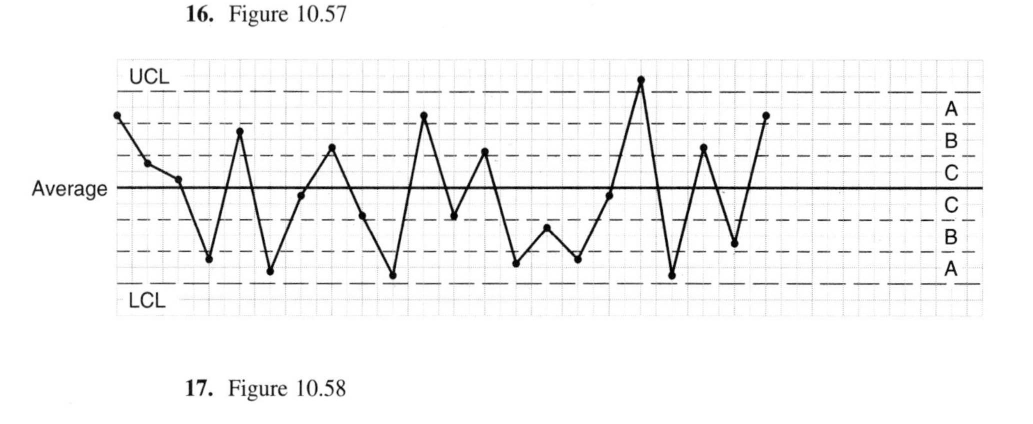

17. Figure 10.58

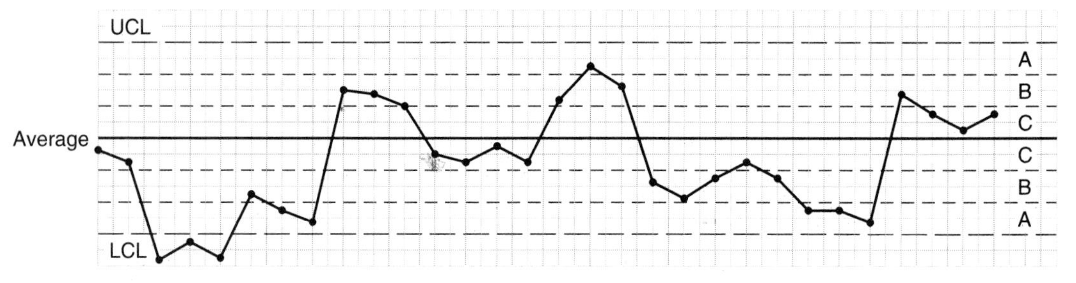

18. Calculate the probability of getting four consecutive points outside the two C zones.

19. Calculate the probability of getting two consecutive points above zone B on the upper half of a control chart.

20. Calculate the probability of getting 13 consecutive points in the two C zones.

21. Calculate the probability of getting six out of seven points above the mean on a control chart.

22. Calculate the probability of getting four consecutive points in a lower C zone *and* the next four consecutive points in the upper B zone.

23. Calculate the probability of getting a point in the lower B zone, the next point in the upper B zone, and the third point back in the lower B zone.

11

PROBLEM SOLVING

OBJECTIVES

- Know the characteristics for problem-solving teams.
- Know the problem-solving sequence.
- Work with the following problem-solving tools: brainstorming, flowcharts, storyboards, cause-and-effect diagrams, Pareto charts, and scatterplots.
- Analyze a problem with histograms.
- Verify incoming quality by comparing histograms and data calculations.
- Apply mistake-proofing concepts.
- Apply problem-solving concepts in the classroom.

Problem-solving methods are used in every stage of quality improvement. A logical, systematic approach to problem solving follows the basic sequence of problem recognition, problem isolation, and problem solution. This chapter investigates various problem-solving tools and methods that can be used effectively in the problem-solving sequence.

11.1 THE PROBLEM-SOLVING SEQUENCE

The sequence follows six steps:

1. Problem recognition
2. Problem definition
3. Problem analysis
4. Choice for action
5. Problem solution
6. Prevention of backsliding

Step 1. Problem recognition. Problem recognition can be as hazy as a feeling that a process can run better or that the situation can somehow be improved. For example, management may believe that the company's market share for a specific product can be increased. Production managers may believe that the operation can run more smoothly and efficiently. Operators may suspect that their phase of the operation can be improved. Problem recognition can also be as obvious as looking at the scrap bin and realizing that some specific part of the process is causing substandard product. Most often, problems are embedded in a process and must be searched out using problem-solving methods.

Some quality improvement programs emphasize that it is management's job to find problems.[1] One approach to this concept is MBWA (management by walking around). Managers at various levels actively seek out problems instead of passively waiting until the problem comes to them. The more comprehensive approach is for management to create an organizational structure that will involve all employees in problem solving. Companies facing today's competition cannot rely on a few people to find and solve problems: Everyone in the organization must be trained to use problem-solving techniques and then shown how their efforts fit into a coordinated, companywide problem-solving structure.

When a company puts quality first in its set of goals and incorporates an ongoing quality process, tools for problem solving are systematically used for problem recognition. Control charts are also helpful in problem recognition. The control chart analysis techniques discussed in Chapter 10 are used for both problem recognition and problem analysis.

Step 2. Problem definition. When a problem is found, it must be clearly defined. A good definitive problem statement is the first step in the problem's solution: It identifies a specific goal so that particular steps can be taken toward achieving that goal. Problems must be carefully analyzed to isolate the root cause. Most problems are complicated by symptoms that hide the root cause of the problem, and time spent trying to eliminate a symptom without attacking the root cause is often time wasted. A good problem definition helps separate symptoms from potential root causes.

Step 3. Problem analysis. Problem analysis involves collecting appropriate data and using those data with the problem-solving tools to suggest various ways of resolving the problem. Control charts and their analyses of potential causes for out-of-control situations are used with flowcharts and cause-and-effect diagrams to track down root causes for problems.

Step 4. Choose appropriate action. The next step in the problem-solving sequence is the choice of the most promising method of solution. There may be several avenues of attack. Broad representation on the problem-solving team pays off at this stage. Each potential course of action must be examined from different perspectives in order to zero in on the best one. Unfortunately, there is no guarantee that the chosen course of action will resolve the problem. The effectiveness of any corrective action should be checked

---

[1]W. Edwards Deming, *Quality, Productivity, and Competitive Position* (Cambridge, MA: Massachusetts Institute of Technology, 1982).

with control charts or other applicable quality tools. If that action didn't work, then another method of solution has to be chosen.

Step 5. Problem solution. The problem solution must be formally recorded along with the problem definition and the pertinent analysis information. It can be important for future reference as well as for the following step.

Step 6. Prevent backsliding. The final phase in the problem-solving sequence is prevention of backsliding. Ensure maintenance of the new methods and techniques that have been incorporated to eliminate the problem.

11.2 TEAMWORK FOR PROBLEM SOLVING

Teamwork is perhaps the most important ingredient embedded in every step of the problem-solving process. The need for total teamwork is not always obvious, but many processes are in a problem situation because teamwork was not used enough when the process was initially created. Total teamwork gives everyone involved in an organization the opportunity to contribute to the resolution of company problems. It is always a challenge for management to organize the problem-solving format in a way that allows and even encourages input from all employees. Involving more people in organized problem-solving efforts increases the likelihood that problems will be eliminated. Process problem-solving teams should be formed with appropriate subteams as needed, depending on the size and complexity of the process.

Problem-solving teams should consist of members from all phases of the process. Supervisors, operators, and personnel from management, purchasing, design, sales, engineering, and quality control should all be involved. Representatives can provide input from their respective points of view for a more complete picture of the problem and for a better chance of solving it. For example, employees at the workstation are often the most knowledgeable about the process and process problems at that point. Given training in statistical process control and permission to track the measurements in their area, they can provide much valuable input to the problem-solving process. Their involvement will help them understand their part of the process more thoroughly, and they will provide positive assistance to the supervisors and engineers who are focusing on the process as a whole. The diverse makeup of the problem-solving team is also important when deciding on a method of problem solution. Both the support and the capability of each department must be carefully considered before the choice is made.

One source of trouble that occurs when a team approach is used in problem solving comes from the different expectations for the team. Management expects the team to be totally goal oriented and to work unselfishly together in achieving the goal. Goal success is total team success. But the importance of individual success is deeply ingrained in all of us (that American individualism) and consequently the team members expect individual recognition along with team success.[2] Too much emphasis on personal success can interfere with teamwork.

---

[2]Karen Bemowski, "What Makes American Teams Tick" (*Quality Progress,* January 1995).

Problem-solving teams have been extremely effective in some companies and a waste of time and effort in others. The success of a team in solving its assigned problem often depends on the degree of internal teamwork. There are some important characteristics of good problem-solving teams[3]:

- The team goal must be clearly understood and accepted by all team members.
- The individual members must know their specific role on the team. It is usually best when they have input on what their role will be. It is also important to ensure that the team work load is evenly distributed among the members.
- Team goals must take precedence over individual goals. The team's success is equally shared by all its members.
- The team must practice good interpersonal communication. Everyone must have an equal chance to express ideas and no disparaging comments should be made about the other members or their ideas. The positive aspects of an idea should always be explored before the negatives.
- The team should choose a leader who can keep the team focused and the communication flowing.
- The team should be following a specific problem-solving sequence.

There have been reports that the failure rate for teams exceeds 50% and that the best chance for team success comes from thorough preparations in forming the team.[4] If the charge to the team is considered too broad, then appropriate data should be gathered that indicates this and a revision should be mutually agreed upon between the team and the manager(s) who initiated the charge. It is also important to identify the "owners" of the process where the problem is (the supervisors and managers) and to discuss with them the charge and potential team action. This will smooth the way for unimpeded action on the problem.

A thorough flowchart of the process that contains the problem should be made. The team members are chosen from different, key process points. They should be good workers and be willing to volunteer to work on the team. The team should be kept to a manageable size of ten or fewer; the number of members depends on the amount of work anticipated. Team members should be given enough leeway in their regular duties to devote the necessary amount of work to their team duties. The team must also receive appropriate training. Training in the use of the needed quality tools immediately comes to mind, but three additional team skills must be developed: communication, trust, and cooperation.[5] Usually the training for these lasts about two days. So, in addition to the previously listed characteristics for problem-solving teams, the formation of the team should include the following:

- Process "owner" involvement.
- A process flowchart.
- Team members chosen from important process points.

---

[3]D. Keith Denton, "Building a Team" (*Quality Progress,* October 1992).

[4,5]James M. Cupello, "The Gentle Art of Chartering a Team" (*Quality Progress,* September 1995).

- Appropriate team training.
- The number of team members ≤ 10.
- Full support for team duties.

Good team leadership is essential for team success. There have been situations in which teams have both a leader to deal with the technical aspects of the charge and a facilitator to ensure that everything runs smoothly. This often leads to problems because it can become unclear as to who the real team leader is. The best approach is to provide the team leader with facilitator training and then the dual leader problem doesn't occur. John Beck and Neil Yeager,[6] directors at the Charter Oak Consulting Group in East Berlin, CT, advocate that the team leader should also receive training in how and when to use the four leadership styles:

- *Style 1 (S1) Directing without dominating.* Take the responsibility to determine what has to be done and assign the tasks to the team members.
- *Style 2 (S2) Problem solving without overinvolvement.* Gather important information from the team members and make decisions with their help.
- *Style 3 (S3) Developing without overaccommodating.* Help the team members complete their assigned responsibilities with suggestions and analysis.
- *Style 4 (S4) Delegating without abdicating responsibilities.* Give the team members the responsibility for completing assigned tasks and for making the related decisions, but maintain enough contact to ensure that the results are satisfactory.

Beck and Yeager stress that "Team leaders have to realize that a team consists of individuals and the group dynamics that surface when these individuals come together as a team . . . they have to know how to lead the group dynamics, but they also need to know how to translate the group effort into individual accountabilities . . . the work ultimately has to be done by individuals who are accountable for their assignments and supported in their efforts . . . the team leader needs to:

- *Start with a team orientation (phase 1).* Form the group by using S1, directing, to tell the team members as much as possible about their mission. Then use S2, problem solving, to focus the group by involving team members in determining the best ways to accomplish the team's objectives. With all of the team members together, be sure everyone is clear about goals and roles.
- *Clarify individual assignments (phase 2).* Use S1, directing, and S4, delegating, to make sure that each team member knows what he or she is empowered to do, what deliverables are expected, and when they are due. Also ask each team member how he or she plans to accomplish the tasks and what he or she needs from the team leader or from the team members to get the work done.
- *Let team members get to work (phase 3).* Use S4, delegating, to give individuals room to accomplish their assignments. Use S3, developing, to check in periodi-

[6]John D. W. Beck and Neil M. Yeager, "How to Prevent Teams From Failing" (*Quality Progress,* March 1996).

cally and offer encouragement. Most important, be responsive when team members need support.

- *Make time for team problem solving (phase 4).* Bring the team together regularly to keep team members informed of progress, to identify and solve problems, and to coordinate each person's efforts. When decisions need to be made, use S2, problem solving, to get input for decisions that the team leader needs to make or use S3, developing, to help team members get input for decisions that they have agreed to make. Always encourage the team to look for ways to work smarter."

11.3 BRAINSTORMING

Brainstorming, developed in the 1930s, is a team technique that generates a large number of ideas on a topic. Its use by a problem-solving team ensures that everyone becomes involved, and with the diverse points of view of the different team members, nothing of importance should be overlooked. There are seven rules for brainstorming:

Step 1. Choose a leader from within the group. The leader is responsible for the following tasks:

- Keeping the group focused on the topic
- Demanding adherence to the brainstorming rules
- Directing the action and individual input
- Starting and finishing the session
- Making sure that everyone clearly understands the topic
- Allowing time for participants to write down their initial ideas

Step 2. Choose an official recorder from within the group. All ideas are to be recorded. The ideas should be listed for group use during the session and provided as minutes in preparation for the next session. Two recorders can be used; one for the official list and one for the action list during the session.

Step 3. As ideas are offered, no comments or disparaging remarks are allowed. Nothing must dampen the spirit of participation or discourage team members from offering ideas on the topic. Quantity of ideas is more important than quality at this point. Occasionally, ideas that sound crazy at first mention are later found to have merit.

Step 4. Call on each person in turn for an idea. Insist on a response at each person's turn. Participants who have no ideas to contribute must say "pass." There are three good reasons for the "pass" routine: First, it forces shy people to "break the ice," which makes it easier for them eventually to contribute. Second, it keeps the pressure on for a contributing response because nobody wants to pass all the time. Third, when each person, in turn, passes, the meeting should end.

Step 5. Provide writing material so that participants can jot down ideas as they think of them. If team members try to remember a particular thought until it is their turn, they may forget their ideas as they mentally react to other ideas presented. The real power of a brainstorming session is in the "hitchhiking" of ideas. One person's idea presented in the group can inspire other ideas. Write the ideas that are presented on a blackboard or screen so that group members can still build on others' ideas after that specific topic has passed.

Step 6. Regulate blurts! Should group members be allowed to blurt out their ideas? Yes and no. If they are allowed to interject ideas indiscriminately, not everyone will contribute. Blurting has a detrimental effect on the more reticent members of the group. This opposes the fourth rule, in which everyone is encouraged to offer ideas. On the other hand, if members must wait their turn to present ideas, the atmosphere becomes a little more subdued; there is a lack of energizing excitement with a resulting loss of creativity. Also, if they are given a chance to think twice about an idea while waiting for their turn, participants may judge their ideas as not worth mentioning, thus defeating the quantity-of-ideas aspect of brainstorming. A good compromise is to proceed systematically in turn but to allow blurts that hitchhike on an idea just presented. Be sure to allow the recorder to catch up before going on to the next team member.

Step 7. Follow brainstorming sessions with evaluating sessions in which the general ideas are sorted by priority. Eliminate no ideas until the problem is solved. Ideas that are considered unlikely or lacking in promise may be set aside, but not erased. Reconsideration from another point of view or further hitchhiking may occur.

Listing group priorities can be done by a weighted voting process. Provide team members with a list of the ideas and instruct them to place a number on each one:

0 This idea will not help in solving the problem.

1 There is a chance that this idea will be useful.

2 This idea should be of some help.

3 This idea will definitely help in solving the problem.

Add the votes. The totals for each idea provide a ranking.

EXAMPLE 11.1 Seven ideas are to be ranked. The votes of the six-member team are recorded in Table 11.1.

Solution

Ties may be given the same rank, as shown in the table with ideas 2 and 7. They may also be arbitrarily ranked in the increasing order, or a show-of-hands vote may establish which is ranked before the other.

TABLE 11.1
Weighted votes for each idea

Idea	Votes	Total	Rank
1	0, 1, 0, 0, 1, 2	4	6
2	2, 3, 1, 3, 2, 2	13	2
3	2, 0, 2, 3, 2, 3	12	3
4	3, 3, 2, 3, 3, 3	17	1
5	2, 1, 3, 2, 1, 2	11	4
6	1, 2, 0, 2, 2, 2	9	5
7	3, 2, 2, 1, 3, 2	13	2

11.4 USING PROBLEM-SOLVING TOOLS

Problem-solving tools involve ways of collecting, organizing, and analyzing information about a process. These tools were introduced in the previous chapters and include the following:

- Brainstorming
- Flowcharts
- Storyboards
- Cause-and-effect diagrams
- Pareto charts
- Graphs, charts, and control charts
- Scatterplots
- Histograms
- Normal curve and capability analysis

When SPC is used in a company, data are systematically gathered at various process points and organized in control charts and histograms. When the charts go out of control and/or the histograms do not have their expected shape, a problem is identified. If the problem can't be easily solved, a problem-solving team is formed to deal with it. More information must be gathered and organized; thus, one of the first tools to be utilized is the process flow chart. The team makes the flow chart and brainstorms for possible root causes of the problem. If that isn't successful, the ideas from the brainstorming sessions are organized in a cause-and-effect diagram. The brainstorming with the cause-and-effect diagram may lead to the root cause or it may show a need for more information, which may be gathered using Pareto charts or scatterplots. Eventually, the team decides on a course of action and the results are checked using control charts, histograms, and/or Pareto charts. A capability analysis is one of the final checks on the solution and indicates whether or not there is more immediate work to be done on the problem.

All processes communicate to the analyst through the data that are gathered. Those data have to be organized so the message is clear; charts and graphs are the tools used to do this. The analyst must learn how to interpret charts and graphs to recognize the existence of a problem and then use the shapes and patterns on those charts and graphs to help solve the problem.

Every chart and graph has an expected shape or pattern. There is also a normal amount of naturally occurring variation from the expected form, so differentiating between normal pattern variation and a trouble-indicating pattern is a learned skill. Many times, the real skill is in spotting the trouble early, before measurements are out of specification.

Histograms for Problem Solving

Histograms are important for spotting trouble with a process, but they are also useful as a problem-solving tool. Measurement data are usually expected to follow a normal distribution with the majority of measurements very close to the average value. As the distance from the average value increases, the number of measurements decreases. The distribution of the data should be symmetric about the mean with approximately the same number of measurements on the high side and the low side.

Some variation is expected within the general pattern of the normal histogram. When a tally histogram is made from a set of measurements, the distribution may look strange initially, but as more data are tallied, the shape of the distribution is expected to look normal. When data do not follow the normal pattern, it is suspected that something out of the ordinary is causing that pattern change. This is the first step in problem recognition using histograms.

Sometimes several histograms are needed for problem solving for the same set of data. By looking at the data in specific groups, more information is gained that can lead to the problem's solution. This concept is demonstrated in the following example.

EXAMPLE 11.2

The control charts and the data values for a run of top adapters are shown in Figures 11.1 and 11.2. The charts were maintained by two machine operators on different shifts. The calculations for the averages are shown at the top of Figure 11.1 using all 40 data values from both charts. Is there a problem with this part of the process?

Solution

The operators marked several out-of-control situations on the control charts. One point is out of control on the range chart, and several are out of control on the $\bar{x}$ chart. Shifts on the $\bar{x}$ chart seem to be the main problem. One major problem that showed up on the second shift on 7/31 is a slow reaction to a trouble indicator. Also, the operator failed to mark the 7:30 P.M. reading out of control (a two-of-three freak pattern). Consequently, the majority of the pieces made during that shift must have been out of specification. Another major problem is that the operators miscalculated the control limits. The narrower control limits show many points out of control and give an earlier warning of trouble with the process. Histograms will be used for further problem analysis and for planning solution steps.

The histogram of all the data values could have a distribution that is crowded right with a longer tail section to the left, as shown in Figure 11.3. The statistical terminology for this shape is "skewed left."

$\bar{R} = \frac{\Sigma R}{k} = \frac{.0175}{40} = .00044$

$\bar{\bar{x}} = \frac{\Sigma \bar{x}}{k} = \frac{8.560}{40} = .214$

$UCL_R = D_4 \times \bar{R} = 2.114 \times .00044 = .0009$

$A_2 \times \bar{R} = .577 \times .00044 = .00025$

$UCL_{\bar{x}} = \bar{\bar{x}} + A_2 \bar{R} = .214 + .00025 = .21425$

$LCL_{\bar{x}} = \bar{\bar{x}} - A_2 \bar{R} = .214 - .00025 = .21375$

Constants

n	A_2	$\tilde{A}_2$	D_4	d_2
3	1.023	1.19	2.574	1.693
4	0.729	—	2.282	2.059
5	0.577	.69	2.114	2.326

Time	0745	0945	1120	1305	1350	1445	4:45	5:45	7:00	8:00	9:30	10:30	9:30	10:30	11:30	12:30	1:15	2:00	2:45	5:00	6:00	7:00	8:00	9:30	10:30
Date/ Shift	7/29	#1					7/29	#2					7/30	#1						7/30	#2				
Top Adaptor	.2134	.215	.214	.2138	.215	.2145	.214	.214	.215	.214	.214	.214	.2146	.2146	.2144	.2146	.2146	.2147	.2149	.215	.215	.215	.215	.215	.215
Machine #15	.2136	.215	.2138	.2143	.214	.214	.215	.2145	.214	.214	.214	.214	.2146	.2146	.2146	.2147	.2148	.2147	.2147	.215	.215	.215	.215	.215	.215
1" Mics	.2138	.2153	.2136	.2143	.214	.2147	.2145	.2145	.215	.214	.214	.214	.2145	.2146	.2146	.2148	.2147	.2147	.2148	.215	.2145	.215	.215	.215	.215
.001 units { +.001	.2136	.2134	.214	.214	.2142	.215	.214	.2145	.2145	.214	.214	.214	.2145	.2146	.2146	.2147	.2146	.2147	.2147	.215	.215	.215	.215	.215	.215
Spec: .215 { −.003	.2147	.214	.214	.2136	.2145	.2143	.214	.2145	.214	.214	.2135	.214	.2146	.2145	.2147	.2146	.2148	.2147	.2147	.215	.2148	.215	.215	.215	.215
$\bar{x}$	.2138	.2145	.2139	.214	.2143	.2145	.2143	.2144	.2145	.214	.2139	.214	.2146	.2146	.2146	.2147	.2147	.2147	.2148	.215	.2149	.215	.215	.215	.215
R	.0013	.0019	.0004	.0007	.001	.001	.001	.0005	.001	.000	.0005	.000	.0001	.0001	.0003	.0002	.0002	0	.0002	0	.0005	0	0	0	0

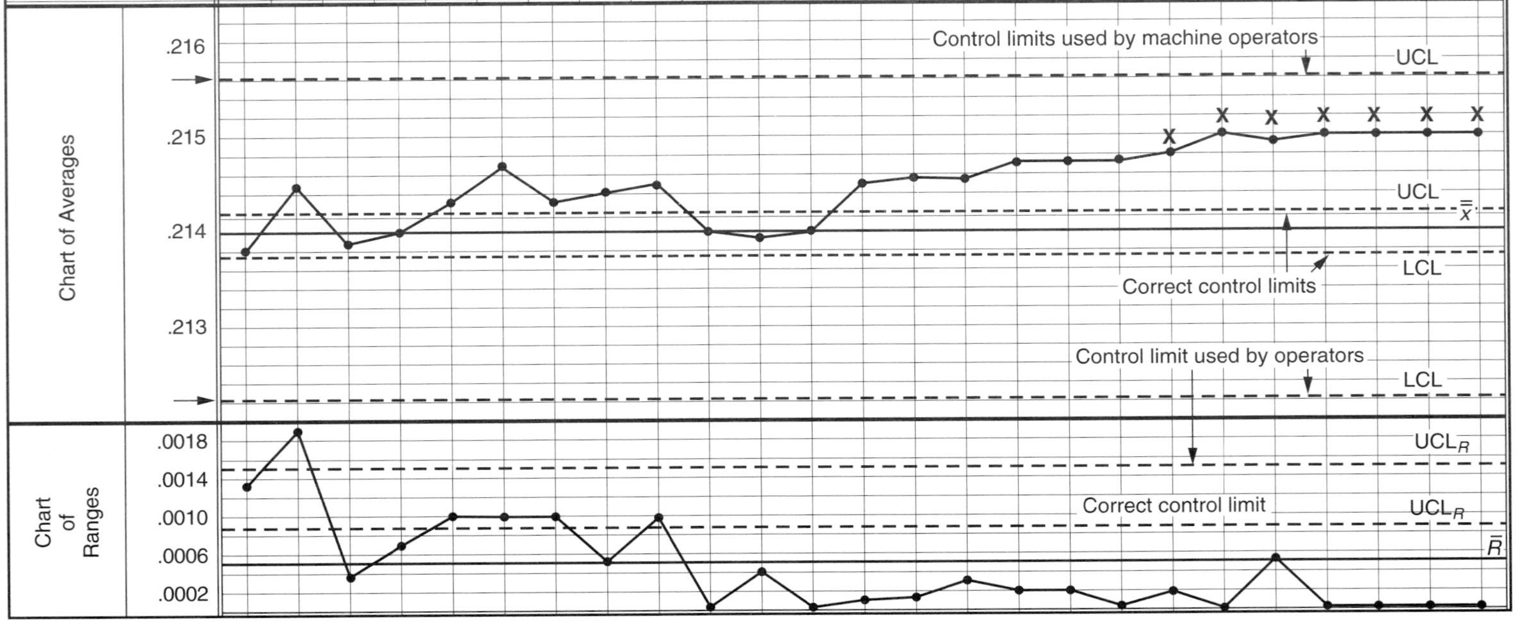

FIGURE 11.1
The $\bar{x}$ and R chart for top adapters, 7/29 to 7/30.

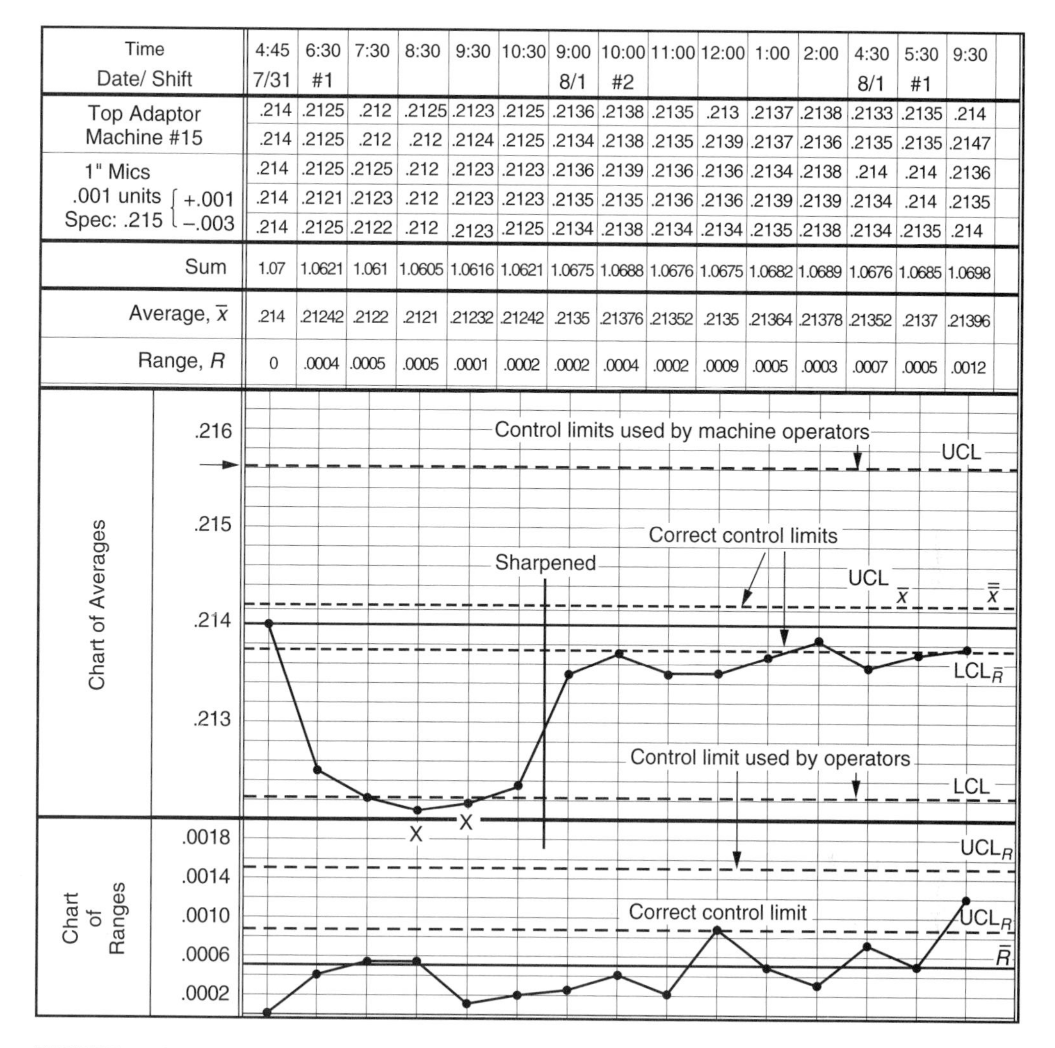

Time	4:45	6:30	7:30	8:30	9:30	10:30	9:00	10:00	11:00	12:00	1:00	2:00	4:30	5:30	9:30
Date/ Shift	7/31	#1					8/1	#2					8/1	#1	
Top Adaptor	.214	.2125	.212	.2125	.2123	.2125	.2136	.2138	.2135	.213	.2137	.2138	.2133	.2135	.214
Machine #15	.214	.2125	.212	.212	.2124	.2125	.2134	.2138	.2135	.2139	.2137	.2136	.2135	.2135	.2147
1" Mics	.214	.2125	.2125	.212	.2123	.2123	.2136	.2139	.2136	.2136	.2134	.2138	.214	.214	.2136
.001 units +.001	.214	.2121	.2123	.212	.2123	.2123	.2135	.2135	.2136	.2136	.2139	.2139	.2134	.214	.2135
Spec: .215 −.003	.214	.2125	.2122	.212	.2123	.2125	.2134	.2138	.2134	.2134	.2135	.2138	.2134	.2135	.214
Sum	1.07	1.0621	1.061	1.0605	1.0616	1.0621	1.0675	1.0688	1.0676	1.0675	1.0682	1.0689	1.0676	1.0685	1.0698
Average, $\bar{x}$	.214	.21242	.2122	.2121	.21232	.21242	.2135	.21376	.21352	.2135	.21364	.21378	.21352	.2137	.21396
Range, R	0	.0004	.0005	.0005	.0001	.0002	.0002	.0004	.0002	.0009	.0005	.0003	.0007	.0005	.0012

FIGURE 11.2
The $\bar{x}$ and R chart for top adapters, 7/31 to 8/1.

The reason for a skewed shape instead of a symmetrical shape is the asymmetric tolerance: $.215^{+.001}_{-.003}$. The target value is .215, but values can be as high as .216 or as low as .212. The distribution of measurements is usually independent of the tolerance, but tolerances can be changed when management considers it permissible to accept a specific nonnormal distribution pattern. In this case, a first analysis may indicate either a symmetric or a skewed pattern. The control chart indicates that the average of all the data is .214, so this machine may actually be working on a symmetric distribution of .214 ± .002. This may indicate a problem. Are the operators trying to use .215 or .214 as a target measurement? Figure 11.4, the histogram of all the data, seems to indicate that the skewed distribution is the correct one for this operation.

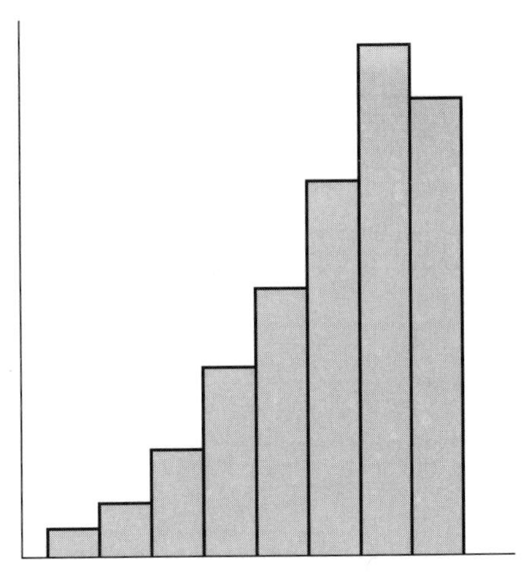

FIGURE 11.3
A histogram with a skewed-left pattern.

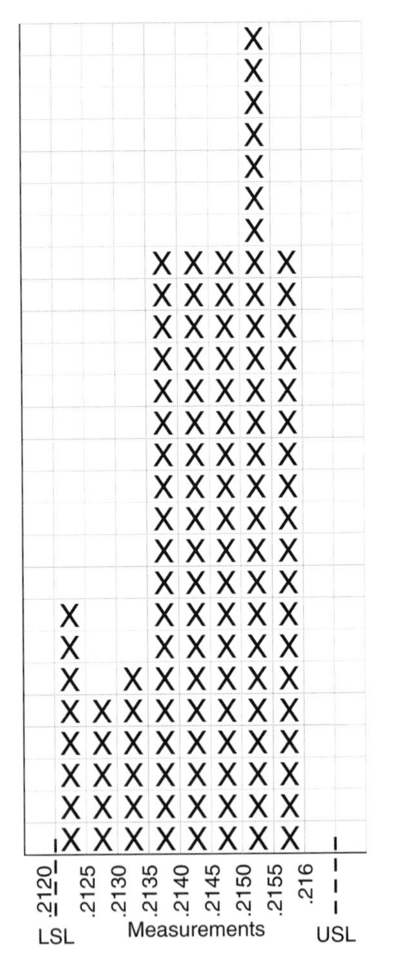

FIGURE 11.4
The histogram for the top adapters in Example 11.2.

The shape of the histogram in Figure 11.4 brings two problems to light. First, the low end of Figure 11.4 does not match the shape of Figure 11.3, which may indicate that some out-of-specification pieces are being forced onto the acceptable range. Second, the histogram shows excessive variation in measurements because virtually all of the tolerance is used and the section in which the bulk of the data occurs (.2135 to .2155) is too wide.

As the work with the normal curve in Chapter 5 showed, the tail sections of the curve predict the percentage of product that is out of specification. Area under the curve corresponds to the percentage of product in that section, a concept that applies to many statistical curves. A statistical curve fitted to the histogram, demonstrated in Figure 11.5, suggests that this machine is regularly producing top adapters that are out of specification. The area in the tail section can be estimated by counting squares. There are about 110 squares under the curve and about 8 squares beyond the specification limits, so approximately 8/110, or 7% of the top adapters are out of specification. This is a very rough esti-

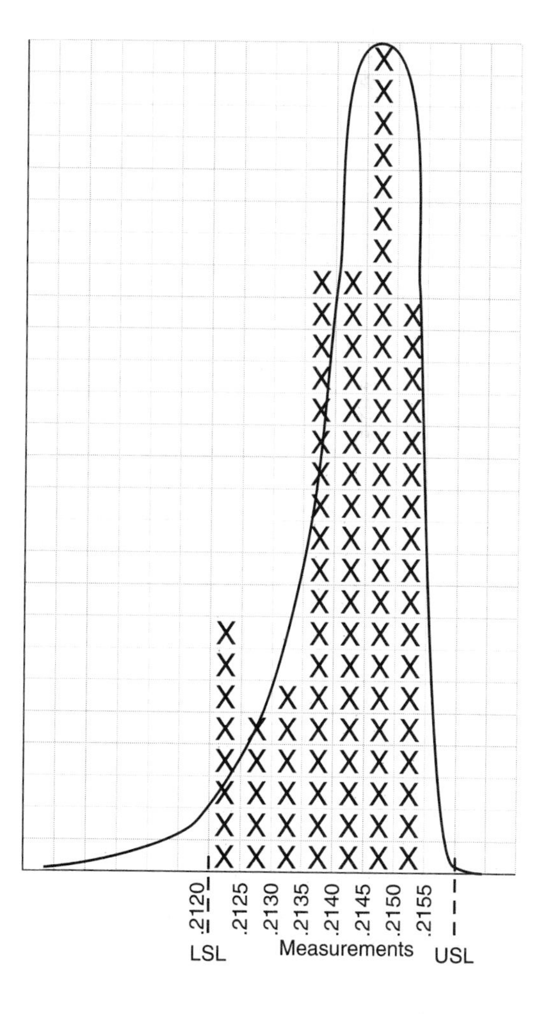

FIGURE 11.5
A skewed-left statistical curve is fitted to the data.

mate, but it is a trouble indicator. As the histogram in Figure 11.4 was being developed from the data, the measurements seemed somewhat streaky. This information may help solve the problem.

The process capability is excellent:

$$s = \frac{\overline{R}}{d_2}$$

$$= \frac{.00044}{2.326}$$

$$= .000189$$

$$6s = 6 \times .000189$$

$$= .001135$$

$$\text{PCR} = \frac{6s}{\text{tolerance}}$$

$$= \frac{.001135}{.004}$$

$$= .28$$

The process uses only 28% of the tolerance. However, problems of excessive variation caused by shifts and out-of-specification products have been identified. The cause of the problems may be attributed to the machine, the operators, or the gauges. The steaky behavior of the data implies that the measurements are grouped somehow. Another view of the data seems warranted.

Figure 11.6 shows histograms of the data separated by operators. The histograms indicate that there could be a problem with the operators. The top adapters produced by the second-shift operator have more variability than those produced by the first-shift operator. Again, the data seemed to come in streaks as these histograms were developed; several sets of successive measurements were the same. Also, the pattern for both operators indicates either a gauge problem with the alternating high and low columns or a mixture of different distributions. The first operator's histogram is bimodal, with two high points, or modes. The second operator's histogram is trimodal, with three high columns. Two or more modes can indicate a mix of two or more distributions. There does appear to be a different distribution of measurements for the second-shift operator on 7/31 that was partially corrected by the tool sharpening.

The data are again separated by operator in Figures 11.7 and 11.8.

The first-shift operator, in Figure 11.7, has more variability on the first day than on the other two days. The control chart indicates an initial setup on that first day, which may account for that difference. The second-shift operator exhibits more trouble with variability, but only on 7/31 does the distribution of measurements have a trouble-indicating pattern. The note on the control chart indicates that a tool was sharpened, and the measurements move back toward the center line after that. The out-of-control X's on the control chart show that the measurements are still too low.

Another problem shows up by comparing histograms on the same-day shifts. Comparing the first shift to the second on 7/29, 7/30, and 8/1 shows a consistent shift to the right each day. This indicates a possible gauge problem, with the second-shift operator reading a value higher than the first-shift operator. This possibility may be tested by running a gauge capability study with the two operators. Another possible cause of the histogram's shift may be some kind of machine creep, in which the measurements gradually get larger as the day wears on. The case does not seem to warrant

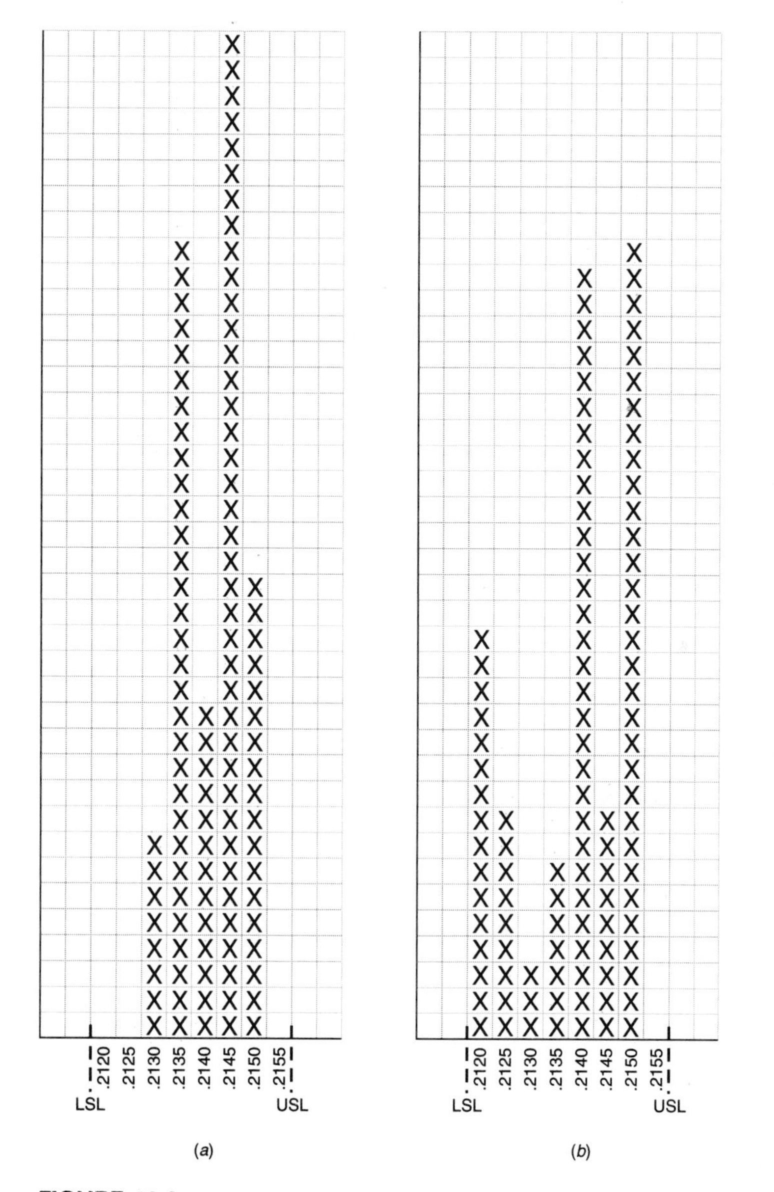

FIGURE 11.6
Data separated by operators: *(a)* first-shift operator; *(b)* second-shift operator.

such an option because there does not appear to be a steady increase in measurements per shift.

The measurements also suggest one more problem. The first shift operator is estimating the measurements to the nearest .0001, whereas the second-shift operator is only occasionally giving them to the nearest .0005. The data can be better interpreted with the more precise measurements. The rule of 10 states that the measuring unit should be one-tenth of the tolerance, or about .0005.

Finally, a major contributor to the total variation is some kind of a day-to-day shift in the measurements. This can be seen by comparing each day's result for the first-shift operator in Figure 11.7 and again for the second-shift operator in Figure 11.8. This may

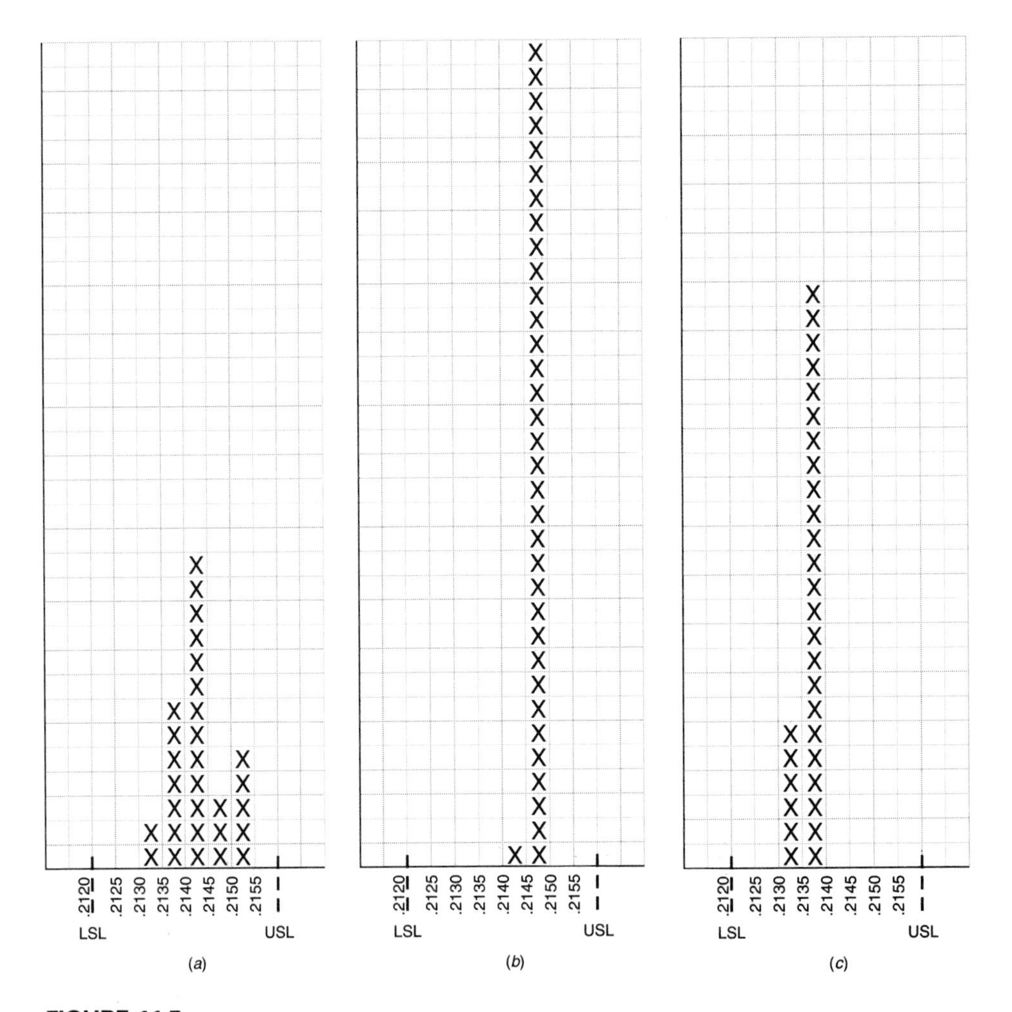

FIGURE 11.7

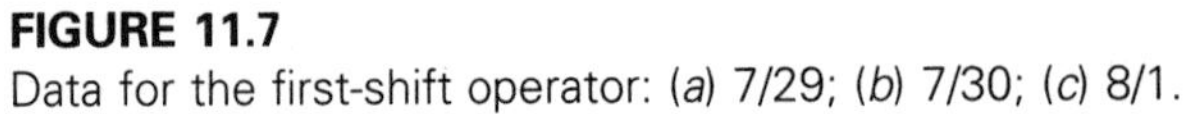
Data for the first-shift operator: (*a*) 7/29; (*b*) 7/30; (*c*) 8/1.

be caused by the day's setup procedure, by something within the machine, or by a gauge problem.

The problem analysis and steps toward solution have been organized in Table 11.2.

Another illustration of the use of histograms for problem analysis is in materials acceptance. Incoming inspection plans are changing drastically. The quality demands on incoming parts and materials are increasing to a point at which the standard attribute

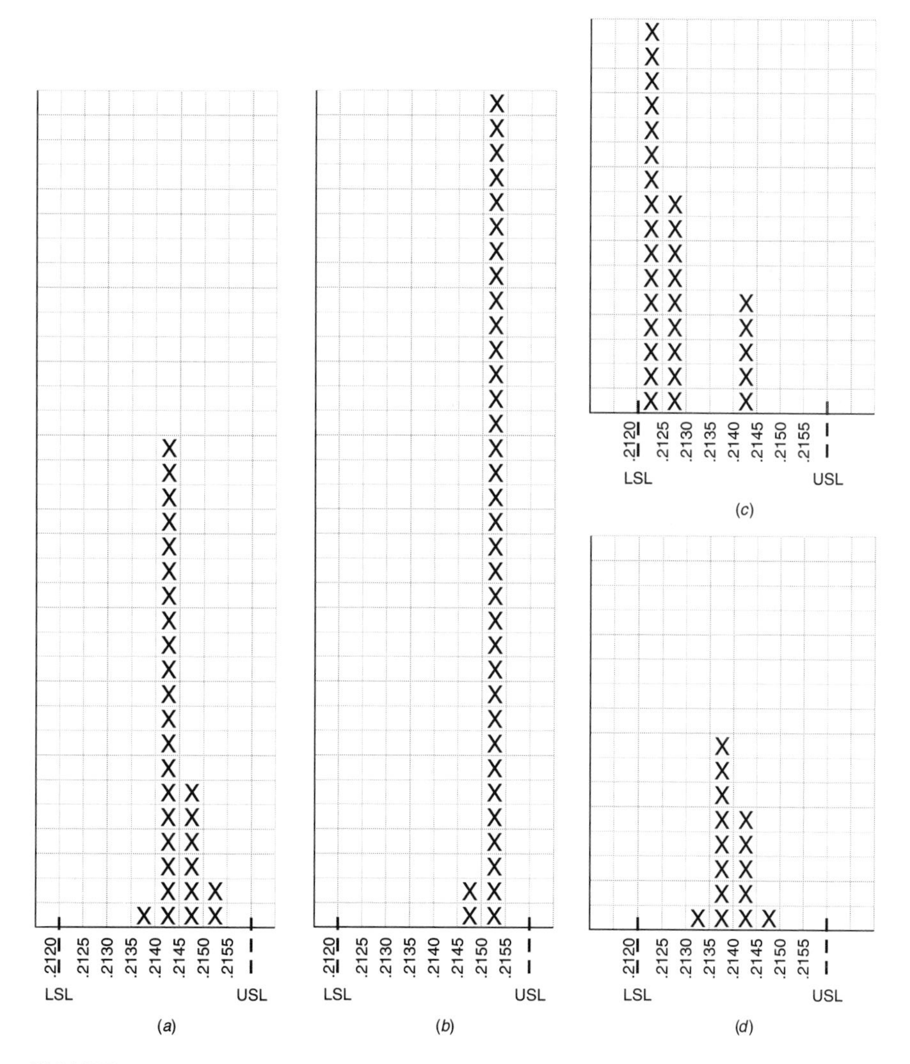

FIGURE 11.8

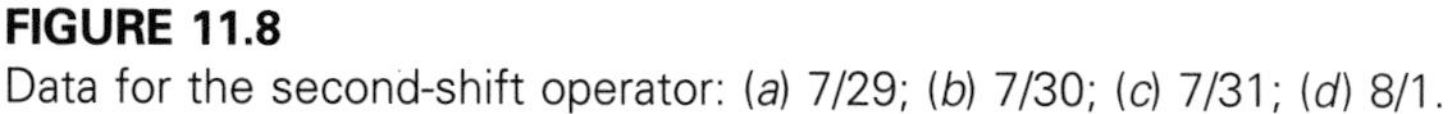
Data for the second-shift operator: (*a*) 7/29; (*b*) 7/30; (*c*) 7/31; (*d*) 8/1.

TABLE 11.2

Trouble Indicators	Source
1. The control chart is out of control.	Control chart
2. The operators may be aiming for a target of .214 instead of .215	Control chart
3. The operators failed to mark all the out-of-control points and miscalculated the control limits.	Control chart
4. Are out-of-specification pieces being accepted on the low end?	Histogram Figures 11.4 and 11.5
5. There is excessive variation.	Histogram Figure 11.4
6. An estimated 7 percent is out of specification.	Histogram Figure 11.5
7. Is there an operator problem? One shows more product variability than the other.	Histogram Figure 11.6
8. There could be a gauge problem.	Histogram Figures 11.6, 11.7, and 11.8
9. There may be a mixture of different distributions.	Histogram Figure 11.6
10. The first operator had more variability on the first day.	Histogram Figure 11.7
11. The second operator had a trouble distribution on 7/31.	Control chart, histogram Figure 11.8
12. The second-shift operator is measuring with less precision.	Data on control chart
13. A day-to-day pattern shift is adding to the product variability.	Histogram Figures 11.7 and 11.8
14. There is a slow reaction to trouble indicators.	Control chart

Problem Number	Action
1, 2, 5, 6, 9, 10	Check the setup, startup, and adjustment procedures.
4, 6, 11	Double-check the output from 7/31.
3, 4, 7, 11, 12	Check on the second-shift operator with respect to training, experience, ability, work habits, and gauge reading.
5, 8, 9, 12, 13	Do a gauge capability study with the two operators.
1, 3, 11, 14	Check the operating procedures. The operator should check the machine log and anticipate when sharpening or adjustment is necessary and then take action at the first sign of trouble. Check on the SPC training for the operators and their supervisors.

inspection process cannot detect unacceptable quality levels unless the sample size is increased drastically. Consequently, quality assurance is becoming a way of life for vendors. Companies with assembly operations rely on vendors for the various bits and pieces that blend together for their final products. The final products are facing world competition, and quality must be inherent.

Verifying Vendor Quality

Companies are now demanding proof of quality from their vendors. They can no longer rely on incoming inspection alone: It is too costly and allows too much chance for accepting poor-quality parts. A shipment that has been rejected because the inspection sample has an unacceptable number of defectives can often be reshipped as is, with perhaps a little reorganization, and stands a good chance of being accepted the second time.

The proof of quality that vendors provide includes copies of the control charts from their shop floor; statements giving mean, standard deviation, and capability information on their product's measurements; and histograms depicting their product's measurement distribution. Vendors are rated according to their own record of quality. A combination of that record and proof of quality is replacing the customary incoming inspection. Receiving companies still verify the vendor claims by using histograms and calculations with sample measurements. Once a vendor is certified, however, the receiving company can decrease its incoming inspection for that vendor to just an occasional audit.

When a lot is chosen for a verification sample, the sample size should be large enough to create a histogram of the measurements that will imply the distribution of the population of measurements: A sample size between 100 and 200 should be sufficient. Measurements are recorded for all pieces in the *random* sample, and a histogram is made for those measurements. That histogram should compare favorably with the histogram of data provided by the vendor. Vendors should be required to provide histograms and control charts of specific critical measurements. Process mean, standard deviation, and capability can be determined from the charts and compared with the manufacturer's claims. They can also be calculated from the data in the sample histogram. The histogram mean, standard deviation, and distribution should be approximately the same as the population mean, standard deviation, and distribution:

Compare and Assess

Vendor's claim	Distribution of measurements from vendor's histograms $\bar{\bar{x}}$ and target measurement s and process capability
Vendor's charts	Distribution of measurements from the data on the vendor's control charts $\bar{\bar{x}}$ and target measurement s and process capability
Sample histogram	Distribution of measurements from the random sample of incoming pieces $\bar{\bar{x}}$ and target measurement s and process capability

There should be close agreement among the three sets of information.

EXAMPLE 11.3

A vendor supplies parts to a manufacturer, whose specifications of a critical measurement are $.750 \pm .0025$ with a process capability, PCR, of .8. The vendor claims that the parts satisfy the requirements and submits a control chart as representative of those available. A copy of the control chart is given in Figure 11.9.

1. Does the control chart support the vendor's claim? Create a histogram of the control chart data and calculate the mean, standard deviation, and process capability from the control chart information.
2. A sample of 200 pieces is randomly chosen from the lot with the control chart. The measurements are recorded in Table 11.3. What conclusions can be made regarding the quality of the incoming pieces? Draw a histogram using sample data and compare it to the vendor histogram formed with control chart data. Also, calculate the mean, standard deviation, and process capability from the sample data.
3. Compare the results for consistency.

Solution

The two histograms shown in Figures 11.10 and 11.11 (the vendor histogram and sample data histogram) match very closely. The normal shape of the distributions and the fact that they are centered at the target measurement indicate that the PCR will give a realistic picture of the process capability. They also indicate that the vendor is honestly reporting the facts about the product's quality.

1. Control chart calculations: The following calculations investigate the vendor's claims:

$$\bar{\bar{x}} = \frac{\Sigma \bar{x}}{N}$$

$$= \frac{19.4998}{26} = 7.4999$$

$$\bar{R} = \frac{\Sigma R}{N}$$

$$= \frac{.046}{26} = .001769$$

$$\text{UCL}_R = D_4 \bar{R}$$

$$= 2.114 \times .001769$$

$$= .0037$$

$$\text{UCL}_{\bar{x}} = \bar{\bar{x}} + A_2 \bar{R}$$

$$= .74999 + (.577 \times .001769)$$

$$= .751$$

$$\text{LCL}_{\bar{x}} = \bar{\bar{x}} - A_2 \bar{R}$$

$$= .74999 - (.00102)$$

$$= .74897$$

The measurements are on target, and the control limits are acceptable. The range chart is in statistical control, but the $\bar{x}$ chart shows instability: Starting with the third point, nine consecutive points are outside the C zones. This is contributing to excessive variability:

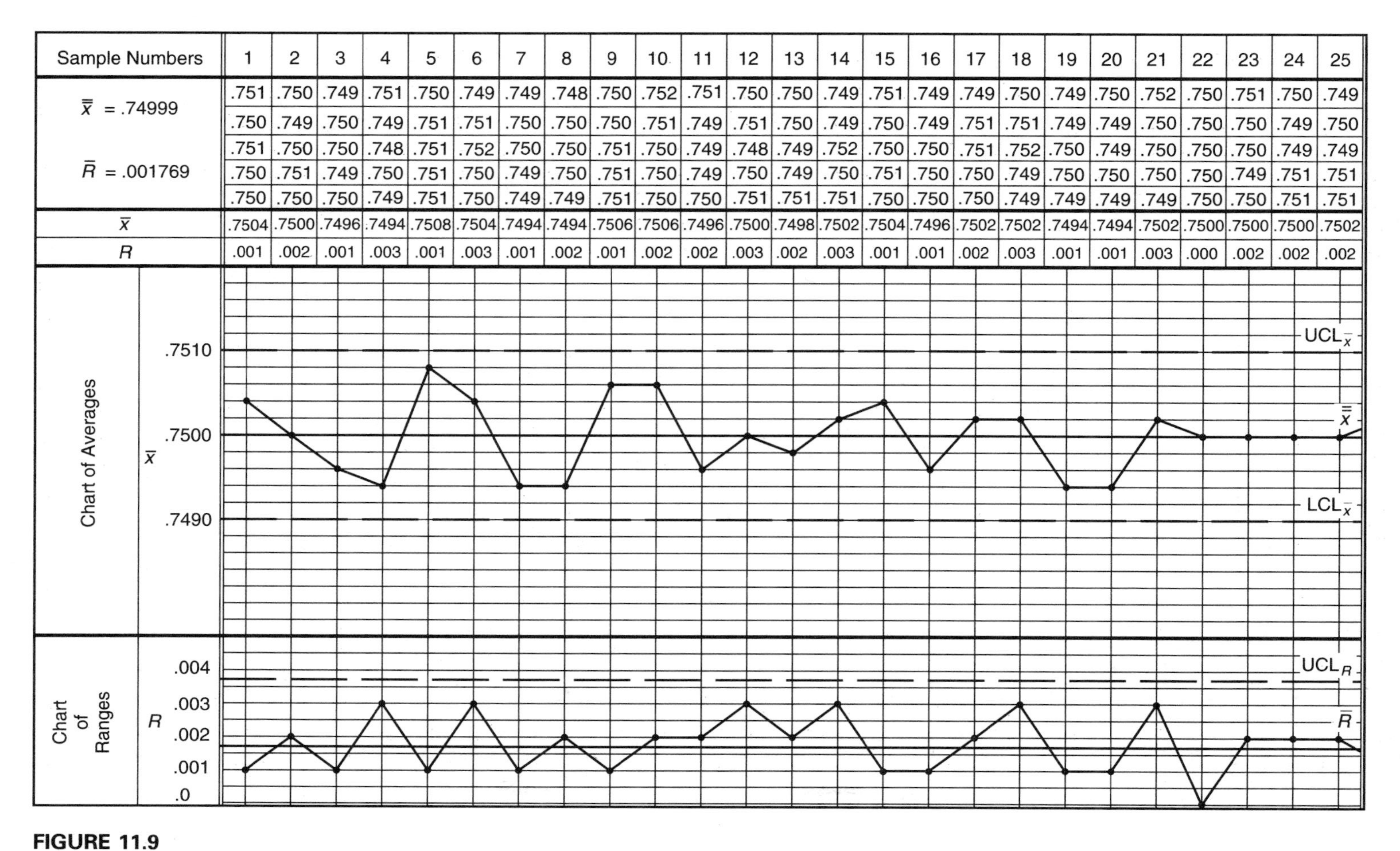

Sample Numbers	1	2	3	4	5	6	7	8	9	10	11	12	13	14	15	16	17	18	19	20	21	22	23	24	25
$\bar{\bar{x}} = .74999$	.751	.750	.749	.751	.750	.749	.749	.748	.750	.752	.751	.750	.750	.749	.751	.749	.749	.750	.749	.750	.752	.750	.751	.750	.749
	.750	.749	.750	.749	.751	.751	.750	.750	.750	.751	.749	.751	.750	.749	.750	.749	.751	.751	.749	.749	.750	.750	.750	.749	.750
$\bar{R} = .001769$	.751	.750	.750	.748	.751	.752	.750	.750	.751	.750	.749	.748	.749	.752	.750	.750	.751	.752	.750	.749	.750	.750	.750	.749	.749
	.750	.751	.749	.750	.751	.750	.749	.750	.751	.750	.749	.750	.749	.750	.751	.750	.750	.749	.750	.750	.750	.750	.749	.751	.751
	.750	.750	.750	.749	.751	.750	.749	.749	.751	.750	.750	.751	.751	.751	.750	.750	.750	.749	.749	.749	.749	.750	.750	.751	.751
$\bar{x}$	.7504	.7500	.7496	.7494	.7508	.7504	.7494	.7494	.7506	.7506	.7496	.7500	.7498	.7502	.7504	.7496	.7502	.7502	.7494	.7494	.7502	.7500	.7500	.7500	.7502
R	.001	.002	.001	.003	.001	.003	.001	.002	.001	.002	.002	.003	.002	.003	.001	.001	.002	.003	.001	.001	.003	.000	.002	.002	.002

FIGURE 11.9
Vendor's control chart for Example 11.3.

TABLE 11.3
Measurements for the random sample of 200 pieces

Measurement	Number of Pieces	Measurement	Number of Pieces
.7524	1	.7498	20
.7522	0	.7496	19
.7520	1	.7494	12
.7518	0	.7492	12
.7516	2	.7490	13
.7514	2	.7488	9
.7512	6	.7486	5
.7510	8	.7484	2
.7508	10	.7482	1
.7506	13	.7480	0
.7504	15	.7478	2
.7502	20	.7476	1
.7500	25	.7474	1

$$s = \frac{\overline{R}}{d_2}$$
$$= \frac{.001769}{2.326}$$
$$= .0007605$$
$$6s = 6 \times .0007605$$
$$= .004563$$
$$\text{PCR} = \frac{6s}{\text{tolerance}}$$
$$= \frac{.00463}{.005}$$
$$= .913$$

The control chart indicates that the process uses 91.3% of the tolerance instead of the 80% that was specified. Again, there is too much variability.

2. Calculations from the random sample of 200 pieces: The 200-sample measurements were entered into a statistical calculator with the following results:

$$\overline{\overline{x}} = .749889$$
$$s = .00080643$$
$$6s = .00483858$$

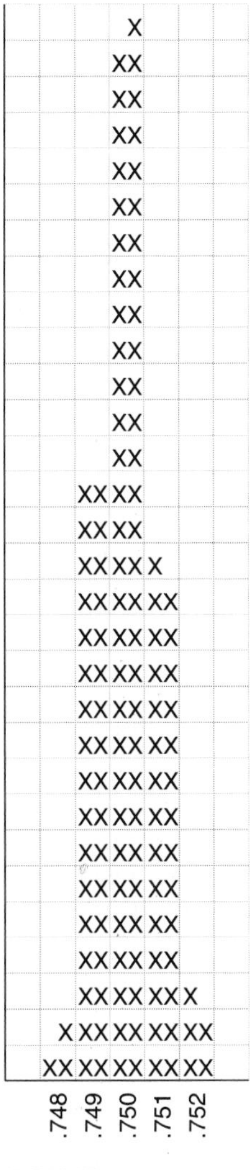

```
          X
         XX
         XX
         XX
         XX
         XX
         XX
         XX
         XX
         XX
         XX
         XX
         XX
      XX XX
      XX XX
      XX XX X
      XX XX XX
      XX XX XX
      XX XX XX
      XX XX XX
      XX XX XX
      XX XX XX
      XX XX XX
      XX XX XX
      XX XX XX
      XX XX XX
      XX XX XX
      XX XX XX X
    X XX XX XX XX
   XX XX XX XX XX
.748 .749 .750 .751 .752
```

FIGURE 11.10
Control chart data for Example 11.9.

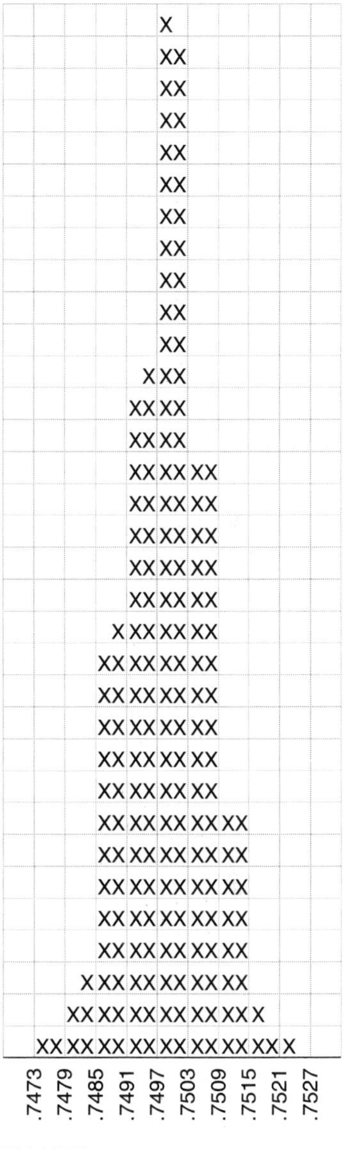

```
                X
               XX
               XX
               XX
               XX
               XX
               XX
               XX
               XX
               XX
               XX
             X XX
            XX XX
            XX XX
            XX XX XX
            XX XX XX
            XX XX XX
            XX XX XX
            XX XX XX
          X XX XX XX
         XX XX XX XX
         XX XX XX XX
         XX XX XX XX
         XX XX XX XX
         XX XX XX XX
         XX XX XX XX XX
         XX XX XX XX XX
         XX XX XX XX XX
         XX XX XX XX XX
         XX XX XX XX XX
       X XX XX XX XX XX
      XX XX XX XX XX XX X
   XX XX XX XX XX XX XX XX X
.7473 .7479 .7485 .7491 .7497 .7503 .7509 .7515 .7521 .7527
```

FIGURE 11.11
Random sample data from Table 11.3.

$$\begin{aligned} PCR &= \frac{6s}{\text{tolerance}} \\ &= \frac{.00483858}{.005} \\ &= .968 \end{aligned}$$

The $\bar{\bar{x}} = .7499$ indicates that the measurement is on target, but the standard deviation of the measurements shows that the measurements use about 97% of the tolerance in the PCR calculation. This is slightly worse than the control chart figure, but not different enough to suspect dishonest charting.

Recommendations The calculations of the control chart and sample data lead to the following recommendations:

1. The measurements on the control chart are not accurate enough. The measurements should be accurate to at least .0002. The vendor should use a more accurate gauge.
2. Neither the operator nor the supervisor recognized the out-of-control situation on the control chart. Check with the vendor on the extent of their SPC training.
3. Either work with the vendor on reducing variability or monitor the vendor's efforts with continued sample checks.

In this instance, the sample calculations were consistent with the vendor information. If the vendor had tried to submit charts that overrated the quality of the product, the analyzed sample would have shown the discrepancy immediately.

When parts are being machined, often several different critical measurements result from a single complex machining operation. The operations may occur in succession, but all operations are on the same machine. It is important to know whether the measurements are all independent, so that each has to be tracked with its own control chart, or whether there is correlation between some of the measurements. If there is good correlation between two measurements, then often just one of the measurements has to be tracked on a control chart. If that one measurement is in control and has good capability, then the same will be true of other measurement. Example 11.4 investigates the possibility of correlation between two top adapter measurements.

EXAMPLE 11.4 The top adapter measurements from Example 11.2 were matched to a second top adapter measurement taken at the same time. The pairs of measurements are listed in Table 11.4. Is there any correlation between the measurements?

Solution

The scatterplot in Figure 11.12 shows that there is *no correlation* between the two measurements, so each will have to be controlled separately. There is no line that the points approach: This set of points has no linear trend.

TABLE 11.4

Measurements for Example 11.4

Date	Time	.215 Measurement	.503 Measurement
7/29	0745	.2134	.5025
		.2136	.5025
		.2138	.5025
		.2136	.5020
		.2147	.5025
	1120	.2140	.5034
		.2138	.5037
		.2136	.5040
		.2140	.5033
		.2140	.5035
	1305	.2138	.5030
		.2143	.5032
		.2143	.5025
		.2142	.5035
		.2136	.5035

FIGURE 11.12
No correlation.

CASE STUDY 11.1

The ACE Spark Plug Company specializes in relatively low-volume aviation plugs. They hired a consultant to train their engineers and workers in SPC, and when the training was completed, they had to select a first application. The company is relatively small with a total work force of approximately 200 people. An initial quality team was formed that consisted of supervisors from four different plug lines: quality control, industrial and ceramic engineers, sales, and receiving.

The first meeting was held to overview the different trouble areas and decide where to start with SPC. After a short brainstorming session, a list of trouble areas was made. Each team member was assigned at least one area to investigate and gather data. They were to meet again the following week.

At the next meeting the data were organized in two Pareto charts, one showing the number of defective items and the other showing the dollar loss associated with each defective item. The A-150 plug line had a high number of returns and resulted in the top priority on the dollar-loss Pareto chart. The second priority was the ceramic insulators that were sourced out to a supplier. The team decided to break up and two new teams were formed. The company's ceramic engineer was to chair a small team to investigate the feasibility of producing in-house ceramic insulators versus working with the supplier to improve quality. The second team was formed around the A-150 supervisor, the industrial engineer, and the quality control representative. They recruited a few more members from the A-150 production line.

The A-150 team met the following week and constructed a flowchart of the A-150 process. They also had an analysis report on the plug returns that they organized in a Pareto chart. The two charts pointed to a problem with the shell that contains the insulator and igniter.

The A-150 team met again a few days later and constructed a cause-and-effect chart for the shell problem. Two measurements were considered critical: the body diameter and the hole diameter. One of the possible causes of the problems, suggested by the machine operator, was that the Acme's they were using were old machines and probably needed replacement. It was decided to chart the critical measurements on an $\bar{x}$ and R chart and a histogram. The next meeting was set for the following week and the work was assigned.

Copies of the two control charts and their histograms were distributed at the beginning of the next meeting. These are shown in Figures 11.13 to 11.17. On Figure 11.13 they noticed that the first point was beyond the control limits for both the $\bar{x}$ and the R charts. The rest of the R chart was in control. The engineer pointed out that R was a little more than half the tolerance, which could indicate that there is excessive variation. A line operator suggested that the high variation would cause the stratification pattern on the $\bar{x}$ chart. A large $\bar{R}$ will spread the control limits on the $\bar{x}$ chart. Someone said that the spread on the histogram verified the excessive variability and questioned the cause of the bimodal pattern. The machinist mentioned that the Acme is a six-spindle machine and that another look at the data by spindle may be helpful. The team discussed the various alternatives for looking at the separate spindles. The quality control representative suggested they construct an x and MR chart for each spindle and analyze the results the following week. When questioned about the data gathering for the $\bar{x}$ and R charts, the machinist said that the samples of six included one from each spindle and were in order on the $\bar{x}$ and R chart. The engineer suggested that the current data be reorganized in separate histograms for a spindle comparison. They reorganized the data to create three histograms with two spindles tracked on each, as shown in Figure 11.16. Spindle 1 uses the right half of the scale and spindle 2 is to the right of center. Spindles 4 and 6 are to the left of center and 3 and 5 are to the left of center with wider spreads. This histogram analysis shows that the spindles on the Acme are inconsistent and appear to be the cause

Date																				
Sample Number	1	2	3	4	5	6	7	8	9	10	11	12	13	14	15	16	17	18	19	20
1	.8712	.8710	.8712	.8712	.8712	.872	.872	.8725	.872	.8725	.873	.872	.872	.8715	.8715	.872	.8735	.872	.8725	.8715
2	.8655	.871	.8706	.8705	.871	.872	.871	.8715	.873	.8715	.8715	.8715	.872	.8715	.872	.872	.8715	.8715	.872	.871
3	.869	.8708	.8705	.872	.8708	.8712	.8705	.8708	.8710	.8709	.8700	.871	.870	.8712	.8712	.8713	.8702	.8701	.8712	.8712
4	.8708	.871	.8708	.8708	.8706	.8706	.8707	.8704	.8708	.8707	.8707	.8706	.8707	.8707	.8709	.8712	.8711	.8711	.8712	.8711
5	.8703	.8706	.8708	.8706	.8704	.8702	.8702	.8703	.8705	.8705	.8705	.8704	.8701	.8705	.8698	.8708	.8704	.8706	.8705	.8692
6	.8708	.8705	.870	.8702	.8702	.8701	.8702	.8703	.8703	.8707	.8705	.8705	.8704	.8708	.8707	.8706	.8703	.8709	.8707	.8707
Sum																				
Average	.86960	.87080	.8707	.8709	.87070	.8710	.8708	.8710	.87130	.87110	.8710	.8710	.87090	.8710	.8710	.87130	.87120	.8710	.87140	.8708
Range	.00570	.00050	.00120	.00180	.0010	.00190	.00180	.00220	.00270	.0020	.00300	.00160	.0020	.0010	.00220	.00140	.00330	.00190	.0020	.00230
Comment																				

Averages: .8720, 18, 16, 14, 12, .8710, 08, 06, 04, 02, .8700, 98, 96 — $UCL_{\bar{x}}$, $\bar{\bar{x}}$, $LCL_{\bar{x}}$

Ranges: .0040, 35, .0030, 25, .0020, 15, .0010, 05, 0 — UCL_R, $\bar{R}$

FIGURE 11.13
The control chart for the body diameter.

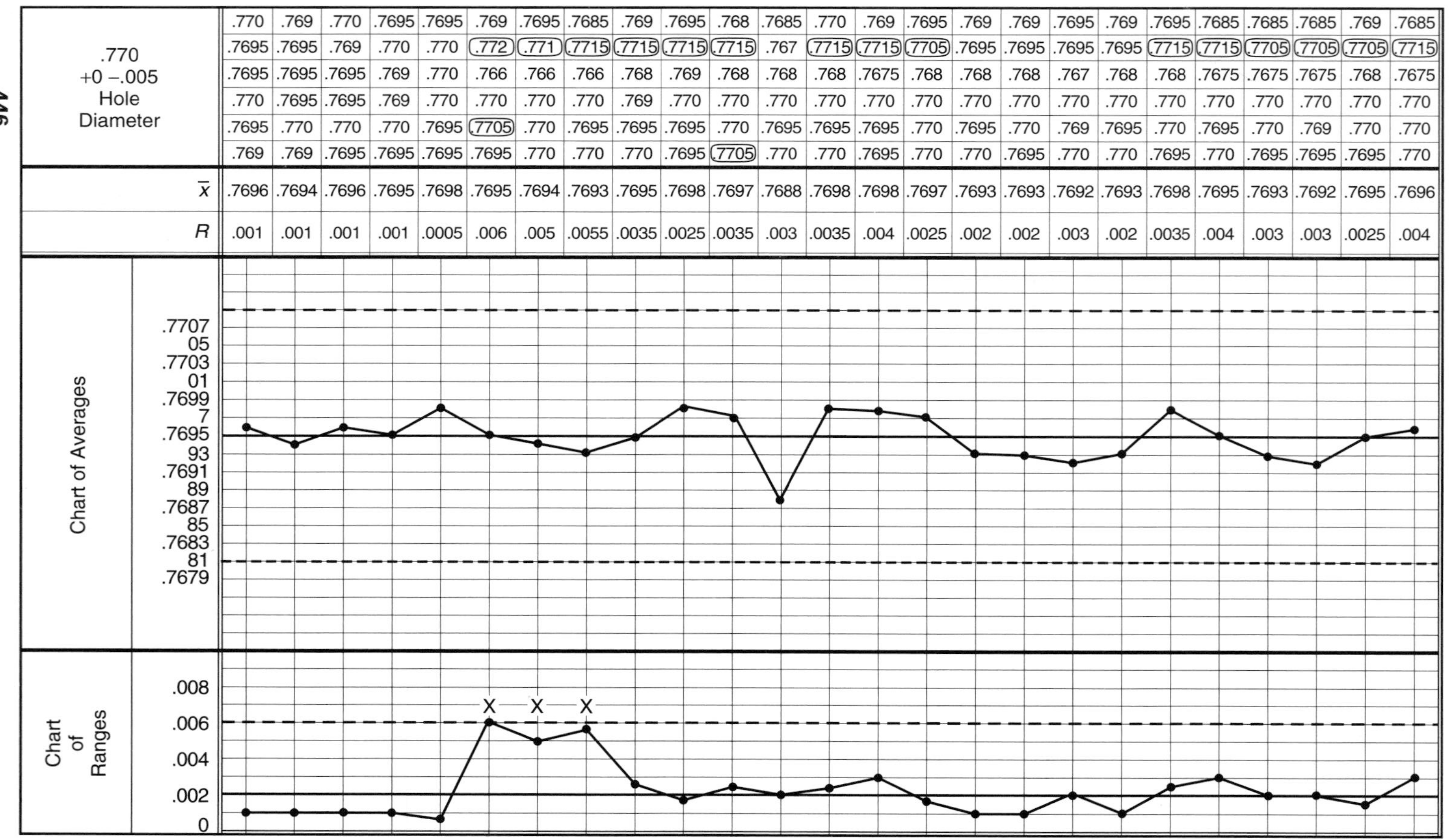

.770 +0 −.005 Hole Diameter													
	.770	.769	.770	.7695	.7695	.769	.7695	.7685	.769	.7695	.768	.7685	.770
	.7695	.7695	.769	.770	.770	.772	.771	.7715	.7715	.7715	.7715	.767	.7715
	.7695	.7695	.7695	.769	.770	.766	.766	.766	.768	.769	.768	.768	.768
	.770	.7695	.7695	.769	.770	.770	.770	.770	.769	.770	.770	.770	.770
	.7695	.770	.770	.770	.7695	.7705	.770	.7695	.7695	.7695	.770	.7695	.7695
	.769	.769	.7695	.7695	.7695	.7695	.770	.770	.770	.7695	.7705	.770	.770
$\bar{x}$	.7696	.7694	.7696	.7695	.7698	.7695	.7694	.7693	.7695	.7698	.7697	.7688	.7698
R	.001	.001	.001	.001	.0005	.006	.005	.0055	.0035	.0025	.0035	.003	.0035

.770 +0 −.005 Hole Diameter												
	.769	.7695	.769	.769	.7695	.769	.7695	.7685	.7685	.7685	.769	.7685
	.7715	.7705	.7695	.7695	.7695	.7695	.7715	.7715	.7705	.7705	.7705	.7715
	.7675	.768	.768	.768	.767	.768	.768	.7675	.7675	.7675	.768	.7675
	.770	.770	.770	.770	.770	.770	.770	.770	.770	.770	.770	.770
	.7695	.770	.7695	.770	.769	.7695	.770	.7695	.770	.769	.770	.770
	.7695	.770	.770	.7695	.770	.770	.7695	.770	.7695	.7695	.7695	.770
$\bar{x}$	.7698	.7697	.7693	.7693	.7692	.7693	.7698	.7695	.7693	.7692	.7695	.7696
R	.004	.0025	.002	.002	.003	.002	.0035	.004	.003	.003	.0025	.004

FIGURE 11.14
The control chart for the hole diameter.

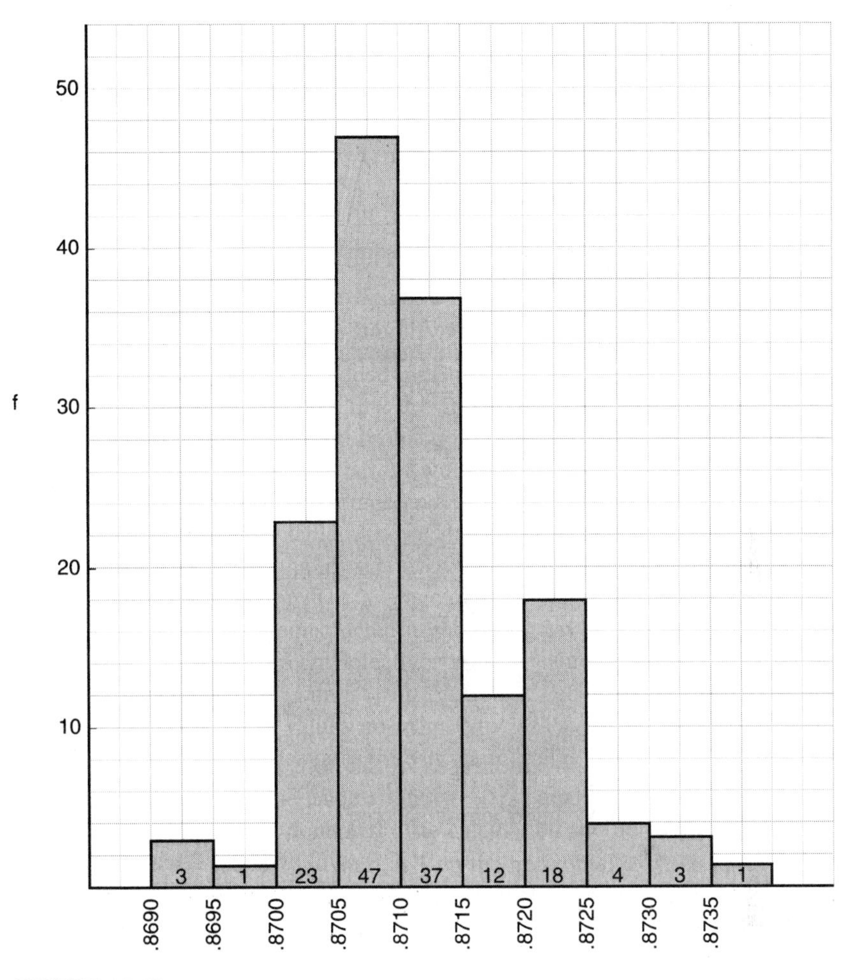

FIGURE 11.15
The histogram for the body diameter data.

of the excessive variation. Assignments to reorganize the data on the other measurement were made and a meeting planned for two days later.

At the next meeting, copies of a histogram that showed the spindle separation for the other measurement were distributed. That histogram is shown in Figure 11.18. Spindles 2 and 3 caused the main problem with the hole diameter and 1, 2, and 5 contributed the most to the variation on the body diameter.

The Acme was immediately scheduled for an overhaul, the work was transferred to another Acme, and a similar analysis was performed on that machine. No evidence of a spindle problem surfaced and the bulk of the problems on the A-150 line were resolved.

The ceramic team continued to pursue both possible solutions to the insulator problem. They organized a small-scale design of experiments project to try to develop an in-house ceramic insulator line. They were running into the same problem as the supplier

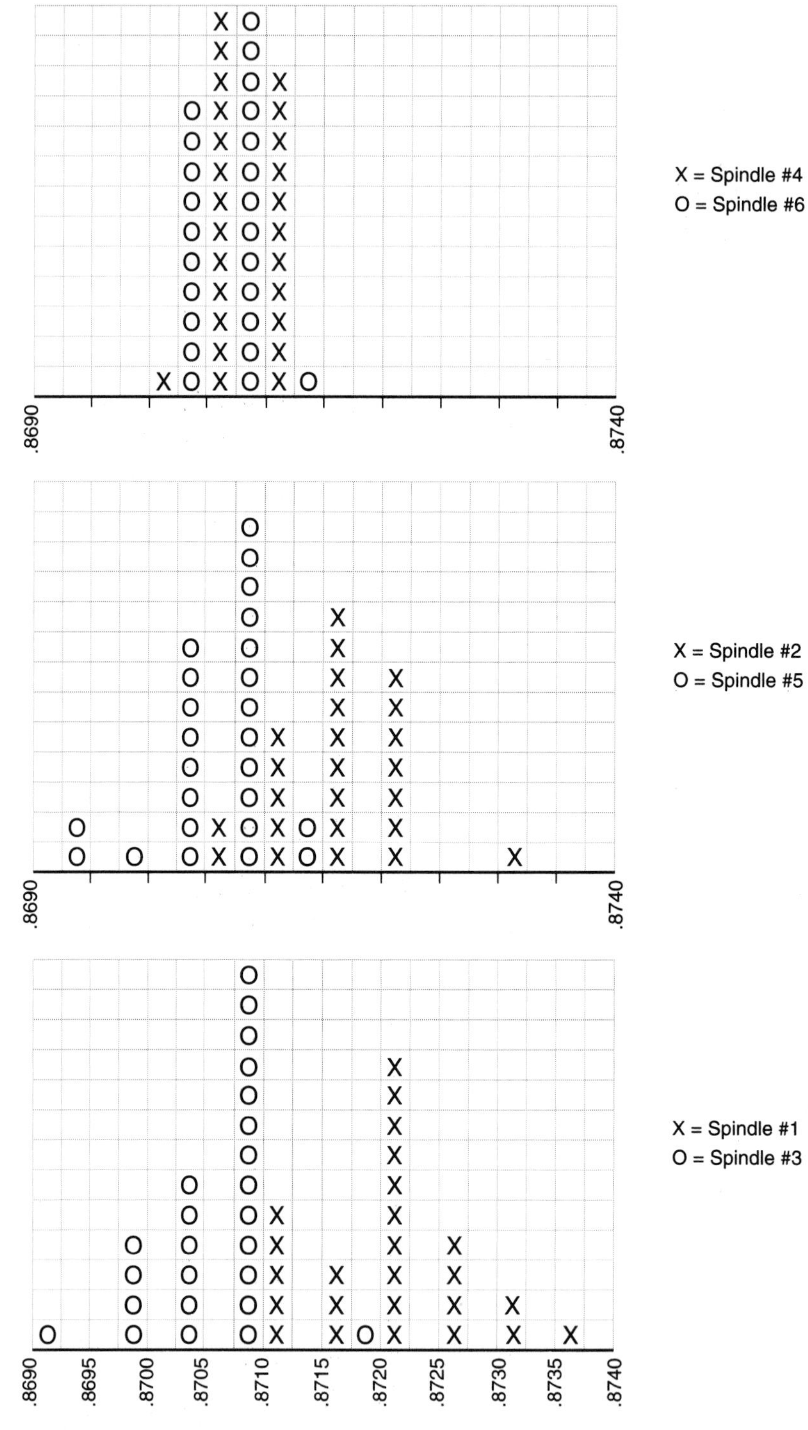

FIGURE 11.16
Histograms showing the spindle breakdown for the body diameter.

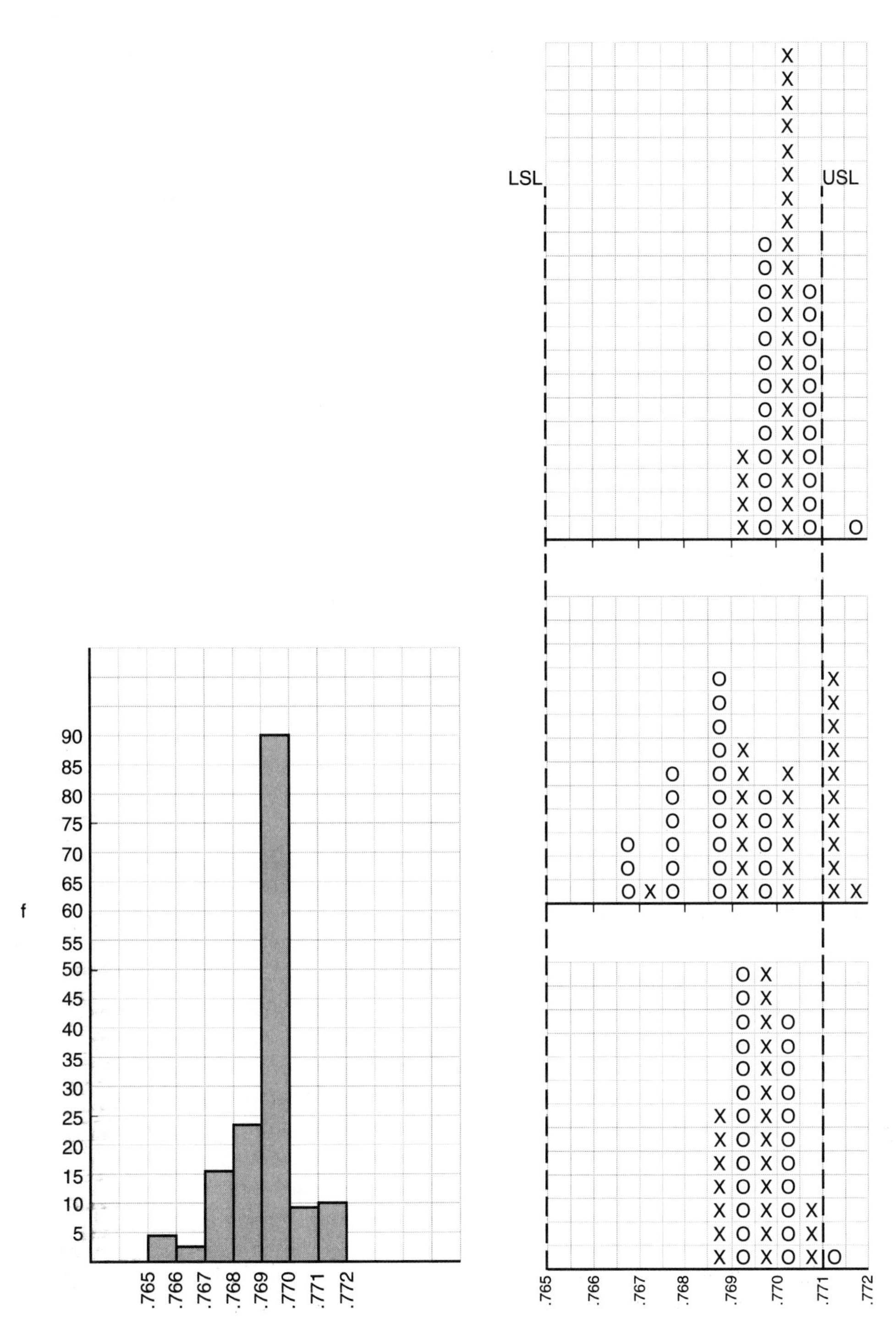

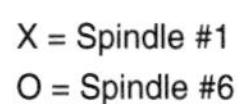

FIGURE 11.17
The histogram for the hole diameter data.

FIGURE 11.18
Histograms showing the spindle breakdown for the hole diameter.

because of the irregular dimension changes that occur when the ceramic is fired. They also continued working with the supplier, exchanging information and ideas for improvement on the insulators.

11.5 MISTAKE PROOFING

Mistake proofing is a process of anticipating errors and preventing them from occurring. This is done by designing products that cannot be manufactured or assembled incorrectly or by designing the process so that the products cannot be made incorrectly. The process utilizes sensors or inspection methods that detect errors that occur within the process as part of a prevention system. Any errors that are found demand immediate attention to prevent them from occurring again.

Each phase of the process should be subject to a complete error analysis. The various types of errors that could occur include the following:

- Lack of standards
- Being forgetful
- Not being attentive
- Lack of training
- Misunderstanding
- Carelessness
- Being startled
- Fear of questioning procedures

How could each type of error occur at each process step and what can be done to prevent that error from occurring? When those questions have been answered and the preventive measures taken, that process step has been mistake-proofed.

The goal of mistake proofing is to prevent errors from occurring and produce 100% defect-free products. When it is applied, it often uses electronic sensors that will either set off alarms or stop the process when errors occur. Mistake proofing is built into a process when potential errors are anticipated in a new process setup. It can be added to an existing process as part of a prevention system when errors or mistakes occur and their root cause is determined.

Mistake proofing can be taught companywide as a basic process that anticipates mistakes, detects them when they are made, and prevents them from occurring again. It can be used in the design stage of a product by making it impossible to assemble the product incorrectly. Anywhere work is being done, the people involved are often aware of mistakes that they have made or that others have made. Training in mistake proofing encourages workers to think of procedures that could be used to prevent those mistakes from happening. It then becomes management's responsibility to respond with using the good ideas and encouraging more thought on ideas that will not work or are economically unsound. There are general sources of errors and workers should be aware of how those sources could cause errors on their job. Is there anything that could startle them? Could forgetfulness or lack of attention be a problem? Is there a chance that misunderstandings or inexperience could lead to errors? Are the standards clear enough? If any of these could lead to errors, how could they be prevented from happening? Answer that and you are mistake proofing the job.

When a company begins to incorporate mistake proofing into its processes, it first prioritizes the potential application stations. The criteria include failure mode effect analysis (FMEA), risk priority numbers (RPNs), scrap records, rework records, and customer complaints (including downstream customers).

Mistake proofing, or Poke-Yoke,[7] builds inspection into the system in various ways. Self-checking puts the responsibility on the persons doing the work. They must inspect their work before they pass it on. They cannot pass on defective pieces or, sometimes, even pieces that are at the minimum or maximum specification values. Successive check systems have both the worker and the receiver check each piece. There may be various mechanical or electronic devices used for the inspection checks. If any defects are detected, the process stops, the defective piece is returned to its sources, and the cause of the defect is determined. If possible, the source of the problem is mistake proofed so the error will not occur again. If the defects seem to occur randomly, they are set aside and the process is allowed to continue. If the problem has a high enough priority, it is given to a problem-solving team. Source inspection concentrates on the process itself and conditions within the process that can cause defects. It stresses the idea that a company does not program its processes to build defects. Any troublesome conditions should be corrected before any defects are made.

When mistake-proofing systems are installed, the detection of an error will result in either a process shutdown or the activation of a warning device. A familiar example of the first would be the safety device on a rotary lawnmower. If the operator lets go of the handle, the engine will stop. In the second case, if you leave your keys in the ignition of a car and then open the door, a warning bell or buzzer will sound.

The actual mistake-proofing checks can be done with various contact devices. If a sensor does not touch a part that is supposed to be installed, either a shutdown or a warning will result. When protection against oversize is needed, a part touching a sensor could trigger a reaction. Contact devices that work well in one area of a plant may very well work in another. The same mistake-proofing ideas and devices may apply in several operations.

Fixed-value methods can be used in mistake proofing too. If a worker is supposed to install eight screws in a cabinet, having the screws set up in groups of eight would protect against forgetting one.

CASE STUDY 11.2

A maker of automobile transmissions ran into an occasional costly error. A speedometer gear was sometimes omitted from a transmission assembly. When that occurred, the customer, a major car company, would reject the lot containing the faulty transmission. They would be sent back, 100% inspected, and reshipped. The problem occurred at an assem-

[7]Shingo, Shigeo, *Zero Quality Control: Source Inspection and the Poke-Yoke System* (Andrew P. Dillon, translator. Preface by Norman Bodek, tr. from JPN) (Productivity Press).

bly point where the worker took the needed gears from adjacent bins, placed them in the case, and bolted the case together. An indented, numbered tray was made to contain all the assembled parts in order. The tray was first filled from the bins, so all the needed parts were accounted for; then they were assembled in the transmission case. The problem was mistake-proofed and there were no further instances of missing gears.

CASE STUDY 11.3

An electronics company makes printed circuit boards with a combination of surface-mount and through-pin transistors and diodes. The automatic-placement machine would stop if one of the through-pin elements had a bent leg that would not align with the hole properly. The downtime of the machine was costly. After brainstorming the problem, it was decided to program the machine to mark the boards that contained a problem and continue the operation. The marked boards were later separated, the through-pin pieces mounted by hand, and then the boards were placed back in line for the solder process. By revising this mistake-proofing action, there was a 35% increase in production.

CASE STUDY 11.4

The spark plug manufacturer has another problem with the Acme's. One of the operations on the shell was a drilling process. When the hole depth was charted on an $\bar{x}$ and R chart, it kept going out of control. An inspection of the data values on the chart revealed that an occasional measurement was too small, well below the specification value. The machinist believed that chips were accumulating and interfering with the drilling operation. They monitored the machine and noticed that the chip buildup was inconsistent. Most of the time the metal chips from the machining operations would fall clear (as they were supposed to), but occasionally they would build up and interfere. A probe was designed and installed near the drilling operation and when a buildup of chips was detected, the sensor activated an alarm. The machinist then removed the accumulation and the short-hole problem was mistake proofed.

11.6 PROBLEM SOLVING IN MANAGEMENT

Management often thinks that SPC and its related problem-solving techniques apply only to the shop floor. Since SPC can be applied in any work situation that produces relevant data, it can be used in the management office too.

CASE STUDY 11.5

Errors in the order entry process can be very costly to a company. Lost orders, miscommunications, incorrect addresses, and inefficient billing are some of the problems that can result in lost customers.

The ABC Corporation uses customer service representatives for the order entry process. There was a general perception that too many costly mistakes were being made. The department manager, responding to heat from "above," held a meeting with all the customer service staff to investigate the problem. The manager had participated in the company's SPC training and proceeded to coach his staff on its implementation. Their first task was to flowchart the process. Since the existing process had evolved over a period of time, it took a good brainstorming session to iron out the details. The order entry flowchart is shown in Figure 11.19.

The first step in the process involves receiving the customer's inquiry. The second, third, and fourth steps involve qualifying the customer and identifying what he or she needs. The fifth transmits the information to the customer. At any point in these first five steps the process can be ended. The inquiry may be from someone who is not a customer, or the product might be obsolete or the quotation might not be acceptable. Once the order is received, additional decisions are required. Step 8 ensures that the customer is included in the company's data base. Step 9 identifies the exact part number and description of the product that the customer has ordered. Step 10 defines the price and delivery. Step 11

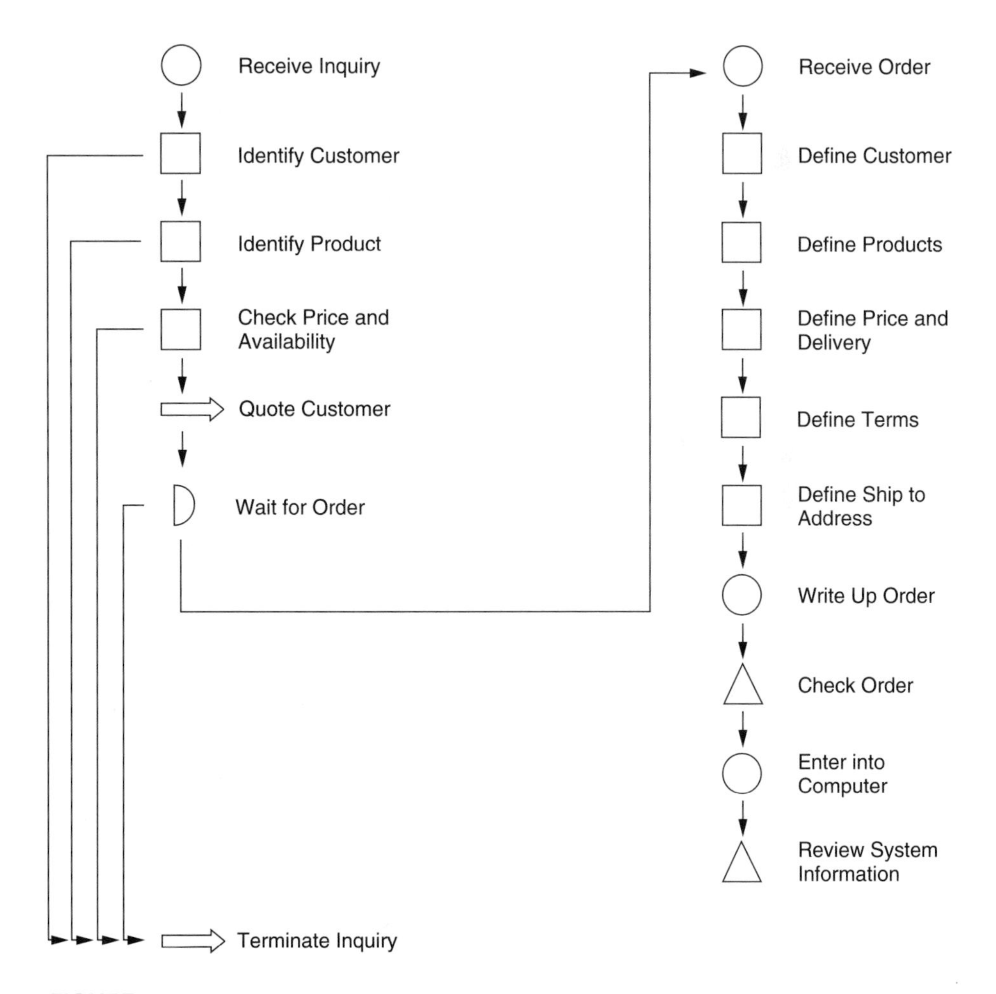

FIGURE 11.19
Customer service order entry process.

defines the payment terms and step 12 defines the ship-to and bill-to addresses. Once this information is complete, the order is written up on an order form. It is then given to the supervisor to be checked and then entered in the system. At the end of each day the orders that have been entered in the system are summarized in a report and checked by the customer service manager as well as managers of other departments involved in the completion of the order.

The department needed to identify the actual errors that were being made to quantify them. Since each step of the process was either done correctly or an error was made, it fit into the go–no-go pattern of a *p* chart. Each sample for the chart consisted of all the orders processed in one day. Since that resulted in different *n* values, the simpler *np* chart could not be used. A tally chart accompanied the *p* chart and classified the type of error made. The tally chart was later reorganized into a Pareto chart for prioritizing.

The *p* chart showed that 14% of the order entries, on the average, were in error. This value was considered excessively high. Everyone in the department realized that improvements must be made. One encouraging thing that showed up on the *p* chart was a downward trend toward the end of the first month. That meant that the percentage of errors was decreasing.

The Pareto analysis showed that most of the errors were made in the shipping address. A new step was put into the process to cross check the address on the order with the customer master list in the computer. The second priority involved completing orders for customers who were suspended due to their past business practices. Another new step was instituted which involved a check with the accounting office on the customer's status at step 2 on the flowchart.

Statistical process control worked very effectively for the department. The downward trend on the *p* chart continued in the following months. When problems were identified, they were mistake-proofed by minor changes in the process. Occasionally reinforced training was needed for the individuals involved. The department continues to monitor the order entry process for further improvement.

CASE STUDY 11.6

The ABC Company sold approximately 20% of its product line to the government. Each government purchase went to the lowest bidder. ABC was in serious competition with one other company, so, being a smaller company, it needed a competitive bidding edge. The government business is extremely important to keep because the high volumes absorb overhead costs and allows the company to be more price competitive in the commercial market.

The sales manager was responsible for setting the bid price. He had estimates from production on bottom-line costs for each product line and the bid history from several previous years. It had always been a "crystal ball" effort to establish the bid price. The ideal price is to be a few cents below the competition so that the bid is won and a maximum income from that specific bid is achieved.

The manager made a scatterplot of the bid history on each product line. The bid price on the vertical axis and the year (coded) on the horizontal. A strong linear trend was implied in most cases. The regression line (line of best fit through the data points) equation was calculated on a statistical calculator that had a linear regression (LR) mode. This gave the manager that additional information for more successful bidding. Figure 11.20 shows the scatterplot with the line of best fit for the winning bid.

The predicted winning bid for June of year 11 is $40.57 using the equation for the line. The estimate from the scatterplot is about $40.50. The production department estimated the cost at about $31. The manager entered a bid of $39.25, slightly below the predicted value. The predicted cost would have allowed a lower bid, but the competitor's history suggested that they would come in higher. The competitor's bid was $41.10 and ABC won the bid the next year.

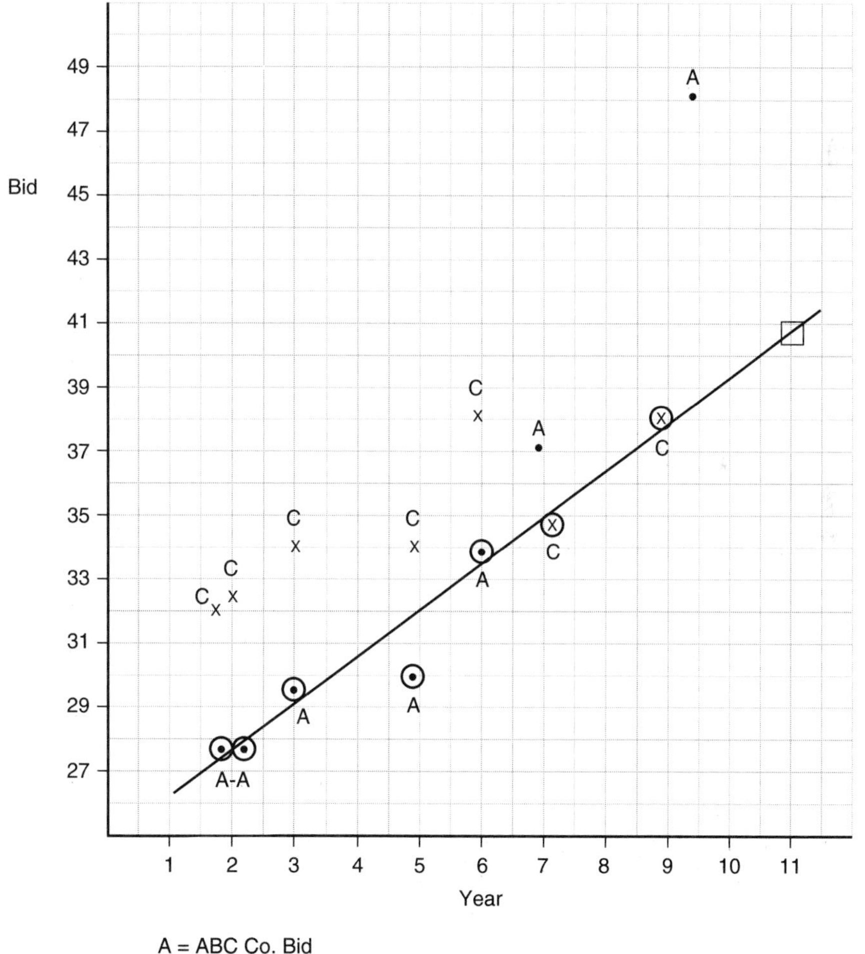

FIGURE 11.20
The bid scatterplot. The successful bid is circled.

11.7 JIT (Just-in-time)

One concept that has developed as a result of problem solving is *just-in-time* (JIT) manufacturing in which production is so efficient that each piece arrives at its next process step just in time, so that no inventories develop. There are two important reasons for JIT manufacturing. First, cutting inventories brings a big cost savings. Second, quality improves because troubles cannot hide in inventory. A defective part taken from a large inventory leaves no trace of the cause of the defect, and the prevention method cannot be applied. This problem is eliminated in JIT manufacturing. When a company wants to adopt JIT manufacturing, inventories should be reduced gradually. In the process of gradual reduction, troubles will arise. Inventory reductions should be stopped until the trouble is resolved, and then further reductions can be made. The logical end to that procedure is zero inventory and single-piece production in which each part arrives at the right place at exactly the right time. There are reports of some manufacturing processes that are choreographed to that extent. One major problem that can develop from JIT manufacturing, however, is an increase in stress when zero inventory is achieved and no leeway exists in product flow.[8] The ideal method is to maintain a small inventory buffer, which will allow some fluctuation in worker rates.

11.8 PROBLEM SOLVING IN THE CLASSROOM

When a student has had more than 12 years in the classroom, it may seem foolish to talk about quality education processes. This attitude is remarkably similar to that of the machine operator who has many years of experience and is told that he/she must learn SPC and apply it on the job. Case Study 11.7 illustrates the application of problem solving in the pre-calculus classroom.

CASE STUDY 11.7

The Pre-Calculus course at a community college had a very low success rate compared with the other mathematics courses at the college. One major reason was that the Pre-Calculus level of mathematics demanded that the students have a good working knowledge of previous mathematics courses. Also, students who were able to get by with a limited amount of work at the more elementary levels, found that their previous learning methods didn't work at the higher level. Warnings by the instructors went unheeded until the students found themselves hopelessly behind and were forced to withdraw.

At the beginning of the Fall '96 semester the instructor utilized a few problem-solving tools. One of the first was the formation of a cause-and-effect diagram for the problem of poor grades. The students contributed their ideas in a brainstorming session and copies of the results were distributed to everyone for future reference. The cause-and-effect diagram is shown in Figure 11.21.

---

[8] Janice M. Klein, "The Human Cost of Manufacturing Reform" (*Harvard Business Review*, March–April 1989).

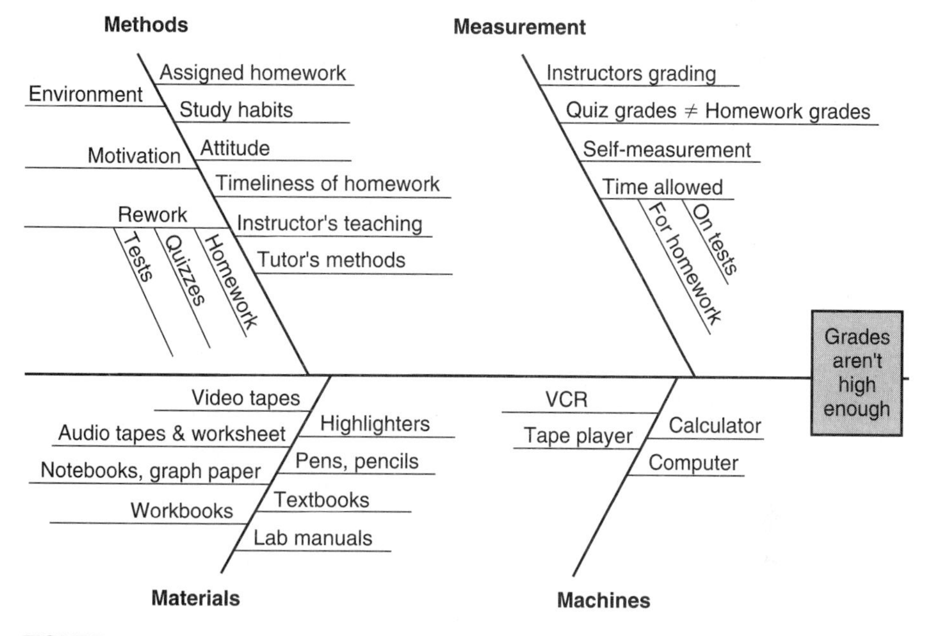

FIGURE 11.21
A cause-and-effect diagram for poor grades.

The instructor then handed out control charts with the instructions that the students check their homework, grade it, and record the grade on the chart. When missed problems are corrected, the improved grade is recorded and the previous grade crossed out. Partial credit is allowed. Quiz grades and test grades are also recorded on the control chart. The chart is similar to the one illustrated in Example 9.6 with the exception that it is a q chart based on the entire homework assignment and not on just a sample from it. The charts had a positive effect on student persistence because they didn't like to see any severe dips on their chart and they reworked the assignments that were poorly done. The charts were also valuable analysis tools when a student needed help. A control chart is illustrated in Figure 11.22.

All students created a learning process flow chart and kept it with their control chart for possible revision. Their first flow chart was generally simple and direct as in Figure 11.23. As some of the students improved their learning skills, their flow charts became more detailed as shown in Figure 11.24.

A few of the students were bothered by frequent errors. They were directed to keep a check sheet or a Pareto chart on the types of errors they were committing, which led to isolating specific repeated errors. A few sessions with the tutors provided refresher lessons on forgotten concepts, and errors decreased.

The overall effect of using the problem-solving tools was that the students became more aware of their responsibility for learning and that learning was a skill that they could improve. Their persistence improved and the overall class success rate improved significantly.

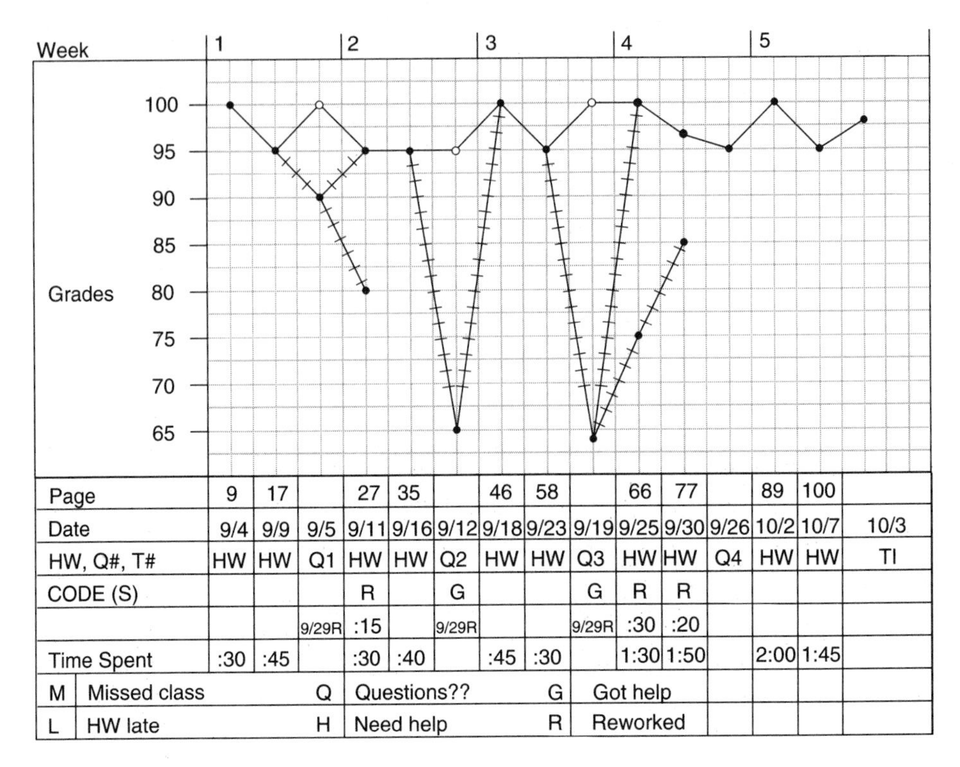

Page	9	17		27	35		46	58		66	77		89	100	
Date	9/4	9/9	9/5	9/11	9/16	9/12	9/18	9/23	9/19	9/25	9/30	9/26	10/2	10/7	10/3
HW, Q#, T#	HW	HW	Q1	HW	HW	Q2	HW	HW	Q3	HW	HW	Q4	HW	HW	TI
CODE (S)				R		G			G	R	R				
			9/29R	:15		9/29R			9/29R	:30	:20				
Time Spent	:30	:45		:30	:40		:45	:30		1:30	1:50		2:00	1:45	

M	Missed class	Q	Questions??	G	Got help
L	HW late	H	Need help	R	Reworked

FIGURE 11.22
A student control chart.

FIGURE 11.23
The initial learning process flowchart.

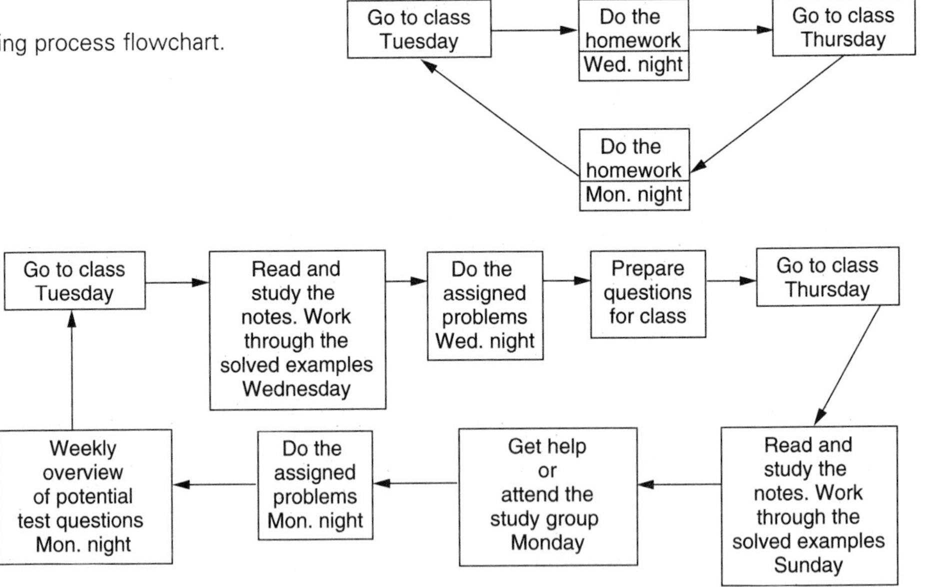

FIGURE 11.24
The more detailed learning process flowchart.

EXERCISES

1. What are the six basic steps in problem solving?
2. Why is the team approach so important in problem solving?
3. What are the basic rules for brainstorming?
4. Why is brainstorming such an effective tool in problem solving?
5. A vendor claims that a specific critical measurement has a mean of .600 with a standard deviation of .0013. The product specifications are .600 ± .005. A random sample is taken from the incoming parts, and the following measurements are recorded:

.596	.598	.600	.602	.599	.598	.599	.601	.600	.600	.602
.603	.601	.604	.598	.598	.599	.597	.596	.599	.598	.601
.604	.600	.600	.602	.601	.601	.602	.599	.600	.600	.598
.597	.598	.598	.600	.602	.597	.599	.600	.601	.600	.599
.599	.594	.598	.595	.596	.600	.600	.602	.598	.599	.599
.598	.602	.605	.601	.600	.596	.601	.600	.600	.599	.600
.599	.601	.600	.598	.600	.598	.600	.602	.600	.598	.600
.599	.599	.600	.600	.598	.599	.597	.598	.601	.602	.601
.597	.598	.600	.600	.601	.600	.598	.600	.601	.600	.601
.599										

   a. What is the PCR, according to the vendor's information?
   b. What are the mean and standard deviation from the random sample?
   c. What is the PCR from the sample information?
   d. Does the histogram of the sample indicate that the product is normally distributed?
   e. Does the sample information indicate that the vendor is supplying honest information?
6. Complete a tree diagram outlining a quality improvement program for a company. Specify the type of company, its size, and any other important details.
7. Analyze the possible causes of the quiz grade dips on Figure 11.22.
8. What are 10 important preparations or characteristics of good problem-solving teams?
9. Why do individual contributions have to be downplayed or deemphasized on problem-solving teams?
10. Why should the "process owner" be consulted by the problem-solving team?
11. Why should the number of members on a problem-solving team be < 11?
12. Give an example of the following leadership methods:
    a. Directing without dominating.
    b. Problem solving without overinvolvement.
    c. Developing without overaccommodating.
    d. Delegating without abdicating responsibilities.
13. What sources of information are available to verify a vendor's honesty and capability? Why is it necessary to do this?
14. Why should two measurements on the same part be checked on a scatterplot?
15. How could the eight types of errors listed in Section 11.5 cost you points on your next examination? How could any of these be mistake-proofed?
16. Analyze the control chart in Figure 11.22.
17. Make your learning process flowchart for each of the courses you are taking.

12

GAUGE CAPABILITY

OBJECTIVES

- Know the effect of accuracy, repeatability, reproducibility, and stability on gauge readings.
- Determine the extent of repeatability and reproducibility variation with a gauge capability study.
- Define maximum deflection and explain the ways accuracy and stability contribute to it.
- Reassess process capability by eliminating the effect of gauge variability.
- Understand the concept of indecisive zones in gauge readings.

Every measuring instrument is subject to variation. To use SPC effectively, gauges must be analyzed to determine the extent of gauge variability. The variation that occurs on a control chart is actually a combination of product variation and gauge variation. Hopefully, the gauge variation is minimized so that the control chart interpretation will reflect primarily process variation.

One descriptive approach to gauge capability involves accuracy and precision. A target analogy best illustrates the two. The measurements shown in Figure 12.1*a* (shots on the target) are precise but not accurate. The measurements shown in Figure 12.1*b* are neither accurate nor precise. Figure 12.1*c* shows measurements that are accurate but not precise. Finally, the measurements in Figure 12.1*d* are both accurate and precise.

A lack of precision reflects an excessive amount of variation in the measurements. A lack of accuracy indicates that the average measurement is off target. The appropriate

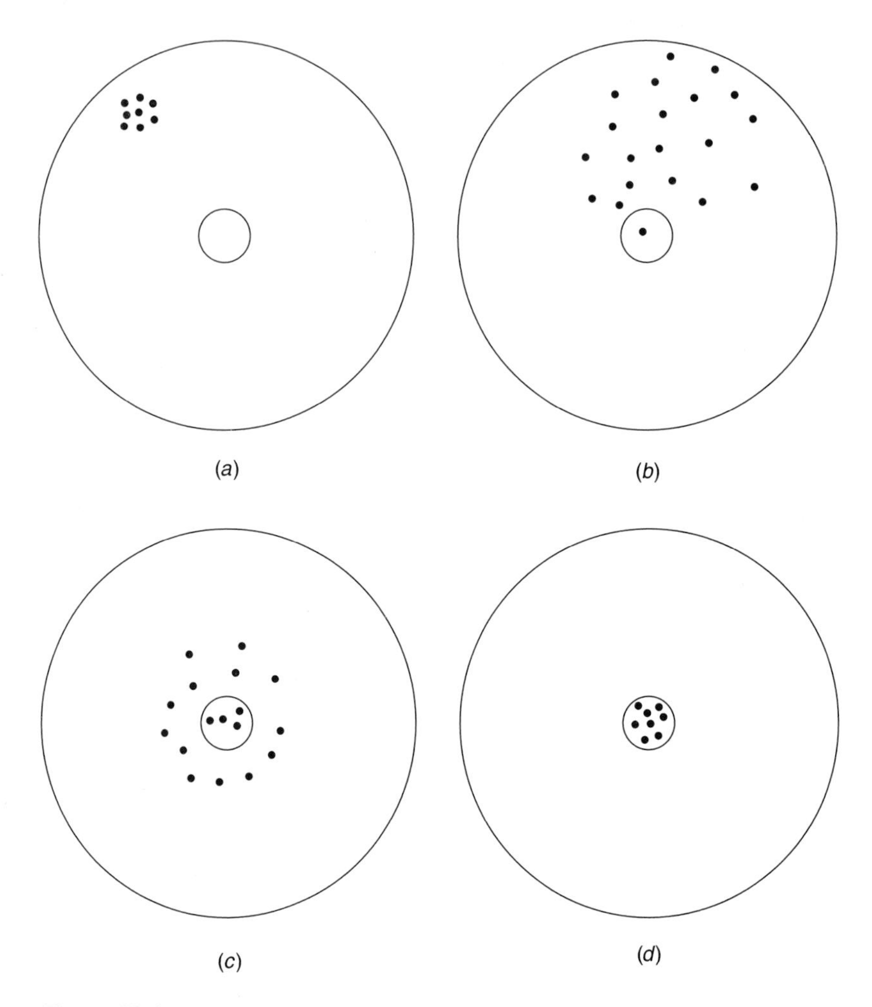

Figure 12.1
(*a*) Measurements are precise but not accurate. (*b*) Measurements are neither accurate nor precise. (*c*) Measurements are accurate but not precise. (*d*) Measurements are both accurate and precise.

corrective steps to either problem can be determined with a gauge capability analysis. A test of gauge capability can measure four individual characteristics and their combined effect.[1]

[1]Robert W. Traver, "Measuring Equipment Repeatability—The Rubber Ruler" (*American Society for Quality Control Annual Convention Transactions,* 1962). Reprinted with the permission of the American Society for Quality Control.

1. Accuracy: The gauge can be calibrated to make the readings more accurate. Accuracy measures the difference between the observed average measurement and the "true" value, although the true value may be unknown in many cases. There is usually some acceptable standard of comparison to determine accuracy of measurements.

2. Repeatability: Measures the consistency of readings of the same item by one person. Poor repeatability usually reflects internal gauge problems. Repeatability analysis can be used for training purposes as well. Given a gauge with very little repeatability variation, a trainee is finished with instruction on gauge use when she or he can match the repeatability standard for the gauge.

3. Reproducibility: Shows the variation in average measurement when different people use the same gauge. An excessive reproducibility value reflects a training problem.

4. Stability: Assesses the difference in average measurement over a long period of time. Gauge wear, upkeep, and periodic calibration are all factors of gauge stability.

The overall accuracy of a gauge is the combined effect of its accuracy, repeatability, reproducibility, and stability.

12.1 PREPARATIONS FOR A GAUGE CAPABILITY STUDY

Before collecting any data, complete the planning. Use the following questions to evaluate the need for the study:

- Is individual gauge capability assessment needed?
- Is training in gauge use in question?
- Is there a gauge problem in some phase of the operation?
- Is this study a preparation for control charting?

Also consider the details of the study procedure:

- How many people will be involved in taking the readings?
- How many like pieces are to be measured?
- How many repeat readings will be made by each individual?

Be sure the gauge is potentially able to do the job. The rule of 10 states that the gauge should be at least one-tenth as accurate as the tolerance of the characteristic that is being measured. If the tolerance is .001, the gauge should read to the nearest .0001. When working with A processes, in which the 6σ spread of the measurements uses less than half of the tolerance, the rule of 10 is amended. In that case more precision is needed and the gauge should be read to one-tenth of the 6σ value.

12.2 THE GAUGE CAPABILITY PROCEDURE

The following directions refer to the gauge capability worksheet. There are 15 steps listed in the procedure, and Figure 12.2 indicates the section on the gauge capability worksheet that corresponds to each step. Example 12.1 works along with the directions. As each step is presented, do the appropriate work on the worksheet in Figure 12.3. Check your results with the completed worksheet in Figure 12.6. You can also use the worksheet in Figure 12.2 as a "roadmap" because it matches the worksheet section with the procedure step.

Step 1. The pieces being measured should be numbered. The worksheet shows room for 10 different pieces.

Step 2. The measurements should be taken in random order.

Step 3. A set of "blind" measurements is best: The individual taking a measurement does not know which piece is being measured, what the previous measurements were on that piece, or what measurements other employees found on that piece.

Step 4. Record all the measurements on the worksheet. There is room on the chart for three sets of measurements by three different operators.

Step 5. After all the measurements are taken, calculate the trial range for each piece for each operator and record the values in columns 4, 8, and 12.

Step 6. Calculate all column totals and record the values in the totals row.

Step 7. Calculate the averages $\bar{x}_A$, $\bar{x}_B$, and $\bar{x}_C$. Transfer the nine trial column totals to the directed boxes at the extension of columns 2, 6, and 10 (follow the arrows). Add the three column totals of each operator for the sum; then divide the sum by 30 for each $\bar{x}$ value. Each operator has taken 30 measurements, so the average measurement is the sum divided by 30.

Step 8. Calculate $\overline{R}$. As directed in the previous step, an average measurement is needed for each operator. The average range, however, is determined from *all* the measurements. The basic calculation is to add all 30 range values and then divide by 30. The worksheet does this a little differently by using the column totals.

Calculate the range column averages by dividing the column total by 10. Do this for columns 4, 8, and 12. Transfer the averages to the box for step 8 (see Figure 12.2), add the three column averages, and divide by 3.

Step 9. Calculate the control limit for the ranges in the box below columns 6, 7, and 8. The box to the left gives the values of the constant D_4.

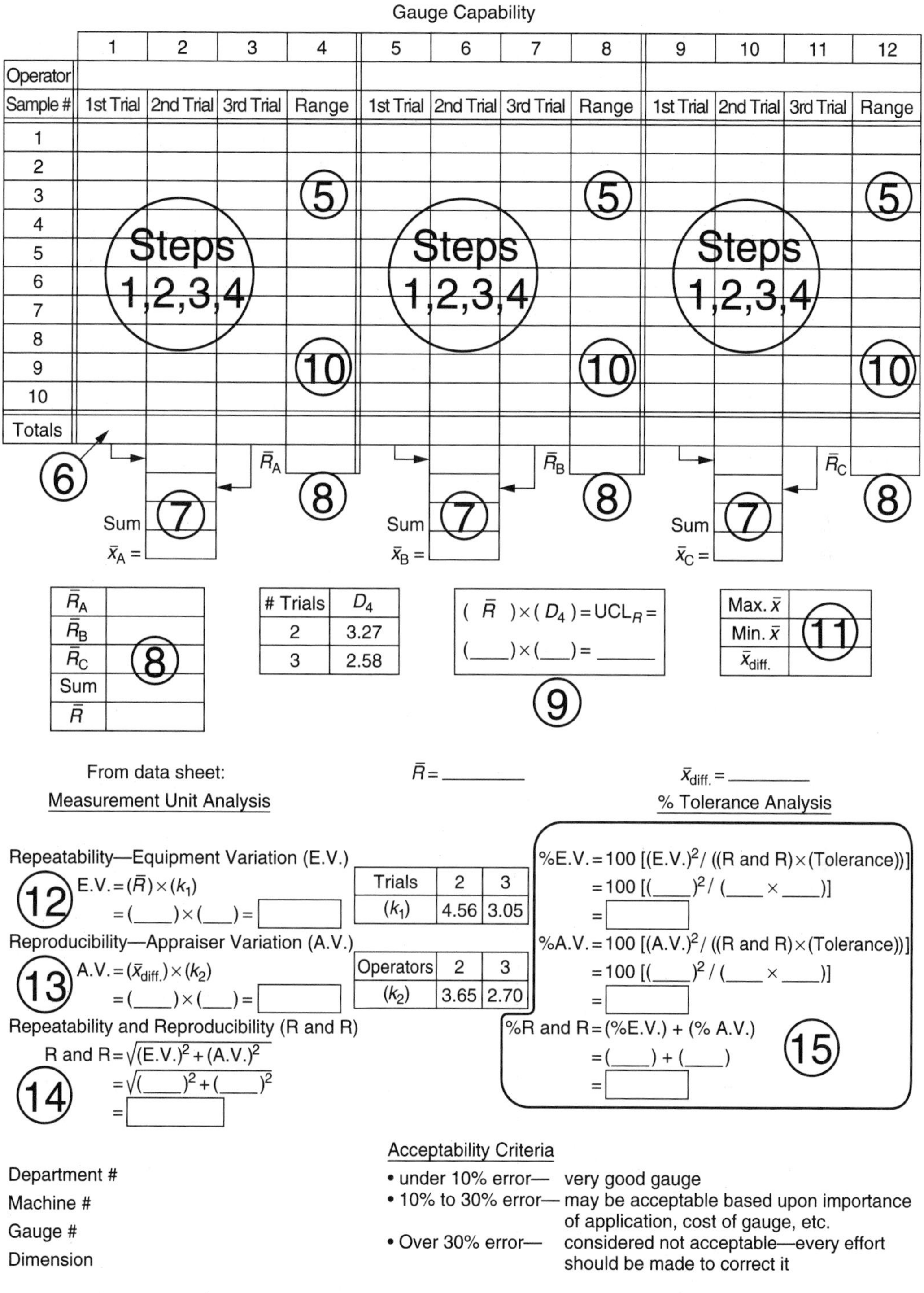

Gauge Capability

	1	2	3	4	5	6	7	8	9	10	11	12
Operator												
Sample #	1st Trial	2nd Trial	3rd Trial	Range	1st Trial	2nd Trial	3rd Trial	Range	1st Trial	2nd Trial	3rd Trial	Range
1												
2												
3				5				5				5
4												
5	Steps 1,2,3,4				Steps 1,2,3,4				Steps 1,2,3,4			
6												
7												
8												
9				10				10				10
10												
Totals												

6

$\bar{R}_A$ 8 — Sum 7 — $\bar{x}_A =$

$\bar{R}_B$ 8 — Sum 7 — $\bar{x}_B =$

$\bar{R}_C$ 8 — Sum 7 — $\bar{x}_C =$

$\bar{R}_A$	
$\bar{R}_B$	
$\bar{R}_C$	8
Sum	
$\bar{R}$	

# Trials	D_4
2	3.27
3	2.58

$(\ \bar{R}\ ) \times (\ D_4\ ) = UCL_R =$

(____)×(____)= _______

9

Max. $\bar{x}$	
Min. $\bar{x}$	11
$\bar{x}_{diff.}$	

From data sheet: $\bar{R} =$ _________ $\bar{x}_{diff.} =$ _________

Measurement Unit Analysis

Repeatability—Equipment Variation (E.V.)

12 $E.V. = (\bar{R}) \times (k_1)$

= (____)×(____)= [ ]

Trials	2	3
(k_1)	4.56	3.05

Reproducibility—Appraiser Variation (A.V.)

13 $A.V. = (\bar{x}_{diff.}) \times (k_2)$

= (____)×(____)= [ ]

Operators	2	3
(k_2)	3.65	2.70

Repeatability and Reproducibility (R and R)

14 R and R $= \sqrt{(E.V.)^2 + (A.V.)^2}$

$= \sqrt{(\_\_\_\_)^2 + (\_\_\_\_)^2}$

= [ ]

% Tolerance Analysis

%E.V. = 100 [(E.V.)²/ ((R and R)×(Tolerance))]

= 100 [(____)²/ (____ × ____)]

= [ ]

%A.V. = 100 [(A.V.)²/ ((R and R)×(Tolerance))]

= 100 [(____)²/ (____ × ____)]

= [ ]

%R and R = (%E.V.) + (% A.V.)

= (____) + (____)

= [ ]

15

Department #

Machine #

Gauge #

Dimension

Acceptability Criteria

- under 10% error— very good gauge
- 10% to 30% error— may be acceptable based upon importance of application, cost of gauge, etc.
- Over 30% error— considered not acceptable—every effort should be made to correct it

Figure 12.2

Sections of the gauge capability worksheet are marked to correspond to the steps of the procedure. (Gauge capability analysis form courtesy of Saginaw Division of General Motors Corporation.)

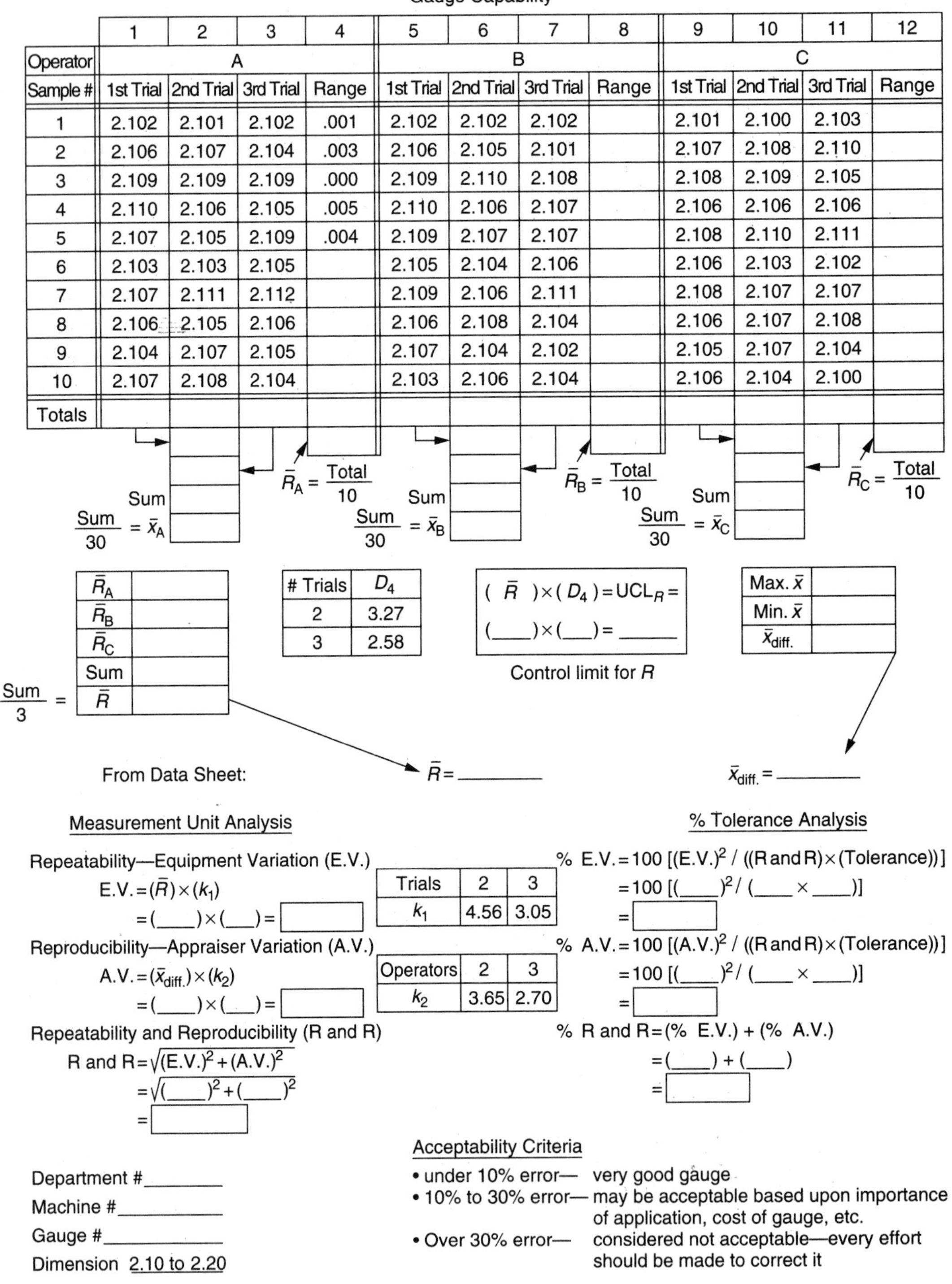

Gauge Capability

	1	2	3	4	5	6	7	8	9	10	11	12
Operator	A				B				C			
Sample #	1st Trial	2nd Trial	3rd Trial	Range	1st Trial	2nd Trial	3rd Trial	Range	1st Trial	2nd Trial	3rd Trial	Range
1	2.102	2.101	2.102	.001	2.102	2.102	2.102		2.101	2.100	2.103	
2	2.106	2.107	2.104	.003	2.106	2.105	2.101		2.107	2.108	2.110	
3	2.109	2.109	2.109	.000	2.109	2.110	2.108		2.108	2.109	2.105	
4	2.110	2.106	2.105	.005	2.110	2.106	2.107		2.106	2.106	2.106	
5	2.107	2.105	2.109	.004	2.109	2.107	2.107		2.108	2.110	2.111	
6	2.103	2.103	2.105		2.105	2.104	2.106		2.106	2.103	2.102	
7	2.107	2.111	2.112		2.109	2.106	2.111		2.108	2.107	2.107	
8	2.106	2.105	2.106		2.106	2.108	2.104		2.106	2.107	2.108	
9	2.104	2.107	2.105		2.107	2.104	2.102		2.105	2.107	2.104	
10	2.107	2.108	2.104		2.103	2.106	2.104		2.106	2.104	2.100	
Totals												

Sum

$\frac{\text{Sum}}{30} = \bar{x}_A$ $\bar{R}_A = \frac{\text{Total}}{10}$

Sum

$\frac{\text{Sum}}{30} = \bar{x}_B$ $\bar{R}_B = \frac{\text{Total}}{10}$

Sum

$\frac{\text{Sum}}{30} = \bar{x}_C$ $\bar{R}_C = \frac{\text{Total}}{10}$

$\bar{R}_A$	
$\bar{R}_B$	
$\bar{R}_C$	
Sum	
$\frac{\text{Sum}}{3} =$ $\bar{R}$	

# Trials	D_4
2	3.27
3	2.58

$(\ \bar{R}\ ) \times (\ D_4\ ) = UCL_R =$

(____)×(___)= ______

Control limit for R

Max. $\bar{x}$	
Min. $\bar{x}$	
$\bar{x}_{diff.}$	

From Data Sheet: $\bar{R}$ = ________ $\bar{x}_{diff.}$ = ________

Measurement Unit Analysis

Repeatability—Equipment Variation (E.V.)

E.V. = $(\bar{R}) \times (k_1)$

= (____) × (___) = [ ]

Trials	2	3
k_1	4.56	3.05

Reproducibility—Appraiser Variation (A.V.)

A.V. = $(\bar{x}_{diff.}) \times (k_2)$

= (____) × (___) = [ ]

Operators	2	3
k_2	3.65	2.70

Repeatability and Reproducibility (R and R)

R and R = $\sqrt{(\text{E.V.})^2 + (\text{A.V.})^2}$

= $\sqrt{(\_\_\_\_)^2 + (\_\_\_\_)^2}$

= [ ]

% Tolerance Analysis

% E.V. = 100 [(E.V.)^2 / ((R and R) × (Tolerance))]

= 100 [(____)^2 / (____ × ____)]

= [ ]

% A.V. = 100 [(A.V.)^2 / ((R and R) × (Tolerance))]

= 100 [(____)^2 / (____ × ____)]

= [ ]

% R and R = (% E.V.) + (% A.V.)

= (____) + (____)

= [ ]

Acceptability Criteria

- under 10% error— very good gauge
- 10% to 30% error— may be acceptable based upon importance of application, cost of gauge, etc.
- Over 30% error— considered not acceptable—every effort should be made to correct it

Department #________

Machine #__________

Gauge #____________

Dimension 2.10 to 2.20

Figure 12.3
Gauge capability worksheet. (Gauge capability analysis form courtesy of Saginaw Division of General Motors Corporation.)

Step 10. Scan the range values in columns 4, 8, and 12 for points out of control. If any out-of-control values occur, take one of two options:

1. Eliminate the out-of-control points from the set of data and recalculate the $\bar{x}$'s, $\bar{R}$'s, and UCL_R's. Be sure to change the calculation divisors as necessary. Remember, every average is the sum divided by the number of values.
2. Have the operator measure that particular item again if only one value is clearly in error. If a few errors have been made, replace all three measurements by that particular operator in each error row. Mix in a few other measurements to keep the test blind (unbiased). Recalculate the necessary $\bar{x}$'s, $\bar{R}$'s, and UCL_R's. The calculation divisors will remain the same because the number of data values has not changed with the replacement(s).

Step 11. Calculate the $\bar{x}$ range. Subtract the smallest $\bar{x}$ value from the largest. This is labeled $\bar{x}_{\text{diff}}$ on the calculation sheet.

Step 12. Complete a repeatability analysis. On the calculation sheet, repeatability is given the symbol E. V. (for equipment variation, the main cause for repeatability errors).

Step 13. Do a reproducibility analysis. Reproducibility is represented on the calculation sheet by A. V. (for appraisor variation, the main cause for reproducibility errors). The k_2 constant is given in the box on the calculation sheet next to the A. V. calculation section.

Step 14. The combined effect of repeatability and reproducibility, called R and R on the calculation sheet is found by adding their squared values and taking the square root of that sum. The $(\text{E. V.})^2$ and $(\text{A. V.})^2$ represent variability measures called the *variance.* When two or more sources of variability are present, the combined effect is found by adding their variances. The square root of the variance is the standard deviation, so the last step, the square root, gives R and R as a standard deviation measure of variability. The constant factor present in the E. V. and A. V. calculation, *k,* makes R and R a measurement value.

Step 15. Compare R and R with the tolerance to determine the percent error relative to the tolerance.

EXAMPLE 12.1 Figure 12.3 is a gauge capability worksheet. Steps 1 through 4 have been completed. Three operators have measured 10 different pieces three times each in the manner described in steps 2 and 3. Do the calculations on Figure 12.3 as the following steps direct. Check your results with Figure 12.6 (presented later in this example).

Solution

Begin at step 5, because steps 1 through 4 have been done. Calculate the range: On chart 1, operator A, piece 1 has a high of 2.102 and a low of 2.101. Subtract for a range of .001.

Piece 2 has a high of 2.107 and a low of 2.104. Subtract for a range of .003. Continue this process for the rest of the range values.

Record the totals for step 6 and do the calculations for step 7:

Calculations for operator A:

21.061	Column 1 total
21.062	Column 2 total
+ 21.061	Column 3 total
63.184	Sum

$$\bar{x}_A = \frac{\text{sum}}{30}$$

$$= \frac{63.184}{30} = 2.1061$$

Calculations for operator B:

21.066	Column 5 total
21.058	Column 6 total
+ 21.052	Column 7 total
63.176	Sum

$$\bar{x}_B = \frac{\text{sum}}{30}$$

$$= \frac{63.176}{30} = 2.1059$$

Calculations for operator C:

21.061	Column 9 total
21.061	Column 10 total
+ 21.056	Column 11 total
63.178	Sum

$$\bar{x}_C = \frac{\text{sum}}{30}$$

$$= \frac{63.178}{30} = 2.1059$$

For step 8, calculate the average range:

$$\bar{R} = \frac{\bar{R}_A + \bar{R}_B + \bar{R}_C}{3}$$

$$= \frac{.0028 + .0032 + .0029}{3}$$

$$= \frac{.0089}{3}$$
$$= .00297$$

Calculate the control limit for step 9. In this case, with three trials, $D_4 = 2.58$:

$$\begin{aligned} \text{UCL}_R &= D_4 \cdot \overline{R} \\ &= 2.58 \times .00297 \\ &= .00766 \end{aligned}$$

Look for out-of-control points (range values larger than .00766) and follow the procedure for step 10. For step 11, calculate the $\bar{x}$ range:

2.1061	Largest $\bar{x}$
− 2.1059	Smallest $\bar{x}$
.0002	$\bar{x}_{\text{diff}}$

At this point, all the boxes on the upper half of Figure 12.3 should be filled.

The repeated measurements for each piece will be normally distributed. To find the repeatability spread for 99% of the measurements on a piece, use a z value of 2.575. This is the z value from the normal-curve table that corresponds to an area of .005 in each tail, which results in a middle area of .99, or 99%. This is shown in Figure 12.4. That z value of 2.575 is combined with a few other necessary numerical constants and is represented

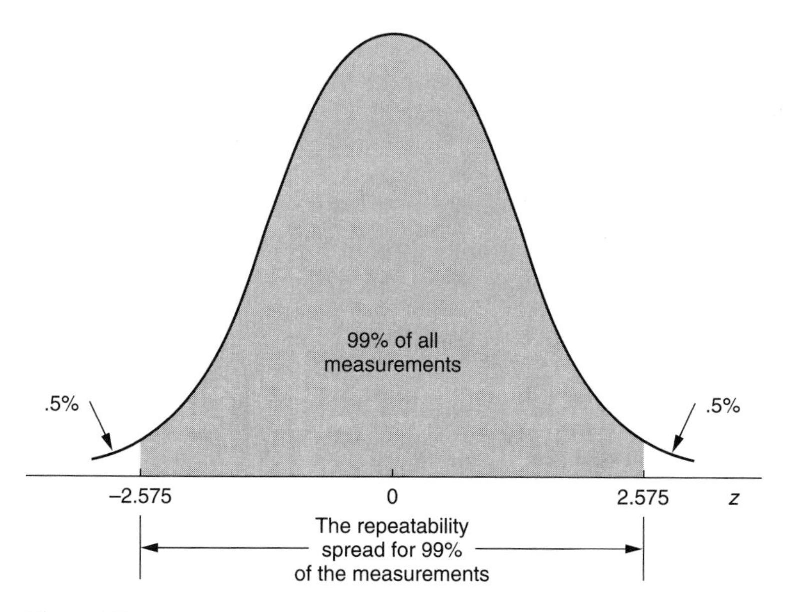

Figure 12.4
The distribution of measurements is normal in Example 12.1.

by k_1 on the calculation sheet. The k_1 constants are given in a box to the right of the E. V. calculation section:

$$\begin{aligned} \text{E. V.} &= \overline{R} \cdot k_1 \\ &= .00297 \times 3.05 \\ &= .00906 \end{aligned}$$

The variation caused by repeatability errors uses .00906 measurement units. Each individual piece has its average measurement, $\overline{x}$. Dividing the .00906 in half gives .0045, so 99% of the time the gauge will measure the piece between $\overline{x} - .0045$ and $\overline{x} + .0045$. This is illustrated in Figure 12.5.

Complete the reproducibility analysis for step 13 as follows:

$$\begin{aligned} \text{A. V.} &= \overline{x}_{\text{diff}} \cdot k_2 \\ &= .0002 \times 2.70 \\ &= .00054 \end{aligned}$$

The standard deviation must be calculated in step 14:

$$\begin{aligned} \text{R and R} &= \sqrt{(\text{E. V.})^2 + (\text{A. V.})^2} \\ &= \sqrt{(.00906)^2 + (.00054)^2} \\ &= \sqrt{.0000824} \\ &= .00908 \text{ measurement units} \end{aligned}$$

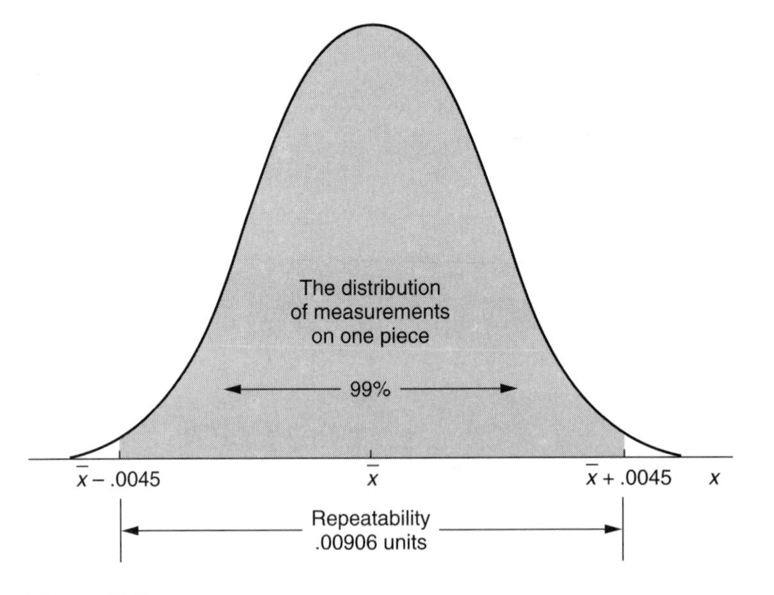

Figure 12.5
The range of measurements on one piece caused by repeatability variation.

(See Figure 12.7 at the end of this example.)

The following is the calculator sequence for R and R:

Enter .00906.

Square it. Either press the x^2 button or do the multiplication: $.00906 \times .00906 = .0000821$.

Store it in the memory: press M.

Enter .00054.

Square it.

Add it to the memory: press M+.

Recall the sum: press MR.

Square root it: press $\sqrt{\ }$.

To determine the percent error relative to the tolerance, first calculate the repeatability percent error:

$$\begin{aligned}
\%\ \text{E. V.} &= \frac{100(\text{E. V.})^2}{(\text{R and R}) \times \text{tolerance}} \\
&= \frac{100(.00906)^2}{.00908 \times .1} \\
&= \frac{100(.0000821)}{.000908} \\
&= \frac{.00821}{.000908} \\
&= 9.04\%
\end{aligned}$$

The calculator sequence for this process is shown here:

Press .00908, $\times$ .1, = .

Store the answer in the memory by pressing the M key.

Enter .00906.

Square it with the x^2 key or by multiplying.

Press $\times$, 100, = .

Divide by MR (the denominator, which was put in memory).

Press = .

Calculate the reproducibility percent error:

$$\begin{aligned}
\%\ \text{A. V.} &= \frac{100(\text{A. V.})^2}{(\text{R and R}) \times \text{tolerance}} \\
&= \frac{100(.00054)^2}{.00908 \times .1}
\end{aligned}$$

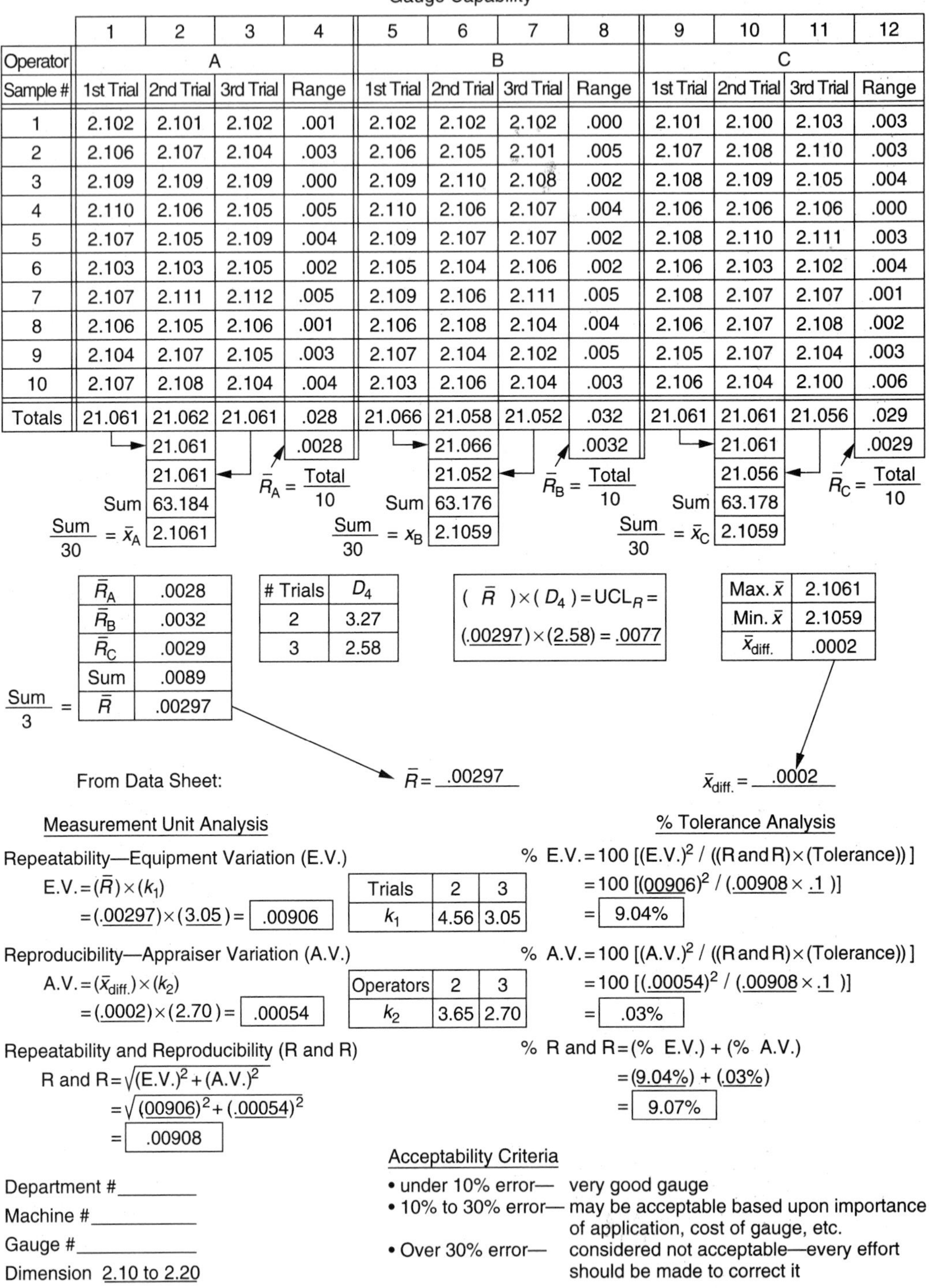

Gauge Capability

	1	2	3	4	5	6	7	8	9	10	11	12
Operator	A				B				C			
Sample #	1st Trial	2nd Trial	3rd Trial	Range	1st Trial	2nd Trial	3rd Trial	Range	1st Trial	2nd Trial	3rd Trial	Range
1	2.102	2.101	2.102	.001	2.102	2.102	2.102	.000	2.101	2.100	2.103	.003
2	2.106	2.107	2.104	.003	2.106	2.105	2.101	.005	2.107	2.108	2.110	.003
3	2.109	2.109	2.109	.000	2.109	2.110	2.108	.002	2.108	2.109	2.105	.004
4	2.110	2.106	2.105	.005	2.110	2.106	2.107	.004	2.106	2.106	2.106	.000
5	2.107	2.105	2.109	.004	2.109	2.107	2.107	.002	2.108	2.110	2.111	.003
6	2.103	2.103	2.105	.002	2.105	2.104	2.106	.002	2.106	2.103	2.102	.004
7	2.107	2.111	2.112	.005	2.109	2.106	2.111	.005	2.108	2.107	2.107	.001
8	2.106	2.105	2.106	.001	2.106	2.108	2.104	.004	2.106	2.107	2.108	.002
9	2.104	2.107	2.105	.003	2.107	2.104	2.102	.005	2.105	2.107	2.104	.003
10	2.107	2.108	2.104	.004	2.103	2.106	2.104	.003	2.106	2.104	2.100	.006
Totals	21.061	21.062	21.061	.028	21.066	21.058	21.052	.032	21.061	21.061	21.056	.029

Operator A: 21.061, 21.062, 21.061; Sum 63.184; $\frac{\text{Sum}}{30} = \bar{x}_A$ 2.1061; $\bar{R}_A = \frac{\text{Total}}{10}$ = .0028

Operator B: 21.066, 21.058, 21.052; Sum 63.176; $\frac{\text{Sum}}{30} = \bar{x}_B$ 2.1059; $\bar{R}_B = \frac{\text{Total}}{10}$ = .0032

Operator C: 21.061, 21.061, 21.056; Sum 63.178; $\frac{\text{Sum}}{30} = \bar{x}_C$ 2.1059; $\bar{R}_C = \frac{\text{Total}}{10}$ = .0029

$\bar{R}_A$	.0028
$\bar{R}_B$	.0032
$\bar{R}_C$	.0029
Sum	.0089
$\frac{\text{Sum}}{3} = \bar{R}$	.00297

# Trials	D_4
2	3.27
3	2.58

$(\ \bar{R}\ ) \times (\ D_4\ ) = UCL_R =$

$(.00297) \times (2.58) = .0077$

Max. $\bar{x}$	2.1061
Min. $\bar{x}$	2.1059
$\bar{x}_{diff.}$	.0002

From Data Sheet: $\bar{R} =$ .00297 $\qquad \bar{x}_{diff.} =$ .0002

Measurement Unit Analysis

Repeatability—Equipment Variation (E.V.)

E.V. $= (\bar{R}) \times (k_1)$

$= (.00297) \times (3.05) =$ .00906

Trials	2	3
k_1	4.56	3.05

Reproducibility—Appraiser Variation (A.V.)

A.V. $= (\bar{x}_{diff.}) \times (k_2)$

$= (.0002) \times (2.70) =$ .00054

Operators	2	3
k_2	3.65	2.70

Repeatability and Reproducibility (R and R)

R and R $= \sqrt{(\text{E.V.})^2 + (\text{A.V.})^2}$

$= \sqrt{(00906)^2 + (.00054)^2}$

$=$ .00908

% Tolerance Analysis

% E.V. $= 100\ [(\text{E.V.})^2 / ((\text{R and R}) \times (\text{Tolerance}))]$

$= 100\ [(00906)^2 / (.00908 \times .1\ )]$

$=$ 9.04%

% A.V. $= 100\ [(\text{A.V.})^2 / ((\text{R and R}) \times (\text{Tolerance}))]$

$= 100\ [(.00054)^2 / (.00908 \times .1\ )]$

$=$ .03%

% R and R = (% E.V.) + (% A.V.)

= (9.04%) + (.03%)

= 9.07%

Department #________

Machine #___________

Gauge #_____________

Dimension 2.10 to 2.20

Acceptability Criteria

- under 10% error— very good gauge
- 10% to 30% error— may be acceptable based upon importance of application, cost of gauge, etc.
- Over 30% error— considered not acceptable—every effort should be made to correct it

Figure 12.6
Gauge capability answer sheet for Example 12.1 (Gauge capability analysis form courtesy of Saginaw Division of General Motors Corporation.)

$$= \frac{100(.000000292)}{.000908}$$

$$= \frac{.0000292}{.000908}$$

$$= .03\%$$

The same calculator sequence can be used here. Simply use .00054 in place of .00906.

The total percent error can be found by adding the individual percent errors:

$$\begin{aligned} \%\ \text{R and R} &= \%\text{E. V.} + \%\text{A. V.} \\ &= 9.04\% + .03\% \\ &= 9.07\% \end{aligned}$$

Check all your calculations on Figure 12.3 with the completed chart in Figure 12.6. The criteria listed on the calculation sheet classify this gauge as very good (with less than 10% error).

The R and R figure calculated at step 14 represents the number of measurement units that account for 99% of the gauge variation of R and R. If the gauge is accurate and the average reading for a piece equals the true reading, then 99% of the readings on that piece will be between

$$\bar{x} - .00454 \qquad \text{and} \qquad \bar{x} + .00454$$

This is pictured in Figure 12.7.

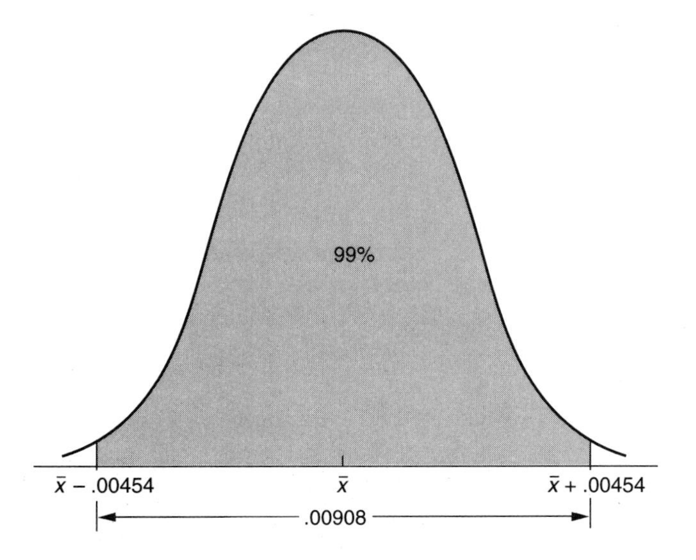

Figure 12.7
Measurement units that account for 99% of the repeatability and reproducibility variation.

EXAMPLE 12.2 Do a gauge capability study with the data in Figure 12.8. Follow steps 1 through 15; the actual work in this example starts at step 5. The chart in Figure 12.2 again shows the section of the chart that corresponds to each step in the procedure. Check your calculations with the completed work in Figures 12.9 and 12.10.

Solution

Complete steps 5, 6, and 7 (steps 1 through 4 have been completed). Check your results with the partially completed chart in Figure 12.9. Do the calculations for steps 8 and 9. Check for out-of-control points in step 10 using the result of step 9 ($\text{UCL}_R = .0061$). Complete the rest of the steps of the procedure.

In this example, the percentage of the tolerance involved in R and R variation is 73.5%, a very high value. This indicates that the gauge is inadequate for the job. Compared with Example 12.1, the big difference is the tolerance demand. The gauge in Example 12.1 varies in thousandths, but the tolerance is one-tenth. In this example, the variation is also in thousandths, but the tolerance is one-hundredth.

12.3 ANALYSIS OF R AND R WITH ACCURACY AND STABILITY: MAXIMUM POSSIBLE DEFLECTION

If a gauge is not accurate, a linear shift occurs either right or left. Figure 12.11 shows a shift to the right in the gauge readings. The *maximum deflection* from the true reading on a piece, illustrated in Figure 12.12, is calculated by

$$\frac{\text{R and R}}{2} + \text{accuracy shift}$$

The (R and R)/2 corresponds to the maximum possible one-way deflection caused by repeatability and reproducibility variation. The range of possible error in measurement units from the true value is determined from the following algebraic sums:

$$-\frac{\text{R and R}}{2} + \text{accuracy shift} \quad \text{to} \quad \frac{\text{R and R}}{2} + \text{accuracy shift}$$

The accuracy shift is a positive number if the gauge is measuring high or a negative number when the gauge is measuring low.

EXAMPLE 12.3 Given that the true reading on piece 2 used in Example 12.1 is 2.103, find:

1. The accuracy shift
2. The maximum possible deflection
3. The 99% range of gauge readings
4. The 99% range of true readings

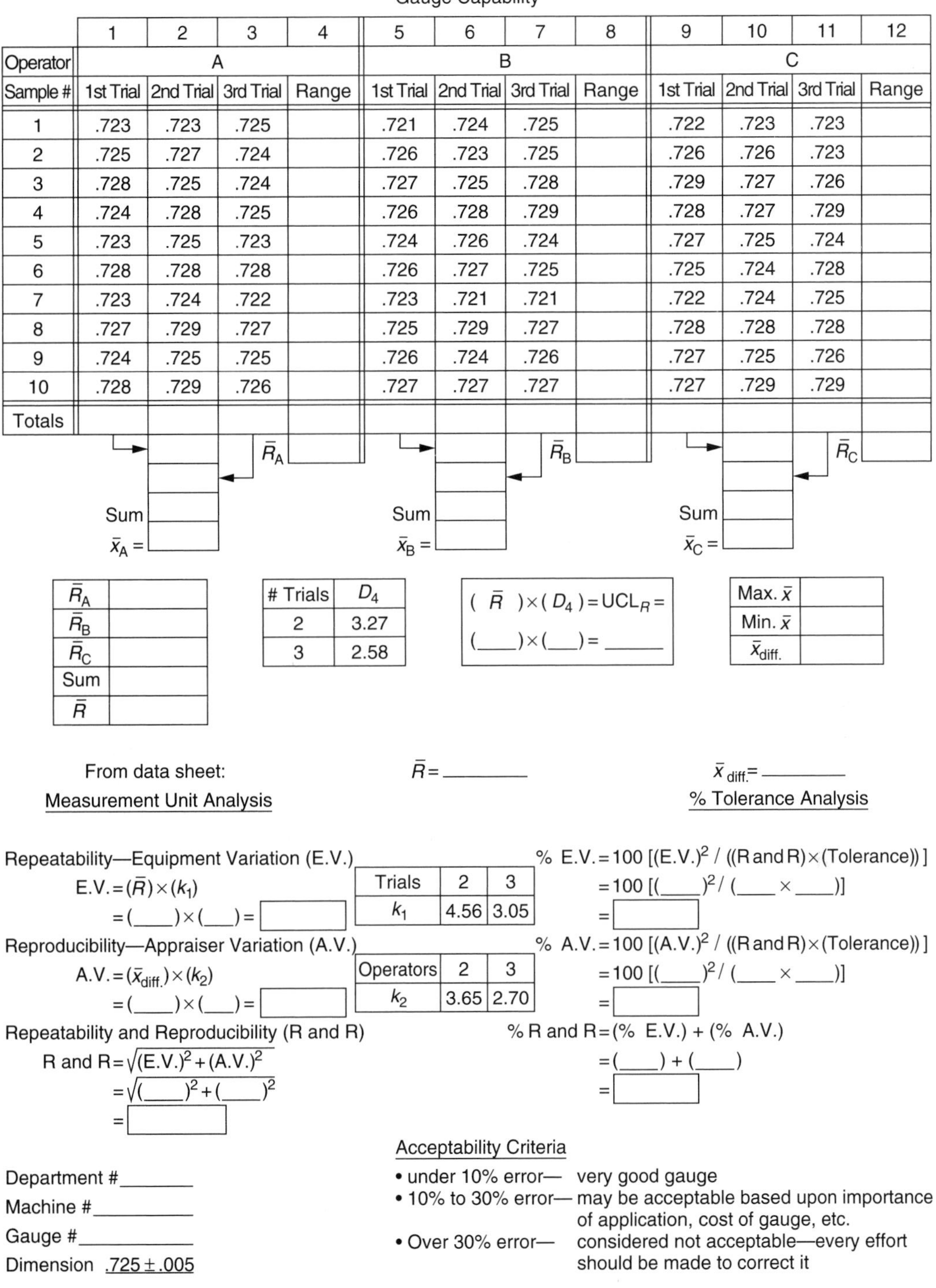

Gauge Capability

	1	2	3	4	5	6	7	8	9	10	11	12
Operator	A				B				C			
Sample #	1st Trial	2nd Trial	3rd Trial	Range	1st Trial	2nd Trial	3rd Trial	Range	1st Trial	2nd Trial	3rd Trial	Range
1	.723	.723	.725		.721	.724	.725		.722	.723	.723	
2	.725	.727	.724		.726	.723	.725		.726	.726	.723	
3	.728	.725	.724		.727	.725	.728		.729	.727	.726	
4	.724	.728	.725		.726	.728	.729		.728	.727	.729	
5	.723	.725	.723		.724	.726	.724		.727	.725	.724	
6	.728	.728	.728		.726	.727	.725		.725	.724	.728	
7	.723	.724	.722		.723	.721	.721		.722	.724	.725	
8	.727	.729	.727		.725	.729	.727		.728	.728	.728	
9	.724	.725	.725		.726	.724	.726		.727	.725	.726	
10	.728	.729	.726		.727	.727	.727		.727	.729	.729	
Totals												

$\bar{R}_A$ ___ $\bar{R}_B$ ___ $\bar{R}_C$ ___

Sum ___ $\bar{x}_A =$ ___ Sum ___ $\bar{x}_B =$ ___ Sum ___ $\bar{x}_C =$ ___

$\bar{R}_A$	
$\bar{R}_B$	
$\bar{R}_C$	
Sum	
$\bar{R}$	

# Trials	D_4
2	3.27
3	2.58

$(\ \bar{R}\ ) \times (\ D_4\ ) = UCL_R =$

$(\_\_\_) \times (\_\_\_) = \_\_\_\_\_\_$

Max. $\bar{x}$	
Min. $\bar{x}$	
$\bar{x}_{diff.}$	

From data sheet: $\bar{R} =$ ________ $\bar{x}_{diff.} =$ ________

Measurement Unit Analysis

Repeatability—Equipment Variation (E.V.)

E.V. $= (\bar{R}) \times (k_1)$

$= (\_\_\_) \times (\_\_\_) =$ ___

Trials	2	3
k_1	4.56	3.05

Reproducibility—Appraiser Variation (A.V.)

A.V. $= (\bar{x}_{diff.}) \times (k_2)$

$= (\_\_\_) \times (\_\_\_) =$ ___

Operators	2	3
k_2	3.65	2.70

Repeatability and Reproducibility (R and R)

R and R $= \sqrt{(E.V.)^2 + (A.V.)^2}$

$= \sqrt{(\_\_\_)^2 + (\_\_\_)^2}$

$=$ ___

% Tolerance Analysis

% E.V. $= 100\ [(E.V.)^2 / ((R\ and\ R) \times (Tolerance))]$

$= 100\ [(\_\_\_)^2 / (\_\_\_ \times \_\_\_)]$

$=$ ___

% A.V. $= 100\ [(A.V.)^2 / ((R\ and\ R) \times (Tolerance))]$

$= 100\ [(\_\_\_)^2 / (\_\_\_ \times \_\_\_)]$

$=$ ___

% R and R = (% E.V.) + (% A.V.)

$= (\_\_\_) + (\_\_\_)$

$=$ ___

Department # ________

Machine # ________

Gauge # ________

Dimension $.725 \pm .005$

Acceptability Criteria

- under 10% error— very good gauge
- 10% to 30% error— may be acceptable based upon importance of application, cost of gauge, etc.
- Over 30% error— considered not acceptable—every effort should be made to correct it

Figure 12.8

Gauge capability worksheet. (Gauge capability analysis form courtesy of Saginaw Division of General Motors Corporation.)

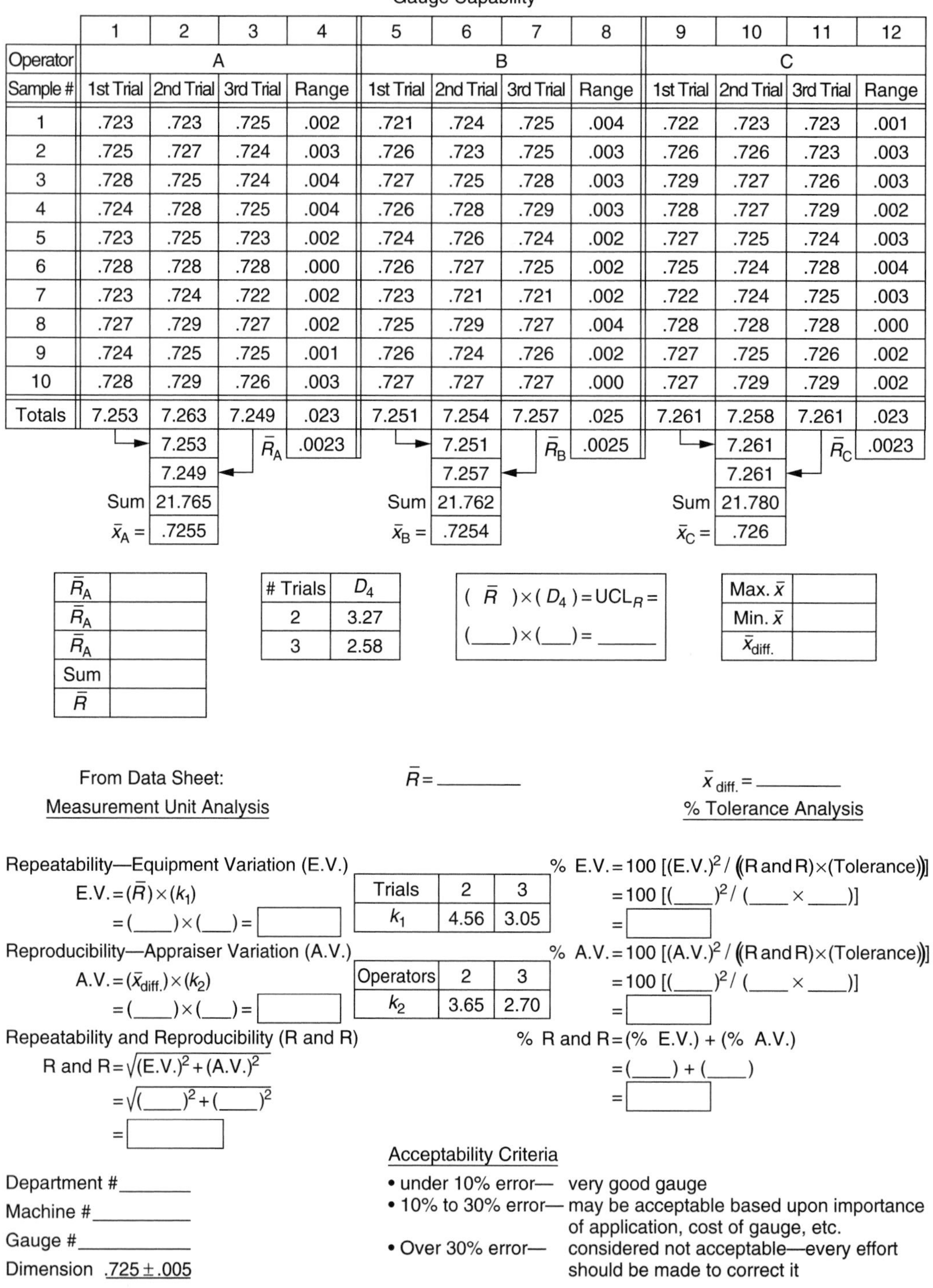

Gauge Capability

	1	2	3	4	5	6	7	8	9	10	11	12
Operator	A				B				C			
Sample #	1st Trial	2nd Trial	3rd Trial	Range	1st Trial	2nd Trial	3rd Trial	Range	1st Trial	2nd Trial	3rd Trial	Range
1	.723	.723	.725	.002	.721	.724	.725	.004	.722	.723	.723	.001
2	.725	.727	.724	.003	.726	.723	.725	.003	.726	.726	.723	.003
3	.728	.725	.724	.004	.727	.725	.728	.003	.729	.727	.726	.003
4	.724	.728	.725	.004	.726	.728	.729	.003	.728	.727	.729	.002
5	.723	.725	.723	.002	.724	.726	.724	.002	.727	.725	.724	.003
6	.728	.728	.728	.000	.726	.727	.725	.002	.725	.724	.728	.004
7	.723	.724	.722	.002	.723	.721	.721	.002	.722	.724	.725	.003
8	.727	.729	.727	.002	.725	.729	.727	.004	.728	.728	.728	.000
9	.724	.725	.725	.001	.726	.724	.726	.002	.727	.725	.726	.002
10	.728	.729	.726	.003	.727	.727	.727	.000	.727	.729	.729	.002
Totals	7.253	7.263	7.249	.023	7.251	7.254	7.257	.025	7.261	7.258	7.261	.023
		7.253	$\bar{R}_A$	.0023		7.251	$\bar{R}_B$	.0025		7.261	$\bar{R}_C$	.0023
		7.249				7.257				7.261		
	Sum	21.765			Sum	21.762			Sum	21.780		
	$\bar{x}_A =$	.7255			$\bar{x}_B =$	.7254			$\bar{x}_C =$	.726		

$\bar{R}_A$	
$\bar{R}_A$	
$\bar{R}_A$	
Sum	
$\bar{R}$	

# Trials	D_4
2	3.27
3	2.58

$(\bar{R}) \times (D_4) = UCL_R =$

(____) × (___) = ______

Max. $\bar{x}$	
Min. $\bar{x}$	
$\bar{x}_{diff.}$	

From Data Sheet: $\bar{R} =$ ________ $\bar{x}_{diff.} =$ ________

Measurement Unit Analysis

Repeatability—Equipment Variation (E.V.)

E.V. = $(\bar{R}) \times (k_1)$

= (____) × (___) = [ ]

Trials	2	3
k_1	4.56	3.05

Reproducibility—Appraiser Variation (A.V.)

A.V. = $(\bar{x}_{diff.}) \times (k_2)$

= (____) × (___) = [ ]

Operators	2	3
k_2	3.65	2.70

Repeatability and Reproducibility (R and R)

R and R = $\sqrt{(E.V.)^2 + (A.V.)^2}$

= $\sqrt{(\_\_\_\_)^2 + (\_\_\_\_)^2}$

= [ ]

% Tolerance Analysis

% E.V. = 100 [(E.V.)^2 / ((R and R) × (Tolerance))]

= 100 [(____)^2 / (____ × ____)]

= [ ]

% A.V. = 100 [(A.V.)^2 / ((R and R) × (Tolerance))]

= 100 [(____)^2 / (____ × ____)]

= [ ]

% R and R = (% E.V.) + (% A.V.)

= (____) + (____)

= [ ]

Department #________

Machine #__________

Gauge #____________

Dimension .725 ± .005

Acceptability Criteria

- under 10% error— very good gauge
- 10% to 30% error— may be acceptable based upon importance of application, cost of gauge, etc.
- Over 30% error— considered not acceptable—every effort should be made to correct it

Figure 12.9
Partially completed worksheet for Example 12.2. (Gauge capability analysis form courtesy of Saginaw Division of General Motors Corporation.)

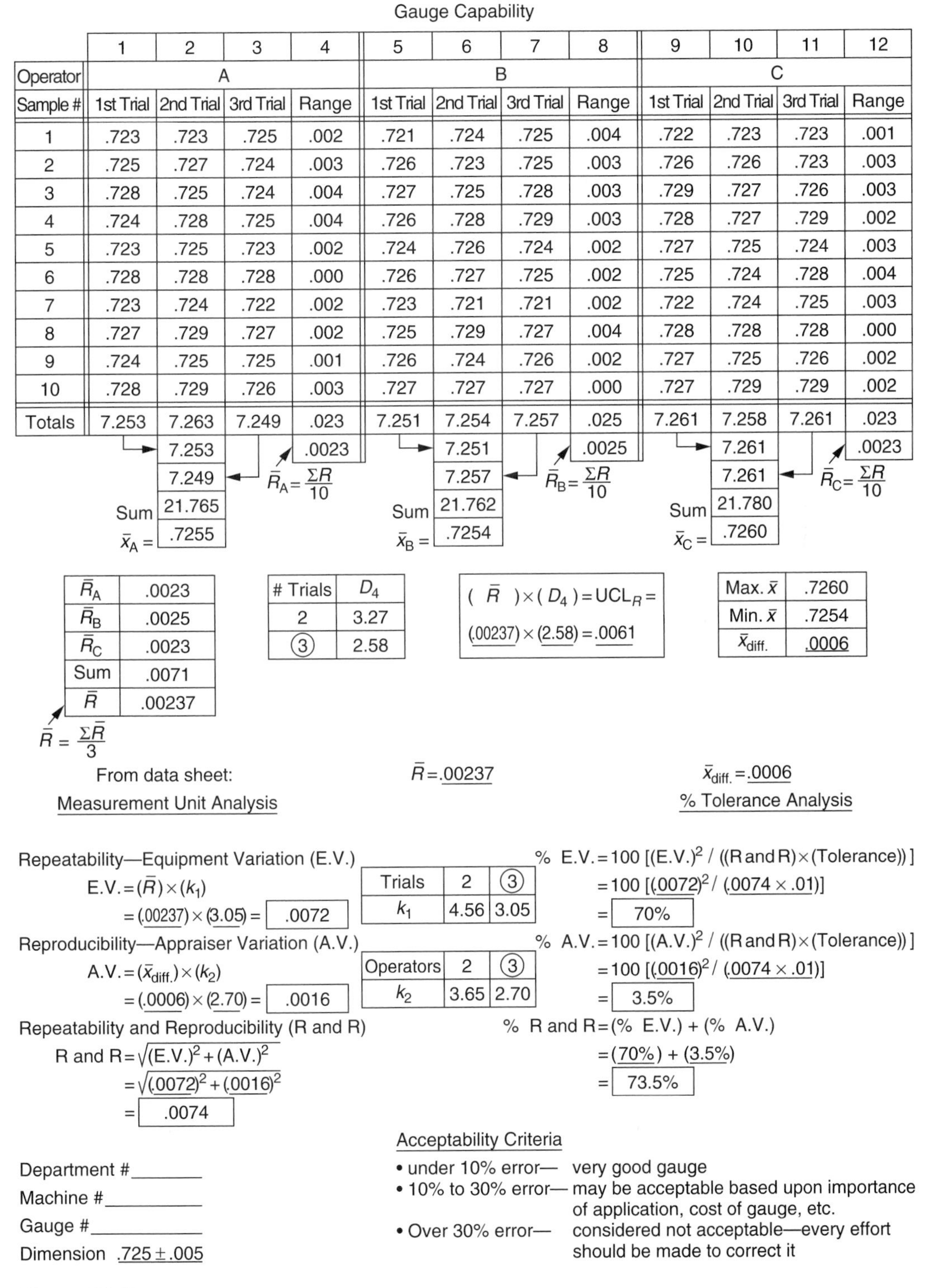

Gauge Capability

	1	2	3	4	5	6	7	8	9	10	11	12
Operator	A				B				C			
Sample #	1st Trial	2nd Trial	3rd Trial	Range	1st Trial	2nd Trial	3rd Trial	Range	1st Trial	2nd Trial	3rd Trial	Range
1	.723	.723	.725	.002	.721	.724	.725	.004	.722	.723	.723	.001
2	.725	.727	.724	.003	.726	.723	.725	.003	.726	.726	.723	.003
3	.728	.725	.724	.004	.727	.725	.728	.003	.729	.727	.726	.003
4	.724	.728	.725	.004	.726	.728	.729	.003	.728	.727	.729	.002
5	.723	.725	.723	.002	.724	.726	.724	.002	.727	.725	.724	.003
6	.728	.728	.728	.000	.726	.727	.725	.002	.725	.724	.728	.004
7	.723	.724	.722	.002	.723	.721	.721	.002	.722	.724	.725	.003
8	.727	.729	.727	.002	.725	.729	.727	.004	.728	.728	.728	.000
9	.724	.725	.725	.001	.726	.724	.726	.002	.727	.725	.726	.002
10	.728	.729	.726	.003	.727	.727	.727	.000	.727	.729	.729	.002
Totals	7.253	7.263	7.249	.023	7.251	7.254	7.257	.025	7.261	7.258	7.261	.023
		7.253		.0023		7.251		.0025		7.261		.0023
		7.249				7.257				7.261		
Sum		21.765				21.762				21.780		
		$\bar{x}_A =$ .7255				$\bar{x}_B =$ .7254				$\bar{x}_C =$ .7260		

$\bar{R}_A = \frac{\Sigma R}{10}$ $\quad \bar{R}_B = \frac{\Sigma R}{10}$ $\quad \bar{R}_C = \frac{\Sigma R}{10}$

$\bar{R}_A$	.0023
$\bar{R}_B$	.0025
$\bar{R}_C$	.0023
Sum	.0071
$\bar{\bar{R}}$	.00237

$\bar{\bar{R}} = \frac{\Sigma \bar{R}}{3}$

# Trials	D_4
2	3.27
(3)	2.58

$(\bar{\bar{R}}) \times (D_4) = UCL_R =$
$(.00237) \times (2.58) = .0061$

Max. $\bar{x}$	.7260
Min. $\bar{x}$	.7254
$\bar{x}_{diff.}$	.0006

From data sheet: $\bar{\bar{R}} = .00237$ $\qquad \bar{x}_{diff.} = .0006$

Measurement Unit Analysis

Repeatability—Equipment Variation (E.V.)

$E.V. = (\bar{\bar{R}}) \times (k_1)$
$= (.00237) \times (3.05) =$ .0072

Trials	2	(3)
k_1	4.56	3.05

Reproducibility—Appraiser Variation (A.V.)

$A.V. = (\bar{x}_{diff.}) \times (k_2)$
$= (.0006) \times (2.70) =$ .0016

Operators	2	(3)
k_2	3.65	2.70

Repeatability and Reproducibility (R and R)

$R \text{ and } R = \sqrt{(E.V.)^2 + (A.V.)^2}$
$= \sqrt{(.0072)^2 + (.0016)^2}$
$=$ .0074

% Tolerance Analysis

$\% \ E.V. = 100\ [(E.V.)^2 / ((R \text{ and } R) \times (\text{Tolerance}))]$
$= 100\ [(.0072)^2 / (.0074 \times .01)]$
$=$ 70%

$\% \ A.V. = 100\ [(A.V.)^2 / ((R \text{ and } R) \times (\text{Tolerance}))]$
$= 100\ [(.0016)^2 / (.0074 \times .01)]$
$=$ 3.5%

% R and R = (% E.V.) + (% A.V.)
= (70%) + (3.5%)
= 73.5%

Department #________
Machine #__________
Gauge #____________
Dimension .725 ± .005

Acceptability Criteria

- under 10% error— very good gauge
- 10% to 30% error— may be acceptable based upon importance of application, cost of gauge, etc.
- Over 30% error— considered not acceptable—every effort should be made to correct it

Figure 12.10
Gauge capability answer sheet for Example 12.2. (Gauge capability analysis form courtesy of Saginaw Division of General Motors Corporation.)

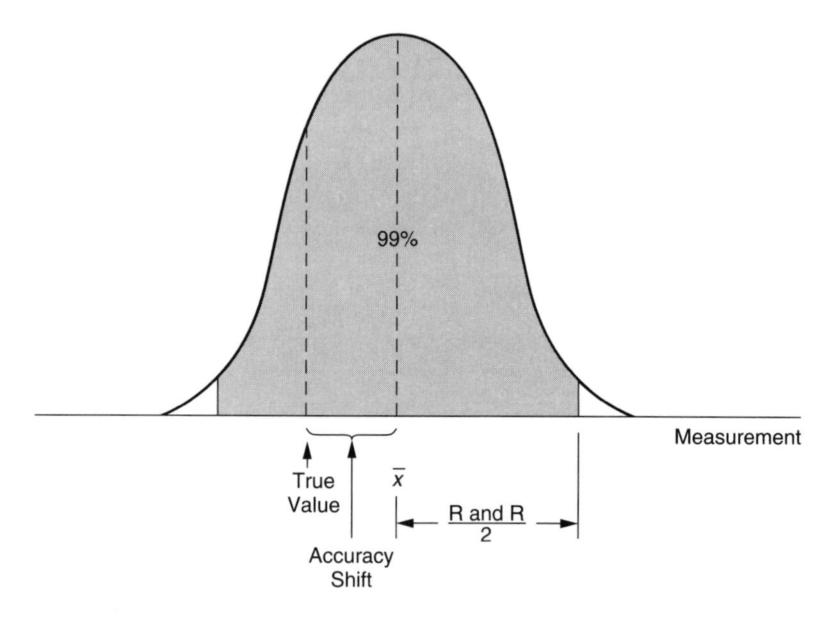

Figure 12.11
A shift in the distribution of gauge readings caused by an inaccurate gauge.

Solution
There are nine measurements of each piece (three measurements by each person), so the sum of the nine readings of piece 2 divided by 9 gives the average measurement:

$$\begin{aligned}\overline{x} &= \frac{\Sigma x}{N} \\ &= \frac{18.954}{9} \\ &= 2.106\end{aligned}$$

2.106	Average reading on the piece
− 2.103	True measurement
.003	Accuracy shift

The gauge variability caused by R and R is .00908. Half is .0045, so 99% of the gauge readings on piece 2 will be in the interval 2.106 ± .0045, or 2.1015 to 2.1105.

The maximum possible deflection from the true measurement is the sum of $\frac{1}{2}$(R and R) and the accuracy shift:

$$.0045 + .003 = .0075$$

The gauge is measuring high because of the accuracy shift, so a deduction of .003 from the gauge reading interval will give the interval in which the true measurement will be

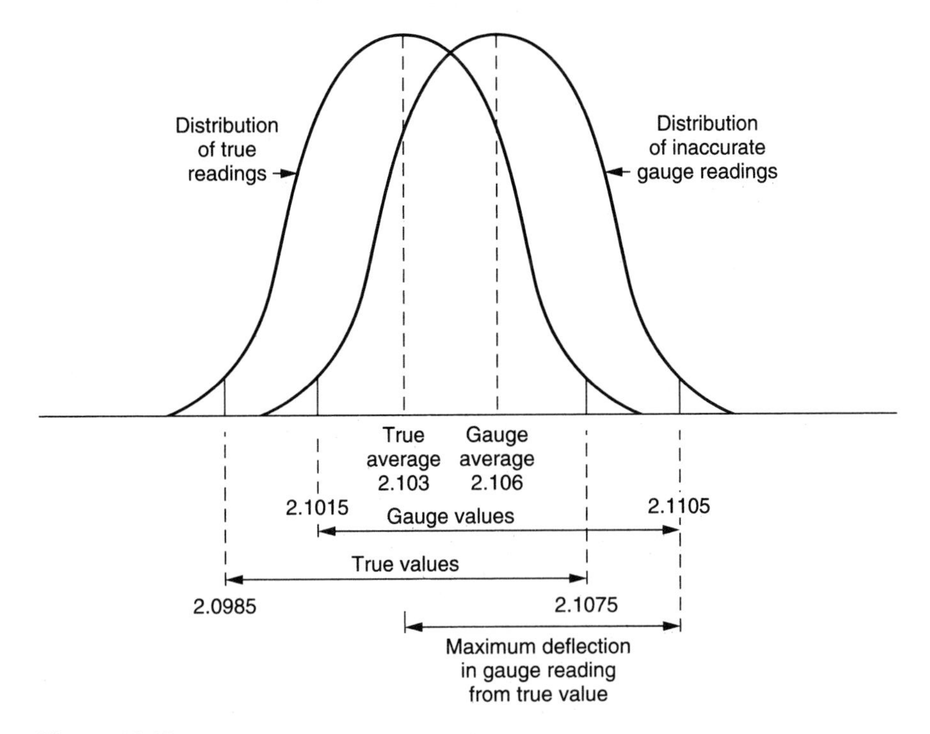

Figure 12.12
R and R variation: the gauge range for 99% of all measurements versus the true range for 99% of all measurements.

99% of the time. This new interval is 2.0985 to 2.1075. The calculations are illustrated in Figure 12.12.

Stability also represents a shift either left or right. The stability shift is added algebraically to the accuracy shift and then combined with R and R, as before, to determine the maximum possible deflection in measurement and the possible range in measurement.

EXAMPLE 12.4 Illustrate the added effect of stability when piece 2 from Example 12.1 is tested two weeks later and $\bar{x} = 2.104$.

Solution

This represents a shift left from the previous readings:

2.106	Previous average value
− 2.104	Latest average value
.002	Stability shift

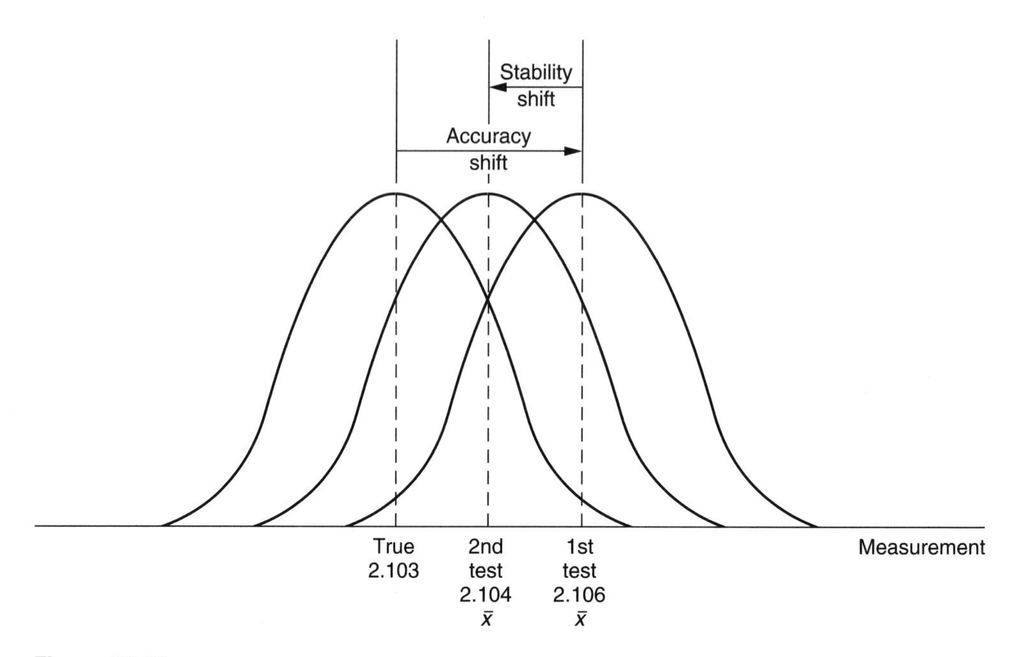

Figure 12.13
The distribution of measurements following an accuracy shift and a stability shift.

The stability shift is –.002 because the measurements shifted left. This is illustrated in Figure 12.13.

In this case the stability shifted in the opposite direction as the accuracy and canceled part of the accuracy shift. The combined accuracy and stability shift becomes

$$+.003 - .002 = +.001$$

The gauge is now reading .001 units high.

The maximum variation from the true value owing to gauge variability is

$$\frac{\text{R and R}}{2} + \text{accuracy shift} + \text{stability shift}$$

Therefore, .0045 + .003 – .002 is .0055 units. The gauge readings on piece 2 vary between 2.104 + .0045 and 2.104 – .0045, or 2.0995 and 2.1085, 99% of the time. The combined effect of accuracy and stability results in the gauge reading .001 units high on any measurement.

EXAMPLE 12.5 Do an accuracy analysis on piece 2 in Figure 12.10. Given that the true measurement of the piece is .727, complete these five steps:

1. Find the average measurement.
2. Find the accuracy shift.
3. Find the 99% range in gauge readings.

4. Find the 99% range in true readings.
5. Find the maximum possible deflection.

Solution

First, find $\bar{x}$:

$$\bar{x} = \frac{\Sigma x}{N} = \frac{6.525}{9} = .725$$

Second, the accuracy shift is the difference between the average and true measurements:

$$\begin{array}{rl} .725 & \text{Average measurement} \\ -\ .727 & \text{True measurement} \\ \hline -\ .002 & \text{Accuracy shift} \end{array}$$

The accuracy shift is a shift to the left on the measurement scale, so it is indicated by a negative number.

Third, half of R and R for the gauge is .0074/2 = .0037. The 99% range for the gauge readings on piece 2 is .725 ± .0037, or .7213 to .7287. For step 4, the 99% range for the true measurement of piece 2 is .727 ± .0037, or .7233 to .7307.

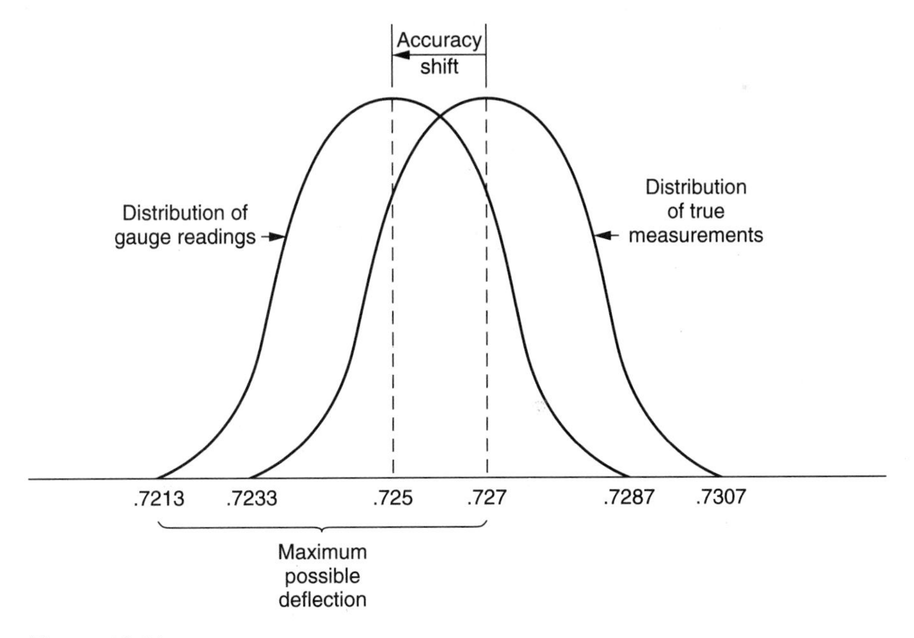

Figure 12.14
The accuracy shift and maximum possible deflection for Example 12.5.

Finally, the maximum possible deflection is

$$\text{Accuracy shift} + \frac{\text{R and R}}{2} = .002 + .0037$$
$$= .0057$$

The maximum possible deflection from the true measurement shifts left because that is the direction in which the combined effect of the two gauge variation factors is the greatest. This is illustrated in Figure 12.14.

12.4 THE ELIMINATION OF GAUGE VARIATION FROM PROCESS VARIATION

The measurement of process variability encompasses the accumulated effects of all sources of variation, including gauge variation, so the elimination of gauge variability allows a truer estimate of process variation. Gauge repeatability errors are generally the largest contributors to gauge variability and can be determined from the gauge capability analysis.

One concept concerning statistical variation states that if a process is in control and satisfies a normal distribution, the total variance caused by a combination of k variation sources is equal to the sum of the individual variances:

$$\sigma^2_{\text{Total}} = \sigma^2_1 + \sigma^2_2 + \cdots + \sigma^2_k$$

Standard deviation σ was introduced in Chapter 3 as a statistical measure of variability. The variance σ^2 is the square of the standard deviation value, so it is also a measure of variability. The concept that variability from several sources is additive should make sense intuitively. As a simple example, consider a lawn mower engine that is secured to a deck with four bolts. If one of the bolts becomes loose, there will be an increase in vibration with some detrimental effect on the various mower parts. If a second bolt loosens, will the vibration increase or decrease? The vibration and its effect will be worse with the two bolts loose than if either bolt were loose alone. The second source of variation causes an increase in total variability.

This concept can also be applied to process capability.[2] The total variance in the process is the sum of the process variance and the gauge variance:

$$\sigma^2_{\text{Total}} = \sigma^2_{\text{Gauge}} + \sigma^2_{\text{Process}}$$

By applying algebra and subtracting from both sides of the equation, the equation changes to a form in which the process variance is equal to the gauge variance subtracted from the total variance:

$$\sigma^2_{\text{Process}} = \sigma^2_{\text{Total}} - \sigma^2_{\text{Gauge}}$$

[2]Robert W. Traver, "Measuring Equipment Repeatability—The Rubber Ruler" (*American Society for Quality Control Annual Convention Transactions,* 1962). Reprinted with permission of the American Society for Quality Control.

The process standard deviation is the square root of the process variance:

$$\sigma_{\text{Process}} = \sqrt{\sigma^2_{\text{Total}} - \sigma^2_{\text{Gauge}}}$$

The total variability, as a standard deviation measurement, is calculated using $\overline{R}$ from the control charts:

$$\sigma_{\text{Total}} = \frac{\overline{R}}{d_2}$$

Gauge variability is mainly caused by gauge repeatability, and the standard deviation for gauge repeatability, σ_{Gauge}, is used as the estimate of gauge variability. The $\overline{R}$ from step 8 in the gauge capability study is used to determine the value of σ_{Gauge}:

$$\sigma_{\text{Gauge}} = \frac{\overline{R}}{d_2}$$

The two standard deviation formulas, σ_{Total} and σ_{Gauge}, look the same; they both equal $\overline{R}/d_2$. However, the two $\overline{R}$ and d_2 values differ. In the σ_{Total} formula, $\overline{R}$ comes from the control chart and d_2 depends on the sample size used in the control chart. For the σ_{Gauge} formula, $\overline{R}$ is the average range from step 8 in the gauge capability procedure and d_2 depends on the number of repeated measurements.

EXAMPLE 12.6

The gauge from the gauge capability study in Example 12.1 is used in a process in which the tolerance is .1. An $\overline{x}$ and R control chart from that process has samples of $n = 5$ and $\overline{R} = .0372$.

1. What is the total process capability?
2. What is the true process capability when the effect of gauge variation is removed?

Solution

For the capability calculations, the sample size on the control chart is $n = 5$. According to Table B.1 in Appendix B, $d_2 = 2.326$:

$$\begin{aligned}\sigma_{\text{Total}} &= \frac{\overline{R}}{d_2}\\ &= \frac{.0372}{2.326}\\ &= .016\end{aligned}$$

$$\begin{aligned}\text{PCR} &= \frac{6\sigma_{\text{Total}}}{\text{tolerance}}\\ &= \frac{6 \times .016}{.1}\\ &= \frac{.096}{.1}\\ &= .96\end{aligned}$$

The process uses 96% of the tolerance.

Second, from Figure 12.6, $\overline{R} = .00297$. Each measurement was taken three times, so the sample size is $n = 3$, and $d_2 = 1.693$:

$$\sigma_{\text{Gauge}} = \frac{\overline{R}}{d_2}$$
$$= \frac{.00297}{1.693}$$
$$= .00175$$
$$\sigma^2_{\text{Process}} = \sqrt{\sigma^2_{\text{Total}} - \sigma^2_{\text{Gauge}}}$$
$$= \sqrt{(.016)^2 - (.00175)}$$
$$= \sqrt{.000253}$$
$$= .0159$$
$$\text{PCR} = \frac{6\sigma_{\text{Process}}}{\text{tolerance}}$$
$$= \frac{6 \times .0159}{.1}$$
$$= .95$$

The conclusion of Example 12.1 indicates that the gauge is classified as very good. The measure of process capability changed from .96 to .95 when the gauge variability was eliminated. This indicates that the gauge has very little effect on the measure of process capability.

EXAMPLE 12.7 The gauge from the capability study in Example 12.2 is used in a process in which the tolerance is .01. A control chart from that process has samples of $n = 5$ and $\overline{R} = .00442$.

1. What is the total process capability?
2. What is the true process capability when the effect of the gauge variation is removed?

Solution

The calculations from the total process capability are as follows:

$$\sigma_{\text{Total}} = \frac{\overline{R}}{d_2}$$
$$= \frac{.00442}{2.326}$$
$$= .0019$$
$$6\sigma_{\text{Total}} = 6 \times .0019$$
$$= .0114$$

$$\text{PCR} = \frac{6\sigma_{\text{Total}}}{\text{tolerance}}$$
$$= \frac{.0014}{.01}$$
$$= 1.14$$

The process uses 114% of the tolerance, a poor process.

From Figure 12.10, $\overline{R}$ = .00237. Use this to calculate the second part of the problem:

$$\sigma_{\text{Gauge}} = \frac{\overline{R}}{d_2}$$
$$= \frac{.00237}{1.693}$$
$$= .0014$$
$$\sigma_{\text{Process}} = \sqrt{\sigma^2_{\text{Total}} - \sigma^2_{\text{Gauge}}}$$
$$= \sqrt{(.0019)^2 - (.0014)^2}$$
$$= \sqrt{.00000165}$$
$$= .00128$$
$$\text{PCR} = \frac{6\sigma_{\text{Process}}}{\text{tolerance}}$$
$$= \frac{6 \times .00128}{.01}$$
$$= \frac{.00768}{.01}$$
$$= .77$$

When gauge variability is eliminated, the process actually uses 77% of the tolerance.

The conclusion of Example 12.2 labels this gauge a very poor one. This example indicates that the gauge is a significant contributor to the measure of process capability. A process thought to use 114% of the tolerance actually uses 77% of the tolerance.

12.5 INDECISIVE GAUGE READINGS

The effect of gauge variability on process measurements is illustrated in Figure 12.15. Pieces whose true measurements are in section A in this figure will always be considered good because the gauge variation is not enough to throw the reading out of specification. The maximum gauge deflection is not large enough to measure a good piece as out of specification. All pieces whose true measurements are in sections B or C will be classified as good or bad according to the effect of gauge variation on that individual measurement. If the gauge variation throws the reading toward the middle, so that it is in specification, the piece is considered good. If the gauge variation throws the reading toward the outside,

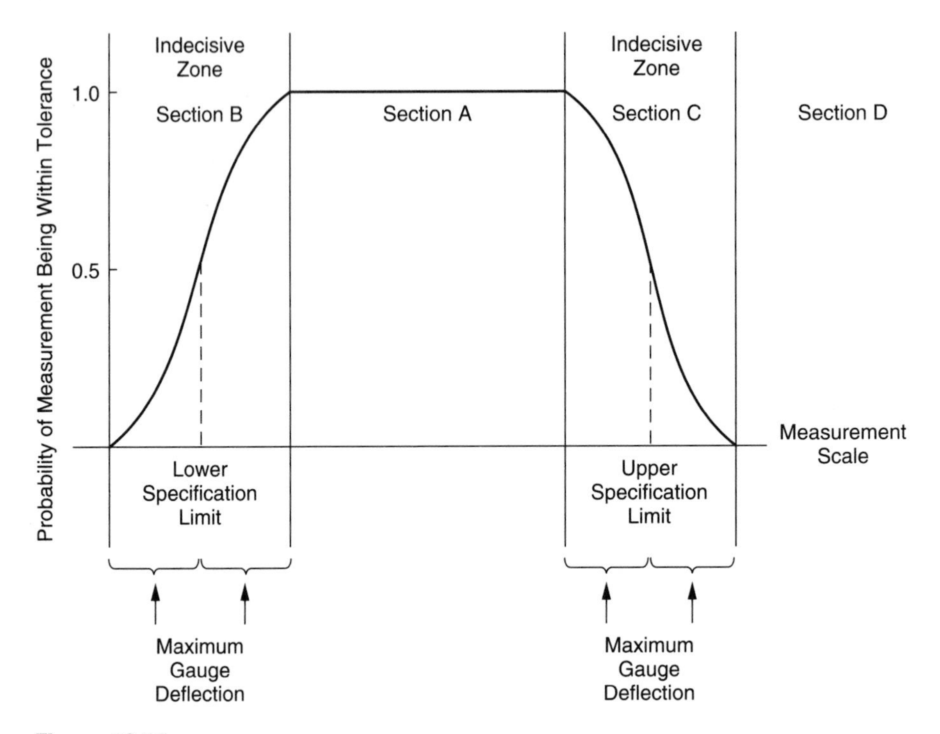

Figure 12.15
Indecisive zones in gauge reading. (From Robert W. Traver, "Measuring Equipment Repeatability—The Rubber Ruler" [*American Society for Quality Control Annual Convention Transactions,* 1962]. Reprinted with permission of the American Society for Quality Control.)

it registers out of specification. Some pieces that are in specification in sections B and C will occasionally be classified as out of specification owing to gauge variability, and some pieces that are out of specification in these areas will be considered good pieces.

A piece whose true measurement falls in section D or E will always measure out of specification because the gauge variation is not large enough to throw the reading into the in-specification section. The width of sections B and C, the indecisive zones, is determined by the amount of gauge variability. The maximum widths in each section will be twice the maximum gauge deflection.

EXERCISES

1. **a.** Complete the gauge capability chart in Figure 12.16.
   b. If the gauge is accurate, what is the maximum possible deflection between the true measurement and the gauge reading?
   c. If the gauge has an accuracy shift left of .003 (that is, if the true measurement is .546, the gauge measurement is .543), what is the maximum possible deflection between the true measurement and the gauge reading?

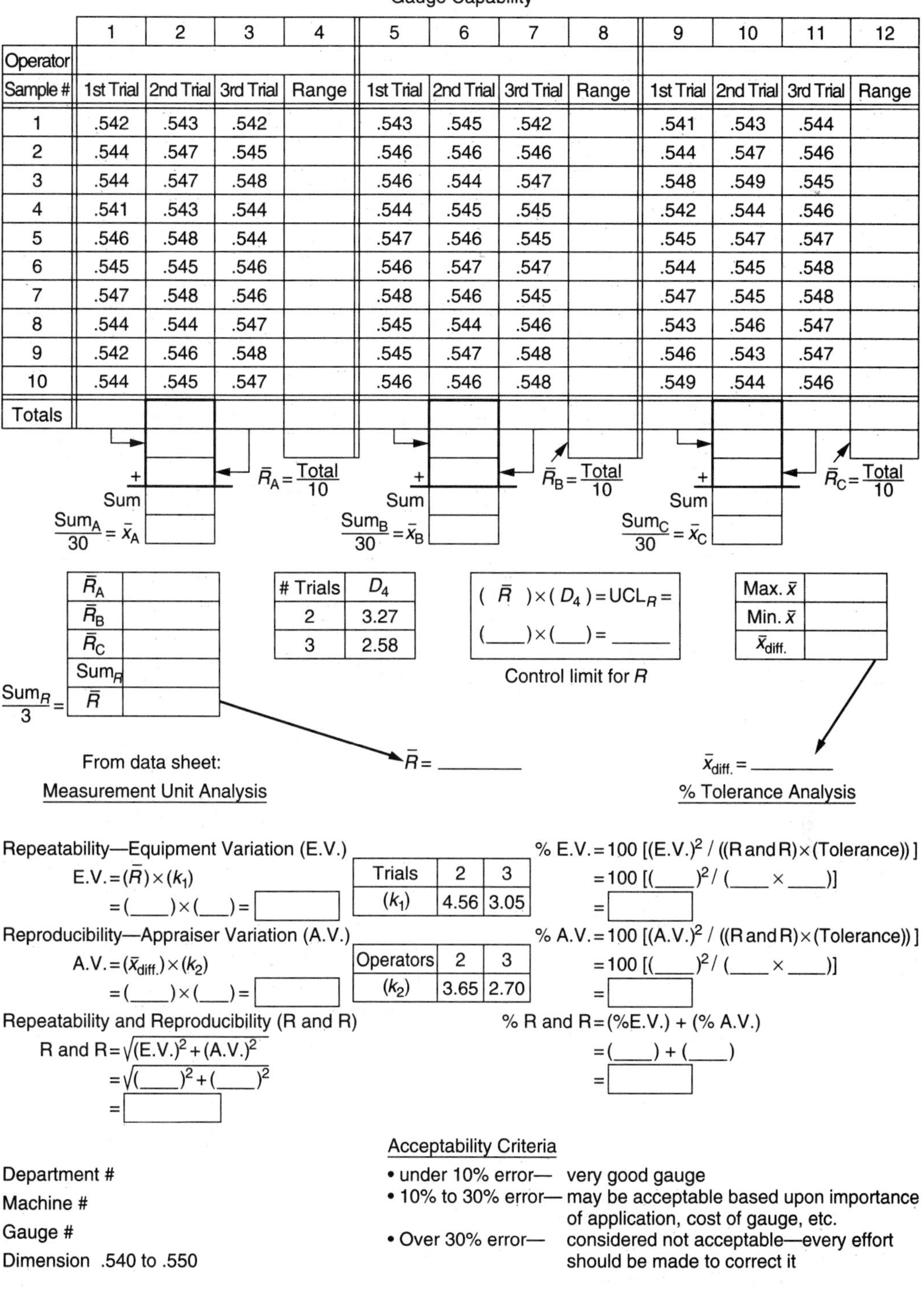

Gauge Capability

	1	2	3	4	5	6	7	8	9	10	11	12
Operator												
Sample #	1st Trial	2nd Trial	3rd Trial	Range	1st Trial	2nd Trial	3rd Trial	Range	1st Trial	2nd Trial	3rd Trial	Range
1	.542	.543	.542		.543	.545	.542		.541	.543	.544	
2	.544	.547	.545		.546	.546	.546		.544	.547	.546	
3	.544	.547	.548		.546	.544	.547		.548	.549	.545	
4	.541	.543	.544		.544	.545	.545		.542	.544	.546	
5	.546	.548	.544		.547	.546	.545		.545	.547	.547	
6	.545	.545	.546		.546	.547	.547		.544	.545	.548	
7	.547	.548	.546		.548	.546	.545		.547	.545	.548	
8	.544	.544	.547		.545	.544	.546		.543	.546	.547	
9	.542	.546	.548		.545	.547	.548		.546	.543	.547	
10	.544	.545	.547		.546	.546	.548		.549	.544	.546	
Totals												

+ Sum

$\frac{Sum_A}{30} = \bar{x}_A$ $\bar{R}_A = \frac{Total}{10}$

$\frac{Sum_B}{30} = \bar{x}_B$ $\bar{R}_B = \frac{Total}{10}$

$\frac{Sum_C}{30} = \bar{x}_C$ $\bar{R}_C = \frac{Total}{10}$

$\bar{R}_A$	
$\bar{R}_B$	
$\bar{R}_C$	
Sum_R	
$\frac{Sum_R}{3} =$ $\bar{R}$	

# Trials	D_4
2	3.27
3	2.58

$(\ \bar{R}\ ) \times (\ D_4\ ) = UCL_R =$

(____) × (___) = ______

Control limit for R

Max. $\bar{x}$	
Min. $\bar{x}$	
$\bar{x}_{diff.}$	

From data sheet: $\bar{R} =$ ________

$\bar{x}_{diff.} =$ ________

Measurement Unit Analysis

% Tolerance Analysis

Repeatability—Equipment Variation (E.V.)

E.V. = $(\bar{R}) \times (k_1)$

= (____) × (___) = [ ]

Trials	2	3
(k_1)	4.56	3.05

% E.V. = 100 [(E.V.)^2 / ((R and R) × (Tolerance))]

= 100 [(____)^2 / (____ × ____)]

= [ ]

Reproducibility—Appraiser Variation (A.V.)

A.V. = $(\bar{x}_{diff.}) \times (k_2)$

= (____) × (___) = [ ]

Operators	2	3
(k_2)	3.65	2.70

% A.V. = 100 [(A.V.)^2 / ((R and R) × (Tolerance))]

= 100 [(____)^2 / (____ × ____)]

= [ ]

Repeatability and Reproducibility (R and R)

R and R = $\sqrt{(E.V.)^2 + (A.V.)^2}$

= $\sqrt{(\_\_\_)^2 + (\_\_\_)^2}$

= [ ]

% R and R = (%E.V.) + (% A.V.)

= (____) + (____)

= [ ]

Acceptability Criteria

- under 10% error— very good gauge
- 10% to 30% error— may be acceptable based upon importance of application, cost of gauge, etc.
- Over 30% error— considered not acceptable—every effort should be made to correct it

Department #

Machine #

Gauge #

Dimension .540 to .550

Figure 12.16

Gauge capability worksheet for exercise 1. (Gauge capability analysis form courtesy of Saginaw Division of General Motors Corporation.)

d. After a week of use, the gauge has a stability shift left of .002. Combine the information with the accuracy shift information from (c) to determine the maximum possible deflection between the true measurement and the gauge reading.

2. An $\bar{x}$ and R chart is made for samples of size $n = 5$ and $\bar{R} = .0073$. The gauge from exercise 1 is used. Calculate the true process standard deviation by removing the effect of the gauge variation.

3. The gauge from the gauge capability study in Figure 12.10 was used on the control chart in Figure 6.19. Eliminate the effect of gauge variability σ_{Gauge} and calculate the process capability σ_{Process}.

13

ACCEPTANCE SAMPLING

OBJECTIVES

- Realize the limitations associated with using operating characteristic (OC) curves and acceptance sampling by attributes.
- Plan a random sample, given the packaging information for an incoming lot.
- Construct and interpret an OC curve.
- Construct and interpret an average outgoing quality (AOQ) curve.
- Read and use the MLT-STD-105D tables.
- Calculate the percentage of defect-free product, given the percent defective of several parts that are used to make the product.

The trend in industry has been to shift the burden of ensuring incoming quality to the vendor. Various quality assurance programs are being developed in which the vendor provides proof of product quality to the purchaser. Vendor documentation of ongoing quality improvement programs is usually required, along with access to control charts at critical process points. The receiving companies also have a vendor-rating system based on the quality history of the vendor.

Some companies, however, continue to use more traditional methods of quality assurance. With current market pressures on quality, acceptance sampling has to be considered a very temporary measure in the effort to attain top-quality input. Incoming inspection is both costly and unreliable. Many different sampling plans are available, but all compromise quality to some extent. The consumer's risk is always involved. The concepts of acceptance sampling are presented here because many companies still voluntarily use it or are forced to use it by government decree (involvement in government contracts may necessitate it). However, the shortcomings of this approach to measuring incoming quality should be apparent.

13.1 THE SAMPLING DILEMMA

Sampling does a good job of accepting very good lots and rejecting very bad lots. Unfortunately, a large area of indecision lies in the middle. The sampling rules in all the formal sample plans are based on probability, but the application of probability predicts the acceptance of lots with substandard quality. This will be demonstrated in the following examples.

Suppose the "lot" is a box of 100 beads. Blue beads are desired, and red beads are considered defective. A random sample of 20 will determine if the lot is acceptable. The maximum percent defective that is allowed is 5%, so if the sample contains more than one defective, the lot will be rejected.

EXAMPLE 13.1 Lot A contains 95 blue beads and 5 red beads; it should be accepted. Find the probability of accepting the lot.

Solution

There are two sampling situations for which the lot will be accepted. Either zero defectives show up in the sample *or* exactly one defective shows up. Use the "or" rule in probability and *add* the probabilities of these two situations to find the probability of acceptance.

Each of the two cases involves the "and" rule of probability because of the 20 successive events when choosing the sample: The 20 probability values are *multiplied.* By the simplest application of the random-sample concept, in which each bead has an equally likely chance to be drawn, each bead will be replaced before the next one is drawn. That way each piece sampled has a probability of .95 (95 chances out of 100) of not being defective:

Case 1: No defectives are drawn.

Case 2: Exactly one defective is drawn.

For case 1, find

$$P\text{ (1st not defective } \textit{and} \text{ 2nd not defective } \textit{and} \ldots \textit{and} \text{ 20th not defective)}$$

$$\begin{aligned} P\text{ (no defectives)} &= .95 \times .95 \times .95 \times \cdots \times .95 \\ &= (.95)^{20} \\ &= .3585 \end{aligned}$$

For case 2, let N represent not defective and D represent defective. Drawing a sample of 20 with one defective can occur in 20 ways. For example, several possibilities are shown here:

The *first* is defective.	$D\,N\,N\,N\,N \ldots N$
The *second* is defective.	$N\,D\,N\,N\,N \ldots N$

The *third* is defective. $N\ N\ D\ N\ N \ldots N$

The *twentieth* is defective. $N\ N\ N\ N\ N \ldots D$

There are 20 ways of getting exactly one that is defective and 19 that are not:

$$P(N) = \frac{95}{100} = .95 \qquad P(D) = \frac{5}{100} = .05$$

$$P(D \text{ and } N \text{ and } N \text{ and } N \text{ and} \ldots \text{and } N) = .05 \times .95 \times .95 \times .95 \times \ldots \times .95$$
$$P(D\ N\ N\ N \ldots N) = .05 \times (.95)^{19}$$
$$= .05 \times .3774$$
$$= .01887$$

Each of the 20 ways will have this same probability because they all have one factor of .05 and 19 factors of .95. Instead of adding the 20 probability values according to the "or" rule, the repeated addition can be done by multiplication:

$$P(\text{one defective}) = 20 \times .01887 = .3774$$

There are 20 ways of getting one defective in 20, and each way has the same probability. Since the lot will be accepted if either the first case or the second case occurs, the addition rule of probability applies:

$$P(\text{accepting the lot}) = .3585 + .3774 = .7359$$

The case 1 probability is .3585, and the case 2 probability is .3774.

The probability of getting either one defective or no defectives in the sample is about .74, so 74 times out of 100 this lot will be accepted, and 26 times out of 100 it will be rejected. The rejected lot will then be either shipped back to the vendor or 100% inspected to sort out the extra defectives.

EXAMPLE 13.2 Lot B contains 97 blue beads and 3 red ones. What is the probability that this acceptable lot will be rejected when a sample of 20 is randomly selected?

Solution

Again, there are two cases in which the lot will be accepted:

Case 1: No defectives show up in the sample.

Case 2: Exactly one defective occurs in the sample.

The "or" rule is used to add the probabilities for each case, and the "and" rule is used to multiply the probabilities for each of the 20 selections in the sample. Each draw has $P(D) = .03$ (3 chances out of 100 for D). For case 1,

$$P\ (\text{no defectives}) = (.97)^{20} = .5438$$

The exponent 20 indicates 20 successive draws. For case 2,

$$\begin{aligned} P\ (\text{one defective}) &= 20 \times (.03) \times (.97)^{19} \\ &= 20 \times .03 \times .5606 \\ &= .3364 \end{aligned}$$

Twenty is the number of ways of getting one defective in 20, and each way has the same "and" probability. Add the probabilities for each case to find the probability of accepting the lot:

$$P\ (\text{accepting the lot}) = .5438 + .3364 = .8802$$

There is an 88% chance that this lot will be accepted and a 12% chance that it will be rejected. Twelve times out of 100 this lot will either be shipped back to the vendor as unacceptable or be 100% inspected.

EXAMPLE 13.3 Lot C contains 90 blue beads and 10 red ones. What is the chance that this bad lot will be accepted?

Solution

The same two cases apply for acceptance: No defectives show up in the sample of 20, or exactly one does:

$P\ (N) = .9$ 90 chances out of 100 for N

$P\ (D) = .1$ 10 chances out of 100 for D

For case 1,

$$\begin{aligned} P\ (\text{no defectives}) &= (.9)^{20} \\ &= .1216 \end{aligned}$$

For case 2,

$$\begin{aligned} P\ (\text{one defective}) &= 20 \times (.1) \times (.9)^{19} \\ &= 20 \times .1 \times .1351 \\ &= .2702 \end{aligned}$$

Combining both cases,

$$P\ (\text{accepting lot 3}) = .1216 + .2702 = .3918$$

This shows a 39% chance of accepting a bad lot. Thirty-nine times out of 100 the bad lot will be accepted because, by chance, not enough of the defectives will show up in the random sample.

Examples 13.1 to 13.3 illustrate that acceptance sampling results in rejecting good lots when too many of the defects in the lot show up in the random sample. There is a nuisance factor if the rejected lot is shipped back to the vendor because the vendor will check it, see that it is acceptable, and ship it back. This adds the cost of two-way shipping as well. If the rejected lot is routed to 100% inspection, additional cost, nuisance factor, and material delays result.

Even worse than rejecting acceptable lots, however, is taking in unacceptable lots, such as lot C. The quality program can be seriously affected by this policy.

EXAMPLE 13.4 Choice of sample size is a major factor in acceptance sampling. To demonstrate the effect of sample size, a sample of 40 will be selected in the bead problem of the first three examples. The *rejection* number will be 3: The lot will be rejected if three or more defectives occur in the sample. This keeps the maximum *acceptance* ratio the same as before: Two out of 40 equals 1 out of 20. The lots from Examples 13.1 and 13.3 will be used:

1. Lot A contains 95 blue beads and 5 red beads. With a sample of 40, what is the chance of rejecting the lot?
2. Lot C contains 90 blue beads and 10 red ones. With a sample of 40, what is the probability of accepting lot C?

Solution

There are now three cases in which the lot will be accepted:

Case 1: No defectives are in the sample.

Case 2: Exactly one defective is in the sample.

Case 3: Exactly two defectives are in the sample:

$$P\ (N) = .95 \qquad \text{95 chances in 100 for } N$$
$$P\ (D) = .05 \qquad \text{5 chances in 100 for } D$$

For case 1 for 40 successive draws,

$$\begin{aligned} P\ (\text{no defectives}) &= (.95)^{40} \\ &= .1285 \end{aligned}$$

For case 2,

$$\begin{aligned} P\ (\text{one defective}) &= 40 \times (.05) \times (.95)^{39} \\ &= 40 \times .05 \times .1353 \\ &= .2706 \end{aligned}$$

For case 3 the number of ways of getting two defectives in 40 pieces is illustrated as follows for several possibilities:

The first two are defective.	$D\,D\,N\,N\,N\,N \ldots N$
The second and third are defective.	$N\,D\,D\,N\,N\,N \ldots N$
The last two are defective.	$N\,N\,N\,N\,N \ldots D\,D$

The two defectives can be in any other positions:

$$N \ldots D\,N\,N \ldots D\,N \ldots N$$

There are 780 ways that the two defects can show up in the sample of 40. The mathematical shortcut for calculating the 780 is called *combinations* and is discussed in Section A.6.2 in Appendix A. Briefly, combinations are randomly ordered groupings that can be calculated using *factorials*. A factorial is the product of descending counting numbers (that is, 3 factorial, or 3!, is $3 \times 2 \times 1 = 6$):

$$C_{40,2} = \frac{40!}{2! \times 38!}$$
$$= 780$$

If 780 is the number of ways of getting two defectives in 40, and if each way has two .05 factors and thirty-eight .95 factors,

$$P\text{ (two defectives)} = 780 \times (.05)^2 \times (.95)^{38}$$
$$= 780 \times .0025 \times .1424$$
$$= .2777$$

The probability of accepting the lot is the sum of the probabilities of the three cases:

$$P\text{ (accepting lot A)} = .1285 + .2706 + .2777$$
$$= .6768$$

There is a 68% chance of accepting lot A and a 32% chance of rejecting it.

There are again three cases for the second part of Example 13.4. For case 1,

$$P\text{ (no defectives)} = (.9)^{40}$$
$$= .0148$$

For case 2 there are 40 ways of getting one defective, and each way has one .1 factor and thirty-nine .9 factors:

$$P\text{ (one defective)} = 40 \times (.1) \times (.9)^{39}$$
$$= .0657$$

For case 3 there are 780 ways of getting two defectives and each way has two .1 factors and thirty-eight .9 factors:

$$P\text{ (two defectives)} = 780 \times (.1)^2 \times (.9)^{38}$$
$$= .1423$$

Combining the cases gives

$$P\text{ (accepting lot C)} = .0148 + .0657 + .1423$$
$$= .2228$$

There is now a 22% chance of accepting the bad lot.

By juggling the sample size, the chance of making an error has been changed. The two errors that were demonstrated are called the *producer's risk* and the *consumer's risk.*

The producer's risk is the chance that a good lot will be rejected. When the sample size was 20 in Example 13.1, the producer's risk was 26% on lot A. However, when the sample size was increased to 40 in Example 13.4, the producer's risk increased to 32%.

The consumer's risk is the chance that a bad lot will be accepted. When the sample size was 20 in Example 13.1, the consumer's risk on lot C was 39%. When the sample size was increased to 40 in Example 13.4, the consumer's risk dropped to 22%.

The two examples showed that the producer's risk and the consumer's risk can be changed by changing the sample size. In general, when the sample size is increased, the consumer's risk will decrease and the producer's risk will increase. The two types of risk will also change when a different acceptance number is used for the sample.

There are three different mathematical approaches to the calculation of the probabilities associated with acceptance sampling. Each results in a different probability distribution. The first is the *hypergeometric probability distribution,* which is used to calculate the chance of getting a specific sample when there is *no replacement* of the pieces that are chosen for the sample.

EXAMPLE 13.5

A box contains 12 parts, 3 of which are defective. If a sample of 4 is taken without replacement, find the probability of getting 2 good parts and 2 defective parts in the sample.

Solution

This is an application of the hypergeometric probability distribution. The 12 parts split into two groups: the 3 defective parts and the 9 good parts. A *success* is defined as choosing 2 of the 3 defective parts *and* 2 of the 9 good parts. The number of successes can be calculated using mathematical combinations: The number of ways of choosing 2 from the 3 defective parts is

$$C_{3,2} = \frac{3!}{2! \times 1!}$$
$$= 3 \text{ ways}$$

The number of ways of choosing 2 from the 9 good parts is

$$C_{9,2} = \frac{9!}{2!\ 7!}$$
$$= 36 \text{ ways}$$

The number of successes is $3 \times 36 = 108$.

The total number of possible samples of 4 parts chosen from the box of 12 is

$$C_{12,4} = \frac{12!}{4! \times 8!}$$
$$= 495$$

The probability of a success is the number of successes divided by the number of possibilities:

$$P\ (2 \text{ good and } 2 \text{ defective}) = \frac{108}{495}$$
$$= .2182$$

More details on combinations and hypergeometric probabilities can be found in Section A.6 in Appendix A.

The second probability distribution that is used in sampling is the binomial distribution. The binomial distribution was used in Examples 13.1 through 13.4 and is also discussed more thoroughly in Section A.6.3 in Appendix A. The binomial distribution is used for sampling *with replacement.* Example 13.5 will be repeated for the case of sampling with replacement. This will show a comparison of the hypergeometric distribution and the binomial distribution.

EXAMPLE 13.6 The same box of 12 parts with 3 defectives is checked by four inspectors. Each inspector checks 1 piece and returns it to the box. Find the probability that two inspectors will check a good part and two inspectors will find a defective part.

Solution

This is an example of sampling with replacement, and the probability is calculated with the binomial probability distribution. Here N is not defective; D is defective:

$$P\ (\text{a good part in a single draw}) = P\ (N)$$
$$= \frac{9}{12}$$
$$= .75$$
$$P\ (\text{a defective part in a single draw}) = P\ (D)$$
$$= \frac{3}{12}$$
$$= .25$$
$$P\ (2 \text{ good and } 2 \text{ defective}) = C_{4,2} \times [P\ (N)]^2 \times [P\ (D)]^2$$
$$= \frac{4!}{2! \times 2!} \times (.75)^2 \times (.25)^2$$
$$= 6 \times .5625 \times .0625$$
$$= .21094$$

A comparison of the two probabilities from Examples 13.5 and 13.6 shows a small difference caused by the different sampling techniques.

The third probability distribution that is used extensively in sampling is the *Poisson distribution.* The Poisson distribution is used in acceptance sampling, and its use is illustrated in Section 13.3.

13.2 RANDOM SAMPLING

It is extremely important to have random samples in any acceptance procedure. The sampling plan must be devised to fit the packaging form for the incoming pieces. If a lot is shipped in several cartons, then a box number has to be part of the sample plan. If the pieces are carefully packed in layers, a layer number must also be included. If each layer has a set number of pieces, a piece number and a well-defined way of counting the pieces in each layer must be included. If the pieces are loose within each layer, then a wire grid should be made to fit the cartons and a piece sampled from a prescribed grid position.

Either a random-number table or a random-number generator on a statistical calculator can be used for selecting the pieces for the sample. If a random number comes up that is not in the proper domain for the numbers that are needed, skip it and go on to the next random number.

EXAMPLE 13.7 A lot of 4000 pieces is shipped in eight boxes. Each box contains five layers, and each layer consists of 10 rows of 10 pieces. A sample of 200 is to be taken. Design a sample plan.

Solution

A three-digit random-number generator on a statistical calculator is used to set up a random sample: The first random number will be used for the piece number (use the *last* two digits, 00 to 99). The second random number will specify the box number with the *first* digit and the layer number with the *last* digit. The sample will be planned as follows for all 200 pieces:

Piece Number	Random Numbers	Piece	Box	Layer
1	585, 811	85	8	1
2	358, 693	58	6	3
3	510, ~~946~~, ~~238~~, ~~726~~, ~~499~~, 293	10	2	3
4	535, ~~636~~, ~~457~~, ~~031~~, ~~982~~, 513	35	5	3

The crossed-out numbers are not in the domain for the box (1 to 8) or the layer (1 to 5). If too many unused random numbers can occur with this method, the plan can be revised to take the box number and layer number separately, if necessary. This method requires only four random numbers for piece 4:

Piece Number	Random Numbers	Piece	Box	Layer
4	535, 636, ~~457~~, 031	35	6	1

13.3 OPERATING CHARACTERISTIC CURVES

Operating characteristic (OC) curves are graphs of the probability of acceptance of a lot for all possible lot levels of percent defective. There are published OC curves for various sampling plans, but their construction is not difficult and will be presented here. The effects of changes in sample size and acceptance number will also be investigated.

The OC curve gives the probability of accepting a lot with a specific percentage of defective items. The dashed line in Figure 13.1 shows that if a lot is 4% defective, it will pass inspection 82% of the time and be rejected 18% of the time. Every OC curve is identified by the sample size n and the maximum acceptance number c. In this case, if the sample of 200 contains 10 or fewer defectives, it will be accepted. In another illustration of the OC curve in Figure 13.1, if the lot is 6% defective, there is a 35% chance that it will be accepted and a 65% chance that it will be rejected.

The OC curve is constructed from a Poisson probability distribution. The Poisson distribution, which is appropriate because it is skewed right to match the outcome of a random sampling situation, is preferred over the binomal distribution because it is better in a situation in which an event can occur in many ways that have low probabilities. This describes the sampling situation quite well. If a lot of 5000 items is 2% defective, there will be 100 defective items in the lot. How many defectives will show up in a sample of 200? The sample will most likely have between 0 and 8 defectives but it may have 13, 28, 54, or any other number, even 100. The chance of getting one of these higher numbers of defectives in the sample is quite low.

The construction of the OC curves is accomplished with the Poisson curves in Figure 13.2. The curved lines on the chart correspond to specific c values. The first construction we will examine is the ($n = 200$, $c = 8$) OC curve. The horizontal axis on the Poisson chart in Figure 13.2 is labeled pn. The proportion of defective items in the lot is given the symbol p, and n is the sample size; pn is their product. The vertical scale gives the probability of acceptance, which is the probability that the sample of 200 will have $c =$

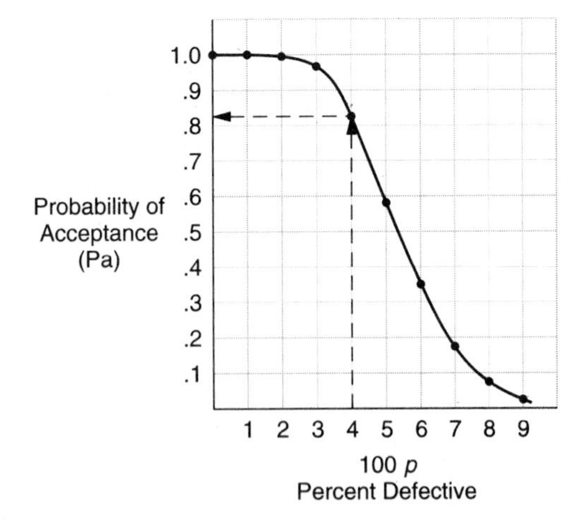

FIGURE 13.1
The ($n = 200$, $c = 10$) OC curve.

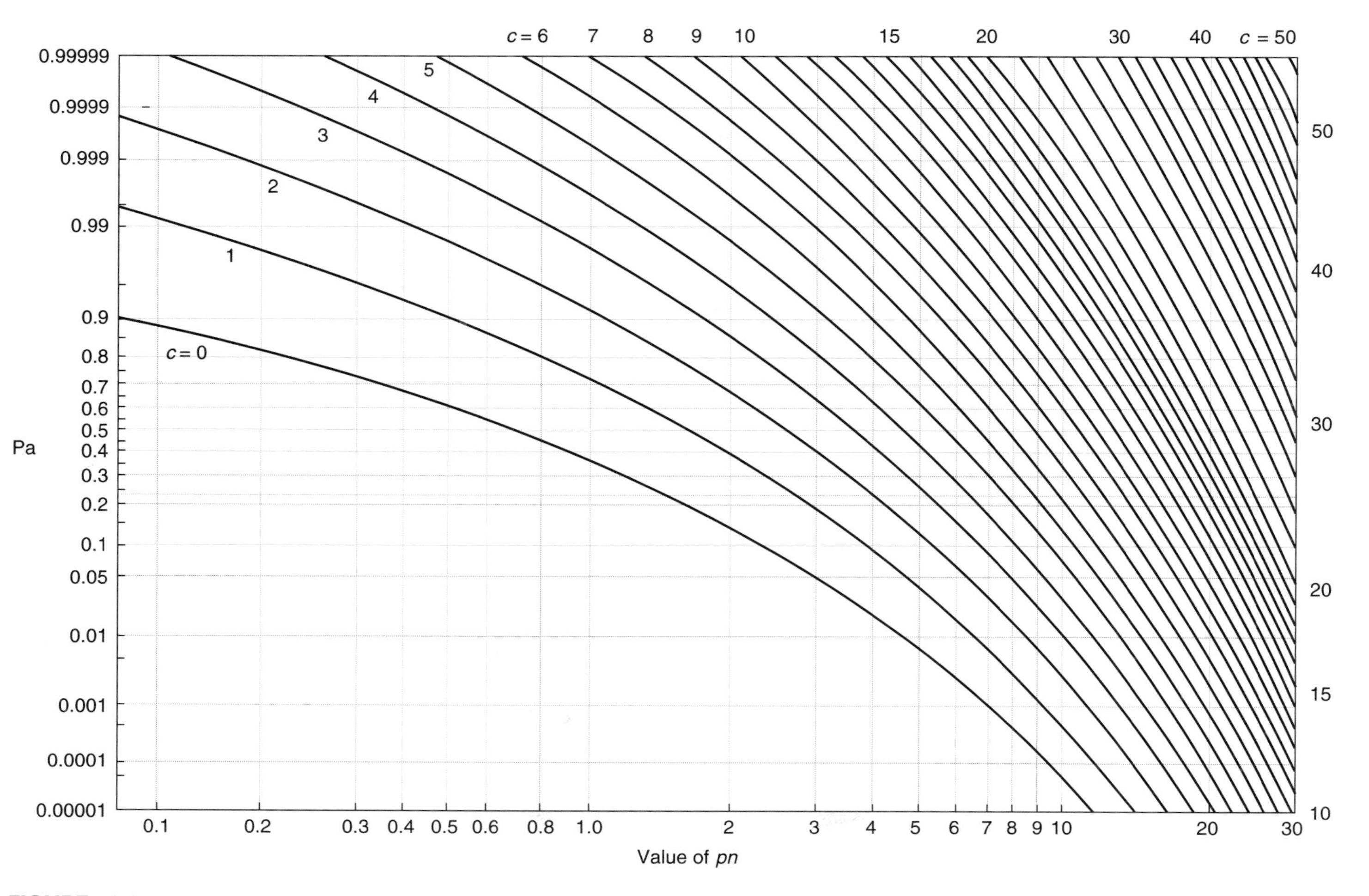

FIGURE 13.2
Poisson probabilities. (Copyright 1956, Western Electric Company, reprinted by permission.)

8 or fewer defective items. The one curved line marked $c = 8$ is used for this construction. A table of values is formed from the Poisson chart in the following way:

1. The p values are chosen from 0 to .09, which will correspond to a lot that has between 0 and 9 percent defective.
2. Multiply each p value by $n = 200$ and enter it in the table.
3. Make a percent column for p and mark it $100p$.
4. Start at the pn value on the horizontal axis in Figure 13.2 and go straight up to the $c = 8$ curve and straight across to the vertical axis on the left. Read the probability-of-acceptance (Pa) value on the vertical axis.
5. Record the Pa value in the table.

For example, if $p = .01$, then $200 \times .01 = 2$. Start at 2; go up to $c = 8$ and across to .9997. If $p = .02$, $200 \times .02 = 4$. Start at 4; go up to $c = 8$ and across to .975. Finally, if $p = .03$, then $200 \times .03 = 6$. Start at 6; go up to $c = 8$ and across to .855. Verify the other table values using the Poisson chart. In Table 13.1, Pa is the probability of acceptance from the vertical axis of the Poisson chart.

The OC curve is constructed with the vertical scale representing the Pa. This scale is labeled from 0 through 1.0. The horizontal axis represents the percentage of defective items in the lot and is labeled from 0 through 9. The horizontal scale is 100 times the p value from the table and is marked $100p$. To graph the values from the table and construct the OC curve, find the $100p$ number on the horizontal scale. Follow the grid upward from that point and mark a point on the graph that is even with the corresponding Pa value. The point should mark the intersection of a vertical line drawn from the $100p$ number and a horizontal line drawn from the Pa value. Do this for each pair of the table values for $100p$ and Pa. The result is Figure 13.3*a*. Then draw a smooth curve through the points. If the shape of the curve is not obvious or easy to draw, graph some in-between points. Choose some inbetween p values, such as $p = .025$ in the fourth line of the table, and find the cor-

TABLE 13.1
Table of values for Figure 13.3

p	np	$100p$	Pa
0	0	0	1.0000
.01	2	1	.9997
.02	4	2	.975
.025	5	2.5	.92
.03	6	3	.855
.04	8	4	.60
.05	10	5	.34
.06	12	6	.15
.07	14	7	.06
.08	16	8	.03
.09	18	9	.006

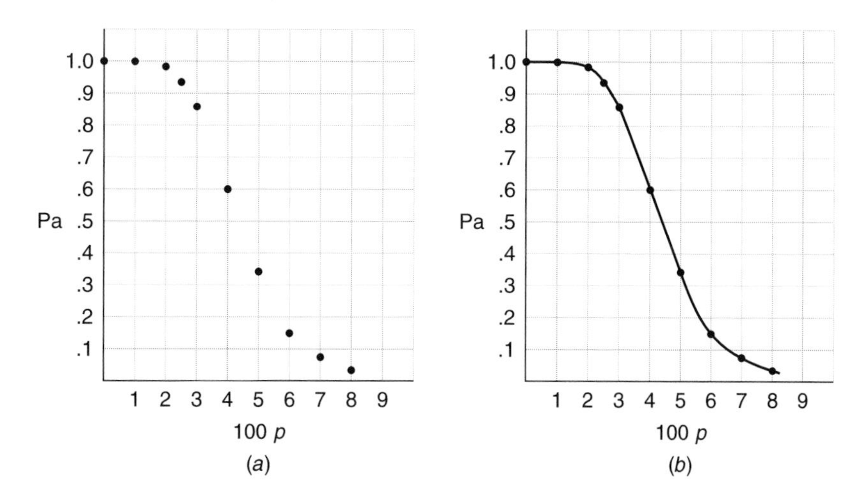

FIGURE 13.3
Construction of the $n = 200$, $c = 8$) OC curve: (*a*) point graphed from Table 13.1; (*b*) a smooth curve drawn through the points.

responding Pa values and plot the points. The in-between point plotting is helpful in the curvy sections of the graph. The curve that is drawn, Figure 13.3*b*, is the OC curve.

Several more OC curves are shown with their corresponding table of values taken from the Poisson chart. Points are also given in Table 13.2 for Figure 13.1. Choose a few points from each table and verify them with the Poisson chart in Figure 13.2, following the curve for the specified *c* value. Then verify the graphed point on the appropriate OC curve. Remember to use the *np* value, not the 100*p* value, on Figure 13.2.

TABLE 13.2
Table of values for Figure 13.1

p	*np*	100*p*	Pa
0	0	0	1.000
.01	2	1	.9999
.02	4	2	.997
.03	6	3	.957
.04	8	4	.816
.05	10	5	.583
.06	12	6	.347
.07	14	7	.176
.08	16	8	.08
.09	18	9	.03

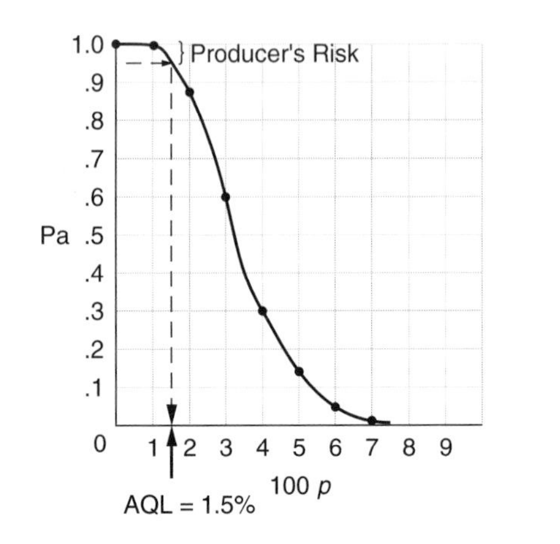

FIGURE 13.4
The ($n = 200$, $c = 6$) OC curve showing the AQL.

TABLE 13.3
Table of values for Figure 13.4

p	np	$100p$	Pa
0	0	0	1.000
.01	2	1	.996
.015	3	1.5	.97
.02	4	2	.88
.025	5	2.5	.78
.03	6	3	.60
.04	8	4	.30
.05	10	5	.14
.06	12	6	.048
.07	14	7	.015

Figure 13.4 shows the ($n = 200$, $c = 6$) OC curve (see Table 13.3 for points). Notice the dashed arrows going from Pa = .95 to the curve and from the curve down to 1.5% defective. These arrows define the AQL, the *acceptable quality level.*

The AQL is used as a measure of quality for acceptance sampling by attributes. There are two definitions of the AQL. The first, set according to military standard MLT-STD-105D, defines the AQL as the maximum defective average allowable. The MLT-STD-105D instructions explain that when a consumer specifies an AQL value, that consumer will accept the great majority of lots with a proportion defective less than or equal to that AQL value. The *Quality Control Handbook* specifies AQL as the proportion defective that is accepted 95% of the time.[1] Example 13.14 in this chapter will show that these two definitions are compatable:

Definition 1: The AQL is the maximum percent defective that is allowed as a process average.

Definition 2: The AQL is the level of quality of a submitted lot that has a 95% chance of being accepted.

The AQL is *not* necessarily the quality level being produced or the quality level being accepted. It is *not* always the quality goal. At the AQL, the producer's risk, or the probability of rejecting the acceptable, is 5%. In Figure 13.4, the AQL is slightly better than the average quality level that is allowed by that particular sample plan.

Figure 13.5 is the ($n = 200$, $c = 4$) OC curve. Its values are given in Table 13.4. The dashed arrows define the IQL, the *indifference quality level:*

Definition: The IQL is the quality level that will be accepted 50% of the time.

[1]Joseph M. Juran, ed., *Quality Control Handbook,* 3d ed. (New York: McGraw Hill, 1974).

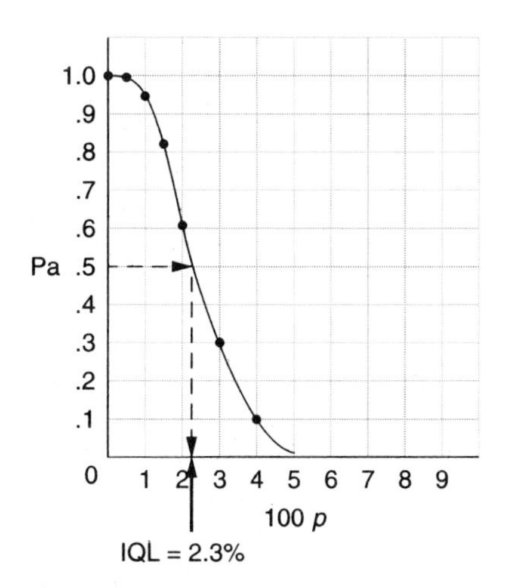

FIGURE 13.5
The ($n = 200$, $c = 4$) OC curve showing the IQL.

TABLE 13.4
Table of values for Figure 13.5

p	np	$100p$	Pa
0	0	0	1.000
.005	1	.5	.996
.01	2	1	.94
.015	3	1.5	.82
.02	4	2	.62
.03	6	3	.30
.04	8	4	.10

Figure 13.6 is the ($n = 100$, $c = 0$) OC curve (see Table 13.5). The dashed lines define the RQL, the *rejectable quality level:*

Definition: The RQL is the level of quality that will be accepted only 10% of the time.

At the RQL, the consumer's risk, or the probability of accepting the unacceptable, is 10%.

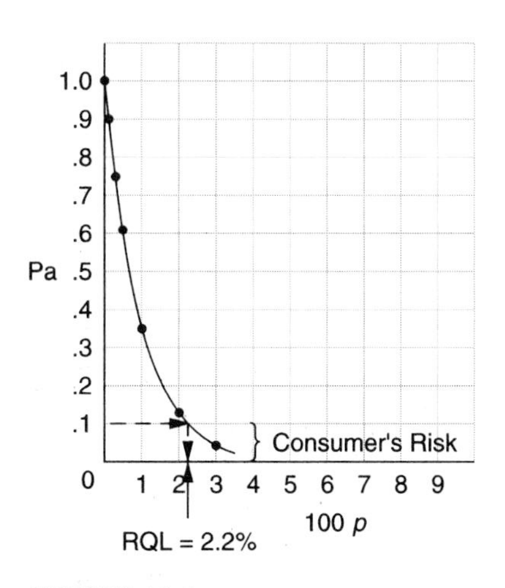

TABLE 13.5
Table of values for Figure 13.6

p	np	$100p$	Pa
0	0	0	1.000
.001	.1	.1	.91
.002	.2	.2	.82
.003	.3	.3	.74
.005	.5	.5	.62
.01	1	1	.35
.02	2	2	.13
.03	3	3	.04

FIGURE 13.6
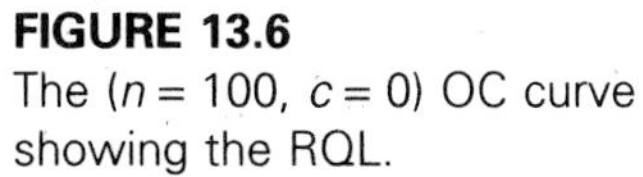
The ($n = 100$, $c = 0$) OC curve showing the RQL.

The interaction of the sample sizes is shown in Figure 13.7 (see Tables 13.3 and 13.6). The acceptance level is kept constant at $c = 6$, and three OC curves are shown together for comparison. The curves show that with increasing the sample size, both the AQL and the RQL values decrease substantially. Furthermore, the steeper curve associated with the larger n value has a greater discriminatory power: It can better discriminate between good and bad lots. The indecisive zone between the AQL and the RQL is only 2.3 percentage units wide:

3.5%	RQL
− 1.2%	AQL
2.3%	

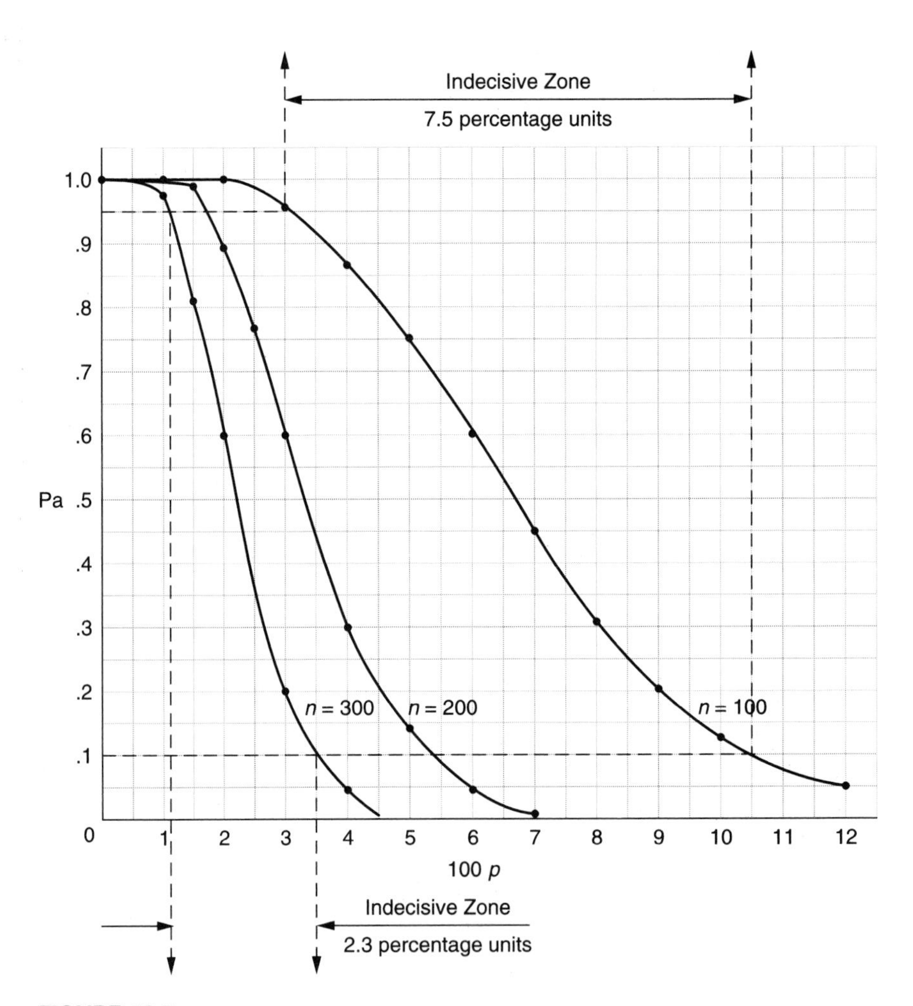

FIGURE 13.7
Three OC curves with a constant $c = 6$ and varying n values. Larger n values create steeper curves, and steeper curves have greater discriminatory power.

TABLE 13.6
Table of values for Figure 13.7

$n = 100, c = 6$				$n = 300, c = 6$			
p	np	$100p$	Pa	p	np	$100p$	Pa
0	0	0	1.000	0	0	0	1.000
.01	1	1	.9999	.01	3	1	.965
.02	2	2	.999	.015	4.5	1.5	.82
.03	3	3	.96	.02	6	2	.60
.04	4	4	.88	.03	9	3	.20
.05	5	5	.75	.04	12	4	.045
.06	6	6	.60				
.07	7	7	.45				
.08	8	8	.32				
.09	9	9	.20				
.10	10	10	.13				
.12	12	12	.05				

This is the area in which acceptable lots have a good chance of being rejected and unacceptable lots have a good chance of being accepted. Compare this zone with the indecisive zone for the less steep ($n = 100$, $c = 6$) OC curve:

10.5%	RQL
− 3.0%	AQL
7.5%	

This one is much wider, which makes it more difficult to discriminate between acceptable and unacceptable lots.

The values for the points on the four curves in Figure 13.8 are given in Tables 13.2 to 13.4 and 13.7. Figure 13.8 shows that changing c while keeping n fixed can have a dramatic effect on the OC curve. The relationship of the grouped curves is quite similar to the effect shown in Figure 13.7. Figure 13.8 shows that the smaller c values decrease both AQL and RQL. Also, the smaller the c value, the steeper the curve, which means that the discriminatory power of the curve is improved. The indecisive zone for $c = 0$ is about 1%, and for $c = 10$ it is about 4.6%:

1.1%	RQL	7.8%	RQL
− .1%	AQL	− 3.2%	AQL
1.0%		4.6%	

The illustrations show that a change in just one variable can lower the AQL and the RQL and increase the discriminatory power as the indecisive zone between the AQL and RQL decreases. This can be done either by increasing the sample size n or by decreasing the acceptance number c. What happens when both change?

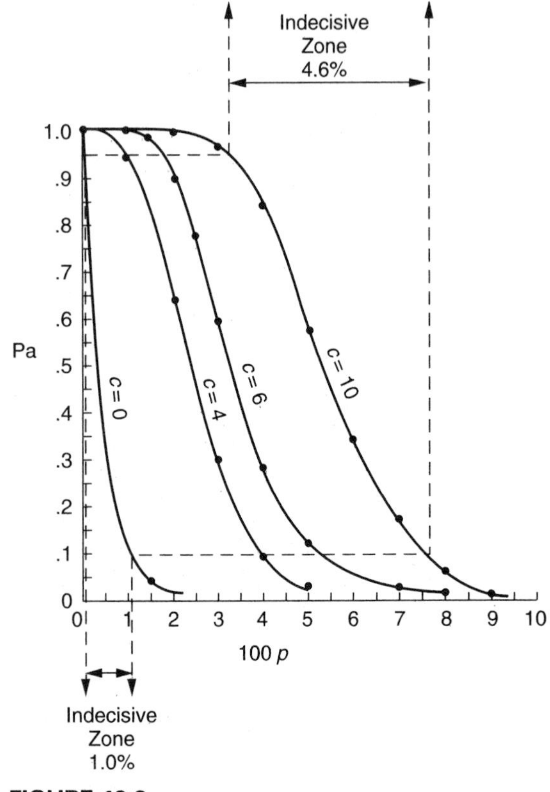

FIGURE 13.8
Four OC curves with constant $n = 200$ and varying c values.

TABLE 13.7
Table of values for the $n = 200$, $c = 0$ curve in Figure 13.8

p	np	$100p$	Pa
0	0	0	1.000
.0005	.1	.05	.9
.001	.2	.1	.82
.005	1	.5	.36
.01	2	1	.12

Figure 13.9 shows the relationship between three curves that have the same c/n ratio (see Table 13.8). The curves are quite similar in shape, although there is a small differential of about .7% in the AQLs and 2% in the RQLs. The greater discriminatory power goes to the curve with the larger n value because it is the steepest and has the shortest percentage span between the AQL and RQL. Maximizing the discriminatory power is one of the considerations in choosing an acceptance procedure, so the use of the c/n ratio can be helpful. Reviewing the other OC curves in Figures 13.6 to 13.8 illustrates that smaller c/n ratios correspond to greater discriminatory power of the curve. In Figure 13.7 the ($n = 300$, $c = 6$) OC curve has the greatest discriminatory power, and it has a c/n ratio of .02, compared to ratios of .03 and .06 for the other two curves. In Figure 13.8 the ($n = 200$, $c = 0$) OC curve has the best discriminatory power. It has a c/n ratio of 0, compared to .02, .03, and .05 for the others. Both figures show the increase in the c/n value as the discriminatory power decreases from curve to curve. Figures 13.6 and 13.8 both have the same c/n ratio of 0. This again illustrates that when the c/n ratio is the same, the higher discriminatory power is in the curve with the larger n value.

Companies establish quality standards for incoming parts and materials, and their acceptance procedures must protect that quality level by ensuring that substandard ship-

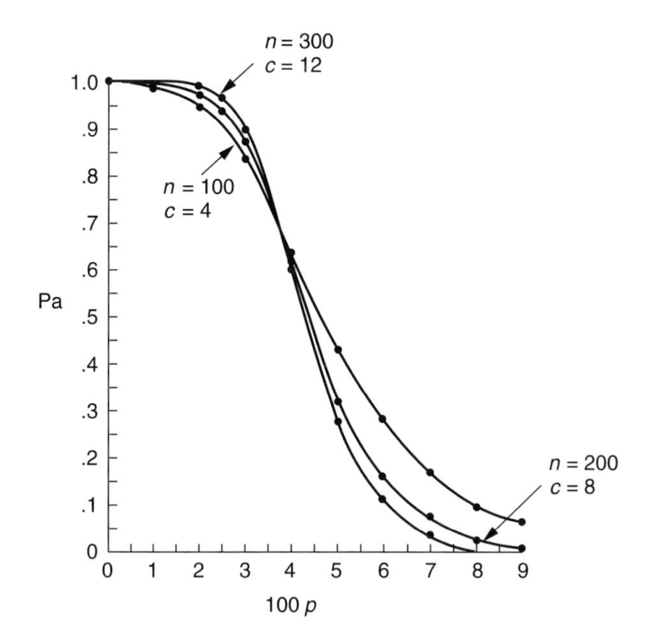

FIGURE 13.9
Three curves with a constant *c/n* ratio of .04.

ments are not accepted. A combination of issues is involved in considerations of acceptance procedures, but quality protection is always a major factor. One rough way to specify that protection is to state the maximum RQL value that can be used. Some companies dodge this issue, however, by specifying a constant percentage of each lot that will be inspected. This is a major error in acceptance procedures because it does not specify a bottom line of quality. Figure 13.10 shows a 10% sampling situation, and a comparison

TABLE 13.8
Table of values for Figure 13.9

$n = 100, c = 4$				$n = 300, c = 12$			
p	np	$100p$	Pa	p	np	$100p$	Pa
0	0	0	1.000	0	0	0	1.000
.01	1	1	.996	.01	3	1	.9999
.02	2	2	.94	.02	6	2	.991
.03	3	3	.81	.025	7.5	2.5	.955
.04	4	4	.62	.03	9	3	.88
.05	5	5	.42	.04	12	4	.60
.06	6	6	.26	.05	15	5	.26
.07	7	7	.16	.06	18	6	.10
.08	8	8	.10	.08	24	8	.006
.09	9	9	.06				

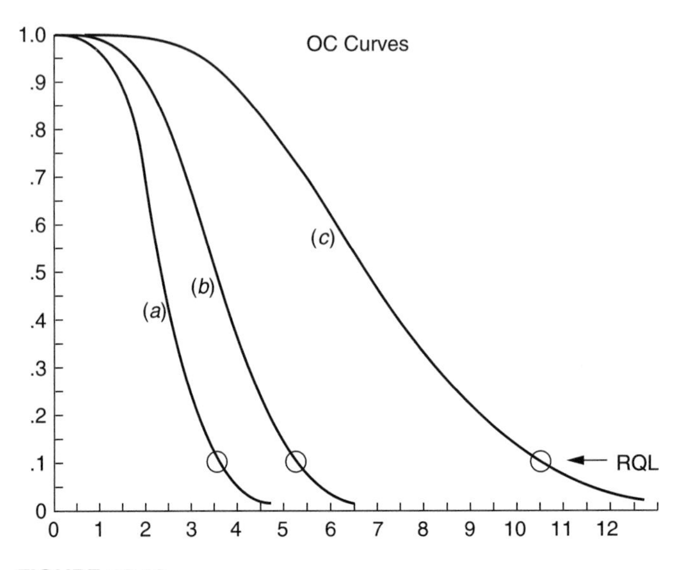

FIGURE 13.10
Three OC curves illustrate the problem with constant percent sampling: (*a*) lot size = 3000, $n = 300$, $c = 6$; (*b*) lot size = 2000, $n = 200$, $c = 6$; (*c*) lot size = 1000, $n = 100$, $c = 6$. These three OC curves illustrate the problem with constant-percentage sampling: They provide no consistent quality estimate.

of the RQL values shows that the variation in incoming quality can be extensive when a constant-percentage approach is used.

Selecting the Best OC Curve

The ideal OC curve, shown in Figure 13.11, is one in which lots within the acceptable quality level are accepted with a probability of 1 and lots that are not within that standard are rejected with a probability of 1. Figure 13.9 illustrates what happens if the c/n ratio is

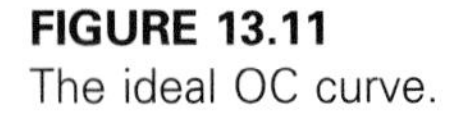

FIGURE 13.11
The ideal OC curve.

Pa
1.0
.9
.8
.7
.6
.5
.4
.3
.2
.1
0
0 1 2 3 4 5 6
100 p

kept constant. If more curves are drawn with the same c/n ratio, such as 24/600 or 40/1000, the sequence of curves will approach the ideal with a vertical line at $100p = 4$. Given a choice of OC curves, choose the one with the smallest c/n ratio and the smallest RQL – AQL difference. That combination gives the best average quality level and a sampling plan with the highest discriminatory power.

Selecting the AQL

Many sampling plans begin with selection of the AQL. In some cases the AQL is chosen arbitrarily. The AQL should represent the maximum percent defective that can be tolerated for the entire population of parts being received. It may be designated by bargaining with the vendor and reflect a compromise between vendor capability and purchaser requirements.

When an assembly operation demands 100% final inspection, the AQL may also be determined on the basis of cost. Let

p = proportion of incoming parts that are defective

C_i = the cost of inspecting one incoming piece

C_r = the cost of dismantling, repairing, reassembling, and testing a unit that failed because of a defective part

The ratio C_i/C_r is the break-even point. Theoretically, if the worst lot has $p < C_i/C_r$, no inspection of incoming materials is necessary. If the best lot has $p > C_i/C_r$, 100% inspection of incoming materials is required.

EXAMPLE 13.8 Let T_i be the time required to inspect a part and T_r the time required for repair:

$T_i = \frac{1}{2}$ minute to inspect each incoming part

T_r = 3 minutes to dismantle and repair a unit with a defective part

Solution

In each case, the times relate directly to cost; R is the rate. For simplicity, assume equal hourly rates:

$$\text{Time} \times \text{rate} = \text{cost}$$

$$T_i \times R = C_i \qquad T_r \times R = C_r$$

$$\frac{C_i}{C_r} = \frac{T_i \times R}{T_r \times R} = \frac{T_i}{T_r} = \frac{.5}{3} = .17$$

If $p < .17$ on the worst lot, no incoming inspection is required. It will be less expensive to repair the defective items when they are found during the final inspection.

Suppose a lot contains $N = 1000$ pieces. At ½ minute per piece for inspection, ½ × 1000 = 500 minutes to inspect the lot. If $p < .17$, say $p = .09$, then there will be .09 × 1000 = 90 defective pieces in the lot. If there is no inspection, there will be 90 repairs to do on the assembled units. At 3 minutes each, 3 × 90 = 270 minutes. This shows that when $p < C_i/C_r$, the cost for repairs is less than the cost of 100% initial incoming inspection.

EXAMPLE 13.9 Suppose T_i is ½ minute to inspect an incoming part and T_r is 20 minutes to dismantle and repair a unit with a defective part. Is inspection or repair more cost-efficient?

Solution

$$\frac{C_i}{C_r} = \frac{T_i \times R}{T_r \times R} = \frac{.5}{20} = .025$$

If $p > .025$ on the best lot, then 100% incoming inspection is necessary. A lot of 1000 parts takes 500 minutes (½ × 1000) to inspect. If $p > .025$, say $p = .04$, then 40 units (.04 × 1000) will have to be dismantled and repaired if no incoming inspection is done. Twenty minutes are needed for each repair, so 40 × 20 = 800 minutes to complete the repairs. In this case, when $p > C_i/C_r$, it is cheaper to complete a 100% incoming inspection and eliminate the defectives before assembly.

Examples 13.8 and 13.9 assume that there is a 100% final inspection, which is contrary to the prevention system of manufacturing. Both examples suggest that some serious work should be done either to improve the vendor's product or to find a better vendor. When incoming quality is at neither of the two extremes illustrated, then sampling will be a cost-effective way to eliminate the poorer quality lots. The AQL of the sampling plan should be less than the break-even ratio C_i/C_r.

13.4 THE AVERAGE OUTGOING QUALITY CURVE

The *average outgoing quality* (*AOQ*) *curve* shows the result of the incoming inspection and sorting of rejected lots. Outgoing quality, in this case, refers to the quality of the parts and materials that go from incoming inspection to manufacturing and assembly. Three assumptions are made for the inspection process:

1. All lots that pass inspection enter production as they are.
2. All lots that are rejected are inspected 100%, and all the nonconforming pieces are removed.
3. All pieces that are removed are replaced by conforming pieces.

Each OC curve has a corresponding AOQ curve. The points on the AOQ curve are determined as follows. For each incoming percent defective, the OC curve indicates what part is accepted for production and what part is rejected for 100% inspection. Only a fraction of the original number of defects makes it to production because of the sorting that occurs with the rejected lots.

EXAMPLE 13.10 Construct an AOQ curve to accompany an ($n = 200$, $c = 4$) OC curve.

Solution

If lots of 3000 are 3% defective, then each time a lot passes inspection, 90 defective pieces ($.03 \times 3000$) go into production. But the OC curve indicates that this happens for only 30% of the defective lots that are submitted. The other 70% of the lots are fully sorted so that *no* defects go into production. If all the lots are considered, an average of 30% of the 90 defective pieces per lot will make it into production. Therefore, $.3 \times 90 =$ 27 defectives per lot, on the average, make it into production. This is done with the percentages in the following table: 30% of the 3% defective ($.3 \times .03 = .009$) go into production. Applying this to the lot size of 3000, 27 defectives ($.009 \times 3000$) per lot, on the average, go into production. The average outgoing percent, $^{27}/_{3000}$ is .9%.

The AOQ curve is constructed in the same way as the OC curves. Label the horizontal axis as incoming percent defective; the table values go from 0 to 5% (see Table 13.9), so set the horizontal scale in units (0, 1, 2, 3, 4, 5). Label the vertical axis as outgoing percent defective. The table values go from 0 to 1.24%, so set the vertical scale in tenths (0, .1, .2,. . . ., 1.4). Plot the points from the paired values. The horizontal value is given first:

0, 0	Mark the corner point.
1, .94	Start at 1 on the horizontal axis, go up to .94, and mark a point.
1.5, 1.23	Start at 1.5 on the horizontal axis, go up to 1.23, and mark a point.

TABLE 13.9
Values for the AOQ curve

Incoming Defective (%)	Accepted from OC Curve (%)	Rejected and Fully Sorted (%)	Defects Outgoing
0	100	0	0%
1.0	94	6	$.92 \times .01 = .0094 = .94\%$
1.5	82	18	$.82 \times .015 = .0123 = 1.23\%$
2	62	38	$.62 \times .02 = .0124 = 1.24\%$
3	30	70	$.30 \times .03 = .009 = .9\%$
4	10	90	$.10 \times .04 = .004 = .4\%$
5	3	97	$.03 \times .05 = .0015 = .15\%$

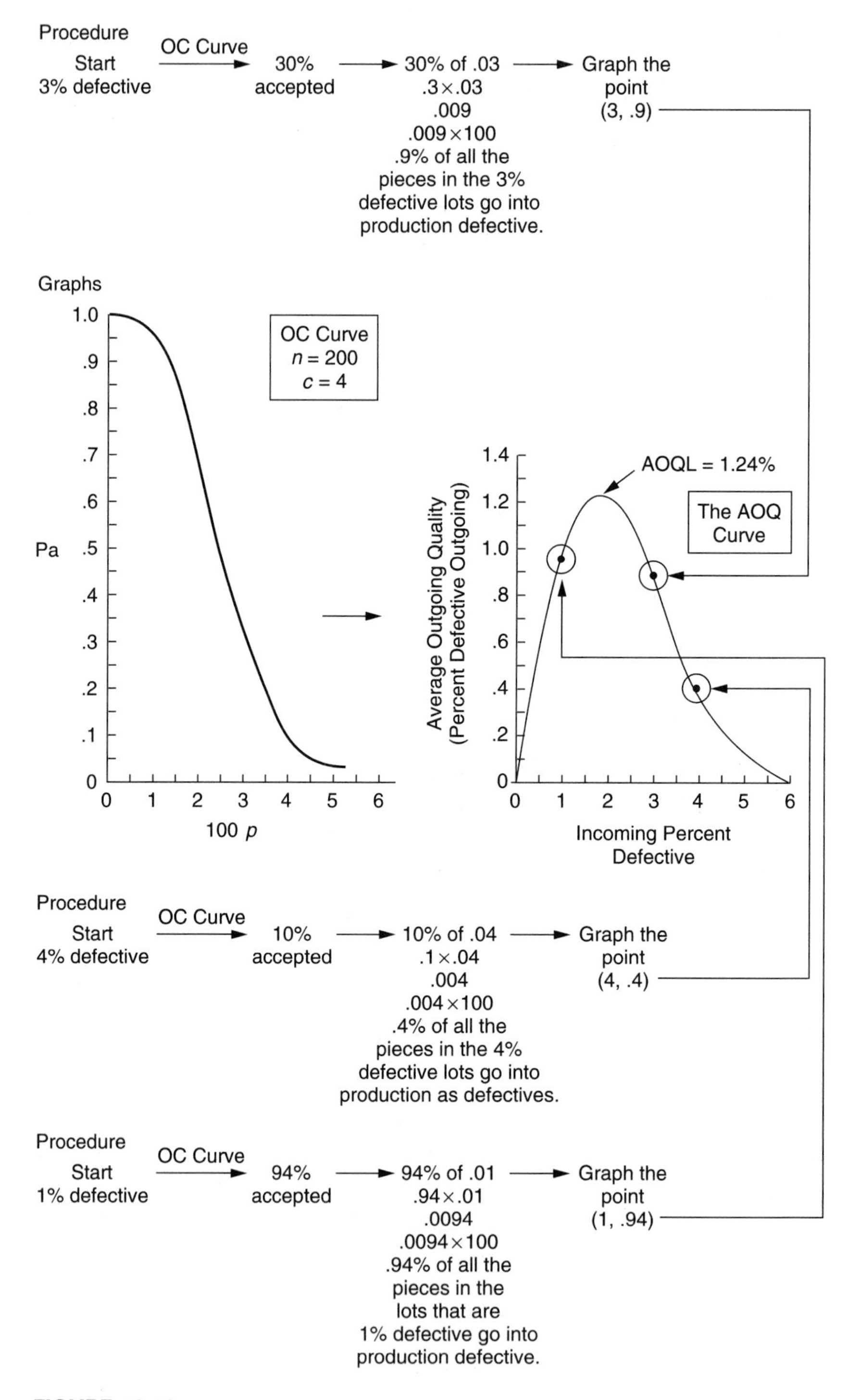

FIGURE 13.12
The graphing sequence for the AOQ curve.

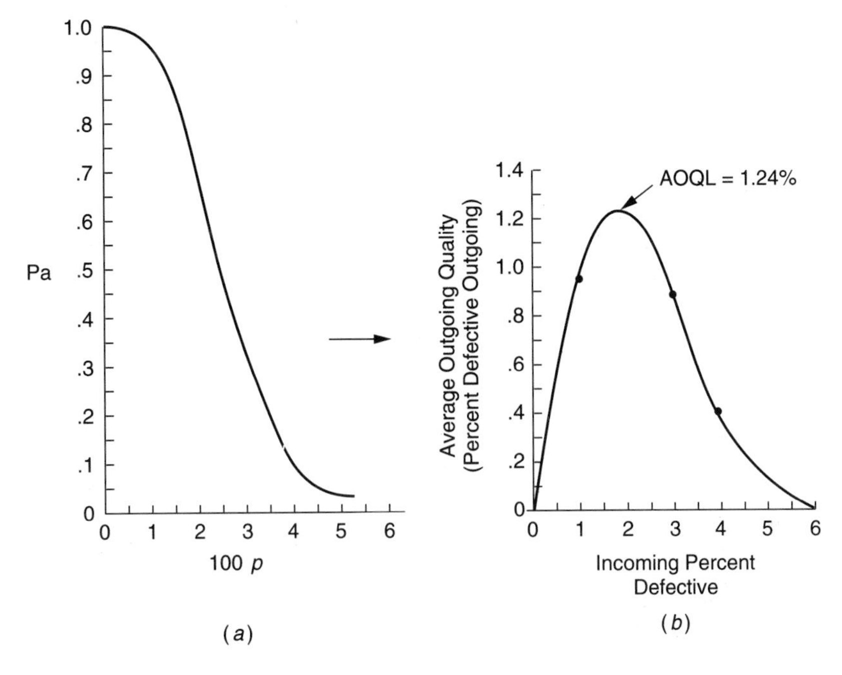

FIGURE 13.13
(*a*) The ($n = 200$, $c = 4$) OC curve; (*b*) the ($n = 200$, $c = 4$) AOQ curve.

The graphing sequence is illustrated in Figure 13.12, and the final result is shown in Figure 13.13. When all the points are graphed, draw a smooth curve through them. The result is the AOQ curve. The highest point on the curve is the maximum percent defective that will go into production, on the average, by this sampling plan. That point is called the AOQL, the *average outgoing quality limit.*

13.5 MLT-STD-105D FOR INSPECTION BY ATTRIBUTES[2]

There are many published sampling plans that can be followed for incoming inspection. Many companies still use these plans because they believe that they are the best available; other companies use them because they are tied into regulations that accompany government contracts. The MLT-STD-105D is presented here in brief form as an example of a standard sampling plan. The complete package of charts and directions can be obtained by writing to the U.S. Government Printing Office.

Some of the variations within the sampling procedure include the classification of defects: critical defects, major defects, and minor defects. Also, inspection is subclassified as normal, tightened, and reduced, and rules dictate when to switch from one classifica-

[2]Naval Publications and Forms Center, 5801 Tabor Avenue, Philadelphia, PA, 19120-5099.

tion to another. This inspection-level classification is determined by the relationship between the lot size and the sample size. Three inspection levels are usually shown—I, II, and III—and level II is used unless otherwise specified. Four additional levels—s-1, s-2, s-3, and s-4—are used when small samples are necessary or when large sampling risks may be taken.

There are three different types of sampling plans—single, double, and multiple—which are illustrated in Figures 13.14 to 13.16. The choice of sampling plan depends on the sample size and the administrative difficulty. Usually the sample size of multiple plans is less than that of double plans. The sample size for double sampling is usually less than for single sampling plans:

d — The number of defective pieces in the sample, d_1 for sample 1 and d_2 for sample 2

$\leq$ — Less than or equal to

$\geq$ — Greater than or equal to

$c < d < r$ — The value for d lies between the values for c and r

The single-sampling diagram in Figure 13.14 demonstrates the sampling decision process that was used with the OC curves. Values for sample size n and acceptance number c are needed for the single-sampling scheme.

The double-sampling scheme shown in Figure 13.15 requires two sets of numbers: n_1, c_1, and r_1 for the first sample and n_2 and r_2 for the second. If d_1 is less than or equal to c_1 in the first sample, accept the lot. If d_1 is greater than or equal to the rejection number r_1, reject the lot. If d_1 is between c_1 and r_1, take a second sample of size n_2. If the total number of defects is less than the rejection number r_2, accept the lot; otherwise, reject it.

Double sampling is preferred sometimes because it appears to give a lot a second chance before rejecting it. Its main attribute, however, is the fact that in some situations it can lead to fewer pieces being sampled overall.

Multiple sampling, as illustrated in Figure 13.16, extends the double-sampling concept to triple sampling and sometimes beyond. Each sampling step except the last

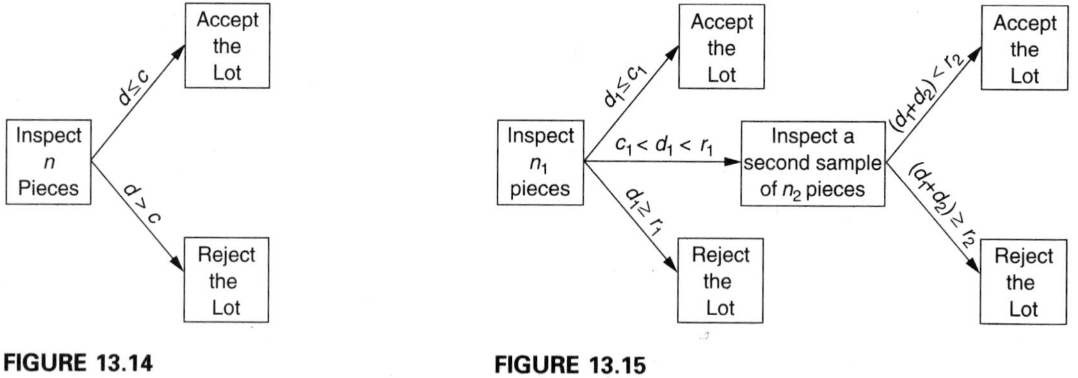

FIGURE 13.14
Single sampling.

FIGURE 13.15
Double sampling.

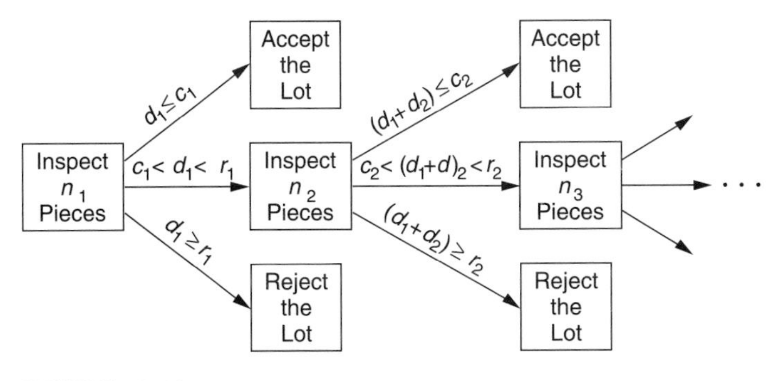

FIGURE 13.16
Multiple sampling.

involves three numbers: sample size n, acceptance number c, and a rejection number r. The last step has just the sample size n and the rejection number r: If the total number of defectives is less than the final rejection number, the lot is accepted.

The various sample sizes are designated by MLT-STD-105D code letters, which are listed in Table 13.10. The code letter and sample size are then used with a master table such as Table 13.11 to determine the acceptance and rejection numbers. Three other master tables are included for comparison: Table 13.12 for tightened inspection, Table 13.13 for reduced inspection, and Table 13.14 for double sampling.

TABLE 13.10
Sample size code letters

	Special Inspection Levels				General Inspection Levels		
Lot or Batch Size	S-1	S-2	S-3	S-4	I	II	III
2–8	A	A	A	A	A	A	B
9–15	A	A	A	A	A	B	C
16–25	A	A	B	B	B	C	D
26–50	A	B	B	C	C	D	E
51–90	B	B	C	C	C	E	F
91–150	B	B	C	D	D	F	G
151–280	B	C	D	E	E	G	H
281–500	B	C	D	E	F	H	J
501–1200	C	C	E	F	G	J	K
1201–3200	C	D	E	G	H	K	L
3201–10,000	C	D	F	G	J	L	M
10,001–35,000	C	D	F	H	K	M	N
35,001–150,000	D	E	G	J	L	N	P
150,001–500,000	D	E	G	J	M	P	Q
500,001 and over	D	E	H	K	N	Q	R

TABLE 13.11
Single-sampling plans for normal inspections (master table)

Sample Size Code Letter	Sample Size	Acceptable Quality																							
		0.010		0.015		0.025		0.040		0.065		0.10		0.15		0.25		0.40		0.65		1.0		1.5	
		Ac	Re	Ac	Re	Ac	Re	Ac	Re	Ac	Re	Ac	Re	Ac	Re	Ac	Re	Ac	Re	Ac	Re	Ac	Re	Ac	Re
A	2																								
B	3	↓		↓		↓		↓		↓		↓		↓		↓		↓		↓		↓		↓	
C	5	↓		↓		↓		↓		↓		↓		↓		↓		↓		↓		↓		↓	
D	8	↓		↓		↓		↓		↓		↓		↓		↓		↓		↓		↓		0	1
E	13	↓		↓		↓		↓		↓		↓		↓		↓		↓		↓		0	1	↑	
F	20	↓		↓		↓		↓		↓		↓		↓		↓		↓		0	1	↑		↓	
G	32	↓		↓		↓		↓		↓		↓		↓		↓		0	1	↑		↓		1	2
H	50	↓		↓		↓		↓		↓		↓		↓		0	1	↑		↓		1	2	2	3
J	80	↓		↓		↓		↓		↓		↓		0	1	↑		↓		1	2	2	3	3	4
K	125	↓		↓		↓		↓		↓		0	1	↑		↓		1	2	2	3	3	4	5	6
L	200	↓		↓		↓		↓		0	1	↑		↓		1	2	2	3	3	4	5	6	7	8
M	315	↓		↓		↓		0	1	↑		↓		1	2	2	3	3	4	5	6	7	8	10	11
N	500	↓		↓		0	1	↑		↓		1	2	2	3	3	4	5	6	7	8	10	11	14	15
P	800	↓		0	1	↑		↓		1	2	2	3	3	4	5	6	7	8	10	11	14	15	21	22
Q	1250	0	1	↑		↓		1	2	2	3	3	4	5	6	7	8	10	11	14	15	21	22	↑	
R	2000	↑		↑		1	2	2	3	3	4	5	6	7	8	10	11	14	15	21	22	↑		↑	

↓ = Use first sampling plan below. If sample size equals or exceeds lot or batch size, do 100% inspection.

↑ = Use first sampling plan above arrow.

Ac = Acceptance number.

Re = Rejection number.

Levels (Normal Inspection)

2.5		4.0		6.5		10		15		25		40		65		100		150		250		400		650		1000	
Ac	Re	Ac	Re	Ac	Re	Ac	Re	Ac	Re	Ac	Re	Ac	Re	Ac	Re	Ac	Re	Ac	Re	Ac	Re	Ac	Re	Ac	Re	Ac	Re
↓		↓		0	1	↓		↓		1	2	2	3	3	4	5	6	7	8	10	11	14	15	21	22	30	31
		0	1	↑				1	2	2	3	3	4	5	6	7	8	10	11	14	15	21	22	30	31	44	45
0	1	↑		↓		1	2	2	3	3	4	5	6	7	8	10	11	14	15	21	22	30	31	44	45		
↑		↓		1	2	2	3	3	4	5	6	7	8	10	11	14	15	21	22	30	31	44	45			↑	
↓		1	2	2	3	3	4	5	6	7	8	10	11	14	15	21	22	30	31	44	45			↑			
1	2	2	3	3	4	5	6	7	8	10	11	14	15	21	22							↑					
2	3	3	4	5	6	7	8	10	11	14	15	21	22			↑		↑		↑							
3	4	5	6	7	8	10	11	14	15	21	22			↑													
5	6	7	8	10	11	14	15	21	22			↑															
7	8	10	11	14	15	21	22			↑																	
10	11	14	15	21	22			↑																			
14	15	21	22			↑																					
21	22			↑																							
		↑																									
↑																											

TABLE 13.12
Single-sampling plans for tightened inspections (master table)

Sample Size Code Letter	Sample Size	Acceptable Quality																					
		0.010		0.015		0.025		0.040		0.065		0.10		0.15		0.25		0.40		0.65		1.0		1.5	
		Ac	Re	Ac	Re	Ac	Re	Ac	Re	Ac	Re	Ac	Re	Ac	Re	Ac	Re	Ac	Re	Ac	Re	Ac	Re	Ac	Re
A	2	↓		↓		↓		↓		↓		↓		↓		↓		↓		↓		↓		↓	
B	3	↓		↓		↓		↓		↓		↓		↓		↓		↓		↓		↓		↓	
C	5	↓		↓		↓		↓		↓		↓		↓		↓		↓		↓		↓		↓	
D	8	↓		↓		↓		↓		↓		↓		↓		↓		↓		↓		↓		↓	
E	13	↓		↓		↓		↓		↓		↓		↓		↓		↓		↓		↓		0	1
F	20	↓		↓		↓		↓		↓		↓		↓		↓		↓		↓		0	1	↓	
G	32	↓		↓		↓		↓		↓		↓		↓		↓		↓		0	1	↓		↓	
H	50	↓		↓		↓		↓		↓		↓		↓		↓		0	1	↓		↓		1	2
J	80	↓		↓		↓		↓		↓		↓		↓		0	1	↓		↓		1	2	2	3
K	125	↓		↓		↓		↓		↓		↓		0	1	↓		↓		1	2	2	3	3	4
L	200	↓		↓		↓		↓		↓		0	1	↓		↓		1	2	2	3	3	4	5	6
M	315	↓		↓		↓		↓		0	1	↓		↓		1	2	2	3	3	4	5	6	8	9
N	500	↓		↓		↓		0	1	↓		↓		1	2	2	3	3	4	5	6	8	9	12	13
P	800	↓		↓		0	1	↓		↓		1	2	2	3	3	4	5	6	8	9	12	13	18	19
Q	1250	↓		0	1	↓		↓		1	2	2	3	3	4	5	6	8	9	12	13	18	19	↑	
R	2000	0	1	↑		↓		1	2	2	3	3	4	5	6	8	9	12	13	18	19	↑		↑	
S	3150					1	2																	↑	

↓ = Use first sampling plan below. If sample size equals or exceeds lot or batch size, do 100% inspection.

↑ = Use first sampling plan above arrow.

Ac = Acceptance number.

Re = Rejection number.

Levels (Reduced Inspection)

2.5	4.0	6.5	10	15	25	40	65	100	150	250	400	650	1000
Ac Re	Ac Re	Ac Re	Ac Re	Ac Re	Ac Re	Ac Re	Ac Re	Ac Re	Ac Re	Ac Re	Ac Re	Ac Re	Ac Re
↓	↓	↓		↓	↓	1 2	2 3	3 4	5 6	8 9	12 13	18 19	27 28
↓	↓	0 1	↓	↓	1 2	2 3	3 4	5 6	8 9	12 13	18 19	27 28	41 42
↓	0 1	↓	↓	1 2	2 3	3 4	5 6	8 9	12 13	18 19	27 28	41 42	↑
0 1	↓	↓	1 2	2 3	3 4	5 6	8 9	12 13	18 19	27 28	41 42	↑	↑
↓	↓	1 2	2 3	3 4	5 6	8 9	12 13	18 19	27 28	41 42	↑	↑	↑
↓	1 2	2 3	3 4	5 6	8 9	12 13	18 19	↑	↑	↑	↑	↑	↑
1 2	2 3	3 4	5 6	8 9	12 13	18 19	↑	↑	↑	↑	↑	↑	↑
2 3	3 4	5 6	8 9	12 13	18 19	↑	↑	↑	↑	↑	↑	↑	↑
3 4	5 6	8 9	12 13	18 19	↑	↑	↑	↑	↑	↑	↑	↑	↑
5 6	8 9	12 13	18 19	↑	↑	↑	↑	↑	↑	↑	↑	↑	↑
8 9	12 13	18 19	↑	↑	↑	↑	↑	↑	↑	↑	↑	↑	↑
12 13	18 19	↑	↑	↑	↑	↑	↑	↑	↑	↑	↑	↑	↑
18 19	↑	↑	↑	↑	↑	↑	↑	↑	↑	↑	↑	↑	↑
↑	↑	↑	↑	↑	↑	↑	↑	↑	↑	↑	↑	↑	↑

TABLE 13.13
Single-sampling plans for reduced inspections (master table)

Sample Size Code Letter	Sample Size	Acceptable Quality																							
		0.010		0.015		0.025		0.040		0.065		0.10		0.15		0.25		0.40		0.65		1.0		1.5	
		Ac	Re	Ac	Re	Ac	Re	Ac	Re	Ac	Re	Ac	Re	Ac	Re	Ac	Re	Ac	Re	Ac	Re	Ac	Re	Ac	Re
A	2																								
B	2	↓		↓		↓		↓		↓		↓		↓		↓		↓		↓		↓		↓	
C	2	↓		↓		↓		↓		↓		↓		↓		↓		↓		↓		↓		↓	
D	3	↓		↓		↓		↓		↓		↓		↓		↓		↓		↓		↓		0	1
E	5	↓		↓		↓		↓		↓		↓		↓		↓		↓		↓		0	1	↑	
F	8	↓		↓		↓		↓		↓		↓		↓		↓		↓		0	1	↑		↓	
G	13	↓		↓		↓		↓		↓		↓		↓		↓		0	1	↑		↓		0	2
H	20	↓		↓		↓		↓		↓		↓		↓		0	1	↑		↓		0	2	1	3
J	32	↓		↓		↓		↓		↓		↓		0	1	↑		↓		0	2	1	3	1	4
K	50	↓		↓		↓		↓		↓		0	1	↑		↓		0	2	1	3	1	4	2	5
L	80	↓		↓		↓		↓		0	1	↑		↓		0	2	1	3	1	4	2	5	3	6
M	125	↓		↓		↓		0	1	↑		↓		0	2	1	3	1	4	2	5	3	6	5	8
N	200	↓		↓		0	1	↑		↓		0	2	1	3	1	4	2	5	3	6	5	8	7	10
P	315	↓		0	1	↑		↓		0	2	1	3	1	4	2	5	3	6	5	8	7	10	10	13
Q	500	0	1	↑		↓		0	2	1	3	1	4	2	5	3	6	5	8	7	10	10	13	↑	
R	800	↑		↑		0	2	1	3	1	4	2	5	3	6	5	8	7	10	10	13	↑		↑	

↓ = Use first sampling plan below. If sample size equals or exceeds lot or batch size, do 100% inspection.

↑ = Use first sampling plan above arrow.

Ac = Acceptance number.

Re = Rejection number.

* = If the acceptance number has been exceeded but the rejection number has not been reached, accept the lot, but reinstate normal inspection.

Levels (Reduced Inspection)*

2.5		4.0		6.5		10		15		25		40		65		100		150		250		400		650		1000	
Ac	Re	Ac	Re	Ac	Re	Ac	Re	Ac	Re	Ac	Re	Ac	Re	Ac	Re	Ac	Re	Ac	Re	Ac	Re	Ac	Re	Ac	Re	Ac	Re
↓		↓		0	1	↓		↓		1	2	2	3	3	4	5	6	7	8	10	11	14	15	21	22	30	31
↓		0	1	↑		↓		0	2	1	3	2	4	3	5	5	6	7	8	10	11	14	15	21	22	30	31
0	1	↑		↓		0	2	1	3	1	4	2	5	3	6	5	8	7	10	10	13	14	17	21	24	↑	
↑		↓		0	2	1	3	1	4	2	5	3	6	5	8	7	10	10	13	14	17	21	24	↑		↑	
↓		0	2	1	3	1	4	2	5	3	6	5	8	7	10	10	13	14	17	21	24	↑		↑		↑	
0	2	1	3	1	4	2	5	3	6	5	8	7	10	10	13	↑		↑		↑		↑		↑		↑	
1	3	1	4	2	5	3	6	5	8	7	10	10	13	↑		↑		↑		↑		↑		↑		↑	
1	4	2	5	3	6	5	8	7	10	10	13	↑		↑		↑		↑		↑		↑		↑		↑	
2	5	3	6	5	8	7	10	10	13	↑		↑		↑		↑		↑		↑		↑		↑		↑	
3	6	5	8	7	10	10	13	↑		↑		↑		↑		↑		↑		↑		↑		↑		↑	
5	8	7	10	10	13	↑		↑		↑		↑		↑		↑		↑		↑		↑		↑		↑	
7	10	10	13	↑		↑		↑		↑		↑		↑		↑		↑		↑		↑		↑		↑	
10	13	↑		↑		↑		↑		↑		↑		↑		↑		↑		↑		↑		↑		↑	
↑		↑		↑		↑		↑		↑		↑		↑		↑		↑		↑		↑		↑		↑	
↑		↑		↑		↑		↑		↑		↑		↑		↑		↑		↑		↑		↑		↑	
↑		↑		↑		↑		↑		↑		↑		↑		↑		↑		↑		↑		↑		↑	

TABLE 13.14
Double-sampling plans for normal inspection (master table)

Sample Size Code Letter	Sample	Sample Size	Cumulative Sample Size	Acceptable Quality																					
				0.010		0.015		0.025		0.040		0.065		0.10		0.15		0.25		0.40		0.65		1.0	
				Ac	Re	Ac	Re	Ac	Re	Ac	Re	Ac	Re	Ac	Re	Ac	Re	Ac	Re	Ac	Re	Ac	Re	Ac	Re
A																									
B	First	2	2																						
	Second	2	4																						
C	First	3	3																						
	Second	3	6																						
D	First	5	5																						
	Second	5	10																					↓	
E	First	8	8																						
	Second	8	16																			↓			
F	First	13	13																					↑	
	Second	13	26																	↓					
G	First	20	20																			↑			
	Second	20	40															↓						↑	
H	First	32	32																	↑				0	2
	Second	32	64													↓								1	2
J	First	50	50															↑				0 ↓	2	0	3
	Second	50	100											↓								1	2	3	4
K	First	80	80													↑				0 ↓	2	0	3	1	4
	Second	80	160									↓								1	2	3	4	4	5
L	First	125	125											↑				0 ↓	2	0	3	1	4	2	5
	Second	125	250							↓								1	2	3	4	4	5	6	7
M	First	200	200									↑				0 ↓	2	0	3	1	4	2	5	3	7
	Second	200	400					↓								1	2	3	4	4	5	6	7	8	9
N	First	315	315							↑				0	2	0	3	1	4	2	5	3	7	5	9
	Second	315	630			↓								1	2	3	4	4	5	6	7	8	9	12	13
P	First	500	500					↑				0 ↓	2	0	3	1	4	2	5	3	7	5	9	7	11
	Second	500	1000									1	2	3	4	4	5	6	7	8	9	12	13	18	19
Q	First	800	800	↓						0 ↓	2	0	3	1	4	2	5	3	7	5	9	7	11	11	16
	Second	800	1600			↑				1	2	3	4	4	5	6	7	8	9	12	13	18	19	26	27
R	First	1250	1250	↑				0 ↓	2	0	3	1	4	2	5	3	7	5	9	7	11	11	16	↑	
	Second	1250	2500					1	2	3	4	4	5	6	7	8	9	12	13	18	19	26	27		

↓ = Use first sampling plan below. If sample size equals or exceeds lot or batch size, do 100% inspection.

↑ = Use first sampling plan above arrow.

Ac = Acceptance number.

Re = Rejection number.

* = Use corresponding single-sampling plan (or alternatively, use double-sampling plan where available).

Levels (NNormal Inspection)

1.5		2.5		4.0		6.5		10		15		25		40		65		100		150		250		400		650		1000	
Ac	Re	Ac	Re	Ac	Re	Ac	Re	Ac	Re	Ac	Re	Ac	Re	Ac	Re	Ac	Re	Ac	Re	Ac	Re	Ac	Re	Ac	Re	Ac	Re	Ac	Re
↓		↓		↓		*		↓		↓		*		*		*		*		*		*		*		*		*	
				*		↑				0	2	0	3	1	4	2	5	3	7	5	9	7	11	11	16	17	22	25	11
										1	2	3	4	4	5	6	7	8	9	12	13	18	19	26	27	17	38	56	57
		*		↑				0	2	0	3	1	4	2	5	3	7	5	9	7	11	11	16	17	22	25	11	↑	
								1	2	3	4	4	5	6	7	8	9	12	13	18	19	26	27	17	38	56	57		
*		↑				0	2	0	3	1	4	2	5	3	7	5	9	7	11	11	16	17	22	25	11	↑			
						1	2	3	4	4	5	6	7	8	9	12	13	18	19	26	27	17	38	56	57				
↑				0	2	0	3	1	4	2	5	3	7	5	9	7	11	11	16	17	22	25	11	↑					
				1	2	3	4	4	5	6	7	8	9	12	13	18	19	26	27	17	38	56	57						
		0	2	0	3	1	4	2	5	3	7	5	9	7	11	11	16	↑		↑		↑							
		1	2	3	4	4	5	6	7	8	9	12	13	18	19	26	27												
0	2	0	3	1	4	2	5	3	7	5	9	7	11	11	16	↑													
1	2	3	4	4	5	6	7	8	9	12	13	18	19	26	27														
0	3	1	4	2	5	3	7	5	9	7	11	11	16	↑															
3	4	4	5	6	7	8	9	12	13	18	19	26	27																
1	4	2	5	3	7	5	9	7	11	11	16	↑																	
4	5	6	7	8	9	12	13	18	19	26	27																		
2	5	3	7	5	9	7	11	11	16	↑																			
6	7	8	9	12	13	18	19	26	27																				
3	7	5	9	7	11	11	16	↑																					
8	9	12	13	18	19	26	27																						
5	9	7	11	11	16	↑																							
12	13	18	19	26	27																								
7	11	11	16	↑																									
18	19	26	27																										
11	16	↑																											
26	27																												
↑																													

EXAMPLE 13.11

Use MLT-STD-105D to determine the sample size and the acceptance number of the following criteria:

- Single sampling
- Tightened inspection
- Inspection level II
- Lot size 6500
- AQL 1.5%

Solution

The code letter is taken from Table 13.10. Find the row that has the lot size 3201 to 10,000. Follow it across to the inspection level II column and read the code letter, L. Single sampling with tightened inspection is shown in Table 13.12. Find the row that corresponds to the code letter L in the left column and follow it across to find the sample size, $n = 200$, and then to the column that corresponds to an AQL value of 1.5. At the intersection of the row and column are two numbers, 5 and 6. Five is the maximum acceptance number, and 6 is the minimum rejection number.

This lot is acceptable if five or fewer defective pieces are found in the sample size of 200. It should be rejected if six or more defective pieces are found in the sample.

EXAMPLE 13.12

Use MLT-STD-105D to determine the sample size and the acceptance number for the following criteria:

- Double sampling
- Normal inspection
- Inspection level II
- Lot size 6500
- AQL 1.5%

Solution

The code letter, L, is found in Table 13.10 as in the previous example. Table 13.14 is used for double sampling and normal inspection. The code letter L in the left column gives the appropriate row. The sample size column indicates that the first sample should be 125 pieces. If a second sample is needed, it should also be 125 pieces. Go across to the column corresponding to an AQL of 1.5. There are four numbers at the intersection of the row and column: 3, 7, 8, and 9. The 3 and 7 in the top row indicate that the acceptance number is 3 and the rejection number is 7. If the number of defectives in the first sample of 125 pieces is between 3 and 7, then a second sample of 125 pieces is taken. The acceptance total becomes 8 and the rejection total becomes 9.

This lot is accepted if 3 or fewer defective pieces are found in the first sample of 125. It is rejected if 7 or more defective pieces are found in the first sample. A second sample of 125 is taken if the first sample has 4, 5, or 6 defective pieces. This lot is accepted if the total number of defective pieces for both samples is 8 or less. If the total number of defective pieces for both samples is 9 or more, the lot is rejected.

EXAMPLE 13.13 Use MLT-STD-105D to determine the sample size and acceptance number for the following criteria. Then analyze the quality protection.

- Single sampling
- Normal inspection
- Inspection level II
- Lot size 6500
- AQL 1.5%

Solution

The code letter L is again obtained from Table 13.10. Single sampling for normal inspection is found in Table 13.11. The row corresponding to the code letter L shows that a sample of 200 is to be taken. Follow the row across to the AQL = 1.5 column. The numbers at the intersection of the row and column are 7 and 8.

If 7 or fewer defective pieces are found in the sample of 200, the lot is accepted. If 8 or more defective pieces are found, the lot is rejected.

The quality protection must be determined from the OC curve (see Table 13.15). The OC curve (Figure 13.17) shows that the actual AQL value is 2% instead of the planned 1.5%. The average proportion defective allowed by this sampling plan is slightly less than 2%. The RQL, rejectable quality level, is 6%.

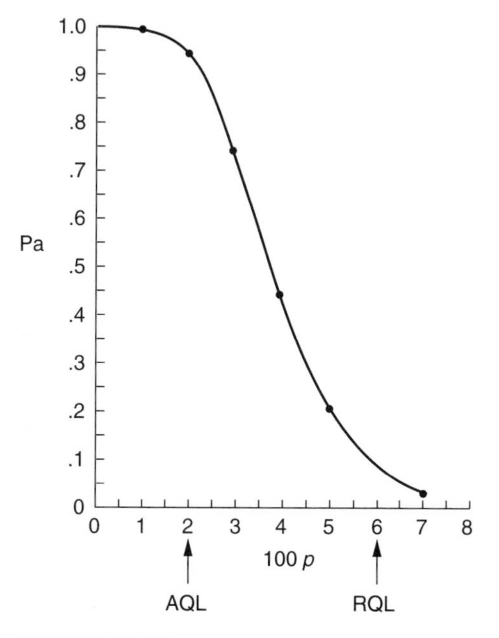

FIGURE 13.17
The ($n = 200$, $c = 7$) OC curve.

TABLE 13.15
Table of values for Figure 13.17

p	np	$100p$	Pa
.01	2	1	.999
.02	4	2	.95
.03	6	3	.74
.04	8	4	.46
.05	10	5	.22
.06	12	6	.09
.07	14	7	.03

13.6 THE AVERAGE PROPORTION DEFECTIVE

One of the problems with the AQL and OC curves is the apparent discrepancy between the quality goal and the quality obtained. The indecisive zone on the OC curve implies that lots with higher proportions defective can be accepted far too often. The following analysis will show that the overall average proportion defective received from a vendor is slightly less than the AQL value on the appropriate OC curve.

EXAMPLE 13.14

For the past year a company received six lots from the vendor whose lot was inspected in Example 13.13, and the incoming inspection records show the following:

Lot 1	Lot 2	Lot 3	Lot 4	Lot 5	Lot 6
$n = 200$	$n = 200$	$n = 200$	$n = 200$	$n = 200$	$n = 200$
$d = 4$	$d = 6$	$d = 2$	$d = 4$	$d = 5$	$d = 3$

Calculate the true incoming quality that is going into production.

Solution

Total number of defects (p) = 4 + 6 + 2 + 4 + 5 + 3 = 24
Total number sampled (N) = 6 × 200 = 1200

$$\bar{p} = \frac{24}{1200}$$

$$= .02 \quad \text{(average proportion defective in the samples)}$$

$$\sigma = \sqrt{\frac{\bar{p}(1 - \bar{p})}{N}}$$

$$= \sqrt{\frac{.02(.98)}{1200}}$$

$$= .004$$

The lot proportion defective p is normally distributed. A variation of the z formula that was introduced in Chapter 5 is used to apply the normal curve to proportions.

$$z = \frac{p - \bar{p}}{\sigma}$$

The horizontal axis of the normal curve is in p units (instead of x, as in Chapter 5), with the average $\bar{p}$ at the center. Calculate the probability that the vendor will ship a lot with 1 percent defective ($p = .01$) by finding the area under the curve from $p = .005$ to $p = .015$. This actually gives the probability that p lies between .005 and .015, but all of those p values will round off to $p = .01$. Similarly, find the probability that the vendor will send a lot with $p = .02$ $(.105 < p < .025)$, $p = .03$ $(.025 < p < .035)$, $p < .005$, and $p > .035$. Figure 13.18 illustrates these sections under the normal curve. These probabilities can then be combined with the information from the OC curve to give a realistic picture of the true incoming quality that makes it to production.

Find the area in the tail of the curve to the left of .005:

$$P(p = 0) = P(p < .005)$$

$$z = \frac{p - \bar{p}}{\sigma}$$

$$= \frac{.005 - .02}{.004}$$

$$= -3.75$$

From Table B.8 in Appendix B, $z = 3.75$ (row 3.7, column 5) gives a tail area of .00009. This is area 1 in Figure 13.19:

$$\text{Area 1} = P(p = 0) = .00009$$

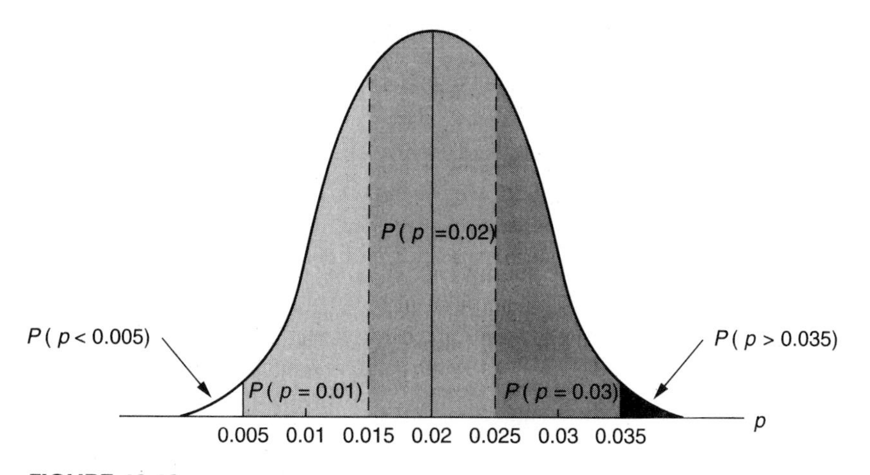

FIGURE 13.18
The normal distribution for p in Example 13.14.

FIGURE 13.19
The probability that $p = .01$.

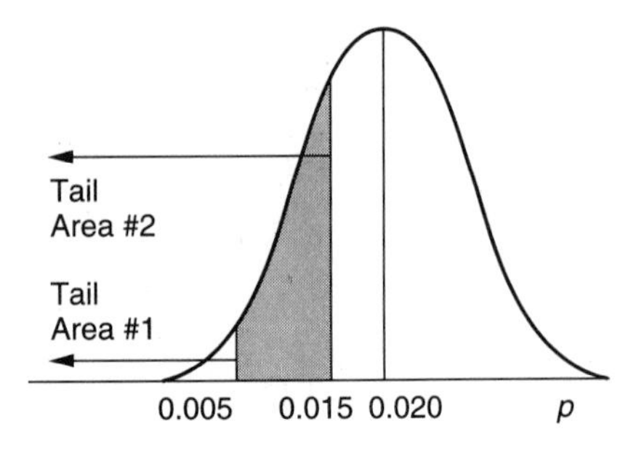

Next, find the area between $p = .005$ and $p = .015$:

$$P(p = .01) = P(.005 < p < .015)$$

$$z = \frac{p - \bar{p}}{\sigma}$$

$$= \frac{.005 - .02}{.004}$$

$$= -1.25$$

A z value of 1.25 has a tail area of .1056, so tail area 2 in Figure 13.19 is .1056:

$$\begin{aligned} \text{Area 2} - \text{Area 1} &= P(p = .01) \\ &= .1056 - .00009 \\ &= .1055 \end{aligned}$$

Find the area between $p = .105$ and $p = .025$:

$$P(p = .02) = P(.015 < p < .025)$$

Because of the symmetry of the normal curve, the area from the center value of $p = .02$ to $p = .015$ is equal to the area from the center to $p = .025$. Remember that the area under one-half of the curve is .5:

$$\begin{aligned} P(p = .02) &= 2 \times (.5000 - \text{area } 2) \\ &= 2 \times (.5000 - .1056) \\ &= 2 \times (.3944) \\ &= .7888 \end{aligned}$$

This is illustrated in Figure 13.20.

The symmetry in Figure 13.18 shows that the area from $p = .025$ to $p = .035$ is equal to the area from $p = .015$ to $p = .005$. Therefore, the probability that $p = .03$ is the same as the probability that $p = .01$. Likewise, the probability that $p = 0$ is the same as the probability that $p > .035$:

$$P(p = .03) = .1055$$
$$P(p > 3.5) = .00009$$

FIGURE 13.20
The probability that $p = .02$.

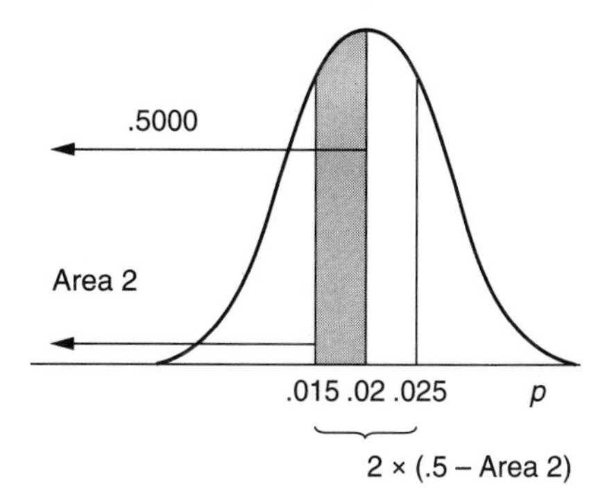

The probability of receiving a lot with a specific proportion defective can be combined with the probability of acceptance of each of those lots from the OC curve to give the true proportion of defectives going into production. The sum of the products in Table 13.16, .9330, is the proportion of the incoming lots that is accepted.

If the rejected lots are 100% sorted, then the products shown in Table 13.16 are the proportions of the incoming lots within each category that are going into production. For example, 7.81% of all the lots going into production are 3% defective.

If the rejected lots are returned to the vendor instead of being sorted, then the products have to be divided by the total, .933, to find the true proportion in each category that is going into production:

$$\frac{.0001}{.933} = .0001 \qquad 0.0\% \text{ of the lots going into production have } p = 0$$

$$\frac{.1054}{.933} = .113 \qquad 11.3\% \text{ of the lots going into production have } p = .01$$

$$\frac{.7494}{.933} = .803 \qquad 80.3\% \text{ of the lots going into production have } p = .02$$

TABLE 13.16
Proportion of incoming lots accepted for each p value

p	$P(p)$	×	P(lot is accepted)	=	Proportion of Incoming Lots Accepted
0	.0001	×	1.0	=	.0001
.01	.1055	×	.999	=	.1054
.02	.7888	×	.95	=	.7494
.03	.1055	×	.74	=	.0781
.04	.0001	×	.46	=	.000046
					.9330

$\frac{.0781}{.933} = .084$ 8.4% of the lots going into production have $p = .03$

$\frac{.000046}{.933} = .000$ 0.0% of the lots going into production have $p = .04$ or larger

Determine the overall percent defective by combining the percentage of the lots going into production with their proportion defective (see Table 13.17). In the first case 1.97% of all the incoming material going into production is defective; in the second case, 1.84% is defective.

TABLE 13.17
Overall proportion defective

With Rejected Lots Returned	With Rejected Lots Sorted
.113 × .01 = .00113	.1054 × .01 = .001054
.803 × .02 = .01606	.7494 × .02 = .014988
.084 × .03 = .00254	.0781 × .03 = .002343
.0197	.0184

This analysis gives a more realistic picture of what production is facing in terms of incoming quality. It also provides data for a cost analysis regarding rejected lots. Is it cheaper to just ship the rejects back to the vendor? Does the fraction defective going into production change enough (.13% in this case) to make the cost and bother of 100% sorting worthwhile?

If the piece in the incoming lot studied in Example 13.14 combines with several different pieces feeding into production, each different piece will have to undergo a similar analysis. Then the "and" rule of probability, applied to the different defect-free proportions, will give the fraction of final product that is defect free (as long as production does not produce any more defects).

EXAMPLE 13.15 Nine different parts received from vendors are assembled in one product. Four of the vendors are submitting parts that are 2% defective, and the other five vendors are producing parts that are 1% defective. If the incoming parts go directly into production, what percentage of the final assembled product is defect free?

Solution

If the parts are 2% defective, they are 98% good. Likewise, the 1% defective parts are 99% good. The "and" rule from probability applies:

$$\begin{aligned} P(\text{good product}) &= P(\text{1st part is good } \textit{and} \text{ 2nd part is good} \\ &\quad \textit{and} \ldots \textit{and} \text{ 9th part is good}) \\ &= (.99)^5 \times (.98)^4 \\ &= .951 \times .922 \\ &= .877 \end{aligned}$$

If no further mistakes are made in the assembly operation, 87.7% of the final product will be defect free.

This concept may be extended: Suppose an assembled product consists of 50 parts and each of the 50 parts is 99% defect free. What percentage of the final product will be free of defects?

$$P(\text{defect-free product}) = (.99)^{50}$$
$$= .605$$

Only 60.5% of the final assembled product will be good. On an even larger scale, consider the automotive industry or another industry in which the final product consists of thousands of parts. Even 99.99% defect-free parts can lead to an excessive number of final products that are defective in one way or another.

13.7 VENDOR CERTIFICATION AND CONTROL CHART MONITORING

In the quest for higher quality, accepting lots that are even 1% defective is becoming unreasonable. As the percent defective is forced down to levels such as .01%, acceptance sampling will become unrealistic owing to the large samples that will be necessary. Even now acceptance sampling is being phased out in favor of vendor certification plans. Each receiving company has its own certification plan, but in general, vendors are being asked for proof of quality: They have to provide evidence of a prevention, not detection, quality control program and of their use of SPC. Some large companies have a vendor rating system for their suppliers. Receiving companies, aiming for higher quality and cost reductions, want to be able to rely on vendor quality assurance and to move the incoming materials directly into production without incoming inspection.

One basic method that can be used by incoming inspection in place of the inappropriate acceptance sampling features chart analysis. Certified vendors are vendors whose previous shipments were honestly and reliably quantified. Receiving companies have to determine the critical measurements on incoming materials. When those materials are purchased from a certified vendor, copies of the vendor's control charts for the critical measurements should be requested. Analysis of the charts can replace incoming inspection. Means, standard deviations, analysis for statistical control, histograms of the data on the charts, and process capability can all be determined. If any of the material is out of specification, it should show up on the control charts for that lot.

For new vendors and vendors that have not been certified and classified, large random samples (n = 100 or more) of incoming materials should be taken and the critical measurements recorded. A histogram of those measurements can give an estimate of the shape of the distribution, the mean, the standard deviation, and the process capability. These estimates can then be compared with the vendor data and the analysis of those data. When the two analyses are compatible, the frequency of checks on that vendor can be reduced until reliability and confidence have been established. Then, as with other certified vendors, there should be just an occasional audit inspection.

If discrepancies do occur between the inspection data and the vendor-submitted control chart data, each lot should be sampled so that an estimate of means, standard deviation, and distribution shape is determined. If differences continue to occur in the two sets of data, on-site inspection and discussion of procedures should take place. If a vendor is not cooperative, a search for a new supplier is in order.

EXERCISES

1. A box contains 20 parts, and 4 of the parts are defective. If a sample of 5 parts is randomly chosen without replacement, find the probability of getting 3 good parts and 2 defectives. Use the hypergeometric probability distribution.
2. The box of 20 parts from exercise 1 is used again. A sample of 5 parts is chosen *with* replacement. Find the probability of getting 3 good parts and 2 defectives. Use the binomial probability distribution.
3. A lot of 2000 pieces arrives in 10 boxes that are uniformly packaged with 5 layers and 40 pieces per layer. Plan a random sample of 100 using a three-digit random-number generator.
4. **a.** Use the Poisson curves in Figure 13.2 to make an ($n = 200$, $c = 3$) OC curve.
   b. Use the ($n = 200$, $c = 3$) OC curve to determine the percentage of lots that will be accepted when the true lot proportion defective is 1%, 2%, and 4%.
   c. What is the AQL?
   d. What is the RQL?
   e. What is the IQL?
5. Repeat exercise 4 for an ($n = 200$, $c = 1$) OC curve.
6. Construct the AOQ curve to accompany the ($n = 200$, $c = 3$) OC curve from exercise 4. What is the AOQL?
7. Construct the AOQ curve to accompany the ($n = 200$, $c = 1$) OC curve from exercise 5. What is the AOQL?
8. Use the MLT-STD-105D sampling plan to determine the sample size and acceptance number for the following criteria:
   a. Single sampling
   b. Normal inspection
   c. Inspection level II
   d. Lot size 2000
   e. The AQL 1%
9. Make an appropriate OC curve for the recommended sample size and acceptance number from exercise 8. Use the OC curve to determine the actual AQL.
10. The product being produced contains 12 parts from vendors. Two of the vendors submit parts that are .5% defective, six vendors submit parts that are 1.5% defective, and the other four submit parts that are 2% defective. If no incoming inspection is used to weed out the defective parts and no further faulty work is done, what percentage of the final product will be defect free?

A

BASIC MATH CONCEPTS AND PROBABILITY

The following list by section numbers is provided for easy reference to specific topics covered in this appendix:

A.1 SIGNED NUMBERS

Signed numbers are necessary when dealing with a numerical scale. A common example of a signed-number scale is a thermometer. It has a reference point of 0 with warmer temperatures indicated by positive numbers and colder temperatures indicated by negative numbers. Similarly, when a scale of numbers is put on a line or gauge, a reference point, 0, is centered between positive numbers to the right (or up) and negative numbers to the left (or down). Signed numbers are occasionally needed in SPC for some gauges and for coding data.

Operations With Signed Numbers

A signed number can be thought of as a combination of direction and a distance:

4 or +4 means 4 units to the right
– 5 means 5 units to the left

When signed numbers are on a number line, the distance is the number of units from 0 (Figure A.1).

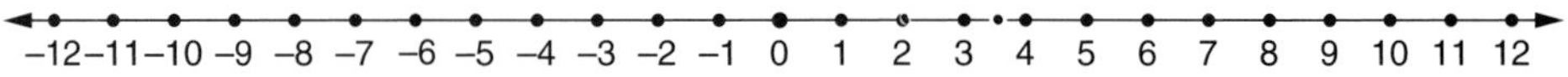

–12 –11 –10 –9 –8 –7 –6 –5 –4 –3 –2 –1 0 1 2 3 4 5 6 7 8 9 10 11 12

FIGURE A.1
A number line.

The absolute value of a number strips away the sign and results in a positive number or zero. The symbol for absolute value is a pair of vertical bars, one on each side of the number:

$\lvert 3 \rvert = 3$	Read, "The absolute value of 3 is 3."
$\lvert -4 \rvert = 4$	Read, "The absolute value of negative 4 is 4."
$\lvert 0 \rvert = 0$	Read, "The absolute value of zero is zero."

Absolute-value terminology is used to describe the rules for operating with signed numbers.

Addition Rules

There are two rules for addition.

Rule 1: If the signs of the two numbers are the same,
(a) add their absolute values and
(b) keep their sign.

This is the addition of two numbers that have the same sign (negative):

$$-5 + (-2)$$

$5 + 2$	(a) Add their absolute values.
-7	(b) Keep the negative sign.

This is the addition of two positive numbers:

$$4 + 3$$

$4 + 3$	(a) Add their absolute values.
7	(b) Keep the positive sign.

This rule is consistent with regular arithmetic addition, so when adding positive numbers, just add them as usual.

Rule 2: If the signs of the two numbers are different (one positive and one negative),
(a) subtract their absolute values and
(b) keep the sign of the one that has the largest absolute value.

This is the addition of two numbers with opposite signs:

$$-9 + 3$$

$9 - 3$	(a) Subtract their absolute values.
-6	(b) Keep the sign of the one with the largest absolute value (the negative sign).

This is the addition of two numbers with opposite signs:

$$12 + (-4)$$

$12 - 4$	(a) Subtract their absolute values.
8	(b) Keep the sign of the largest in absolute value (the positive sign).

The addition of signed numbers is always a three-step process:

1. Classify the problem as an addition or subtraction problem.
2. Do the appropriate addition or subtraction of absolute values.
3. Affix the proper sign (positive or negative).

EXAMPLE A.1 Do the following additions by the three step-process.

Solution

$-12 + 3$	Classify as addition with opposite signs.
$12 - 3$	Subtract the absolute values.
-9	Keep the negative sign.
$-8 + (-7)$	Classify as addition with the same signs.
$8 + 7$	Add the absolute values.
-15	Keep the minus sign.
$4 + (-9)$	Classify as addition with opposite signs.
$9 - 4$	Subtract the absolute values.
-5	Keep the negative sign.

$-6 + 13$	Classify as addition with opposite signs.
$13 - 6$	Subtract the absolute values.
7	Keep the positive sign.
$-4 + (-8)$	Classify as addition with the same signs.
$4 + 8$	Add their absolute values.
-12	Keep the negative sign.
$12 + 8$	Just add. There is no need to follow the signed-number
20	rule when it's a familiar problem from arithmetic.

Subtraction

The addition rules are also used for subtraction.

Rule: (a) Change the subtraction sign to addition.
(b) Change the sign of the number on the right.
(c) Use the appropriate addition rule.

Here the subtraction is changed to addition and the 5 is changed to –5:

$$-8 - 5$$
$$-8 + (-5)$$

$8 + 5$	The signs are the same, so add the absolute
-13	values and keep the negative sign.
$9 - (-4)$	The subtraction is changed to addition, and
$9 + 4$	the –4 is changed to 4.
13	Just add.
$-5 - (-8)$	The subtraction is changed to addition, and
$-5 + 8$	the –8 is changed to 8.
$8 - 5$	For addition with opposite signs, subtract the
3	absolute values and keep the positive sign.

When a mixture of addition and subtraction of signed numbers is encountered, the classification step of the three-step addition process is very important. If the operation is subtraction, it is changed to addition by the two-step process shown above. Then the addition has to be classified and the appropriate rule applied.

EXAMPLE A.2 Follow the addition and subtraction procedures for the following problems.

Solution

$4-(-7)$	Classify as subtraction of signed numbers.
$4+7$	Change the problem to addition and change the – 7 to 7.
11	Add.
$-5+3$	Classify as addition with opposite signs.
$5-3$	Subtract their absolute values.
-2	Keep the negative sign.
$-8-(-2)$	Classify as subtraction of signed numbers.
$-8+2$	Change the problem to addition and change the –2 to 2.
$8-2$	For addition with opposite signs, subtract the absolute
-6	values and keep the negative sign.
$-9-4$	Classify as subtraction of signed numbers.
$-9+(-4)$	Change the problem to addition and change the 4 to –4.
$9+4$	For addition with the same signs, add the absolute
-13	values and keep the negative sign.
$8+(-13)$	Classify as addition with opposite signs.
$13-8$	Subtract the absolute values.
-5	Keep the negative sign.
$12-5$	Just subtract. Again, if a familiar arithmetic operation
7	is encountered, just do it as usual.

Multiplication and Division

The sign rules are the same for both multiplication and division.

Rule 1: If one of the two numbers is negative, the answer is negative.

These all have negative answers:

$$8 \times (-3) = -24 \qquad -9 \times 6 = -54$$

$$\frac{50}{-5} = -10 \qquad \frac{-32}{16} = -2$$

Rule 2: If both of the numbers are negative, the answer is positive.

These have positive products or quotients:

$$-7 \times (-4) = 28$$

$$\frac{-48}{-8} = 6$$

EXAMPLE A.3 Classify the following operations and apply the appropriate rule.

Solution

-9×3	Classify as multiplication with one negative number.
-27	The product is negative.
$-8 + 17$	Classify as addition with opposite signs.
$17 - 8$	Subtract the absolute values.
9	Keep the positive sign.
$\frac{24}{3}$	Just divide. Another familiar problem from arithmetic.
8	
$-7 \times (-6)$	Classify as multiplication; both values are negative.
42	The product is positive.
$-9 - 15$	Classify as subtraction.
$-9 + (-15)$	Change the operation to addition, and change the 15 to −15.
$9 + 15$	For addition with the same signs, add the absolute values
-24	and keep the negative sign.
$\frac{-36}{-12}$	Classify as division; both values are negative.
3	The quotient is positive.
$\frac{55}{-5}$	Classify as division with one negative value.
-11	The quotient is negative.

A.2 VARIABLES

A variable is a symbol that represents some number. Variables are often used in formulas that determine a specific value of interest. Perhaps a familiar example is the formula for the area of a rectangle:

$$\text{Area} = \text{length} \times \text{width}$$

That formula can be written in a shorter form using variables:

$$\text{Area} = LW$$

In the short form, the variable L stands for the length of the rectangle and W represents the width of the rectangle. When two variables are placed next to each other, their values are to be multiplied.

The symbols that are used in statistics for variables are primarily letters, but sometimes letters with subscripts are used. For example, x is often used to represent a measurement. When there is more than one measurement, they can be represented by different letters such as x, y, and z or by letters with subscripts such as x_1, x_2, and x_3.

One of the most commonly used formulas in statistics is the average-value formula: Add all the values, then divide the sum by the number of values added. In variable form, the formula can be written in different ways:

$$\bar{x} = \frac{x + y + z}{3}$$

$\bar{x}$, read "x bar," is the symbol for the average. The formula is specific for three different measurements represented by x, y, and z.

$$\bar{x} = \frac{x_1 + x_2 + \cdots x_n}{n}$$

This formula is more general because it applies to any number of values n.

The second formula uses subscripted variables where x_1, read "x one," represents the first measurement, x_2, read "x two," represents the second measurement, and x_n, read "x sub n" or just "x, n," stands for the nth, or last, measurement. If there were 30 measurements to average, the n value would be 30; if there were four measurements, the n value would be 4. The three dots mean to keep adding all the other measurements that lie between the x_2 and the x_n values, specifically, x_3, x_4, and all others up to x_{n-1}.

Another formula, equivalent to the previous one, uses the Greek letter Σ (capital sigma) as an addition indicator:

$$\bar{x} = \frac{\Sigma x}{n}$$

The formula is read, "x bar equals summation x divided by n." Σx, summation x, means to add all the x measurements, and the denominator in the formula indicates that there are n measurements to average. This is the preferred formula because it is short and concise.

Operations

The formulas in statistics, as well as in other applications, contain precise directions for handling the numerical values represented by the variables. The operations are indicated by both symbol and position, and the order in which the operations are to be performed follows a strict algebraic hierarchy.

Addition

Addition is indicated by either a "+" or a "Σ" symbol:

$x + y + z$	Add the three values.
Σx	Add all the x values.
$x_1 + x_2 + x_3 + \cdots + x_{15}$	Add the 15 x values.

Subtraction

Subtraction uses the "–" symbol:

$c - b$	Subtract the b value from the c value.

Multiplication

Multiplication has many indicators. When variables are not used, the ×, adjacent parentheses, a number next to a parenthesis, and a raised dot all mean multiplication:

$$4 \times 5 = 20 \qquad (8)(5) = 40$$

$$7(9) = 63 \qquad 6 \cdot 5 = 30$$

When variables are used, the × may be confusing as a multiplication indicator because x is often used as a variable. Adjacent positions, raised dots, and parentheses are all alternative multiplication indicators:

xy	Multiply the x value by the y value.
$5x$	Multiply the x value by 5.
$(3)(x)$	Multiply the x value by 3.
$7 \cdot c$	Multiply the c value by 7.
$b(8)$	Multiply the b value by 8.

Division

Division is sometimes indicated by the "÷" symbol, but more often by the fraction bar:

$x \div y$	Divide the x value by the y value.
$\frac{c}{d}$	Divide the c value by the d value.

Powers

Powers are indicated by a raised number on the right side of the base value. A power indicates a repeated multiplication of the base value, and the exponent, or raised number, indicates how many factors are multiplied:

3^4, read "three to the fourth power," means $3 \cdot 3 \cdot 3 \cdot 3$:

$$3^4 = 3 \cdot 3 \cdot 3 \cdot 3 = 81$$

Second and third powers have the special names of square and cube:

5^2, read "five squared," means $5 \cdot 5$:

$$5^2 = 5 \cdot 5 = 25$$

6^3, read "six cubed," means $6 \cdot 6 \cdot 6$:

$$6^3 = 6 \cdot 6 \cdot 6 = 216$$

x^2, read "x squared," means $x \cdot x$.

There are several algebraic properties associated with powers, but the only ones needed at this time are:

Rule 1: For multiplication of powers with the same base, keep the base the same and add the exponents.

$$x^2 \cdot x^3 = x^5$$
$$2^4 \cdot 2^3 = 2^7$$
$$5^2 \cdot 5 = 5^3 \qquad (5 \text{ is the same as } 5^1)$$

Writing 5 is the same as writing 5^1. Also, $x^3 \cdot x^5$ means $xxx \cdot xxxxx$, but that can be written as x^8. Similarly, $2^4 \cdot 2^3$ means $(2 \cdot 2 \cdot 2 \cdot 2) \cdot (2 \cdot 2 \cdot 2)$, and that can be written as 2^7.

The cancellation rule from fractions leads to the division rule for powers:

$$\frac{6}{8} = \frac{2 \cdot 3}{2 \cdot 4} = \frac{3}{4}$$

Although the middle step is usually omitted, it is the "official" concept: Cancel common factors. Applied to powers,

$$\frac{x^3}{x^5} = \frac{1 \cdot x^3}{x^2 \cdot x^3} = \frac{1}{x^2} \qquad \frac{x^8}{x^5} = \frac{x^3 \cdot x^5}{1 \cdot x^5} = \frac{x^3}{1} = x^3$$

Rule 2: For division of powers with the same base, keep the base the same and subtract the exponents. If the larger power is in the numerator, the resulting power is in the numerator. If the larger power is in the denominator, the resulting power is in the denominator (retain a 1 in the numerator).

The following examples illustrate this rule:

$$\frac{x^6}{x^2} = x^4 \qquad \frac{3^2}{3^5} = \frac{1}{3^3} = \frac{1}{27}$$

$$\frac{y^4}{y^5} = \frac{1}{y} \qquad \frac{2^7}{2^2} = 2^5 = 32$$

Rule 3: When a product or quotient is raised to a power, the exponent applies to each factor within the parentheses.

See the following examples:

$$(ab)^3 = a^3b^3 \qquad (4x)^2 = 4^2 \cdot x^2 = 16x^2 \qquad \left(\frac{2}{3}\right)^4 = \frac{2^4}{3^4} = \frac{16}{81} \qquad \left(\frac{2x}{5}\right)^3 = \frac{2^3 \cdot x^3}{5^3} = \frac{8x^3}{125}$$

Roots

Roots are the opposite of powers. The symbol for a root is called a radical sign, $\sqrt{}$. A square root is indicated by the $\sqrt{}$ alone. Any other root has a root index: $\sqrt[3]{}$ indicates a cube root, and $\sqrt[4]{}$ signals a fourth root. The square root of 16, $\sqrt{16}$ means the number whose square is 16. Because $4^2 = 16$, $\sqrt{16} = 4$.

The root and power have a cancelling effect when they match:

$$\sqrt{4^2} = 4$$

$$\sqrt{x^2} = x \quad \text{for positive values of } x$$

$$\sqrt[3]{y^3} = y$$

Square roots of numbers that are not perfect squares can be done with a calculator. Virtually all calculators have a $\sqrt{}$ key. If some other root is needed, such as cube root, a scientific calculator is needed.

$$\sqrt{49} = \sqrt{7^2} = 7 \qquad \sqrt[3]{8} = \sqrt[3]{2^3} = 2$$

$\sqrt{12}$
Press 12 $\sqrt{}$ on the calculator
$= 3.46410$

As with powers, there are several algebraic properties of radicals, but just the multiplication and division properties are needed here:

$$\sqrt{a} \cdot \sqrt{b} = \sqrt{ab} \qquad \frac{\sqrt{a}}{\sqrt{b}} = \sqrt{\frac{a}{b}}$$

$$\sqrt{4} \cdot \sqrt{9} = \sqrt{36} \qquad \frac{\sqrt{144}}{\sqrt{9}} = \sqrt{\frac{144}{9}} = \sqrt{16}$$

The two rules can be verified by evaluating the roots on the numerical examples:

$$\sqrt{4} \cdot \sqrt{9} = \sqrt{4 \cdot 9} = \sqrt{36} = 6$$

or

$$\sqrt{4} \cdot \sqrt{9} = 2 \cdot 3 = 6$$

$$\frac{\sqrt{144}}{\sqrt{9}} = \sqrt{\frac{144}{9}} = \sqrt{16} = 4$$

or

$$\frac{\sqrt{144}}{\sqrt{9}} = \frac{12}{3} = 4$$

The radical rules can be used to simplify algebraic expressions containing radicals:

$$\sqrt{2} \cdot \sqrt{x} = \sqrt{2x}$$

$$\frac{\sqrt{15x}}{\sqrt{3}} = \sqrt{\frac{15x}{3}} = \sqrt{5x}$$

$$\sqrt{\frac{6x}{25}} = \frac{\sqrt{6x}}{\sqrt{25}} = \frac{\sqrt{6x}}{5}$$

A.3 ORDER OF OPERATIONS

When more than one operation has to be performed, it is necessary to know the correct sequence. For example, the calculation $4 + 3 \cdot 2$ equals 14 if the addition is done first or 10 if the multiplication is done first. The calculation should have just one answer, so the proper sequence must be followed.

The order of operations sequence has four steps.

1. If there are any grouping symbols, do the operations within the grouping symbols.
2. Evaluate all powers and roots.
3. Do the multiplication and division as they occur in order, left to right.
4. Do the addition and subtraction in order from left to right.

Order of operations with special statistical quantities involves Σx^2, $(\Sigma x)^2$, and expressions such as $\Sigma(x - 5)^2$.

Σx^2	Square all the x values, then add all the squared values.
$(\Sigma x)^2$	Add all the x values first, then square the sum.
$\Sigma(x - 5)^2$	Subtract 5 from all the x values first, square each difference, then add all the squared values.

In each of the preceding statistical quantities, the Σ sign acts as a grouping symbol around the sum of the quantities to its right.

EXAMPLE A.4 Use the order of operations to evaluate the following:

$$4 + 3 \cdot 2, \qquad (5 + 4) - \frac{15 + 6}{3}, \; \frac{3 + 18 - 2 \cdot 3}{3 + 2}, \qquad \text{and} \qquad \sqrt{3^2 + 4^2}$$

Solution

$4 + 3 \cdot 2$	Do the multiplication first, then add.
$4 + 6$	
10	
$(5 + 4)^2 - \frac{15 + 6}{3}$	Add inside the grouping symbols (both the parentheses and the fraction line).
$9^2 - \frac{21}{3}$	Square the 9.
$81 - \frac{21}{3}$	Divide by 3.
$81 - 7$	Subtract.
74	
$\frac{3 + 18 - 2 \cdot 3}{3 + 2}$	The division line is a grouping symbol, so all the operations in both the numerator and denominator must be done first before the division can be performed. The multiplica tion in the numerator and the addition in the denominator are done first. The addition and subtraction are done as they occur, left to right.
$\frac{3 + 18 - 6}{5}$	
$\frac{21 - 6}{5}$	
$\frac{15}{5}$	Divide.
3	
$\sqrt{3^2 + 4^2}$	The $\sqrt{}$ is a grouping symbol, so all the operations must be completed inside before the square-root step. Square first.
$\sqrt{9 + 16}$	Add.
$\sqrt{25}$	Take the square root.
5	

EXAMPLE A.5 Given the measurements 4, 2, 7, and 3, calculate $\sqrt{4\Sigma x^2 - (\Sigma x)^2}$.

Solution

There are actually four grouping symbols: The $\sqrt{}$, the parentheses, and the two Σ signs.

Calculate the summations first:

$$\begin{aligned}\Sigma x &= 4 + 2 + 7 + 3 \\ &= 16 \\ (\Sigma x)^2 &= (16)^2 \\ &= 256 \\ \Sigma x^2 &= 16 + 4 + 49 + 9 \\ &= 78\end{aligned}$$

Substitute the values into the original expression. Multiply first, then subtract. The square root is done with a calculator: Press 56 $\sqrt{\ }$.

$$\begin{aligned}\sqrt{4 \cdot 78 - 256} &= \sqrt{312 - 256} \\ &= \sqrt{56} \\ &= 7.48333\end{aligned}$$

EXAMPLE A.6 Given the measurements 2, 4, 5, and 9, find $\Sigma(x - 5)^2$.

Solution

The subtraction in the parentheses is done first (second column). The differences are then squared (third column):

x	$x - 5$	$(x - 5)^2$
2	−3	9
4	−1	1
5	0	0
9	4	16
		26

The last step is to find the sum of the squared values (add the third column):

$$\Sigma(x - 5)^2 = 26$$

A.4 INEQUALITIES

The algebraic properties of inequalities are not needed for the work in this text, but the language of inequalities is used. When quantities are not equal, a statement indicating which is larger or smaller can be made using an inequality sign.

The symbol ">" is read "greater than."

The symbol "<" is read "less than."

Both symbols will always point to the smaller number:

$4 > -1$ Read, "4 is greater than negative 1."

$7 < 12$ Read, "7 is less than 12."

Two less than statements or two greater than statements can be combined for a "between" statement:

$-6 < 4 < 10$ Read, "4 is between negative 6 and 10."

$8 > 5 > 2$ Read, "5 is between 8 and 2."

When a variable is used with an inequality, a set of numbers is described and a picture of that set of numbers can be drawn using a number line. For example, $x > 4$ is illustrated by a line that extends infinitely to the right. There is no "next number" after 4 because of the infinitely many decimal values (4.1, 4.01, 4.0001, and so on), so the graph, or picture, illustrates this with an open circle. This is shown in Figure A.2.

FIGURE A.2
The graph of $x > 4$ and $x \leq 3$.

When a set of numbers has a definite beginning, a different symbol is used: ≥, read "greater than or equal to," and ≤, read "less than or equal to." A solid endpoint is used when graphing the "or equal to" inequalities:

$x \leq 3$ Read, "x is less than or equal to 3."

This is also illustrated in Figure A.2.

$0 \leq p \leq 1$ Read, "p lies between 0 and 1, inclusive" (including the 0 and the 1).

See Figure A.3.

FIGURE A.3
The graph of $0 \leq p \leq 1$.

EXAMPLE A.7 Interpret the following with a statement and a graph: $x > 2.5$, $8.3 < x < 12.2$, $x < 3.25$, and $6 \geq x \geq 1.7$.

Solution

	Statement	Figure
$x > 2.5$	x is greater than 2.5	Figure A.4
$8.3 < x < 12.2$	x is between 8.3 and 12.2	Figure A.5
$x < 3.25$	x is less than 3.25	Figure A.6
$6 \geq x \geq 1.7$	x is between 6 and 1.7, inclusive	Figure A.7

See Figures A.4 through A.7.

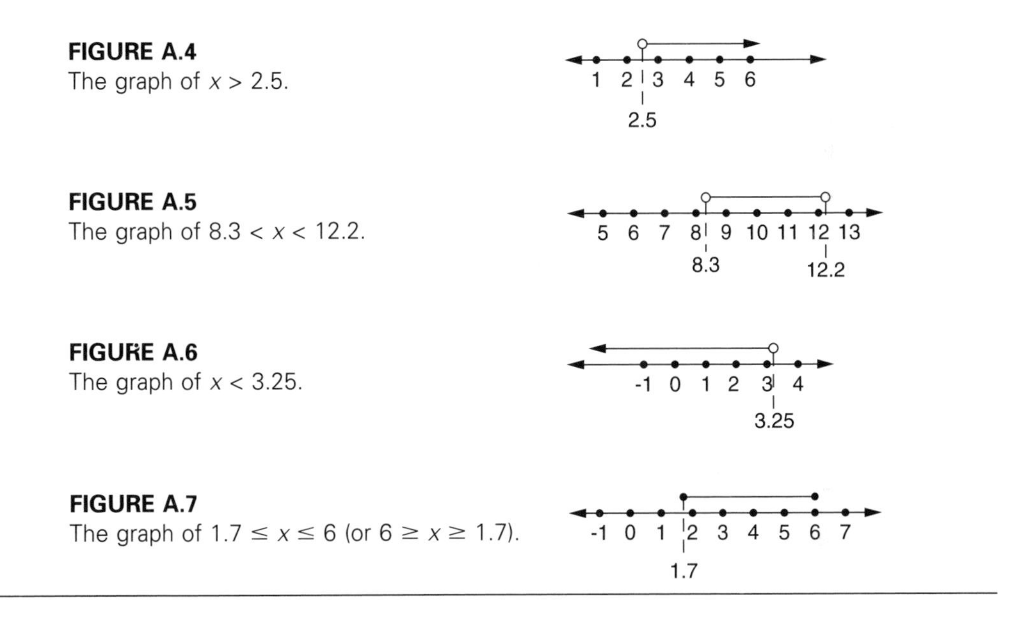

FIGURE A.4
The graph of $x > 2.5$.

FIGURE A.5
The graph of $8.3 < x < 12.2$.

FIGURE A.6
The graph of $x < 3.25$.

FIGURE A.7
The graph of $1.7 \leq x \leq 6$ (or $6 \geq x \geq 1.7$).

A.5 USING THE STATISTICAL CALCULATOR

When a calculator is put in the statistics mode, data are usually entered with the Σ key. Some calculators use a different data entry key such as the M+ key.

After all the data have been entered, the values of the mean and standard deviation can be found by pressing the appropriate keys. Usually there are two standard deviation keys: σ_n and σ_{n-1}. The two versions refer to the divisors in the formula.

The first formula is used for the population standard deviation, when all the measurements in the population are used in the calculation:

$$\sigma_n = \sqrt{\frac{\Sigma(x - \overline{x})^2}{n}}$$

The second formula is used to find the sample standard deviation and is the formula used in SPC. It is technically the best estimate of the population standard deviation, based on sample information:

$$\sigma_{n-1} = \sqrt{\frac{\Sigma(x - \bar{x})^2}{n - 1}}$$

Other calculator key variations include

σ_n	σ_{n-1}
σ	s
s_n	s_{n-1}
	$\hat{\sigma}$

EXAMPLE A.8 Use a statistical calculator to find the standard deviation for the set of data 8, 10, 11, 15, 16, 18.

Solution

1. Put the calculator in the statistical mode.
2. Enter the data: 8 Σ, 10 Σ, 11 Σ, . . . , 18 Σ. In some calculators, the number of data entries will show on the display when the Σ key is pressed.
3. When all the data are entered, press $\bar{x}$ for the mean, $\bar{x} = 13$, and s for the standard deviation, $s = 3.899$. These may be second function keys for which the 2nd key or the INV key must be pressed first.

A.6 PROBABILITY

When a set of counts or measurements can be separated into two distinct groups, the concepts of probability can be applied. Let one group be classified as the successes and the other as failures:

$$P(\text{success}) = \frac{\text{number of successes}}{\text{number of successes} + \text{failures}} = \frac{\text{number of successes}}{\text{number of possibilities}}$$

The following are some applications of probability. First, a coin is tossed. Let a head be a success and a tail be a failure:

$$P(\text{success}) = \frac{1 \text{ success}}{2 \text{ possibilities}}$$

$$= \frac{1}{2}$$

Second, if a bag contains 3 red beads, 4 white beads, and 5 blue beads and one bead is drawn from the bag, find the probability that is a red bead. Let a red bead be a success and the white and blue beads be failures:

$$P(\text{red}) = \frac{3 \text{ red}}{12 \text{ beads}} = \frac{3}{12} = \frac{1}{4}$$

Finally, a routing tray contains 88 good parts and 12 defective parts. One part is chosen from the tray. Find the probability that it is a good part. Let a good part be a success and a bad part be a failure:

$$P(\text{good}) = \frac{88 \text{ good parts}}{100 \text{ parts}} = \frac{88}{100} = \frac{22}{25}$$

In each case, the probability represents the chance or likelihood of a success. Probability is defined as a fraction, but as with any other fraction, it can also be expressed as a decimal or a percentage. In the preceding example, the language of probability indicates that the probability of drawing a red bead is ¼ or .25, or that there is a 25% chance of drawing a red bead.

Probability values are restricted to numbers between 0 and 1 inclusive:

$$0 \leq P(\text{success}) \leq 1$$

This restriction is easily demonstrated. In a bag of 12 red, white, and blue beads, find the $P(\text{green})$. There are no green beads, so

$$P(\text{green}) = \frac{0 \text{ successes}}{12 \text{ possible beads}} = \frac{0}{12} = 0$$

This is an impossible situation; there are no successes in the set of possibilities.

When one bead is drawn from this same bag, find the probability that it is not orange:

$$P(\text{not orange}) = \frac{12 \text{ successes}}{12 \text{ possible beads}} = \frac{12}{12} = 1$$

This is a sure thing because every element in the set of possibilities is a success.

A.6.1 COMPOUND PROBABILITY

The "Or" Probability When more than one possible success can occur in a probability calculation, it is described using the word "or." The probability that either event A or event B occurs, $P(A \text{ or } B)$, is equal to the sum of the individual probabilities:

$$P(A \text{ or } B) = P(A) + P(B)$$

EXAMPLE A.9 A bag contains beads of several colors: 4 red, 2 white, 5 black, 1 green, and 3 blue beads.

1. If one bead is drawn from the bag, what is the probability that it is either red or white?
2. If one bead is drawn from the bag, what is the probability that it is either a black, green, or blue bead?

Solution

For the first problem,

$$P(\text{red or white}) = P(\text{red}) + P(\text{white}) = \frac{4}{15} + \frac{2}{15} = \frac{6}{15} = \frac{2}{5}$$

For the second,

$$P(\text{black or green or blue}) = P(\text{black}) + P(\text{green}) + P(\text{blue}) = \frac{5}{15} + \frac{1}{15} + \frac{3}{15} = \frac{9}{15} = \frac{3}{5}$$

The one necessary condition for the "or" rule is that the events are *mutually exclusive*; that is, no two events can occur simultaneously. For example, when one bead is drawn, it cannot be both green and blue; it must be one or the other.

There are situations in which different events can occur simultaneously. When that happens, simply adjust the additional probabilities to exclude the simultaneous occurrences. Make sure that no event is counted more than once:

$$P(A \text{ or } B) = P(A) + \text{adjusted } P(B)$$

EXAMPLE A.10 Draw one card from a standard deck of 52 cards. Find the probability of getting either a 9 or a heart.

Solution

The two events could occur simultaneously because a 9 of hearts could be drawn. The adjusted "or" rule must be used. Adj P signifies adjusted probability:

$$P(9 \text{ or heart}) = P(9) + \text{adj } P(\text{heart})$$
$$= \frac{4}{52} + \frac{12}{52}$$

The numerator 4 is the 4 nines and the numerator 12 is the remaining 12 hearts. The 9 of hearts was counted in the first term:

$$P(9 \text{ or heart}) = \frac{16}{52} = \frac{4}{13}$$

The terms could be written in either order, but the second term must always be adjusted accordingly. The main idea is to make sure that none of the successful events are counted more than once. If the hearts are taken first, there are 13 successes; the nines would then be adjusted to three additional successes because the 9 of hearts was already counted among the 13 hearts of the first term.

EXAMPLE A.11 Find the probability of drawing either a face card or a seven.

Solution

These events are mutually exclusive; none of the success cards occur simultaneously. No adjustment is needed, so the first version of the "or" rule is used. There are 12 face cards: the Jack, Queen, and King of each suit.

$$P(\text{face card or } 7) = P(\text{face card}) + P(7)$$
$$= \frac{12}{52} + \frac{4}{52}$$

$$= \frac{16}{52}$$

$$= \frac{4}{13}$$

EXAMPLE A.12 Find the probability of getting an 8 or a diamond or a 10.

Solution

These events can occur simultaneously, so the adjusted probability rule must be used:

$$P(8 \text{ or diamond or } 10) = P(8) + \text{adj } P(\text{diamond}) + \text{adj } P(10)$$

$$= \frac{4}{52} + \frac{12}{52} + \frac{3}{52}$$

The numerators signify the 4 eights, the remaining 12 diamonds (8 of diamonds is in the first term) and the remaining 3 tens. The 10 of diamonds is in the second term:

$$P(8 \text{ or diamond or } 10) = \frac{19}{52}$$

If the order of the terms is changed, the adjustments change accordingly:

$$P(8 \text{ or } 10 \text{ or diamond}) = P(8) + P(10) + \text{adj } P(\text{diamond})$$

$$= \frac{4}{52} + \frac{4}{52} + \frac{11}{52}$$

The last term is adjusted to eliminate the 8 and 10 of diamonds, which were counted in the first two terms:

$$P(8 \text{ or } 10 \text{ or diamond}) = \frac{19}{52}$$

The result is the same.

"And" Probabilities When successive events are described, the "and" is used in the probability description. The probability of an "and" situation is found by multiplying the individual probabilities:

$$P(A \text{ and } B) = P(A) \times P(B)$$

EXAMPLE A.13 A coin is tossed three times. Find the probability of getting three heads.

Solution

The three tosses of the coin are three successive events, so the "and" rule applies. A success consists of getting a head on the first toss *and* a head on the second toss *and* a head on the third toss. Let H represent a head:

$$\begin{aligned} P(H \text{ and } H \text{ and } H) &= P(H) \times P(H) \times P(H) \\ &= \frac{1}{2} \times \frac{1}{2} \times \frac{1}{2} \\ &= \frac{1}{8} \end{aligned}$$

The "and" rule can be verified by using the basic definition of probability on the set of all possible outcomes. Let H represent a head, T represent a tail, and adjacent letters represent "and" (HT means a head *and* then a tail). The following are the possible outcomes of tossing a coin three times:

$$HHH \quad HHT \quad HTH \quad THH \quad TTH \quad THT \quad HTT \quad TTT$$

The probability is

$$P(3H) = \frac{1}{8}$$

This result agrees with the solution in Example A.13.

The "and" rule applies when there is only one way for a success to occur. When a success can happen in more ways than one, a combination of "and" and "or" rules applies.

EXAMPLE A.14 A coin is tossed three times. Find the probability of getting two heads and one tail.

Solution

The basic probability rule is the "and" rule because of the three successive events. The "or" rule is used to add the probabilities of all the different ways the success can occur. The event ($2H$ and $1T$) can occur three ways: The tail could be on the first toss *or* the second toss *or* the third toss:

$$\begin{aligned} P(2H \text{ and } 1T) &= P(THH) \text{ or } P(HTH) \text{ or } P(HHT) \\ &= \left(\frac{1}{2} \times \frac{1}{2} \times \frac{1}{2}\right) + \left(\frac{1}{2} \times \frac{1}{2} \times \frac{1}{2}\right) + \left(\frac{1}{2} \times \frac{1}{2} \times \frac{1}{2}\right) \\ &= 3\left(\frac{1}{2} \times \frac{1}{2} \times \frac{1}{2}\right) \end{aligned}$$

Each way has the same "and" probability. Multiply by 3, the number of ways, instead of the repeated addition:

$$P(2H \text{ and } 1T) = \frac{3}{8}$$

This combination of "and" and "or" rules can be verified by applying the basic definition of probability to the set of possible outcomes. There are eight possible outcomes (*HHH*, *HHT*, and so on) and three of them have two heads and one tail:

$$P(2H \text{ and } 1T) = \frac{3}{8}$$

EXAMPLE A.15 A tray contains eight good parts and two defective parts. A part is chosen at random, inspected, and replaced by four different inspectors. Find the probability that two of the inspectors found a defective part.

Solution

On each single inspection the probability of getting a good part is $P(G) = 8/10 = .8$. The probability of getting a defective part is $P(D) = 2/10 = .2$.

There are four successive events, so the basic probability rule that applies is the "and" rule. A success can occur in several ways, so the "or" rule applies too. The probabilities for each different success must be added, but because each of the different successes has the same probability, multiplication by the number of ways can be used instead of the repeated addition. Two good parts and two defective parts can be chosen in the following ways: *GGDD* or *GDGD* or *DGGD* or *DGDG* or *DDGG* or *GDDG:*

$$\begin{aligned} P(2G \text{ and } 2D) &= P(G \text{ and } G \text{ and } D \text{ and } D) \times 6 \text{ ways} \\ &= .8 \times .8 \times .2 \times .2 \times 6 \\ &= .1536 \end{aligned}$$

Adjusting the "And" Rule When the "and" rule is used, the successive events must be *independent*: The probability of each successive event is not affected by the preceding events. When tossing a coin, each toss is independent of all the previous tosses. When sampling with replacement, each selection is independent of the previous selections. However, if the successive sampling is done without replacement, the probability of successive events is affected by what happened in the previous events. In this case, adjust the probabilities accordingly. When successive events are not independent,

$$P(A \text{ and } B) = P(A) \times \text{adj } P(B)$$

EXAMPLE A.16 If two successive cards are drawn from a standard deck (no replacement), find the probability of drawing two kings.

Solution

Use an adjusted probability:

$$P(\text{king and king}) = P(\text{king}) \times \text{adj } P(\text{king})$$
$$= \frac{4}{52} \times \frac{3}{51}$$

The second factor is adjusted because three kings remain, with 51 cards from which to choose:

$$P(2 \text{ kings}) = \frac{12}{2652}$$
$$= \frac{1}{221}$$

Another variation of the "and" rule includes both adjusting factors and determining the number of ways that a success can occur. This is the most inclusive form because all possible variations are shown. The overall version of the "and" rule is

$$P(A \text{ and } B) = P(A) \times \text{adj } P(B) \times \text{number of ways}$$

The thought sequence is as follows:

1. These are successive events, so the "and" rule applies.
2. Is adjustment needed?
3. Can a success occur in more than one way?
   a. If all the probabilities are the same, multiply by the number of ways.
   b. If the probabilities are different, add all the different products.

EXAMPLE A.17 Draw three cards from a standard deck of cards without replacement. Find the probability of getting two aces and one face card.

Solution

1. There are three successive events, so the "and" rule applies.
2. Is adjustment needed? Yes, it is, because the cards are not replaced.
3. Is there more than one success? Yes, the face card may be drawn either first or second or third (*FAA* or *AFA* or *AAF*). The probabilities are all the same, so multiply by the number of ways:

$$P(2 \text{ aces and 1 face card}) = P(A) \times \text{adj } P(A) \times \text{adj } P(F) \times \text{number of ways}$$
$$= \frac{4}{52} \times \frac{3}{51} \times \frac{12}{50} \times 3$$

The numerators represent 4 aces on the first draw, 3 aces in the remaining 51 cards, and 12 face cards in the remaining 50 cards. The 3 signifies the three ways this can happen:

$$P(\text{2 aces and 1 face card}) = \frac{1}{13} \times \frac{1}{17} \times \frac{6}{25} \times 3$$
$$= \frac{18}{5525}$$

The adjustments will be different if the factors are rearranged, but the product of the three fractions will remain the same. The only change is that the numerator factors would occur in a different order. This is why the "× 3 ways" shortcut can be used. For example, the *FAA* product is $^{12}/_{52} \times {}^{4}/_{51} \times {}^{3}/_{50}$. Both the numerator product, $12 \times 4 \times 3$, and the denominator product, $52 \times 51 \times 50$, are the same as before.

A.6.2 COUNTING WITH PERMUTATIONS AND COMBINATIONS

The Number of Ways In the previous examples, the number of ways that a success could occur was determined by listing all the possibilities. A more efficient method is needed for complicated situations.

Permutations, $P_{N,n}$ The number of *permutations* refers to the number of arrangements of N objects when n of them are used. As a numerical example, $P_{7,3}$ is the number of arrangements of seven objects when three of them are used. If the objects are letters, $P_{7,3}$ gives the number of different three-letter "words" that can be formed when no letter occurs more than once in any word.

EXAMPLE A.18 How many three-letter words can be formed from the seven letters *A, B, C, D, E, F,* and *G* if no letter occurs more than once in any word?

Solution

There are three decisions to make when forming a word:

_______	and	_______	and	_______
Choose the first letter (7 choices)		Choose the second letter (6 choices)		Choose the third letter (5 choices)

When the letters are chosen to form a word, there are seven choices for the first letter *and* six choices for the second letter *and* five choices for the third letter:

$$P_{7,3} = 7 \times 6 \times 5$$
$$= 210 \text{ words}$$

EXAMPLE A.19 How many seven-letter words can be formed from the seven letters if the letters cannot be repeated in any word?

Solution

There are seven decisions to make when forming a word:

____	and	____	and	____	and	____	and	____	and	____	and	____
First letter		Second letter		Third letter		Fourth letter		Fifth letter		Sixth letter		Seventh letter

When filling in the blanks to form a word, there are seven choices for the first letter, six choices for the second letter, and so on:

$$\begin{aligned} P_{7,7} &= 7 \times 6 \times 5 \times 4 \times 3 \times 2 \times 1 \\ &= 5040 \text{ words} \end{aligned}$$

Factorials A *factorial* is the product of a number and all the counting numbers descending from it to 1. The product of the descending counting numbers from 7 to 1 is 7!, read "7 factorial,":

$$7! = 7 \times 6 \times 5 \times 4 \times 3 \times 2 \times 1 = 5040$$

The factorial can be thought of as a mathematical shorthand:

$$\begin{aligned} 4! &= 4 \times 3 \times 2 \times 1 = 24 \\ 12! &= 12 \times 11 \times 10 \times 9 \times \cdots \times 1 = 479{,}001{,}600 \end{aligned}$$

There is one special definition of convenience:

$$0! = 1$$

This definition allows the general use of factorials in mathematical formulas without the need for special exceptions.

The formula for the permutations of n things taken from a population of N things is

$$P_{N,n} = \frac{N!}{(N-n)!}$$

The denominator, $(N - n)!$, eliminates the lower factors by cancellation when only part of the population, N, is used. If the formula is applied to Example A.30,

$$\begin{aligned} P_{7,3} &= \frac{7!}{(7-3)!} = \frac{7!}{4!} = \frac{7 \times 6 \times 5 \times 4 \times 3 \times 2 \times 1}{4 \times 3 \times 2 \times 1} \\ &= 7 \times 6 \times 5 = 210 \end{aligned}$$

For simpler writing, factorials can be canceled in total:

$$P_{7,3} = \frac{7 \times 6 \times 5 \times 4!}{4!} = 7 \times 6 \times 5 = 210$$

For Example A.31, the 0! = 1 definition is needed:

$$P_{7,7} = \frac{7!}{(7-7)!} = \frac{7!}{0!} = \frac{5040}{1} = 5040$$

Many scientific and statistical calculators have a factorial key, $x!$, for easy calculations of factorials. All the factorial calculations can be calculated by one of two procedures:

1. Repeatedly multiply the descending factors.
2. Use the factorial key, $x!$, on the calculator. On some calculators, $x!$ is a second function and the 2nd or INV key has to be pressed first.

For 6!, the calculator sequence is

6 $x!$ The display shows 720.

or

6 2nd $x!$ The result is 720.

Use whichever of these two sequences is appropriate on your calculator.

For the permutation

$$P_{8,2} = \frac{8!}{(8-2)!} = \frac{8!}{6!} = \frac{8 \cdot 7 \cdot 6!}{6!} = 8 \cdot 7 = 56$$

the calculator sequence is 8 $x!$ ÷ 6 $x!$ = for 56.

The *Mississippi* Problem Another variation of the arrangement problem involves repeated letters. For an easy first example, how many different "words" could be formed by rearranging the six letters in the word *kisses*? There are six letters, so 6! gives the total number of rearrangements. However, some of the rearrangements do not form a new word. In fact, any rearrangement of the three *s*'s in the word *kisses* does not give a new word. There are 3! ways of rearranging the three *s*'s in every different word, and that has to be eliminated from the 6! total. It is eliminated by division:

$$\frac{6!}{3!} = \frac{6 \times 5 \times 4 \times 3!}{3!}$$
$$= 6 \times 5 \times 4$$
$$= 120 \text{ words}$$

On the calculator 6 $x!$ ÷ 3 $x!$ = gives the answer.

How many "words" can be formed by rearranging the letters in *ABBBBCC*?

$$\frac{7!}{4! \times 2!}$$

There are seven letters. The 4! eliminates the arrangements of the four B's, and the 2! eliminates the arrangements of the two C's:

$$\frac{7!}{4! \times 2!} = \frac{7 \times 6 \times 5 \times 4!}{4! \times 2 \times 1}$$
$$= \frac{7 \times (3 \times 2) \times 5}{2 \times 1}$$
$$= 7 \times 3 \times 5$$
$$= 105$$

On the calculator: 7 $x!$ ÷ 4 $x!$ ÷ 2 $x!$ = is 105.

Now solve the *Mississippi* problem. How many different words can be formed by rearranging the letters?

$$\frac{11!}{4! \cdot 4! \cdot 2!} = \frac{11 \times 10 \times 9 \times 8 \times 7 \times 6 \times 5 \times 4!}{4! \times 4 \times 3 \times 2 \times 1 \times 2 \times 1}$$
$$= 34{,}650$$

There are four s's, four i's, and two p's in the 11-letter word. 11! gives the total number of rearrangements of the eleven letters and the two 4!'s and the 2! eliminate repeated words. On the calculator: 11 $x!$ ÷ 4 $x!$ ÷ 4 $x!$ ÷ 2 $x!$ = gives 34,650.

The Two-Letter Word Problem Determining the number of words that can be formed by rearranging the letters of a set that contains repetitions of just two different letters is an application of the *Mississippi* problem.

How many words can be formed by rearranging *ssssff*? With two f's and four s's,

$$\frac{6!}{2! \times 4!} = \frac{6 \times 5 \times 4!}{2 \times 1 \times 4!}$$
$$= 3 \times 5$$
$$= 15$$

On the calculator, 6 $x!$ ÷ 4 $x!$ ÷ 2 $x!$ = gives 15.

The choice of letters in this problem may have given you a hint at the eventual usefulness of all these "word games." The two-letter word problem applies directly to the probability calculations illustrated in Example A.15 and to probabilities associated with random sampling. In Example A.15, the number of ways that two good parts and two defective parts can occur was determined by listing all the possibilities. Using the two-letter word approach,

$$\frac{4!}{2! \times 2!} = \frac{4 \times 3 \times 2 \times 1}{2 \times 1 \times 2 \times 1} = 6$$

Now the more complicated "number of ways" can be easily calculated using the word game concepts.

EXAMPLE A.20 A box of parts contains 40 good parts and 10 bad parts. If eight parts are randomly removed from the box, fine the probability of getting five good parts and three bad ones.

Solution

1. This is an "and" problem because of the eight successive events. The probabilities are multiplied.
2. Adjustments must be made because the parts are not replaced.
3. There is more than one way to draw five good parts and three bad parts from the box, so the number of ways must be calculated.

The number of ways of drawing five good parts and three bad parts, *GGGGGBBB,* can be determined by using the two-letter word problem:

$$\frac{8!}{3! \times 5!} = \frac{8 \times 7 \times 6 \times 5!}{3 \times 2 \times 1 \times 5!}$$
$$= 8 \times 7$$
$$= 56$$

On the calculator, 8 $x!$ ÷ 5 $x!$ ÷ 3 $x!$ = gives 56.

$$P(5G \text{ and } 3B) = \frac{40}{50} \times \frac{39}{49} \times \frac{38}{48} \times \frac{37}{47} \times \frac{36}{46} \times \frac{10}{45} \times \frac{9}{44} \times \frac{8}{43} \times 56$$

Adjust the "and" probabilities for the 5 G's and the 3 B's:

$$P(5 \text{ good and } 3 \text{ bad}) = .147$$

If a set of objects is withdrawn from a population in an *ordered* fashion, the number of ways that it can be done is found by using *permutations.* Orderly withdrawal without replacement, a sampling concept, is closely associated to the word games. The number of ways the sample can be drawn from the population and the number of words that can be made by choosing and rearranging letters is the same; both are calculated by permutations.

EXAMPLE A.21 Given the population (A, B, C, D, E, F, G, H):

1. How many three-letter words can be formed by choosing and rearranging three letters at a time?

2. Eight parts are in a bin with letter tags on them. How many ways can three parts be removed from the bin for inspection?

Solution

For the first problem, to form a word, there are eight choices for the first letter, seven choices for the second letter, and six choices for the third letter of the word (8 *and* 7 *and* 6):

$$\begin{aligned} P_{8,3} &= 8 \times 7 \times 6 \\ &= \frac{8!}{(8-3)!} \\ &= \frac{8!}{5!} \\ &= 336 \text{ words} \end{aligned}$$

For the second problem, to form the sample, there are eight choices for the first part, seven choices for the second part, and six choices for the third part (8 *and* 7 *and* 6):

$$8 \times 7 \times 6 = P_{8,3} \ldots$$

This uses exactly the same analysis for 336 ways.

Combinations When objects are withdrawn from a population *without regard to order,* the number of possible groups that can be formed is calculated by *combinations.* In Example A.21, part 2, the order in which the parts were chosen really was not important from a sampling point of view. Whether parts *A, C,* and *G* or parts *C, G,* and *A* were sampled did not matter because they formed the same sample.

Another everyday example that illustrates the difference between permutations and combinations occurs with clubs or organizations. Electing three officers from the group—president, vice president, and secretary—is a *permutation* because the order in which the three are chosen is important: who is elected president, and so on. Choosing a committee of three from the group to work on a project, on the other hand, is a *combination* because the order in which the three are chosen is not important. The committee consisting of Jack, Jane, and Phyllis is the same as the committee of Jane, Phyllis, and Jack.

The formula for calculating combinations starts with the permutation formula and eliminates the rearrangements by division. This is the same concept that was used in the *Mississippi* problem. The formula for permutations,

$$P_{N,n} = \frac{N!}{(N-n)!}$$

includes all the rearrangements of the n objects in the sample. In the formula for combinations, however, the $n!$ eliminates the rearrangements of the n objects in the sample:

$$C_{N,n} = \frac{N!}{(N-n)! \cdot n!}$$

EXAMPLE A.22 A club consists of 12 members.

1. How many slates of officers could be formed if the offices of president, vice president, and secretary are to be filled?
2. How many committees of three could be formed to plan the Christmas party?
3. If the club consists of nine women and three men, how many ways could the men and women be seated in a row of 12 chairs (for example, *M, W, W, W, M, M, W, W, W, W, W, W*).

Solution

Problem 1 is a permutation because the order of the three people on the slate is important:

$$\begin{aligned} P_{12,3} &= \frac{12!}{(12-3)!} \\ &= \frac{12!}{9!} \\ &= \frac{12 \times 11 \times 10 \times 9!}{9!} \\ &= 1320 \end{aligned}$$

On the calculator, the sequence 12 $x!$ ÷ 9 $x!$ = gives 1320.

Problem 2 is a combination because the order in which the committee is chosen is not important:

$$\begin{aligned} C_{12,3} &= \frac{12!}{(12-3)! \cdot 3!} \\ &= \frac{12!}{9! \times 3!} \\ &= \frac{12 \times 11 \times 10 \times 9!}{3 \times 2 \times 1 \times 9!} \\ &= 220 \end{aligned}$$

On the calculator, 12 $x!$ ÷ 9 $x!$ ÷ 3 $x!$ = is 220.

The third problem is a two-letter word problem:

$$\begin{aligned} \frac{12!}{3! \cdot 9!} &= \frac{12 \times 11 \times 10 \times 9!}{3 \times 2 \times 1 \times 9!} \\ &= 220 \end{aligned}$$

The two-letter word problem has the same calculation as a combination.

A.6.3 THE BINOMIAL PROBABILITY DISTRIBUTION

Binomial probability refers to the possible ways two specific events can occur and the probabilities associated with those ways. The two events will be classified as *s*, success, and *f*, failure. Any binomial outcome can be categorized in this general fashion. A systematic look at the binomial distribution of successes and failures involves the number of repeated trials, the possible outcomes from the repeated trials, and a simplified algebraic form to represent each outcome. See Table A.1.

Letters placed together represent an "and" situation: *sf* means a success followed by a failure. Similarly, *fss* means a failure on the first trial *and* a success on the second trial *and* a success on the third trial.

TABLE A.1
Binomial outcomes for repeated trials

Number of Trials	Possible Outcomes
1	*s* or *f*
2	*ss* or *sf* or *fs* or *ff*
3	*sss* or *ssf* or *sfs* or *fss* or *ffs* or *fsf* or *sff* or *fff*
4	*ssss* or *sssf* or *ssfs* or *sfss* or *fsss* or *ssff* or *sffs* or *ffss* or *sfsf* or *fsfs* or *fssf* or *fffs* or *ffsf* or *fsff* or *sfff* or *ffff*

The algebraic form simplifies the situation. In the case of four trials, *ssss* can be written in exponent form as s^4. Several ideas that were discussed previously come together here.

Repeated trials form an "and" situation:

- "And" leads to a multiplication.
- A repeated multiplication is a power.
- The number of ways of getting a particular number of successes and failures is the two-letter word problem.
- The two-letter word problem is solved by combinations.

Three successes and one failure can occur in four ways:

$$\text{Algebraic form:} \quad 4s^3f$$

Two successes and two failures can occur in six ways:

$$\text{Algebraic form:} \quad 6s^2f^2$$

The "or" situation leads to addition. See Table A.2.

TABLE A.2
The binomial expansion

Number of Trials	Algebraic Form for the Possible Outcomes
1	$s + f$
2	$s^2 + 2sf + f^2$
3	$s^3 + 3s^2f + 3sf^2 + f^3$
4	$s^4 + 4s^3f + 6s^2f^2 + 4sf^3 + f^4$
5	$s^5 + 5s^4f + 10s^3f^2 + 10s^2f^3 + 5sf^4 + f^5$
6	$s^6 + 6s^5f + 15s^4f^2 + 20s^3f^3 + 15s^2f^4 + 6sf^5 + f^6$
7	$s^7 + 7s^6f + 21s^5f^2 + 35s^4f^3 + 35s^3f^4 + 21s^2f^5 + 7sf^6 + f^7$

The algebraic form shows all possible outcomes in a condensed form. In the seven-trials case, 21 s^5f^2 indicates that five successes and two failures can occur in 21 different ways. If the seven-trials case were written in the long form shown in Table A.1, there would be 21 "words" with 5 *s*'s and 2 *f*'s.

If *s* represents the probability of a success in a single trial and *f* represents the probability of a failure, the probability of any repeated-trial event can be calculated by choosing the appropriate term from the algebraic form. The algebraic form is referred to as the *binomial expansion.* When probabilities are substituted for the variables *s* and *f,* the result is a *binomial probability distribution.*

EXAMPLE A.23 If a single die is tossed four times, find the probability of getting a 3 twice (two 3's and two other numbers).

Solution

- There are four trials because the die is tossed four times.
- A success is getting a 3 when the die is rolled.
- A failure is getting a 1, 2, 4, 5, or 6 when the die is rolled.
- $s = P(\text{success}) = 1/6$
- $f = P(\text{failure}) = 5/6$
- The binomial distribution term for getting two successes is $6s^2f^2$. Calculation of the combination gives six ways, and the *s* and *f* are squared to signify two successes and two failures:

$$C_{4,2} = \frac{4!}{2!2!}$$

$$= 6 \text{ ways}$$

$$P(2 \text{ threes in 4 tosses}) = 6s^2f^2$$

$$= 6\left(\frac{1}{6}\right)^2\left(\frac{5}{6}\right)^2 = \frac{6 \times 1 \times 25}{36 \times 36}$$

$$= .1157$$

$$P(2 \text{ threes in 4 tosses}) = .1157$$

EXAMPLE A.24 A box contains 48 good parts and 12 defective parts. If 10 parts are inspected and returned at different times, find the probability that 4 of the inspected parts were defective.

Solution

There are 10 repeated trials because 10 parts are inspected. A success is getting a defective part because the problem is set up in terms of defective parts:

$$s = P(\text{success}) = P(\text{defective})$$

$$= \frac{12}{60}$$

$$= .2$$

A failure is getting a good part:

$$f = P\ (\text{failure}) = P\ (\text{good})$$

$$= \frac{48}{60}$$

$$= .8$$

The binomial distribution term that represents four successes and six failures is determined by matching the exponents to the number of successes and failures. The coefficient (number of ways) is determined by a combinations calculation. The formula $C_{10,4}s^4f^6$ indicates a combination for 10 trials with 4 successes times two power factors that represent 4 successes and 6 failures:

$$C_{10,4} = \frac{10!}{(10-4)!4!}$$

$$= \frac{10!}{6! \cdot 4!}$$

$$= \frac{10 \times 9 \times 8 \times 7 \times 6!}{4 \times 3 \times 2 \times 1 \times 6!}$$

$$= 210$$

The calculator sequence 10 $x!$ ÷ 6 $x!$ ÷ 4 $x!$ = gives 210:

$$P(4 \text{ defectives}) = 210 \times (.2)^4 \times (.8)^6$$

On the calculator use the sequence .2 x^y 4 = × .8 x^y 6 = × 210 = to get the answer:

$$P(4 \text{ defectives}) = .088$$

EXAMPLE A.25 An assembly line produces a continuous product that is 3% defective. If 100 pieces are randomly checked, find the probability that:

1. Exactly three are defective.
2. None are defective.

Solution

The probabilities stay the same for each piece chosen in the sample [P(defective) = .03]. This makes it a binomial application.

1. $P(3 \text{ defective}) = P(3 \text{ defective and } 97 \text{ good})$
$$= C_{100,3} \times s^3 \times f^{97}$$
$$= \frac{100!}{3! \times 97!} \times (.03)^3 \times (.97)^{97}$$
$$= \frac{100 \times 99 \times 98 \times 97!}{3 \times 2 \times 1 \times 97!} \times .000001406$$
$$= 161{,}700 \times .000001406$$
$$= .227$$
2. $P(0 \text{ defective}) = C_{100,0} \times s^0 \times f^{100}$
$$= 1 \times 1 \times (.97)^{100}$$
$$= .048$$

The Simple Random Sample Drawing a random sample is not really simple at all. It has to be carefully planned. Two basic procedures can be considered:

- When a single element is chosen for a sample, every element in the population must have an equal chance of being chosen. This concept, used in the introductory samples of Chapter 11, allows the possibility that an individual item will be chosen more than once for a particular sample because each item is returned to the population before the next is drawn. The probability model for this is the binomial probability distribution.
- When a sample of size n is chosen from a population, every possible sample of that size must have an equal chance of being chosen. Choosing a set of items from a population, without regard to order, is related to the concept of combinations; the combinations formula gives the number of different sets or samples of size n that could be drawn from a population of size N. The probability model that was previously used for this case was the "and" situation with adjusted prob-

abilities (no replacement), and the number of ways was determined by combinations.

A.6.4 THE HYPERGEOMETRIC PROBABILITY DISTRIBUTION

The probability model that can be used for no-replacement sampling involves the use of combinations within the basic probability formula:

$$P(\text{success}) = \frac{\text{number of successes}}{\text{number of possibilities}}$$

The number of successes and the number of possibilities are calculated using combinations.

EXAMPLE A.26 From a bag that contains 5 red and 4 blue marbles, find the probability of drawing a sample that contains 3 red marbles and 1 blue marble (no replacement).

Solution

A success consists of getting 3 red from the available 5 red marbles *and* getting 1 blue from the available 4 blue marbles. The number of possible successes is $C_{5,3} \times C_{4,1}$. The total number of possibilities consists of choosing any 4 marbles from the available nine, which is $C_{9,4}$:

$$P(\text{success}) = \frac{C_{5,3} \times C_{4,1}}{C_{9,4}}$$

$$C_{5,3} = \frac{5!}{3!2!} = \frac{5 \times 4 \times 3!}{3! \times 2 \times 1} = 10$$

$$C_{4,1} = \frac{4!}{1!3!} = \frac{4 \times 3!}{1 \times 3!} = 4$$

$$C_{9,4} = \frac{9!}{4!5!} = \frac{9 \times 8 \times 7 \times 6 \times 5!}{4 \times 3 \times 2 \times 1 \times 5!} = 126$$

$$P(\text{success}) = \frac{10 \times 4}{126}$$

$$= \frac{40}{126}$$

$$= .317$$

This is an illustration of the use of the hypergeometric probability distribution. The hypergeometric distribution is used for sampling *without replacement.*

Example A.26 can also be solved using the adjusted "and" probabilities;

$$P(\text{3 red and 1 blue}) = \frac{5}{9} \times \frac{4}{8} \times \frac{3}{7} \times \frac{4}{6} \times 4 \text{ ways} = .317$$

EXAMPLE A.27 A box contains 40 good parts and 10 bad ones. If a sample of 8 parts is taken, find the probability that the sample contains 5 good parts and 3 bad ones. This is sampling without replacement, so use the hypergeometric distribution.

Solution

A success is choosing 5 good parts from the available 40 good parts ($C_{40,5}$) *and* 3 bad parts from the available 10 bad parts ($C_{10,3}$). The total number of successes is

$$C_{40,5} \times C_{10,3}$$

The "and" signifies multiplication. The total number of possibilities involves choosing any 8 parts from the available 50, which is $C_{50,8}$:

$$P(\text{5 good and 3 bad parts}) = \frac{C_{40,5} \times C_{10,3}}{C_{50,8}}$$

Calculate each combination:

$$C_{50,8} = \frac{50!}{8!42!} = 536{,}878{,}650$$

$$C_{40,5} = \frac{40!}{5!35!} = 658{,}008$$

$$C_{10,3} = \frac{10!}{3!7!} = 120$$

$$P(\text{5 good and 3 bad}) = \frac{658{,}008 \times 120}{536{,}878{,}650}$$

$$= .147$$

Compare this result with that of Example A.20. The same problem is done by the two different methods, and the same probability results.

Factorial numbers increase in size rapidly and extend beyond the whole-number display capability of calculators at about 14!. When that happens, the calculator automatically changes into scientific notation, and a split-number format shows up in the calculator display:

14! shows 8.7178291 10 — This means that the decimal point is really 10 places to the right, which would give approximately 87,178,291,000.

50! shows 3.0414093	64	The decimal point is 64 places to the right, which would give a 65-digit number.
	70!	This extends beyond the calculator capability and shows an error message, E.

The preceding calculation can be done entirely within the calculator without having to write down the intermediate results. Do the combination in the denominator first and store it in the memory. Then do one of the combinations from the numerator, divide it by the quantity in the memory, and place the result back in the memory. Do the last combination in the numerator and multiply it by the quantity in the memory. That completes the calculation. This procedure is shown here for calculators:

$$50\ x! \div 8\ x! \div 42\ x! = \text{M}$$
$$40\ x! \div 5\ x! \div 35\ x! = \div\ \text{MR} = \text{M}$$
$$10\ x! \div 3\ x! \div 7\ x! = \times\ \text{MR} = \quad (\text{ display will show } .147)$$

EXAMPLE A.28 A lot of 60 pieces is 5% defective. If a sample of 6 pieces is drawn, find the probability of getting 1 defective piece in the sample. This is sampling without replacement, so use the hypergeometric distribution.

Solution

If 5% of the 60 pieces are defective, there are 3 (.05 × 60) defective pieces in the lot. A success is choosing 5 good pieces from the 57 good ones ($C_{57,5}$) *and* choosing 1 bad piece from the 3 bad ones ($C_{3,1}$). The "and" signifies multiplication:

$$C_{57,5} \times C_{3,1}$$

The total number of possibilities involves choosing any 6 from the 60 pieces, $C_{60,6}$:

$$P(1 \text{ defective}) = \frac{C_{57,5} \times C_{3,1}}{C_{60,6}}$$

Calculate each part of the formula. The calculator sequence is shown on the right:

$$C_{60,6} = \frac{60!}{6! \cdot 54!} = \quad 60x! \div 54x! \div 6x! = \text{M}$$

$$C_{57,5} = \frac{57!}{5! \cdot 52!} = \quad 57x! \div 5x! \div 52x! = \div\ \text{MR} = \text{M}$$

$$C_{3,1} = \frac{3!}{1! \cdot 2!} = \quad 3x! \div 2x! \div 1x! = \times\ \text{MR} = \ P(1 \text{ defective}) = .251$$

EXERCISES

1. Evaluate:

a. $-12 + 7$ **b.** $-8 - 5$ **c.** $9 - 17$

d. $7 - (-11)$ **e.** $\frac{-32}{4}$ **f.** $\sqrt{36}$

g. $-16 + (-3)$ **h.** $-8 - (-6)$ **i.** $\sqrt{\frac{100}{90}}$

j. $17 + (-9)$ **k.** $-\frac{40}{-10}$ **l.** $4 + 3 \times 5$

m. $(-12)(-3)$ **n.** $5(-9)$ **o.** $(5 + 2 + 3)^2 + \frac{20 + 2}{4}$

p. $-4(5)$ **q.** $\frac{48}{-6}$ **r.** $\sqrt{49 + 9}$

2. Use the given numerical values to evaluate the algebraic expressions:

$$x_1 = 4 \qquad y = 12$$
$$x_2 = 9 \qquad w = 3$$
$$x_3 = 2 \qquad z = 5$$
$$x_4 = 7$$

a. $\sum x$ **b.** z^2w **c.** $\frac{y + w}{w}$

d. wy **e.** $4w^2$ **f.** $(w + z)^2$

g. $\sqrt{yw}$ **h.** $\sqrt{x_1} \times \sqrt{x_2}$ **i.** $3z$

j. $\frac{w^2}{y}$ **k.** w^4 **l.** $\frac{\sqrt{y}}{\sqrt{w}}$

m. $\sum x^2$ **n.** $(\sum x)^2$ **o.** $\sum (x - 2)^2$

3. Seven percent of a line's output must be reworked, and 2% of the output must be scrapped.

a. If one piece is randomly chosen from the line, what is the probability that it is a good piece?

b. If one piece is chosen, what is the chance that it must be either scrapped or reworked?

c. If two pieces are chosen, what is the probability that the first must be reworked and the second is acceptable?

d. If three pieces are taken from the line, what is the chance of getting one in each category (good, scrap, rework)?

4. Evaluate:

a. $6!$ **b.** $P_{8,3}$ **c.** $0!$ **d.** $\frac{10!}{6!4!}$

e. $C_{9,4}$ **f.** $C_{5,5}$ **g.** $\frac{12!}{7!}$ **h.** $C_{11,3}$ **i.** $P_{4,4}$

5. A bin contains 50 good pieces and 10 defective pieces. If a sample of 4 is taken, what is the probability that 3 out of the 4 are good pieces?

6. A line is producing 5% defectives. If 10 pieces are randomly chosen with replacement, what is the probability that 2 are defective? Use the appropriate binomial distribution term.
7. A box contains 30 good parts and 10 bad parts. If a sample of 9 parts is taken, find the probability of getting:
   a. Nine good parts
   b. Six good parts and three bad parts

B

CHARTS AND TABLES

B.1 FORMULAS AND CONSTANTS FOR CONTROL CHARTS

Average and Range Charts: $\bar{x}$ and R

The sample size n is less than or equal to 10; n is usually three to five consecutive pieces.

The $\bar{x}$ chart:

- Center line: $\bar{\bar{x}} = \dfrac{\Sigma \bar{x}}{k}$ for k samples
- Upper control limit: $UCL_{\bar{x}} = \bar{\bar{x}} + (A_2 \times \bar{R})$
- Lower control limit: $LCL_{\bar{x}} = \bar{\bar{x}} - (A_2 \times \bar{R})$

The R chart:

- Center line: $\bar{R} = \dfrac{\Sigma R}{k}$
- Upper control limit: $UCL_R = D_4 \times \bar{R}$
- Lower control limit: $LCL_R = D_3 \times \bar{R}$

The standard deviation for all measurements: $s = \dfrac{\bar{R}}{d_2}$

TABLE B.1
Constants for an $\overline{x}$ and R chart

Sample size n	A_2	D_3	D_4	d_2
2	1.880	0	3.267	1.128
3	1.023	0	2.574	1.693
4	0.729	0	2.282	2.059
5	0.577	0	2.114	2.326
6	0.483	0	2.004	2.536
7	0.419	0.076	1.924	2.704
8	0.373	0.136	1.864	2.847
9	0.337	0.184	1.816	2.970
10	0.308	0.223	1.777	3.078

Median and Range Charts: $\tilde{x}$ and R

The sample size n is most often an odd number less than 10, usually three or five consecutive pieces.

The $\tilde{x}$ chart:

- Center line: $\overline{\tilde{x}} = \dfrac{\Sigma \tilde{x}}{k}$ for k samples
- Upper control limit: $\text{UCL}_{\tilde{x}} = \overline{\tilde{x}} + (\tilde{A}_2 \times \overline{R})$
- Lower control limit: $\text{LCL}_{\tilde{x}} = \overline{\tilde{x}} - (\tilde{A}_2 \times \overline{R})$

The R chart:

- Center line: $\overline{R} = \dfrac{\Sigma R}{k}$
- Upper control limit: $\text{UCL}_R = D_4 \times \overline{R}$
- Lower control limit: $\text{LCL}_R = D_3 \times \overline{R}$

TABLE B.2
Constants for an $\tilde{x}$ and R chart

Sample size n	$\tilde{A}_2$	D_3	D_4	d_2
3	1.187	0	2.574	1.693
5	0.691	0	2.114	2.326
7	0.509	0.076	1.924	2.704
9	0.412	0.184	1.816	2.970

The standard deviation for all measurements: $s = \dfrac{\overline{R}}{d_2}$

Average and Standard Deviation: $\overline{x}$ and s

The sample size n is usually three or more.

The $\overline{x}$ chart:

- Center line: $\overline{\overline{x}} = \dfrac{\Sigma \overline{x}}{k}$ for k samples
- Upper control limit: $UCL_{\overline{x}} = \overline{\overline{x}} + (A_3 \times \overline{s})$
- Lower control limit: $LCL_{\overline{x}} = \overline{\overline{x}} - (A_3 \times \overline{s})$

The s chart:

- Center line: $\overline{s} = \dfrac{\Sigma s}{k}$
- Upper control limit: $UCL_s = B_4 \times \overline{s}$
- Lower control limit: $LCL_s = B_3 \times \overline{s}$

The standard deviation for all measurements: $\hat{\sigma} = \dfrac{\overline{s}}{C_4}$

TABLE B.3
Constants for an $\overline{x}$ and s chart

Sample size n	A_3	B_3	B_4	C_4
2	2.659	0	3.267	.7979
3	1.954	0	2.568	.8862
4	1.628	0	2.266	.9213
5	1.427	0	2.089	.9400
6	1.287	0.030	1.970	.9515
7	1.182	0.118	1.882	.9594
8	1.099	0.185	1.815	.9650
9	1.032	0.239	1.761	.9693
10	0.975	0.284	1.716	.9727

Individual Measurement and Moving Range: x and MR

Chart consecutive pieces.

Center line: $\overline{x} = \dfrac{\Sigma x}{N}$ for N consecutive pieces

Upper control limit: $\text{UCL}_x = \bar{x} + E_2 \times \overline{MR}$

Lower control limit: $\text{LCL}_x = \bar{x} - E_2 \times \overline{MR}$

n = number of consecutive measurements used to calculate MR

MR = range of each consecutive set of n measurements

$$\overline{MR} = \frac{\Sigma MR}{(N - n) + 1}$$

TABLE B.4
x and *MR* constants

n	E_2	D_3	D_4
2	2.659	0	3.267
3	1.772	0	2.574
4	1.457	0	2.282
5	1.290	0	2.114

n = number of measurements used to calculate MR

Attributes Charts

***P* Charts** A p chart works for large samples, $n > 20$. The number of nonconforming pieces, np, should be approximately 5:

$$p = \frac{np}{n} = \frac{\text{number nonconforming}}{\text{total number in sample}}$$

- Center line: $\bar{p} = \dfrac{\Sigma np}{\Sigma n}$
- Upper control limit: $\text{UCL}_p = \bar{p} + 3 \times \sqrt{\dfrac{\bar{p}(1 - \bar{p})}{\bar{n}}}$
- Lower control limit:

  $\text{LCL}_p = \bar{p} - 3 \times \sqrt{\dfrac{\bar{p}(1 - \bar{p})}{\bar{n}}}$

- When sample sizes differ: $\bar{n} = \dfrac{\Sigma n}{k}$ for k samples

***np* Charts** Use an np chart: with a constant sample size of 20 or more when $\overline{np}$ is approximately 5.

- Center line: $\overline{np} = \frac{\Sigma\, np}{k}$ for k samples
- Upper control limit: $UCL_{np} = \overline{np} + 3 \times \sqrt{\overline{np}\,(1 - \bar{p}\,)}$
- Lower control limit: $LCL_{np} = \overline{np} - 3 \times \sqrt{\overline{np}\,(1 - \bar{p}\,)}$

***c* Charts** For a c chart, the inspection unit should be large enough that $\bar{c} \geq 5$, where c is the number of nonconformities per inspection unit. Inspection units should be of constant size.

- Center line: $\bar{c} = \frac{\Sigma c}{k}$ for k units
- Upper control limit: $UCL_c = \bar{c} + 3 \times \sqrt{\bar{c}}$
- Lower control limit: $LCL_c = \bar{c} - 3 \times \sqrt{\bar{c}}$

***u* Charts** With a u chart, n should be large enough that $\bar{u} \geq 5$, where

$$u = \frac{c}{n} = \frac{\text{number of nonconformities}}{\text{sample size}}$$

Use with a sample size of $n > 20$; n may vary.

- Center line: $\bar{u} = \frac{\Sigma c}{\Sigma n}$
- Upper control limit: $UCL_u = \bar{u} + 3 \times \sqrt{\frac{\bar{u}}{\bar{n}}}$ $\bar{n}$ = average sample size
- Lower control limit: $LCL_u = \bar{u} - 3 \times \sqrt{\frac{\bar{u}}{\bar{n}}}$

TABLE B.5
Critical values of T for setup approval

n	3	4	5	6	7	8	9	10
Critical T	0.885	0.529	0.388	0.312	0.263	0.230	0.205	0.186

TABLE B.6
Modified $\overline{x}$ and R control chart factors

g	Subgroup Size: 1 ($\overline{R}$ based on moving range of 2)				2				3				4				5			
	A_{2F}	D_{4F}	A_{2S}	D_{4S}	A_{2F}	D_{4F}	A_{2S}	D_{4S}	A_{2F}	D_{4F}	A_{2S}	D_{4S}	A_{2F}	D_{4F}	A_{2S}	D_{4S}	A_{2F}	D_{4F}	A_{2S}	D_{4S}
1	NA	NA	237	128	NA	NA	167	128	NA	NA	8.21	14	NA	NA	3.1	13	NA	NA	1.8	5.1
2	12.0	2.0	20.8	16	8.49	2.0	15.7	15.6	1.6	1.9	2.7	7.1	0.8	1.9	1.4	3.5	0.6	1.7	1.0	3.2
3	6.8	2.7	9.6	15	4.8	2.7	6.8	14.7	1.4	2.3	1.9	4.5	0.8	1.9	1.1	3.2	0.6	1.8	0.8	2.8
4	5.1	3.3	6.6	8.1	3.6	3.3	4.7	8.1	1.3	2.4	1.6	3.7	0.8	2.1	1.0	2.9	0.6	1.9	0.8	2.6
5	4.4	3.3	5.4	6.3	3.1	3.3	3.8	6.3	1.2	2.4	1.5	3.4	0.8	2.1	1.0	2.8	0.6	2.0	0.7	2.5
6	4.0	3.3	4.7	5.4	2.8	3.3	3.3	5.4	1.2	2.5	1.4	3.3	0.8	2.2	0.9	2.7	0.6	2.0	0.7	2.4
7	3.7	3.3	4.3	5.0	2.7	3.3	3.1	5.0	1.1	2.5	1.3	3.2	0.8	2.2	0.9	2.6	0.6	2.0	0.7	2.4
8	3.6	3.3	4.1	4.7	2.5	3.3	2.9	4.7	1.1	2.5	1.3	3.1	0.8	2.2	0.9	2.6	0.6	2.0	0.7	2.3
9	3.5	3.3	3.9	4.5	2.5	3.3	2.7	4.5	1.1	2.5	1.3	3.0	0.8	2.2	0.9	2.5	0.6	2.0	0.7	2.3
10	3.3	3.3	3.7	4.5	2.4	3.3	2.6	4.5	1.1	2.5	1.2	3.0	0.8	2.2	0.8	2.5	0.6	2.0	0.7	2.3
15	3.1	3.5	3.3	4.1	2.2	3.5	2.3	4.1	1.1	2.5	1.2	2.9	0.8	2.3	0.8	2.4	0.6	2.1	0.6	2.2
20	3.0	3.5	3.1	4.0	2.1	3.5	2.2	4.0	1.1	2.6	1.1	2.8	0.7	2.3	0.8	2.4	0.6	2.1	0.6	2.2
25	2.9	3.5	3.0	3.8	2.1	3.5	2.1	3.8	1.1	2.6	1.1	2.7	0.7	2.3	0.8	2.4	0.6	2.1	0.6	2.2

g = number of groups used to compute stage 1 control limits.

B.2 THE *G* CHART

TABLE B.7
The *G* chart

Number of Measurements	Recommended Number of Groups
10	4
20	5
30	6
40	7
50	8
70	9
100	10
130	11
180	12
230	13
300	14
350	15
430	16
520	17
640	18
750	19
900	20

For any other number of measurements, choose the closest table value and use the corresponding recommended number of groups.

B.3 THE NORMAL DISTRIBUTION TABLE (TAIL AREA)

TABLE B.8
The normal distribution table

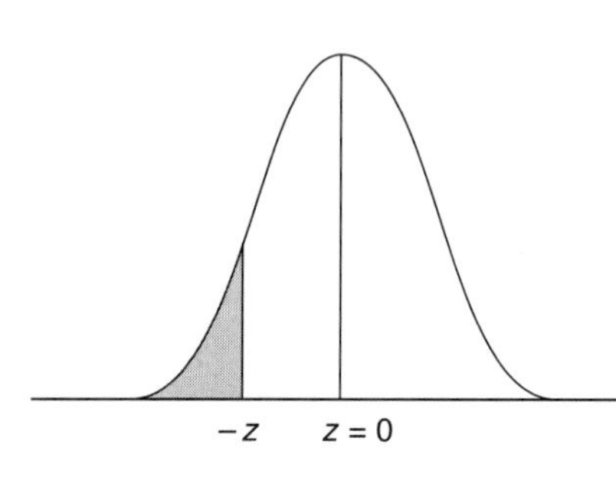

A negative z gives the left tail area.
Area = probability that a measurement is less than x.

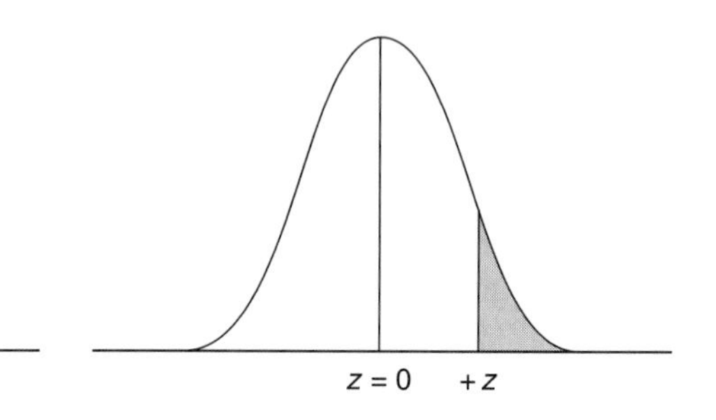

A positive z gives the right tail area.
Area = probability that a measurement is greater than x.

$$z = \frac{x - \bar{\bar{x}}}{s}$$

$\lvert z \rvert$	0	1	2	3	4	5	6	7	8	9
0.0	.5000	.4960	.4920	.4880	.4840	.4801	.4761	.4721	.4681	.4641
0.1	.4602	.4562	.4522	.4483	.4443	.4404	.4364	.4325	.4286	.4247
0.2	.4207	.4168	.4129	.4090	.4052	.4013	.3974	.3936	.3897	.3859
0.3	.3821	.3783	.3745	.3707	.3669	.3632	.3594	.3557	.3520	.3483
0.4	.3446	.3409	.3372	.3336	.3300	.3264	.3228	.3192	.3156	.3121
0.5	.3085	.3050	.3015	.2981	.2946	.2912	.2877	.2843	.2810	.2776
0.6	.2743	.2709	.2676	.2643	.2611	.2578	.2546	.2514	.2483	.2451
0.7	.2420	.2389	.2358	.2327	.2297	.2266	.2236	.2206	.2177	.2148
0.8	.2119	.2090	.2061	.2033	.2005	.1977	.1949	.1922	.1894	.1867
0.9	.1841	.1814	.1788	.1762	.1736	.1711	.1685	.1660	.1635	.1611
1.0	.1587	.1562	.1539	.1515	.1492	.1469	.1446	.1423	.1401	.1379
1.1	.1357	.1335	.1314	.1292	.1271	.1251	.1230	.1210	.1190	.1170
1.2	.1151	.1131	.1112	.1093	.1075	.1056	.1038	.1020	.1003	.0985
1.3	.0968	.0951	.0934	.0918	.0901	.0885	.0869	.0853	.0838	.0823
1.4	.0808	.0793	.0778	.0764	.0749	.0735	.0721	.0708	.0694	.0681
1.5	.0668	.0655	.0643	.0630	.0618	.0606	.0594	.0582	.0571	.0559
1.6	.0548	.0537	.0526	.0516	.0505	.0495	.0485	.0475	.0465	.0455
1.7	.0446	.0436	.0427	.0418	.0409	.0401	.0392	.0384	.0375	.0367
1.8	.0359	.0351	.0344	.0336	.0329	.0322	.0314	.0307	.0301	.0294
1.9	.0287	.0281	.0274	.0268	.0262	.0256	.0250	.0244	.0239	.0233
2.0	.0228	.0222	.0217	.0212	.0207	.0202	.0197	.0192	.0188	.0183
2.1	.0179	.0174	.0170	.0166	.0162	.0158	.0154	.0150	.0146	.0143
2.2	.0139	.0136	.0132	.0129	.0125	.0122	.0119	.0116	.0113	.0110
2.3	.0107	.0104	.0102	.0099	.0096	.0094	.0091	.0089	.0087	.0084
2.4	.0082	.0080	.0078	.0075	.0073	.0071	.0069	.0068	.0066	.0064

TABLE B.8 *(continued)*

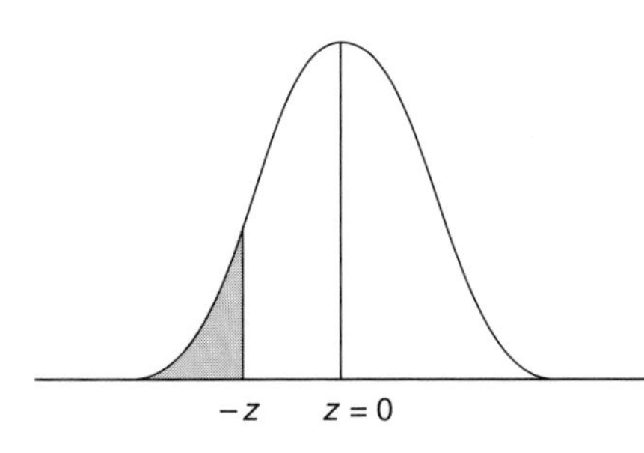

A negative z gives the left tail area.
Area = probability that a measurement is less than x.

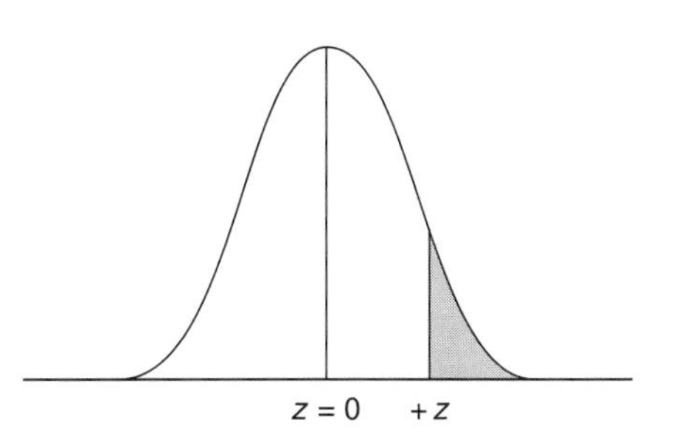

A positive z gives the right tail area.
Area = probability that a measurement is greater than x.

$$z = \frac{x - \bar{\bar{x}}}{s}$$

$\lvert z \rvert$	0	1	2	3	4	5	6	7	8	9
2.5	.0062	.0060	.0059	.0057	.0055	.0054	.0052	.0051	.0049	.0048
2.6	.0047	.0045	.0044	.0043	.0041	.0040	.0039	.0038	.0037	.0036
2.7	.0035	.0034	.0033	.0032	.0031	.0030	.0029	.0028	.0027	.0026
2.8	.0026	.0025	.0024	.0023	.0023	.0022	.0021	.0021	.0020	.0019
2.9	.0019	.0018	.0018	.0017	.0016	.0016	.0015	.0015	.0014	.0014
3.0	.00135	.00131	.00126	.00122	.00118	.00114	.00111	.00107	.00104	.00100
3.1	.00097	.00094	.00090	.00087	.00084	.00082	.00079	.00076	.00074	.00071
3.2	.00069	.00066	.00064	.00062	.00060	.00058	.00056	.00054	.00052	.00050
3.3	.00048	.00047	.00045	.00043	.00042	.00040	.00039	.00038	.00036	.00035
3.4	.00034	.00032	.00031	.00030	.00029	.00028	.00027	.00026	.00025	.00024
3.5	.00023	.00022	.00022	.00021	.00020	.00019	.00019	.00018	.00017	.00017
3.6	.00016	.00015	.00015	.00014	.00014	.00013	.00013	.00012	.00012	.00011
3.7	.00011	.00010	.00010	.00010	.00009	.00009	.00008	.00008	.00008	.00008
3.8	.00007	.00007	.00007	.00006	.00006	.00006	.00006	.00005	.00005	.00005
3.9	.00005	.00005	.00004	.00004	.00004	.00004	.00004	.00004	.00003	.00003
4.0	.00003									

B.4 THE NORMAL DISTRIBUTION TABLE (CENTER AREA)

TABLE B.9
Normal probability distribution: Area from the center

|z|	0	1	2	3	4	5	6	7	8	9
0.0	.0000	.0040	.0080	.0120	.0150	.0199	.0239	.0279	.0319	.0357
0.1	.0398	.0439	.0479	.0519	.0558	.0598	.0638	.0678	.0718	.0758
0.2	.0793	.0832	.0871	.0910	.0948	.0987	.1026	.1064	.1103	.1141
0.3	.1179	.1217	.1255	.1293	.1331	.1368	.1406	.1443	.1480	.1517
0.4	.1554	.1591	.1628	.1664	.1700	.1736	.1772	.1808	.1844	.1879
0.5	.1915	.1950	.1985	.2019	.2054	.2088	.2123	.2157	.2190	.2224
0.6	.2258	.2291	.2324	.2357	.2389	.2422	.2454	.2486	.2518	.2549
0.7	.2580	.2612	.2642	.2673	.2704	.2734	.2764	.2794	.2823	.2852
0.8	.2881	.2910	.2939	.2967	.2996	.3023	.3051	.3079	.3106	.3133
0.9	.3159	.3186	.3212	.3238	.3264	.3289	.3315	.3340	.3365	.3389
1.0	.3413	.3438	.3461	.3485	.3508	.3531	.3554	.3577	.3599	.3621
1.1	.3643	.3665	.3686	.3708	.3729	.3749	.3770	.3790	.3810	.3830
1.2	.3849	.3869	.3888	.3907	.3925	.3944	.3962	.3980	.3997	.4015
1.3	.4032	.4049	.4066	.4082	.4099	.4115	.4131	.4147	.4162	.4177
1.4	.4192	.4207	.4222	.4236	.4251	.4265	.4279	.4292	.4306	.4319
1.5	.4332	.4345	.4357	.4370	.4382	.4394	.4406	.4418	.4430	.4441
1.6	.4452	.4463	.4474	.4485	.4495	.4505	.4515	.4525	.4535	.4545
1.7	.4554	.4564	.4573	.4582	.4591	.4599	.4608	.4616	.4625	.4633
1.8	.4641	.4649	.4656	.4664	.4671	.4678	.4686	.4693	.4700	.4706
1.9	.4713	.4719	.4726	.4732	.4738	.4744	.4750	.4756	.4762	.4767
2.0	.4773	.4778	.4783	.4788	.4793	.4798	.4803	.4808	.4812	.4817
2.1	.4821	.4826	.4830	.4834	.4838	.4842	.4846	.4850	.4854	.4857
2.2	.4861	.4865	.4868	.4871	.4875	.4878	.4881	.4884	.4887	.4890
2.3	.4893	.4896	.4898	.4901	.4904	.4906	.4909	.4911	.4913	.4916
2.4	.4918	.4920	.4922	.4925	.4927	.4929	.4931	.4932	.4934	.4936
2.5	.4938	.4940	.4941	.4943	.4945	.4946	.4948	.4949	.4951	.4952
2.6	.4953	.4955	.4956	.4957	.4959	.4960	.4961	.4962	.4963	.4964
2.7	.4965	.4966	.4967	.4968	.4969	.4970	.4971	.4972	.4973	.4974
2.8	.4974	.4975	.4976	.4977	.4977	.4978	.4979	.4980	.4980	.4981
2.9	.4981	.4982	.4983	.4983	.4984	.4984	.4985	.4985	.4986	.4986

TABLE B.9 ***(continued)***

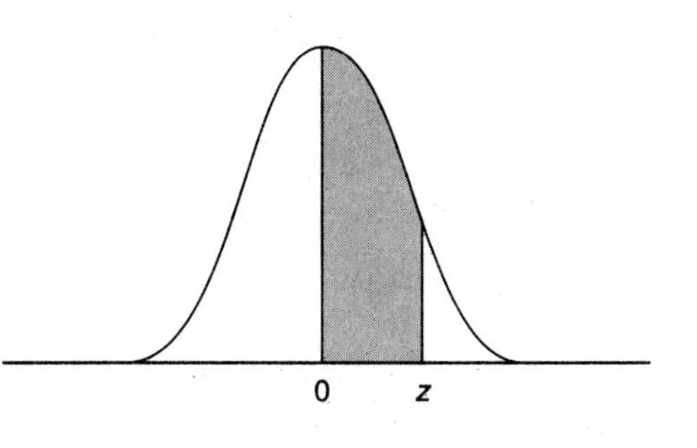

$\lvert z \rvert$	0	1	2	3	4	5	6	7	8	9
3.0	.4987	.4987	.4987	.4988	.4988	.4989	.4989	.4989	.4990	.4990
3.1	.4990	.4991	.4991	.4991	.4992	.4992	.4992	.4992	.4993	.4993
3.2	.4993	.4993	.4994	.4994	.4994	.4994	.4994	.4995	.4995	.4995
3.3	.4995	.4995	.4996	.4996	.4996	.4996	.4996	.4996	.4996	.4997
3.4	.4997	.4997	.4997	.4997	.4997	.4997	.4997	.4997	.4998	.4998
3.5	.4998	.4998	.4998	.4998	.4998	.4998	.4998	.4998	.4998	.4998
3.6	.4998	.4999	.4999	.4999	.4999	.4999	.4999	.4999	.4999	.4999
3.7	.4999	.4999	.4999	.4999	.4999	.4999	.4999	.4999	.4999	.4999
3.8	.4999	.4999	.4999	.4999	.4999	.4999	.4999	.5000	.5000	.5000

B.5 THE NORMAL DISTRIBUTION TABLE (LEFT AREA)

TABLE B.10
Normal probability distribution: Total area to the left

$\lvert z \rvert$	0	1	2	3	4	5	6	7	8	9
0.0	.5000	.5040	.5080	.5120	.5150	.5199	.5239	.5279	.5319	.5357
0.1	.5398	.5439	.5479	.5519	.5558	.5598	.5638	.5678	.5718	.5758
0.2	.5793	.5832	.5871	.5910	.5948	.5987	.6026	.6064	.6103	.6141
0.3	.6179	.6217	.6255	.6293	.6331	.6368	.6406	.6443	.6480	.6517
0.4	.6554	.6591	.6628	.6664	.6700	.6736	.6772	.6808	.6844	.6879
0.5	.6915	.6950	.6985	.7019	.7054	.7088	.7123	.7157	.7190	.7224
0.6	.7258	.7291	.7324	.7357	.7389	.7422	.7454	.7486	.7518	.7549
0.7	.7580	.7612	.7642	.7673	.7704	.7734	.7764	.7794	.7823	.7852
0.8	.7881	.7910	.7939	.7967	.7996	.8023	.8051	.8079	.8106	.8133
0.9	.8159	.8186	.8212	.8238	.8264	.8289	.8315	.8340	.8365	.8389
1.0	.8413	.8438	.8461	.8485	.8508	.8531	.8554	.8577	.8599	.8621
1.1	.8643	.8665	.8686	.8708	.8729	.8749	.8770	.8790	.8810	.8830
1.2	.8849	.8869	.8888	.8907	.8925	.8944	.8962	.8980	.8997	.9015
1.3	.9032	.9049	.9066	.9082	.9099	.9155	.9131	.9147	.9162	.9177
1.4	.9192	.9207	.9222	.9236	.9251	.9265	.9279	.9292	.9306	.9319
1.5	.9332	.9345	.9357	.9370	.9382	.9394	.9406	.9418	.9430	.9441
1.6	.9452	.9463	.9474	.9485	.9495	.9505	.9515	.9525	.9535	.9545
1.7	.9554	.9564	.9573	.9582	.9591	.9599	.9608	.9616	.9625	.9633
1.8	.9641	.9649	.9656	.9664	.9671	.9678	.9686	.9693	.9700	.9706
1.9	.9713	.9719	.9726	.9732	.9738	.9744	.9750	.9756	.9762	.9767
2.0	.9773	.9778	.9783	.9788	.9793	.9798	.9803	.9808	.9812	.9817
2.1	.9821	.9826	.9830	.9834	.9838	.9842	.9846	.9850	.9854	.9857
2.2	.9861	.9865	.9868	.9871	.9875	.9878	.9881	.9884	.9887	.9890
2.3	.9893	.9896	.9898	.9901	.9904	.9906	.9909	.9911	.9913	.9916
2.4	.9918	.9920	.9922	.9925	.9927	.9929	.9931	.9932	.9934	.9936
2.5	.9938	.9940	.9941	.9943	.9945	.9946	.9948	.9949	.9951	.9952
2.6	.9953	.9955	.9956	.9957	.9959	.9960	.9961	.9962	.9963	.9964
2.7	.9965	.9966	.9967	.9968	.9969	.9970	.9971	.9972	.9973	.9974
2.8	.9974	.9975	.9976	.9977	.9977	.9978	.9979	.9980	.9980	.9981
2.9	.9981	.9982	.9983	.9983	.9984	.9984	.9985	.9985	.9986	.9986

TABLE B.10 *(continued)*

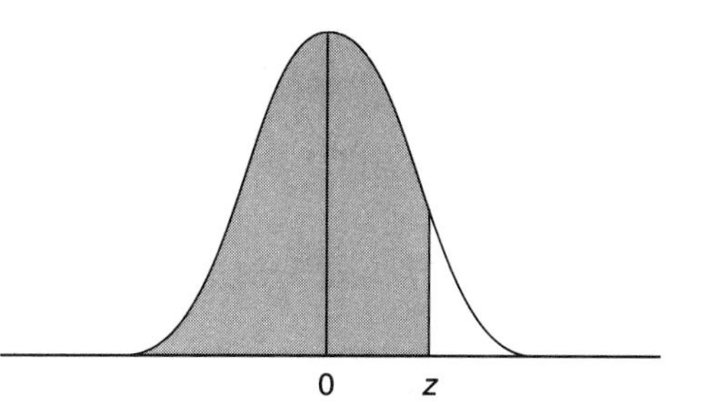

$\lvert z \rvert$	0	1	2	3	4	5	6	7	8	9
3.0	.9987	.9987	.9987	.9988	.9988	.9989	.9989	.9989	.9990	.9990
3.1	.9990	.9991	.9991	.9991	.9992	.9992	.9992	.9992	.9993	.9993
3.2	.9993	.9993	.9994	.9994	.9994	.9994	.9994	.9995	.9995	.9995
3.3	.9995	.9995	.9996	.9996	.9996	.9996	.9996	.9996	.9996	.9997
3.4	.9997	.9997	.9997	.9997	.9997	.9997	.9997	.9997	.9998	.9998
3.5	.9998	.9998	.9998	.9998	.9998	.9998	.9998	.9998	.9998	.9998
3.6	.9998	.9999	.9999	.9999	.9999	.9999	.9999	.9999	.9999	.9999
3.7	.9999	.9999	.9999	.9999	.9999	.9999	.9999	.9999	.9999	.9999
3.8	.9999	.9999	.9999	.9999	.9999	.9999	.9999	1.0000	1.0000	1.0000

B.6 PROCESS CAPABILITY

$$\text{PCR} = \frac{6s}{\text{tolerance}}$$

$$\text{Cpk} = \left| \frac{\text{Nearest specification limit} - \bar{x}}{3s} \right|$$

TABLE B.11
The process capability grid

Process Classification	Process Capability	Control Chart
A	PCR $\leq$ .5 Cpk $\geq$ 2	Recommended
B	.51 $\leq$ PCR $\leq$ .7 1.42 $\leq$ Cpk $\leq$ 1.99	Recommended
C	.71 $\leq$ PCR $\leq$ .9 1.11 $\leq$ Cpk $\leq$ 1.41	Required
D	PCR $\geq$ .91 Cpk $\leq$ 1.1	Required

C

GLOSSARY OF SYMBOLS

$A_2, \tilde{A}_2$	Control chart factors
AOQ	Average outgoing quality
AOQL	Average outgoing quality limit
AQL	Acceptable quality level
A.V.	Appraiser variation (reproducibility)
c	The number of defects for control charting; the acceptance number in sampling
$\bar{c}$	The average number of defects
$C_{n,r}$	Combinations of n objects using r of them
C_{pk}	A process capability measure
d_2, D_3, D_4	Control chart factors
E.V.	Equipment variation (reproducibility)
f	Frequency
***G* chart**	A chart for determining the optimum group size for frequency distributions or histograms
LCL	Lower control limit
LSL	Lower specification limit
MR	Moving range
$\overline{MR}$	The average moving range value
μ	The mean of a population

n	Sample size
N	Lot size; the number of values in a population
np	The number of defective pieces
OC	Operating characteristic
p	The proportion of defective items
$\bar{p}$	The average proportion of defective items
$100p$	The percentage of defective items
PCR	A process capability measure
$P_{n,r}$	The permutation of n objects using r of them
$P(A)$	The probability that event A occurs
R	Range
$\bar{R}$	Average range
R & R	Repeatability and reproducibility in gauge variation
s	Sample standard deviation; the standard deviation of a large sample used as an estimate of the population standard deviation.
s^2	Sample variance
UCL	Upper control limit
USL	Upper specification limit
u	The average number of defects per unit in a sample
$\bar{u}$	The average of the u values
x	A measurement value
$\bar{x}$	The average measurement
$\bar{\bar{x}}$	The average of the $\bar{x}$'s, or the grand average
$\tilde{x}$	The median
$\bar{\tilde{x}}$	The average of the medians
z	The scale value for the normal curve
Σ	The Greek letter sigma (uppercase), which indicates "the sum of"; Σx is the sum of the x values
σ	The Greek letter sigma (lowercase), which is the symbol for the standard deviation of a population of measurements
$\hat{\sigma}$	The standard deviation of a large sample used as an estimate of the population standard deviation
σ_{n-1}	The calculator key for s
σ^2	The variance of a population of values

D

LAB EXERCISES

SPC LAB EXERCISES

Chapter 1

1. Make a detection quality model for one of your courses.

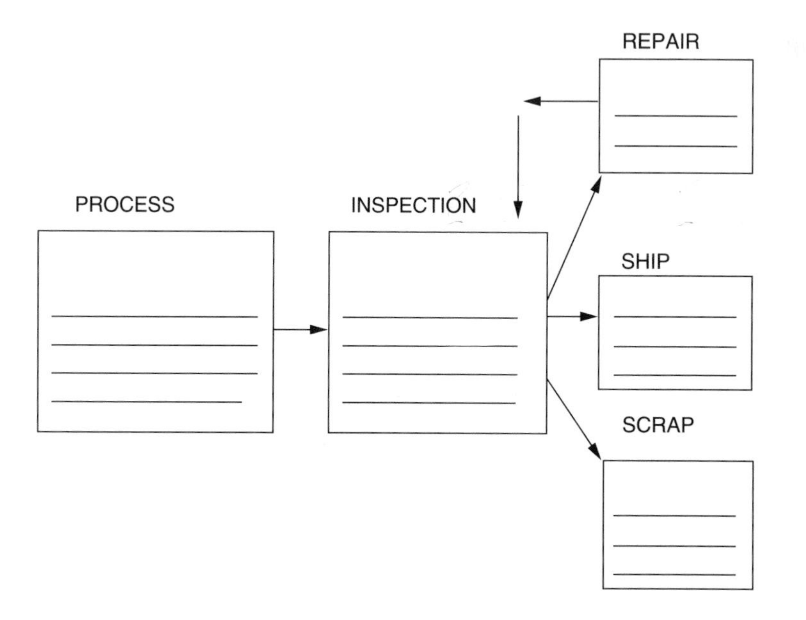

2. Make a prevention quality model for one of your courses.

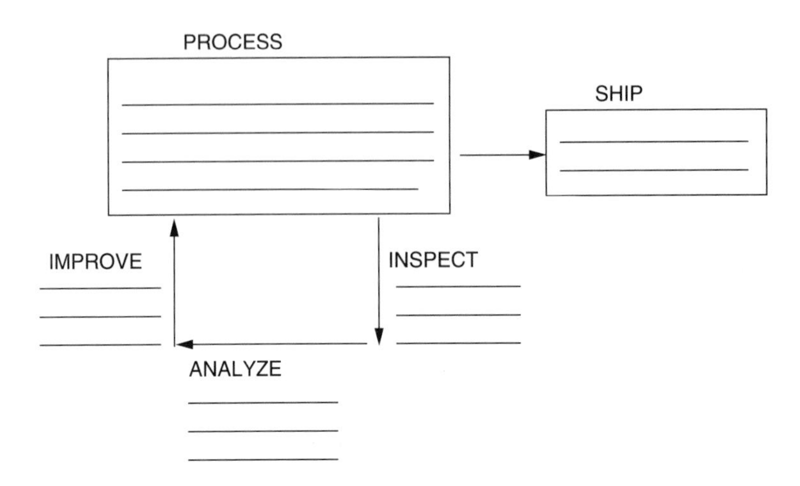

3. Show how the list of process improvements is interactive as a result of using SPC. Complete the "leads" diagram.

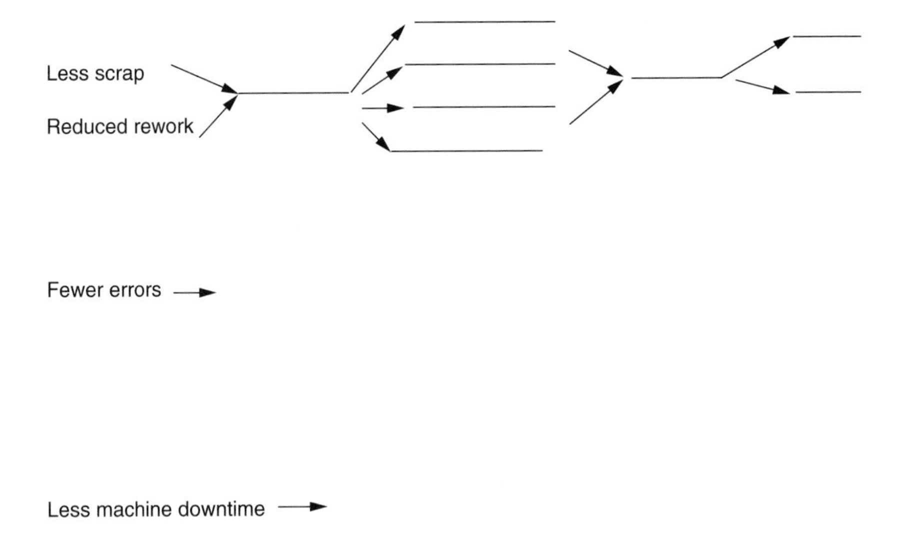

Chapter 2

Adapt Deming's 14 points for management to your management of your college career.

1. Constancy of purpose:
2. The new philosophy:
3. Cease mass inspection:
4. End price alone as deciding factor:
5. Constantly improve the system:

6. Modern training methods:
7. Modern methods of supervision:
8. Drive out fear:
9. Break down barriers between areas:
10. Eliminate numerical goals:
11. Eliminate quotas:
12. Remove barriers to doing good work:
13. Institute a program of education and training:
14. Create a management structure for the 13 Points:

Chapter 3

1. Use the various measuring instruments provided by the instructor and determine for each:
   a) Accuracy:
   b) Maximum possible error:
2. Use a measuring device with appropriate accuracy to measure the lengths of the paper clips in a box. Each team member should measure at least 50 clips.
   a) Make a list of the measurements. Group them according to the person doing the measuring. (Save this information for Lab 4).
   b) Make a sequence of tally distributions similar to Figures 3.1 to 3.4. (First 30, First 50, First 80, First 100, All).
3. a) Compare the distributions for each box with respect to
   (1) Location (2) Spread (3) Shape
   b) Calculate μ and σ for each box using the histogram from 2b and the frequency button on a statistical calculator.
   c) What percentages of the measurements are beyond?
   (1) $\mu \pm 2\sigma$ (2) $\mu \pm 3\sigma$
4. Plan a random sample of 30 paper clips from the list in 2a using random numbers (from a statistical calculator or from a random number table). Calculate $\bar{x}$ and s for your sample to compare with μ and σ.

Chapter 4

1 Make a stemplot for each person's measurements from Lab 3 and determine
   a) Position b) Spread c) shape
2. Compare the individual stemplots to see if any two would have a bimodal shape if they were combined into one.
3. Based on the paper clip specifications given to you by your instructor, re-analyze each person's stemplot with respect to
   a) Position b) Spread c) Shape
4. Combine all the stemplots from Lab 3 into one and compare it with the tally histogram from Lab 3.
5. Make a frequency distribution and a histogram of all the measurements. Use the number of groups recommended by the G chart in Table B7. Compare it with the stemplot and the tally histogram from Lab 3 with respect to
   a) Position b) Spread c) Shape
6. Based on your experience in Labs 3 and 4, create a flowchart that could be used for the two labs next semester.
7. Make a cause-and-effect diagram for the problem "The individual stemplot distributions differ too much."

Chapter 5

The bottom of this sheet contains 30 samples, $n = 5$, of the outside diameter of sparkplug shell casings. The specifications are .743 to .757. The histogram of the 150 measurements is included.

1. Use the frequency button on a statistical calculator and the given frequency histogram to determine, for the 150 measurements,
   a) $\bar{\bar{x}} =$ b) $s =$
2. Clear your statistical registers and enter the 30 $\bar{x}$ values. Find
   a) $\bar{\bar{x}} =$ b) $s_{\bar{x}} =$
3. Use a colored marker to plot the 30 $\bar{x}$ values on the given histogram (superimpose the $\bar{x}$ histogram on the data histogram).
   a) $\bar{\bar{x}}$ for data = b) $\bar{\bar{x}}$ for the sample means =
   c) s for data = d) $s_{\bar{x}}$ for sample means =
   e) Observations:
4. If the same set of 150 data values were rearranged in different samples of $n = 5$, what will happen? Use the second list of $\bar{x}$ values to determine
   a) $\bar{\bar{x}} =$ b) $s_{\bar{x}} =$
5. According to the Central Limit Theorem, what is the theoretical value of $s_{\bar{x}}$?
6. Analyze the data histogram with respect to the specification limits.
   a) Position:
   b) Spread:
   c) Shape:
7. What percent of the data values are out of specification?
8. Use the normal curve and the s value for the 150 data values to determine the percentage of the population that is out of specification.

Sample measurements and $\bar{x}$ values:

.751	.754	.752	.758	.755	.753	.746	.751	.746	.742	.746	.750
.753	.751	.749	.750	.745	.748	.748	.749	.750	.747	.748	.752
.750	.751	.757	.753	.747	.747	.750	.750	.752	.749	.750	.754
.748	.750	.750	.751	.750	.749	.744	.743	.748	.745	.750	.751
.750	.752	.749	.750	.751	.749	.748	.749	.751	.749	.747	.751
.7504	.7516	.7514	.7524	.7496	.7492	.7472	.7484	.7494	.7464	.7482	.7516

.756	.749	.755	.753	.752	.755	.752	.749	.750	.749	.748	.757
.754	.750	.750	.752	.750	.751	.750	.749	.749	.751	.749	.753
.756	.750	.751	.751	.750	.751	.751	.750	.753	.752	.749	.751
.751	.753	.754	.759	.751	.752	.750	.747	.753	.750	.750	.751
.752	.752	.754	.755	.753	.749	.748	.752	.750	.750	.749	.750
.7538	.7508	.7528	.7540	.7512	.7516	.7502	.7494	.7510	.7504	.7490	.7524

.754	.750	.747	.755	.749	.750
.752	.749	.748	.748	.751	.753
.748	.749	.749	.750	.750	.748
.748	.748	.744	.747	.751	.751
.751	.747	.746	.748	.749	.749
.7506	.7486	.7468	.7496	.7500	.7502

$\overline{x}$ values for the same data rearranged in different samples of $n = 5$:

.7450	.7496	.7516	.7498	.7504	.7476
.7484	.7496	.7458	.7492	.7512	.7508
.7522	.7518	.7512	.7522	.7504	.7506
.7518	.7514	.7516	.7508	.7506	.7504
.7500	.7502	.7498	.7492	.7482	.7478

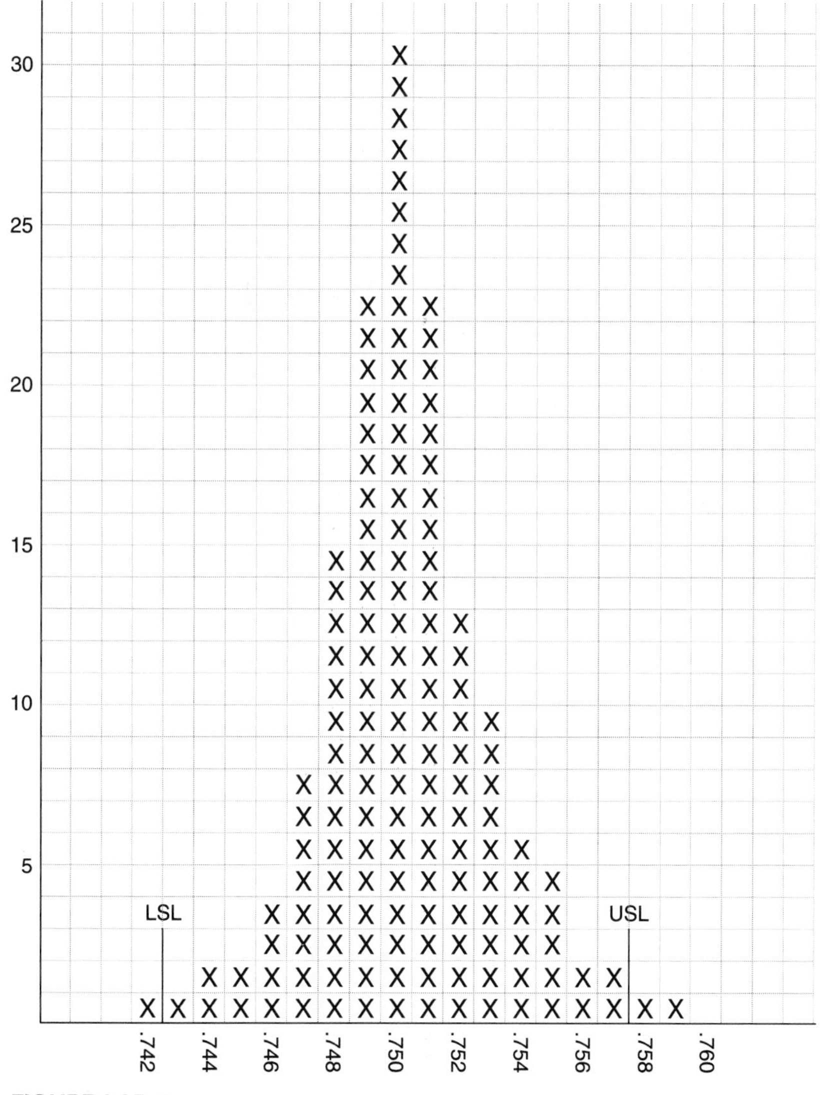

FIGURE LAB 1

Chapter 6

For the control chart on the back of the lab sheet,

1. Complete the remaining $\bar{x}$ and R calculations.
2. Complete the chart by plotting the $\bar{x}$ and R values and completing the graph.
3. Calculate $\bar{\bar{x}}$ and $\bar{R}$ and draw the solid lines. Check for reasonableness. You should see a balance between points above the average line and points below.
4. Calculate the control limits and draw the dashed lines. They should be evenly spaced on the $\bar{x}$ chart.
5. Mark any points that are out of control with an X: points beyond the control limits, shifts, runs and indicate any trends with an arrow.
6. Eliminate all out-of-control data, recalculate averages and control limits, reanalyze for additional points out-of-control, and repeat this step if necessary.
7. Draw a picture of the distribution of data: either a stemplot or tally histogram.
8. Do a complete capability analysis: visually from #7, PCR, C_p, C_{pk}, percent out-of-specification.
9. Keep this chart for further analysis with Chapter 10.

Part/Asm. Name	Operation	Specification .810 – .840	Chart No.
Part No.	Department	Gage	Unit of Measure
Parameter	Machine	Sample Size/Frequency	Zero Equals

Date																										
Time																										
Operator																										
Sample Measurements	1	.830	.825	.820	.814	.826	.828	.830	.828	.835	.838	.829	.831	.826	.824	.827	.829	.828	.819	.815	.818	.827	.831	.835	.838	.826
	2	.820	.822	.812	.811	.830	.825	.832	.834	.836	.830	.824	.820	.818	.833	.819	.834	.820	.826	.829	.824	.819	.836	.820	.836	.820
	3	.835	.831	.815	.817	.818	.830	.819	.822	.825	.830	.827	.822	.820	.828	.832	.818	.815	.817	.816	.825	.832	.820	.836	.835	.822
	4	.827	.810	.813	.822	.822	.817	.827	.828	.824	.834	.828	.823	.820	.827	.822	.827	.817	.822	.820	.817	.826	.833	.837	.823	.824
	5																									
Sum																										
Average $\bar{X}$		.828	.822	.815	.816	.824	.825	.827	.828	.830	.833	.827	.824	.821	.828	.825	.827	.820	.821	.820	.821					
Range R		.015	.021	.008	.011	.012	.013	.013	.012	.012	.008	.005	.011	.008	.009	.013	.016	.013	.009	.014	.008					

1 2 3 4 5 6 7 8 9 10 11 12 13 14 15 16 17 18 19 20 21 22 23 24 25

Averages

$\bar{X}$

.840
.835
.830
.825
.820
.815

Ranges

R

.030
.020
.010
0

FIGURE LAB 2

Chapter 7

For the control chart on the back of the lab sheet,

1. Complete the remaining $\overline{x}$ and s calculations.
2. Complete the chart by plotting the $\overline{x}$ and s values and completing the graph.
3. Calculate $\overline{\overline{x}}$ and $\overline{s}$ and draw the solid lines. Check for reasonableness.
4. Calculate the control limits and draw the dashed lines.
5. Mark any points that are out of control with an X: points beyond the control limits, shifts, runs and indicate any trends with an arrow.
6. Eliminate all out-of-control data, recalculate averages and control limits, re-analyze for additional points out-of-control, and repeat this step if necessary.
7. Draw a picture of the distribution of data: either a stemplot or tally histogram.
8. Do a complete capability analysis: visually from #7, PCR, C_p, C_{pk}, percent out-of-specification.
9. Keep this control chart for further analysis with Chapter 10.

Part/Asm. Name	Operation	Specification .600 to .640	Chart No.
Part No.	Department	Gage	Unit of Measure
Parameter	Machine	Sample Size/Frequency	Zero Equals

		1	2	3	4	5	6	7	8	9	10	11	12	13	14	15	16	17	18	19	20	21	22	23	24	25
Date																										
Time																										
Operator																										
Sample Measurements	1	.615	.625	.628	.630	.638	.628	.607	.630	.632	.617	.630	.604	.605	.625	.621	.634	.618	.630	.639	.635	.619	.624	.611	.625	.607
	2	.610	.606	.614	.632	.621	.610	.620	.621	.610	.626	.610	.607	.603	.619	.634	.620	.616	.611	.637	.620	.630	.610	.618	.614	.634
	3	.608	.604	.610	.619	.624	.611	.614	.628	.618	.621	.605	.612	.606	.625	.625	.627	.627	.617	.624	.618	.628	.620	.615	.622	.623
	4	.607	.609	.632	.619	.621	.611	.615	.629	.628	.628	.615	.617	.622	.631	.628	.615	.627	.626	.636	.615	.615	.618	.624	.615	.618
	5																									
Sum																										
Average $\overline{X}$		.610	.611	.621	.625	.626	.615	.614	.627	.622	.623	.615	.610	.609	.625	.627	.624	.622	.621	.634	.622					
		.0036	.0096	.0106	.0070	.0081	.0087	.0054	.0041	.0099	.0050	.0108	.0057	.0088	.0049	.0055	.0083	.0058	.0086	.0068	.0089					

Averages

$\overline{X}$: .635, .630, .625, .620, .615, .610

S: .0020, .0015, .0010, .0005, 0

FIGURE LAB 3

Chapter 8

1. A process has specifications .645 inches to .655 inches. Calculate the maximum value that the process standard deviation should be to be able to use a precontrol chart.
2. Set up a precontrol chart based on the above specifications.
3. Use the following list of measurements, by column, as sequential measurements (as needed) and indicate your precontrol decisions.

.648	.648	.646	.647	.650	.651	.643	.651	.652	.650
.646	.651	.652	.652	.649	.648	.650	.651	.652	.650
.649	.650	.651	.650	.651	.646	.654	.653	.650	.653
.650	.649	.646	.654	.650	.648	.653	.652	.649	.649

4. An A process has a standard deviation of .0016 and specification limits of .385 to .405. Set up a modified precontrol chart for tight control.
5. Use the first six columns of the data given in #3 and make an *x and MR* chart with $n = 4$ using the blank chart on the back of the lab sheet.
6. Calculate the control limits and mark any out-of-control points with an X.
7. Calculate C_{pk} from the in-control points on the *x and MR* chart.

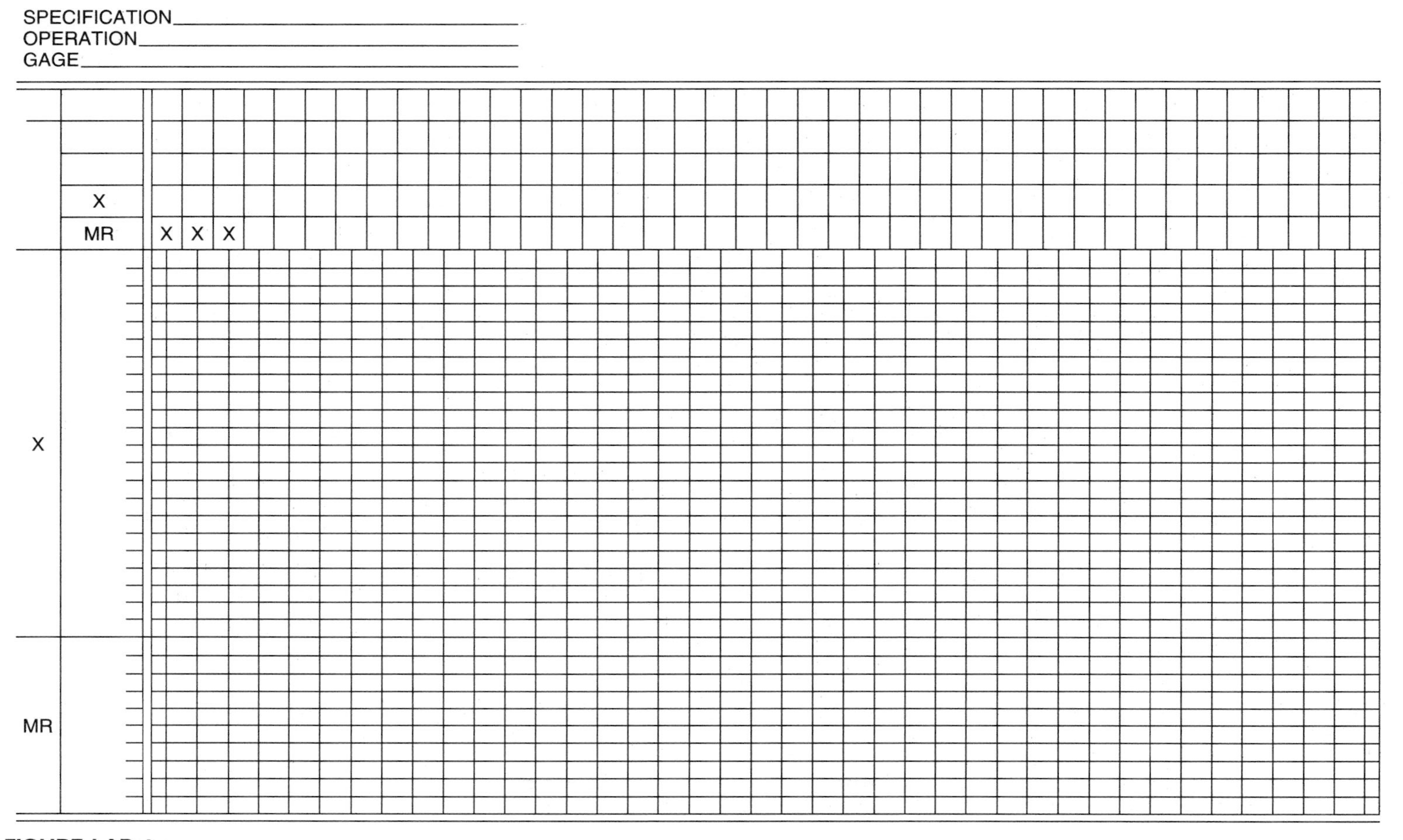

FIGURE LAB 4

Chapter 9

For the control charts on the back of the lab sheet,

1. Complete Chart A as a p chart and Chart B as a np chart.
2. Do the remaining p calculations for Chart A, plot the p values, and draw the broken-line graph.
3. Calculate $\overline{p}$ and the control limits. Draw the solid and dashed lines.
4. Analyze for points out-of-control: mark them with an X.
5. What is the capability?
6. Plot the np values on Chart B.
7. Calculate $\overline{np}$ and the control limits. Draw the solid and dashed lines.
8. Analyze for points out-of-control: mark them with an X.
9. What is the capability?
10. How do the two charts compare?
11. Save the control charts for further analysis in Chapter 10.

A

Discrepancies																									
Sample Size (n)	400	400	400	400	400	400	400	400	400	400	400	400	400	400	400	400	400	400	400	400	400	400	400	400	400
Number (np. c)	12	11	9	8	7	5	4	2	0	6	3	4	2	13	12	3	11	12	1	3	0	5	4	3	5
Proportion (p.u)	.03	.0275	.0225	.02	.0175	.0125	.01	.005	0	.015	.0075	.01	.005	.0325											
Date																									

	1	2	3	4	5	6	7	8	9	10	11	12	13	14	15	16	17	18	19	20	21	22	23	24	25
P																									

B

Discrepancies																									
Sample Size (n)	400	400	400	400	400	400	400	400	400	400	400	400	400	400	400	400	400	400	400	400	400	400	400	400	400
Number (np. c)	12	11	9	8	7	5	4	2	0	6	3	4	2	13	12	3	11	12	1	3	0	5	4	3	5
Proportion (p.u)																									
Date																									

	1	2	3	4	5	6	7	8	9	10	11	12	13	14	15	16	17	18	19	20	21	22	23	24	25
nP																									

FIGURE LAB 5

Chapter 10

1. In Example 6.15, the range chart was out of control and the control limits were recalculated after the out-of-control data was removed. Analyze the new control chart in Figure 6.32.
2. Case study 7.1 produced a control chart that was out of control. Analyze the revised chart in Figure 7.15.
3. Case study 9.1 produced the *p* chart with the variable control limits. The points beyond the control limits and any shifts are easy to spot, but the other patterns are more difficult. One way to identify freak patterns and instability would be to label the points with their zone letter. Then the zone letters would easily identify these out-of-control patterns. At each control limit count the lines above and below the average line and divide the number of lines by 3 for an approximate number of lines per zone. Label the zone letters in figure 9.4 and identify any out-of-control patterns.
4. Reinterpret the control chart from Lab 6.
5. Reinterpret the control chart from Lab 7.
6. Reinterpret the control chart from Lab 8.

Chapter 11

At this point in your studies, final examinations are not too far away. Utilize teamwork to develop a process by which everyone can be successful in all their courses.

1. Form teams of 4 to 6 members.
2. Follow the problem solving sequence.
3. Use brainstorming and any other problem solving tools that are helpful.

Chapter 12

Do a gauge capability study on the gauge provided by your instructor. Follow the gauge capability procedure in Section 12.2 of the text.

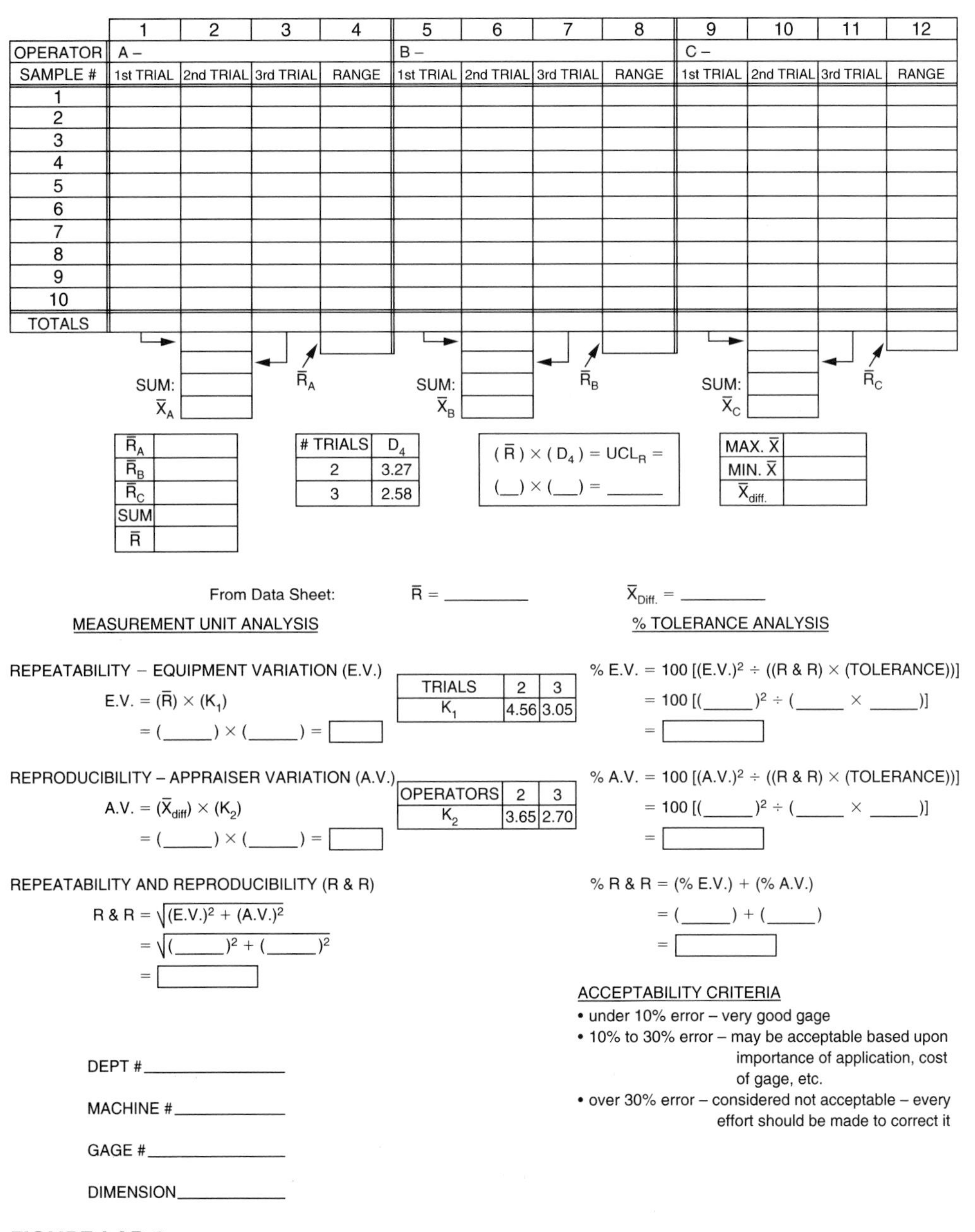

GAGE CAPABILITY

	1	2	3	4	5	6	7	8	9	10	11	12
OPERATOR	A –				B –				C –			
SAMPLE #	1st TRIAL	2nd TRIAL	3rd TRIAL	RANGE	1st TRIAL	2nd TRIAL	3rd TRIAL	RANGE	1st TRIAL	2nd TRIAL	3rd TRIAL	RANGE
1												
2												
3												
4												
5												
6												
7												
8												
9												
10												
TOTALS												

SUM: $\bar{X}_A$ $\bar{R}_A$ SUM: $\bar{X}_B$ $\bar{R}_B$ SUM: $\bar{X}_C$ $\bar{R}_C$

$\bar{R}_A$	
$\bar{R}_B$	
$\bar{R}_C$	
SUM	
$\bar{R}$	

# TRIALS	D_4
2	3.27
3	2.58

$(\bar{R}) \times (D_4) = UCL_R =$

$(\_\_) \times (\_\_) = \_\_\_\_\_\_$

MAX. $\bar{X}$	
MIN. $\bar{X}$	
$\bar{X}_{diff.}$	

From Data Sheet: $\bar{R} =$ ________ $\bar{X}_{Diff.} =$ ________

MEASUREMENT UNIT ANALYSIS

REPEATABILITY – EQUIPMENT VARIATION (E.V.)

E.V. $= (\bar{R}) \times (K_1)$

$= (\_\_\_\_) \times (\_\_\_\_) = \square$

TRIALS	2	3
K_1	4.56	3.05

REPRODUCIBILITY – APPRAISER VARIATION (A.V.)

A.V. $= (\bar{X}_{diff}) \times (K_2)$

$= (\_\_\_\_) \times (\_\_\_\_) = \square$

OPERATORS	2	3
K_2	3.65	2.70

REPEATABILITY AND REPRODUCIBILITY (R & R)

R & R $= \sqrt{(E.V.)^2 + (A.V.)^2}$

$= \sqrt{(\_\_\_\_)^2 + (\_\_\_\_)^2}$

$= \square$

% TOLERANCE ANALYSIS

% E.V. = 100 [(E.V.)^2 ÷ ((R & R) × (TOLERANCE))]

= 100 [(____)^2 ÷ (____ × ____)]

$= \square$

% A.V. = 100 [(A.V.)^2 ÷ ((R & R) × (TOLERANCE))]

= 100 [(____)^2 ÷ (____ × ____)]

$= \square$

% R & R = (% E.V.) + (% A.V.)

= (____) + (____)

$= \square$

ACCEPTABILITY CRITERIA

- under 10% error – very good gage
- 10% to 30% error – may be acceptable based upon importance of application, cost of gage, etc.
- over 30% error – considered not acceptable – every effort should be made to correct it

DEPT # __________

MACHINE # __________

GAGE # __________

DIMENSION __________

FIGURE LAB 6

Chapter 13

1. Construct an OC curve for $n = 250$ and $c = 9$.
   a) Check the following pair of values on the Poisson probability graph in Figure 13.2 in the text.

Pa	pn	100p
.99	4.2	
.95	5.5	
.9	6.3	
.8	7.5	
.6	9.1	
.3	12	
.1	15	

   b) Complete the third column in the table above by calculating

$$100\,p = \frac{(pn) \times 100}{n} = \frac{(pn) \times 100}{250} \text{ for each } pn \text{ table value.}$$

   c) Graph the ordered pairs ($100p$, Pa) and draw the OC curve.
   d) Interpret the OC curve for the following percentage defective:

% defective	Probability of Acceptance (Pa)
1%	
3%	
3.2%	
4%	
5%	

2. Vendor Auditing: the sample of $n = 250$ pieces were measured and 6 defective pieces were found. The histogram for the measurements had a normal shape, their mean was 4.74 inches, and their standard deviation was .0108 inches. The specifications are 4.72 inches to 4.78 inches.
   a) What percentage of the sample is defective?
   b) What percentage of the lot is out of specification? What is the probability that this lot is accepted?
   c) According to the sample plan, what percentage of the sample can be defective for the lot to be accepted?

ANSWERS TO EXERCISES

Chapter 1

1. An operator is responsible for his/her part of the process and is required to make the necessary changes to keep it in statistical control.
3. Statistical control indicates that there is no special cause variation present.
5. Variation that is inherent in a process: it can't be eliminated; only reduced.
7. Special cause variation is variation that can be attributed to a specific source and can be eliminated by "fixing" the source.
9. Decreased labor costs for fixing defective items leads to increased profits.
11. To make knowledgeable decisions about process changes based on SPC interpretation.
13. No. Only appropriate action to the statistical signals provided by SPC can lead to high quality.
15. Inspection methods may not find all the poor quality items.
17. The company is paying someone to make a defective item, pays someone else to try and find it, and then pays someone else to try and fix it. Warranty costs may be involved if the defective item ends up in the hands of the customer.
19. Nobody can be exempt from performing quality work.
21. The workers must know how SPC can be applied to their specific jobs.
23. They may have to provide "coaching" to their workers and they have to present SPC based process changes to management.
25. Worker turnover: new jobs within the company and new workers being hired.
27. A process is capable when the resulting product or service is within specifications or meets customer expectations.
29. A data gathering sheet that categorizes problems or defects.

31. A chart with the entire process diagrammed showing each step from start to finish.
33. As quality problems are found within the process they are eliminated and the process is being improved continuously.

Chapter 2

1. Management problems include
   a) Management techniques based on detection.
   b) Adversarial relationship with employees.
   c) Underestimating the abilities of the workforce.
   d) Not understanding the importance of quality.
   e) Not knowing or applying management techniques that can lead to quality products and services.
3. (1) SPC and other statistical applications. (2) Appropriate management methods.
5. A history of worker exploitation and union reaction. Mutual distrust.
7. Suppliers should be
   a) Actively involved in a quality program such as Deming's 14-point program.
   b) Aggressively using SPC.
   c) Aiming for improved quality and decreased costs.
   d) Working closely with their customers toward a single-source goal.
9. Modern methods of supervision include
   a) The supervisors should thoroughly know all the jobs in their area of responsibility.
   b) Be a quality coach. Help the workers produce quality work by
      1) Providing assistance and suggestions as needed.
      2) Reinforcing procedures and techniques as needed.
      3) Recognizing when additional training is needed.
      4) Reporting any barriers to quality or efficient production to management and continually press for their resolution.
      5) Recognize (with appropriate praise) when good work is being done.
11. SPC leads to high quality by showing where variation can be decreased. High quality leads to high productivity because fewer defects are made (more good products). Production costs are lower because there are fewer defects (workers are not paid to make or fix defective items).
13. Work quotas are used to specify the minimum quantity of work expected, but they become the quantity goal and as such, they actually hinder production. All workers understand variation in the sense that, if left on their own, their output would vary from day to day. But they also know that management, ignoring the variation concept, would expect them to match their best output every day with appropriate penalties when they don't. Consequently workers never exceed a quota because that would cause management to raise it. Management's goal should be to maximize the quality output of each individual through effective supervision. Supervisors should know what their workers are capable of and encourage them to achieve it (within acceptable variation.)

15. The most immediate result of the quality process is the elimination of many inspector positions as the company changes form the detection system to the prevention system. People won't work effectively with the quality process if they or their friends are faced with a job loss.
17. 15% is worker based and 85% is management based.
19. Walter Shewhart in 1924.
21. POC is the price of conformance, usually 3% or 4% of the sales dollar. PONC is the price of nonconformance, usually 20% or more of the sales dollar.
23. When inspection uncovers a problem, it's fixed and the process is improved.
25. Middle management must be included in TQM because decisions made at that level may conflict with the quality process or program. Also, middle management and support areas should have the same quality and work standards as the rest of the company.
27. The quality of the tool must be considered. If the cheaper tool breaks or wears out prematurely and has to be replaced, it is more costly than paying the extra for the better tool.
29. If the customer is completely satisfied, then quality has been achieved and the customer will stay with the company. It is five times as costly to recruit a new customer than to keep a current one.
31. The Baldrige Award was designed to recognize world-class quality management and to provide a framework for the development of a TQM system.
33. Variation is present in every process.
    There are two distinct types of variation: Special cause and Common cause.
    Special cause variation can be attributed to a specific source and can be eliminated.
    Common cause variation is imbedded in the process that management created and can never be eliminated, only reduced by improving the process.
    When variation is reduced, quality is increased.
    Management must be constantly directing efforts to improve processes and decrease variation.
35. Eventually the competition will have a better quality product if the quest for quality is stopped at "good enough".
37. Improve two-way communication at all levels.
    Actively seek suggestions and then provide positive feedback.
    Recognize good work.
    Define roles and responsibilities, especially with empowerment.
    Emphasize training and retraining as needed.
    Have an open-door policy.
39. The only way to improve the process is to decrease the common cause variation and that can only be achieved with a management directed change in the process.
41. One section of a company can't have different quality standards than another. Quality must pervade the entire company and one way to ensure this is to initiate TQM.
43. Top management must provide the leadership, initiate the quality process, and maintain constant pressure to develop the quality process.
45. $\frac{1}{.06} = 16.7$ is the multiplier.
    40 complaints $\times$ 16.7 = 668 customers.
    15% of 668 = 100 customers switched to a competitor.

47. 86% of the workforce is in services.

49. a) $\frac{1}{.03}$

b) $\frac{1}{.03} \times 24 = 800$ dissatisfied customers.

c) 40% of 800 = 320 customers lost.

Chapter 3

1. a) (1) .0001 inch (2) .001 mm
b) (1) .00005 inch (2) .0005 mm
c) (1) 1.5 inch (2) .6 mm
d) (1) 1.50 inch (2) .63 mm
e) (1) 1 inch (2) 1 mm

3. a) 14.3 inch (Add first, then round off)
b) Minimum is 14.2685 inches, maximum is 14.3684 inches. The actual sum should lie between these two extremes.

5. An acceptable range of measurements for that dimension.

7. Every adjustment results in a new distribution with the center of the distribution changed. The result of all the different distributions is an increase in the variability of the measurements as shown in Figure 3.10.

9. a) Target = 3.5
b) Tolerance is 3.2 to 3.8
c) Not acceptable because its not on target, the spread is too wide indicating parts out of specification, and the shape is not normal.

11. Collection, Organization, Analysis, and Interpretation of data.

13. 1) Measures of central tendency
2) Measures of dispersion

15. 3:11 3:44 5:27 6:10 7:06 7:48 7:52 8:50 9:18
9:28 9:51 9:55 (two consecutive samples)

17. a) 40 b) 234 c) −5 d) 24 e) 3 f) 40 − 19 = 21
g) 3 h) 11.5

19. a) −3.2 b) −16.8 c) 14.7 d) −2.5 e) −4
f) −13.7 g) 1.2 h) −12.2 i) 7.84 j) −5.29

21. a) 31.5 b) 6.3 c) 6.4 d) 3.0 e) 1.159
f) 1.296 g) 1.344 h) 1.68
i) Mean, median, range, population standard deviation, sample standard deviation
j) σ^2, s^2

23. $\frac{.87}{2.059} = .423$

25. a) 27.86 b) 1.83 c) 1.86 d) 3.337 e) 3.460

27. There are two modes: 27 and 28; $\tilde{x} = 28$

29. a) 26 to 29.72 b) 24.14 to 31.58 c) 22.28 to 33.44
d) 32 and 24; $\frac{2}{28} \times 100 = 7\%$

31. a) 4.1 4.0 3.7 4.2 4.4
    b) .5 1.2 .4 .6 .5
    c) $\frac{\sum \bar{x}}{k} = \frac{20.4}{5} = 4.08$ $\frac{\sum x}{N} = \frac{81.6}{20} = 4.08$
    d) $\bar{R} = .64$ $s = .311$

33.

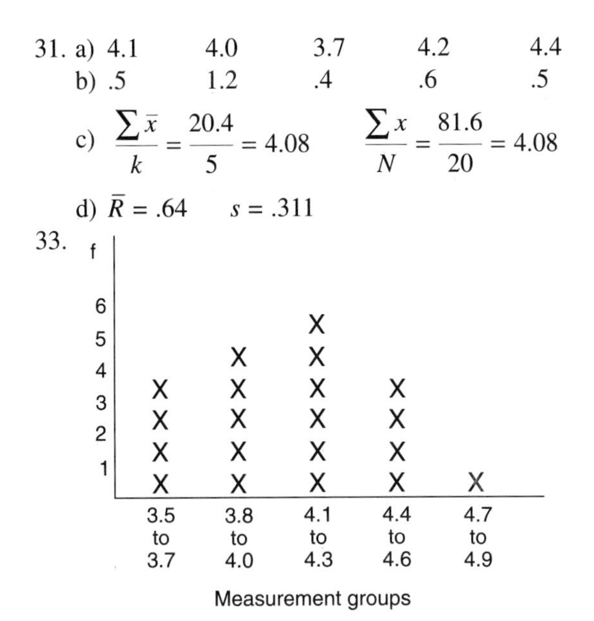

35. a) 3.458 to 4.702
    b) 3.147 to 5.013
    c) 100%
    d) Only 4.8. (1/20) × 100 = 5%
37. There are two modes: 3.9 and 4.2.
39. 4.1 to 4.3
41. They are basically the same shape.
43. A random sample gives the best chance for the sample to fairly represent the population.
45. Some combination of the design engineer, the manufacturer and the customer estimates the amount of variation that can be tolerated without sacrificing too much quality.

Chapter 4

1. a) 11 b) 9 c) 14
3. Use a group range ≥.0083. Using a group range of .0085, the groups are
   5.7850 to 5.7935
   5.7935 to 5.8030
   5.8020 to 5.8105
   5.8105 to 5.8190
   5.8190 to 5.8275
   5.8275 to 5.8360
   5.8360 to 5.8445
5. 1. Set the alarm.
   2. Get up when the alarm goes off.
   3. Stay up.

4. Bathroom routine.
5. Get dressed.
6. Inspect.
7. Eat breakfast.
8. Clean up.
9. Organize and pack materials.
10. Go to school or work.

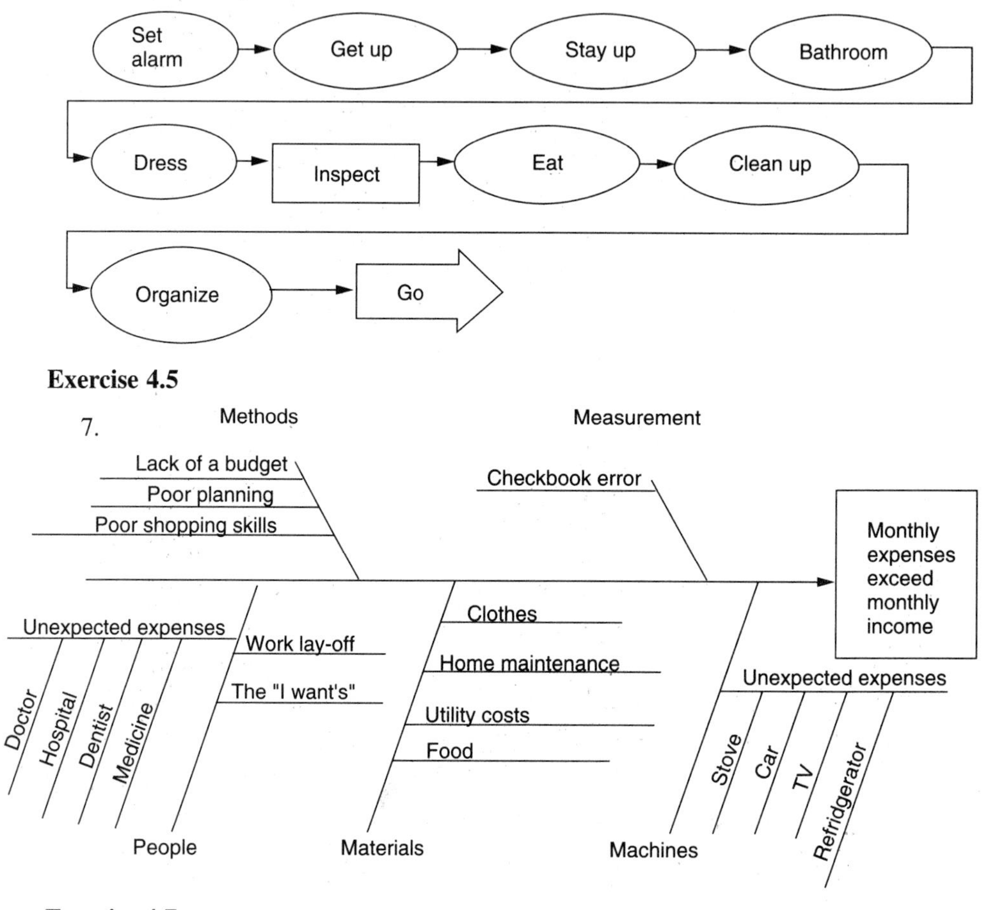

Exercise 4.5

7.

Exercise 4.7

9. Line	Defectives	%	Cum.%
5	920	36.8	36.8
4	855	34.2	71.0
2	350	14	85.0
6	170	6.8	91.8
1	100	4	95.8
7	60	2.4	98.2
3	45	1.8	100
	2500		

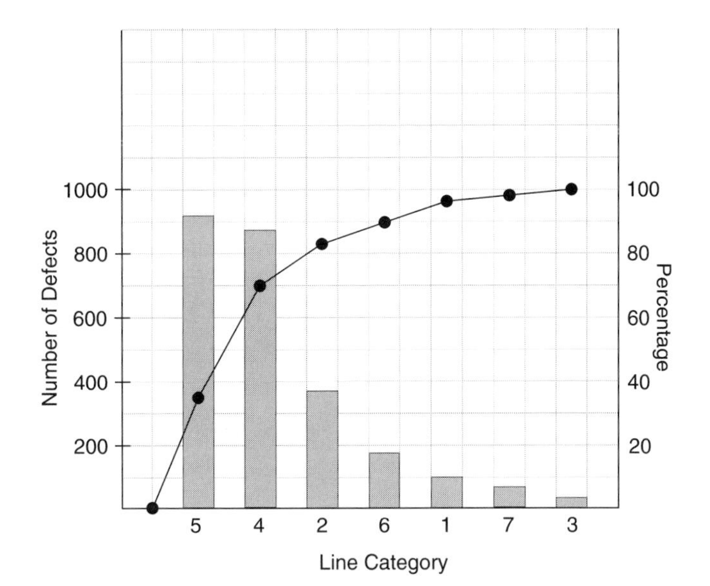

Exercise 4.9

11. a)

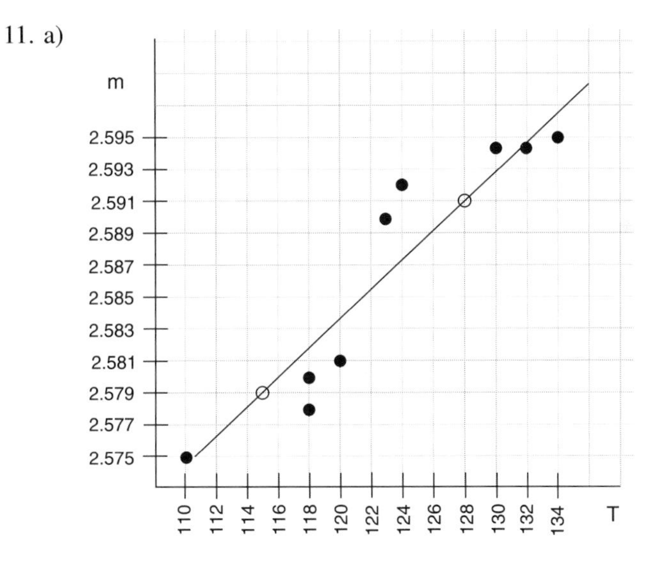

Exercise 4.11

b) Two points on the line are (128, 2.591) (115, 2.579)
c) 2.579
d) $r = .94$ good correlation

13. The G chart recommends 9 or 10 groups. Since there are only 13 different data values, the logical choice would be either 13 individual groups or 7 groups with 2 measurements per group. The 7 groups matches the split stemplot in Exercise 12.

Data Value	Frequency
.283	1
.284	3
.285	4
.286	7
.287	4
.288	8
.289	11
.290	12
.291	11
.292	6
.293	6
.294	4
.295	3

15. a)

```
1.7   3 4 4 3 4 3
1.7   6 7 9 9 8
1.8   1 4 1 2 1 1 3 4 1 1 3
1.8   7 6 7 5 5 9 7 7 6 6 5 7 6 7
1.9   3 4 0 0
1.9   8 5
```

b) The position is good, 1.85 is in the middle. The spread is too wide, several measurements are smaller than the Lower Specification Limit, LSL = 1.75. There is one measurement larger than the USL = 1.95. The shape is not normal, perhaps even bimodal.

c) The position is off; the center is larger than the target value. The spread is good. Even though the distribution crowds the USL = 2.00, the spread on both ends would be well within specifications if the position were adjusted. The shape is still not normal.

17. For 72 data values the G chart recommends 9 groups, but with 12 different data values, either 12 individual groups or the 7 groups shown in the split stemplot of Exercise 16 would be best.

Measurement	Frequency
.55	1
.56	2
.57	3
.58	11
.59	8
.60	9
.61	8
.62	9
.63	4
.64	7
.65	9
.66	1

Measurement Group	Frequency
.54 to .55	1
.56 to .57	5
.58 to .59	19
.60 to .61	17
.62 to .63	13
.64 to .65	16
.66 to .67	1

19. Inspector___Smith Date__9/9/99 Time__3:05 PM

    Item ___TI-3X SOLAR Calculator

    Cover: Appearance__________ Fit_______

    Reference Card__________

    Print: Legible__________ Color coded__________

    Keys: Size__________ Obvious abbreviations__________

    Case: Finish__________ Warped__________

    Printout: Size__________ Clarity__________

    Solar cell: Amount of light needed____________________

    Functions tested: __________ __________ __________

    __________ __________ __________

Chapter 5

1. 24/30 or 4/5 or .8 or 80%
3. 20/280 or 1/14 or .071 or 7.1%
5. a) B9: Center area = .4854, Tail area = .5 − .4854 = .0146
      B8: Tail area = .0146
   b) B9: Center area = .3212, Tail area = .5 − .3212 = .1788
      B8: Tail area = .1788
   c) B9: Center area = .5000, Tail area = .5 − .5000 = .0000
      B8: Tail area = .00000 (one more decimal place in accuracy)
7. a) $z = -1.96$, Left tail area = .0250
   b) $z = .76$, Left area = .7764
   c) $z = -1.88$, Left tail area = .0301

9. a) $z = -1.35$, Left tail area = .0885
   b) $z = 2.03$, Right tail area = .0212
   c) $z_1 = -2.03$, $z_2 = 2.03$, $2 \times$ Center area = 2(.4788) = .9576
   d) $z_1 = 1.35$, $z_2 = 2.03$, Center A_1 − Center A_2 = .4788 − .4115 = .0673
11. $z_1 = 4.8$, Tail A_1 = .00000, $z_2 = -3.2$, Tail A_2 = .00069
    $\text{Area}_1 + \text{Area}_2$ = .00069 or .069%
13. a) $\bar{\bar{x}} = 1.450$
    b) $s_{\bar{x}} = \dfrac{\sigma}{\sqrt{n}} = \dfrac{.0175}{\sqrt{5}} = .00783$
    c) 1.427 and 1.473
15. a) Between .691 and .892 ($\mu \pm 3\sigma$)
    b) Between .7255 and .7945 ($\mu \pm 3s_{\bar{x}}$)
17. $s_{\bar{x}} = \dfrac{s}{\sqrt{n}} = \dfrac{.025}{\sqrt{5}} = .01118$ UCL = 3.50 + 3(.01118) = 3.53
19.

Group	Probability
1.84 to 1.90	.06
1.90 to 1.96	.16
1.96 to 2.02	.24
2.02 to 2.08	.28
2.08 to 2.14	.12
2.14 to 2.20	.10
2.20 to 2.26	.04

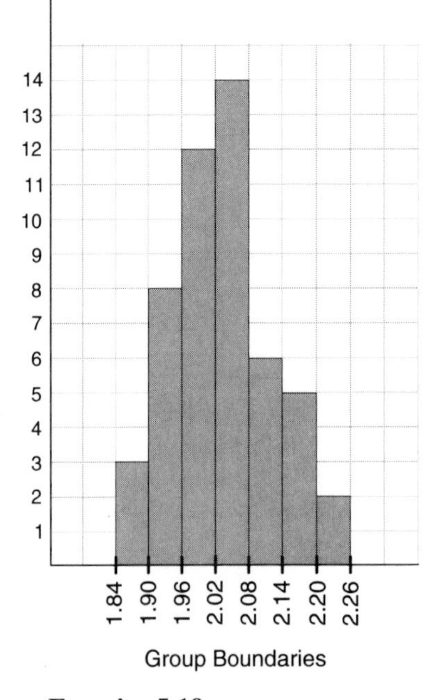

Exercise 5.19a

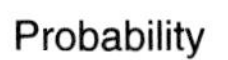
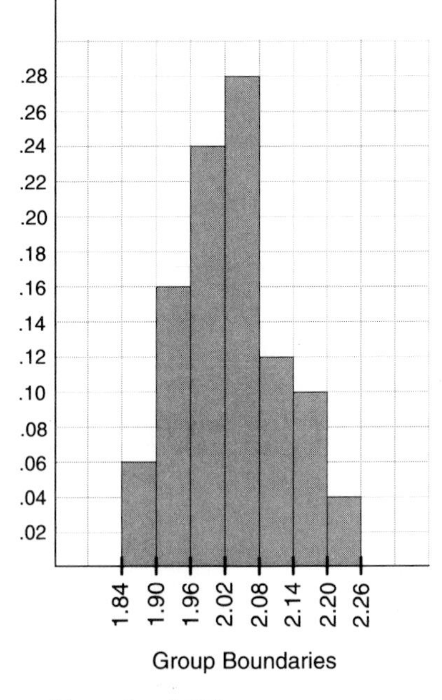

Exercise 5.19b

21. a) 2.302 and 2.398 $\mu \pm 3\sigma$
    b) (i) 2.350

    (ii) $\dfrac{\sigma}{\sqrt{n}} = \dfrac{.016}{\sqrt{5}} = .0072$

    (iii) 2.328 and 2.372 $\left(\bar{\bar{x}} \pm \dfrac{3\sigma}{\sqrt{n}}\right)$

    c) $z_1 = -1.88$, $z_2 = 1.88$, Tail area_1 = Tail area_2 = .0301
    6.02% out of specification.
23. a) The distribution of sample means, $\bar{x}$, from any large population will approach the normal distribution as the sample size, n, increases. $\bar{\bar{x}} = \mu$ and $s_{\bar{x}} = \dfrac{\sigma}{\sqrt{n}}$.
    b)

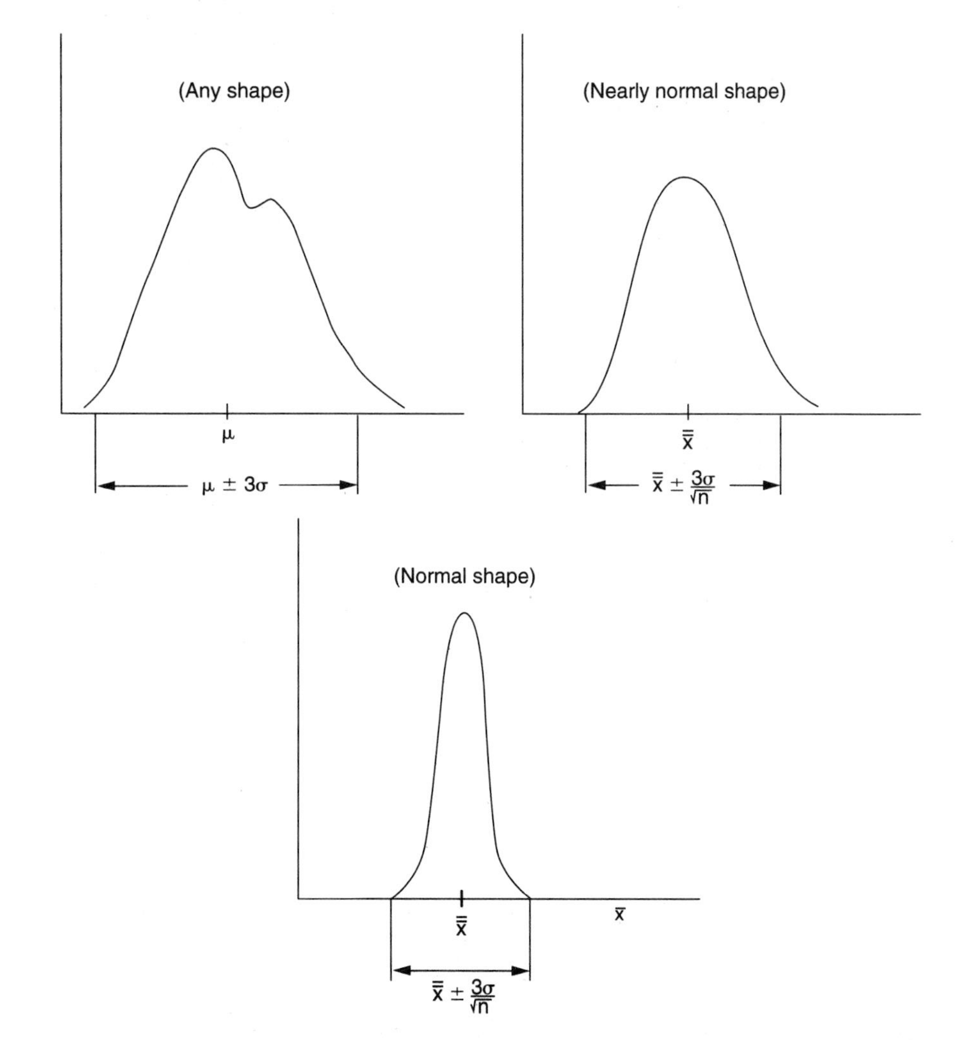

25. a) 2.250
    b) $s_{\bar{x}} = \frac{\sigma}{\sqrt{n}} = .0188$
    c) z = ±2, 95.46%
    d) z = ±3, 99.73%
    e) $\pm 2 = \frac{x - 2.25}{.042}$, x = 2.116 and x = 2.334
    f) 2.1936 to 2.3064

Chapter 6

1. Special cause variation and common cause variation.
3. 1) Select a measurement to chart.
   2) Eliminate obvious sources of variation.
   3) Check the gauges.
   4) Make a sample plan.
   5) Set up the chart and the process log.
   6) Set up the tally histogram.
   7) Collect the data and plot the points.
   8) If the chart is a continuation chart, analyze as the chart develops. If it is a new chart, calculate the averages and control limits when the chart is completed and analyze. Take action on any special cause variation.
   9) Calculate the capability.
   10) Monitor the process when it is capable.
   11) Continue to look for ways to improve the process and decrease common cause variation.
5. a) $s = .00466$
   b) $UCL_R = .022$ $LCL_R = 0$
   c) $UCL_{\bar{x}} = 4.732$ $LCL_{\bar{x}} = 4.718$
7. Special cause variation.
9. The process log may contain information that will lead to the cause of an out-of-control situation.
11. It allows the on-going analysis of the process to begin sooner. Even though there are no control limit lines, shifts and runs can be spotted relative to the target line.
13. Any analysis of the $\bar{x}$ chart is meaningless if the R chart is out of control. If there is a problem with variation, the average sample value is of little importance. Any adjustment toward the target value is a waste of time when variation is out of control.
15. There are no out-of-control points. $s = .0006615$, $3s = .001985$, $6s = .00397$
    a) $C_{pk} = .45$ 222% of the tolerance.
    b) PCR = 1.985 198.5% of the tolerance.
    c) $z_1 = -1.36$, $Area_1 = .0869$, $z_2 = 1.66$, $Area_2 = .0485$
       13.54% out of specification.
17. $\bar{\bar{x}} = .30084$, $\bar{R} = .01868$, $UCL_{\bar{x}} = .314$, $LCL_{\bar{x}} = .287$, $UCL_R = .043$

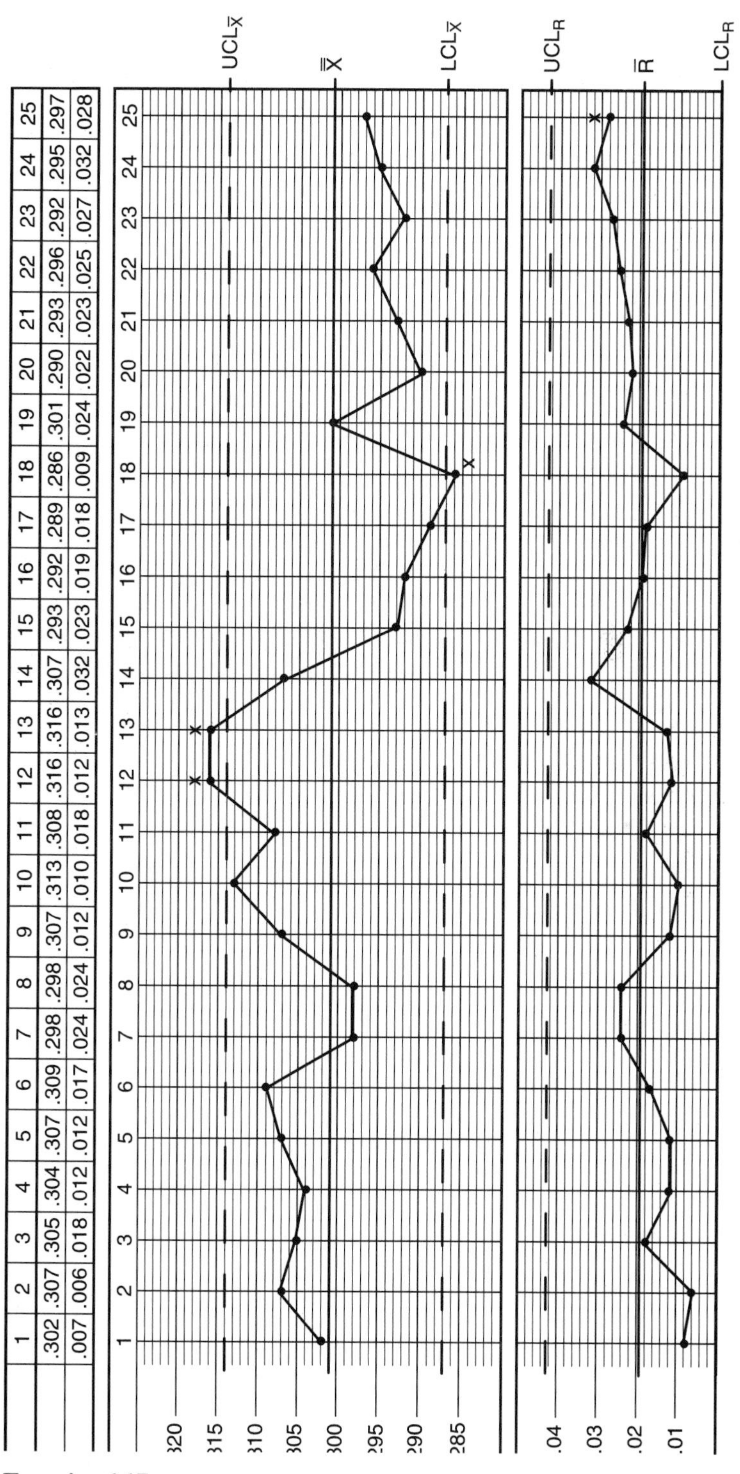

Sample	1	2	3	4	5	6	7	8	9	10	11	12	13	14	15	16	17	18	19	20	21	22	23	24	25
$\bar{X}$	.302	.307	.305	.304	.307	.309	.298	.298	.307	.313	.308	.316	.316	.307	.293	.292	.289	.286	.301	.290	.293	.296	.292	.295	.297
R	.007	.006	.018	.012	.012	.017	.024	.024	.012	.010	.018	.012	.013	.032	.023	.019	.018	.009	.024	.022	.023	.025	.027	.032	.028

Exercise 6.17

19. Eliminate sample numbers
    12, 13, and 18 (Beyond the control limits on the $\bar{x}$ chart)
    25 (Shift on the R chart)

    New $\bar{R} = .01833$ $s = .008902$ New $\bar{\bar{x}} = .30029$

    PCR = 1.07 or 107% of the tolerance is used. $C_p = .93$ $C_{pk} = .93$

    $z_1 = -2.84$ Tail $A_1 = .0023$ $z_2 = 2.78$ Tail $A_2 = .0027$

    .005 or .5% is out of specification.

21.

Sample Number	Comment
4	Is this a shift? This is the 5th consecutive point below $\bar{\bar{x}}$.
6	The official shift begins here (count sample #25 from Figure 6.8). Investigate for cause. The sample measurements and the averages are reasonably close to the target so 100% inspection shouldn't be necessary.
9	A sample average below the LCL. Investigate. 100% inspection until the problem is identified. Also, 100% inspection on all pieces made since the last sample.
11	A shift is noted on the Range chart. The shift is below $\bar{R}$, perhaps variation has decreased.
14	We're on target. There seems to be an increasing trend.
20	The trend on the $\bar{x}$ chart continues, investigate. The range values are staying low, I'll have to recalculate the control limits.

23. *p*, *np*, *c*, *u*, *q*, *nq* charts.

25. a) Shift (first seven), also a downward trend.
    b) Shift from #6 to the last value.
    c) A point beyond the control limits at #4, 8, and 9.

27. a)

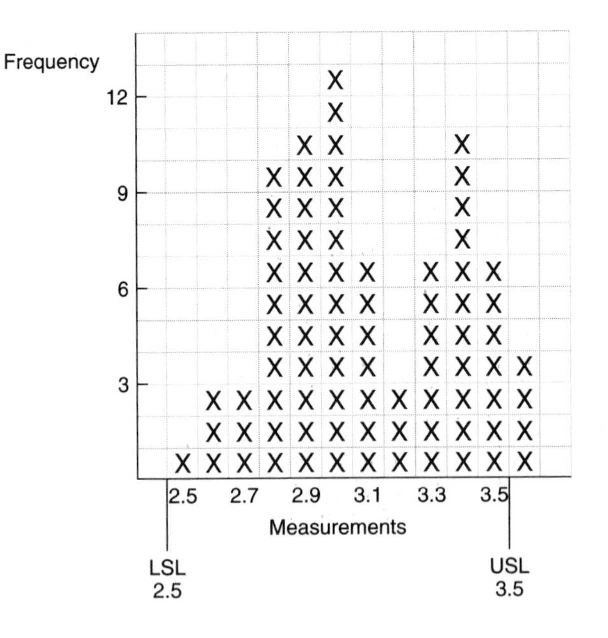

d) $\bar{\bar{x}} = 3.1$ $\bar{R} = .31$ $UCL_R = .707$
$UCL_{\bar{x}} = 3.33$ $LCL_{\bar{x}} = 2.87$

e) Samples 3–6 are beyond the UCL.
8–10 are part of a shift pattern.
17–20 are below the LCL.
New $\bar{\bar{x}} = 3.04$ New $\bar{R} = .278$
New $UCL_R = .63$ New $UCL_{\bar{x}} = 3.27$ New $LCL_{\bar{x}} = 2.84$
Eliminate one more point, #7 is beyond the new control limit.

f) Using the remaining 8 points, $\bar{R} = .2625$, $\bar{\bar{x}} = 3.0125$, $s = .1275$
PCR = .765, $C_p = 1.31$, $C_{pk} = 1.27$ The capability of the remaining points is fairly good, but too many out-of-control points occurred on the chart to be able to give much credibility to the capability calculation.

g) For the in-control data, $z_1 = -4.02$, Tail $A_1 = .00000$
$z_2 = 3.82$, Tail $A_2 = .00007$ .007% out-of-specification.

29. 1) The relative size of the index numbers PCR, C_p, and C_{pk}.
2) The percentage of the tolerance used can be found from the three index numbers.
3) The standard deviation of the data distribution, s, can be determined from $\bar{R}$, and then $\bar{\bar{x}} \pm 3s$.
4) The percentage of population data that is out-of-specification can be calculated.

Exercise 6.27b

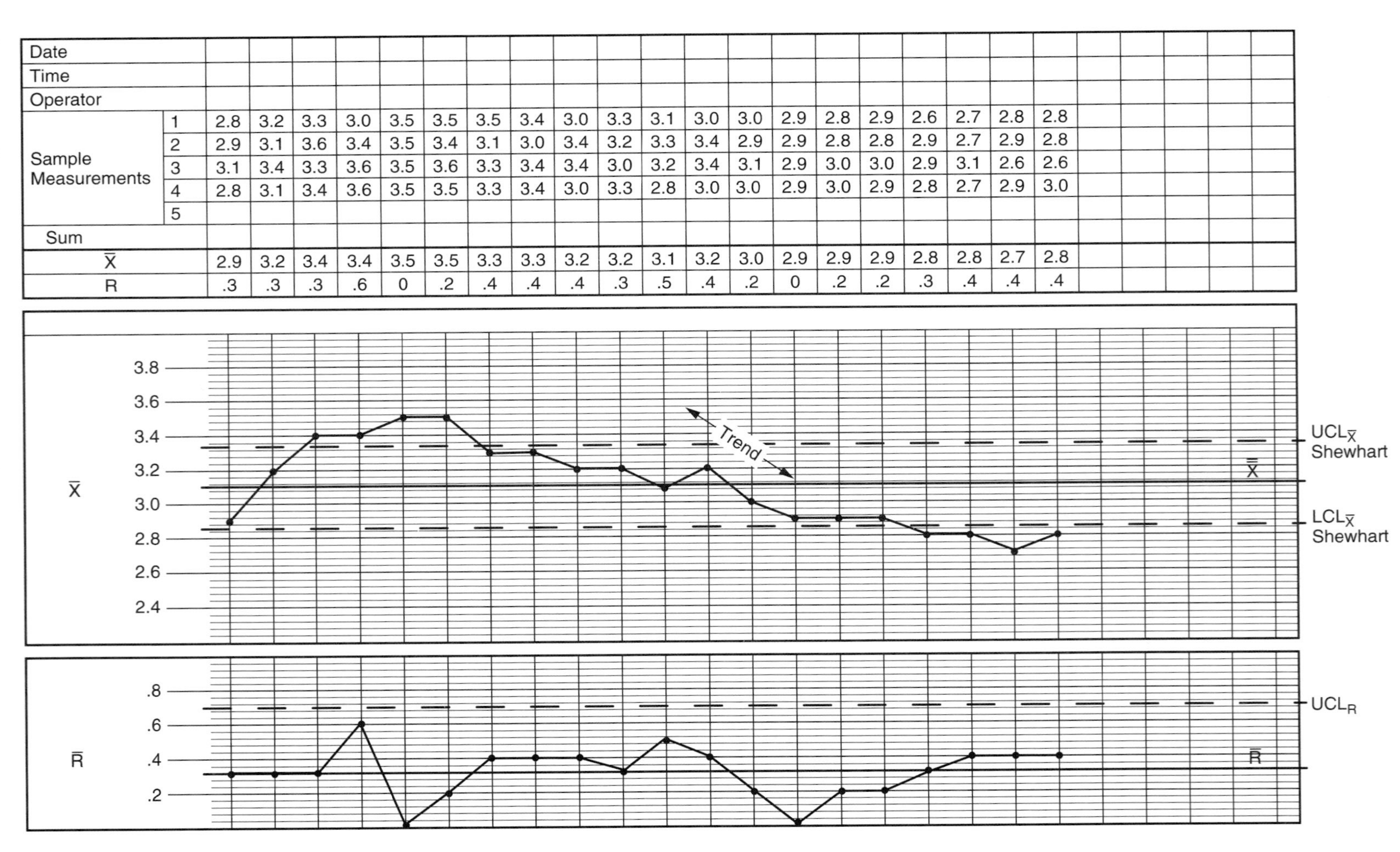

Date																					
Time																					
Operator																					
Sample Measurements	1	2.8	3.2	3.3	3.0	3.5	3.5	3.5	3.4	3.0	3.3	3.1	3.0	3.0	2.9	2.8	2.9	2.6	2.7	2.8	2.8
	2	2.9	3.1	3.6	3.4	3.5	3.4	3.1	3.0	3.4	3.2	3.3	3.4	2.9	2.9	2.8	2.8	2.9	2.7	2.9	2.8
	3	3.1	3.4	3.3	3.6	3.5	3.6	3.3	3.4	3.4	3.0	3.2	3.4	3.1	2.9	3.0	3.0	2.9	3.1	2.6	2.6
	4	2.8	3.1	3.4	3.6	3.5	3.5	3.3	3.4	3.0	3.3	2.8	3.0	3.0	2.9	3.0	2.9	2.8	2.7	2.9	3.0
	5																				
Sum																					
$\bar{X}$		2.9	3.2	3.4	3.4	3.5	3.5	3.3	3.3	3.2	3.2	3.1	3.2	3.0	2.9	2.9	2.9	2.8	2.8	2.7	2.8
R		.3	.3	.3	.6	0	.2	.4	.4	.4	.3	.5	.4	.2	0	.2	.2	.3	.4	.4	.4

Chapter 7

1. a) From the specifications:

$$\text{Range scale} \approx \frac{Tolerance}{\#\,spaces} = \frac{.03}{10} = .003, \text{ use } .005 \text{ units per line.}$$

$$\tilde{x} \text{ scale} \approx \frac{3 \times Tolerance}{n \times \#\,spaces} = \frac{3 \times .03}{5 \times 25} = .00073, \text{ use } .001 \text{ units per line.}$$

From the samples:

$$\text{Range scale} \approx \frac{2 \times R_R}{\#\,lines} = \frac{2 \times .019}{10} = .0038, \text{ use } .005 \text{ units per line.}$$

$$\tilde{x} \text{ scale} \approx \frac{2 \times R_{\tilde{x}}}{\#\,lines} = \frac{2 \times .015}{25} = .0012, \text{ use } .001 \text{ units per line.}$$

b) The $\bar{x}$ chart setup is the same as the $\tilde{x}$ chart in part (a).

$$s \text{ chart} \approx \frac{2 \times R_s}{\#\,lines} = \frac{2 \times .0078}{10} = .00156, \text{ use } .002 \text{ units per line.}$$

$\bar{x}$ *or* $\tilde{x}$ chart	2.757 ______
start at the	2.756 ______
Midspecification	2.755 ______
in the	2.754 ______
middle	2.753 ______

R chart	.015 ______	s chart	.006 ______
start at	.010 ______	start at	.004 ______
0 at the	.005 ______	0 at the	.002 ______
bottom	0 ______	bottom	0 ______

3. The R chart has virtually the same pattern as the s chart. The analysis is the same on both charts, both the R and the s chart are in-control $\hat{\sigma} = \frac{\bar{R}}{d_2} = \frac{14.8}{2.326} = .000636.$

The $\bar{R}$ was calculated from the same 20 samples as $\bar{s}$ (after the out-of-control data was eliminated). PCR = .64 which compares with the previous value of .65.

$\bar{R} = \frac{399}{25} = \underline{15.88}$

$\bar{\bar{X}} = \frac{\Sigma \bar{x}}{k} = \frac{605.4}{25} = \underline{24.2}$

$UCL_R = D_4 \bar{R} = 2.114\ (15.88) = 33.6$

$UCL_{\bar{X}} = \bar{\bar{X}} + A_2\bar{R} = 24.2 + .577\ (15.88) = 33.4$

$LCL_{\bar{X}} = \bar{\bar{X}} - A_2\bar{R} = 24.2 - 9.2 = 15.0$

Specification : 1.3280 to 1.3340 (−20 to 40 coded)

Sample Numbers	1	2	3	4	5	6	7	8	9	10	11	12	13	14	15	16	17	18	19	20	21	22	23	24	25
Coded 0 = 1.3300	29	29	28	28	22	33	33	33	27	28	38	36	25	26	20	16	29	17	21	29	24	27	20	14	08
	30	19	27	27	26	36	27	20	21	21	29	27	25	35	31	19	29	04	28	16	10	25	21	20	30
	28	25	14	14	08	32	20	21	31	17	25	24	11	20	19	23	05	27	16	19	30	21	18	29	16
	37	21	18	18	27	27	33	27	30	34	26	26	27	25	30	16	19	26	17	33	16	27	26	27	09
	29	35	34	34	24	28	21	33	27	25	27	21	36	24	35	25	21	16	21	15	21	30	31	22	00
Average $\bar{X}$	30.6	25.8	24.2	24.2	21.4	31.2	26.8	26.8	27.2	25.0	29.0	26.8	24.8	26.0	27.0	19.8	20.6	18.0	20.6	22.4	20.2	26.0	23.2	25.2	12.6
Range R	9	16	20	20	19	9	13	13	10	17	13	15	25	15	16	9	24	23	12	18	20	9	13	9	30

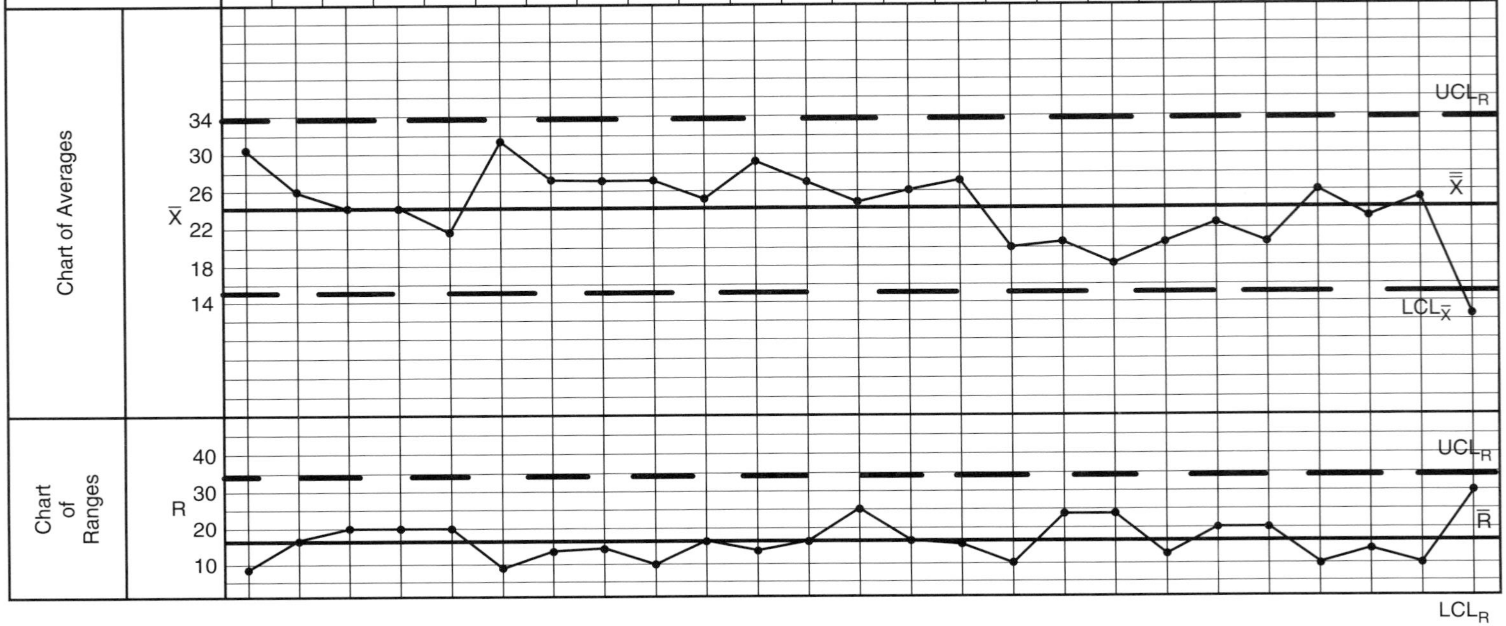

Exercise 7.3

5. $\bar{x}$ 71.41 68.81 70.55 $\bar{\bar{x}} = 70.26$ $D_{4f} = 1.9$ $D_{4s} = 3.2$
R 2.16 5.86 4.26 $\bar{R} = 4.093$ $A_{2f} = .8$ $A_{2s} = 1.1$
Initial control limits: $UCL_R = D_{4f} \times \bar{R} = 7.78$ No R values to eliminate.
$UCL_{\bar{x}}$ 73.53 $LCL_{\bar{x}} = 66.99$ No data points to eliminate.
Final control limits: $UCL_R = 13.098$, $UCL_{\bar{x}} = 74.76$, $LCL_{\bar{x}} = 65.76$

7. $\bar{\bar{x}} = \dfrac{\sum coded\bar{x}}{N} = \dfrac{-.0025}{21} = -.0001$ $\bar{R} = .0015$

$UCL_R = .0039$, $UCL_{\bar{x}} = .0014$ $LCL_{\bar{x}} = -.0016$

9. The two charts have the same pattern. Both are in-control.

$\bar{s} = 14.9$ $UCL_s = 31$ $\hat{\sigma} = \dfrac{14.9}{.94} = 15.9$

For R, $\hat{\sigma} = \dfrac{37.04}{2.326} = 15.9$ Both charts will have the same capability.

11. a) $\bar{\bar{x}} = 299.75$ $\bar{R} = .0189$
$UCL_R = .04$, $UCL_{\bar{x}} = .3128$ $LCL_{\bar{x}} = .2867$
Eliminate out-of-control data assuming the problem has been fixed.
New $\bar{\bar{x}} = .2987$, New $\bar{R} = .0164$ (Eliminated #'s 10, 11, 18–20)
$\hat{\sigma} = .007051$ PCR = .846 $C_p = 1.18$ $C_{pk} = 1.12$
$z_1 = -3.36$, Tail $A_1 = .00039$, $z_2 = 3.73$, Tail $A_2 = .00010$,
Total out-of-specification = .00049 or .049%

b. $\bar{\bar{x}} = .30084$, $\bar{s} = .0084$, $UCL_{\bar{x}} = .3145$, $LCL_{\bar{x}} = .2872$, $UCL_s = .019$

New $\bar{\bar{x}} = .30029$, New $\bar{s} = .0085$, (Eliminate #'s 12, 13, 18, 25)
$\hat{\sigma} = .00923$, PCR = 1.02, $C_p = .98$, $C_{pk} = .77$
$z_1 = -2.98$, Tail $A_1 = .0014$, $z_2 = 2.91$, Tail $A_2 = .0018$
Total out-of-specification = .0032 or .32%
When compared to the $\bar{x}$ and R chart in Exercises 17 and 19 in Chapter 6, the same four samples were out-of-control and the capability analyses were in close agreement.

Specification:

Hole diameter $.770 \pm {}^{+.002}_{-.005}$

		1	2	3	4	5	6	7
Coded measurements		−.0005	0	.0010	.0015	.0015	−.0005	−.0005
		−.0005	0	.0015	.0015	.0015	−.0005	.0015
		−.0010	.0020	.0015	−.0030	.0005	−.0005	.0015
Coded	$\bar{X}$	−.0007	.0007	.0013	0	.0012	−.0003	.0008
	R	.0005	.0020	.0005	.0045	.0010	0	.0020

Body diameter $.871 \pm .002$

		1	2	3	4	5	6	7
Coded measurements		−.0055	−.0005	0	.0005	.0010	.0010	.0010
		0	0	.0005	.0005	.0005	.0005	0
		−.0004	.0010	.0020	.0005	.0010	.0005	.0010
Coded	$\bar{X}$	−.0020	.0002	.0008	.0005	.0008	.0007	.0007
	R	.0055	.0015	.0020	0	.0005	.0005	.0010

Hole depth $.900 \pm .005$

		1	2	3	4	5	6	7
Coded measurements		−.0015	−.0010	−.0015	−.0010	−.0015	−.0020	−.0010
		−.0020	−.0010	.0010	0	.0015	0	−.0010
		−.0020	−.0025	−.0020	0	−.0025	−.0020	−.0040
Coded	$\bar{X}$	−.0018	−.0015	−.0015	−.0003	.0018	−.0013	−.0080
	R	.0005	.0015	.0010	.0010	.0010	.0020	.0030

Chart of Averages

$\bar{X}$: .0016, .0012, .0008, .0004, 0, −.0004, −.0008, −.0012, −.0016, −.0020

$UCL_{\bar{X}}$, $\bar{\bar{X}}$, $LCL_{\bar{X}}$

Chart of Ranges

R: .005, .004, .003, .002, .001, 0

UCL_R, $\bar{R}$

Exercise 7.7

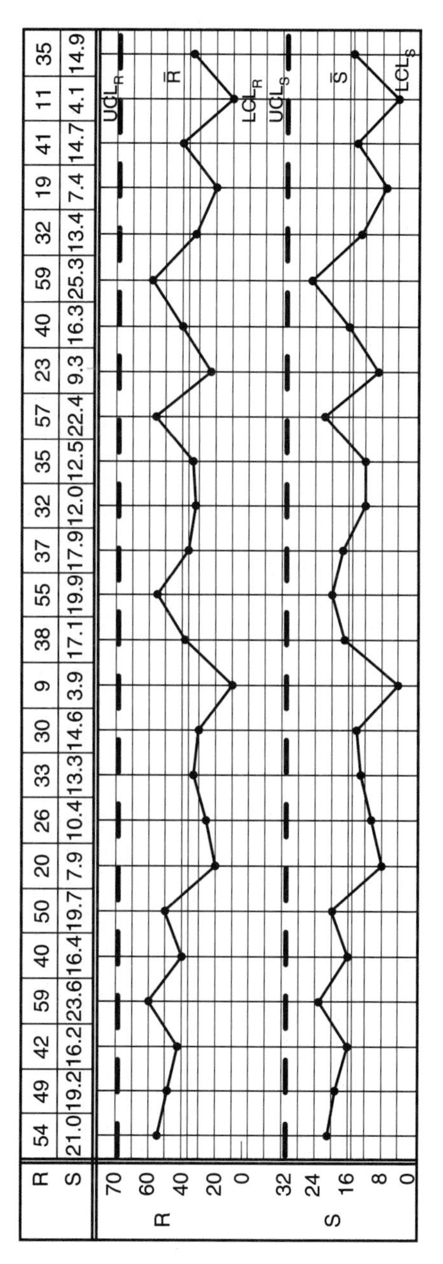
R
S
54 49 42 59 40 50 20 26 33 30 9 38 55 37 32 35 57 23 40 59 32 19 41 11 35
21.0 19.2 16.2 23.6 16.4 19.7 7.9 10.4 13.3 14.6 3.9 17.1 19.9 17.9 12.0 12.5 22.4 9.3 16.3 25.3 13.4 7.4 14.7 4.1 14.9
R
70 60 40 20 0
S
32 24 16 8 0
UCL_R
$\bar{R}$
LCL_R
UCL_S
$\bar{S}$
LCL_S

Exercise 7.9

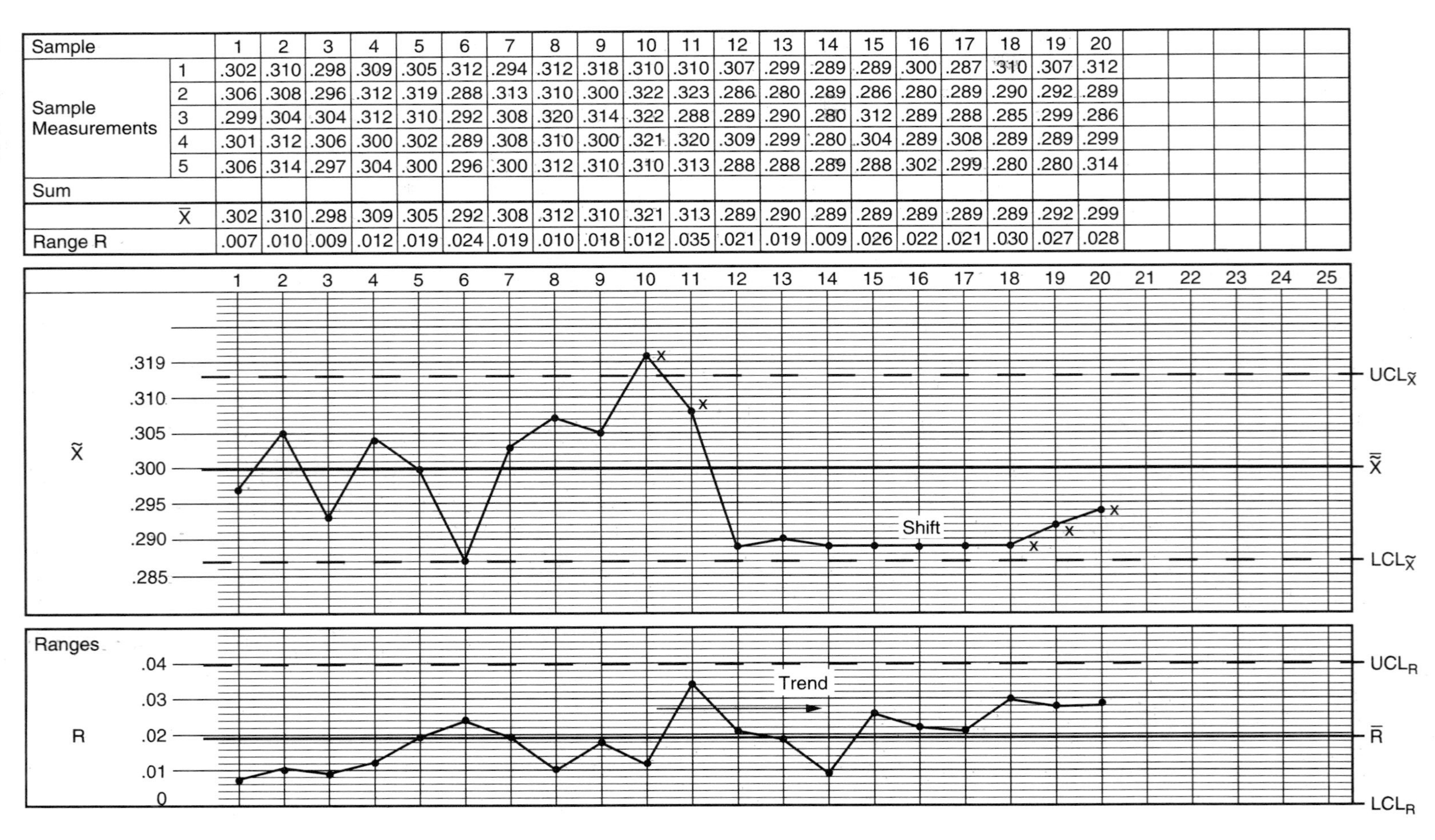

Sample		1	2	3	4	5	6	7	8	9	10	11	12	13	14	15	16	17	18	19	20
Sample Measurements	1	.302	.310	.298	.309	.305	.312	.294	.312	.318	.310	.310	.307	.299	.289	.289	.300	.287	.310	.307	.312
	2	.306	.308	.296	.312	.319	.288	.313	.310	.300	.322	.323	.286	.280	.289	.286	.280	.289	.290	.292	.289
	3	.299	.304	.304	.312	.310	.292	.308	.320	.314	.322	.288	.289	.290	.280	.312	.289	.288	.285	.299	.286
	4	.301	.312	.306	.300	.302	.289	.308	.310	.300	.321	.320	.309	.299	.280	.304	.289	.308	.289	.289	.299
	5	.306	.314	.297	.304	.300	.296	.300	.312	.310	.310	.313	.288	.288	.289	.288	.302	.299	.280	.280	.314
Sum																					
	$\overline{X}$	.302	.310	.298	.309	.305	.292	.308	.312	.310	.321	.313	.289	.290	.289	.289	.289	.289	.289	.292	.299
Range R		.007	.010	.009	.012	.019	.024	.019	.010	.018	.012	.035	.021	.019	.009	.026	.022	.021	.030	.027	.028

Exercise 7.11a

Exercise 7.11b

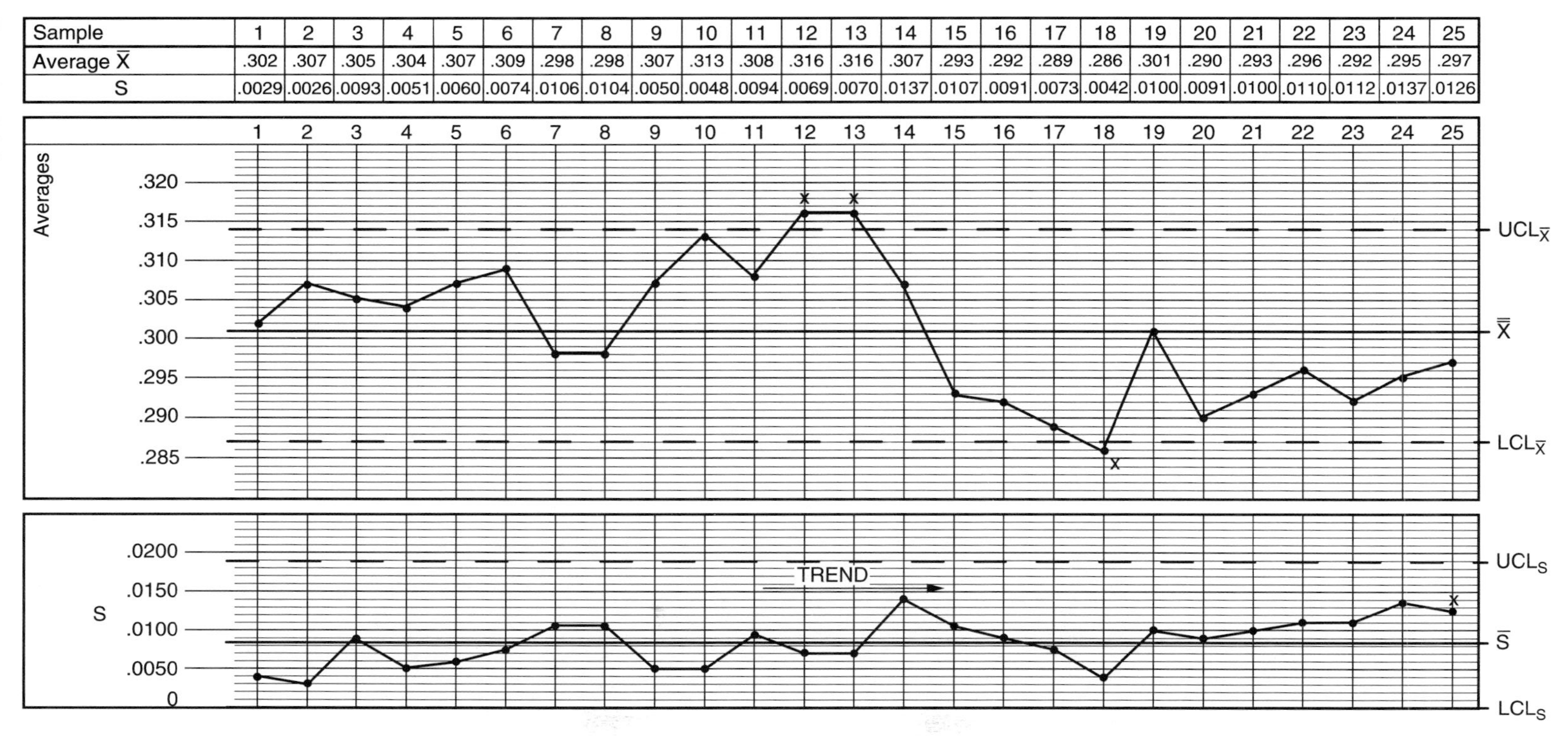

Sample	1	2	3	4	5	6	7	8	9	10	11	12	13
Average $\overline{X}$	.302	.307	.305	.304	.307	.309	.298	.298	.307	.313	.308	.316	.316
S	.0029	.0026	.0093	.0051	.0060	.0074	.0106	.0104	.0050	.0048	.0094	.0069	.0070

Sample	14	15	16	17	18	19	20	21	22	23	24	25
Average $\overline{X}$	.307	.293	.292	.289	.286	.301	.290	.293	.296	.292	.295	.297
S	.0137	.0107	.0091	.0073	.0042	.0100	.0091	.0100	.0110	.0112	.0137	.0126

13. a) Coded values

Coded values	−1	−3	1	2	−2
	3	0	0	−1	0
	−3	−1	0	1	0
	0	−1	1	−1	−3
Range	6	3	1	3	3

b) coded $\bar{x} = \frac{-8}{20} -.4$ Actual $\bar{x} = .315 - .4 \times .001 = .3146$

c) $\bar{x} = \frac{6.292}{20} -.3146$

d) Coded $\bar{R} = \frac{16}{5} = 3.2$ Actual $\bar{R} = 3.2 \times .001 = .0032$

Actual range: .006 .003 .001 .003 .003

$\bar{R} = \frac{.016}{5} = .0032$

e) $s = \hat{\sigma} = \frac{\bar{R}}{d_2} = \frac{.0032}{2.059} = .00155$

f) Coded values

Coded values	4	2	6	7	5
	8	5	5	4	5
	2	4	5	6	5
	5	4	6	4	2
Range	6	3	1	3	3

Coded $\bar{x} = \frac{92}{20} = 4.6$ Actual $\bar{x} = .310 + 4.6 \times .001 = .3146$

Coded $= \bar{R} = \frac{16}{5} = 3.2$ Actual $\bar{R} = 3.2 \times .001 = .0032$

$s = \hat{\sigma} = \frac{.0032}{2.059} = .00155$

15.

c) (1) The $\bar{x}$ and s chart contains no out-of-control points, so no adjustment is needed. $\hat{\sigma} = \frac{.00059}{.8862} = .0006658$

PCR = 1.985 or 198.5% of the tolerance. $C_p = .50$

$C_{pk} = .45$ $\frac{1}{C_{pk}} = 2.22$ or 222% of the tolerance

(2) $z_1 = -1.35$ Tail $A_1 = .0885$
$z_2 = 1.65$ Tail $A_2 = .0495$ 13.8% out-of-specification.

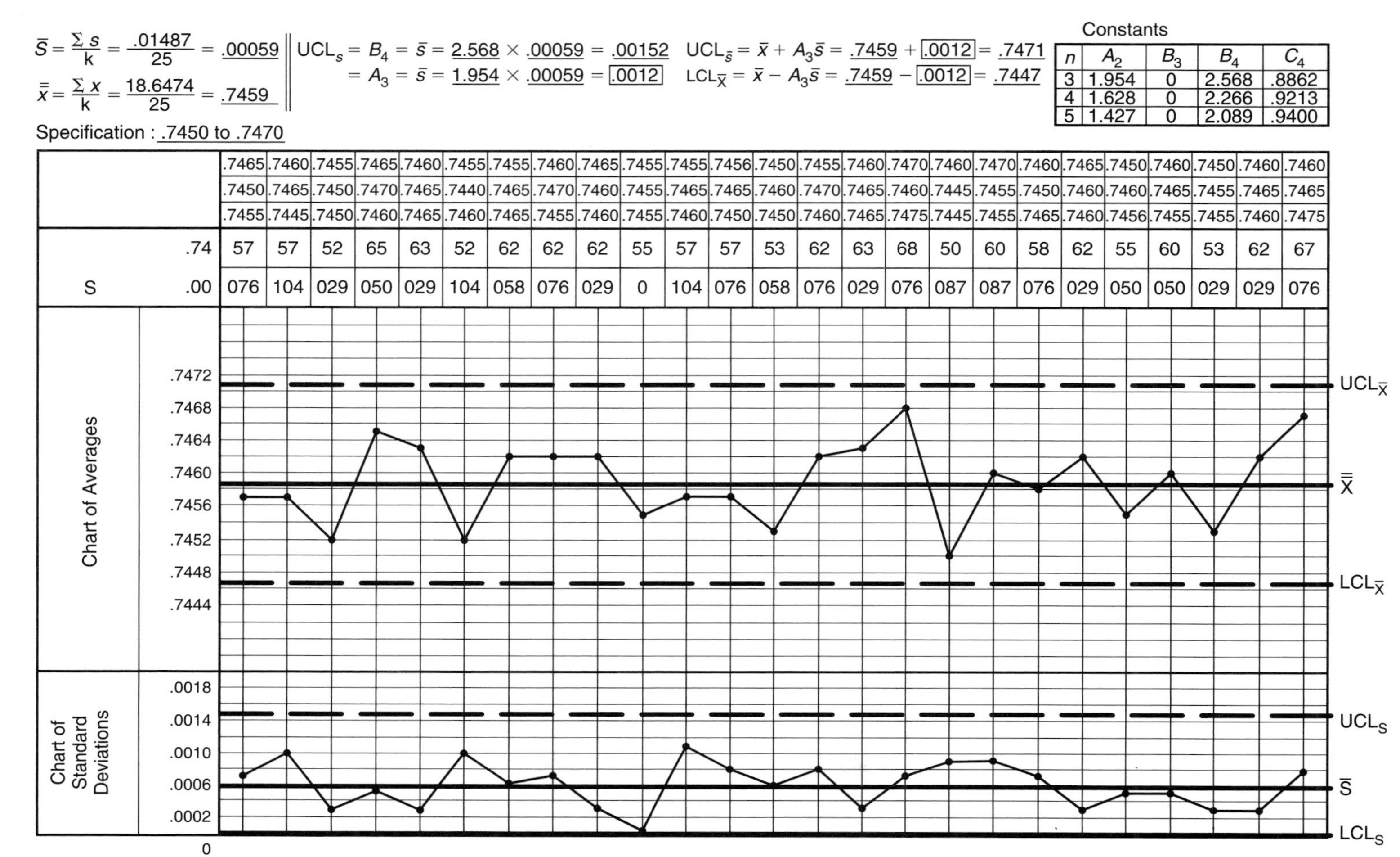

$\bar{S} = \frac{\Sigma s}{k} = \frac{.01487}{25} = \underline{.00059}$

$\bar{\bar{x}} = \frac{\Sigma x}{k} = \frac{18.6474}{25} = \underline{.7459}$

$UCL_s = B_4 = \bar{s} = \underline{2.568} \times \underline{.00059} = \underline{.00152}$

$= A_3 = \bar{s} = \underline{1.954} \times \underline{.00059} = \boxed{.0012}$

$UCL_{\bar{s}} = \bar{x} + A_3\bar{s} = \underline{.7459} + \boxed{.0012} = \underline{.7471}$

$LCL_{\bar{X}} = \bar{x} - A_3\bar{s} = \underline{.7459} - \boxed{.0012} = \underline{.7447}$

Constants

n	A_2	B_3	B_4	C_4
3	1.954	0	2.568	.8862
4	1.628	0	2.266	.9213
5	1.427	0	2.089	.9400

Specification : .7450 to .7470

	1	2	3	4	5	6	7	8	9	10	11	12	13	14	15	16	17	18	19	20	21	22	23	24	25
	.7465	.7460	.7455	.7465	.7460	.7455	.7455	.7460	.7465	.7455	.7455	.7456	.7450	.7455	.7460	.7470	.7460	.7470	.7460	.7465	.7450	.7460	.7450	.7460	.7460
	.7450	.7465	.7450	.7470	.7465	.7440	.7465	.7470	.7460	.7455	.7465	.7465	.7460	.7470	.7465	.7460	.7445	.7455	.7450	.7460	.7460	.7465	.7455	.7465	.7465
	.7455	.7445	.7450	.7460	.7465	.7460	.7465	.7455	.7460	.7455	.7460	.7450	.7450	.7460	.7465	.7475	.7445	.7455	.7465	.7460	.7456	.7455	.7455	.7460	.7475
.74	57	57	52	65	63	52	62	62	62	55	57	57	53	62	63	68	50	60	58	62	55	60	53	62	67
S .00	076	104	029	050	029	104	058	076	029	0	104	076	058	076	029	076	087	087	076	029	050	050	029	029	076

Exercise 7.15

Chapter 8

1. Tolerance = .945 − .925 = .020 $\frac{.020}{4}$ = .005 units per yellow zone.

RED I YELLOW I GREEN	IYELLOW I RED
.925 .930	.940 .945

The measurements for the start-up procedure are omitted in this exercise.

MEASUREMENTS		DECISION
Start-up procedure (5 consecutive measurements in the GREEN zone).		
A	B	
.932		OK
.935		OK
.942	.935	OK
.938		OK
.927	.933	OK
.937		OK
.936		OK
.918		Stop and check for trouble. Start-up procedure is needed.
.937		OK
.927	.929	Stop and check for process shift. Start-up procedure is needed.
.938		OK
.929	.942	Stop and check for a variability increase. Start-up procedure is needed.
.931		OK
.943	.944	Stop and check for a process shift. Start-up procedure is needed.
.944	.949	Stop and check for a process shift. Start-up procedure is needed.

3. Two measurements in the same yellow zone indicate a shift problem.
5. Variation is excessive. There will be pieces that are out of specification.
7. a) Tolerance = .027 − .021 = .006 $\frac{.006}{10}$ = .0006, use .0005 per line

   The chart is in control. $UCL_x - LCL_x = .0301 - .0193 = .0108 = 6s$.

   $PCR = \frac{6s}{Tolerance}$ = 1.80 or 180% of the tolerance.

$\overline{MR} = \frac{\Sigma MR}{N-1} = \frac{.0591}{30} = .00204$

$\overline{X} = \frac{\Sigma \overline{x}}{N} = \frac{.7408}{30} = .0247$

$UCL_{MR} = 3.267 \times .00204 = .0067$

$2.659 = \overline{MR} = 2.659 = .00204 = \boxed{.00512}$

$UCL_X = \overline{X} + 2.659\overline{MR} = .0247 + \boxed{.0054} = .0301$

$LCL_X = \overline{X} + 2.659\overline{MR} = .0247 - \boxed{.0054} = .0193$

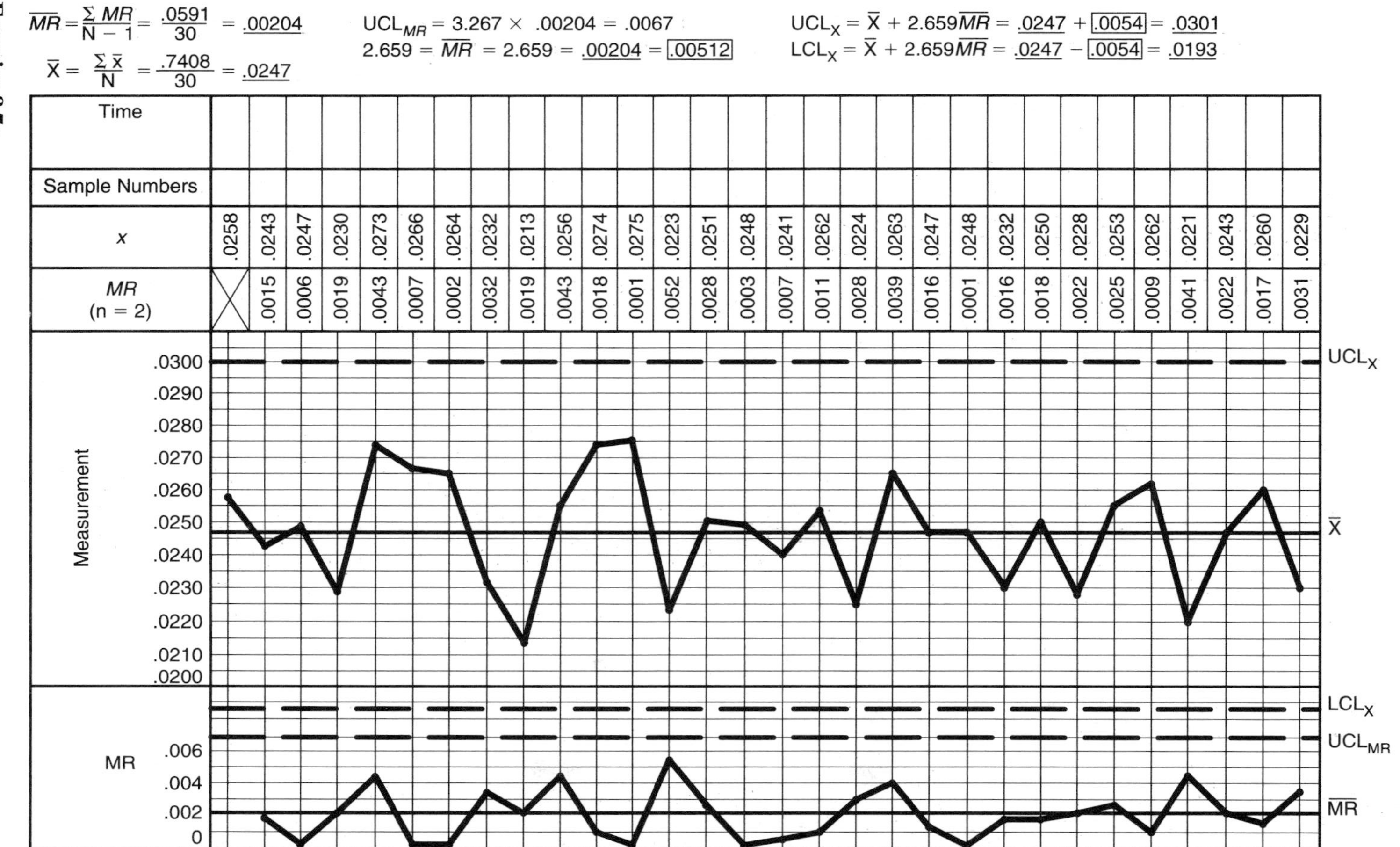

Exercise 8.7a

b) The chart is in control. $UCL_x - LCL_x = .0303 - .0191 = .0112 = 6s$.
PCR = 1.87 or 187% of the tolerance.

9. The standard deviation for the sample of 30 data values is $s = .0066$ and $6s = .0396$
Tolerance = .945 − .925 = .020
$.88(.020) = .0176$
Since $6s$ is not less than 88% of the tolerance, it is NOT recommended that pre-control be used.

11. There is no relationship between the $\bar{x}$ values and the specification limits. The Central Limit Theorem indicates that the distribution of $\bar{x}$ values will be narrower than that of the individual x values and the specification limits on an $\bar{x}$ chart will give a false sense of security. Since individual measurements are charted on an x and MR chart, specification limits can be included. The person interpreting the chart must be cautioned that both analyses are important: out-of-control analysis with respect to the control limits AND out-of-specification analysis with respect to the specification limits.

Chapter 9

1. $\bar{p} = \dfrac{370}{10000} = .037 \qquad 1 - \bar{p} = .963 \qquad UCL_p = .065 \qquad LCL_p = .009$

c) The chart is in control.
d) The process capability is $\bar{p} = .037$

Exercise 8.7b

$$\overline{MR} = \frac{\Sigma MR}{N-2} = \frac{.0879}{28} = .00314 \qquad UCL_{MR} = D_4\overline{MR} = 2.574 \times .00314 = .0081$$

$$\overline{X} = \frac{\Sigma x}{N} = \frac{.7408}{30} = .0247 \qquad UCL_X = \overline{X} + E_2\overline{MR} = .0247 + 1.772 \times .00314 = .0303$$

$$LCL_X = \overline{X} - E_2\overline{MR} = .0191$$

(n=3)

Sample	1	2	3	4	5	6	7	8	9	10	11	12	13	14	15
X	.0258	.0243	.0249	.0230	.0273	.0266	.0264	.0232	.0213	.0256	.0274	.0275	.0223	.0251	.0248
MR	—	—	.0015	.0018	.0043	.0043	.0007	.0034	.0051	.0045	.0061	.0019	.0052	.0052	.0028

Sample	16	17	18	19	20	21	22	23	24	25	26	27	28	29	30
X	.0241	.0252	.0224	.0263	.0247	.0248	.0232	.0250	.0228	.0253	.0262	.0221	.0243	.0260	.0229
MR	.0010	.0011	.0028	.0039	.0039	.0016	.0016	.0018	.0022	.0025	.0034	.0041	.0040	.0039	.0031

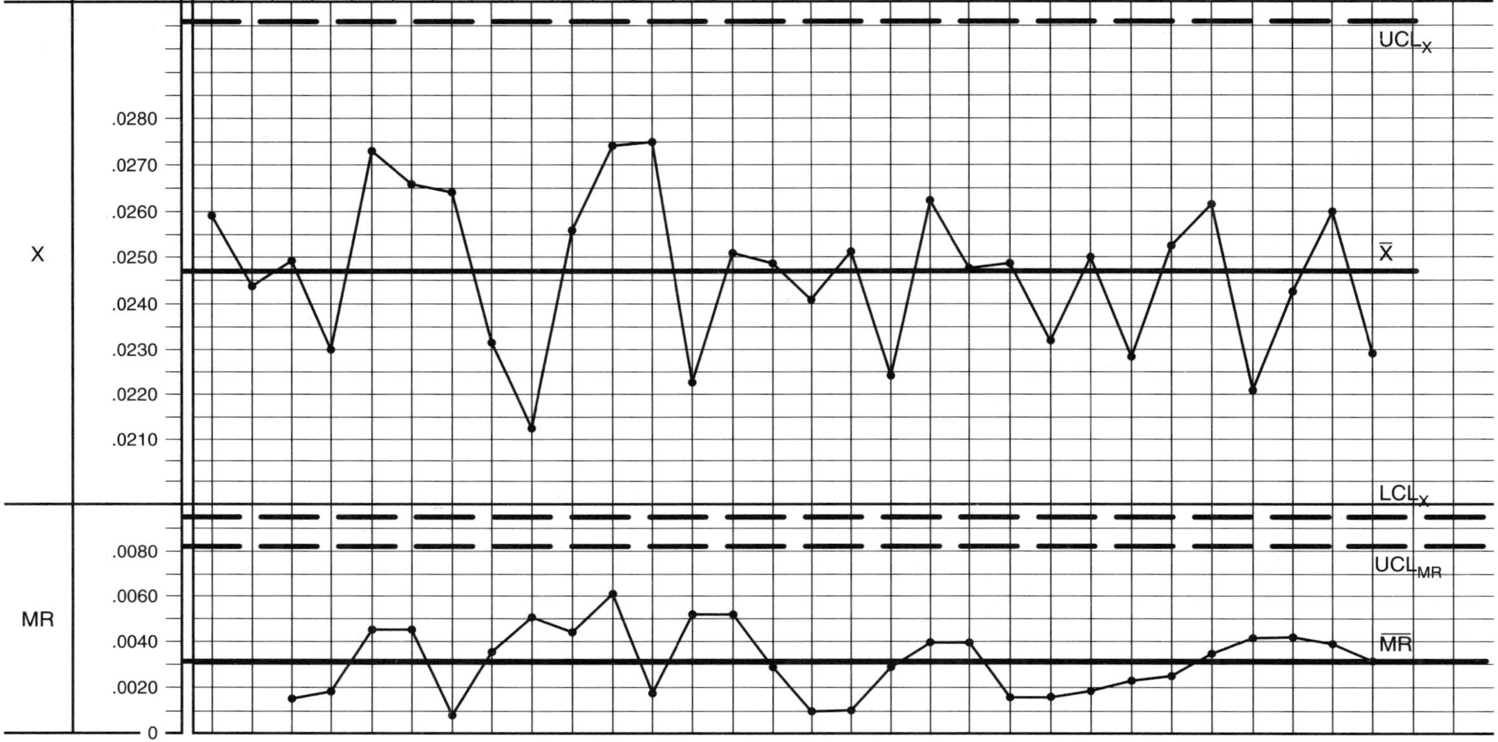

X: UCL_X, $\overline{X}$, LCL_X

MR: UCL_{MR}, $\overline{MR}$

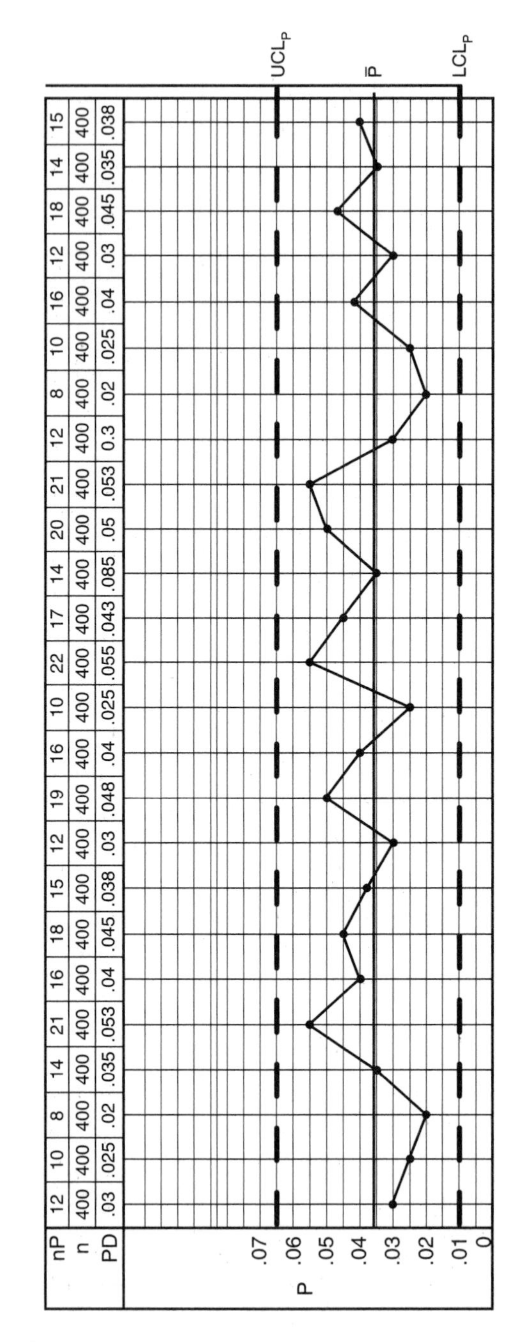

Exercise 6.17

3. $\overline{c} = 3.5$ $\quad UCL_c = 9.1 \quad LCL_c = 0$

c) The process is in statistical control, however there seems to be some problem with variation.

d) The process capability is $\overline{c} = 3.5$ defects per item.

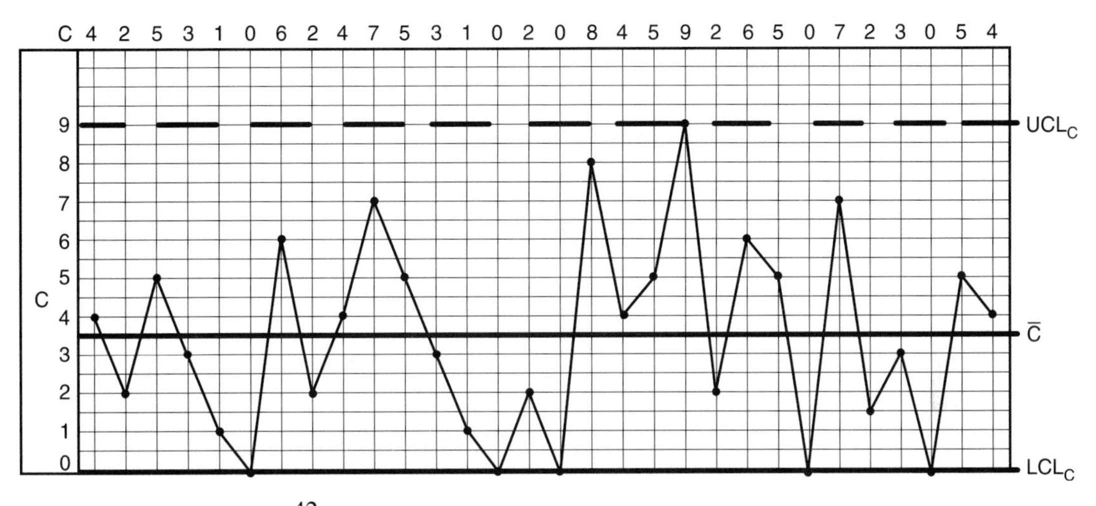

7. $\overline{np} = \dfrac{42}{13} = 3.23$ nonconforming items per sample of 40. $\overline{p} = .081 \quad 1–\overline{p} = .919$

$UCL_{np} = 8.4 \quad LCL_{np} = 0$

There is one point out of control. With that point removed, assuming the source of the problem was eliminated, a recalculation gives

$\overline{np} = 2.5 \qquad \overline{p} = .0625 \qquad 1 - \overline{p} = .9375$

$UCL_{np} = 7.1$ There is one additional point out-of-control. A second recalculation gives $\overline{np} = 2 \quad \overline{p} = .05 \quad UCL_{np} =- 6.1$

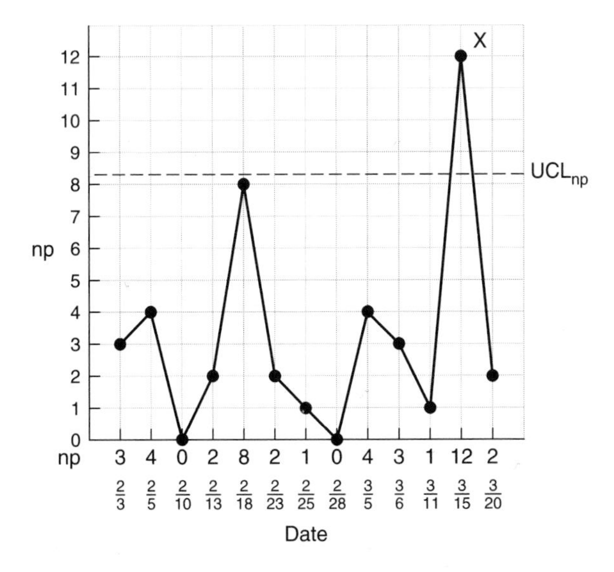

The remaining points are now in control. The capability is now $\overline{np}$ = 2 nonconforming items per sample of 40.

Chapter 10

1.

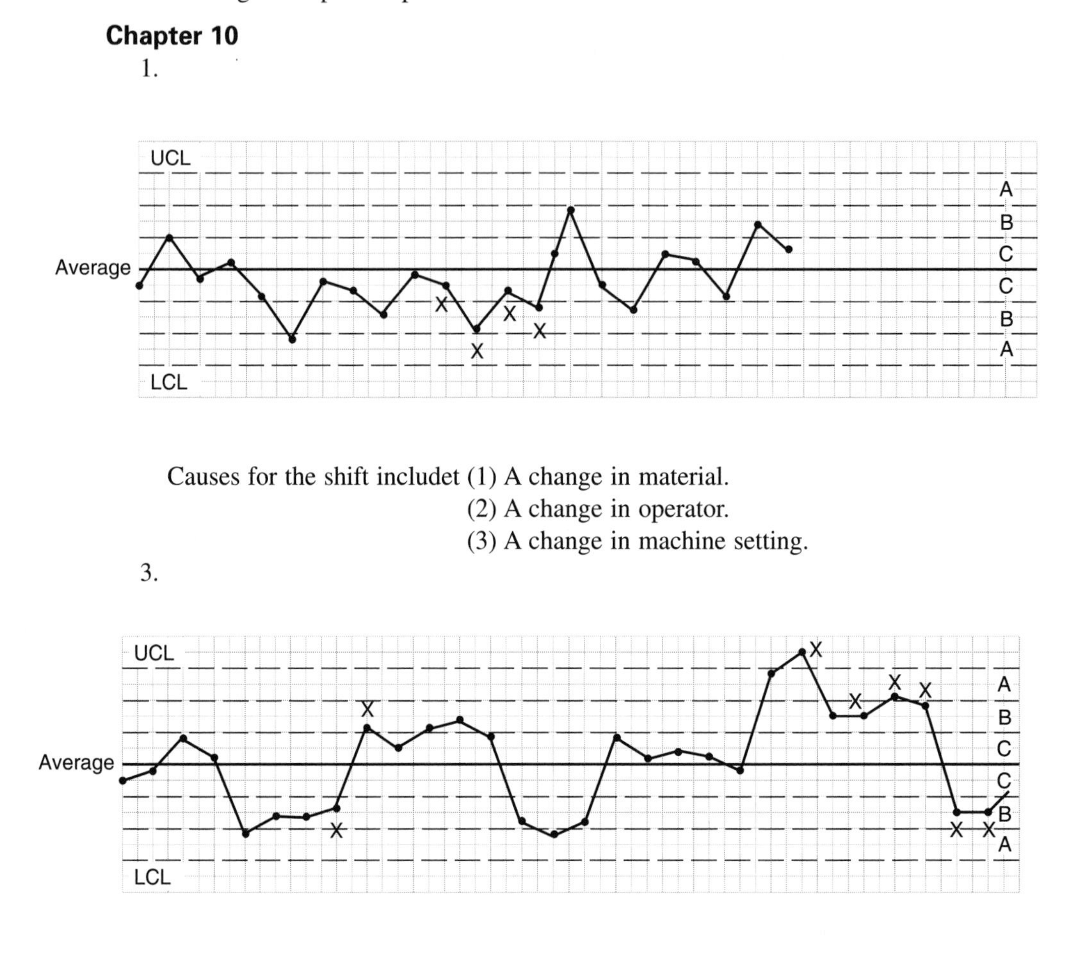

Causes for the shift includet (1) A change in material.
(2) A change in operator.
(3) A change in machine setting.

3.

From left to right the x's indicate a four-of-five freak pattern, stable mixture, a freak, another four-of-five, and another stable mixture pattern. There is also an indication of bunching. Causes include

1) Two different populations are being measured.
2) Different operators or inconsistent work by one operator.
3) Variation is material.

5.

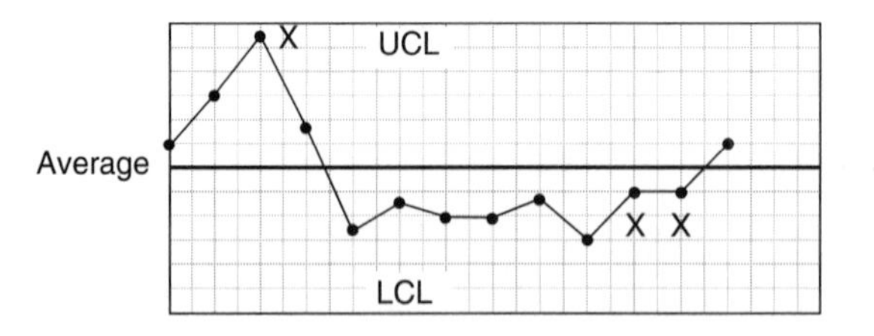

There is a freak followed by a shift. Causes include

1) A tool problem followed by an adjustment.
2) A change in material.
3) A change in operator.

7.

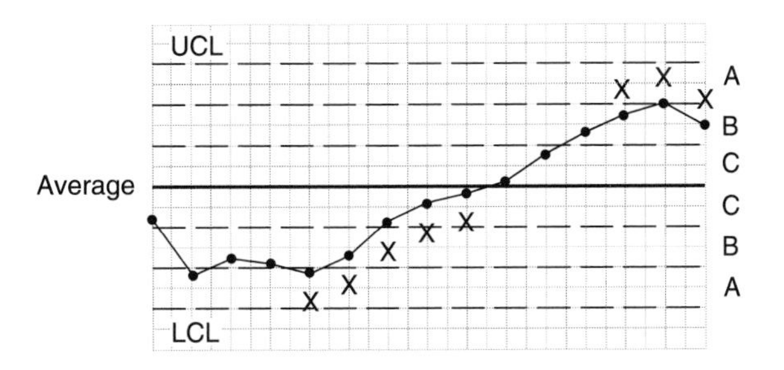

There is a shift followed by a run with four-of-five freak patterns imbedded in each. Causes include

1) An adjustment followed by tool wear.
2) A machine fixture is gradually loosening causing an increase in variability
3) There is a gradual change in a gauge.

9.

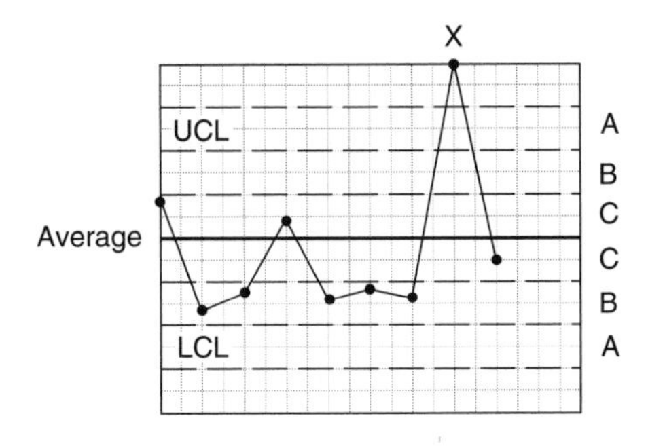

A single freak can be caused by (1) A measurement error.
(2) A plotting error.
(3) A calculation error.

11.

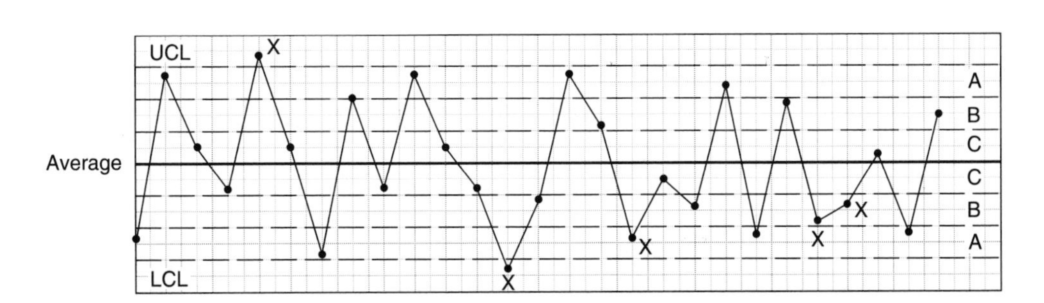

The overall pattern is that of an unstable mixture. The x's indicate two freaks and stable mixture patterns. Causes include

1) Fluctuations in measurements or variation due to loose fixtures.
2) Inconsistent work due to poor training or other operator problems.
3) Different populations of measurements are merging (suppliers, operators).

13.

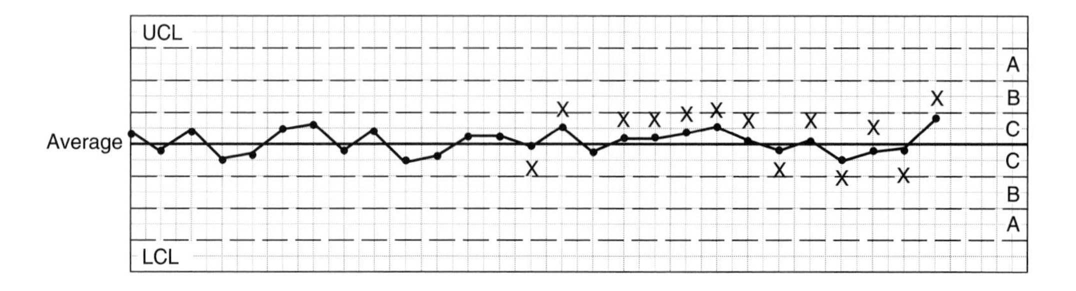

The stratification pattern can be caused by

1) Dishonest measurement.
2) Selective sampling.
3) A decrease in variation and control limits have to be recalculated.

15.

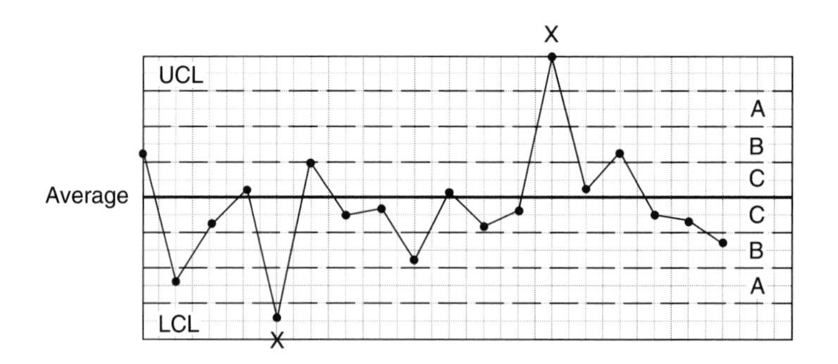

The two freaks can be due to

1) Mistakes in measurement.
2) Errors in calculation.
3) Charting errors.

17.

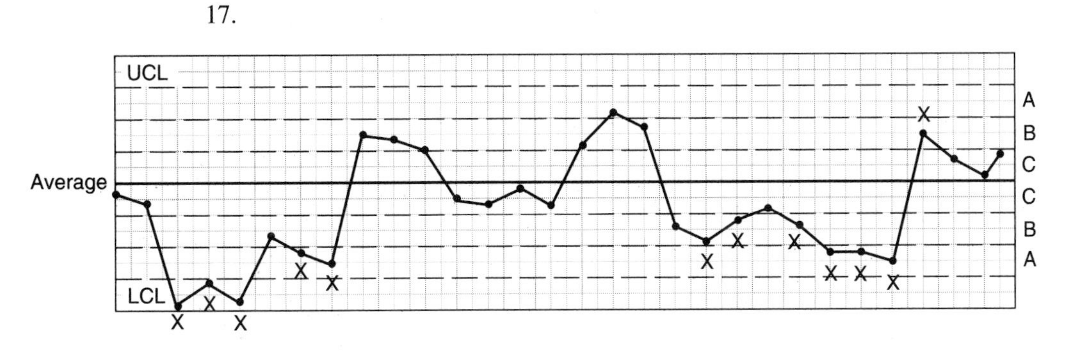

There is a shift which contains 3 freaks and a four-of-five freak pattern. The next shift begins with a stable mixture pattern, then a four-of-five freak pattern, and then ends with another stable mixture pattern. This may be due to

1) Different operators.
2) Adjustment problems or inconsistent work habits.
3) Material changes.

19. $(.0227)^2 = .000515$

21. $(.5)^6 \times (.5) \times 7 = .05469$

6 above 1 below 7 ways

23. .1587 z > 1

−.0228 z > 2

.1539 P(B)

P(lower B, then upper B, then lower B) = $(.1539)^3 = .00251$

Chapter 11

1.
1) Problem recognition.
2) Problem definition.
3) Problem analysis.
4) Choice for action.
5) Problem solution.
6) Prevention of backsliding.

3.
1) Choose a group leader.
2) Record all suggestions.
3) No disparaging remarks are allowed.
4) Keep everyone in the group involved in the generation of ideas.

5. a) PCR $= \frac{.0078}{.01} = .78$ or 78% of the tolerance.
   b) $\bar{x} = .59953 \quad s = .0018934$
   c) PCR $= \frac{.01136}{.01} = 1.136$ or 114% of the tolerance.
   d) The distribution may be bimodal and not normal. It also crowds the specification limit on one end and overlaps the specification limit on the other.
   e) This sample casts serious doubts on the reliability of the vendor's information.

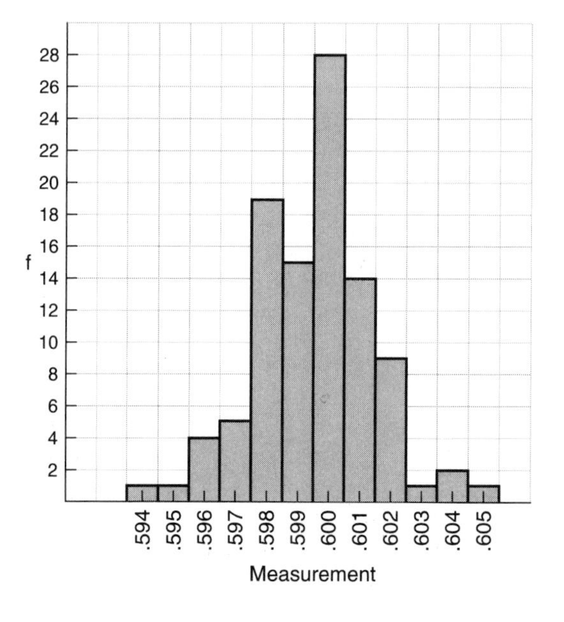

7. The help and re-work on the homework assignments may not have been finally reworked by the student alone to ensure that the student could do the work in a testing situation.
9. Too much emphasis on individual contributions will interfere with teamwork. Ideas are not openly shared or discussed when team members are concerned about someone else getting credit for their ideas.
11. Too many team members are hard to organize. Any recognizable imbalance in team member duties will have a negative impact on team success.
13. • The vendor's claim
    • Vendor information such as control charts and histograms.
    • The mean, standard deviation and histogram of a random sample of incoming material.

    It is necessary to know the quality incoming in order to assess the quality outgoing. Also, vendors may be reluctant to report any problems with a shipment.

15. 1) Lack of standards: You don't know how to do some of the examination problems.
If you don't set high standards for yourself, your preparation will most likely be inadequate. Aim for that A or B and keep your preparation work consistent with your goal.
2) Being forgetful: You forgot a formula that is needed for one of the problems.
Improve your organization. Reinforce your learning by reworking things you have a tendency to forget.
3) Not being attentive: You vaguely remember the instructor doing a problem in class that was similar to one on the examination that you're stuck on. Work on being an aggressive learner. Follow up on problems that are demonstrated in class. Find similar problems in the homework assignment.
4) Lack of training: You can only get part way through an examination problem that is similar to one that you did on a homework assignment. Don't skip any of the assigned homework. Rework a sample of homework problems at a later time to ensure that you know how to do them.
5) Misunderstanding: The examination question asks for one procedure and you do another.
Reread the question to make sure you are doing what is asked for.
6) Carelessness: You lose points for careless errors doing problems that you know how to do.
Check your answers and give a reasonableness check as well.
7) Being startled: You lose points for omitting part of a procedure because you were startled during the examination.
If you get distracted in the middle of a problem, do a logical trace from the beginning to the point where you were distracted.
8) Fear of questioning procedures: You didn't know how to do a problem on the examination because you didn't understand the procedure when it was demonstrated.
Either ask the questions during class (you're probably asking the question that several other students would like to ask), or make a note to investigate it later.

Chapter 12

1.

b) Maximum possible deflection $= \frac{.0096}{2} = .0048$

c) Maximum possible deflection $= .003 + .0048 = .0078$

d) Maximum possible deflection $= .002 + .003 + .0048 = .0098$

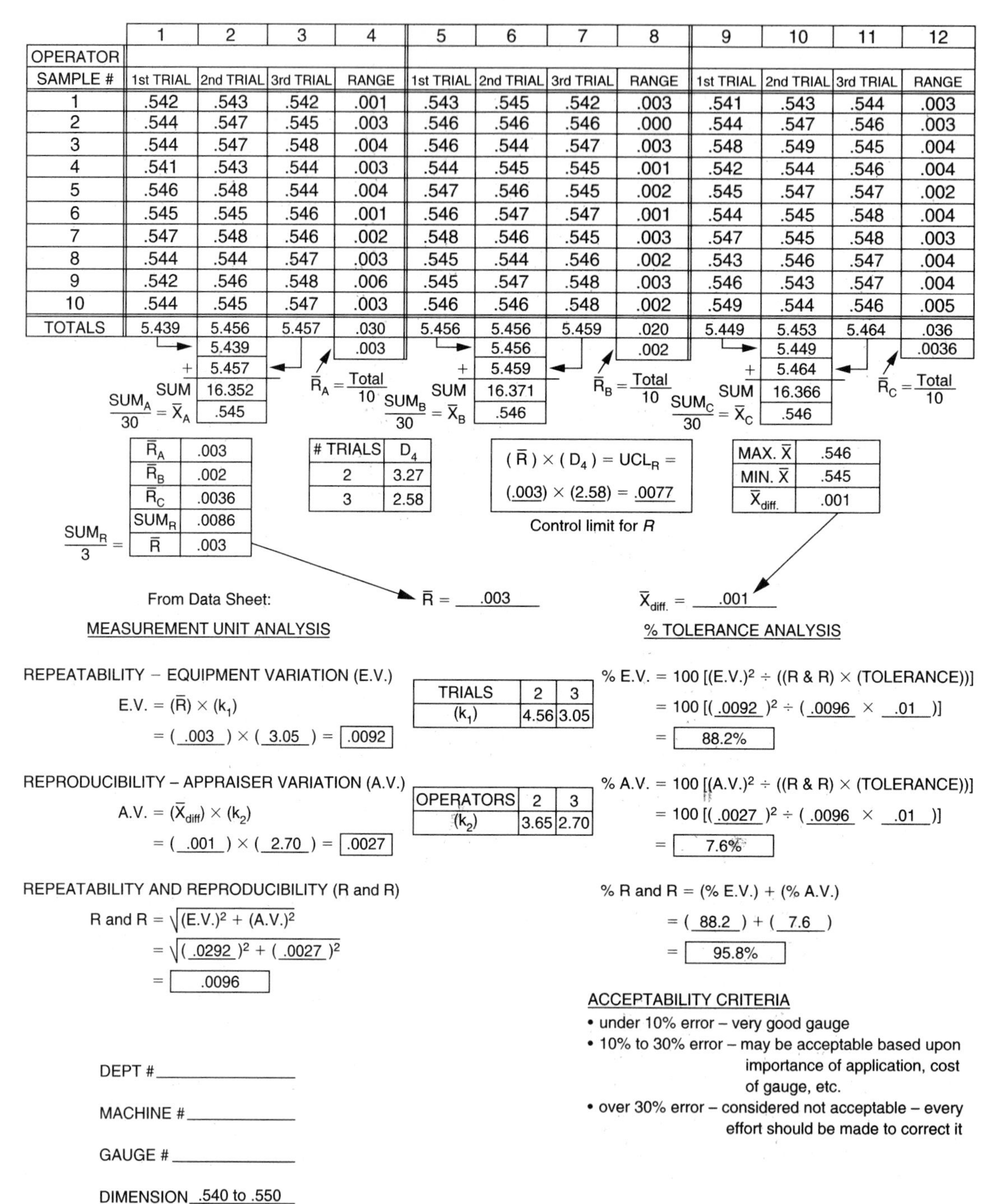

GAGE CAPABILITY

	1	2	3	4	5	6	7	8	9	10	11	12
OPERATOR												
SAMPLE #	1st TRIAL	2nd TRIAL	3rd TRIAL	RANGE	1st TRIAL	2nd TRIAL	3rd TRIAL	RANGE	1st TRIAL	2nd TRIAL	3rd TRIAL	RANGE
1	.542	.543	.542	.001	.543	.545	.542	.003	.541	.543	.544	.003
2	.544	.547	.545	.003	.546	.546	.546	.000	.544	.547	.546	.003
3	.544	.547	.548	.004	.546	.544	.547	.003	.548	.549	.545	.004
4	.541	.543	.544	.003	.544	.545	.545	.001	.542	.544	.546	.004
5	.546	.548	.544	.004	.547	.546	.545	.002	.545	.547	.547	.002
6	.545	.545	.546	.001	.546	.547	.547	.001	.544	.545	.548	.004
7	.547	.548	.546	.002	.548	.546	.545	.003	.547	.545	.548	.003
8	.544	.544	.547	.003	.545	.544	.546	.002	.543	.546	.547	.004
9	.542	.546	.548	.006	.545	.547	.548	.003	.546	.543	.547	.004
10	.544	.545	.547	.003	.546	.546	.548	.002	.549	.544	.546	.005
TOTALS	5.439	5.456	5.457	.030	5.456	5.456	5.459	.020	5.449	5.453	5.464	.036
		5.439		.003		5.456		.002		5.449		.0036
	+	5.457			+	5.459			+	5.464		
	SUM	16.352			SUM	16.371			SUM	16.366		
	$\frac{SUM_A}{30} = \bar{X}_A$	.545			$\frac{SUM_B}{30} = \bar{X}_B$	.546			$\frac{SUM_C}{30} = \bar{X}_C$	.546		

$\bar{R}_A = \frac{Total}{10}$ $\bar{R}_B = \frac{Total}{10}$ $\bar{R}_C = \frac{Total}{10}$

$\bar{R}_A$	.003
$\bar{R}_B$	.002
$\bar{R}_C$	.0036
SUM_R	.0086
$\frac{SUM_R}{3} =$ $\bar{R}$	.003

# TRIALS	D_4
2	3.27
3	2.58

$(\bar{R}) \times (D_4) = UCL_R =$

$(.003) \times (2.58) = .0077$

Control limit for *R*

MAX. $\bar{X}$	.546
MIN. $\bar{X}$	.545
$\bar{X}_{diff.}$	.001

From Data Sheet: $\bar{R}$ = .003 $\bar{X}_{diff.}$ = .001

MEASUREMENT UNIT ANALYSIS

REPEATABILITY – EQUIPMENT VARIATION (E.V.)

E.V. = $(\bar{R}) \times (k_1)$

= (.003) × (3.05) = .0092

TRIALS	2	3
(k_1)	4.56	3.05

REPRODUCIBILITY – APPRAISER VARIATION (A.V.)

A.V. = $(\bar{X}_{diff}) \times (k_2)$

= (.001) × (2.70) = .0027

OPERATORS	2	3
(k_2)	3.65	2.70

REPEATABILITY AND REPRODUCIBILITY (R and R)

R and R = $\sqrt{(E.V.)^2 + (A.V.)^2}$

= $\sqrt{(.0292)^2 + (.0027)^2}$

= .0096

% TOLERANCE ANALYSIS

% E.V. = 100 [(E.V.)² ÷ ((R & R) × (TOLERANCE))]

= 100 [(.0092)² ÷ (.0096 × .01)]

= 88.2%

% A.V. = 100 [(A.V.)² ÷ ((R & R) × (TOLERANCE))]

= 100 [(.0027)² ÷ (.0096 × .01)]

= 7.6%

% R and R = (% E.V.) + (% A.V.)

= (88.2) + (7.6)

= 95.8%

ACCEPTABILITY CRITERIA

- under 10% error – very good gauge
- 10% to 30% error – may be acceptable based upon importance of application, cost of gauge, etc.
- over 30% error – considered not acceptable – every effort should be made to correct it

DEPT #________

MACHINE #________

GAUGE #________

DIMENSION .540 to .550

Exercise 12.1

3. $$\sigma_{Gauge} = \frac{.00237}{1.693} = .0014$$

$$\sigma_{Total} = \frac{.00393}{2.059} = .0019$$

$$\sigma_{Process} = \sqrt{(.0019)^2 - (.0014)^2} = .0013$$

$$PCR = \frac{6 \times .0013}{.01} = .78 \text{ or } 78\% \text{ of the tolerance.}$$

Chapter 13

1. $$P(3 \text{ good and } 2 \text{ defective}) = \frac{C_{16,3} \times C_{4,2}}{C_{20,5}} = \frac{560 \times 6}{15504} = .217$$

3. Take the first digit from the first random number for the box number (0 = box #10). Use the last digit for the layer number (additional random numbers may be needed for a digit in the 1 to 5 range). After the box and layer are determined, choose another random number and use the last two digits for the piece number. The following random numbers illustrate the procedure.

RANDOM NUMBER				BOX	LAYER	PIECE
141	448	411		1	1	11
box	layer	piece				
625	472	934		6	5	34
254	235			2	4	35
854	731			8	4	31
956	144	699	029	9	4	29

This procedure is continued until the 100 sample pieces are identified, then the list is reordered by box, then by layer in each box and finally by piece within each layer.

5.

p	np	100p	Pa
0	0	0	1
.002	.4	.2	.92
.005	1	.5	.72
.01	2	1	.36
.015	3	1.5	.18
.02	4	2	.09
.025	5	2.5	.04
.03	6	3	.02
.04	8	4	.005

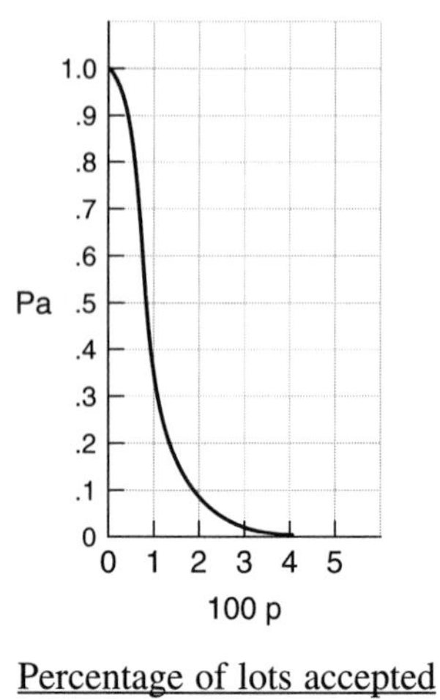

b)

Lot proportion defective	Percentage of lots accepted
1%	36%
2%	9%
4%	.5%

c) AQL = .2%

d) RQL = 1.9%

e) IQL = .8%

7.

100p	× Pa	= AOQ	(100p, AOQ)
0	1	0	(0, 0)
.2	.92	.18	(.2, .18)
.5	.72	.36	(.5, .36)
1	.36	.36	(1, .36)
1.5	.18	.27	(1.5, .27)
2	.09	.18	(.2, .18)
2.5	.04	.10	(2.5, .10)
3	.02	.06	(3, .06)

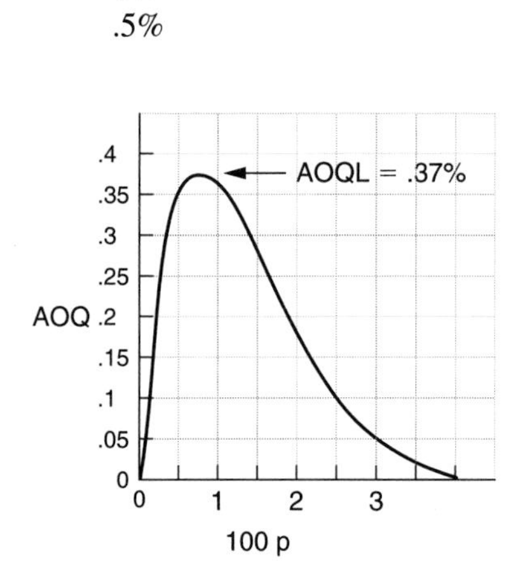

9.

p	np	100p	Pa
0	0	0	1
.01	1.25	1	.97
.015	1.88	1.5	.88
.02	2.5	2	.75
.03	3.75	3	.47
.04	5	4	.26
.05	6.25	5	.13
.06	7.5	6	.07
.07	8.75	7	.04
.08	10	8	.01

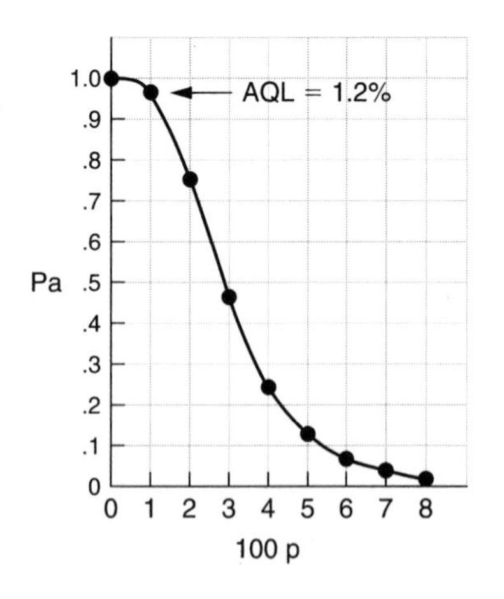

Appendix A

1. a) −5 b) −13 c) −8 d) 18 e) −8 f) 6 g) −19 h) −2 i) 3.33 j) 8 k) 4 l) 19 m) 36 n) −45 o) 105.5 p) −20 q) −8 r) 7.6158
3. a) P(good) = 91%
   b) P(scrapped or reworked) = 9%
   c) P(reworked, then good) = .07 × .91 = .064 or 6.4%
   d) P(good and scrap and rework) = .91 × .02 × .07 × 6 = .0076 or .76%
5. P(3 good and 1 defective) = $\dfrac{C_{50,3} \times C_{10,1}}{C_{60,4}} = \dfrac{19600 \times 10}{487635} = .402$
7. a) P(9 good) = $\dfrac{C_{30,9}}{C_{40,9}} = .0523$

   b) P(6 good and 3 defective) = $\dfrac{C_{30,6} \times C_{10,3}}{C_{40,9}} = .26$

INDEX

L

M

N

O

P

Q

R

S

T

U

V

W

X

Y

Z